1st International Scientific and Practical Conference "Innovative Technologies in Environmental Engineering and Agroecosystems" (ITEEA 2021)

E3S Web of Conferences Volume 262 (2021)

Nalchik, Russia
18-19 March 2021

Editor:

R.Z. Abdulkhalikov

ISBN: 978-1-7138-3007-8

Printed from e-media with permission by:

Curran Associates, Inc.
57 Morehouse Lane
Red Hook, NY 12571

Some format issues inherent in the e-media version may also appear in this print version.

For additional information, please contact EDP Sciences – Web of Conferences
at the address below.

EDP Sciences – Web of Conferences
17, Avenue du Hoggar
Parc d'Activité de Courtabœuf
BP 112
F-91944 Les Ulis Cedex A
France

Phone: +33 (0) 1 69 18 75 75
Fax: +33 (0) 1 69 28 84 91

contact-edps@webofconferences.org

Additional copies of this publication are available from:

Curran Associates, Inc.
57 Morehouse Lane
Red Hook, NY 12571 USA
Phone: 845-758-0400
Fax: 845-758-2633
Email: curran@proceedings.com
Web: www.proceedings.com

TABLE OF CONTENTS

PREFACE

We are pleased to present this collection of scholarly articles of the 1st International Scientific and Practical Conference "Innovative Technologies in Environmental Engineering and Agroecosystems" (ITEEA 2021). The conference was held in Kabardino-Balkarian State Agrarian University, Nalchik, Russian Federation, on March 18 - 19, 2021.

The key sections of the Conference were:

Section 1: Innovative technologies, engineering systems and energy in the agro-industrial complex.

Section 2: Basic and applied research in zooengineering and veterinary medicine.

Section 3: Highly productive agroecosystems and environmental safety.

Section 4: Innovative technologies in environmental engineering

The overall objective of ITEEA 2021 is to provide a platform to analyze the interrelation between environmental engineering, agricultural pollution and agroecosystems. It is expected to open the way to elimination of environmental pollution and ecological crisis as well as to protection of green environment for human existence.

The Conference has received 302 manuscripts, of which 106 submissions have been accepted by our reviewers (well-known international experts in the field of agroengineering). Full papers submitted to the conference were peer-reviewed by two - tree reviewers per manuscript before being selected for publication. In case when two experts had disagreements, the paper was sent to another two reviewers for evaluation.

We would like to express our high appreciation to the organizers, members of the program committees and reviewers of ITEEA 2021 for their hard work. Without their participation, it would be impossible to hold ITEEA 2021 successfully and ensure high quality of papers of the conference proceedings. We also would like to express our gratitude to the unbelievable number of participants for contributing their research results to the Conference. Special acknowledgements go to E3S Web of Conference team. We also hope to meet again in the upcoming conferences.

Statement of Peer review

In submitting conference proceedings to *Web of Conferences*, the editors of the proceedings certify to the Publisher that

1. They adhere to its **Policy on Publishing Integrity** in order to safeguard good scientific practice in publishing.
2. All articles have been subjected to peer review administered by the proceedings editors.
3. Reviews have been conducted by expert referees, who have been requested to provide unbiased and constructive comments aimed, whenever possible, at improving the work.
4. Proceedings editors have taken all reasonable steps to ensure the quality of the materials they publish and their decision to accept or reject a paper for publication has been based only on the merits of the work and the relevance to the journal.

Title, date and place of the conference

ITEEA 2021 - 1st International Scientific and Practical Conference
"Innovative Technologies in Environmental Engineering and Agroecosystems"
(Nalchik, Russian Federation, March 18 - 19, 2021)

Proceedings editor(s):

R. Z. Abdulkhalikov

Date and editor's signature

03.05.2021

Analysis of the reclamation canal condition and cleaning methods

Khamzat Abdulmazhidov*

Federal State Budgetary Educational Institution of Higher Education "Russian State Agrarian University-Moscow Timiryazev Agricultural Academy", Timiryazevskaya street, 49, 127550 Moscow Russian Federation

Abstract. The operating quality of reclamation drainage and irrigation systems depends on the condition of their main elements. The main elements of reclamation systems are canals with various profiles, ranging from rectangular to parabolic. The most widespread in land reclamation are canals with a trapezoidal profile. At the same time, it is possible to note that the natural production characteristics of reclamation systems are quite diverse, this issue is especially relevant for geometric parameters of the canals. In this case, the stable slopes of canals, restoring their profile and cleaning from sediments and silting can only be provided by special complexes of machines. Knowledge of defect types and causes in reclamation canals allows them to be prevented and eliminated in a timely manner. One of the main tasks in this case is to ensure the stability of canal slopes.

1 Introduction

The high-quality functioning of reclamation systems is directly related to the condition of drainage or irrigation canals. During the flood period, it becomes necessary to discharge excess water; for this, canals that meet operating requirements are required, that is, the canals shall have design geometric dimensions. During a drought period, moisture shall be maintained in compliance with the water supply regime, in this case, various structures on the canals are used to save water, for example, locks, their closing makes it possible to keep moisture in the reclamation system. In addition, locks are an element of water level regulation in a constantly changing water situation, depending on natural and seasonal conditions. During operation, the initial design cross-section of the canals changes, sediments, siltation and vegetation appear, in some cases, when the frequency and timing of current and major repairs is violated, the appearance of small shrubs is observed. These factors affect the capacity of canals, respectively, and system operation as a whole. At first glance, economically, periodic cleaning of reclamation canals as part of current and major repairs in comparison with other types of agricultural production (for example, grain growing) does not bring economic benefits, however, various disasters and the elimination of their consequences confirms the need for cleaning and renewal of canals. Such cataclysms include

* Corresponding author: hamzat72@mail.ru

floods during the flood period and fires during dry seasons. Standard sizes of drainage canals are shown in table 1.

Table 1. Standard sizes of canal cross-sections in the earthen bed of drainage systems

Canal name	Construction depth, m	Canal width, m	Ratio of slope
1. Canals made by general construction machines			
Conductive Regulator Catchwater and intercepting canals	from 0.8 to 1.5 from 1.5 to 2.5 from 1.5 to 3.0	0.4; 0.6; 0.8 0.6; 0.8; 1.0 0.6; 0.8; 1.0; 1.5	1.0; 1.5; 2.0 1.5; 2.0; 2.5 1.5; 2.0; 2.5
2. Canals made by specialized machines			
Regulator	from 0.8 to 1.0 from 1.0 to 1.2from 1.2 to 1.7	0.25; 0.25; 0.40; 0.6 0.25; 0.40; 0.6	1.0 1.0; 1.5

The costs incurred in the process of reclamation systems, agricultural production and forest areas renewal in general after periodic floods or after fires (for example, fires in the European part of the Russian Federation in 2010) are disproportionately high compared to what has to be implemented during current and major repairs of drainage systems. Routine repair of reclamation canals means their cleaning, including the bottom and slopes, from vegetation, sediment and silting; elimination of landslides and defects, formation of elements for slope securing. The bottom is deepened up to the initial design marks, developed sediments are distributed along the canal [1].

Overhaul of the canals is accompanied by canal deepening and widening, slope planning, soil leveling and silting removed from the canal; in some cases, it is allowed to change the canal route. The most time-consuming operations in the overhaul are slope securing with concrete slabs, a plank wall and fascines. Overhaul of canals can be performed using special canal cleaning machines, and in cases where the canal depth exceeds 2 m, using single-bucket general construction excavators with special working bodies. Routine repair and maintenance of open canals is most efficiently performed by special canal cleaners with active working bodies of continuous action (milling or rotary type). Termination of current repairs and overhaul of reclamation canals of the drainage network result in overgrowth and change in the canal profile (Fig. 1), which also results in a violation of its capacity.

Fig. 1. Drainage canal condition during long-term operation without current and major repairs

High-quality use of reclamation lands is achieved with modern and efficient performance of a wide range of repair and renewal operations using special machines of high quality and productivity. At present, the most optimal method for cleaning and renewal of reclamation canals is a complex mechanized method, implying implementation of all basic and auxiliary operations using machines and mechanisms. This method combines machines and mechanisms so that when they are fully loaded, the minimum cost of works is ensured at their maximum productivity. It is also worth mentioning that some sections of the canals still require the manual labor of canal repairmen, it applies to those sections where the use of special machines is impossible, for example, sections under bridges and near locks, etc. [2].

Another important component in the formation of a complex mechanized method is the number of machines in a complex. On the one hand, a large number of machines should be selected and interconnected for full load with the lowest cost of work at maximum productivity, on the other hand, a large number of machines of a particular complex is already economically unprofitable. There is no single special machine that performs all types of cleaning and renewal operations, it also follows that a significant increase in the number of machines and their standard sizes in one complex can provide high-quality cleaning, but there will also be significant costs. Thus, the optimal value of the number of machines within one complex for cleaning works of reclamation systems becomes a very important task. A certain scope of canal cleaning and renewal operations calls for a certain mechanical equipment from the enterprise or economy, which is determined by the ratio of the balance cost of the fleet of machines and mechanisms B_m to the cost of repair and renewal operations C_r, performed during a year:

$$\text{М}_\text{о} = \frac{\text{Б}_\text{м}}{\text{С}_\text{г}}$$

When determining the composition of the optimal complex of machines, it is important to take into account both the technical and operational and technical and economic machine parameters, the first group of indicators includes productivity, and the second group is the cost of cleaning operations. In general, the performance of a canal cleaning machine is the volume of production (i.e., sediment and silt volume) produced by the machine per unit of time. By integrating the units of sediment and silt volumes over time, performance can be determined. Thus, it is important to understand the scope of land reclamation farm cleaning operations [3].

The use of high-performance powerful machines can be effective for large volumes of operations, and, such machines are mainly used in the construction of reclamation system canal. The scope of performed cleaning operations is not high; moreover, the distribution of sediments along the canal length is extremely uneven. Their highest number is observed at the junctions with other canals, as well as at the junctions of the drainage pipes with the canal.

The complex of canal cleaning machines can contain main and auxiliary machines. To determine the number of main machines at the preliminary stage, the planned intensity of the studied type of operations is calculated:

$$I_i = \frac{V_i K_{\text{нер}}}{T_i}$$

where V_i is a scope of studied operation type; $K_{\text{нер}}$ is a coefficient of uneven performance of operations in the studied period; T_i is a deadline of operations. The required number of machines in this case is determined by the formula:

$$n_i = \frac{I_i}{\Pi_i}$$

where T_i is the design performance of machines under specific operating conditions.

Fig. 2. Reclamation canal in the process of cleaning and renewal

An important characteristic for maintenance of reclamation canals is the frequency of repair operations. Here the cycles of cleaning operations are considered, they are divided into the overhaul period, the overhaul cycle and the overhaul cycle structure. The operation of canals and their structures between two next scheduled repairs refers to the overhaul period, the operation of networks and structures between two next major overhauls refers to the overhaul cycle. The overhaul cycle structure implies alternating repairs of a system or a separate canal [4,5].

As part of the maintenance of the canals of the drainage network, cleaning operations are performed annually, the specific volumes of sediment and siltation for the drainage network are low and range from 0.015 to 0.025 m^3. However, these values increase over time and can range from 0.1 to 0.3 m^3.

Mechanical cleaning of canals is complicated by the fact that reclamation systems include a large number of various standard sizes of canals, various maintenance and repair operations. It is often necessary to deal with canals made in a cut, a semi-cut-and-fill or in a fill. It is reflected in the designs of canal cleaning machines. Some inconveniences when performing mechanized canal cleaning are caused by canal structures such as bridges, crossings, outlets, etc.

2 Materials and Methods

The formation of machine complexes for canal cleaning is usually performed with selection of the main leading machine. In this study, a canal cleaner with a bucket moving during operation on rigid guides along the canal axis is taken as the leading machine. The operating equipment, including a double telescopic boom, rigid guides made of two channels and a bucket on rollers, is installed in the form of a side hitch to a caterpillar tractor. The machine, moving along the canal berm, works positionally to clean the bottom of the canal from sediments and silting. The first version of the working body, a bucket, has a rectangular profile and is designed to clean the bottom of canals, 0.4 and 0.6 m wide, with fixed slopes. Canals with fixed slopes are usually present on drainage systems in Kaliningrad region and Belarus. The second version of the working body, a bucket, has a trapezoidal profile and is designed to clean the bottom and slope parts adjacent to the bottom. Such profiles of drainage canals are most common and have large lengths in the European part of the Russian Federation [6,7,8].

If a canal cleaner with a bucket on rigid guides is considered as the main machine of one of many possible complexes, then other operations, such as mowing of canal slopes, collecting and moving sediment and silt removed from the canal, can be performed by a canal trowel and a single-bucket excavator based on a pneumatic-wheeled tractor with a bulldozer blade.

It was necessary to ensure the lowest traction resistance during cleaner bucket operation. To determine the traction forces, reduced models of rectangular and trapezoidal profile buckets were designed and manufactured. Laboratory works to determine traction resistance with models of working bodies was performed on a soil tray, which is a metal container with soil and rigid guides on its walls. A trolley with a model can move along rigid guides, excavating the ground, with characteristics as close as possible to the soil-alluvial environment in natural production conditions. Digging efforts were recorded by special sensors and transmitted to a computer. The obtained experimental data were processed and then recalculated to working bodies in full size. Traction forces as a function of optimization were determined depending on the digging depth (thickness of chips removed by the bucket).

3 Results and Discussion

The obtained valuesof traction forces for the model and recalculated for the bucket in full size turned out to be quite sufficient for the hydraulic system of the base tractor in combination with a four-speed pulley block. The use of a trapezoidal section bucket results in certain changes in operational characteristics: firstly, the bucket capacity increases, respectively, the bucket mass with soil also increases, all this can affect the machine stability during operation; secondly, an increase in the bucket mass with soil means an increase in loads, requiring strength calculations of the working equipment structures; thirdly, the working equipment hydraulic drive is to be checked.

The stable operation of the canal cleaner is ensured through the use of a tracked base tractor, as well as the use of a counterweight with a variable reach. Preliminary strength calculations in Inventor Pro show that the safety factor for steel structures is within acceptable limits.

Fig. 3. Testing the working body model of a canal cleaner in laboratory conditions

 https://doi.org/10.1051/e3sconf/202126201001

In recent years, the volume of maintenance and repair operations on reclamation canals has become very significant, it requires cleaning machines of high productivity, mobility and maneuverability, as well as not causing problems with transportation from one system to another or from one canal to another. In this sense, the most demanded machines will be continuous action machines, in comparison with cyclic action machines, they have high technical and operational indicators. However, a canal cleaner with a bucket movement along the canal axis has a high quality of cleaning operations, especially when it comes to ensuring the required slope. Laboratory tests with a model bucket and field studies with a life-size working body showed the high quality of cleaning operations and productivity.

4 Conclusions

1. The analysis of the studies showed that as the leading machine of the cleaning complex for reclamation drainage canals, up to 2 m, a canal cleaning machine with a bucket longitudinal along the canal axis can be used with the greatest efficiency.

2. Calculations of canal cleaner stability were made taking into account the stability factor of 1.4 adopted by Rostekhnadzor. The stability calculations include all design changes in the working equipment that could result in an increase in weight. An analysis of graphical dependencies of the bucket capacity on its reach showed that when using a tracked tractor as a base with reach of about 5-6 meters, it makes sense to focus on the bucket capacity of 0.25 ... 0.3 m^3, and for reach up to 3-4 meters, the capacity can be increased to 0.40 ... 0.45 m^3.

3. The first main parameter in the design of working equipment is the depth of the canal to be cleaned. The second main parameter of the bucket is its width, this parameter can be considered initially selected, as it is related to the width along the bottom of most drainage canals. Therefore, the minimum bucket width can be taken equal to 400 mm, in cases where the width along the bottom is 600 mm, a replacement bucket of the same width can be used. For canals with a bottom width of 800 mm, a 400 mm wide double pass bucket can be used. In the process of laboratory tests and field studies, the main advantage of the canal cleaner with the bucket movement along the canal axis was confirmed - the high quality of cleaning operations. Tests have also identified ways to increase leading canal cleaner performance by improving bucket design and matching bucket length and height with its width. Technological studies have revealed the optimal dimensions of buckets (width, length and height).

4. As experimental studies demonstrated, the traction forces at soil digging with canal cleaner buckets were determined, they are converted to working bodies in full size and may well be implemented by the hydraulic drive of the base machine.

References

1. N.N. Dubenok, Yu.G. Yanko, A. F. Petrushin, R. V. Kalinichenko, Modern Problems of Remote Sensing of the Earth from Space, **16(3)**, 96 (2019)
2. I.P. Kruzhilin, N.N. Dubenok, M.A. Ganiev, K.A. Rodin, S.D. Fomin, IOP Conference Series: Earth and Environmental Science, 341(1), 012100 (2019)
3. N.N. Dubenok, A.V. Gemonov, A.V. Lebedev, IOP Conference Series: Earth and Environmental Science, 548(8), 082014 (2020)
4. F. Abdrazakov, A. Rukavishnikov, O. Miheeva, C. Churkina, M. Yarmashevich, IOP Conference Series: Materials Science and Engineering, 883(1), 012209 (2020)
5. F.K. Abdrazakov, A.A. Rukavishnikov, A.V. Povarov, Y.E. Trushin, E3S Web of Conferences, 140, 09009 (2019)

6. N. Martinova, K. Shavazov, N. Telovov, S. Toigambayev, S. Yusupov, IOP Conference Series: Earth and Environmental Science, 614(1), 012145 (2020)
7. A.A. Rukavishnikov, Materials of the VII Intramural International Scientific and Practical Conference Saratov State Agrarian University Named after N. I. Vavilov, 233 (2019)
8. Yu.M. Kosichenko, Mssu Bulletin, **7**, 114 (2016)

Analysis of drainage canal defects and review of canal cleaner designs

Khamzat Abdulmazhidov*

Federal State Budgetary Educational Institution of Higher Education "Russian State Agrarian University-Moscow Timiryazev Agricultural Academy", Timiryazevskaya street, 49, 127550 Moscow, Russian Federation

Abstract. The drainage system elements, namely, the canals, lose their initial design characteristics over time. In general, the canals with unstable profiles are often in the reclamation construction. The problems of unstable profiles of the canals are, primarily, related to the physical and mechanical properties of soils, formation of internal mechanical, as well as chemical and biological processes occurring by the action of natural conditions. Maximum stability is characteristic for canals formed in cohesive organic soils, as well as in various kinds of peat with their thickness considerably exceeding the depth of the canal. With insignificant peat thickness above mineral soils, the stability of the drainage canal slopes will be rather insignificant. Wood peats are the most stable among organic soils. The canals laid in sapropelic soils characterized by the natural flow have rather low stability. That is why, the required stability in wet soils can be achieved by removing the excess moisture. The slopes of the canals made in various clays and loams with their large thickness considerably exceeding the depth of the canal have relatively high stability.

1 Introduction

Stability of the canals laid in the soils of different origin with various mechanical strength Ensuring the stability of the drainage canals and elimination of their deformations to a certain extent require the formation of slight slopes with shallow depth, however, such solution leads to widening the canal. And this is one of the deficiencies related to a reduction of the drained (reclaimed) area. The drainage canals of open streams made in various soil conditions and operated for a long period of time, with complex and various deformations, often become non-functioning. The most common kinds of canal deformations include: peat absorbing; slope destruction; underwater and slope erosion; bottom silting and accumulation of mud; canal silting and accumulation of mud; overgrowth of bottom, slopes and berms; destruction of benches, dump banks, conical depressions and mouths of canals; destruction of facilities.

Loss of function of the drainage canals can occur under the influence of natural and artificial operation conditions. The negative natural operation conditions include: overgrowth of bottom and slopes of drain ditches with grassland vegetation and small shrubs, and small

* Corresponding author: hamzat72@mail.ru

trees in some cases; slope destruction due to periodical freezing and melting of top soil; sedimentation in the flood period; landslide developments and soil settlement; significant silting of closed drainage connected to the canal, etc. [1, 2].

The artificial operation conditions that have negative influence on the drainage canal operation include: non-compliance with technical requirements during the engineering and construction of hydro-reclamation systems; steep slopes; disturbed canal bottom slope. Significant adverse impact on drainage canals is made by non-compliance with technical documentation of the system by business companies in general; crossing over the canals by tractors and other agricultural machinery; clogging of the canal with stones, weeds and garbage (Fig. 1).

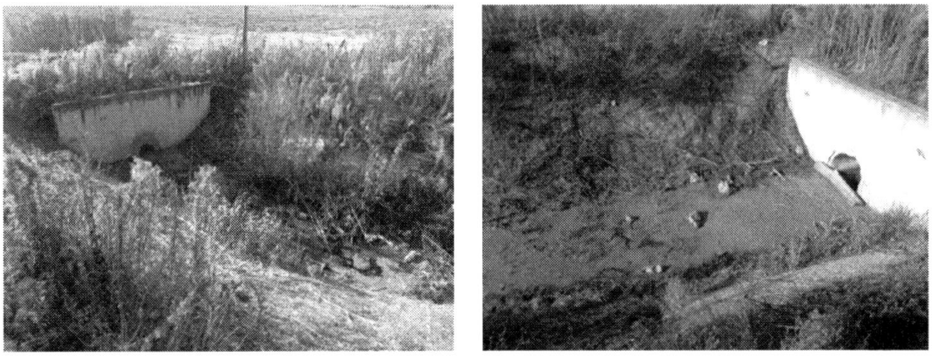

Fig. 1. Condition of the drainage canal and its facilities in long-term operation without current and capital repairs

The initial structural cross-section of the canal disappears, namely, it becomes smaller as a result of an accumulation of vegetation on the canal slopes and bottom, in addition, the bed roughness significantly increases, all these reduce the cross-section throughput and facilitate intense silting. Vegetation in the canal leads to increase in the roughness coefficient from allowable values of 0.003 to high values of about 1.7 and more. The all round overgrowth of the canal bed with hydrophilious vegetation is usually observed in the areas with low water flow rate [3, 4].

The most common kind of canal deformation is soil creep in case of melting and degradation of slopes by surface waters. The soil creep and degradation of slopes are also caused by the presence of the mineral layer or the water-saturated soil layer under the upper coherent soil layer in the slope base. As a result of creeping, the slope becomes more flat, correspondingly, the canal width increases.

Such phenomena as slope erosion and silting even in the small areas can impair the operation of the whole drainage system. These processes usually occur during the spring flood period. In the other periods, the canal erosion and silting may occur but not so catastrophic. This is also due to the canal crossing the loose soils, canal turns of large radius of rounding and drastic slope change.

The intense silting of the canal bottom is usually observed in the areas where the water flow rate drastically decreases. The deposited silt particles rise the water level and impair the system operation.

Draining the peaty and mineral soils leads also to their absorbing. When the peat layer is drained, soil absorbing can reach 40%, and in clay soils, it can be within 10% of the bed depth nearby the channel. The negative impact of soil absorbing consists in decrease in the transverse and longitudinal profiles of the canal. To avoid soil absorbing, the soil excavated from the canal should be placed in a significant distance from the canal bench during canal construction. Moreover, it is necessary to provide the bench width for various service equipment passing.

The drain ditches are the part of shallow network which depth varies within 0.8 and 1.7 m. This parameter makes primarily about 1.2 m.

Restoration and repair of drainage canals require the selection of certain types of machines which controls are designed for operation with the drain ditches having specific geometry. Such parameter of the canal as the width along the bottom is important during development of the operating equipment of the canal cleaners. For most of the drainage canals, this parameter is within 0.2 and 0.6 m. The width along the bottom of 0.2 m is characteristic for canals made by plow-type ditchers. In the territory of the Russian Federation, such equipment has not been virtually used in making the long-length canals.

The percentage of the canal width along the bottom for various zones is given in Table 1.

Table 1. Width along the bottom of open ditchers

Width along the bottom	Russian Federation (Smolensk region)	Belarus
0.2…0.25	none	26.0%
0.4	87.0%	49.0%
0.6	10.0%	19.0%
over 0.6	3.0%	6.0%

From the point of view of ensuring the stability of drainage canal slopes and their mechanical cleaning, such parameter as the ratio of slope is important. For most elements of the drainage network, the ratio of slope is within 1:1 and 1:1.5. The canals with slighter slopes make 1…4%. This is explained by that the steeper slopes are less stable, and the performance of the canals with slight slopes is low. With the average deep of the canal being equal to 1.5 m, the width along the top will be within 5…6 m. The low values of the canal width along the top make it possible to use the cleaning machine operating according to the saddle scheme.

For quality functioning of drainage systems, it is necessary to take into account that it is much easier to prevent a canal deformation than to correct it, that's why the regular maintenance and control over the condition of all elements of the drainage network are required. The most common canal deformation, the overgrowth of the bottom and slopes with vegetation, is a significant deformation caused by natural factors. The process of overgrowth of the slopes develops so quickly that the effect from the performed cleaning works is brought to a minimum. Grassland vegetation on the slopes should be cut down. It is desirable to perform this process before blossoming at the end of May - in the beginning of June taking into account that there is few water in the canals during this period. All the cut vegetation should be taken out of the canal.

The low water flow rate in the canal facilitates the growth of vegetation, and the high flow rate exceeding the critical value can lead to erosion. In the soils with low erosion resistance, it makes sense to maintain the bottom inclination. It is known that the scouring action of the water flow will be stronger if to reduce the section, in view of this, to avoid erosion, all obstructions compressing the water flow should be removed from the canal. To avoid erosion, the areas with weak slopes and bottom should be strengthened by gravel, fagot structures, rock blankets or plank structures. The wet slopes should be strengthened by stone and porous concrete slabs, and the dry slopes can be grassed.

The measures to fight against sedimentation are brought to the activities against bed erosion, catchment of rain and melt waters at water discharge or construction of additional catch drains and creation of uniform non-silting flow rates. Also in places where sedimentation is possible, for example, in the areas of descending inclination of the canal bottom, the setting reservoirs are arranged.

2 Materials and Methods

The materials and methods for eliminating the defects of the drainage canals and the ways to implement them offered in this article are mainly related to the review of design of different kinds of canal cleaners and to the selection of the optimal complex on the basis of their performance indicators. During the 1960s and 1970s, special canal cleaners started being developed. To create special-purpose equipment for canal cleaning, the common construction bucket excavators with various kinds of detachable operating equipment were used. The operating equipment of the bucket excavator, a dragline with side attachment, was mainly used. Such design made it possible to clean the canal bottom with the bucket moving longitudinally along the canal axis. But the dragline could not qualitatively clean the sediments off the canal bottom in the framework of regular repairs. There appeared new canal cleaners with new kinds of the operating equipment that differed by the operation mode (cyclic and continuous), method of attachment (with side attachment, with rear attachment), type of base machine (crawler or pneumatic-tyre tractor), arrangement of base chassis and operating unit (cantilever-type, saddle-type, in-the-canal operation scheme), operating unit design (single bucket, multibucket, cutting-type, scraper, rotary) [5].

The scopes of cleaning works on the reclamation drainage systems is though less than when laying new canals but not remain rather extensive. This requires the creation and use of high-performance machines. The operation of canal cleaners should not create difficulties in transportation, as well as in relocation from one object to the other. In addition to the other properties, the specialized canal cleaner should have high-quality performance. In terms of performance indicators, the continuous-type machines have advantages before the intermittent-type canal cleaners. However, it cannot be always stated in such unambiguous way. The operating equipment aggregated with general-purpose tractors will also have a certain advantage, such decision makes it possible to reduce the cost of use of canal cleaners. If the business company has various kinds of base machine (crawler or pneumatic-tyre tractor), the use of specialized replaceable working equipment obviously reduces the cost of cleaning operations in contrast to the necessary manufacture of base machine [6, 7].

The canal cleaners with cantilever-type attachment of the operating equipment are most widely used, their advantages are quite significant. Such scheme provides for multi-purpose use of machines, allows for bypassing various obstacles located in the canal bed (this is rather difficult operation for saddle-type scheme), performing cleaning operations in any kinds of deformations. The disadvantage of such scheme are the difficulties in ensuring stability of the base machine.

The saddle-type scheme has its advantages and disadvantages: firstly: the stability is ensured irrespective of arrangement of the operating equipment elements as the center of gravity will be always within the support contour; secondly, the machine can transfer along the benches. However, the canal cleaners operating according to the saddle-type scheme are cumbersome with due regard to that the gauge of tracks of wheels should be more than the canal width along the top.

In the cases when driving along the benches is impossible due to tree planting, pasture fencing and other facilities on the canals, the canal cleaners operating within the canal contour with wheel or floating travel device may compete with the basic kinds of machines. The in-the-canal cleaners have not been widely used as they can move only over rather flat slopes and along the wide bottom, as well as with the high soil bearing capacity under the carrying wheels, and this is a rare case for the drainage systems.

Therefore, the cantilever-type scheme of attachment of operating units is the most advantageous that is agreeable to experience.

The multibucket canal cleaners have mainly the traversal digging scheme. The quality and performance characteristics of this kind of continuous-type machines are rather high even at low power costs. These machines are especially efficient for capital repairs when it is required to restore the initial geometrical profile. The positive characteristics of the multibucket canal cleaners can include: separate cleaning both of the bottom and the slopes; operability even in the conditions of wood and stone inclusions; adaptability to work with the wet soil not sticking to the operating equipment elements, as the operating equipment includes the forced cleaning mechanisms; operability in the watered canals; possible cleaning when moving both forwards and backwards. One of the disadvantages of such machines is shouldering the lower parts of the slopes that leads to the impaired geometrical parameters of the canal cross-section.

The multibucket canal cleaners have become most widely used in Belarus where the open-cut drain is a common practice, in the Russian Federation, the use of such machines is limited due to a small number of shallow drainage network and its short length. Of course, there were the attempts to create multibucket canal cleaners for operation in the canals with the depth of to 3 m. However, such structures were rather cumbersome and were not widely implemented.

The creation of canal cleaners with the scraper-type operating unit was limited by manufacture of pilot samples which testing identified the deficiencies that consisted in the limited use (only on peat soils), quick wear of the chain-drive operating unit and soil pelting of the base machine.

The cutting-type canal cleaners have relatively wide range of products and are mounted on the base crawler tractors of drawbar categories 3.0…6.0. Such machines are used for regular repairs when only the bottom deposits are to be cleaned off. The feature of the cutting-type canal cleaners is high rotation of the cutter that cuts the soil layer 10-30 cm thick and spreads it in the distances of 10-15 m from the edge of the canal without dike formation. The most efficient cutting-type canal cleaners work on the wetlands with the water level of 15-20 cm in the canals. Such conditions occur during cleaning operations of the drainage system. However, with higher water levels, the cutting-type canal cleaner performance decreases. Also, it is rather difficult to operate the cutting-type canal cleaners in the mineral dry soils with rocky inclusions, such operation leads to heavy and quick wear of blades with cutting elements of the cutter. The main disadvantage of the cutting-type canal cleaners consists in poor quality of works and impossibility to reach the initial design dimensions of the canal. Hence, it follows that the cutting-type canal cleaners must not be used as the base machines of complexes for regular repairs of canals.

The next kind of machines used for canal cleaning, the bucket-type canal cleaners, are rather common and still manufactured at the plants of the Russian Federation and Belarus. Irrespective of that they are cyclic-type machines and their performance fall short of that of the intermittent-type machines, and are efficiently used by the land-reclamation companies. When there are no special-purpose machines, standard construction bucket excavators are often used. Some land-reclamation companies use the replaceable extended buckets for bucket excavators. The main advantages of such machines include simple design and multipurposeness that provide for performing cleaning operations in various conditions. As of today, it can be stated that the bucket canal cleaners together with the bucket excavators still take the leading position in the use of complexes of machines during repair operations on the reclamation systems. Their efficient use is related to specific volume of the sediments. Statement that the bucket excavators are efficient on the drainage systems does not refer the non-full swing bucket excavators with rear attachment of the operating equipment to the standards pneumatic-tyre tractors (for examples, analogues of EO-2621 excavator). Such machines can be used for small work scopes and in particular cases when the frequent point-to-point operation is not required.

The bucket excavators with replaceable extended reclamation buckets for traversal digging are the most efficient from the group of machines being considered. The following characteristics of the excavators are the most critical during cleaning operations: insignificant operating weight to avoid additional deformation of the canal bed; full-circle operating equipment to ensure full operating cycle with soil unloading in the sufficient distance from the edge of the canal; crawler attachment to ensure minimum specific pressure on the soil and stable operation during soil and sediments digging, moreover, the point-to-point motion trajectory of the machine is always parallel to the canal axis.

The low performance of the bucket cleaners and excavators is not their main disadvantage. The work quality is more important, i.e. significant overdigs during repair, required additional planning, installation of fixtures in particular cases, all these considerably increased the cost of the cleaning operations performed. But all the same, simple design, reliability and multipurposeness are the prerequisites for development of the bucket cleaners.

The main area in development of the bucket canal cleaners is to make the bucket move along the longitudinal axis of the canal in contrast to the traversal digging machines. The dragline with side attachment was such first machine, but it has not been widely used due to its cumbersome design.

Berkenheger firm offered interesting design decisions that consisted in mounting of the operating equipment, the backhoe, on the powered chassis capable of functioning both according to the saddle-type and in-the-canal scheme. These schemes were further developed, in addition to that the full-circle wheel machine resting upon the slope could move along the canal axis, the sediments could be removed both using the backhoe along the canal axis. Such system is quite operable on the shallow network systems, but there appear difficulties in the bucket unloading in significant distances from the edge of the canal when cleaning the deep canals.

Often, it is necessary to preserve the slope strengthening structures, as well as to ensure the turf integrity on the canal slopes during cleaning operations. Such requirements can be complied with only when the canal cleaners with the bucket moving longitudinally along the canal axis are used. The most efficient and promising in this case can be the crawler-tractor based machine, drawbar category 3, with the bucket moving longitudinally along the canal axis on rigid guides. The longitudinal motion of the bucket is provided by the used rigid transverse beam made of beam channels along which the bucket moves by the action of hydraulic cylinders and four-fold high-speed block pulley [8, 9].

3 Results and Discussion

Irrespective of the relatively low performance, the canal cleaner with the bucket on rigid guides ensures high quality of the performed cleaning operations, this is confirmed by the field tests conducted. The strictly straight line motion of the bucket with the regulation of chip thickness and the use of replaceable buckets of various width corresponding to the width along the bottom of the most of drainage canals, as well as with the function of soil and sediments development from under water make the machine demanded and necessary for cleaning and recovery in the framework of regular and capital repairs. The canal cleaner is capable of working on the canals clogged with wood and stone inclusions. The use of this crawler-tractor based machine with the use of extended tracks makes it possible to ensure stable operation of the machine in various conditions.

Considering that the bearing capacity of the slopes can appear to be considerably lower than that of the benches, the strict flotation requirements are established for attachments of in-the-canal machines. In the operating equipment of the canal cleaner, it is recommended to use the trapezoidal cross-section bucket (Fig. 2) duplicating the lower levels of the canal that will clean not only the bottom from sediments and siltage but also bottom-related parts of the slopes to prevent the cut part from creeping.

 https://doi.org/10.1051/e3sconf/202126201002

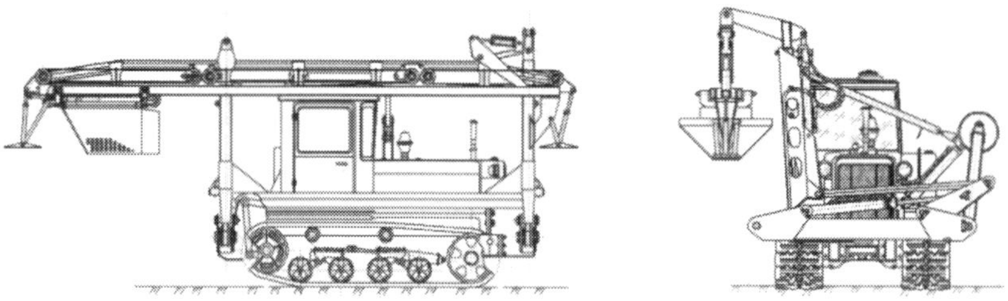

a *b*

Fig. 2. Canal cleaner with the bucket on rigid guides: *a* – side view, with the trapezoidal cross-section bucket - front view

Such advantages allow to consider the canal cleaner with the bucket on rigid guides as a leading machine of particular complex.

It is obvious that the highest effect will be achieved with the comprehensive use of canal cleaners with passive and active operating units.

4 Conclusions

1. The conducted analysis of defects of drainage canals shows that their current condition is ambiguous. The canals require repairs with the use of various kinds of machines.
2. The review of the canal cleaners allows for stating that there are various kinds of machines with the operating units suitable for the corresponding operation conditions.
3. The multibucket canal cleaners have the best performance.
4. The canal cleaners with the bucket-type operating units are the most universal.
5. By the quality of the conducted cleaning operations, the canal cleaners with the bucket moving longitudinally along the canal axis are the most promising. Moreover, the efficiency of such machines increases with the use the buckets of different width and profile according to the parameters of the cleaned canal.

References

1. N.N. Dubenok, Yu.G. Yanko, A.F. Petrushin, R.V. Kalinichenko, Modern Problems of Remote Sensing of the Earth from Space, **16(3)**, 96 (2019)
2. F. Abdrazakov, A. Rukavishnikov, O. Miheeva, C. Churkina, M. Yarmashevich, IOP Conference Series: Materials Science and Engineering, 883(1), 012209 (2020)
3. F.K. Abdrazakov, A.A. Rukavishnikov, A.V. Povarov, Y.E. Trushin, E3S Web of Conferences, 140, 09009 (2019)
4. N. Martinova, K. Shavazov, N. Telovov, S. Toigambayev, S. Yusupov, IOP Conference Series: Earth and Environmental Science, 614(1), 012145 (2020)
5. Y.N. Nikolskii, I.P. Aidarov, C. Landeros-Sanchez, V.V. Pchyolkin, Irrigation and drainage, **68 (5)** (2019)
6. S.-H. Cheng, L. Wang, Y.-B. Wang, Z.-Z. Wang, B.A. Engel, Irrigation and drainage, **68 (5)** (2019)
7. A. Singh, Irrigation and drainage (2020)
8. H. Ojaghlou, T. Sohrabi, H. Javani, F. Abbasi, Irrigation and drainage (2020)
9. A.L. Buber, I.G. Bondarik, A.A. Buber, Irrigation and drainage, **69 (1)** (2020)

Increasing the efficiency of use of wheeled harrow units in regions of risk farming

A. E. Slepenkov*, *E. S.* Polikutina, *S. V.* Shchitov, *E. E.* Kuznetsov, and *Z. F.* Krivutsa

Far Eastern State Agrarian University, Polytechnicheskaya street, 86, Blagoveshchensk

Abstract. Crop yields largely depend on the quality of the soil prepared for sowing. At the same time, one of the most important field operations is harrowing, with the main technological parameter of harrowing is the depth of processing of the fertile soil layer. Failure to do not have the necessary depth of processing leads to a decrease in yields, and the excess - to additional energy costs, and therefore to increase the cost of a unit of the resulting products. As you know, the main parameter influencing the depth of processing is the weight that falls on the working organ of the harrow. This article proposes a device that has obtained a patent for the intellectual property of the Russian Federation, allowing to regulate the weight of the working organ of the harrow, due to its optimal redistribution within the machine-tractor unit (MTU), and the theoretical and experimental results of the experimental harrowing unit are provided. The results suggest that the proposed solution is highly effective, which provides as an opportunity for quality pre-planting treatment, and contributes to the compliance with the necessary agro-technological requirements for the harrowing process.

1 Introduction

The analysis of existing disc harrows has shown that currently produced both domestic and foreign designs are focused, as a rule, on light and medium low-clay soils. Moreover, if we analyze the main indicator that ensures the quality of field work, it is the depth of cultivation of the fertile layer. At the same time, this parameter directly depends on a number of indicators, including the vertical load on the working body (disk), its geometric parameters, as well as the position and spatial orientation on the field surface [1,2].

A number of scientific works are devoted to the issues of determining the basic geometric parameters of disk harrows [3,4]. The works of scientists are devoted to the substantiation of the relationship between the radius of curvature of the disk and its diameter.

In the studies, an analytical relationship was obtained between the diameter of the disc and the depth of cultivation with a harrow, which makes it possible to determine the required diameter of the working body (disc), which allows to optimize the depth of the cultivated soil layer. Perspective calculations of the parameters of spherical disks are given in the works of the authors [5,6].

* Corresponding author: slepen555@mail.ru

One of the problems when working with harrows with spherical disc working bodies is their sticking with soil at high humidity. To eliminate this problem, in study [4] it is proposed to use a spherical disc with slots. The author explains this by the fact that by reducing the area of contact with the soil, the specific pressure of the working body increases, thereby improving the process of its self-cleaning.

The quality of the disc harrows is largely determined by the rotational speed of the disc harrow. The author in [7] proposes to make asymmetric cuts for these purposes, which will ensure sliding cutting. To reduce clogging of discs with soil in study, it is proposed to ensure rotation of adjacent discs with different angular velocities.

At the same time, as shown by the analysis, all the proposed ways to improve the quality of harrowing work are aimed mainly at structures intended for work on sandy loam and chernozem soils with a low degree of moisture, which is not always acceptable for other soils.

From the above, it can be concluded that the vertical load on the working body is a fundamental parameter that ensures the burrowing capacity of the disc harrow. The analysis and review of existing designs of disc harrows showed that the vertical load on one disc ranges from 500 ... 600H to 1300 ... 1400H, depending on the width and weight of the harrow, while they can be successfully operated on light and medium low-stone soils.

The land resources of the Amur Region are mainly represented by clay and heavy loamy soils with a maximum stickiness coefficient at high humidity. It should be noted that all work related to soil preparation for sowing and sowing in the region takes place in the presence of a solid underlying layer in the form of permafrost. At the same time, the climatic features of the region are such that, simultaneously with temperature fluctuations during the spring field work, uneven soil melting occurs, depending on the direction to the horizon and the slope of the terrain.

Thus, it can be concluded that in the conditions of the Amur Region it is necessary to use harrows that are capable of changing the value of the vertical load on the working body as quickly as possible, which will ensure the proper quality of the work performed.

In the studies carried out in [8,9], it is noted that to ensure the optimal track depth, it is necessary to automatically reload the harrow working body due to the redistribution of weight in the links of the machine-tractor unit.

Consequently, the goal of the presented work is to create and test a reliable, highly efficient device with low cost, which allows to automatically adjust the load on the working body in order to improve the quality of soil preparation for sowing with the absolute fulfillment of agrotechnological requirements for machine and tractor units.

2 Material and Methods

As an example of the aforementioned device capable of automatically redistributing the adhesion weight as part of the MTA, it is proposed to use the adhesion weight regulator of the harrow unit for which RF patent No. 196181 was obtained [10].

For automatic adjustment of the tillage depth, it is proposed to install on the tractor a regulator of the coupling weight of the harrow unit, which allows it to maintain the required depth of tillage, depending on the agrotechnological requirements (Figure 1-2).

 https://doi.org/10.1051/e3sconf/202126201003

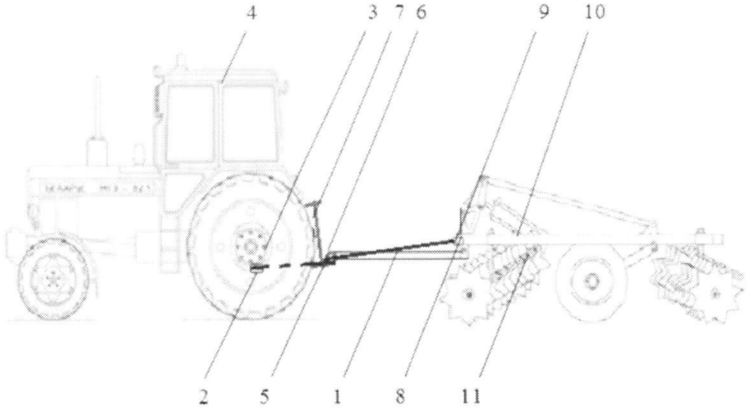

Fig. 1. Schematic diagram of the harrow unit coupling weight regulator:
1-cable power connection; 2-support bracket; 3- unit for installing a cable power connection; 4-tractor body ; 5-towing device; 6-transverse lower link; 7- tractor hitch assemblies; 8-mount brackets; 8 - the cross beam of the harrow; 9 - harrow frame; 11- disc harrow tools

Fig. 2. General view of a tractor with a harrow unit coupling weight regulator

In more detail, the general structure and operation of the regulator of the adhesion weight of the harrow unit is given in study [11].

As an object for research on automatic adjustment of the working depth, a harrow unit was taken as a part of a tractor of class 1.4 of MTZ type and a heavy disc harrow BDT-3, on which a regulator of the adhesion weight of the harrow unit was installed.

Experimental studies were carried out according to the general and private methods recommended by the All-Russian Scientific Agroengineering Center (VIM-ARSAC). At the same time, an on-board measuring complex installed on a tractor, reading devices and devices, specialized programs for mathematical calculation, experiment modeling and well-known methods of statistical and variance-regression analysis were used. During the experimental studies, the on-board measuring complex measured and recorded the following parameters: theoretical and operating speed, distance traveled, experience time, slipping, tractive effort, additional transmitted load. In the process of conducting experimental studies, the depth of processing by the disc working bodies of the harrow was also measured.

The processing of the data obtained during the experiment was carried out by the well-known methods of mathematical statistics using information technologies [12].

3 Results

Conducted on the basis of the Far Eastern State Agrarian University, the city of Blagoveshchensk, the Amur region, the Russian Federation and production organizations of the region, theoretical and experimental studies on the use of the adhesion weight regulator of the harrow unit showed the validity and correctness of the chosen method of increasing the traction-coupling properties of the machine-tractor unit [6-11].

As a result of the theoretical studies, mathematical dependences were obtained, which made it possible to evaluate the effect of the regulator of the coupling weight of the harrow unit on the operating parameters of the harrow. The proposed device allows you to adjust the vertical load on the disk working body, thereby allowing you to adjust the depth of processing of the fertile layer. This issue is directly relevant for regions in which spring work takes place at a time when the depth of soil thawing in areas of the same arable field is different.

Let us consider the case when it becomes necessary to reduce the working depth with the harrow, using the working diagrams in Figures 3 and 4.

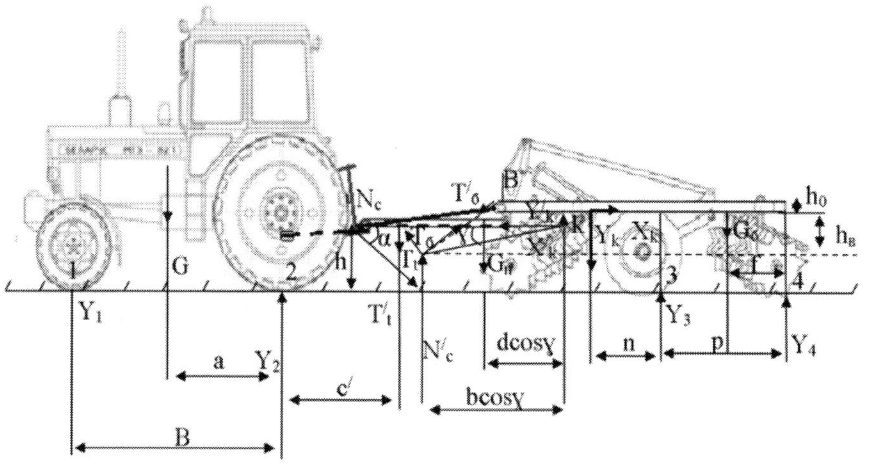

Fig. 3. Scheme for determining the vertical components of surface reactions during the operation of the adhesion weight regulator - lowering the hitch

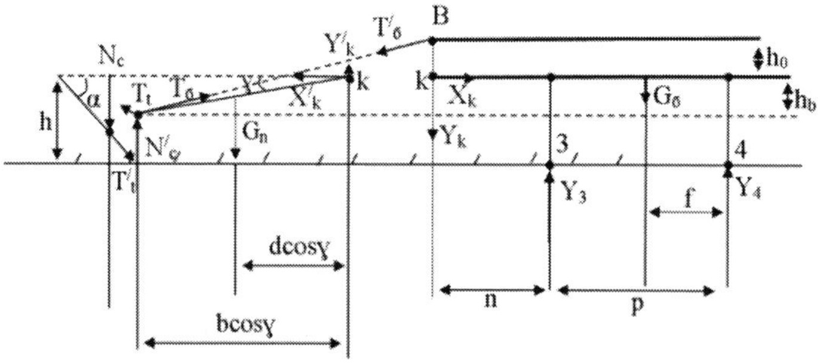

Fig. 4. Working scheme for determining and calculating the vertical components of the surface reactions during the operation of the adhesion weight regulator - lowering the hitch

Where $G, G_n, G_б$ – is the weight of the tractor, hitch, harrow frame, respectively, H; Y_1, Y_2, Y_3, Y_4 – force vertical reactions of the surface at the corresponding points, H; $\overline{N_c}=\overline{N_c^{/}}$ - force reaction of the internal connection at points C, H; $\overline{X_k}=\overline{-X_k^{/}}$ and $\overline{Y_k}=\overline{-Y_k^{/}}$ - force reactions of the hinge at the point K,H; B – longitudinal wheelbase of the tractor, m; a - distance from the center of gravity of the tractor to the rear propulsion support of the tractor, м; c – distance from the rear propulsion support of the tractor to point C, м; b – the length of the hitch, m; d - distance from the center of gravity of the hitch to the hinge K, м; p – is the distance from the points of application of vertical surface reactions to the harrow frame, m; n- is the distance from the hinge K to point 3 of the application of the vertical reaction of the surface to the section and frame of the harrow, м; *f*– Distance from the center of gravity of the harrow to the point of application of the vertical reaction of the surface to the section and frame of the harrow at point 4, м.; h – Distance from the soil surface to the sample, m;

As a result of the theoretical studies, by constructing the equations of the equilibrium of the parts of the composite structure, a mathematical dependence was obtained, which makes it possible to determine the load on the working body of the harrow, depending on the design of the harrow unit and the features of the redistributing device, and to substantiate the parameters for expanding of its technological characteristics:

$$Y_3=\frac{G_{\delta}f}{P}+\frac{G_n(B-d)(n+P)}{BP}+F\left(\frac{\cos d}{\sin(d+\theta)}-\frac{1}{\sin\theta+\cos\theta \operatorname{tg} d}\right)\frac{\cos\theta h_B(n+P)}{B\cos\gamma P}+\frac{F\cos\theta\cos d}{\sin(d+\theta)}\cdot\frac{h}{P}+\frac{F(\cos\theta\frac{h_0}{P}+\sin\theta(\frac{n+p}{P}))}{\sin\theta+\cos\theta \operatorname{tg} d}. \quad (1)$$

Analyzing expressions (1), it should be noted that the proposed device allows you to increase the load on the harrow sections due to the use of a redistributing device.

To identify additional dependencies of the effect of the grip weight regulator of the harrow unit on the efficiency of the harrowing process and experimental confirmation of the theoretical results obtained, experimental studies were carried out in real operating conditions. The experiments were carried out in field production conditions, the soil background of which, according to its characteristics, corresponded to the agrotechnical requirements. In terms of its properties (mechanical composition), the soil was a heavy loam with a high degree of viscosity.

According to the obtained experimental data after their processing, the dependences presented in Figures 5 and 6 were built.

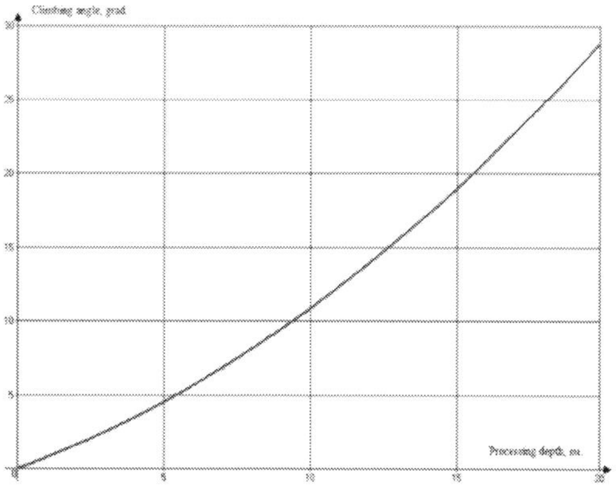

Fig. 5. Dependence of the tillage depth on the lifting angle of the device at the fixed values of the speed of the unit (V = 7 km / h) and the width of the disc harrow (b = 2.5 m) (built on the basis of the data of the third factorial experiment)

Analysis of the data obtained (Figure 5) allows us to conclude that with a decrease in the angle of lowering of the device, the depth of soil cultivation decreases.

During the operation of the harrow unit, it was found that the working depth also depends on the speed and width of the harrow unit. This relationship is visually visible on the research results presented in Figures 6 and 7.

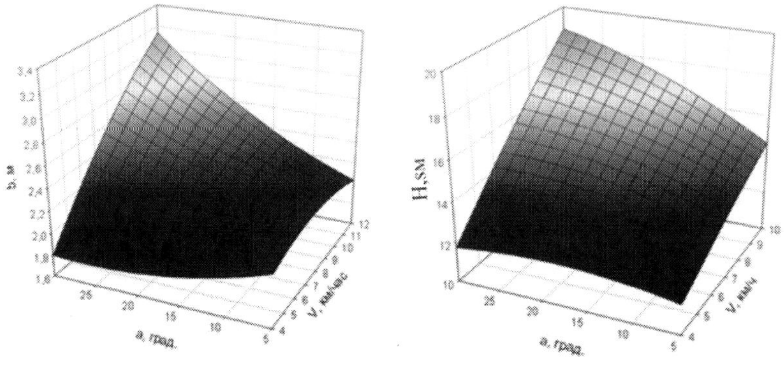

Fig. 6. Response surfaces (3D models) of the working depth and width of the disc harrow depending on the speed of the disc harrow and the lowering angle of the hitch (with b = 2.5 meters and h = 16 cm fixed at the zero level.)

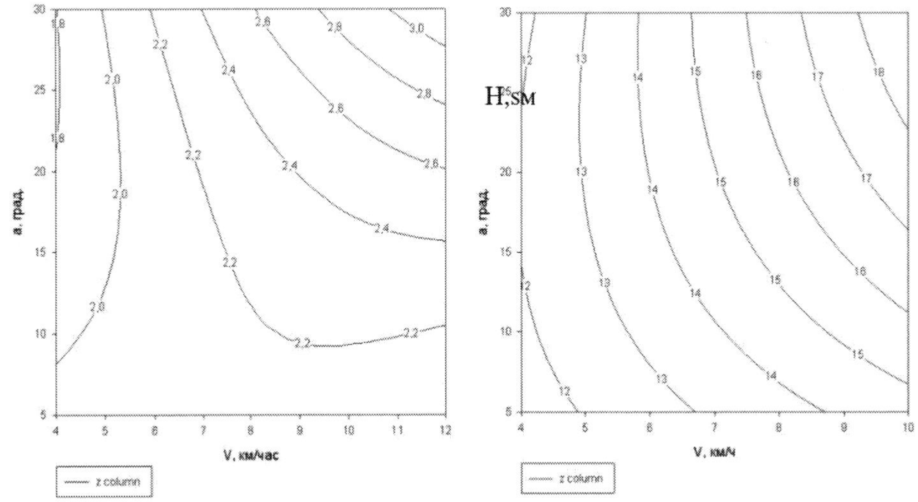

Fig. 7. Cross-section of the response surface (3D -model) of the dependence of the working depth and the width of the disc harrow on the speed of the disc harrow and the lowering angle of the traverse of the hitch (with b = 2.5 meters and h = 16 cm fixed at the zero level.)

4 Discussion

Conducted theoretical and experimental studies have shown that the use of the proposed device allows you to adjust the depth of soil cultivation by regulating the load on the working body.

With a maximum tillage depth of 20 cm, the angle of the traverse of the tractor attachment to the device was 26-30 degrees, the width of the disc harrow was 2.2-2.8 meters, the speed of the unit was 9-10 km / h.

However, the results obtained in terms of comparing related indicators in previous studies [13-15] on this issue allow us to conclude that in the modern period of development of agricultural science, the proposed solution is the most rational, which undoubtedly distinguishes the proposed methodological and technical solutions, forming a new conceptual line in the environment of applied science, in particular, dedicated to the study of interaction in the field of knowledge "man-machine-nature" in agricultural production in the conditions of risky farming areas, dedicated to the study of improving the efficiency of soil preparation for sowing ..

5 Conclusion

Based on the foregoing, it can be concluded that the proposed device, the regulator of the adhesion weight of the harrow unit, is a highly efficient design that implements the original ideas and has a design novelty intended for automatic regulation of the depth of soil cultivation. The research materials are used in the agricultural production technology used by «Soyuz» in the Oktyabrsky District and a number of other peasant farms in the Amur Region.

The introduction of the obtained results into production made it possible to increase the efficiency of the process of preparing the soil for sowing when using the harrow MTA in the technology of cultivating agricultural crops. Methodological and mathematical solutions are massively used in the educational process of the federal state budgetary educational institution of higher education "Far Eastern State Agrarian University".

Reference

1. M. G. Bekker, *Introduction to Terrain-Vehicle Systems*, 1, 520 (1973)
2. M. Behnassi, M. Syomiti Muteng'e, G. Ramachandran, K. N. Shelat, *Vulnerability of Agriculture, Water and Fisheries to Climate Change: toward Sustainable Adaptation Strategies*, 336 (2014)
3. D. Wulfsohn, T. R. Way, Advances in Soil Dynamics, 3, 209 (2019)
4. G.N. Sinyakov, I.M. Panov, *Theory and calculation of tillage machines*, 311 (1977)
5. J. Bulinski, H. Niemczyk, Ann.Warsaw Agr. Univ. Agr., 50, 3 (2007)
6. V.I. Belyaev, M. Fruhauf, T. Mainel, Europa Regional, **1(4)**, 13 (2004)
7. F.M. Kanarev, *Rotary tillage machines and implements*, 142 (1983)
8. E. E. Kuznetsov, *The use of the multi-axis energy devices of class 1.4*,153 (2013)
9. E.E. Kuznetsov, S.V. Shchitov, *Improving the efficiency of the use of mobile energy resources in the technology of cultivation of agricultural crops*, 272 (2017)
10. Shchitov S.V., Kuznetsov E.E., Pat. for utility model No. 196181 Rus. Federation applicant and patent holder Far East state. agr. university. declared 2019, registered 09.25.2019, 5, 10 (2019)
11. A.E. Slepenkov, O.A. Kuznetsova, Actual problems of science and technology, Issue VI, Samara, 21 (2019)
12. Milan Janić, *Advanced Transport Systems: Analysis, Modeling, and Evaluation of Performances*, 408 (2014)

13. Keshav Maharjan, Joshi Lall, Prakash Niraj, *Climate Change, Agriculture and Rural Livelihoods in Developing Countries: Series: Advances in Asian Human-Environmental Research*, 176 (2013)
14. Armando Mendes, L. D. G. Soares da Silva, Azevedo Santos Emiliana, M Jorge, *Efficiency Measures in the Agricultural Sector: With Applications*, 197 (2013)
15. Eric Lichtfouse, Sustainable Agriculture Reviews, Series: Sustainable Agriculture Reviews, **15**, 407 (2015)

E3S Web of Conferences **262**, 01004 (2021) https://doi.org/10.1051/e3sconf/202126201004
ITEEA 2021

Optimization of the width of the transportation corridor of tractor trains for agricultural purpose

A. N. Kushnarev*, *E. E.* Kuznetsov, *S. V.* Shchitov, *V. G.* Evdokimov, and *E. I.* Reshetnik

Far Eastern State Agrarian University, Polytechnicheskaya street, 86, Blagoveshchensk

Abstract. The main works related to the production and transportation of crop production take place in the Amur region in conditions when the surface of the movement has a weak carrying capacity, which imposes additional requirements on transport units to increase their traction-chain properties.So the issue of exporting crop products from the fields is important and relevant both for peasant farms (PF) and for private farms. The use of multi-link tractor trains (MTU) in the region is limited by a significant deterrent, such as the increase in the width of the transport corridor in turns. Why the issue of increasing the efficiency of the ICC should be considered both in the direction of increasing traction-chain properties, and at the same time optimizing the width of the transport corridor when it moves. The proposed article provides theoretical, experimental research and experience with the use of the designer and scientific novelty, designed to enhance traction-clutch properties and reduce the width of the MTU corridor in an arbitrary turn.

1 Introduction

It is possible to increase the efficiency of using wheeled mobile power-tractors as part of tractor-transport units by increasing the volume (mass) of the transported cargo [1]. The main limiting factor in this case is the insufficient traction and coupling properties of tractors with a wheeled chassis during the transportation of agricultural goods from the fields. This issue is especially acute in small farms, where the main energy source is universal row-crop tractors with a wheel arrangement of 4K2. The analysis of the known methods of increasing the efficiency of using these tractors showed that it is possible to solve this problem for improving their traction and coupling properties [2, 3].

The works of a significant number of researchers are devoted to the improvement of the traction-coupling properties of wheeled tractors [4,5]. At the same time, the methods considered and proposed by them to increase the traction and coupling properties may not always be acceptable, since their application does not take into account changes in the natural production and road conditions of the region where they are operated.

In the conditions of the Amur Region, the main transportation operations associated with the transportation of goods from the fields occur during the period when the main

* Corresponding author: leha.kushnarev.79@gmail.com

amount of precipitation falls. Consequently, the transportation of goods occurs along field dirt roads, which have a significant number of sections associated with a change in rectilinear movement, that is, with turns having a weak bearing surface.

Another limiting factor in the use of multi-link tractor-transport trains is the small width of the carriageway and, therefore, when cornering, they cannot withstand the optimal width of the transport corridor, which affects traffic safety.

In connection with the above, in the conditions of the road network of the Amur region and regions with a small width of the carriageway of agricultural roads, in order to increase the efficiency of the use of tractor trains, it is necessary to follow the path of increasing the traction and coupling properties of wheeled energy vehicles with simultaneous optimization of the width of the transport corridor [6]

In earlier studies, it is noted that the optimization of the transport corridor can be carried out by shifting the attachment point of the towing device of the power unit and the trailer, depending on the direction of rotation.

In the studies carried out in the works, it is noted that the displacement of the attachment point of the trailer to the energy device contributes to an increase in the traction-coupling properties of tractors due to the redistribution of the coupling weight inside the machine-tractor unit.

Thus, taking into account the hypothesis formulated, the aim of the work is to conduct a patent search and create a reliable, high-performance device with a low cost, which allows to automatically increase the traction-coupling properties of an energy vehicle at a turn, while optimizing the width of its transport corridor.

2 Material and Methods

An example of the above-mentioned device is the use of a device of a new design, which, due to the displacement of the attachment point of the trailer and the tractor, can achieve a redistribution of the coupling weight, and, consequently, an increase in its traction and coupling properties, while simultaneously optimizing the transport corridor.

For this purpose, it is proposed to install on the tractor a hydro-regulated towing device with an inventive level and novelty [7], which will allow achieving a significant and effective result (Figure 1-2).

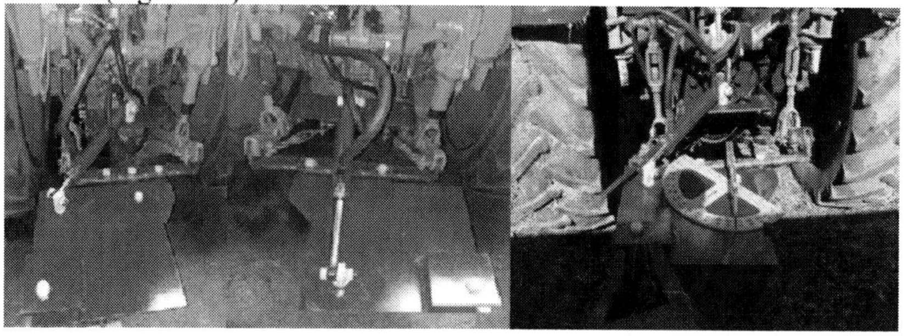

Fig. 1. Experimental hydraulically adjustable towing device

Fig. 2. Connection diagram of a trailer with a tractor through a hydraulic adjustable towing device

As an object for the study of traction and coupling properties and an experiment to optimize the width of the corridor, a wheeled tractor was taken, as it is the most often used in small farms in the agricultural sector of the region, with an experienced hydro-adjustable towing device.

The studies were carried out according to the recommended general and specific methods using specialized programs for mathematical calculation, experiment modeling and methods of regression analysis [8]. At the same time, the following parameters were measured: adhesion weight, turning radius, turning angle of the guide wheels, changing the attachment point of the trailer to the tractor, the length of the hydraulic cylinder rod outlet.

To measure the above parameters, specialized instruments and equipment were used. The processing of experimental data obtained in the course of research was carried out by the well-known methods of mathematical statistics using information technologies.

3 Results

Theoretical and experimental studies on the use of a tractor with a hydraulically controlled towing device, carried out on the basis of the institution of higher education, the Far Eastern State Agrarian University, Amur Region, the city of Blagoveshchensk, the Russian Federation and the peasant farms of the region, have shown the validity and correctness of the chosen method of increasing the traction and coupling properties of a wheeled power plant.

It was found that the movement of the coupling point of the tractor and the trailer made it possible to reload the tractor wheels in a turn, which were in conditions of low adhesion qualities.

In general, the width of the transport corridor formed by the tractor and the trailer will be (Figure 3)

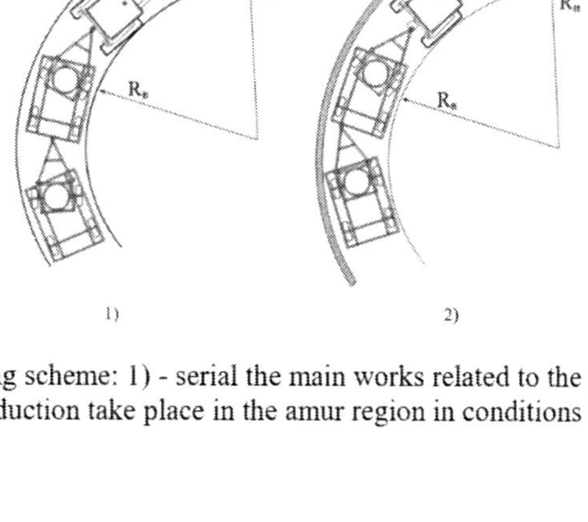

Fig. 3. Tractor train turning scheme: 1) - serial the main works related to the production and transportation of crop production take place in the amur region in conditions when the surface of the

MTU, 2) - experimental the main works related to the production and transportation of crop production take place in the amur region in conditions when the surface of the MTU

Where $R_н$ - outer turning radius of TTU, m; $R_в$ - inner turning radius of MTU, m; H- width of the transport corridor of serial the main works related to the production and transportation of crop production take place in the amur region in conditions when the surface of the MTU, m; $H_э$- the width of the transport corridor of the experimental the main works related to the production and transportation of crop production take place in the Amur region in conditions when the surface of the MTU, m.

A formula for forming the width of the transport corridor is proposed:

$$H = R_н - R_в , \qquad (1)$$

As can be seen from formula (1), the inner turning radius of the trailer has a great influence on the width of the transport corridor.

When the hydraulically adjustable towing device is operating, the attachment point of the trailer to the power unit moves in the longitudinal plane along a circle, the length of the arc of which is determined by the formula:

$$l = \int_{\varphi_1}^{\varphi_2} 2R d\varphi = 2R\varphi\Big|_{\varphi_1}^{\varphi_2} = 2R(\varphi_2 - \varphi_1) \qquad (2)$$

where R - the turning radius of the hydraulically adjustable towing device relative to the point relative to the attachment point to the tractor, м; φ_2 and φ_1 - angles of rotation of the hydraulically adjustable towing device.

During the operation of the hydraulically adjustable towing device, the attachment point of the trailer to the tractor in the horizontal-transverse direction is displaced by the following value B:

$$B = \int_{\varphi_1}^{\varphi_2} 2R d\varphi = 2R\varphi|_{\varphi_1}^{\varphi_2} = 2R(\varphi_2 - \varphi_1)cos\alpha. \qquad (3)$$

As a result of displacement of the attachment point of the trailer to the tractor, the inner turning radius of the trailer increases by the amount:

$$R_{нэ} = R_{вэ} + B \qquad (4)$$

In this case, the transport width of the corridor for the experimental the main works related to the production and transportation of crop production take place in the Amur region in conditions when the surface of the MTU with the installed device will be equal to:

$$H_э = R_н - R_{нэ} = R_н - (R_{вэ} + B) \qquad (5)$$

The analysis of formula (5) showed that when using the proposed device, the width of the transport corridor decreases in comparison with the serial the main works related to the production and transportation of crop production take place in the amur region in conditions when the surface of the MTU.

To confirm the obtained theoretical studies, experimental studies were carried out in real operating conditions.

The experiments were carried out on a dirt road used for transport goods from the fields in the location of the farm. The results of the studies carried out in the form of 3 D-models and response surfaces were built using the specialized program SigmaPlot v.11.0, while two factors were varied - the length of the hydraulic cylinder rod outlet $h_{пр}$ and the angle of deflection of the device $a_{ш}$,

Figure 4.

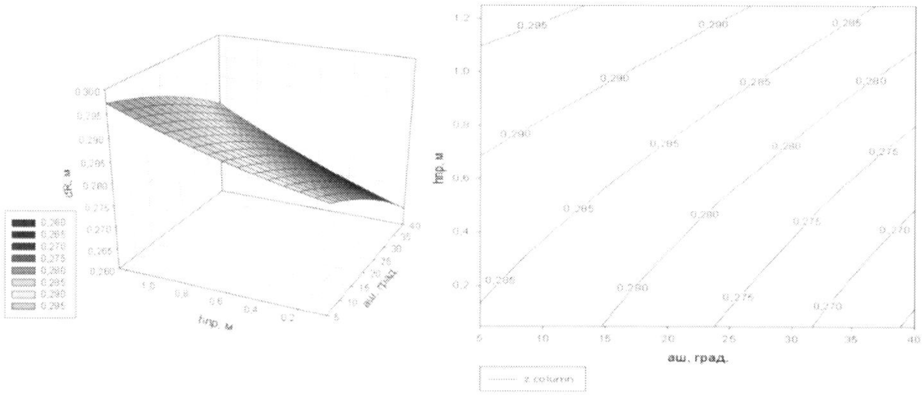

Fig. 4. The surface and cross-section of the response surface dR depending on changes in the parameters of the hydraulically controlled towing device

As a result of the analysis of variance (Figure 4), it was found that the use of the proposed device allows you to adjust the turning radius of the tractor transport train due to:

- ensuring the angle of rotation of the plan-washer installed between the tractor and the first trailer, within up to 40 degrees;

-correction of the angle of rotation of the drawbar of the trailer within the range of up to 30 degrees with the length of the output of the hydraulic cylinder rod up to 0.25 m.

The data obtained allow us to conclude that the proposed device allows you to optimize the width of the transport corridor.

4 Discussion

Conducted theoretical and experimental studies carried out with a class 1.4 wheeled tractor and a 2PTS-4 trailer showed that the use of the proposed device allows to reduce the width of the the main works related to the production and transportation of crop production take place in the amur region in conditions when the surface of the MTU transport corridor in a turn by 7-12% compared to the serial the main works related to the production and transportation of crop production take place in the amur region in conditions when the surface of the MTU.

The introduction of the obtained results of research and development work into the technology of transport and technological support used in "Soyuz" in the Oktyabrsky District of the Amur Region made it possible to save energy costs up to 10 MJ / tkm, which in total amounted to more than 4 million rubles when processing and harvesting crops from a sown area of 400 hectares.

The data obtained as a result of the carried out theoretical and experimental studies

The conducted theoretical, experimental studies and the obtained experimental data are within the confidence interval, which indicates their reliability, and their analysis in terms of comparing the related indicators in the previously conducted works of the authors [9,10,11,12] on this issue allow to conclude that in the modern period of development of agricultural science, the presented solution is the most rational, which undoubtedly highlights the proposed methodological and technical solutions, forming a new conceptual line in the environment of applied science, in particular devoted to the study of interaction in the field of knowledge "man-machine-nature" when carrying out transport operations in agricultural production.

5 Conclusion

Based on the foregoing, it can be concluded that the proposed tug-loading device is a highly efficient design that implements original ideas and has a design novelty, designed to maintain the optimal width of the transport corridor of a transport train. The research materials are widely used in the transport and technological support of the crop production technology used by "Soyuz" in the Oktyabrsky District and a number of other peasant farms in the Amur Region.

The implementation of the results obtained in production has made it possible to increase the efficiency when using class 1.4 wheeled tractors in transport operations and to obtain significant economic benefits.

Based on the results of the work, the team of authors responsibly declares that there is no conflict of interest, information data and results were obtained as a result of the activities of the entire research group. At the same time, the spent material resources are not attracted funds and belong personally to the co-authors.

References

1. N.V. Aldoshin, A.S. Pehutov, Mechanization and electrification of agriculture, **2**, 26 (2012)
2. S.V. Shchitov, P.V. Tikhonchuk, I.V. Bumbar, Z.F. Krivuca, SamuiloVv, A.V. Yakimenko, O.P. Mitrokhina, Journal of Mechanical Engineering, **41(2)**, 31 (2018)
3. E.E. Kuznetsov, S.V. Shchitov, *Increasing the efficiency of the use of mobile energy resources in the technology of cultivation of agricultural crops*, 272 (2017)
4. J. Bulinski, H. Niemczyk, Ann.Warsaw Agr. Univ. Agr, **50**, 3 (2007)
5. M.D. Gholkar, V.M. Salokhe, A. Keen, *The Effect of axle load and tyre inflation pressure on the tractive performance of a two wheel drive tractor on soft clay paddy field*, 1 (2009)
6. A.N. Kushnarev, E.E. Kuznetsov, S.V. Shchitov, Technics and equipment for the village, **6(276)**, 14 (2020)
7. S.V. Shchitov, E.E. Kuznetsov, Pat. for invention No. 2728162 Rus. Federation applicant and patent holder Far East state. agr. University, Application No. 2018114603, app. 13.07.2018, registered 13.07.2018, publ. 07/28/2020 Bul. No. 22.10
8. G.G. Maslov, O.N. Didmanidze, V.V. Tsibulevsky, *Optimization of parameters and operating modes of machines by methods of experiment planning*, 292 (2007)
9. Armando Mendes, L. D. G. Soares da Silva, Azevedo Santos Emiliana, M Jorge, Efficiency Measures in the Agricultural Sector: With Applications, 197 (2013)
10. Eric Lichtfouse, Sustainable Agriculture Reviews, Series: Sustainable Agriculture Reviews, 15, 407 (2015)
11. J. Bulinski, H. Niemczyk, Ann.Warsaw Agr. Univ. Agr., **50**, 3 (2007)
12. N.F. Skuryatin, E.V. Soloviev, S.V. Soloviev, A.V. Bondarev, *Methods for optimizing the design and operational parameters of tractor transport and technological units*, 129 (2020)

Amaranth bran powder is a promising raw material for the confectionery industry

Svetlana Khasanova, and *Zinaida* Skobelskaya

Moscow State University of Food Production, Volokolamskoe sh., 11, 125080 Moscow, Russia

Abstract. The research is devoted to the creation of a new raw material - a product of processing amaranth seeds of the species Amaranthus L. (variety "Krepysh", selection of FNTSO). Developed amaranth bran powder, which is a waste product in the production of oil from seeds. The bran contains 19.6% proteins, including water-soluble ones; 17.9% dietary fiber; 46.9% highly dextrinated starch; 1.01% of macro- and microelements, including deficient selenium. The selenium content in the studied batches of amaranth bran is 0.053 mg% per 100 g of dry weight, which allows us to meet the average daily human need for selenium by 70%. The new technology includes infrared processing, bran cooling and fine grinding. Infrared processing allows targeted action on protein and carbohydrate complexes, increasing the biological value of amaranth bran powder, imparting increased moisture absorption properties to the product, and maintaining high quality indicators. Amaranth bran powder was stored at a temperature of 18 ± 2^{0}C, relative air humidity 75 ± 1^{0}C, in the presence of photocatalytic air purifiers "Tiokraft M-400". A new product - amaranth bran powder is recommended for use in the production of sugar and flour confectionery products

1 Introduction

In modern conditions, the most important task for the enterprises of the confectionery industry is to increase the output of competitive products. For this, in particular, new raw materials. [1]

In the confectionery industry of our country, special attention is paid to enrichment with functional ingredients and biologically active ingredients. The development of such products is carried out in accordance with the concept of healthy nutrition, which is aimed at maintaining and strengthening the health of the population, as well as preventing diseases caused by unbalanced nutrition. [2].

The development of plant-based powders provides an expansion of raw materials for the industry and an increase in the range of finished products. [3,4,5]

The work was carried out at the Federal State Budgetary Educational Institution of Higher Education "Moscow State University of Food Production". The research was carried out in order to develop a technology for obtaining a new raw material for the confectionery industry - amaranth bran powder, which is a secondary product in the production of amaranth oil using short-wave infrared radiation and a shock-pin disintegrator.

2 Methods and types of research on the properties of amaranth bran powder and products based on it

In the work, we used a batch of bran from seeds of amaranth of the species Amaranthus L., variety "Krepysh", selection of the Federal State Budgetary Scientific Institution "Federal Scientific Center for Vegetable Growing. bran is a fat-free seed part of amaranth with a rich chemical composition, so it can also be used in the confectionery industry. [6,7,8].

Such processing of amaranth bran allows a gentle and targeted effect on protein and carbohydrate complexes, to increase its biological value, to impart moisture absorption properties to the product and maintain high quality indicators.

The thermally treated and ground powder of amaranth bran was stored in accordance with the current regulatory and technical requirements for bulk products: temperature 18 ± 2 ±C, relative air humidity Ψ = 75 ± 1°C.

Powder of amaranth bran was stored under conditions with additional air purification by medical photocatalytic installations "Tiokraft-M400".

3 Analysis of the main results of research on the properties of amaranth bran powder and products based on it

The composition of the native amaranth bran is presented in Table 1.

The studied bran contains 19.6% proteins, including soluble ones; 17.9% dietary fiber; 46.9% highly dextrinated starch; 1.01% of macro and microelements, including deficient selenium. The selenium content in the test batch of bran is 0.053 mg per 100 g of dry matter. This is enough to meet the daily requirement of a person by 70%. A new product - amaranth bran powder is recommended for use in the production of sugar and flour confectionery products.

Table 1. Physicochemical indicators of the quality of amaranth bran

Main factors	Quality indicators of native amaranth bran	
	from literary sources	experimental data
Chemical composition:		
Proteins, %	19,2	19,6±1,2
Fats, %	0,6	0,4±0,1
Starch, %	45,5	46,9±1,3
Mono- and disaccharides, %	2,8	0,9±0,1
Alimentary fiber, %	15,8	17,9±1,2
Ash, %	3,0	1,04±0,2
Sodium, mg	8,2	10,1±1,2
Potassium, mg	535	528,0±5,2
Calcium, mg	145	242±2,5
Magnesium, mg	45	35,9±1,2
Phosphorus, mg	76	66±2,5
Iron, mg	35	26,0±1,1
Selenium, mg	0,108	0,050±0,01
Tocopherol equivalent, mg	1,70	1,70±0,1
Vitamin C, mg	8,0	8,0±0,5
Energy value, kcal	233,00	227,00±4,0
Physical indicators:		
Average particle size, mcm	240	300
Moisture content, %	13,10	13,26
Expiration date, months	6,0	6,0

A tendency towards intensive soil pollution is observed near large cities, and since amaranth belongs to plants with cumulative properties, experimental batches of Amaranth L. seed bran were studied for the content of heavy metals by inductively coupled plasma mass spectrometry (MS-IS) as a source of ions with a quadrupole mass analyzer and a discrete-dynode detector. The choice of the method is due to the high accuracy of determining the concentration of elements and individual isotopes in raw materials - at a level from hundredths of nanograms to hundreds of milligrams per liter. The results are shown in Table 2.

The content of cadmium, mercury and lead is within the normalized values; iodine, bismuth and arsenic were not found. The results of the study indicate the food safety of the new raw materials recommended for the confectionery industry and the favorable state of the soil conditions of the growing region.

Table 2. Toxic elements in amaranth bran

	Element	Content in the studied maranth bran, μg / g	Maximum Allowable Concentration in food bran according to TR TS 021/211, mg / g, no more
1	Bi	Not found	Unacceptable
2	As	Not found	0,2
2	I	Not found	0,04
3	Cd	0,02±0,003	0,2
4	Hg	<0,0036	0,03
5	Pb	0,05±0,008	1,0

The native amaranth bran has a solid structure, a specific smell, and is difficult to grind. Its use in the production of confectionery products requires special training.

After heat treatment at the UTZ-4 unit with an infrared power drive, the bran particles become brittle and are effortlessly crushed to the required size on impact-pin disintegrators. Intense IR radiation of the UTZ-4 installation has the ability to influence the vibrations of atoms in biological polymers, cause resonance and destruction of molecules, causing thermal destruction of almost any grain raw material. [9]

The choice of a disintegrator unit for grinding amaranth bran is due to a number of advantages over other mills: high productivity (one unit replaces 4–5 eight-roll mills), compactness, high dispersion (the unit allows achieving a dispersion of 94%).

Heat treatment at UTZ-4 also contributes to the disruption of the integrity of protein-starch matrices in the endosperm of amaranth seeds, an increase in the availability of proteins to digestive enzymes, partial destruction and dextrinization of starch. After heat treatment and grinding of raw materials, moisture absorption properties increase by 1.8 times. [10,11]

The developed powder of amaranth bran is characterized by the following quality indicators: mass fraction of moisture 6.3 ± 0.1%; dispersion not less than 90.0% (average particle size 20-30 microns).

The principle of operation of the photocatalytic cleaner "Tiokraft M-400", in the presence of which samples of amaranth bran powder were stored, is based on the acceleration of chemical reactions with the combined action of a catalyst of titanium dioxide and ultraviolet radiation. The air is forced into the purifier, passes through ultraviolet lamps, which impart quantum energy of light to it. Then the air enters the lattice with a nanocrystalline catalyst - titanium dioxide, where organic molecules are oxidized to carbon dioxide and water (Figure 1).

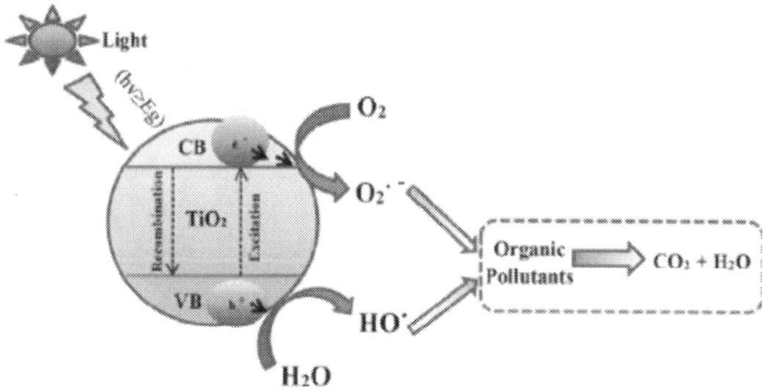

Fig.1 Principle of operation a photocatalytic air purifier works "Tiokraft M-400"

It was revealed that under such storage conditions it is possible to maintain a satisfactory microbiological state of amaranth bran powder for all standardized indicators in accordance with the requirements of the technical regulation of the Customs Union 022/2011 for 30 days (table 3).

Table 2. Microbiological parameters of amaranth bran powder during storage

№	Product	Storage conditions	Indicators		
			NMAFAM, CEF/g, no more	**CB, g/sm ^{3}, no more**	**Mycelial mushrooms, CEF/g, no more**
1	Food bran (Norms according to TR CU 021/211)	t=18±2°C, ϕ=75%	$5x10^{3*}$ $5x10^{4**}$	0,1	50
2	maranth bran powder. 1st day of storage	t=18±2°C, ϕ=75%	$1,1x10^3$	Not found	40
	Amaranth bran powder. 1st day of storage	t=18±2°C, ϕ=75% photocatalytic cleaners	$1,1x10^3$	Not found	40
3	Amaranth bran powder. 30th day of storage	t=18±2°C, ϕ=75%	$3,4x10^3$	Not found	55
	Amaranth bran powder. 30th day of storage	t=18±2°C, ϕ=75% photocatalytic cleaners	$1,3x10^3$	Not found	45

*Notes: * for baby food; ** for adult food.*

To study the moisture absorption properties, samples of the powder and native amaranth bran were stored in three different rooms with low (56%), medium (66%), and high (83%) relative air humidity. The results are shown in Figure 2.

The powder actively absorbs moisture from the ambient air: the higher the relative humidity during storage, the more intense the sorption. On the 10th day of storage, equilibrium humidity was established in all samples. The mass fraction of moisture in the native bran during storage at a relative humidity of φ = 56% practically did not change; at φ = 66% and 83% changed insignificantly. In amaranth bran at φ = 56%, the mass fraction of moisture after 10 days of storage increased by 41%; at φ = 66% - by 75%; at φ = 83% - by 80%.

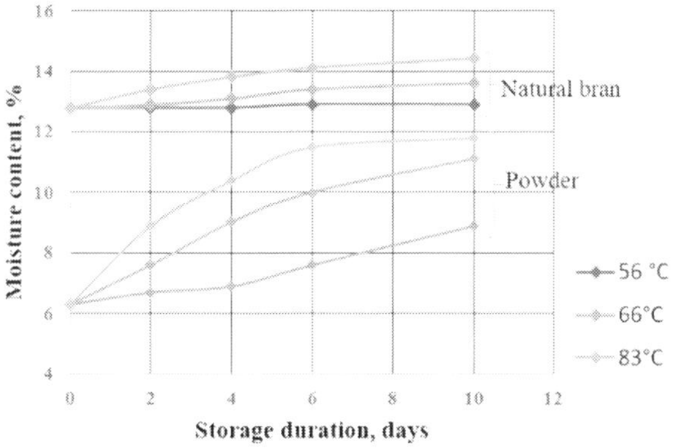

Fig.2. Dynamics of moisture sorption during storage of native bran and amaranth bran powder

Amaranth bran powder was used in the development of the technology of sweets with fondant bodies. The introduction of new raw materials into the recipe to improve the quality of sweets is a promising direction in the development of the confectionery industry. [12,13].

Based on the results of the study, it was revealed that amaranth bran powder allows to increase the fraction of the fraction with crystal sizes of 10-15 microns, improve organoleptic characteristics. The bodies of the powdered chocolates had a uniform color, a uniform shiny surface, and a fine-crystalline structure.

On the 70th day of storage, candies with amaranth bran powder had an organoleptic rating higher than the control ones: sugar - by 16%, fruit - 13%, dairy - 8% (relative%).

4 Conclusions

The basis for the creation of new raw materials for the confectionery industry is the use of a product of processing amaranth seeds - bran, followed by its processing with short-wave infrared radiation and grinding to a powdery state. Amaranth bran powder has a rich chemical composition and high moisture absorption properties. The composition of amaranth bran, its availability and low cost, makes it possible to recommend the use of this raw material in the production of sweets with fondant bodies, fillings for caramel, chocolate bars and flour confectionery

References

1. A. Kochetkova, Foods and Raw Materials, **20** (2020)
2. O. E. Bakumenko, *Fortified food technology for target groups. Scientific foundations and technology* (2013)
3. E. B. Alexeenko, Storage and processing of agricultural raw materials, **116** (2019)
4. I. A. Nikitin, International Journal of Pharmaceutical Research, **971** (2020)
5. Krasina I.B. American Journal of Biochemistry and Biotechnology, **99** (2014)
6. M. S. Gins, Bulletin AAS, **15** (2008)
7. M. Gins, Retirement, **330** (2018)
8. V. V. Kolpakova et al., Storage and processing of agricultural raw materials, **44** (2011)
9. V. V. Kirdyashkin, A.A. Andreeva, D.S. Bakhtina et al, Bakery products, **42** (2020)
10. S. D. Khasanova, Z. G. Skobelskaya, Confectionery production, **16** (2004)

11. S. D. Khasanova, Z. G. Skobelskaya, Confectionery production, **40** (2005)
12. I. B. Krasina, U. F. Berestova et al, News of universities. Food technology, **62** (2006)
13. S. D. Khasanova, Z. G. Skobelskaya, A.U. Lebezova, *Modification of the properties of amaranth bran for the confectionery industry* (2017)

A priori estimates of the mass of the burnt materials in rooms of buildings in the integral mathematical model of the initial stage of the fire

Victoria Pavlidis, *Yury* Fedorov, *Marina Chkalova*, and *Rayslu* Suleymenova

Orenburg State Agrarian University, Institute of Risk Management and Complex Safety, Chelyuskintsev str., 18, 460014 Orenburg, Russia

Abstract. The integral mathematical model of the initial stage of a fire in the premises of buildings focuses on the problem of estimating the mass of burnt materials assuming that the data on the rate of combustion of combustible materials are not complete. A priori estimates of the mass of burnt materials are found, which do not use data on the combustible material combustion rate. To derive these estimates, a special representation of the modulus of the increment (fall) of oxygen density in a room with a fire in the initial stage was preliminarily found. Two approaches to the derivation of a priori estimates of the mass of burnt materials have been implemented - using the upper and lower estimates of the modulus of the increment (fall) of the oxygen density. A comparison of a priori estimates obtained with different approaches is performed. Research can be useful at assessment of ecological situation and consequences of the fire.

1 Introduction

One of the main objectives of technical researches in agriculture is production risk mitigation.

The grain processing and grain-collecting facilities pose a significant threat for the population and environment. Modern technologies for receiving, storing and processing grain cause the accumulation of a significant amount of finely dispersed combustible organic materials at elevators, flour and feed mills. The explosive and fire-hazardous dust-air mixtures may occur in free spaces of technological, transport and aspiration equipment, as well as the production facilities and premises [1, 2]. In addition, the fires and explosions at the grain storage and processing facilities can be caused by the self-heating process, which may develop into spontaneous combustion and malfunction of electrical equipment.

Therefore, fires at the grain processing plants pose a very serious danger, as they are often accompanied by explosions leading to death and injury of people, as well as to significant material damage. It is known that successful evacuation from the premises of public buildings in the event of a fire that is not accompanied by an explosion, as well as other actions to eliminate the danger to people in the event of a fire are possible only at the initial stage of fire development, until conditions incompatible with human life are reached [3, 4, 5].

The development of new, as well as the analysis and modernization of the existing mathematical models of fire (MMF) is a topical direction of modern engineering and applied research since the calculations of the dynamics of dangerous factors of fire (DFF) are the basic elements of fire risk (FR) assessment.

In the article under consideration, estimates of the mass of burnt materials are given for an integral mathematical model of the initial stage of a fire in the premises of buildings.

An integral mathematical model of a fire in the premises of buildings was developed by Yu.A. Koshmarov and his students [6, 7]. The integral mathematical model of the initial stage of a fire, which can be significantly simplified, corresponds to the generally accepted concept of a safe mass evacuation of people in the event of a fire in its initial stage [5, 6, 7, 8]. The theoretically possible assumptions for simplifying the model, as well as a simplified system of independent ordinary differential equations, to which the fire equations are reduced under such assumptions, were taken from [5] (equations (15) - (19) on page 407 of that paper). Earlier in [9, 10] for such a fire model, we found a priori estimates of the increments (falls) of such dangerous fire factors (DFF) as an increase in the concentration of toxic gases and a decrease in the oxygen concentration.

In this article, estimates of the mass of burnt materials are given for an integral mathematical model of the initial stage of a fire in the premises of buildings. To do this, in a simplified fire model, we have selected an equation for the average volumetric $r(t)$ oxygen density, kg/m^3

$$V\frac{dr}{dt} = -hy(t)(k + k_1 r), (1)$$

where V is the volume of the room on fire (m^3); h - coefficient of completeness of combustion; function $y(t)$ – burnout rate (kg/s); k - stoichiometric coefficient for oxygen (determines the amount of oxygen required for the combustion of a mass of combustible material), kg/kg; t - time (s);

$$k_1 = \frac{Q_н^p(1-j)+i_э}{c_p r_0 T_0} (2)$$

[5] (equation (18), p.407). In formula (2) $Q_н^p$ - heat of combustion, J/kg; j - coefficient of heat loss; $i_э$ - enthalpy of gasification products of combustible material, J/kg; c_p – isobaric heat capacity of an ideal gas, J/kg K; r_0, T_0 –the average volumetric density and temperature of the gaseous environment in the room before the start of the fire. The listed parameters at this stage of the fire are considered constant.

This model considers a continuous solution $r(t)$ to equation (1) with the initial condition

$$r(0) = r_0, (3)$$

where r_0 is the oxygen density in the room before the start of the fire. After separating the variables in (1) and integrating by s O$[0,t]$, we get:

$$\int_0^t (k + k_1 r)^{-1} dr = -hV^{-1} \int_0^t \psi(s) ds \; . (4)$$

The solution $r(t)$ of the Cauchy problem (1), (3) can be represented by the exponent of function $M(t)$ on the right-hand side of (4)

$$M(t) = \int_0^t \psi(s) ds \; , (5)$$

expressing the law of change in the mass of combustible materials burnt by the time t .

If the speed of burning $\psi(t)$ is known, then $M(t)$ is found using the formula (5). Function $M(t)$ may have different appearance depending on conditions of burning and spreading mud process of the fire. For example, in [5], [7] the function $M(t)$ is presented in the form $M(t) = At^n$, where $A = \frac{\pi}{3} \psi_{уд} v_л^2$, $n = 3$ if the process of fire propagation over the surface of a solid combustible material (SCM) is circular; $A = \frac{b}{2} \psi_{уд} v_л$, $n = 2$ if the spread of fire over the surface of the SCM is linear; $\psi_{уд}$ - specific mass rate of burnup (kg/m^{2}s), $v_л$ - linear speed of flame propagation (m/s), b – width of the flame front (m), and the flame propagates in one direction.

If the data on the burn-up rate $\psi(t)$ is incomplete, then the function $M(t)$ is unknown. Therefore, in this article, the following problem is considered: for the integral mathematical model of the initial stage of a fire in the premises of buildings, find a priori estimates of the mass of burnt materials $M(t)$ under conditions when the data on the burnout rate is $\psi(t)$ not complete, and all other necessary parameters of the model are known.

2 Methods for a priori estimation of the mass of burnt materials with the incomplete data on the burn-up rate

To estimate $M(t)$, we first find estimates $|\Delta r|$, where $\Delta r = r(t) - r_0$. We calculate the integral on the left-hand side of (4), then by potentiating we find

$$(k + k_1 r(t))(k + k_1 r_0)^{-1} = \exp\left(-k_1 hV^{-1} M(t)\right). \qquad (6)$$

We expand in Maclaurin's series the function $\exp\left(-k_1 hV^{-1} M(t)\right)$ in (6)

$$\exp(-k_1 hV^{-1} M(t)) = \sum_{m=0}^{\infty} (-1)^m \frac{\left(k_1 hV^{-1} M(t)\right)^m}{m!}, \qquad (7)$$

which converges absolutely for all $k_1 hV^{-1} M(t) \geq 0$ [11] and transform the right-hand side of (7)

$$\exp(-k_1 hV^{-1}M(t)) = 1 - k_1 hV^{-1}M(t)\sum_{j=0}^{\infty}(-1)^j \frac{(k_1 hV^{-1}M(t))^j}{(j+1)!}. \quad (8)$$

Let us denote by $S(t)$ the sum of the series converging absolutely for the series $k_1 hV^{-1}M(t) \neq 0$

$$S(t) = \sum_{j=0}^{\infty}(-1)^j \frac{(k_1 hV^{-1}M(t))^j}{(j+1)!}. \quad (9)$$

Then formula (8) takes the form of

$$\exp(-k_1 hV^{-1}M(t)) = 1 - k_1 hV^{-1}M(t)S(t). \quad (10)$$

Let us replace the right-hand side of equation (6) with the formula (10)

$$\frac{k + k_1 r(t)}{k + k_1 r_0} = 1 - k_1 hV^{-1}M(t)S(t). \quad (11)$$

Finally, in (11), we transfer the unit to the left-hand side and simplify the resulting difference. As a result, we find a representation of the increment Δr at the point $t = 0$

$$\Delta r \equiv r(t) - r_0 = -bM(t)S(t), \quad (12)$$

$$b = hV^{-1}(k + k_1 r_0). \quad (13)$$

It is more convenient to consider further $|\Delta r|$, since $\Delta r < 0$:

$$|\Delta r| = bM(t)|S(t)|. \quad (14)$$

Sing-alternating series (9) converges pursuant to the Leibniz criterion for sufficiently small $M(t)$ (in a sufficiently small neighborhood of the point $t = 0$), which are determined by the inequality

$$k_1 hV^{-1}M(t) < 2, \text{ i.e., } M(t) < 2V(k_1 h)^{-1}, \quad (15)$$

(the sequence of absolute values of the terms of series (9) is infinitesimal and decreases monotonically) and its sum $S(t)$ satisfies the estimate

$$0 < S_2 \leq S(t) \leq S_1, \quad (16)$$

where, $S_1 = 1$, $S_2 = 1 - k_1 hM(t)(2V)^{-1}$ are the first and second partial sums of series (9) [11].

Considering (14) and properties $S(t)$ (16), we conclude that

$$|S(t)| = S(t), \ |\Delta r| = bM(t)S(t), \quad (17)$$

From (16) and (17), we derive estimates of the modulus of the increment (fall) of oxygen density $|\Delta\rho|$ at the point $t = 0$ in the integral mathematical model of the initial stage of the fire under the constraint (15)

$$\beta M(t)\left(1 - k_1 h M(t)(2V)^{-1}\right) \le |\Delta\rho| \le \beta M(t). \quad (18)$$

Inequalities (18) allow to give estimates $M(t)$.

Pursuant to GOST 12.1.004-91, the normal oxygen content in the air is 0.226 kg/m3, and the ratio $\rho_0 / \rho_0 = 0{,}23$ [12]. A twofold decrease in this value leads to disturbances in the activity of the human pulmonary and cardiovascular systems, and the threefold decrease – to cessation of respiration and cardiac arrest [5]. It should be noted that with a drop in oxygen density $\rho(t)$, the modulus of increment (drop) of oxygen density $|\Delta\rho|$ increases. Therefore, we will consider the initial stage of the fire, at which the modulus of the increase (decrease) in the oxygen density does not exceed a certain critical value $|\Delta\rho|$ $d_{кр}$

$$|\Delta\rho| \le d_{кр}, \quad (19)$$

$$d_{кр} = |\Delta\rho|_{кр} = |\alpha\rho_0 - \rho_0| = (1-\alpha)\rho_0,\ 0 \le d_{кр} \le \rho_0, \quad (20)$$

α – the a priori selected coefficient that determines the critical value $d_{кр}$, $0 < \alpha < 1$. For instance, if the density drop is allowed by a factor of two, then $\rho(t)$

$$d_{кр} = \rho_0 / 2. \quad (21)$$

3 Analysis of two approaches to assessment of the burnt material mass

Let us utilize and compare two approaches to estimation of $M(t)$, which are based on upper and lower estimates of $|\Delta\rho|$ in inequality (18). In (18), we first choose the right inequality, i.e., an upper $|\Delta\rho|$ bound, and adopt an additional, stronger than (19), constraint

$$|\Delta\rho| \le \beta M(t) \le d_{кр}. \quad (22)$$

Assessment $M(t)$ comes down to the solution of system linear inequalities (15) and (22)

$$\beta M(t) \le d_{кр},\ M(t) < 2V(k_1 h)^{-1}, \quad (23)$$

Substituting instead of β and $d_{кр}$ their expressions from (13) and (20), respectively, we find the solution to the system of inequality (23) and, therefore, the estimate $M(t)$ in the first approach

$$0 \le M(t) \le V(1-\alpha)\left(h(k + k_1\rho_0)\right)^{-1}\rho_0. \quad (24)$$

In the second approach to the estimate $M(t)$, we compose a system of the left inequality (18) (lower $|Dr|$ bound) and constraint (15), assuming $|Dr|$ J $d_{кр}$:

$$bM(t)\left(1- k_1hM(t)(2V)^{-1}\right) \text{J } d_{кр},\; M(t) < 2V(k_1h)^{-1}. \quad (25)$$

Having substituted in (25) instead of b its expression from (13), we will have (25) as

$$M^2(t) - \frac{2V}{k_1h}M(t) + \left(\frac{2V}{k_1h}\right)^2 \frac{k_1}{2}\frac{d_{kp}}{(k+k_1r_0)} \, i \; 0, M(t) < 2V(k_1h)^{-1} \quad (26)$$

Taking for the $d_{кр}$ value (21) and finding the zeros M_1 and M_2 the quadratic function on the left-hand side of the first inequality in (26), we find the solution to the system of inequalities (26) as a set of the following inequalities

$$0 \text{ J } M(t) < M_1,\; M_2 < M(t) < 2V(k_1h)^{-1}. \quad (27)$$

where expressions for zero M_1 and M_2 have the appearance of

$$M_1 = V(k_1h)^{-1}\left(1- \sqrt{k(k+k_1r_0)^{-1}}\right),\; M_2 = V(k_1h)^{-1}\left(1+ \sqrt{k(k+k_1r_0)^{-1}}\right),$$

and their borders, considering the inequality $0 < k(k+k_1r_0)^{-1} < 1$,

$$0 < M_1 < V(k_1h)^{-1},\; V(k_1h)^{-1} < M_2 < 2V(k_1h)^{-1}.$$

Based on the conditions of this model and the meaning of $M(t)$, we choose the first of inequalities (27) as an estimate of $M(t)$ in the second approach

$$0 \text{ J } M(t) < V(k_1h)^{-1}\left(1- \sqrt{k(k+k_1r_0)^{-1}}\right). \quad (28)$$

Estimate (28) is more natural and accurate as compared to (24), since in deriving estimate (24) we used an additional artificial, stronger than (19), constraint (22).

4 Conclusions

1. In the integral mathematical model of the initial stage of a fire in the premises of buildings, the problem of estimating the mass of burnt materials assuming that the data on the rate of combustion of combustible materials are not complete is considered. A priori estimates (24) and (28) are found for the mass of burnt materials, which do not use data on the rate of combustion of combustible materials, and all other necessary parameters of the model are known.
2. To derive a priori estimates (24) and (28) of the mass of burnt materials, a special representation (17) was previously found for the modulus of increment (fall) of oxygen density in a room with a fire in the initial stage – a representation using the sum of an alternating series (9).
3. Two approaches have been implemented to derive a priori estimates of the mass of burnt materials based on estimate (18) from above and below the modulus of increment

(fall) of oxygen density. A comparison of a priori estimates obtained with different approaches is performed.

4. Research can be useful at an assessment of ecological situation and consequences of the fires.

References

1. V.D. Pavlidis, Collection of scientific papers of the VII International Scientific and Practical Conference dedicated to the 60th anniversary of the creation of the first scientific unit in the Republic of Belarus in the field of prevention and elimination of emergencies and fires. In 2 parts, 245 (2016)
2. V.D. Pavlidis, Bulletin of Michurinsk branch of the Russian University of Cooperation, **4**, 36 (2013)
3. M. Haghani, Safety Science, 128, 104745 (2020)
4. A. D'Orazio, L. Grossi, D. Ursetta, G. Carbotti, L. Poggi, International Journal of Safety and Security Engineering, **10(1)**, 1 (2020)
5. S.A. Kolodyazhny, I.I. Pereslavtseva, Bulletin of KGASU, Mathematical modeling, numerical methods and program complexes (in construction), **4 (30)**, 403 (2014)
6. Yu.A. Koshmarov, V.V. Rubtsov, Moscow Institute of Fire Safety of the Ministry of Internal Affairs of Russia, 90 (1998)
7. Yu.A. Koshmarov, Academy of State Fire Service of the Ministry of Internal Affairs of Russia, 118 (2000)
8. I.I. Pereslavtseva, S.A. Yaremenko, Bulletin of civil engineers, **5 (40)**, 122 (2013)
9. Yu.I. Fedorov, V.D. Pavlidis, V.A. Urban, Ye.V. Yakovleva, Safety in technosphere, **2 (83)**, 37 (2020)
10. Yu.I. Fedorov, V.D. Pavlidis, Scientific and technical bulletin of the Volga region, **2**, 76 (2021)
11. L. D. Kudryavtsev, Mathematical Analysis, **1**, 590 (1970)
12. GOST 30494-2011 Residential and public buildings

https://doi.org/10.1051/e3sconf/202126201007

Development and research of metal-cladding additive to lubricants

Victor Balabanov*

Russian State Agrarian University- Moscow Timiryazev Agricultural Academy, 127550 Moscow Timiryazevskaya 49, Russia

Abstract. The article indicates that one of the directions of increasing the wear resistance of rubbing parts of machines and equipment, reducing friction losses and temperatures in the friction zone is the use of special metal-plating additives to lubricants. The Engineering Center of the Timiryazev Academy has developed an experimental metal-cladding additive under the code name "MP-21" based on a mixture of fatty acids (oleic, stearic, linoleic) and plastic metal salts of the same fatty acids. The tribotechnical properties of lubricating compositions with various additives in recommended concentrations were investigated on a modernized Timken-mashine friction machine in a steady state using samples from real rolling bearing assemblies. It has been established that the developed metal-cladding additive "MP-21", along with the additive Renom Engine, provides maximum antiwear properties and allows to reduce the wear of rubbing samples in comparison with other investigated additives from 30% to several.

1 Introduction

The current stage of Russia's economic development is characterized by the growth of domestic and imported automotive and agricultural equipment with high mileage and service life. The operation of such equipment leads to a significant increase in the cost of its maintenance: an increase in the consumption of fuel and lubricants and spare parts, an increase in the volume of repair and adjustment work, etc. [1-2].

One of the directions of increasing the wear resistance of rubbing parts of machines and equipment, reducing friction losses and temperatures in the friction zone is the use of special metal-plating additives to lubricants [1-2].

At present, many such additives are known that implement metal-coating of rubbing surfaces during operation and implement the effect of selective transfer during friction. They contain a mixture of fatty acids, salts or powders of plastic metals, glycerin and a number of other components [1-10].

Numerous studies note that the process of cladding rubbing joints with ductile metals is carried out as a result of recovery and deposition of ductile metals from salts in the presence of surfactants contained in the lubricant composition. When used in a lubricant

* Corresponding author: vbalabanov@rgau-msha.ru

composition, a composite protective film of plastic additives with high antiwear and antifriction properties is formed on the friction surfaces [1-10].

In this regard, the development and research of such additives is a modern and urgent task for increasing the durability of various types of machinery and equipment.

2 Materials and methods

The Engineering Center of the Timiryazev Academy (ECTA) has developed an experimental metal-cladding additive under the code name "MP-21" based on a mixture of fatty acids (oleic, stearic, linoleic) and plastic metal salts of the same fatty acids.

The tribotechnical properties of lubricating compositions with various additives in recommended concentrations were studied at the Russian State Agrarian University - Moscow Timiryazev Agricultural Academy (RSAU-MTAA) on a modernized friction machine of the Timken-mashine brand in a steady state using samples from real rolling bearing assemblies: an inner race and rollers with a diameter of 8 mm, rolling bearing type 7202, made of ball bearing steel.

The design of the modernized Timken-Mashine friction machine is shown in Figure 1, and a general view in Figure 2.

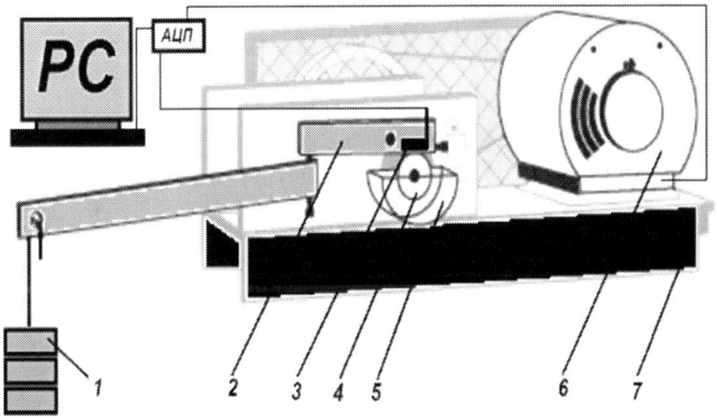

Fig. 1. Scheme of the modernized Timken-Mashine: 1 - cargo; 2 - load mechanism; 3 - upper roller sample with a temperature sensor; 4 - lower annular sample; 5 - a cuvette for lubricant; 6 - electric motor; 7 - bed; ADC - analog-to-digital converter; PC - personal computer with software

The laboratory tribotechnical tests were determined by the structural capabilities of the friction machine: specific pressure in the contact zone p = 5 MPa, sliding speed v = 1.83 m/s, lubricant was supplied to the friction zone by spreading, the time of one test was t = 3600 s.

Fig. 2. General view of the modernized Timken-Machine

The principle of operation of the Timken-Mashine installation is based on the implementation of a rotary motion according to the "ring-roller" scheme with the friction of the upper roller sample 3 against the lower annular sample 4 (Figures 3-4).

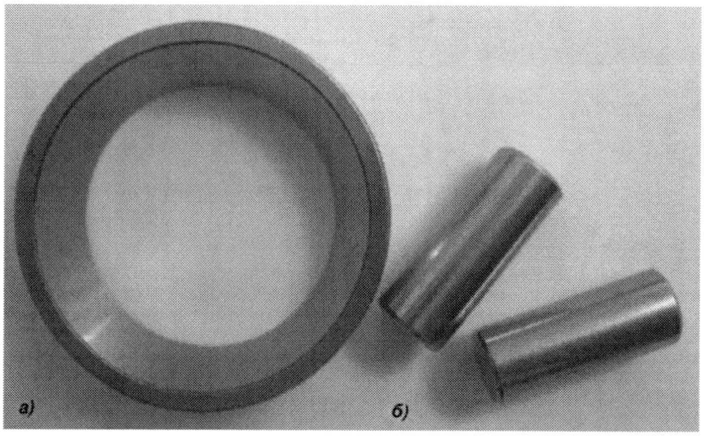

Fig. 3. General view of the samples for tribological testing on a Timken-Mashine: a - lower annular, b - upper roller

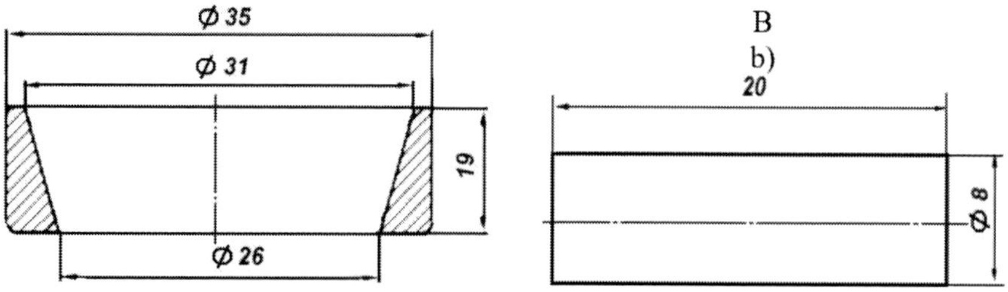

a)

Fig. 4. Sketches of samples for the Timken-Mashine: a - lower annular, b - upper roller

The composition of the studied lubricating compositions is presented in the table 1.

Table 1. The composition of the studied lubricating compositions [1-10]

Name	Manufacturer	Type of additive	Content, %
Resurs	Firm «VMPAauto»	Metal-clad	5,0
Motor Healer	JSC «Fine Metal Powders»	Metal-clad	5,0
MP-21	ECTA	Metal-clad	5,0
Valena	Kuskovo grease plant	Metal-clad	0,6
MKF-18NT	Firm «Tribotechnology»	Metal-clad	5,0
Kremnezem	GOSNITI	Geomodifier	1,0
Bemit	GOSNITI	Geomodifier	1,0
DLCf	Firm «Tribotechnology»	Geomodifier	1,0
Renom Engine	Firm «Autohimproject»	Metal-clad	5,0

The results of tests of the same structural materials on pure Lukoil-Super 10W40 oil without additives were taken as a comparison base.

3 Results and discussion

The results of laboratory tribotechnical wear tests are presented in Figure 5.

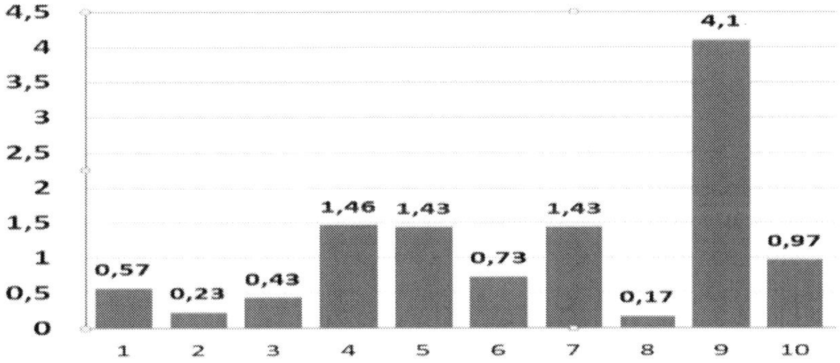

Fig. 5. Wear of samples (I, mg) during tribotechnical tests of various lubricating compositions: 1 – Resurs; 2 – Motor Healer; 3 – «MP-21»; 4 – Lukoil-Super 10W40 (base); 5 – Valena; 6 – Kremnezem; 7 – Bemit; 8 – Renom Engine; 9 – DLCf; 10 – MKF-18NT

The wear recovery of the test samples was determined at the end of each test by the weighing method on an WA-21 analytical balance (Mechaniki zaklady precyzyjnej, Poland) with a weighing accuracy of up to 0.0001 g.

Metal-cladding additive to lubricants contributes to the filling of microdefects (scuffs) on the crankshaft journals with plastic metals with subsequent implementation during operation in rubbing joints of selective transfer during friction (wear-free effect).

4 Conclusions

1. The test result indicates that the use of the additive ensures the formation of a protective composite film of plastic metals contained in the additive on the rubbing surfaces, which reduces the wear rate of steel samples by 8.5 times.

2. It has been established that the developed metal-cladding additive "MP-21", along with the additive Renom Engine, provides maximum antiwear properties and allows to

reduce the wear of rubbing samples in comparison with other investigated additives from 30% to several times.

Acknowledgments

The article was prepared with the support of the Ministry of Education and Science of Russia within the framework of agreement No. 075-15-2021-032 dated March 23, 2021 on the provision of a grant in the form of a subsidy for the creation and development of an engineering center on the basis of an educational institution of higher education and (or) a scientific organization in within the framework of the federal project "Development of infrastructure for research and training" of the national project "Science and Universities".

References

1. D.N. Garkunov, *Wear and tear*, 616 (2001)
2. D.N. Garkunov, V.I. Balabanov, Heavy mechanical engineering, **2**, 18 (2000)
3. V.I. Balabanov, G.K. Potapov, Diagnostics, reliability and repair of machines: Collection of scientific works of MGAU, 92 (1995)
4. V.I. Balabanov, Vestnik mashinostroeniya, **8**, 14 (2001)
5. V.I. Beklemyshev, I.I. Makhonin, A.F. Letov, Herald of mechanical engineering, **10**, 51 (2004)
6. V.I. Balabanov, V.I. Beklemyshev, I.I. Makhonin, V.K. Filippov, Rural machine operator, **11**, 40 (2005)
7. A.F. Sinelnikov, V.I. Balabanov, *Automotive oils, fuels and technical fluids*, 155 (2007)
8. V.I. Balabanov, *Nanotechnology and nanopreparations for automotive machinery*, 77 (2008)
9. V.F. Fedorenko, M.N. Erokhin, *Nanotechnology and nanomaterials in the agricultural sector*, 312 (2011)
10. V.I. Balabanov, V.Yu. Bolgov, *Car additives and inclusions*, 224 (2011)

E3S Web of Conferences 262, 01008 (2021) https://doi.org/10.1051/e3sconf/202126201008
ITEEA 2021

Motor vehicle unmanned driving technology enhancement

Larisa Zhuravleva

Russian State Agrarian University - Moscow Timiryazev Agricultural Academy, Timiryazevskaya str., 49, 127434 Moscow, Russia

Abstract. The relevance of the introduction of unmanned vehicles, the improvement of their control systems is increasing every year due to the ever-increasing volume of cargo transportation. The undoubted advantages of the developed direction are the reduction of the influence of the human factor, management errors, and, consequently, reduction of the number of accidents, as well as reduction of the transportation costs. The research was aimed to develop the algorithmic, programmatic control of an unmanned vehicle when transporting goods along a given route, in particular, when unmanned vehicles move in a convoy behind a pilot vehicle with a given distance along an arbitrary trajectory. The article discusses the control algorithm, technical means of the control system and the technical vision tracking system. The article presents the results of experimental studies of the movement of an unmanned vehicle based on a serial KamAZ 6520 vehicle.

1 Introduction

The importance of the transport industry can hardly be overestimated. The total freight turnover in Russia is increasing and automobile transport remains the leading mode of transport. The development of transport infrastructure, innovative technologies and technical means comprise the basis for further social-and-economic development, the increase in the volume of foreign and domestic trade, as well as the reduction in the cost of transportation.

Many automotive giant manufacturers and scientific institutions are studying the problem of creating unmanned vehicles capable of moving and performing the assigned technological tasks either without direct human participation or with the minimum human participation in control.

Currently, all world companies conduct the research and work on the creation of unmanned vehicles. The most famous automakers engaged in this area are General Motors, Ford, Mercedes Benz, Volkswagen, Audi, BMW, etc. [1-4].

There is an interesting development of the Volvo company. This is the Safe Road Trains for the Environment (SARTRE) system installed on vehicles moving along the road in a convoy one after another [5]. A convoy of several cars follows a car with a driver, repeating its movement, copying the trajectory with a distance of 6 m.

The control system (New Energy and Industrial Technology Development Organization, NEDO) was also developed to move four trucks in a convoy in Japan to save fuel [6].

The most difficult task is to create a so-called intelligent control system enabling a vehicle to perform a task, analyze the external conditions and internal processes in the vehicle systems and make decisions under conditions of uncertainty.

The goal is to develop the algorithmic, programmatic control of an unmanned vehicle for its autonomous operation when driving along a given route under conditions of uncertain road conditions and situations.

A vehicle control system with a set of functions ensuring the unmanned movement along the route using the navigation system and movement in a convoy behind a car with a driver is needed.

2 Materials and methods

Considering the movement of a vehicle, the following basic elements of the digital infrastructure should be provided for:

- a linear network of satellite base stations along the road;
- a control center;
- a satellite receiver;
- the communication and information transfer means.

The neural network technology has proven itself in the best way, it is often used in intelligent control systems. For instance, a neural network model that tracks a sequence of images captured during movement is considered in [7].

The work [8] considers the possibility of building trajectories of movement of robotic vehicles when bypassing statically located and moving obstacles. The advantages of such a system are adaptability and high interference immunity.

The authors [9-14, 18-21] considered control systems for mobile robotic means based on the use of semantic and neural networks for pattern recognition, issues of digitalization of the road infrastructure.

Recently, the most widespread are control systems based on fuzzy logic consisting in calculating the optimum required behavior or actions based on the analysis of the current state of the controlled object.

Pursuant to the tasks set, we will present a fuzzy logic unmanned vehicle control system capable of making the real-time decisions. The control system principle is shown in Fig. 1.

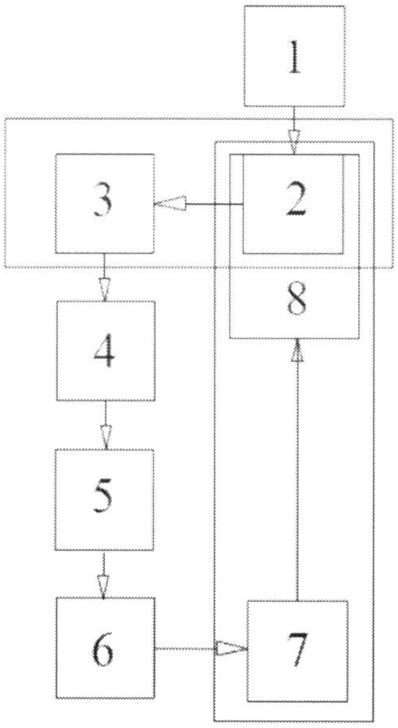

Fig. 1. Control system principle: 1-task, 2-database, 3-control algorithm, 4-actuating mechanism, 5-control object, 6-external environment, 7-measuring and information tools, 8- data generalization and decision-making mechanism

Functioning of intellectual control system requires the following:

- technical vision data about the recognizable objects of the environment, their mutual location in relation to an unmanned vehicle;
- data about the state of the internal system of the vehicle, parameters of operation, movement.

The capabilities of the control system should provide for the following modes:

- Mode of direct vehicle control.
- Mode of remote control of a vehicle using a computer.
- Mode of unmanned vehicle control using technical vision.
- Mode of movement in the convoy behind the pilot vehicle.

The simulation diagram of the control system of an unmanned vehicle when driving in a column is shown in Figure 2.

When driving behind a pilot vehicle, in the process of modeling, the coordinates of the location x_0, y_0 are determined. The block of motion simulation, considering the position of the unmanned vehicle and the coordinates of the pilot vehicle, determines the direction and distance to it d. The task of the fuzzy controller is to determine at each moment of time the linear and angular velocities of the unmanned vehicle.

The input signals for the fuzzy controller are the parameters of the distance d between the unmanned vehicle and the pilot vehicle and the angle α between the direction of movement and the straight line that determines the distance between them.

Since the speed of the pilot vehicle, which sets the mode and trajectory of movement, and relative to which the action of the unmanned vehicle is performed, has a known unknown value, it is necessary for the linear speed controller to proportionally change the speed of movement.

Thus, the controller's task is to determine the required speed of the unmanned vehicle and change it in accordance with the change in the speed of the pilot vehicle, respectively, the task is to timely change or regulate the acceleration.

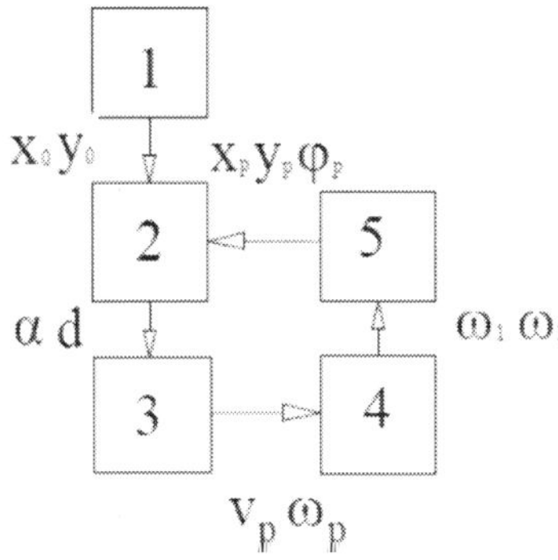

Fig. 2. Simulation scheme of the control system 1-Block of the tracking system of technical vision, analysis of beacon signals, 2-block of motion simulation, 3-fuzzy controller, 4-block of analysis of angular velocities of driving wheels, 5-block of simulation of kinematics and dynamics of a vehicle

If the unmanned vehicle and the pilot vehicle have different speeds and the relative speed has some value, some acceleration value will be determined at the output of the controller.

When turning, the input variable of the controller is the to-the-object direction angle, determined by the built-in vision system. The output variable of the controller is the angular velocity of the unmanned vehicle.

Let us represent the model of motion in the form of a system of equations:

$$\begin{cases} \dot{x}_0 = v_0 \cos\varphi_0 \\ \dot{y}_0 = v_0 \sin\varphi_0 \end{cases} \tag{1}$$

where v_0 is the speed of the pilot vehicle;

φ_0 is the angle of the direction of movement of the pilot vehicle;

x_0, y_0 are the coordinates of the position of the pilot vehicle.

Determination of the distance and direction angle is as follows:

$$\begin{cases} d = \sqrt{(x_0 - x_p)^2 + (y_0 - y_p)^2} \\ \alpha = \mathrm{arctg2}[(x_0 - x_p), (y_0 - y_p)] - \varphi_p \end{cases} \tag{2}$$

where α is the angle of direction to the object;

d is the distance between the unmanned vehicle and the pilot vehicle;

x_0 , y_0 are the coordinates of the position of the pilot vehicle;

x_p , y_p, φ_p are the coordinates of the position of the unmanned vehicle.

Simulation of the movement of an unmanned vehicle behind a pilot vehicle with movement parameters corresponding to the parameters of the pilot vehicle, repeating its trajectory with a certain specified distance, ensuring traffic safety was carried out in the MatLab program using the Simulink application [15].

For practical confirmation of the reliability of theoretical studies, an experimental model of a system with an unmanned vehicle based on Kamaz 6520 was developed and created for movement along a given trajectory of cargo transportation.

The methodology for studying the characteristics of movement along a curved trajectory in accordance with GOST R 52302-2004 "Road vehicles. Handling and stability. Technical requirements Test methods" [16]. And also GOST R 52280-2004 Trucks. General technical requirements (with Amendment No. 1) [17].

The practical study included the movement of vehicles in a convoy one after another and the determination of the characteristics of the movement and trajectory of the unmanned vehicle in the mode of movement behind the pilot vehicle at specified speeds.

The program for controlling the position of servomotors was developed in the MPLab IDE v5.4 program, using the Assembler language.

The set of equipment comprised: a computer control and decision-making unit, a communication system, a set of antennas, a navigation and orientation system.

Following actuators were installed on the unmanned vehicle: linear actuators, electric power steering, and servo drives.

Vision system: SLR 1 light detecting and ranging system; AXIS P1346 video camera and ultrasonic sensors.

The measuring equipment included linear displacement sensors of the actuators and a potentiometer.

3 Results and discussion

A number of experiments consisted of determining the trajectory of movement of an unmanned vehicle in the mode of following each other at a certain distance. The value of the distance between the unmanned vehicle and the object in front of the pilot vehicle was taken to be 10 m. The experiments showed the operability of the control system.

The next study was aimed to determine the trajectory of an unmanned vehicle when driving behind a pilot vehicle along an arbitrary trajectory at a speed of 60 km/h. At the initial moment of time on the remote computer, the system was switched to automatic mode.

After a series of experiments that showed the efficiency and safety of the system, the unmanned vehicle was launched along a given trajectory of the track section.

Graph 3 shows the deviation of the actual trajectory from the one set when making turns. After a number of adjustments, the movement of the unmanned vehicle showed satisfactory results.

After additional experiments with arbitrary movement, the performance of an unmanned vehicle was demonstrated with an acceptable quality of the work process and recommendations were given.

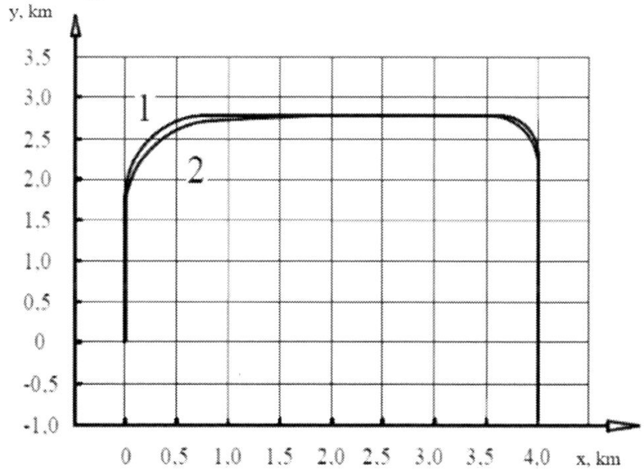

Fig. 3. Movement trajectory: 1-set, 2-actual

4 Conclusions

As a result of the analysis, the prospects of using the fuzzy logic method were substantiated. An algorithm for controlling unmanned vehicles, a mathematical model was developed and computer simulation of movements was carried out.

The equipment was selected and experimental studies were carried out to confirm the theoretical studies.

References

1. Audi: company website, http://www.audi.com
2. BMW: company website, http://www.bmw.com
3. Google: company website, http://www.google.com
4. Volkswagen: company website, http://www.volkswagenag.com
5. Volvo Trucks - self-driving road caravan vehicles, http:/interdalnoboy.com
6. Japan launches a road train with robotic trucks on the road, http:/www. nanonewsnet.ru
7. Fundamentals of artificial neural networks, http://www.intuit.ru
8. V. M. Lokhin, I. M. Makarov, S.V. Manko, M. P. Romanov, Extreme robotics, 73 (1999)
9. V. V. Kruglov, *Fuzzy logic and artificial neural networks* (2001)
10. Zh.L. Loryer, *Systems of artificial intelligence* (1991)
11. V. M. Lokhin, M. P. Romanov, Extreme robotics, 1 (2014)
12. D.A. Pospelova. *Models and methods. Artificial Intelligence* (1990)
13. Neurocybernetics, neuroinformatics, neurocomputers, http://www.bigrouplabs.ru.
14. Fundamentals of artificial neural networks, http://www.intuit.ru.
15. Matlab: Company website, http://matlab.ru
16. GOST R 52302- 2004 "Road vehicles. Handling and stability. Technical requirements Test methods".
17. GOST R 52280-2004 Trucks. General requirements (With Amendment 1)
18. T. Peter, Int. J. of Applied Mathematics and Computer Sciences, **3** (2012)
19. J. Hu, D. Zhao, F. Zhu, Proc. 14th International IEEE Conference on Intelligent Transportation Systems (ITSC) (2011)
20. A.I. Diveev and E.A. Sofronova, Proceedings of European Control Conference (ECC-2013) (2013)
21. G.B. Castro, J.C. Martini, A. Hirakawa, Proc. of 17th International Conference on Intelligent Transportation Systems (2014)

Synergy of a complex of complex technologies of the future agro-industrial complex

Igor Bakin[1,*], *Viktor* Panfilov[2], and *Anatoliy* Popov[1]

[1]Kemerovo State University, Krasnaya Str., 6, 650043 Kemerovo, Russia
[2]Russian State Agrarian University – Moscow Timiryazev Agricultural Academy, Timiryazevskaya Str., 49, 127550 Moscow, Russia

Abstract. Considering a synergistic approach to the development of complex technologies for the production of the agro-industrial complex is considered. An analysis of the prospects for the unification of manufacturing and processing industries is given. The features of the structure of simple, small, large and complex technological systems are described. The influence of the stages of digitalization in the new synergetic picture of the agricultural complex is revealed. To effectively use the existing capacities of agrarian and food technologies and predict their further development, a qualitative change in the state of the entire complex system of the complex is considered, to identify control parameters and the least stable modes. The development of technological processes is considered taking into account the synergistic effects of interaction. Bifurcation transitions on the way of obtaining products and system-forming factors of the organization of complex technological production systems are described. The influence of internal and external relations on the structure of small and large systems of processing and food industries is analyzed.

1 Introduction

A characteristic trend in the development of modern technological civilization is the development and creation of extremely complex systems [1]. This trend extends to the technologies of the agro-industrial complex (AIC). The prospect of such technologies is seen in the unification of manufacturing and processing technologies, that is, the creation of end-to-end branching technologies for food production. The integration of industrial forms of production is explained by the complication of a set of interrelated elements - tools of labor, means and procedures of management, as well as functional relations [1].

Complex systems are characterized by a hierarchical structure and include simple, small and large systems. Simple systems in technology are separate various processes of transformation of the food environment. Small technological systems are autonomous groupings of simple systems in which there is a stochastic interaction of individual processes. Small systems - production areas, implementing local technologies. The integration of small systems into large ones is characterized by the emergence of separate agricultural processing technologies. They have completely new features. Complex systems have an even more

* Corresponding author: bakin@kemsu.ru

complex integrity than their constituent large systems. They are characterized by openness, as well as energy, material and informational connections with the environment. Examples of large systems in agriculture are modern technologies for growing, harvesting and storing crop products and modern technologies for producing livestock products, and in the processing and food industries, these are technological lines. Complex technologies represent the next system-forming level of development of productive forces, in comparison with large systems.

Analysis and forecast of the development of complex technologies in the agro-industrial complex allows the branch of scientific knowledge - synergetic. From the standpoint of synergy, a complex technology in the agro-industrial complex is an open, non-linear, non-equilibrium system of technological processes for the production and processing of agricultural products in order to obtain food. This system has two distinct zones of bifurcation: the transition from agricultural production to storage and the transition from storage to processing. Both zones are characterized by significant changes in the material and energy backbone factors.

Important to adhere by that complexity is an essential characteristic of the evolution of artificial systems, that is, each subsequent level is more complex than the previous one. Moreover, the development of a complex system is an increase in orderliness and is expressed in the qualitative improvement of the system, an increase in the level of its organization. At the same time, it should be borne in mind that the modern methodology of scientific research, based on the systems approach, cybernetics and synergetic, is still in its infancy.

A systematic methodology that provides methods for understanding the complex holistic process of food production is becoming an urgent direction in solving promising scientific and technological problems of the agro-industrial complex. Linear thinking, characteristic of the study of simple, small and even large systems, can be unacceptable in a non-linear complex reality. This kind of thinking can lead to failure in exploring complex systems and making diagnoses in order to improve and develop complex technology. The fact is that, unlike the traditional practice of organizing large systems, the development of complex systems solves a number of the following issues. When creating complex end-to-end technologies, the mechanisms of phenomena and the regularities of individual technological processes of transforming the initial food environment in simple, small and large systems are considered to a lesser extent, and to a greater extent - the regularities of organization and functioning, taking into account the influence of the external environment. When scaling agricultural technologies, one should focus more on positive links that give rise and development of the entire system.

The purpose of the study is to show the decisive role of system-forming factors in the organization of complex technologies of the future agro-industrial complex of Russia at the present stage of development, to describe the bifurcation transitions of complex technological systems of the agro-industrial complex, to identify the control parameters of the organization of complex technological production systems.

2 System-forming factors of complex technological systems

A synergistic approach to the organization of complex agro-industrial complex technologies requires a search and formulation of system-forming factors. The system-forming factor is the factor that forms the system. The search for a system-forming factor is the main problem in organizing a complex technology, since, having found it, it is possible to delimit the system from the environment.

The search for the backbone factor of a complex technology can be carried out in two directions. The first is followed by specialists from certain branches of the agro-industrial complex. Agricultural specialists investigate the specifics and nature of the system-forming factors in each technology that produces products of plant and animal origin, as well as their storage. Specialists of the processing industries, whose technologies disassemble agricultural products into anatomical parts and ensure their storage, also follow the first path. Food production specialists follow the same path, who in their technologies, in accordance with the recipe, collect the anatomical parts of plants and animals into a finished food product and store it until sale [2]. Thus, each agro-industrial complex technology is provided with its "own" system-forming factor.

In the second variant, the backbone factor determines the general functions of the development of all technologies of a complex system. The possible development of the system is determined by the duration of the processes and its part - the future. Improving the efficiency of a complex agro-industrial complex system is the goal of the future. The mutual reinforcement of large technologies connected into a whole is manifested in the form of a synergistic effect of the functioning of complex systems. The interaction of the backbone factors of different technologies also leads to the development of a complex, integral technology of processing industries.

3 Evolution of internal backbone factors

The connection between the elements of a large technological system (technological processes) is provided by internal system-forming factors. The same internal system-forming factor exists in a complex technology between its elements. These internal connections can be of the following types: interaction, structure, functioning, management, and development.

For the conditions of industrial production, the concept of multilateral connections between cyber-physical systems and people was formulated in 2011 in the form of the postulates of Industry 4.0 technology, put forward as a national strategy for the development of science and technology in Germany [3]. The development of sophisticated technologies has implied vertical linkages of individual enterprises along the entire product chain and horizontal links of production processes and machines within enterprises. Interaction and control connections were based on digital integration of production processes through adaptive, evolutionary and self-organizing networks [4]. Co-creation of products based on new technologies and communications gave an advantage to small and medium-sized enterprises with limited resources, which alone lacked the capabilities and skills to do this. The synergistic effect of innovation in manufacturing and operating systems contributes to faster, customized production and optimization of the manufacturing process. Integration benefits include fine-tuning the product, improving product quality, traceability and process control. Vertical links between retailers and agricultural producers enable a faster response to new consumer demands through a range of technological innovations that improve the functionality, properties and quality of the supplied products [4].

Considering the development of the system of manufacturing and processing industries, experts of the international community in 2017 [5] formulated a forecast for sustainable development of the agro-industrial complex until 2030. It was based on the analysis of the main problems and trends in the development of food systems. However, after a short time, the development of technologies led to the manifestation of competition in a complex nonlinear system, which was expressed in the integration of various structures and the emergence of new attractors described in 2018 in the study of the food technology industry FoodTech.

The distribution and implementation of online platforms in the areas of technological solutions in agricultural production (FoodScience, AgTech) and the globalization of the food distribution system as a whole (Delivery & Retail, Food Service, Media) became a new system-forming factor in the development of technologies for the future of the agro-industrial complex. The change in the spatio-temporal dynamics of the parameters of the system was

https://doi.org/10.1051/e3sconf/202126201009

reflected in the global FoodNet market and the Russian sector of agricultural and processing industries in the form of the concept of a digital platform for the food and personalized food market "Foodnet" within the framework of the National Technological Initiative (NTI), under the national project "Science". Scientific and technological innovations in the processing industries, or otherwise "end-to-end" technologies, have changed the structure of traditional industries based on related high-tech products and services.

Connections of interaction, management and development of complex nonlinear systems of the agro-industrial complex with the introduction of digital technologies have determined the evolutionary development of the system. One of the effects of the transition to information and communication technologies was a change in the structure of production in the agro-industrial complex, in which productivity growth is not associated with an increase in agricultural areas and volumes. According to Rosselkhozbank experts, the introduction of digital agricultural technologies in the world will increase the profitability of the agro-industrial complex by more than 500 billion rubles in the next five 5 years. However, for the Russian agricultural sector, the change in the production paradigm is much slower, the number of modern technologies being introduced is up to 5% of enterprises. The forecast scenario of the Ministry of Agriculture of the Russian Federation assumes an increase in the Russian market of digital technologies in the agro-industrial complex from 360 billion rubles fivefold by 2026. The economic effect is expected to be achieved using the Internet of things, the introduction of information and production technologies, developments in the field of biotechnology and new materials.

The influence of management ties can be traced by analyzing the dialectics of the development of Russian manufacturing and processing technologies of the system complex "Agricultural and food technology" [1], where the leading role is assigned to the internal state policy and regulation in the sphere of the agro-industrial complex. A possible scenario for sustainable development of the agro-industrial complex with state support, developed by the Higher School of Economics, involves the expansion of the domestic market by reducing external imports and balanced regulation of the industry. Overcoming import dependence is associated with developments in the field of "precision" agriculture, agro telematics and robotization, information and biotechnology.

Smart products are considered as systems consisting of natural objects, artificial subsystems and controllers [6]. Their integrity is ensured by the internal interaction between the elements of the system and external relations with consumers, as well as a certain autonomy.

The development of "precision" and "smart" agriculture is aimed at creating links between automated enterprise management systems, systems for accounting and certification of products. The implementation of a modern complex intellectual system was the national digital platform of the Ministry of Agriculture of Russia. Its goal is to ensure the comprehensive digitalization of the agro-industrial complex, the use of digital services on one site, and the creation of a common industry data base. The creation of the system is aimed at standardizing and digitalizing technologies, increasing productivity and environmental friendliness in the structure of agricultural processing industries. A separate area of the platform is the task of control (monitoring) and labeling of agricultural products. The use of technologies of unmanned aerial and near-earth space systems in agriculture provides accurate mapping and accounting of farmland, monitoring of the condition and control of crops. Analysis of information arrays using Big Data and artificial intelligence (AI) is used in the segment of "smart" farming to predict crop yields based on data on climate change, soil, plant appearance and other parameters.

Management connections are highlighted at the stages of implementation of small automation and robotization systems for "smart" agriculture. Global manufacturers of agricultural equipment (CaseIH and CNH Industrial (USA), in cooperation with the company-developer of navigation systems Trimble Inc. (TRMB); Japanese Yanmar Holdings Co., Ltd .; John Deere; Cognitive Technologies (Russia), etc.) developed unmanned tractor models of various functions for the mass market. The emergence of new technology is aimed at increasing the intensity of production, simplifying the functions of farm management and, in general, replacing industrial agricultural technologies.

The internal connections of the structure of the system determine the use of digital twins, when based on the flows of digital information about the change or use of the product, a response and control strategy is developed [7].

The destruction of linear connections from producer to consumer through retail chains or food enterprises occurred with the emergence of digital online platforms (Wolt, Deliveroo, etc.). New technologies have led to the connection of production, preparation and delivery of food products directly to the consumer, taking into account his preferences and diet. The forecasts for the growth of the Russian food technology market (Foodtech) are estimated at 20-30% per year. Links in the fields of medicine, food technology, nutrition, dietetics, marketing and logistics have propelled the emergence of the personalized nutrition industry. An interdisciplinary field of science - nutrigenomics, which studies the relationships between interactions between human genes, environmental factors, such as nutrition, health status, has given rise to new technologies in the field of specialized products [8].

The qualitative restructuring of processes in the agro-industrial complex system is conditioned by the system-forming factor that arose during the massive introduction and use of nanotechnology. This governing mechanism provides new functions and connections in food and feed processing and production, packaging, transportation, analysis and control methods, and other areas.

New technologies for treating plants using pesticides and herbicides have emerged with the advent of nano-encapsulation and controlled release methods on an ultra-small scale, usually from 1 to 100 nm [9]. Nanoscale delivery systems are applied to nutrient growth regulators directly in plant tissue. The introduction of agricultural nanotechnology solves the problem of increasing yields and improving varieties, ensuring their sustainability.

Technologies for nanoencapsulation of valuable compounds (vitamins, minerals, antimicrobial drugs, drugs, dyes, antioxidants, probiotics, etc.) have been proposed to be used to provide the body with them for long periods of time [10]. Systems for the delivery and release of particles have been developed under certain environmental factors.

The synergistic effect of nanotechnology processes with biotechnology and information technology has manifested itself in the development of nanobiosensors for the detection of pathogens and contaminants in food systems. Nano-biosensors are used in agriculture to determine a wide range of materials, including fertilizers, herbicides, pesticides, insecticides, pathogens, moisture, soil pH, etc. [11].

The introduction of nanoparticles into packaging materials ensures the preservation, extension of the shelf life and the quality of the product. The bonding of particles is carried out at the level of the packaging material, its part or inside the container. Active protection of food products from external conditions is provided by the absorption of various substances (oxygen, moisture, etc.) or their release (antioxidants, carbon dioxide, antimicrobial agents, etc.) [12]. Communication of information exchange with the external environment of intelligent packaging systems is based on the functions of nanosensors embedded in packaging materials used to receive and transmit data on the quality and safety of packaged food.

A new qualitative change in biological characteristics with a change in the bonds and systems of nanoparticles underlies the technology of genetically modified production [10]. According to research by the Russian Agricultural Bank, the market for products obtained using genetic modification will amount to $ 9.7 billion by 2025. Technological solutions of genetic engineering have given perspectives for adaptive areas of crop breeding, genetically engineered modification of farm animals, and synthetic biology technologies.

Bioproduct Engineering encompasses the related fields of biotechnology, modeling and food systems. Studying the relationships between the chemical composition and properties of raw materials, processing technologies allows you to create products with specified properties and purposes. The development of new types of organic products is based on connections from the molecular level to the macro-level of manufacturing enterprises and the mega-level of distribution networks [13].

4 Evolution of external backbone factors

In addition to internal connections, a complex technology has external connections that contribute to the emergence, formation and development of such technology. But external connections can also be systemic. These links are aimed at destabilizing and losing the stability of the system. In this case, the backbone connections are unable to maintain the adaptability of a complex technology. The destruction of the system occurs when external factors prevail. The manifestation of a resonance of external disturbances and internal connections is possible, as a result of which negative consequences are manifested in processing technologies. At the same time, even a small fluctuation in the functioning of technology can serve as an impetus for its development in a completely new, innovative direction.

The sequence of technological advances in the Fourth Industrial Revolution led to the development of new types of organizations and structures. The links between the real (physical) and digital world are shrinking and blurring [14]. Digital technologies are becoming the decisive system-forming link of globalization. The previously listed systemic factors (climate change; growth in the world population and megalopolises; decrease in biological, water and raw materials; impoverishment of the population, etc.) are being changed by digital technologies, which are becoming a global instrument of economic impact.

New technological solutions arise when a qualitative transformation of the system parameters, or otherwise, when a bifurcation event occurs. The future development of technologies is a probabilistic parameter of the ambiguity of the development of the system [1].

Considering the qualitative restructuring of the food market in the areas of Food Service, Delivery & Retail and Media, one can trace the evolutionary impact of individual fluctuations on the transformation of a complex system. An example of the adaptation of the system of services and public catering products is the development of the previously popular organizational structures "home kitchen", "kitchen factories" into the format of the "dark kitchen" restaurant business. Well-known technological approaches to the processing of raw materials into semi-finished products have been developed into high-margin production based on online services and interaction with food delivery aggregators. Products are developed only for delivery to the consumer, from a limited range of raw materials, but according to a varied menu that meets the needs of the buyer. Cost optimization is achieved with less floor space and less overhead. The rapid growth of this public catering sector occurred during a pandemic and imposed restrictions on the catering industry. New food production and delivery format continues to scale and improve efficiency.

Global factors such as climate change and world population growth mean that traditional farming methods may not be able to meet the expected growth in food demand. In this regard, alternative agro-industrial development and production strategies arise. New technologies of cell agriculture are aimed at the production of renewable agricultural products in bioreactors using cells of microorganisms, animals or plants [13].

For the system of the national agro-industrial complex, destabilizing effects are the consequences of the global financial and economic crisis and the increasing world instability [14]. Technologies for the extensive development of domestic agro-industrial production cannot be competitive on the world food markets, given the dumping conditions of global corporations. To maintain the stability of the system, a balanced regulation of domestic markets for agricultural raw materials and food is required, in accordance with the mechanisms for ensuring food security of the Russian Federation.

External factors of direct and indirect government intervention in the activities of enterprises are expressed in the environmental regulation of production. The imposed restrictions, on the one hand, reduce competitiveness and market advantages [15], on the other hand, enterprises introduce environmental innovations.

The transition to "green" technologies includes the creation of new environmentally friendly products, reducing energy costs, reusing products and raw materials, reducing environmental pollution throughout the entire production cycle through technological improvements [16]. Consumers prefer sustainable products based on a newly formed mindset and a desire for technological advancement.

The introduction of innovative industrial technologies in the agriculture sector [17] is aimed at: automating data collection and monitoring along the entire value chain; increasing the flexibility and intensity of production facilities; reducing the number of errors and costs; improving product quality and the ability to meet customer needs.

5 Discussion

The evolution of agro-industrial complex technologies is an example of development from simple to complex, from lower forms of organization to higher, that is, developing, technologies of the agro-industrial complex acquire an increasingly complex organization. In the new synergistic picture of the agro-industrial complex, the priority will be:

- behind the mechanism for creating interdependent elements that make up a developing integral system (complex technology);
- behind a coordinated innovative flow in space and time of several processes, which manifests itself during their addition;
- behind the increasing complexity of technology, which is an open, nonlinear, no equilibrium system.

Water is a system-forming substance of a large system for the production of biological raw materials and its processing to products, the amount of which, the state and forms of interaction with other substances and components of biological objects determine the development of living organisms. The energy factor is heat. A rational amount of heat energy ensures the rational development of organisms of biological origin, their storage before processing, as well as the formation of the quality of finished products. The information factor, the most complex one, is presented as a subjective perception by a person of the state of the first two factors and his management of these factors. The information factor controls the previous two and its subjectivity has long been an obstacle to the development of complex systems. Modern technology development requires a person to be completely withdrawn from the decision-making system for the development and functioning of less complex systems for the production of agricultural products, their storage and processing into food. At the same time, ensuring the stability of the exchange of matter and energy requires maximum isolation of the agricultural production system from the influence of natural factors (weather conditions, etc.) Thus, the unity of three closed systems can be organized: production of agricultural products, storage and processing.

6 Conclusions

1. Modern technologies of the agro-industrial complex are on the verge of accelerated development due to the introduction of digital technologies. There has been a transformation of traditional raw materials processing technologies with the advent of digital products and services. The internal communications of the system, under the influence of digitalization, have provided new functions of cooperation and competition.

2. Digital transformation, on the one hand, destroys existing processes and connections, on the other hand, generates innovation, the creation of new values and business ecosystems. New technologies are more focused on external communications and networking. The digital transformation of complex agro-industrial complex systems is critical, giving new meaning to the value chain of agricultural products.

References

1. V. A. Panfilov, Food Processing: Techniques and Technology, **50(4),** 642 (2020)
2. A. L. Maytakov, E. G. Vinogray, L. N. Beryazeva, Advances in Social Science, Education and Humanities research, **273,** 238 (2019)
3. M. Bernhard, H. Otthein, *Industry 4.0 and Urban Development – The Case of India. acatech Materialien* (2015)
4. M. J. Oltra-Mestre, V. Hargaden, P. Coughlan, Creat Innov Manag, 1 (2020)
5. The future of food and agriculture: Trends and challenges, http://fao.org
6. T. Dickopf, A. Hristo, P. Müller, J.Göbel, S.Forte, Procedia CIRP, **84,** 538 (2019)
7. D. M. Maio, G.D. Kapos, N. Klusmann, In: 4th IEEE International Symposium on Systems Engineering, ISSE (2018)
8. L. R. Ferguson, M. P. Barnett, Nutrition & Dietetics, **69,** 198 (2012)
9. A. Gogos, K. Knauer, T.D. Bucheli, Food Chem., **60(39),** 9781 (2012)
10. L. Rashidi, K. Khosravi-Darani, Food Sci. Nutr., **51(8),** 723 (2011)
11. C. Bartolucci, In Nanotechnology in Agriculture and Food Science (2017)
12. G. Nathalie, P. Stéphane, M.L. Jose, E. Yolanda, G. Carole, In Nanotechnology in Agriculture and Food Science (2017)
13. G. Gubser, S. Vollenweider, D. Eibl, R. Eibl, Eng Life Sci., 1 (2020)
14. J.-P. Dal Pont, In Process Industries, 2 (2020)
15. D. Lee, International Journal of Environmental Research and Public Health, **17,** 2274 (2020)
16. J. Álvarez Jaramillo, J. W.Zartha Sossa, G. L. Orozco Mendoza, Business Strategy and the Environment, **28,** 512 (2019)
17. G. Büchi, M. Cugno, R. Castagnoli, Technological Forecasting and Social Change, 150 (2020)

Development of self-pollinated maize lines based on the teosinte collection of the N.I. Vavilov institute of plant industry (VIR)

Safar Appaev[1,*], *Alan* Kagermazov[1], *Azamat* Khachidogov[1], *Murat* Bizhoev[1], and *Eduard* Khatefov[2]

[1]Institute of Agriculture - Branch of the Kabardino-Balkarian Scientific Center of the Russian Academy of Sciences, 224, Kirov Street, 360004 Nalchik, Kabardino-Balkarian Republic
[2]Federal Research Centre the N.I. Vavilov All-Russian Institute of Plant Genetic Resources, 42, 44, B. Morskaya Street, 190000 St. Petersburg, Russia

Abstract. The decline in the genetic polymorphism in maize is a cause of concern for many breeders. VIR maize collection represents the world's maize variety over the past 100 years. The results of the study of maize quantitative properties based on 169 samples from the VIR collection created with the teosinte are shown. KB 595 hybrid, which belongs to the late-season group and is prone to multi-cob was used as a standard. Sowing and evaluation of phenotypic traits were carried out for 2 years in the foothills of Kabardino-Balkaria, in conditions of sufficient soil moisture, in the optimal time. All measurements were carried out on 10 plants with average values of 2 years and an LSD0.05. The results of the research allocated a core collection of 21 samples that promising for hybrid maize breeding and are characterized by significant height and leafiness of the stem, as well as a tendency to form 2 or more cobs on one stem. It has been suggested that the involvement of the selected core collection of 21 samples in thybrid maize breeding programs will significantly expand its genetic polymorphism and increase grain yield in hybrid combinations.

1 Introduction

Maize is one of the most common crops in world agriculture. Among the main grains, maize is the second largest crop, second only to wheat [1].

The state of and prospects for the development of modern crop production are evaluated by the degree to which the objective of providing food to the world's population has been achieved. Agricultural production will have to produce as much food as it has produced during the entire period of human existence [2]. Taking into account the evolution of maize from teosinte for more than 6,000 years, it should be noted that maize has achieved great genetic diversity through its genome selection and has become one of the food sources for many peoples on our planet [3.4].

*Corresponding author: kbniish2007@yandex.ru

The challenge of increasing food production directly depends on an increase in gross food grain and feed grains. Maize as the most productive grain fodder crop has an important role in the solution of the grain problem. The level and quality of maize grains largely depend on the implementation of new high-yield varieties and hybrids into production, as well as the use of research-based technology for their cultivation [5]. In the Russian Federation, technologies for the deep processing of grains are becoming increasingly common, with more end products, including starches with different thermodynamic properties [6.7].

All this indicates that the expansion of maize cultivation is an obvious need, that maize should be given special attention, as a crop that produces high yields and enables the full provision of livestock with succulent and concentrated feed, and industry with raw material for processing [8].

The development of high-yield hybrids is possible only if there are inbred lines that meet the requirements of the modern commercial maize grain and seed production market. Therefore, it is important to develop and study the source material for a number of economic attributes, such as: grain and green mass yield, speed of moisture recovery during grain maturation, resistance to the lodging and overmature of the root in the field, quantity characteristics of plant and cob, resistance to diseases and pests [9].

The trend in the reduction of genetic polymorphism of hybrid maize is currently a matter of concern [10, 11, 12]. It is known that the vast majority of inbred maize lines widespread among breeders are from only three American varieties [13, 14, 15]. The narrowing of the genetic base and the use of closely related genotypes contributes to the rapid spread of diseases to which varieties, lines and hybrids are susceptible, reducing the growth rate of crop yields. In this regard, there is an urgent need to find and involve in breeding programs new source material that concentrates on favorable characteristics. With the widespread use of interline hybrids, teosinte is a repository of germplasm and can serve as a starting material for the development of self-pollinated lines. One of the best opportunities for breeders to expand the genetic polymorphism of the initial material is to involve in the breeding process the sources and donors of valuable maize characteristics, concentrated in the VIR maize collection, which represents the world's maize diversity over the last 100 years [11].

Selecting the source material by origin and maturity group, reaction to the CMS, grain type, its biochemical composition and physiological characteristics during its maturation, structure of quantitative characteristics of plant and cob, resistance to various stress factors of the environment, the breeder can quickly develop a valuable starting material for the hybrid maize breeding. The use of haploinduction techniques can accelerate the production of inbred dihaploid lines within 2 years. For example, plant height is one of the most important morphological characteristics correlated with other indicators - cob initiation height, leafiness, lodging and productivity. Furthermore, plant height and cob attachment height are important in industrial maize technology, as well as in the selection of parental pairs at hybridization sites. The hybrid plants height often significantly depends on the parental plants height [16]. The number of leaves on a plant can be used to indicate the early or late maturity of hybrids, and to specify the ways in which the lines and the hybrids obtained by crossing them can be used.

The purpose of the research is to study the economically valuable morphological characteristics of the maize lines obtained using the Chalco teosinte from the VIR collection as the starting material for hybrid maize breeding.

2 Materials and methods

The material for the study was 169 middle-late (FAO 400-550) maize lines developed using Chalco teosinte from the VIR collection. For estimating plant height and cob fixation height, number of leaves and cobs a hybrid of the KBNIISH selection - KB 595 st from the FAO late maturity group was chosen as a standard.

The selection estimation of the VIR maize collection was carried out in 2018-2019. in accordance with the following methods: Dospekhov B.A. "The methodology of field experience" [17], Guidelines for the study and maintenance of corn collection samples, approved by the VASKhNIL VIR [18]. The description of morphological characteristics was performed according to "A wide unified classifier of CMEA and the international classifier of CMEA of the species *ZEA MAYS* L." [19].

The research was carried out at the research and production site NPU No.1 (village Nartan, foothill zone) of the Institute of Agriculture of the Kabardino-Balkarian Scientific Center of the Russian Academy of Sciences. The NPU (research and production site) No.1 territory is located within the foothill zone of the North Caucasus and lies on the Urvan-Nalchik watershed.

The soil of the experimental site is represented by ordinary carbonate chernozem of a heavy loamy variety. The agrochemical composition of the soil of the experimental site (according to Chirikov) is represented by the following indicators: pH-7.2; labile P_2O_5 - 9.8 mg/100 g of soil; exchangeable K_2O -7.2 mg/100 g of soil; humus (according to Tyurin) - 4.4%. The arable horizon contains 3.9-4.2% humus, 18-27 mg of nitrogen, 27-34 mg of labile phosphorus and 230-250 mg of exchange potassium.

During the years of research the climate of the zone was moderately hot. The amount of active temperatures was 3000-3200^{0}C with moderate moisture (moisture coefficient - 0.5-0.9), hydrothermal coefficient 0.9-1.2. In early April, the average daily air temperature passed through + 10^{0}C. On average, the frost stopped at the end of the second decade of April. The frost-free period lasted until the end of October and amounted 190 days. Up to 17.5 days with strong winds were recorded per year. The amount of precipitation for the year was 615 mm, of which 75.8% (416 mm) were for the growing season.

The site area was 7.84m^2, the location of the sites was systematic. The row spacing was 0.7 m, the plant density was 50-60 thousand plants per 1 ha. According to the breeding program, all the necessary phenological observations for all the main phases of plant development were carried out on the studied samples during the growing season. The morphological characteristics of maize (plant height and attachment of economically suitable cobs, number of leaves on the main stem, number of cobs on a plant) were measured on 10 typical plants of each sample.

3 Research results

The analysis of the research results showed that all of the studied lines belong to the middle-late and late maturity groups according to FAO 400-550. From the studied 169 samples of the VIR collection, a selection of 21 best biometric genotypes was formed (Table 1). The main estimation criteria were: plants height, cob fixation height, number of leaves and number of cobs on a stem. None of the samples showed any tillering at standard planting density, despite the fact that the lines were developed on the basis of teosinte, which is characterized by a high degree of tillering. It should be noted that the selection of 21 samples included all samples with a multi-cob tendency. Indicator variation ranged from 1.0 to 2.45 cobs on average over 2 years (Fig. 1).

Table 1. Biometric characteristics of the matured maize plants of the VIR collection (2019-2020, foothill zone).

No.	Line name	Plant height, m			Attachment height of economically suitable cob, m			The number of leaves on the main stem, pcs.			Number of economically suitable cobs on one plant, pcs.		
		2019	2020	2-year average	2019	2020	2-year average	2019	2020	2-year average	2019	2020	2-year average
1	KB 595 st	2.21	2.15	2.18	0.88	0.82	0.86	15.4	16.0	15.5	1.4	1.6	1.5
2	MP 151 A yellow	2.42	2.37	2.40	1.18	1.08	1.13	17.0	17.0	17.0	2.1	1.9	2.0
3	MP 153 V yellow	2.67	2.64	2.66	1.43	1.35	1.39	16.0	16.5	16.25	1.9	2.2	2.15
4	MP 75 B yellow	3.01	2.88	2.95	1.61	1.54	1.58	16.2	15.4	15.9	2.3	2.6	2.45
5	MR 16616-5	2.79	2.75	2.77	1.11	1.08	1.10	15.0	16.0	15.5	1.1	1.3	1.25
6	MR 16616-8	2.41	2.44	2.42	1.29	1.22	1.26	14.8	15.6	15.2	2.2	2.3	2.25
7	MR 16616-24	2.71	2.63	2.67	1.23	1.25	1.24	15.8	16.4	16.1	2.2	2.1	2.15
8	MRxArg1	2.51	2.57	2.54	1.17	1.14	1.16	14.5	14.8	14.7	2.2	2.3	2.25
9	1120 MP-213	2.79	2.77	2.78	1.59	1.49	1.55	16.4	16.8	16.6	2.4	2.3	2.35
10	MP 135 D	2.55	2.52	2.54	1.08	1.10	1.09	14.9	15.3	15.1	1.1	1.1	1.1
11	MR 16616-17	2.61	2.56	2.59	1.21	1.15	1.18	16.1	15.9	16.0	1.0	1.0	1.0
12	MR 16616-25 yellow	2.53	2.56	2.55	1.01	1.08	1.05	15.2	15.7	15.5	1.1	1.4	1.25
13	MP 16616-33 yellow	2.89	2.77	2.83	1.37	1.29	1.33	16.2	16.8	16.6	1.1	1.1	1.1
14	I 627401	2.63	2.54	2.59	1.09	1.15	1.12	15.5	16.3	15.9	1.4	1.7	1.55
15	I 627403	2.79	2.69	2.74	1.31	1.25	1.28	16.3	16.8	16.6	1.5	1.3	1.4
16	KBN 16-93-1	2.91	2.88	2.90	1.21	1.18	1.20	15.5	15.9	15.7	1.3	1.3	1.3
17	102AMP x Nart37	2.91	2.85	2.88	1.51	1.44	1.48	15.8	16.3	16.1	1.5	1.8	1.65
18	1084 MP-71	2.59	2.56	2.58	1.18	1.19	1.19	15.8	16.7	16.3	1.3	1.6	1.45
19	1087 MP-76	2.79	2.74	2.77	1.19	1.17	1.18	16.0	16.5	16.3	1.2	1.4	1.3
20	1103 MP-113	2.78	2.74	2.76	1.17	1.12	1.15	15.8	16.3	16.1	1.5	1.4	1.45
21	1105 MP-121	2.91	2.85	2.88	1.57	1.47	1.52	16.2	16.7	16.5	2.0	2.3	2.15
22	1113MP-150	2.69	2.65	2.67	1.17	1.24	1.21	15.7	16.5	16.4	1.5	1.7	1.6
22	Average			2.68			1.25			16.01			1.54
LCD_{05}				0.21			0.17			1.1			0.11

Biometric measurements of the quantitative characteristics of maize lines showed that plant height allocated such lines as: MP 153 V yellow, MP 75 B yellow, MR 16616-5, MR16616-24, 1120 MP-213, MR 16616-17, MR 16616-33 yellow, I 627401, I 627403, KBN 16-93-1102A, MP x Nart37, 1084MP-71, 1087MP-76, 1103MP-113, 1105MP-121, 1113MP-150. The variation ranged from 2.40 to 2.95m, which is higher than the LCD=0.21cm.

On average for 2 years in the selection (without a standard), the number of samples with a multi-cob value above 2.0 was 8 samples (MP 151 A yellow, MP 153 V yellow, MP 75 B yellow, MR 16616-8, M 16616-24, MPxArg, 11120 MP-213, 1105 MP-121) and according to the number of leaves on the stem, the sample MP 151 A yellow. showed a significant excess over the standard.

Fig. 1. Polymorphism based on the phenotypic trait of multi-cob in some of the studied maize lines from the VIR collection (the leftmost is hybrid KB 595 - standart).

4 Conclusions

The breeding material of maize samples developed on the basis of teosinte from the VIR collection is of practical interest and makes it possible to expand the genetic polymorphism of the starting material for the breeding of hybrid maize. Based on the studied material presented by the VIR maize collection, a selection of 21 samples was created that represent a breeding value as a starting material for improving maize for a number of economically valuable characteristics. Valuable genotypes characterized by a high stem (MP 153 V yellow, MP 75 B yellow, MR 16616-5, MR 16616-24, 1120 MP-213, MR 16616-17, MR 16616-33 yellow, I 627401, I 627403, KBN 16-93-1102A, MP x Nart37, 1084MP-71, 1087MP-76, 1103MP-113, 1105MP-121, 1113MP-150), a large number of leaves on the stem (MP 151 A yellow) and a tendency to multi-cob (MP 151 A yellow, MP 153 V yellow, MP 75 B yellow, MP 16616-8, MR 16616-24, MPxArg, 11120 MP-213, 1105 MP-121), which will later be used in the breeding programs of Institute of Agriculture of the Kabardino-Balkarian Scientific Center of the Russian Academy of Sciences, on the creation of maize hybrids with high grain yield were allocated.

References

1. D. A. Fasoula, V. A. Fasoula, Maydica, **49** (2005)
2. Perspective resource-saving technology for the production of corn for grain, **72** (2010)
3. M. B. Hufford, et al, Teosinte as a model system for population and ecological genomics. Trends in Genetics, **606**
4. M. B. Hufford, et al, *Inferences from the historical distribution of wild and domesticated maize provide ecological and evolutionary insight*
5. I. M. Chilashvili, Scientific journal KubSAU, **1** (2012)
6. E. B. Khatefov, M. R. Gonikova, V. I. Khoreva, V. G. Goldstein, L. P. Nosovskaya, L. V. Adikaeva, Works on applied botanics, genetics and breeding, **56** (2020)
7. L. A. Wasserman, A. G. Filatova, E. B. Khatefov, V. G. Goldstein, I. G. Plashchina, **74** (2021)
8. A. M. Kagermazov, A. V. Khachidogov, Bulletin of the agro-industrial complex of Stavropol, **57** (2019)

9. E. B. Khatefov, Seed productivity of tetraploid corn and ways to increase it in the conditions of Kabardino-Balkaria, **391**
10. B. Gouesnard et al, Theor Appl Genet, **2165** (2017)
11. S. Kanagarasu et al, African Journal of Biotechnology, **5723** (2012)
12. M. C. Romay et al, Genome Biol, (2013)
13. P. S. Schnable et al, Science, **326** (2009)
14. E. Anderson, Heterosis, **124** (1952)
15. M. B. Hufford, X. Xu, J. van Heerwaarden, T. Pyhajarvi, J.-M. Chia, et al, Nature Genetics, **808** (2012)
16. P. P. Domashnev, Selection of corn, **208** (1992)
17. B. A. Dospekhov, The methodology of field experience (with the basics of statistical processing of research results) (2012)
18. G. E. Shmaraev, G. V. Matveeva, *Guidelines for the study and maintenance of corn collection samples* (1985)
19. V. G. Kukekov, *A wide unified classifier of CMEA and the international classifier of CMEA of the species Zea mays* (1977)

The efficiency of using different fertilization systems in dryland and irrigated crop rotations on ordinary carbonate chernozems of the Central Ciscaucasia

Tamara Bizhoeva, *Ruslan* Bizhoev*, *Asiyat* Sarbasheva, *Olga Batyrova*, and *Aslanbek* Kushkhabiev

Institute of Agriculture - Branch of the Kabardino-Balkarian Scientific Center of the Russian Academy of Sciences, 224, Kirova Street, 360004 Nalchik, Kabardino-Balkarian Republic

Abstract. The search for the rational use of fertile irrigated land in the steppe zone of Central Ciscaucasia and the determination of factors for the sustainable production of crop products, provided that soil fertility is maintained, remained as valid today as ever. The relevance of research is also increasing due to changes in climatic characteristics in specific agro-landscape conditions, including, an increase in average daily and annual air temperatures by 2 - 50C. The novelty of research is the establishment of the most efficient fertilization systems in dryland and irrigated conditions to increase winter wheat yields, to increase the productivity of 1 ha of crop area, as well as to maintain and reproduce soil fertility and to increase the productivity of irrigated crop rotation. The article contains the research results on the use of different fertilization systems and their influence on winter wheat yields and crop rotation productivity in dryland and irrigated agriculture on ordinary carbonate chernozems in specific agro-landscape, soil and climatic characteristics of the Central Ciscaucasia.

1 Introduction

Intensive use of arable land and accelerated mineralization of organic matter, together with a sharp decrease in the use of organic and mineral fertilizers, irrational use of irrigation water and an insufficient share of perennial grasses in crop rotation leads to a general decline in soil fertility and, as a result, a decline in crop yields and crop rotation efficiency [1]. In this regard, farming systems are focused on evidence-based crop rotation systems that reduce production costs, improve soil fertility, regulate the food, water and air regimes of the soil, without requiring additional material costs, which is very important for producing environmentally friendly crop products and reducing their cost [2].

*Corresponding author: kbniish2007@yandex.ru

2 Materials and methods

Scientific research was carried out in a long-term stationary field experiment, located on the experimental field of the Institute of Agriculture of the Kabardino-Balkarian Scientific Center of the Russian Academy of Sciences in the village Krasnoarmeyskoye, Tersk region, Kabardino-Balkarian Republic, founded in 1948 by P. E. Prostokov, professor of the Department of Agrochemistry, North Ossetian Agricultural Institute. From the moment of setting up to the present time, the fertility of the ordinary carbonate chernozem is monitored in the experiment with long-term irrigation and systematic use of fertilizers in the agricultural landscape system of agriculture. The research is coordinated by the Geographical Network of Experiments with Fertilizers and Other Agrochemicals in the Russian Federation, the experiment is included in the Register of Long-term Experiments and registered under No. 037, and is conducted according to classical methods. R&D reports are annually submitted to the Geographic Network, which maintains an information database of long-term stationary experiments for more than 70 years at the federal level.

This article presents the results of research on winter wheat. Experiment scheme: alternation of 4 crops in a crop rotation - winter wheat, maize, sunflower and alfalfa for 1 and 2 years of use with different mineral and organo-mineral fertilizers. Estimated mineral and organo-mineral fertilization systems are designed to preserve and reproduce soil fertility and to increase the productivity of irrigated crop rotation.

The number of options in the experiment is 100, the number of plots is 200. The site area is 165.6 m^2, with a site width of 6.9 m and a length of 24.0 m; the accounting area of the site varies from 33.6 to 73.2 m^2, depending on the culture and method of accounting.

Fertilizers were manually applied according to the experiment scheme in autumn for the main plowing. The following forms of fertilizer were used: ammonium nitrate, simple and double super phosphate, potassium salt, semi-superior manure. In recent years, ammonium nitrate, ammonium phosphate and nitro-ammonium phosphate were used because of the lack of required fertilizers such as super phosphate and potassium fertilizers and their cost.

Yield data from the experiment are reduced to 14% (maize grain, winter wheat, peas), 16% (alfalfa hay) and 12% (sunflower seeds) of standard humidity.

The prevailing climatic characteristics of the studied growing seasons are presented in Table 1.

The studies were performed in accordance with the guidelines for conducting research in long-term experiments with fertilizers [3.4.5], field experience methodology [6], agrochemical research technique [7] and methodological and organizational bases of conducting agroecological monitoring in intensive agriculture [8]. The methodology of the All-Russian Research Institute of Agrochemistry named after D.N. Pryanishnikov was used in selecting plant and soil samples. Agrochemical analyses of soil and plants were performed by standard methods, according to the GOST. The research results were statistically evaluated [6].

Table 1. Characteristics of the meteorological conditions of the crop growing seasons in 2017-2020.

Year	Month	Air temperature, °C		Monthly precipitation, mm		Relative humidity, %	
		current	long-term average annual	current	long-term average annual	current	n average annual
2017	September	20.7	17.3	13.7	36.1	63	74
	October	11.1	10.5	41.2	30.1	75	82
	November	4.9	3.9	23.6	23.5	86	88
	December	2.1	- 0.9	40.7	21.0	90	90
2018	January	- 1.5	- 3.1	27.9	17.9	91	89
	February	2.1	- 2.4	40.4	17.4	90	86
	March	5.1	2.5	92.0	28.0	79	84
	April	11.9	10.7	16.2	43.5	67	74
	May	19.4	16.3	87.1	63.5	65	72
	June	22.3	20.6	42.8	77.9	59	69
	July	25.9	23.1	34.0	59.9	65	67
	August	22.7	22.5	43.6	46.9	65	69
	September	19.9	17.3	8.4	36.1	65	74
2019	January	- 0.1	- 3.1	13.1	17.9	84	89
	February	1.9	- 2.4	8.6	17.4	83	86
	March	6.6	2.5	36.4	28.0	70	84
	April	11.1	10.7	50.4	43.5	67	74
	May	18.8	16.3	72.6	63.5	70	72
	June	25.9	20.6	28.9	77.9	60	69
	July	24.9	17.3	39.0	36.1	67	74
	August	24.9	10.5	26.9	30.1	76	82
	September	18.1	3.9	16.8	23.5	88	88
	October	13.2	10.5	26.9	36.1	67	74
	November	4.1	3.9	16.8	30.1	76	82
	December	2.9	- 0.9	11.4	23.5	88	88
2020	January	1.6	- 3.1	6.2	17.9	81	89
	February	2.8	- 2.4	13.1	17.4	75	86
	March	7.5	2.5	23.1	28.0	71	84
	April	10.5	10.7	15.9	43.5	70	74
	May	17.9	16.3	105.5	63.5	75	72
	June	23.5	20.6	55.6	77.9	64	69
	July	26.9	23.1	32.0	59.9	63	67
	August	23.2	22.5	73.9	46.9	61	69
	September	19.7	17.3	0	36.1	70.3	74

3 Results and discussion

During the autumn of 2018 (from September to December inclusive), 51.2 mm of precipitation fell at a rate of 110.7 mm, which did not even allow winter wheat to grow in autumn, despite the optimal planting time. The scarcity of soil moisture was apparently not enough for wheat seed germination and smooth and timely sprouts. No snow cover was formed and was observed for only 1-3 days. The sprouts emerged slowly from late January through February and developed poorly.

The yield of Cheget winter wheat in the option 1 - without fertilizer and irrigation (control) - after 69 years of the experiment averaged 18.6 dt/ha (Table 2), and for three different predecessors: alfalfa, maize, sunflower, it was almost identical, which is statistically proven.

Table 2. Crop yields and crop rotation productivity in the experience of 1948 in 2018, dt/ha.

Water availability	Fertilizer options	Cheget winter wheat	Terek maize	SPK sunflower	Kevsala alfalfa two-year-old	Productivity 1 ha, dt.g.u.
Without irrigation	1. Without fertilizers	18.6	21.9	10.9	31.6	17.4
	2. Recommended mineral fertilization system $N_{69}P_{63}K_{45}$	51.3	32.0	14.0	53.8	33.1
	3. Recommended organo-mineral fertilization system manure 8 t/ha + $N_{44}P_{42}K_{24}$	58.4	36.7	15.6	64.6	38.0
Irrigation, 75-80% minimum moisture-holding capacity	4. Without fertilizers	20.0	29.6	10.7	60.8	23.1
	5. Recommended mineral fertilization system $N_{69}P_{63}K_{45}$	48.1	43.8	13.6	82.7	38.7
	6. Recommended organo-mineral fertilization system manure 8 t/ha + $N_{44}P_{42}K_{24}$	55.6	49.0	16.2	105.5	45.3
	7. Estimated mineral fertilization system $N_{141}P_{68}K_{60}$	55.7	47.7	15.8	96.8	44.0
	8. Estimated organo-mineral fertilization system manure 15 t/ha $+N_{87}P_{36}K_7$	61.1	62.0	17.6	118.6	52.6
	LCD_{05} particular differences	2.64	4.10	2.60	2.82/4.40	
	LCD_{05} for irrigation	0.84	1.30	1.16	1.81/1.97	
	LCD_{05} for fertilizers and interaction.	1.07	1.44	1.84	2.28/3.10	
	Experiment accuracy, S_x, %	3.46	3.69	6.40	2.95/1.85	

Recommended fertilization systems: mineral and organo-mineral in non-irrigated conditions (options 2 and 3), in 2018 provided, on average, a higher yield of Cheget winter wheat (51.3 and 58.4, respectively) than wheat yield in similar irrigated crop options (options 5 and 6) by 3.2-2.8 dt/ha, which is mathematically reliable, and is explained by the fact that soil fertility in dryland options is undoubtedly higher than in irrigated options, and the weather conditions favorable for winter wheat made it possible to realize the advantage of a higher humus and nutrient content in the root layer of the soil to produce a higher yield.

The estimated fertilization systems in the irrigated crop rotation (options 7 and 8) provided an increase in crop yield compared to the recommended fertilization systems of 5.5-7.6 dt/ha, which exceeded the LSD05 by 2.5-2.7 times.

The highest grain yield of Cheget winter wheat 60.0-63.2 dt/ha, depending on the predecessor, is obtained in option 8 on the estimated organo-mineral fertilization system aimed at reproduction of soil fertility. In 2018, a large yield of winter wheat in this option was achieved on the alfalfa predecessor, which is mathematically reliable (Table 2).

Not only weather conditions have contributed to this yield, but also better timely soil preparation, planting of crops in optimal time and compliance of other agricultural practices.

It was found that with an optimal combination of accompanying factors, the efficiency of the studied fertilization systems increased more than 2 times. Thus, in the dryland crop rotation in 2018, mineral and organo-mineral fertilization systems achieved crop rotation productivity of 33.1 and 38.0 dt.g.u/ha with an average of 40 years of research (1970-2016) - 32.1 and 34, 4 dt.g.u/ha. At the same time, the productivity of crop rotation without fertilization and irrigation was equal to 17.4 dt.g.u/ha with average productivity of dryland unfertilized crop rotation of 17.2dt.g.u/ha.

The climatic characteristics of the 2009 growing season were also unfavorable for the growth, development and production of all field crops. The sprouts emerged slowly from late January through February and developed poorly. From January to March 2019, precipitation fell only 58.1 mm, which was also less than the long-term average annual amount (63.3 mm) with temperatures significantly exceeding long-term ones by 3-40C. From April to August, 208.6 mm of precipitation fell, with a norm of 291.7 mm. Temperatures during the summer months were from 2.4 to 5.3o higher than the long-term average annual temperature, which contributed to the water scarcity for all crops. The relative air humidity in all months of growing season was less than the annual average by 2-14%. June 2019 was particularly unfavorable for climatic characteristics (Table. 1).

Such changes in climatic characteristics indicate a tightening of dryland cultivation conditions in comparison with previous decades. Low precipitation and high air temperatures are not only responsible for soil and air drought, but also for reduced crop yields.

The yield of Cheget winter wheat in option 1, without fertilizer and irrigation, after 70 years of the experiment averaged 19.8 dt/ha. In option 4, without fertilizers in irrigation, its average yield was 21.4 dt/ha (Table 3). In the non-irrigated crop rotation of the experiment in 2019, the influence of predecessors on winter wheat yields: alfalfa, maize, sunflower, was almost the same, the changes in yield indicators for all studied fertilization systems did not exceed the smallest significant difference.

Recommended fertilization systems: mineral and organo-mineral in non-irrigated conditions (options 2 and 3) provided, on average, lower winter wheat yields in 2019 than wheat yields in similar irrigated crop options (options 5 and 6) by 0.7 and 2.1 dt/ha, respectively, which is mathematically reliable (Table 3). Based on a comparative estimation of the soil fertility in these options under dryland and irrigated conditions, when the soil fertility in the dryland options, undoubtedly better, weather conditions unfavorable to winter wheat did not allow to realize the advantage of sufficient humus and nutrients in the root soil layer to produce a higher yield, since the decisive factor in its formation in an extremely dry year was the provision of plants with moisture.

The estimated fertilization systems in irrigated crop rotation, options 7 and 8, provided an increase in crop yield compared to the recommended fertilization systems of only 1.7-1.8 dt/ha, which exceeded the LCD05 for fertilizers by 2.3 times.

The highest grain yield of Cheget winter wheat is 38.7-39.2, i.e. 1.5 times less than in the previous year, depending on the predecessor, produced in option 8 on the estimated organo-mineral fertilization system aimed at reproduction of soil fertility. In 2019 the winter wheat yield in this option was almost the same for all studied predecessors, the difference in yield did not exceed the least significant difference for this factor (Table. 3).

Table 3. Crop yields (dt/ha) and productivity of 1 hectare of crop rotations in the experiment of 1948, 2019.

Water availability	**Fertilizer options**	**Cheget winter wheat**	**Terek maize**	**SPK sunflower**	**Kevsala alfalfa two-year-old**	**Productivity 1 ha, dt.g.u.**
Without irrigation	1.Without fertilizers	19.8	0	0	2.3/8.4	6.5
	Recommended mineral fertilization system $N_{69}P_{63}K_{45}$	30.6	0	0	23.9/20.2	11.4
	3.Recommended organo-mineral fertilization system manure 8 t/ha + $N_{44}P_{42}K_{24}$	35.1	0	0	29.6/21.8	13.1
Irrigation, 75-80% minimum moisture-holding capacity	1 Without fertilizers	21.4	26.5	6.2	40.6/35.4	19.6
	2 Recommended mineral fertilization system $N_{69}P_{63}K_{45}$	31.3	36.7	12.9	70.0/79.5	30.5
	3 Recommended organo-mineral fertilization system manure 8 t/ha + $N_{44}P_{42}K_{24}$	37.2	57.5	15.7	82.0/93.8	36.6
	4 Estimated mineral fertilization system $N_{141}P_{68}K_{60}$	33.0	40.2	13.7	75.0/100.9	33.6
	5 Estimated organo-mineral fertilization system manure 15 t/ha +$N_{87}P_{36}K_{7}$	39.0	64.8	18.5	89.1/118.6	45.6
LCD_{05} particular differences		1.82	2.27	1.04	1.99/4.40	
LCD_{05} for irrigation (maize-predecessor)		0.58	Not substantive	-	0.96/1.97	
LCD_{05} for fertilizers and interaction.		0.74	1.14	-	1.41/3.10	
Experiment accuracy, S_x, %		2.03	1.75	1.94	1.31/3.39	

The importance of grain crop yields for crop rotation and crop rotation productivity in non-irrigated conditions in different years confirms their dependence on the amount of precipitation during the growing season and the year as a whole.

Without irrigation, the efficiency of the used fertilization systems is low. With the systematic use of the recommended mineral fertilizers N69P63K45 (option 2) to the crop rotation culture in 2019, the crop rotation productivity was increased by 4.9 dt.g.u./ha to control or 75.4%. The systematic use of the organo-mineral fertilization system (option 3), manure 8 t/ha + N44P42K24, provided the increase in the productivity of dryland crop rotation 6.6 dt.g.u/ha to control or 101.5%.

The productivity of 1 hectare of crop rotation without fertilizers under irrigation (option 4) in 2019 was 19.6 dt.g.u. Irrigation increased the efficiency of the recommended mineral (option 5) and organo-mineral (option 6) fertilization systems and increased the annual average productivity of irrigated crop rotation by 55.6 and 86.7%, respectively, compared to the option without fertilizers in irrigation, and 3.7 and 4.6 times compared to the control, without fertilizer or irrigation. The use of estimated fertilization systems, mineral and organo-mineral (to preserve and reproduce soil fertility and to obtain the productivity of irrigated crop rotation 52-53 dt.g.u/ha) increased the productivity of 1 ha in irrigation by 71.4% and 132.7% respectively compared to the option 4, without fertilizer in irrigation (Table 5) and 4.2 and 6.0 times compared to control.

In 2019, the productivity of crop rotation without irrigation was significantly (almost 3 times) less than the long-term average annual productivity.

In 2020, the research results also confirmed the dependence of crop yields and productivity of non-irrigated crop rotation, primarily on the amount of precipitation during the growing season and the year as a whole, and then on the conditions of mineral nutrition.

From January to March 2020, precipitation fell 42.4 mm, which is 20.9 mm less than the long-term precipitation for the same period, at temperatures exceeding long-term ones. From April to August, 282.9 mm of precipitation fell, with a norm of 291.7 mm. The temperatures during the summer months were 0.7-3.8o higher than the average long-term temperature, which contributed to the development of moisture deficiency for spring crops.

The yield of Yuzhanka winter wheat in 2020 with fertilization without irrigation was relatively high, as against the other crop rotations, which was achieved due to the biological characteristics of the crop and its ability to efficiently use of available soil moisture.

In option 1, without fertilizer or irrigation, after 71 years of the experiment (Table 4), yields averaged 18.2 dt/ha and for three different predecessors: alfalfa, maize, sunflower, it was almost the same, which is statistically proven. In option 4, without fertilizers in irrigation, its average yield was 19.1 dt/ha.

Recommended fertilization system: mineral in non-irrigated conditions (option 2) in 2020, provided, on average, a lower yield of Yuzhanka winter wheat of 32.9 dt/ha than wheat yield in a same version of irrigated crop rotation (option 5 - 34.3 dt/ha) by 1.4 dt/ha. The estimated fertilization systems in irrigated crop rotation, options 7 and 8 (33.1 and 39.7 dt/ha), provided an increase in crop yield compared to the recommended fertilization systems of 1.2-3.0 dt/ha.

The highest yield of Yuzhanka winter wheat grain is 38.0-39.7 dt/ha, depending on the predecessor, produced in option 8 from the estimated organo-mineral fertilization system aimed at reproduction of soil fertility (Table 4). In 2020, high yield of winter wheat in this option was achieved by the maize predecessor, which is mathematically relevant.

Table 4. Crop yields and crop rotation productivity in the experiment of 1948 in 2020, dt/ha.

Water availability	**Fertilizer options**	**Yuzhanka winter wheat**	**Karat maize**	**SPK sunflower**	**Kevsala alfalfa two-year-old**	**Productivity 1 ha, dt.g.u.**
Without irrigation	1. Without fertilizers	18.2	0	0	30.8/30.5	17.2
	2. Recommended mineral fertilization system $N_{69}P_{63}K_{45}$	32.9	0	0	46.3/55.9	30.2
	3. Recommended organo-mineral fertilization system manure 8 t/ha + $N_{44}P_{42}K_{2}$	31.8	0	0	53.9/67.9	33.0
Irrigation, 75-80% minimum moisture-holding capacity	4. Without fertilizers	19.1	22.4	7.8	47.5/70.1	19.9
	5. Recommended mineral fertilization system $N_{69}P_{63}K_{45}$	34.3	29.9	10.2	71.3/85.1	28.3
	6. Recommended organo-mineral fertilization system manure 8 t/ha + $N_{44}P_{42}K_{24}$	36.7	40.1	16.2	80.8/119.2	36.2
	7. Estimated mineral fertilization system $N_{141}P_{68}K_{60}$	33.1	35.6	15.6	77.0/108.7	32.9
	8. Estimated organo-mineral fertilization system manure 15 t/ha +$N_{87}P_{36}K_{7}$	39.7	42.9	10.2	85.5/142.7	38.4
	LCD_{05} particular differences	1.89	3.47	1.31	2.82/4.40	
	LCD_{05} for irrigation	0.59	2.65	0.59	1.81/1.97	
	LCD_{05} for fertilizers and interaction.	0.78	1.33	0.93	2.28/3.10	
	Experiment accuracy, S_x, %	2.11	5.13	3.06	2.95/1.85	

The efficiency of the applied fertilization systems in non-irrigated conditions is low, especially in years with insufficient water availability. Thus, the efficiency of the applied fertilization systems in 2020 with the systematic use of recommended mineral fertilizers for crops ($N_{69}P_{63}K_{45}$ on average for the year of rotation) was 13.0dt.g.u./ha to control, or 75.5% higher. The systematic use of the organo-mineral fertilization system (option 3), manure 8 t/ha + $_{44}P_{42}K_{24}$, provided an increase the productivity of dryland crop rotation of 15.8 dt.g.u/ha to control or 91.8% higher.

The productivity of 1 hectare of crop rotation without fertilizer in irrigation (option 4) in 2020 was 19.9 dt.g.u/ha. Irrigation increased the efficiency of the recommended mineral (option 5) and organo-mineral (option 6) fertilization systems (28.3 and 36.2 dt.g.u/ha) (Table 5) and increased the average annual productivity of irrigated crop rotation in these options by 65 and 110%, respectively, compared to the control (17.2 dt.g.u/ha).

The use of estimated fertilization systems, mineral and organo-mineral (options 7 and 8 - 32.9 and 38.4 dt.g.u./ha) In order to maintain and reproduce soil fertility and to increase the productivity of irrigated crop rotation, the productivity of 1 hectare in irrigation increased by 91% and 123%, respectively.

4 Conclusion

1. As a result of scientific research in 2018-2020 it was identified that the efficiency of the use of different mineral and organo-mineral fertilization systems in conditions of natural humidification and irrigation on ordinary carbonate chernozem of the Central Ciscaucasia differed and depended largely on water availability and moisture reserves in the soil.

2. With an optimal combination of factors, the efficiency of the studied fertilization systems increased in dryland crop rotation by 1.8-2.2 times. Thus, in the dryland crop rotation, in 2018, the mineral and organo-mineral fertilization systems achieved crop rotation productivity of 33.1 and 38.0, and in 2020 - 32.5 and 35.9 dt.g.u./ha, respectively, with average productivity over 40 years of the study (1970- 2016) - 32.1 and 34.4dt.g.u./ha. At the same time, the productivity of crop rotation without fertilizers and irrigation was 17.4 and 17.2dt.g.u./ha, respectively, while the average productivity of the dryland inconvenient crop rotation was 17.2 dt.g.u/ha

3. Under unfavorable climatic characteristics, the efficiency of the studied fertilization systems has decreased more than 3 times. Thus, in the dryland crop rotation in 2019, with the use of mineral and organo-mineral fertilization systems, productivity of the crop rotation of 11.4 and 13.1 dt.g.u./ha was achieved with an average of 32.1 and 34.4 dt.g.u./ha in 40 years of research (1970-2016). At the same time, the productivity of crop rotation without fertilization and irrigation was only 6.5 dt.g.u./ha with average productivity of 17.2dt.g.u. /ha.

4. The estimated fertilization systems in irrigated crop rotation, aimed at maintaining and reproducing soil fertility and increasing the productivity of irrigated crop rotation, have increased the productivity of irrigated crop rotation in comparison with recommended fertilization systems in different years by 2.2-9.0 dt.g.u./ha.

References

1. V. M. Novikov, Agriculture, **13** (2015)
2. V. M. Bizhoev, Agrochemistry, **37** (1988)
3. *Guidelines for the Geographic Network of Fertilizer Experiments* (1970)

4. Guidelines for conducting research in long-term experiments with fertilizers, Experimental technique and soil analysis, **1** (1975)
5. Guidelines for conducting research in long-term experiments with fertilizers, Program and methods of soil research, **2** (1983)
6. B. A. Dospekhov, *Field experiment methodology (with the basics of statistical processing of research results* (1985)
7. F.A. Yudin, Agrochemical research technique, **11, 16, 100** (1971)
8. *Methodological and organizational bases of conducting agoecological monitoring in intensive agriculture* (1991)

Productivity of sudan grass and sorghum-sudangrass hybrids depending on seeding rates and planting methods in the steppe dryland zone of the Kabardino-Balkarian Republic

Aslan Kushkhov*, *Natalya* Berbekova, and *Alena* Zhurtova

Institute of Agriculture - Branch of the Kabardino-Balkarian Scientific Center of the Russian Academy of Sciences, 224, Kirova Street, Nalchik, 360004, Kabardino-Balkarian Republic

Abstract. The article contains the results of three-year field research in the steppe dryland zone of the republic, the best options of grass sorghum cultivation, which increase the gross yield of green forage and hay were developed. Optimum planting standards for dry steppe conditions, planting and harvesting methods of Sudan grass and sorghum-Sudangrass hybrids were identified to produce a high, more energy-intensive green and dry forage of the studied crops. The quality and feed qualities of fodder for production and direct feeding to animals are determined. The crop growing time after hay harvest and quantity for feeding after harvesting are given.

1 Introduction

Feed production is the most labor-intensive and complex process of the agrarian economy. The most important challenge in the intensification of agricultural production in the dryland zone of the Kabardino-Balkarian Republic is the development of a sustainable feed base. Expanding the range of feed crops adapted to specific soil and climatic conditions will better meet the needs of animals for high-quality feed, improve land management and also significantly reduce the adverse climate effects.

In the zone of unstable humidification of the republic, research on promising varieties and hybrids of sorghum crops to identify signs of drought resistance, prematurity, high feed productivity and productive longevity to take full advantage of the favorable bioclimatic potential of local conditions for the production of selected feed products was carried out.

The studied varieties and hybrids have a number of valuable qualities: they grow quickly after cutting or feeding, contrary to other annual feed grasses, which allows them to be more widely used in a green conveyor; with increased yields, they produce high-quality hay, which is superior in protein and fat content to other feed crops, second only to legumes.

The development of the proposed agronomic practices is motivated by the importance of growing feed crops that increase the productivity of the arable land of the dry steppe zone,

*Corresponding author: kbniish2007@yandex.ru

and by the implementation of feed technology that stabilizes the gross yield in different weather years.

The goal of the research is to improve the elements of grass sorghum cultivation technology by applying optimal timing and seeding rates, planting techniques and the use of varieties and hybrids with different indicators of the biochemical composition of green forage used for the preparation of high-quality and varied feeds.

The task of the field research is to study the growth and development of Sudan grass and sorghum-Sudangrass hybrids with different seeding rates and techniques, to estimate their economic characteristics, and to develop some techniques for the cultivation of Sudan grass and sorghum-Sudangrass hybrids for green forage and hay; to study the varietal differences and chemical composition of studied grasses to identify the most productive varieties and hybrids for production in the unstable humidification zone.

The novelty of the research is that for the first time in the dryland steppe zone of the republic the possibility of obtaining high yields of green and dry forage of grass sorghum with different technological direction of preparation of quality feed is identified. For the first time, on the chernozem soils of the dry steppe zone of the Kabardino-Balkarian Republic, a comprehensive study of various grass sorghum hybrids was performed and their reaction to the implementation of various seeding rates and planting methods, cutting and use regimes were identified.

The research provides science and production with a range of Sudan grass and sorghum-Sudangrass hybrids, optimally selected to the specific soil and climate zone that produces high-quality feeds with high productivity and nutritional value.

The research results were tested and implemented in various farms in the steppe zone of the republic on an area of 85 ha.

2 Materials and methods

The research was conducted for three years (2015-2017) at the experimental site of the laboratory of technology for the cultivation of field crops of the Institute of Agriculture of the Kabardino-Balkarian Scientific Center of the Russian Academy of Sciences, Tersky district, village Opytnoye Soils are Ciscaucasian carbonate chernozem. The predecessor is winter wheat (2015), grain maize (2016), peas (2017). The humus content in the soil is 3.5%; soil solution reaction is neutral: ph - 6.5-7.1; labile phosphorus content - 2.0-2.5; exchange potassium - 30-40 mg/100g of soil, average annual precipitation is 435 mm with fluctuations from 320 to 545 mm over the studied years, the thickness of the humus horizon is 30-35 cm. Over the years of research, precipitation fell: 2015 - 335.5 mm; 2016 - 545.7 mm; 2017 - 430.5 mm, of which in the autumn-winter months, respectively: 142.5 mm; 170.4 mm; 164.0 mm. Over three years of the experiments, the growing season of 2016 was the wettest - 545.7 mm, compared to 435 mm of average long-term indicators. Meteorological observations over the years of the study are shown in Table No. 1. The choice of the site, the size and location of the plots, the replicates were formed according to the "Methodology of field experiment", Dospekhov B.A. M.: 1979; the scheme of the field experiment, seeding rate, planting methods are presented in Table No. 2.

The following seeds were used in the research

Sudan grass, Shirokolistnaya variety, originator is the Institute of Agriculture of the Kabardino-Balkarian Scientific Center of the Russian Academy of Sciences

Sorghum-Sudangrass hybrid Navigator, originator is Stavropol NAC

Sorghum-Sudangrass hybrids, originator is Agroplasma LLC, Krasnodar

Sorghum-Sudangrass hybrid Solaris, Agroplasma LLC, Krasnodar

Soil preparation technology: autumn plowing at 25-27 cm with application of the main mineral fertilizer - P20K90, presowing cultivation on depth of 8-10 cm; 6-8 days before seeding introduce soil herbicide Doual, Avangard - 1,5 l/ha. Solid planting was carried out with C3-3.6 seed, raw planting in 70 cm with SUNP-8 seed with selection of seeding plates - 2 mm. During the growing season, 2-3 intercultivations were carried out with a cultivator with the introduction of nitrogen fertilization - 60 kg of active agent/ha. The plantings were carried out according to the years of research April 24, 2015, May 02, 2016, May 08, 2017.

Phenological observations were made in terms of age of incineration, tillering tube exit, ear formation, dairy and wax ripeness. Green forage accounting was carried out during the heading of panicles phase of sorghum-Sudan hybrids, and the aftergrass was counted as the maximum crop growth.

Table 1. Meteorological data in research years (data of the Kuyan agricultural and meteorological post, village Opytny village, Tersky district of the Kabardino-Balkarian Republic).

No.	Month decade		Rainfall, mm						max, min		
			2015		2016		2017		2015	2016	2017
			by decades	total	by decades	total	by decades	total	max min	max min	max min
1	January	1	8.8	12.9	6.9	48.4	7.5	23.8	7.5	11.50	7.5
		2	0		41.1		-20.5		-25.0	-21.0	-20.5
		3	4.1		30.4						
2	February	1	4.3	12.2	4.1	19.6	21.0	1.3	13.5	21.0	21.0
		2	7.7		15.5		-21.0		-8.0	-6.5	-21.0
		3	0.2		0						
3	March	1	13.7	41.7	4.1	33.4	22.9	18.3	16.0	18.0	22.9
		2	4.7		10.7		-4.0		-5.0	-5.0	-4.0
		3	23.3		18.6						
4	April	1	5.1	14.2	1.2	65.2	26.5	40.5	29.5	29.0	26.5
		2	3.9		35.9		-3.5		0.5	2.0	-3.5
		3	5.2		28.1						
5	May	1	35.8	104.5	27.9	81.3	27.5	79.5	32.0	28.0	27.5
		2	34.7		4.7		6.5		7.5	8.0	6.5
		3	14.0		48.7						
6	June	1	26.8	61.3	139.2	151.9	35.0	69.6	35.0	33.5	35.0
		2	5.2		10.4		12.0		11.0	10.5	12.0
		3	29.3		2.3						
7	July	1	0.5	19.8	77.8	136.2	40.5	34.3	40.0	35.0	40.5
		2	3.5		33.1		13.5		14.0	13.5	13.5
		3	15.8		25.3						
8	August	1	19.4	37.9	2.2	6.7	41.5	23.6	41.0	38.5	41.5
		2	4.5		4.5		16.0		11.5	16.0	16.0
		3	14.0		0						
9	September	1	0	0	8.5	44.9	35.5	13.7	37.5	32.0	35.5
		2	0		2.2		-1.0		7.0	6.0	-1.0
		3	0		34.2						
10	October	1	21.3	37.5	4.2	67.1	23.5	41.2	27.5	25.0	23.5
		2	15.8		21.2		-2.5		-2.0	-1.5	-2.5
		3	0.5		41.7						
11	November	1	0	14.1	0.3	9.1	14.0	13.6	18.5	19.5	14.0
		2	14.1		8.5		-5.0		-3.0	-7.5	-5.0
		3	0		0.3						
12	December	1	5.4	26.8	5.3	23.9	13.0	40.7	16.0	10.0	13.0
		2	0.8		18.6		-7.5		-10.0	-16.0	-7.5
		3	20.6		0						

Table 2. Field experience scheme, seeding rates, planting methods for grass sorghum

No.	Experimental option	Planting methods and seeding rates	
		solid planting 15 cm thousand pcs./ha	**band planting 70 cm thousand pcs./ha**
1	Sudan grass, Shirokolistnaya variety	1400	800
2	Sorghum-Sudangrass Navigator hybrid	1200	650
3	Sorghum-Sudangrass Sabantuy hybrid	1200	650
4	Sorghum-Sudangrass Solaris hybrid	1200	650

Scientific research on the study of these crops is carried out in a four-pole crop rotation, the location of plots in the experiment is systematic, and the placement of three tiers is by freights (Table. 2). The experiment is one-factor, the area of the plots is 36 m2, the repetition is three-fold.

Biometric measurements and inventories identified the following characteristics: plant height, tilling capacity, number of plants per square meter, percentage of leaves and stems in total mass. All measurements were made before the green forage of each bite was harvested. There were two hay harvest during the growing season. The first hay harvest of Sudan grass and sorghum-Sudangrass hybrids was carried out on 55-60 days after the sprouting, or 5-6 days before the start of the ear formation, and the second was carried out 40-45 days after the first hay harvest. From each plot, two test sheaves of 5 kg were selected to determine the yield of hay, the structure of the crop (percentage of leaves and stems) and the energy content of green and dry forage.

The Sudan grass Shirokolistnaya variety was used as the standard. Statistical processing of experimental data was performed according to B.A. Dospekhov (1985).

3 Research results

One of the most significant indicators giving an initial estimate of the studied hybrids is their morphological characteristics. The plant height varies considerably from year to year due to different water availability and growing medium temperature (Table. 3).

Table 3. Analysis of the height of plants studied in the experiment, cm

No.	Experimental option	Hay harvests	Years of research							
			2015		2016		2017		Three-year average	
			Solid planting	Band planting	Solid planting	Band planting	Solid planting	Band planting	Solid planting	Band planting
1	Sudan grass Shirokolistnaya (standard)	1	135	142	215	235	186	205	178	194
		2	95	102	185	195	156	146	145	147
2	Sorghum-Sudangrass Navigator hybrid	1	144	170	265	278	242	255	217	234
		2	123	134	202	226	189	194	172	185
3	Sorghum-Sudangrass Sabantuy hybrid	1	152	174	275	284	246	264	224	240
		2	128	132	215	235	200	210	181	192
4	Sorghum-Sudangrass Solaris hybrid	1	148	166	270	276	232	247	216	229
		2	121	134	225	227	204	193	183	185

On average, in three years of research, plant heights in solid planting in the second hay harvest ranged from 95 to 185cm in the Sudan grass population, with the highest being the sorghum-Sudan Sabantuy hybrid in band planting - 174-284 cm.

One of the significant characteristics of plants is tilling capacity. Comparing the number of stems, it should be noted that before the second hay harvest in all variants, this number is greater than in the first harvest (Table.4).

Table 4. Analysis of tilling capacity depending on the hay harvest and planting method, (pcs/plant) by years of research

No.	Experimental option	Hay harvests	Years of research							
			2015		2016		2017		Three-year average	
			Solid planting	Band planting	Solid planting	Band planting	Solid planting	Band planting	Solid planting	Band planting
1	Sudan grass Shirokolistnaya (standard)	1	3.2	4.8	5.4	6.2	4.4	4.9	4.4	5.3
		2	4.6	6.3	6.2	7.0	5.1	5.0	5.3	6.1
2	Sorghum-Sudangrass Navigator hybrid	1	4.3	5.1	7.2	8.8	6.0	6.9	5.8	6.9
		2	4.9	5.9	7.8	7.3	6.3	6.5	6.3	6.6
3	Sorghum-Sudangrass Sabantuy hybrid	1	4.1	5.6	7.9	9.3	6.0	6.8	6.0	7.2
		2	4.6	6.1	9.3	10.1	6.5	6.1	6.8	7.4
4	Sorghum-Sudangrass Solaris hybrid	1	3.8	4.9	7.6	9.1	6.9	7.3	6.1	7.1
		2	4.4	5.1	8.8	10.0	6.4	6.8	6.5	7.3

As can be seen from Table 4, over the years of research, the tilling capacity of the investigated plants varies from year to year and from harvest to harvest. Compared to the Sudan grass Shirokolistnaya variety (standard), sorghum-Sudangrass hybrids have a larger tilling capacity of 1.5-2 times, while after the second harvest, according to this indicator the sorghum-Sudangrass Solaris and Sabantuy hybrids can be distinguished - 10.0-10.1 respectively.

The quality of the green and dry forage of plants depends on the size of the leaves in total mass and on the genotype of the variety or hybrid, as well as on the weather conditions of plant development. In years with the abundant precipitation, the leafiness of sorghum-Sudangrass hybrids and Sudan grass (in percentage terms) will be lower than in dryland years. In humid years, the plant height reaches 2.5-2.8m, in dryland years - 130-170cm, therefore internodes are longer and the number of leaves per plant is equal. Accordingly, the percentage of stems in the total plant mass in a wet year will be higher and leaves will be lower (Table. 5).

As can be seen from Table 5, the lowest green forage yield in years of the research was obtained in 2015 due to soil and air drought, which hindered normal plant development.

It should be noted that the yields of sorghum-Sudangrass hybrids exceed the standard for years of the research and for hat harvests. On average over the years of research, there is an advantage of first hay harvest in all experiments, due to the high temperatures in mid-summer and lack of moisture to fully develop the second hay harvest (Table. 6).

Table 5. Appearance of plants of Sudanese grass and sorghum-Sudanese hybrids, depending on sowing and stingray, %, Leafiness of Sudan grass and sorghum-Sudangrass hybrids, depending on the planting and hay harvest method, %

No.	Experimental option	Hay harvests	Years of research							
			2015		2016		2017		Three-year average	
			Solid planting	Band planting	Solid planting	Band planting	Solid planting	Band planting	Solid planting	Band planting
1	Sudan grass Shirokolistnaya (standard)	1	34.2	38.3	31.6	33.8	33.6	34.5	33.1	35.5
		2	38.3	41.1	34.5	34.6	36.2	35.8	36.3	37.1
2	Sorghum-Sudangrass Navigator hybrid	1	43.1	51.3	35.8	38.2	40.2	41.8	39.7	43.7
		2	46.2	53.2	37.6	41.6	41.6	43.6	41.8	46.1
3	Sorghum-Sudangrass Sabantuy hybrid	1	46.1	53.6	37.3	39.6	43.1	44.6	42.1	45.9
		2	49.5	54.2	39.8	42.3	45.3	45.2	44.8	47.2
4	Sorghum-Sudangrass Solaris hybrid	1	44.5	48.3	36.8	39.9	4.01	43.2	40.4	43.8
		2	45.2	51.3	38.6	43.6	42.3	45.6	42.0	46.8

Table 6. Yield of green forage of Sudan grass and sorghum-Sudangrass hybrids by year and harvests, dt/ha

No.	**Variants option**	**Hay harvests**	**Years of research**						**Yield of green forage, average for three years, dt/ha**		**Yield of air-dry forage, average for three years, dt/ha**	
			2015		**2016**		**2017**					
			Solid planting	**Band planting**	**Solid planting**	**Band planting**	**Solid planting**	**Band planting**	**Solid planting**	**Band planting**	**Solid planting**	**Band planting**
1	Sudan grass Shirokolistnaya(standard)	1	70.3	78.6	232.3	258.1	139.6	155.8	147.4	161.4	36.8	41.0
		2	63.8	69.8	215.6	240.6	122.3	141.2	133.9	150.4	33.5	37.6
	For 2 hay harvests		134.1	148.4	447.9	498.7	261.9	297.0	281.3	311.8	70.3	78.6
2	Sorghum-Sudangrass Navigator Navigator hybrid	1	92.3	116.6	261.9	278.9	142.5	166.3	165.6	187.3	41.4	46.8
		2	83.6	107.0	245.8	256.3	131.2	151.2	153.5	171.5	38.3	42.8
	For 2 hay harvests		176.8	223.6	507.7	353.2	273.7	317.5	319.1	358.8	79.7	89.6
3	Sorghum-Sudangrass Navigator Sabantuy hybrid	1	96.3	124.3	268.3	286.8	153.6	169.8	172.7	193.6	43.1	48.4
		2	83.8	112.5	251.6	263.9	141.2	152.3	158.8	176.2	39.7	44.0
	For 2 hay harvests		180.1	236.8	519.9	550.7	845.5	322.1	331.5	369.8	82.8	92.4

Table 6. Continued.

4	Sorghum-Sudangrass Navigator Solaris hybrid	1	94.6	118.6	245.6	273.8	149.3	164.3	163.2	185.6	40.8	46.4
		2	82.8	108.3	230.1	252.7	138.7	149.8	150.5	170.2	37.6	42.5
	For 2 hay harvests		177.4	226.9	475.7	526.5	288.0	314.1	313.7	355.8	78	88.9

Table 7. Productivity and nutritional value of Sudan grass and sorghum hybrids (average for 2015-2017)

No.	Variants option	Crop yield, dt/ha				Digestible protein kg/ha				Feed units dt/ha				Carotene dt/ha			
		green forage		dry forage		green forage		dry forage		green forage		dry forage		green forage		dry forage	
		15 cm	70 cm	15 cm	70 cm	15 cm	70 cm	15 cm	70 cm	15 cm	70 cm	15 cm	70 cm	15 cm	70 cm	15 cm	70 cm
1	Sudan grass Shirokolistnaya (standard)	281.3	311.8	70.3	78.6	78.7	87.3	45.7	51.1	73.2	81.1	36.5	40.8	16.9	18.7	1.49	1.57
2	Sorghum-Sudangrass Navigator hybrid	319.1	358.8	79.7	89.6	63.8	71.7	51.8	58.8	82.9	93.3	41.4	46.6	19.1	21.5	1.59	1.79
3	Sorghum-Sudangrass Sabantuy hybrid	331.5	369.8	82.8	92.4	66.3	74.0	53.8	60.1	86.9	96.1	43.0	48.1	19.9	22.2	1.66	1.85
4	Sorghum-Sudangrass Solaris hybrid	313.7	355.8	78.4	88.9	62.7	71.6	51.0	57.8	81.5	95.5	40.7	46.2	18.8	21.3	1.57	1.78

Alyzing the productivity of the investigated options over all years of research, sorghum-Sudangrass hybrids are above the standard for all investigated indicators and can be recommended for cultivation in the dry steppe zone for green forage and hay (Table 7).

The largest share in the structure of the cost of livestock products is the feed cost. Consequently, the cultivation of sorghum-Sudanicgrass hybrids with lower mass cost is one way of reducing production costs (Table 8).

Table 8. Cost-effectiveness of green forage sorghum-Sudangrass hybrid production Navigator, Sabantuy and Solaris

No.	Indicators	Variety, hybrid			
		Sudan grass Shirokolistnaya (standard)	Sorghum-Sudangrass Navigator	Sorghum-Sudangrass Navigator Sabantuy hybrid	Sorghum-Sudangrass Navigator Solaris hybrid
1	Crop yield feed units dt/ha	81.1	93.3	96.1	95.5
2	Prime cost 1 dt. feed unit/ ha (rubles)	148.80	129.34	125.57	126.36
3	Cost 1 dt. feed unit/ ha (rubles)	210.00	210.00	210.00	210.00
4	Cost cost feed units (rubles)	17.031	19.593	20.181	20055
5	Expenses Per 1 ha (rubles)	12.068	12.068	12068	12068
6	Net operating income	4963	7525	8113	7987
7	Profitability level, %	1.41	1.62	1.67	1.66

According to the information in Table 8, it can be concluded that sorghum-Sudangrass hybrids exceed the standard in terms of gross green forage charge. The cost of feed units for sorghum-Sudangrass hybrids is lower - 125.59-129.34 rubles. compared to the standard - 148.80 rubles. The profitability of Sudan grass is 1.41%, while that of sorghum-Sudangrass hybrids is 1.62-1.67%. At the same cost, the net profit increases almost twofold.

4 Conclusion

The research carried out in 2015 - 2017 on the productivity of sorghum-Sudangrass hybrids at the experimental site of the Institute of Agriculture of the Kabardino-Balkarian Scientific Center of the Russian Academy of Science in the dry-steppe zone of the Kabardino-Balkarian Republic leads to the following conclusions:

1. In the unstable moisture zone, the studied sorghum-Sudangrass hybrids Navigator, Sabantuy and Solaris have proved promising, capable of producing high yields of green and air-dry forage, well-eaten by livestock animals.
2. It is recommended to carry out crops in a band manner (60-70 cm), which makes it possible to combat weed vegetation, to apply mineral fertilizers in the form of a feed during intercropping, and to irrigate on raws.
3. Recommended planting standards for sorghum-Sudangrass hybrids Navigator, Sabantuy and Solaris for band planting - 600-700 thousand pcs. /ha, with solid planting - 1100-1200 thousand pcs./ha.

In order to obtain stable yields of 30-35 t/ha of silage in steppe areas with sufficient moisture, it is recommended to apply mineral fertilizers at a dose of N90P120K60 kg/ha of fractions: phosphorus potassium for plowing, nitrogen for feeding.

References

1. A. V. Alabushev, N. A. Kovtunova, E. A. Shishova, Fodder production, **11,** 33 (2015)

2. A. V. Alabushev, *Technological methods of cultivation and use of sorghum* (2007)
3. E. I. Bazarev, *Agroenergetika* (1987)
4. A. B. Volodin, M. P. Zhukova, Feed Production, **4**, 11 (2002)
5. A. B. Volodin, Feed production, **4**, 16 (2015)
6. A. B. Volodin, R. I. Kapustin, Yu. P. Danilenko, *Recommendations for the cultivation of sorghum for grain, silage and green forage in the Stavropol Territory* (2015)
7. B. A. Armor, *Field experiment technique* (1985)
8. *Recommendations for the cultivation of sugar sorghum* (2013)

Assessment of yield and quality of winter barley grain in southern Russia

Khamid Malkanduev, *Rustam* Shamurzaev, and *Aminat* Malkandueva

Branch of the Kabardino-Balkarian Scientific Center of the Russian Academy of Sciences, 224, Kirova Street, 360004 Nalchik, Kabardino-Balkarian Republic

Abstract. A new high yielding winter barley of Madar variety is presented. One of the main conditions to obtain high yields is the use of new varieties adapted to the climatic conditions of a particular region. Therefore, the objective of our research, when creating the variety, was to increase productivity by improving adaptive properties and elements of productivity of plants and heads. As a result of joint work with breeders of FSBSI of the National Center of Grain named after P.P. Lukyanenko a new variety of winter barley Madar for forage direction was created, with a potential yield of more than 80 dt/ha. For the years of the comparative testing at FSBSI of the National Center of Grain named after P.P. Lukyanenko by predecessors sunflower, winter wheat and white mustard, yields of the variety were 118.7 dt/ha; 87.5 and 68.4 dt/ha respectively. In the tests on seeding time, yields in the variety ranged from 65.3 to 81.7 dt/ha. The new variety is resistant to lodging, winterhardiness and drought resistance. The variety is suitable for cultivation according to all technologies; it can be cultivated in all zones of Russia. The estimated economic effect of the new variety is an increase in yields from 3 to 8dt/ha.

1 Introduction

Winter barley is the most important fodder, food and industrial crop, which has a significant share in the structure of sown areas of grain crops in the Russian Federation. It is widely cultivated all around the world and has an area of about 60 million ha. In the Russian Federation, the Southern and North Caucasian Federal Districts are the main cultivating regions [1].

Growth in grain production remains a major problem for the development of Russia AIC, including the North Caucasus, and winter barley is one of the most productive crops in the area. It makes better use of autumn-winter moisture reserves, uses them more sparingly, gives a much higher yield than the spring one, making cultivation of this crop more profitable [2].

The main challenge in creating a new variety is not only to use new and better specimens for hybridization, but also to create a local breeding material that is adaptable to specific conditions. Only in this way can work on the further improvement of breeding methods and the development of new varieties [3].

Winter barley has several advantages over spring barley and other grain crops. It is capable of producing grain yields that are on average by 1.6 times exceeding the yield of spring barley, and in some years by 1.9-3.0 times. Increasing grain yields is an important condition for crop production. There are two main lines for solving this problem. First, the creation of new high yielding varieties and second, the improvement of their cultivation technology [4].

The role of the variety in increasing crop yields is indisputable. New varieties approved for use in a region have higher yields or other agronomic characters. The rapid introduction of new high yielding varieties in production will lead to a minimum of 20-30% increase in product yields [5].

A significant role in grain balance is attributed to barley as a particularly valuable culture of versatile use (fodder, beer, cereal, green feed, hay, etc.), which occupies the second place. In the structure of sown areas, it is allocated between 15% and 30% [6].

In the study of the gene pool of the source material, it is important to choose the right selection criteria, taking into account the conditions of the environment. For this purpose, it is recommended to test varieties on different background and under different environmental conditions. Such studies enhance knowledge of ecological and geographical variability and species response rates, allowing to distinguish forms with agronomic characters [7].

Extensive technologies cannot be discarded with limited material resources since on fertile soils, well-fertilized predecessors, under normal weather conditions and timely and high-quality operations in the North Caucasus region, winter crops are capable of grain yielding more than 3 t/ha [8].

Taking into account the clear advantage of winter barley in terms of crop yields over spring barley and the observed climate warming, in order to increase gross grain production at the same cost, it is advisable in the southern region of the Russian Federation to increase the seeding area up to 1 million ha. The proposed production of new highly adaptive barley varieties and advanced crop agronomic techniques will be more conducive to a successful resolution of this issue [9].

Drought is an important factor limiting crop production in arid and semi-arid environments. Therefore, the use of high yielding drought-resistant varieties is an effective approach to reduce its harmful effects. Growing barley in drylands requires a reliable assessment of the variability of drought-resistance among breeding varieties. These varieties can also be used in barley breeding programs to increase drought-resistance [10.11.12].

2 Materials and Methods

Ecological varieties were carried out according to the methodology of the State Commission for the Testing and Protection of Breeding Achievements, the method of ecological grading of grain crops of FSBSI of the National Center of Grain named after P.P. Lukyanenko, the field experiment technique (B.A. Dospekhov, 1985) on the basis of the Institute of Agriculture KBSC RAS, in the steppe zone of the Kabardino-Balkarian Republic, in the conditions of the Krasnodar and Stavropol territories, the Republic of Crimea [13,14,15].

The registered area of the plots is 25 m², repetition is 3-fold, the predecessor - peas, the seeding rate is 4.0-4.5 million seedlings per 1 ha. Mineral fertilizers at a dose of $N_{60}P_{60}K_{30}$ were used for basic tillage. Feeding with a dose of N_{30} was carried out in the tube exit phase. The seeding was carried out by Klen-1.5 seed planter. The harvesting was carried out with a Terrion-2010 combine.

3 Results and discussion

The Madar fodder winter barley is obtained by intra-specific hybridization and individual selection from a combination of 386-2/Parallelum 1633, 386-2(330-1/Kozyr), 330-1 (Radikal/Birgit). The variety plant height reaches 110 cm, resistance to lodging is high - 5 points, the number of grains in the head - 59 pcs. Winterhardiness is 100%, drought resistance is high - 5 points. Grain unit - 684 g/l, weight of 1,000 grains - 38.9 g, protein content - 10.3%. The growing season is 250 days. The variety has average field resistance to mildew, net blotch, has tolerance to leaf rust, susceptible to smut types.

In the ecological test of the Institute of Agriculture of the Kabardino-Balkarian Scientific Center of the Russian Academy of Sciences (steppe zone of the Kabardino-Balkarian Republic, 2018-2020), the average yield of the Madar variety was 51.4 dt/ha or exceeding the Samson standard of 7.7 dt/ha. The Insar variety exceeded the Samson standard by 2.3 dt/ha. In the Madar variety, on average for three years, it was obtained 5.4 dt/ha more compared to Insar variety. The maximum yield by varieties was obtained in 2018, 54.7 dt/ha were obtained for the Madar variety (Table 1).

Table 1. Yield of winter barley varieties in Kabardino-Balkarian Republic steppe zone (according to the Institute of Agriculture of the Kabardino-Balkarian Scientific Center of the Russian Academy of Sciences, the predecessor was a pea, 2018-2020)

Variety/line	**Yield, dt/ha**				
	2018	**2019**	**2020**	**Medium**	**± to the standard**
Samson, (st)	48.3	37.0	45.8	43.7	-
Madar	54.7	45.7	53.8	51.4	+7.7
Insar	49.3	42.5	46.1	46.0	+2.3
LSD $_{05}$	2.9	3.2	3.0		

Under production conditions (Kabardino-Balkarian Republic foothill zone, Nartan village, 2018-2020) 50.0 dt/ha were obtained from an area of more than 150 ha for the Madar variety.

In conditions of Krasnodar territory in 2015-2017 the new Madar winter barley variety exceeded the standard Samson variety by 9.3-18.8 dt/ha, Rubezh variety by 2.5-12.1 dt/ha. An even higher yield of 92.2 dt/ha was obtained at the North Kuban Experimental Station. The maximum yield in the Krasnodar territory for the Madar winter barley variety was obtained in 2015, which amounted to 98.2 and 98.8 dt/ha (Table 2).

Table 2. Yields of winter barley varieties in the Krasnodar territory (according to the National Center of Grain named after P.P. Lukyanenko, 2015-2017)

Variety/line	**Yield, dt/ha**				
	2015	**2016**	**2017**	**Medium**	**± to the standard**
Samson, st	86.6	66.0	67.3	73.3	-
Rubezh	94.8	67.5	78.2	80.1	+6.8
Madar	98.2	70.2	79.5	82.6	+9.3
Madar (North Kuban Agricultural Experimental Station)	98.8	88.2	89.2	92.2	+18.9
LSD $_{05}$	3.2	3.5	3.3		

According to the grain quality, the Madar variety in the conditions of Krasnodar territory on average for the period from 2015 to 2017 was at the level of the Samson standard. The protein content of grains in Madar variety ranged from 9.6% to 10.8%, the weight of 1,000 grains from 35.8 to 40.7 g, at 10.2% to 10.5% and 40.5 g to 41.7 g in Samson variety, respectively. By weight of 1,000 grains, the Madar winter barley variety, on average, was inferior to the Samson variety by 2.9 g. Compared to the best similar variety of winter barley Rubezh, the Madar variety, on average for three years, was at its level, both in protein content and in weight of 1,000 grains (Table 3).

Table 3. Grain quality of winter barley varieties (according to the National Center of Grain named after P.P. Lukyanenko, 2015-2017)

Variety	**2015**		**2016**		**2017**		**average**	
	Protein, %	**Weight of 1,000 grains, g**	**Protein, %**	**Weight of 1,000 grains g**	**Protein, %**	**Weight of 1,000 grains, g**	**Protein, %**	**Weight of 1,000 grains, g**
Samson, st	10.5	41.7	10.5	40.5	10.2	43.2	10.4	41.8
Rubezh	9.7	39.1	10.3	35.4	10.7	38.4	10.2	37.6
Madar	9.6	40.7	10.6	35.8	10.8	40.2	10.3	38.9

Under conditions of insufficient moisture in the Kabardino-Balkarian Republic, the protein content in the grain of the Madar variety varied from 12.2% to 12.7%, while in the Samson standard these indicators varied from 11.3% to 12.5%. On average over three years, the Madar variety exceeded the standard by 0.6%. Among the studied varieties in terms of protein content, the Insar variety is notable - 12.9% (Table 4).

Table 4. Grain quality of new winter barley varieties of the Institute of Agriculture of the Kabardino-Balkarian Scientific Center of the Russian Academy of Sciences (Kabardino-Balkarian Republic steppe zone, predecessor was peas, 2018-2020)

Variety	**2018**		**2019**		**2020**		**average**	
	Protein, %	**Weight of 1,000 grains, g**	**Protein, %**	**Weight of 1,000 grains, g**	**Protein, %**	**Weight of 1,000 grains, g**	**Protein, %**	**Weight of 1,000 grains, g**
Samson, st	11.9	40.3	12.5	41.2	11.3	48.5	11.9	43.3
Insar	11.4	41.0	12.8	36.4	14.4	38.9	12.9	38.8
Madar	12.2	42.7	12.5	38.5	12.7	46.0	12.5	42.4

When assessing the grain quality of the Madar variety in the Krasnodar territory and Kabardino-Balkaria, it can be noted that on average, in terms of protein content and weight of 1,000 grains, the variety had advantages in the conditions of Kabardino-Balkarian Republic. In terms of protein content, the excess was 2.2% and in terms of weight of 1,000 grains was 3.5 g.

In the study of a Madar winter barley variety at the various plots of the Russian Federation, positive results were obtained. Thus, when tested in the conditions of the Crimea for the Madar variety, the yield exceeded the standard Kondrat variety by 2.5 dt/ha, the excess by the weight of 1,000 grains was 2.1 g. In terms of lodging and drought resistance, the variety was at the standard level. The Madar variety exceeds the Kondrat variety in height by 4.6 cm (Table 5).

Table 5. Comparative assessment of winter barley varieties at variety plots (2018-2020)

Variety	**Crop yield, dt/ha**	**Plant height, cm**	**Forage 1,000 grains, g**	**Lodging resistance, point**	**Lodging resistance, point**
Republic of Crimea					
Kondrat, st	57.4	89.2	34.2	5	5

Table 5. Continued.

Madar	59.7	93.8	36.3	4.5	4
LSD 05					
Stavropol Territory					
Average standard	50.3	83.3	38.7	4	4
Madar	56.5	91.7	39.7	4	4
LSD 05					

For the Stavropol region in terms of yields, Madar variety exceeded the standards on average years of research by 6.2 dt/ha, by weight of 1,000 grains per 1.0g, by plant height by 8.4cm. In terms of lodging and drought resistance, the Madar winter barley variety was at the standard level.

In terms of disease resistance, the Madar winter barley variety has an average field resistance to mildew, net blotch, has tolerance to leaf rust, susceptible to smut types (Table 6).

Table 6. Disease incidence, %
(according to the National Center of Grain named after P.P. Lukyanenko, 2015-2017)

Causal agent agent	Madar			Samson, st			Indicator variety		
	2015	2016	2017	2015	2016	2017	2015	2016	2017
In competitive variety testing									
Rust types:							Mikhailo		
leaf rust	0	0	0	0	0	0	0	0	5
Smut types:									
							Sekret		
head smut	0	0	0	0	0	0	0	0	0
kernel smut	0	0	0	0	0	0	0	0	0
							Dobrynya 3		
mildew	5	0	15	0	0	10	15	20	30
							Kondrat		
root rot	10	15	15	30	50	30	30	50	50
With artificial infection:									
							Mikhailo		
leaf rust	20	30	30	20	40	30	50	50	70
									Sekret
head smut	15.8	12.0	9.7	16.5	12.5	10.4	22.4	17.2	10.4
kernel smut	4.5	4.7	4.0	4.0	3.8	4.2	6.7	7.2	5.4
							Dobrynya 3		
mildew	50	30	55	50	30	60	50	50	70
							Romans		
net blotch	50	30	50	70	50	50	80	60	80

The Madar winter barley is resistant to root rots, on average over three years, the susceptibility was 10-15%, with 30-50% for the Samson standard and Kondrat indicator variety. The susceptibility to leaf rust of the Madar variety was at the level of 20-30%, while the Mikhailo indicator variety was 50-70%. On average, during the years of research, the susceptibility of Madar variety with mildew at the level of 30-55%, with the susceptibility of

Dobrynya 3 by 50-70%, and the Samson standard by 30-60%. The susceptibility to the net blotch of the new Madar variety varied between 30 and 50 % during the research period, with the Romans indicator variety by 60-80 % and the Samson standard by 50-70 %.

The bush of the Madar winter barley variety has an intermediate shape, the stem is medium, hollow, strong. The head is cylindrical, of medium length, straw-yellow, the density of the segments of the head is 16–18. Glume is small, narrow, nervation is weak. The leaf is not pubescent, with a weak wax bloom during tillering, dark green in color, intermediate type. The color of the stem nodes is brownish, the ears are sickle-shaped, the tongue is ordinary. The awns are long, half-pressed, straw-yellow in color. The grain is medium in size, the base of the grain is bare, the character of the seta at the base of the grain is short. The main morphological feature of the new variety, which makes it possible to distinguish it from other varieties is a very weak anthocyanin coloration of the ears, a strong wax bloom on the spathe. Tilling capacity of the variety is 2.8, resistance to lodging - 5 points, ability to shed and brittleness of the head, the degree of flagging of the head - 1 point, resistance to germination at the root - 5 points, resistance to soaking and damping off - 5 points. The variety is highly resistant to autumn and spring frosts (5 points). The critical freezing temperature for the Madar winter barley variety is -14C. The grain has good thrashing capacity.

The most important distinctive biological characteristics of the variety are ear formation two days earlier than the standard Samson variety, possessing an elastic, strong straw, and high lodging resistance. The optimum seeding rate for the Madar variety is 140-160 kg/ha.

4 Conclusion

The search and attraction of created sources in breeding programs aimed at increasing the productivity and adaptability of winter barley will allow to develop a new breeding material for highly productive varieties.

The main advantages of the new variety Madar are expressed in an increase in yield over the standards by 6-18 dt/ha. In terms of grain quality indicators, the new variety exceeds the standards or is at their level.

The estimated economic effect from the use of the Madar variety is on average 3-8 dt/ha higher compared to the standards.

The tests performed in the conditions of the North Caucasus (6) region make it possible to recommend the Madar variety for wide use in breeding programs and for implementation into production. The variety is included in the State Register of Breeding Achievements of the Russian Federation in 2020 and is protected by a patent. The variety differs in high yields, resistance to lodging, higher productivity of plants and heads, average frost resistance. Suitable for cultivation in the southern regions of Russia.

References

1. N. V. Repko, K. V. Smirnova, K. V. Sukhinina, A. S. Koblensky, Scientific Journal of KubSAU, **1** (2016)
2. Kh. A. Malkanduev, N. V. Serkin, A. Kh. Malkandueva, R. I. Shamurzaev, V. V. Nesterenko, News of Kabardino-Balkarian Scientific Center of RAS, **39** (2020)
3. A. A. Bisharev, V. A. Shevchenko, I. A. Kalyakulina, M. A. Dyuldina, Izvestia RAS SamSC, **612** (2017)
4. A. A. Bisharev, V. A. Shevchenko, V. A. Zheleznikova, I. A. Kalyakulina, M. A. Dyuldina, Izvestia RAS SamSC. 478 (2018)

5. N. V. Repko, K. V. Smirnova, A. S. Koblensky, Scientific Journal of KubSAU, **1** (2015)
6. E. G. Filippov, A. A. Dontsova, D. P. Dontsov, A. A. Bulanova, Agrarnyi vestnik Urala, **75** (2017)
7. I. S. Sysenko, S. I. Novoseletsky, O. E. Patseka, Scientific Journal of KubSAU, **1** (2016)
8. A. S. Ereshko, V. B. Khronyuk, N. V. Repko, Vestnik agrarnoj nauki Dona, 57 (2012)
9. Yu. P. Pryadun, Izvestia Orenburg State Agrarian University, 84 (2019)
10. Mahdi Zare, African Journal of Biotechnology, 15975 (2012)
11. Mehdi Feizi, Mahmood Solouki, Behzad Sadeghzadeh, BaratAli Fakheri, Seyed Abolghasem Mohammadi, Bioscience Journal, 1518 (2020)
12. Saeideh Dorostkar, Hassan Paknyiat, Mahmood Ahmadi Kordshooli, Raziyeh Ghorbani, Massumeh Aliakbari, Neda Sobhanian, Reza Valiloo, Science Research, **26** (2016)
13. *Methodology of state variety testing of agricultural crops,* 2 (1989)
14. N. I. Dzhelali, *Guidelines for ecological variety testing of grain crops* (1985)
15. B. A. Dospekhov, *Field experiment technique* (1985)

Impact of main tillage methods on weed infestation of crops in the winter maize, wheat, dried peas rotation

Khasan Tarchokov, *Magomed* Chochaev, and *Julietta* Tutukova

Institute of Agriculture - Branch of the Kabardino-Balkarian Scientific Center of the Russian Academy of Sciences, 224, Kirova Street, 360004 Nalchik, Kabardino-Balkarian Republic

Abstract. The paper presents ways of suppressing weed plants, including a complex of agricultural practices aimed at reducing their harmful effects and further distribution. Many methods to control weed vegetation are known in the field cultivation technology. They differ in terms of the amount of expired costs and labor, economic and technical efficiency. As a result of our research, the impact of some tillage methods on the effectiveness of weed suppression on the main field crops in selected Kabardino-Balkaria natural climatic conditions is identified. The goal of the research is to determine the effectiveness of main tillage methods on the Ciscaucasian carbonate chernozems in different zones of the republic. In research conducted in 2015-2018 in the Kabardino-Balkaria steppe zone in crop rotation, with an alternation of maize-winter wheat-pea crops in the suppression of weed plants, a soil treatment system is allocated when the soil treatment on 12-14cm is combined with plowing to a depth of 25-28 cm with the use of related chemical herbicides of various spectrum of actions.

1 Introduction

One of the main problems of the agricultural industry worldwide is the harmful effects of the weed-field community on agricultural crops. Obviously, without proper science-based control, it is meaningless to undertake activities aimed at increasing and maintaining soil fertility, protecting crops from known pathogens, pests [8, 9]. The effectiveness of different methods of weed suppression on agricultural crops in our earlier studies has been thoroughly studied and found application in agricultural products of different forms of ownership of Kabardino-Balkaria [1, 2].

Multiple studies have found that nonmoldboard tillage with high impact protects the soil from deflation and water erosion processes that contribute to moisture conservation in the soil [3, 4]. At the same time their high economic efficiency is proved, but its technical efficiency in terms of decreasing the number of weed plants is observed.

It should be noted that at present, there are still problems of soil cultivation aimed at the principles of resource and energy conservation. Therefore, on the Ciscaucasian carbonate chernozems of heavy granulometric composition, we conducted researches, which were necessary due to the fact that recently (not only in Kabardino-Balkaria) agricultural vehicles and tractors of new modernization, new chemical preparations of various selectivity and environmental friendliness have entered the market.

2 Materials and Methods

The research was carried out in the steppe zone of Kabardino-Balkaria (Opytny village, Tersky district of the Kabardino-Balkarian Republic) (Table 1). The average annual temperature was 10.1°C.

Table 1. Meteorological indicators for research years (According to the Kuyan agrometeorological post, Opytny village, Tersky district of the Kabardino-Balkarian Republic)

Year	Air temperature, °C	Rainfall, mm	Relative air humidity, %
2015	12.6	12.6 382.9	70.3
2016	14.0	571.1	72.6
2017	11.3	387.0	73.8
2018	11.4	470.0	73.3
Long-term average annual data (standard)	10.1	410.0	77.0

The amount of precipitation (452.7 mm) and relative air humidity (72.5%) during the years of the research were practically within the normal range (long-term average annual data).

The humus content in the arable soil layer was in the range of 3.0-3.5%; P_2O_5 (according to Machigin) - 0.14-0.27%; K_2O (according to Machigin) - 0.2-2.6%; PH - 6.8-7.0%. Average annual precipitation - 340-400ml [10].

In the experiments, maize seeds zoned for the steppe zone of Kabardino-Balkaria were used: Kabardinskaya 3812 variety - a hybrid population (originator of Institute of Agriculture of the Kabardino-Balkarian Scientific Center of the Russian Academy of Sciences), belongs to the late-ripening group (FAO-600), grain yield on the irrigation reaches 9.5 t and more, silage mass - 55.0-60.0t. It responds positively to a high agricultural background.

Peas. Aksaysky usatiy-55 variety (originator - Donskoy zonal Research Institute of Agriculture), belongs to the mid ripening group of varieties with a growing season of 85-90 days, leafless. Seeds are spherical, smooth with a sign of non-shedding, yellow. The weight of 100 seeds is 17-235g with a protein content in the grain of 25-27% and a maximum grain yield of 4.5-4.9 t/ha. Resistance to lodging, shedding and drought are high.

Winter wheat. Yuzhanka variety, bred by the Institute of Agriculture of the Kabardino-Balkarian Scientific Center of the Russian Academy of Sciences, has a high yield (6.5-7.9 t/ha) on condition of irrigation and good predecessors (peas, soybeans, perennial grasses). Resistant to lodging, pests and diseases.

The experiment scheme consisted of three types of main tillage for seeding maize, winter wheat and peas:

1. Cultural tillage (25-28cm)
2. Minimum tillage (12-14cm)
3. Shallow tillage (6-8cm).

Main tillage methods. In order to suppress some species of perennial weeds (Johnson grass, species of sow thistles, bindweed, etc.) after the winter wheat and peas were harvested against the background of light primary tillage with tools such as LDG, Roundup + Banvel herbicides were used at dosages of 5.0 and 0.1 l/ha. The preparations were applied to vegetative weeds with an average height of 10-15 cm with a working solution consumption of 250-300 l/ha in the summer-autumn period (August-September).

On peas areas for "blind" harrowing in the spring Prometrin (Gezvgard) at a dose of 3.5 l/ha and Bazagran at a dose of 3.5 l/ha were used in the phase of 3-5 leaves of the crop. Maize areas were treated in the phase of 3±5 leaves of the crop. The winter wheat was treated in the tillering phase in spring with Banvel (0.2 l/ha) + a mixture of falcon and Bi-58 (1.2 l/ha), and karate (0.2 l/ha) was introduced during the appearance of flag leaf.

For winter wheat and corn areas, $N_{90}P_{60}K_{60}$ was applied, including N_{70} as a top dressing, nitrogen fertilizers were not applied on pea crops and $P_{90}K_{60}$ was used for cultivation in the spring before seeding.

In the experiments, the weed infestation of crops was determined by quantitative-specific and qualitative-weight methods. [5, 6] The yield of the tested crops was determined on a plot basis, after which it was brought to 100% purity and standard moisture, calculated in tons per hectare planted area.

The agricultural technique of crop cultivation in crop rotation was in line with the recommended parameters for production crops. The by-products of maize, winter wheat and peas were crushed in the field and planted in the soil. The tillage was carried out with a DT-75 tractor with an agricultural tools (BDT-3.0 - for grinding of shredded residues after grain maize); disc cultivators (LDG-10) for treatment of the stern after winter wheat and peas, PN-4-35 plows, KPS-4.0, KPN-5.6 cultivators; SZ-3.6 seeders were also used The experimental material was statistically processed by the dispersion analysis according to Dospekhov [6].

3 Results and Discussion

Various main tillage methods in the short crop rotation (corn-winter wheat-peas) on average slightly changed practically the soil moisture with its agrophysical properties over the period 2015-2018. Thus, the amount of productive moisture in the 0-100 cm soil layer for surface and minimum tillage was 152.0 and 150.0 mm, respectively. In the cultural plowing variant, these figures were slightly higher - 160.5 mm. The soil density of 0-30 cm soil layer on average in spring ranged from 1.09 to 1.12 g/cm^3 in the variant with cultural tillage, minimal and shallow methods of main tillage.

The determination of the weed infestation of experimental plots prior to the use of herbicides in the post-harvest period showed that the main weeds of field crops under the conditions of the research were Johnson grass (Jardium Jalepense Pers.J), species of sow thistles (Jonchis arvensis.J.), and bindweed (Jonchis arvensis.J.) (Convolvulus arvense.j.). The young gramineous-dicotyledonous weed plants were a part of the redroot pigweed (Amazantus vulgaris.J.), field mustard (Jinapis arvensis.J.),regweed (Amdrozia artemisia-folia.J.), etc.

As can be seen from Table 2, the different types of main tillage in the crop rotation had a corresponding impact on the level of crop infestation.

Table 2. The effectiveness of the main tillage methods and herbicides on crops in the short rotation (the number of weeds, pcs/m^2 on average over the years of research)

Terms of accounting for weed infestation by crops	**Main tillage methods**		
	1$^{+)}$	**2$^{+)}$**	**3$^{+)}$**
Maize (on average for 2015-2018)			
Summer-autumn before the use of the herbicide mixture (average for 2014-2016)	52-7$^{*)}$	57-12$^{*)}$	69-19$^{*)}$
Before the spring cultivation of the plow	95-2	110-4	140-4
Before the grain harvesting	12	22	24

Table 2. Continued

Dried peas (on average for 2015-2017)			
Before chemical weeding of crops in spring	68-12	85-20	90-27
2 weeks after chemical weeding	25-2	30-3	35-3
Before the grain harvesting	12-1	11-2	10-1
Dried winter wheat			
Before chemical weeding of the stubble (summer-autumn 2016-2018)	78-15	83-19	85-20
In spring during the renewal of vegetation (2016-2018)	8	12	16
Before harvesting	17	20	27

+) Note:
1. Cultural tillage (25-28 cm)
2. Minimum tillage (12-14 cm)
3. Shallow tillage (6-8 cm)
*) The first number is the young gramineous-dicotyledonous group of weeds; the second - from the group of perennial weeds (mainly species of sow thistles and bindweed).

Thus, in maize crops planted after winter wheat, the smallest amount of weed plants before the use of chemical weeding in the summer-autumn period was recorded on a variant with cultural plowing - 52 plants of young weeds and 7 - perennial ones. For the variants of minimum (12-14 cm) and surface tilling, the values were 57-69 and 12-19 young and perennial weed species, respectively. Before the first spring cultivation of the plow and before the corn harvesting, the efficiency of weed suppression increases rapidly. On the background of cultural plowing, they do not exceed 25.0pcs./m^2 for a small amount (2.0 plants per 1 sq.m.) of perennials. Before harvesting maize grain, cultivated plowing for its crops in combination with chemical weeding of stubble from under winter wheat in summer-autumn period and during the growing season, practically cleared the crops of perennials. A small amount of young weeds (12.0-24.0pcs./m$^{2)}$ were underdeveloped and had no negative impact on the crop. The pea crops following the late harvesting of drain maize were not treated with a sustainable herbicide of a general destructive nature. Therefore, our research on this culture was aimed at determining the effectiveness of the use of Prometrin (a soil preparation) and Bazagran for seedlings in the 3-5-leaf phase of the crop in weed control in the context of various forms of main tillage. As can be seen from the given data in Table 2, the main tillage methods had an effect on the degree of weed infestation of crops. In the variant with cultural tilling before the use of herbicides, the amount of young weeds decreased from 85-90 to 68, perennials - from 20.0-27.0 to 12.0 plants per 1m^2.

The suppression efficiency of weeds is sharply increased when chemical weeding is applied against the background of various main tillage methods, where the total amount of weeds was only 27 versus 80 or 66.7% of their suppression. In variants with minimal and shallow methods of main tillage, these indicators were 58.3 and 68.0%. By the harvest period, the level of weed infestation of pea crops continued to decline, and was limited to between 11.0 and 13.0 plants per m^2 of area.

On the crops of winter wheat, placed on pea stubble for grain, the initial amount of weeds before various methods of main tillage and the use of chemical weeding in the summer-autumn period was 78.0-85.0 plants of young weeds and 15-20 pcs/m^2 of perennials. In the spring, during the resumption of vegetation, there is a high degree of suppression of both young and perennial weeds from the combined action of main tillage + general destructive herbicides used in summer-autumn period and during the Banvel vegetation period - 0.2l/ha.

However, to the greatest extent, the technical efficiency of herbicides can be seen against the background of cultural tillage, where the total amount of weeds in spring during the renewal of the growing season did not exceed 8.0 versus 12.0 and 16.0 plants per 1m^{2}against the background with minimal and shallow main tillage methods.

In the technology of suppressing weeds on crops of agricultural plants, along with the quantitative and species composition, the determination of their quantitative-weight (g/m^2) composition of the weed-field community is of paramount importance [5, 3]. As can be seen from Table 3, in our research for the harvesting period, the total amount of weed plants per 1m^2 was between 12.0 and 23.0 for all crops.

Table 3. Impact of different main tillage methods on weed infestation of crops$^{+)}$ and carrying out of nutrients in the link of short crop rotation before harvesting

Crop and time of accounting for weed infestation of crops	**Main tillage methods**		
	1*)	**2*)**	**3*)**
Maize (average for 2014-2016)	23-170$^{+)}$	27-210$^{+)}$	29-230$^{+)}$
Peas (average for 2015-2017)	25-150	29-170	33-190
Winter wheat (average for 2016-2018)	12-110	15-140	20-155
Total weight of weeds (average for crops); g/m^2	430	520	575
Nutrients carried out of the soil by weeds; kg/ha:			
Nitrogen	67	79	83
Phosphorus	25	30	36
Potassium	86	106	109

Note:
1. Cultural tillage (25-28 cm)
2. Minimum tillage (12-15 cm)
3. Shallow tillage (6-8 cm)

$^{+)}$ The numerator is the amount; denominator - their wet aboveground weight in g/m^2

In variants with minimal and shallow main tillage methods, they occupied the limits of 15.0-29.0 and 20.0-29.0 pics./m^2, respectively.

However, when considering these data by determining their weight per unit of planted area, there is a clear and visible difference in efficiency of weed suppression compared to the different main tillage methods. Thus, in our experiments, the largest amount by weight of aboveground wet weight (g/m^2) was recorded in the third variant (shallow tillage to a depth of 6-8 cm) - 575.0 g/m^2. In second place is the variant where the minimum (12-15 cm) tillage is carried out - 520.0 g/m^2.

In the variant with cultural tillage (25-28 cm), these indicators did not exceed 430.0 g/m^2. The reviewed data also show that this mass of 430.0g/m^2 carries out the smallest amount of nitrogen (67.0 kg/ha), phosphorus (25.0 kg/ha) and potassium (86.0 kg/ha) from the soil. The variants with minimum and shallow main tillage were: nitrogen 79.0 and 83.0 kg/ha, phosphorus 30.0 and 36.0 kg/ha and potassium 106.0 and 109.0 kg/ha, respectively.

These data indicate that even with a smaller number of weeds before harvesting, but with a greater mass, they remove a greater amount of nutrients from the soil.

Along with a change in the weed infestation of crops, various main tillage methods in combination with chemical weeding and without it, influenced the processes of the formation of the productivity of experimental crops (Table 4).

According to the research results, the highest yields of maize grain (on average for 2015-2018) were obtained on the variant with cultural tilling (25-28 cm) - 7.8 t/ha. These results exceed the data obtained on the variants with the minimum (12-15 cm) and shallow (6-8 cm) main tillage methods by 1.3 and 3.8 t/ha compared to the grain yield formulated under the condition of maize placement against the background of cultural plowing.

On the peas and winter wheat areas, grain productivity against the background of cultural tilling was also the highest, occupying the limits of 3.2 and 4.5 t/ha versus 2.8 and 3.0 t/ha against the background of minimal and shallow tillage methods.

Table 4. The impact of the main tillage methods and chemical weeding on the grain productivity of crops in the link of the short crop rotation

Main tillage methods	**Yield (t/ha)**		
	maize (2015- 2018)	**peas (2015-2017)**	**winter wheat (2016-2018)**
Cultural tillage (25-28cm)	7.8-8.3+)	3.2-4.5+)	4.5-5.9+)
Minimal (12-15 cm)	6.5-7.0	2.8-3.7	4.3-5.6
Shallow (6-8 cm)	4.0-6.3	2.0-3.0	3.0-4.6
HCP-05	0.30	0.05	0.4

+) Note: backgrounds of chemical weeding: the first number - no herbicide; the second - with the use of herbicides

The efficiency of main tillage methods is improved and characterized by the higher productivity of the studied crops, while the weeds are suppressed with the aid of chemical weeds. However, the grain productivity of winter wheat in variants with cultural tillage and minimal main tillage was almost identical (4.3-4.5 t/ha on a herbicide-free) against 5.6-5.9 t/ha on a herbicide. With shallow tillage, these rates reliably decrease to 3.0 and 4.6 t/ha.

The average crop yield of maize on a herbicidal background for all studied main tillage methods was 7.2 compared to 6.1 t/ha without the use of herbicides. The same pattern applies to pea crops (3.7 and 2.7 t/ha) and winter wheat crops (4.3 and 3.4 t/ha).

4 Conclusion

1. Weed control techniques for maize, winter wheat and peas (short rotation phase) should consist of basic and preharvest tillage, use of herbicides and their combinations in different periods - summer-autumn, growing season, which makes it possible to reduce the pollution of crops with a high technical effect.

2. The use of cultural (25-28 cm) and minimum (12-15 cm) tillage as the main tillage produces the highest yield of the seasonal crop rotation. With further minimization (shallow 6-8 cm) of main tillage for maize, peas and winter wheat, grain yields are reduced by 2.0; 1.5 and 2.9 t/ha, respectively.

3. In order to increase the technical and economic efficiency of chemical weeding in fields with the placement of maize, peas and winter wheat with mixed type of weed infestation of crops, it is necessary to use general-purpose herbicides in the summer-autumn period after the harvesting of peas and winter wheat.

4. Cultural tilling provided that winter wheat is placed over dried peas does not have a reliable advantage in the formation of grain productivity compared to the data obtained in the minimum (12-15cm) main tilling method of the Ciscaucasian carbonate chernozems of heavy granulometric composition.

References

1. P.N. Yaroslavskaya, *Peculiarities of weed control on oilseeds with anti-erosion soil cultivation. Scientific bases of soil protection agriculture in the Krasnodar Territory*, (1998)
2. Kh.Sh. Tarchokov, *Technology for growing high yields of peas in Kabardino-Balkaria*, (2020)
3. Kh.A. Malkanduev, Kh.Sh. Tarchokov, M.M. Chochaev et al, *Agrotechnology of a new generation in an adaptive landscape farming system for various natural and climatic zones of the Kabardino-Balkar Republic* (2020)
4. Kh.Sh. Tarchokov, D.A. Tutukova, *Rolling the soil and harvest of field crops* (2020)
5. A.V. Fisyunov, *Weed Control Handbook* (1984)
6. A.V. Armor, *Field experiment technique* (1985)
7. Kh.Sh. Tarchokov, *Agriculture*, **37** (2014)
8. Kh.Sh. Tarchokov et al., *Agrotechnology of a new generation in an adaptive landscape farming system for various natural and climatic zones of the KBR* (2020)
9. *Agrotechnological assessment, design of adaptive landscape systems of agriculture and agricultural technologies, edited by Acad. IN AND. Kiryushina, acad. A.L. Ivanova* (2005)
10. K.N. Kerefov, B.Kh. Fiapshev, *Soil regions of Kabardino-Balkaria and their agricultural characteristics* (1968)

Integrated protection system against progressive corn diseases

Anzor Shabatukov, and *Liudmila* Khromova

The branch of the RAS (Russian Academy of Sciences) Kabardino-Balkarian Research Center, Kirov str., 224, 360004, Nalchik, Russia

Abstract. A biologized system of integrated protection against pathogens in corn crops is an important feature to reduce pesticide load. The use of new generation chemical fungicides with a low consumption rate and a wide spectrum of action against dominant infections, as well as their manufacturing application, will contribute to the improvement of corn agrocenosis. Due to the predicted, and subsequently planned yield, the phytosanitary situation will help in the development and implementation of effective protective measures to obtain biologically cleaner production of corn grain. Seed disinfection only solves the problem of seed infection. The appearance and harmfulness of boil smut also occurs during the growing season. Therefore, the transmission of primary infection occurs by plant residues and airborne droplets, which suggests the possibility of a systemic fungicide response to effectively contain not only the causative agent of boil smut, but also other types of infectious diseases on vegetative and reproductive bodies of corn. Investigations carried out in 2018-2020 to reduce the harmfulness of blister smut and fusarium head smut showed that seed infections are caused by a pronounced prevalence of diseases of roots, stalks and ears, such as fusarium, boil smut, alternaria, bacterial leaf burn, etc. The resistance degree of cross-breeds to dominant diseases depends on the infection terms, weather and climatic conditions and which corn body is more susceptible to these pathogens. Considering the possibility of infection penetration during the post-emergent period of corn development it is required to use system fungicides with a low consumption rate and prolonged action. In the Kabardino Balkarian Republic, the dominant corn disease is boil smut. It is known that the harmfulness of boil smut is higher with cob forms than with stem, leaf forms of infection, but in 2020 only the cob form of the disease was recorded, the dry and hot weather was the reason for this. In the steppe zone of the Kabardino Balkarian Republic, it is more expedient to use effective fungicides in the developed complex protection of corn crops against progressive harmful bodies.

1 Introduction

The use of increasing volumes of chemicals and plant protection products must be put on a strictly scientific basis. Therefore, the development and implementation of new features of the integrated protection system will become an urgent requirement to obtain more environmentally friendly corn grain products. The development of such a system of protective measures in the harmful pathocomplex response is an important area of research for corn production.

The object of research is a new cross-breed of Terek corn bred by ISKh KBNTs RAN.

The relevance of research involves monitoring of diseases as a system for monitoring the dynamics of the phytosanitary condition during the period of intensive growth and development of corn.

Ensuring plant protection in the period between the emergence of seedlings and their development will prevent or reduce the pathologies incidence, thereby increasing the yield of seed and industrial crops of corn.

The **aim of the research** is to develop an ecologically safer and more economically viable system for the integrated protection of corn against diseases in modern land husbandry.

The investigation objectives are as follows:

– to determine effective fungicides in the boil smut response and other dominant phytophages on experimental crops of corn;

– to reveal the influence of growth regulators and agrochemicals on the resistance of corn cross-breeds to unfavorable environmental factors and infectious diseases.

The **scientific novelty** of the research lies in the fact that for the first time in the arid steppe zone of the Kabardino-Balkarian Republic, a comparative effectiveness assessment of the chemical and biological fungicides used in the progressive pathogens response to corn diseases has been given.

One of the conditions for obtaining high yields of corn grain is the integrated protection of crops against pests, diseases and weeds using plant protection chemicals. It is known that bacterial preparations, along with the production of antibiotics are capable of triggering induced systemic resistance of plants, which can be successfully used in biological plant protection, which in some cases makes it possible to replace chemical agents. The problem of boil smut is completely solvable both by breeding for resistance and by reducing the sources of infection in the post-emergent period through the use of fungicides with a long period of protective action. Thus, the investigations carried out confirm the positive role of treating crops with Baileton fungicide, SP in protecting corn crops against diseases and increasing grain yield.

Agrotechnical measures associated with obtaining seeds that are practically clear of seed infections and suppress their activity during the growing season require constant investigation and improvement of plant protection methods.

2 Materials and methods

Field experiments were set up according to existing techniques [1,2]. Humus reserves in the A + B horizon reach 256 tons/ha [3]. At the same time, the dependence of the appearance of progressive diseases on the main weather predictors and on the protective measures taken was revealed. At the same time, the biological effectiveness of the tested biofungicides against the indicated diseases was determined [4].

The use of plant protection products and their assessment were carried out in accordance with the "Methodological guidelines for state testing of fungicides, antibiotics and disinfectants of agricultural seeds" [5,6]. To clarify the hectare consumption rate of fungicides, microbiopreparations, agrochemicals, the corresponding catalogs were used [7, 8].

The following chemical fungicides were used as tested plant protection products: Baileton, SP (0.5 kg/ha), Abakus Ultra, SE (1.75 l/ha), Prozaro, KE (1 l/ha). The following biofungicides were tested: Orgamika S, Zh (1.4 l/ha); Pseudobacterin 3, Zh (0.7 l/ha); Fitolavin, VRK (2l/ha); Rizoplan, Zh (1.0 l/ha); BisolbiSan, Zh (2.0 l/ha); Trichotsin, SP

(0.04 kg/ha); Alirin B, TAB. (80 tab /ha); Gamair, TAB. (80 tab/ha). Chemical fungicides were used in a mixture with the Agromaster (18:18:18) agrochemical with a consumption rate of 2.5 kg/ha and the Maxifol Dynamics growth regulator with a consumption rate of 1.0 l/ha [9]. Plants were treated with preparations using the Comfort – OE-12U shoulder sprayer. The total area of the plot was 56m^2 accounting plot area was 14m^2. During the growing season, observations and checks were carried out: the emergence of seedlings, flowering of panicle and corn cob, the total number of plants at the plot. In the phase of the grain full ripeness the percentage of attacked plants was calculated and the biological effectiveness of the preparations used against boil smut and other diseases was determined. The corn cobs were harvested manually in the full ripeness phase. The grain yield was taken into account by weighing, the data were processed by the method of variance analysis.

3 Results and discussion

One of the conditions for obtaining high yields of corn grain is the system of integrated protection of crops against pests, diseases and weeds using plant protection chemicals.

Experimental attempts laid down in the NPU No. 2 ISKh KBNTs RAN in 2018-2020 confirm the positive role of treatment with chemical fungicides: Baileton, SP; Abakus Ultra, SE and Prozaro, KE in the protection of corn against boil smut and other diseases and, as a result, an increase in grain yield. The agricultural technology of producing corn seeds clear of seed infections or suppressing their activity during the growing season requires constant investigation and improvement of the integrated protection system. High efficiency of fungicide Baileton, SP application against boil smut was revealed in the treatment of experimental plots in the phenophase of 7-8 corn leaves. As a result of calculating the weighted average percentage of attacked plants and determining the biological effectiveness, it was found that these values were 1.1% and 91.1%, respectively.

In 2018-2020. on experimental crops of corn, boil smut was a progressive disease. The frequency of he bud form occurrence of boil smut was 2–3 times higher than the leaf and stem forms of this disease.

On the crops of the Terek corn cross-breed, 3 fungicides of chemical origin with different active ingredients were used: Baileton, SP; Abakus Ultra, SE and Prozaro, KE. The tests of fungicides in the boil smut response on the crops of the Terek corn cross-breed made it possible to give a comparative assessment of their biological and economic efficiency.

From the data in the table it can be seen that the option 2 shows the lowest percentage of the prevalence of boil smut – 1.1%, which is 1.3% and 1.9% lower than in options 3 and 4, respectively (Table 1).

Option 2 (Baileton, SP + Agromaster (18:18:18) + Maxifol Dynamics (st.)) shows the highest biological efficiency (91.1%), which is 10.6% and 15.5% higher than in options 3 and 4. The yield was higher in option 2 and amounted to 4.1 ton/ha, which is 0.9 ton/ha more than the control and by 0.3 ton/ha and 0.6 ton/ha in options 3 and 4.

Table 1. Comparative assessment of the biological and economic effectiveness of fungicides in the corn boil smut response (steppe zone, Terek cross-breed, 2018-2020)

№	Fungicide names	Application rate (l/ha, kg/ha)	Average % of prevalence	Biological efficiency,%	Crop yield, (ton/ha)	Addition yield, ton/ha
1	control	-	12.3	-	3.2	-
Table 1. Continued						

2	Baileton, SP +Agromaster (18:18:18) + Maxifol Dynamics (st.)	0.5 +2.5+1.0	1.1	91.1	4.1	0.9
3	Abakus Ultra, SE + Agromaster (18:18:18) Maxifol Dynamics	1.75 +2.5+1.0	2.4	80.5	3.8	0.6
4	Prozaro, KE +Agromaster (18:18:18) + Maxifol Dynamics	1.0 +2.5 +1.0	3.0	75.6	3.5	0.3
-	HCP $_{05}$					0.1

Based on the investigation results, it should be noted that all the studied tank mixtures of fungicides with agrochemicals (options 2, 3 and 4) on the crops of the Terek corn cross-breed showed a positive result in the boil smut response. Among them, the most effective in reducing harmfulness and increasing biological and economic efficiency was option 2, where a tank mixture of Baileton, Agromaster (18:18:18) and Maxifola Dynamics (st.) was used.

The obtained results of the effectiveness of plant protection products will be used in the development of an improved system, integrated protection of varieties and cross-breeds of field crops.

The experimental sowing of Terek F_1 corn was sprayed with biofungicides after the first irrigation, since a significant concentration of pathogens - polyphages - is concentrated in the upper soil layer. Plant diseases were recorded after 10, 20 and 30 days according to their symptoms, and then the causative agents of root and foot rot of different etiology were identified in laboratory conditions. At the same time, the dependence of the boil smut appearance, root and foot rot, mold fungi on weather predictors and ongoing agrotechnical measures was revealed. At the same time, the biological effectiveness of the tested biofungicides against the indicated diseases was determined .

As shown by the Table 2, all tested biofungicides in terms of biological effectiveness against root and foot rot of fungal and bacterial origin had low effectiveness. In the fusarium root rot response, variants 3 (Pseudobacterin 3, Zh), 4 (Fitolavin, VRK), 5 (Rizoplan, Zh) distinguished, the values of which were above 50% and amounted to 41.3%; 52.9%; 47.8%; 47.9%; 41.7% respectively.

In the bacterial root and foot rot response, the efficiency was 50% higher in options 3 (Pseudobacterin 3, Zh), 4 (Fitolavin, VRK), 5 (Rizoplan, Zh), 6 (BisolbiSan, Zh) and amounted to 51.7% ; 78.4%; 64.3%; 69.5% respectively. In the causative agent response of Alternaria root rot, the following options have distinguished: 6 (BisolbiSan, Zh), 7 (Trichocin, SP), 8 (Alirin B, tab.), where the biological effectiveness was important: 53.4%; 51.5%; 52.6% respectively. A representative of mold fungi – gray rot was comparatively controlled by only 2 preparations: Pseudobacterin-3, Zh - 50.3% (option 3) and Alirin-B, tab. – 55.2% (option 8).

Biofungicides indicated in Table 2 had low biological effectiveness against the causative agent of boil smut and therefore the costs of purchasing preparations and spraying will not be justified.

According to the results of three years of investigations, it can be noted that, in the steppe arid zone, the species composition of pathogens was identified on corn crops, and the dominant phytopathogens were identified among them. The most progressive corn disease is boil smut as a result of poor resistance of sown corn cross-breeds.

Table 2. Scheme for experiments set up on a biologized system of corn protection against diseases (Terek F_1, steppe zone, KBR), 2018-2020

No.	Experimental option	Consumption rate (l/ha, kg/ha, tab. /ha)	Biological efficiency,%		
			Root and foot rot	Cob mold	Boil smut
1	Control (out of treatments)	-	-	-	-
2	Orgamika S, Zh	1,40	39,80	31,10	13,40
3	Pseudobacterin-3, Zh	0,70	41,30	36,50	10,50
4	Fitolavin, VRK	2,00	52,90	38,80	16,40
5	Rizoplan, Zh	1,00	47,80	37,80	8,30
6	BisolbiSan, Zh	2,00	47,90	42,20	9,70
7	Trikhotsin, SP	0,04	33,60	34,00	14,80
8	Alirin B, Tab.	80,00	41,70	36,40	10,40
9	Gamair, Tab.	80,00	37,30	35,80	6,30

Note: spraying was carried out the next day after the first watering in phenophase of 5-6 corn leaves

4 Conclusion

Based on the three year investigation results, the following conclusions can be drawn:

1. The use of a tank mixture of agrochemical fungicides and a growth regulator was effective, since the increase in grain yield according to the options was the following: from 0.3 ton/ha (Prozaro, KE) to 0.9 ton/ha (Baileton, SP).

2. On experimental crops of the Terek corn, the high biological effectiveness of the Baileton, SP fungicide in the boil smut response was determined, where the biological effectiveness was 91.1%.

3. A comparative assessment of the biological effectiveness of 8 biofungicides showed that against the types of root and foot rot, fungal bacterial etiology, the following preparations were effective: Fitolavin, VRK; BisolbiSan, Zh and Rizoplan, Zh, where the efficiency was higher than others and met 52.9%; 47.9%; 47.8%.

4. All studied microbiological preparations operated poorly against boil smut, since their effectiveness was lower than 16.4%.

References

1. V .F. Peresypkin et al., *Workshop on the methodology of experimental work in plant protectio* (1989)
2. S. M. Pospelova, *Workshop on plant protection* (1988)
3. A. L. Ivanov, L. N. Molchanov, H. Sh.Tarchokov, M. M. Chochaev, *Features of the adaptive-landscape farming system of the Kabardino-Balkarian Republic under total* (2013)
4. A. I. Chumakov, T. I. Zakharova, *Harmfulness of crop diseases* (1990)
5. K. V. Novozhilov, *Guidelines for state testing of fungicides, antibiotics and seed dressing agricultural crops* (1985)
6. M. K. Khokhryakov, V. I. Potlaychuk et al., *Identifier of crop diseases* (1984)
7. *AgroMaster Modern agrochemicals. Effective plant nutrition* (2015)

8. List of pesticides and agrochemicals permitted for use on the territory of the Russian Federation (2020)
9. A. Kh. Shabatukov, L. M. Khromova, Bulletin of the agro-industrial complex of Stavropol, **3**, 78 (2019)

The role of innovations in the development of agriculture in Russia

Evgeniy Kharitonov[1,*], *Karina* Krikun[1], and *Alexander* Nesmyslenov[2]

[1]Federal State Budgetary Educational Institution of Higher Education "Kuban State Agrarian University named after I.T. Trubilina", Kalinina St., 13, 350044 Krasnodar, Russia
[2]Federal State Budgetary Scientific Institution (FSBSI) "Povolzhskiy Research Institute of Economics of the Organization of Agroindustrial Complex", Shekhurdina st. 12, 410010 Saratov, Russia

Abstract. This article is intended to substantiate the role of innovations in the development of agriculture in Russia. The authors of the article focused on identifying the importance of innovative techniques and technologies in the modern development of the agricultural industry. The current directions of innovative developments in the agricultural sector in Russia have been identified and investigated. The conditions for achieving effective implementation of digital land use projects are presented; "Smart field"; "Smart garden; Smart farm. The experience and level of development of innovative techniques and technologies in foreign countries are considered. An up-to-date list of both Russian and foreign manufacturers of innovative developments, the buyers of which are domestic agricultural organizations, is presented. The factors that hinder the active implementation of innovations in the activities of agricultural entities in Russia have been identified. A diagram of the relationship between the components of the process of creating and applying innovative technologies has been developed. Also the author's schematic position of the role and place of innovations in the process of creating and applying innovative technologies is presented. As a result of the research, there were given recommendations for eliminating the identified problems in the field of introducing innovative techniques and technologies in agricultural production.

1 Introduction

The use of innovative digital technologies for the agricultural sector is necessary to increase the efficiency of the functioning of agricultural production entities, especially private forms of ownership. The use of innovative digital technologies contributes to a radical change in the quality of manufactured products, and also makes it possible to correctly predict the level of marketability of products in a rapidly transforming economy.

The practice of foreign agricultural manufactures, as well as the experience of Russian agricultural entities, show that the use of innovative technologies in the production process makes it possible to determine the optimal combination of soil-agrotechnical and

* Corresponding author: istor-polit@kubsau.ru

organizational-territorial conditions that contribute to significantly increase yields, productivity, labor productivity throughout the entire period of agricultural production, minimize direct production costs, preserve soil fertility and positively influence environmental safety.

At the present time, the research institute of the Russian Agricultural Academy, together with the Ministry of Agriculture of Russia and agricultural universities of the country, are developing highly productive breeds of animals, birds, fish, as well as high-yielding varieties and hybrids of agricultural crops. It should be noted that foreign scientific structures have the most successful experience in creating resource-saving and environmentally friendly equipment. So, in modern conditions, among Russian agricultural producers, there is a tendency to an increase in demand for the use of foreign innovative techniques and technologies.

Despite the positive effect of the use of innovative technologies, at present there remains an acute problem in their use among Russian agricultural entities. The main reason for this is the long-term ineffective investment climate, which was a typical source for the current period, of a low level of provision of agricultural entities with innovative equipment, techniques and technologies, which emphasizes the relevance of the scientific article.

2 Materials and Methods

The theoretical and methodological basis of our research is the scientific development of foreign and domestic scientists on the use of innovative techniques and technologies in agricultural production in the context of modern world politics.

The scientific article examines the experience of using innovation in agriculture, both foreign and domestic agricultural producers. In the scientific article, we used research methods: monographic, comparison, index, graphic, calculation and analytical techniques and other methods based on statistical data from 2017 to 2019.

3 Results and Discussion

The use of innovative techniques and technologies in agriculture is an innovative activity of an economic entity. Note that at the present time, innovative activity does not have the character of a one-time stage, especially for the agricultural sector, the specificity of this activity is in its cyclical nature. Carrying out innovative activities, the management of an agricultural organization should focus on the simplest innovations when choosing scientific developments for their introduction into the production process, in view of the fact that complicated innovations are often unfeasible and require scientific improvements, which leads to high financial costs and requires a long time for their effective use [4].

We emphasize that the use of innovative techniques and technologies plays an important role in the functioning of an agricultural organization. In research of Kovalchuk M. D. in the development of an integral model for assessing the economic efficiency of forms of ownership of agricultural entities, the "level of innovative development" indicator is 0.3, which suggests that the author assesses the role of this indicator in the activities of an agricultural entity on 30% of the total number of recorded indicators, which is a fairly high proportion [1, 5–6].

The development of agriculture in Russia in recent years in the context of an embargo on the import of a number of products has identified some sectoral problems that need to be addressed, including through the introduction of innovations. The transition of agricultural production to the level of a high-tech industry is due to the relevance of ensuring the country's food security in the context of modern world politics. There is an urgent need to create a favorable innovation and investment climate for the introduction of innovative techniques and technologies in the agricultural sector to provide the Russian population with high-quality and safe agricultural products.

The current level of innovative development of Russian agricultural organizations is of serious concern due to the lack of scientific and practical knowledge in the application of modern innovative agricultural technologies, difficulties in predicting the cost of agricultural products and the required number of innovative techniques and technologies for the effective operation of agricultural organizations [2–3]. At the present time, innovative developments in the agricultural sector in Russia are carried out in the following areas: innovations for the efficient production of agricultural products; innovations to improve the processes of public administration in the field of agriculture [7-8].

Innovation for efficient agricultural production focuses on:

- the formation and continuous replenishment of the information base on the areas of agricultural land, on sown areas by types and varieties of agricultural crops, yield, number and breeds of farm animals, the state of agricultural machinery and equipment;
- involvement of agricultural enterprises in resource-saving and environmentally friendly production;
- minimization of production costs of agricultural enterprises, optimization of land use, involvement of unused agricultural land in economic circulation;
- reduction of the share of material costs in the structure of the unit cost of agricultural products;
- reducing the labor intensity of agricultural production.

To this date, the following directions for the introduction of innovative techniques and technologies in the production of agricultural products have been identified in Russia: digital land use; smart field; smart garden; smart farm. Further, we will explore each of the above directions in more detail.

The main goal of digital farming is the creation and implementation of an innovative intelligent system for management, planning and use of agricultural land, the functionality of which is based on geoinformation technologies and the latest methods of computer modeling. The successful implementation of digital land use can only be recognized if:

- by 2024, the share of agricultural enterprises that have implemented an innovative system of planning and optimization of agricultural landscapes in their activities will be about 50%;
- the level of profitability of production of agricultural enterprises will reach 50%;
- by 2024, it is necessary to achieve complete digitization of the planning and cartographic material;
- by 2024, it is necessary to carry out a complete delineation in the information bases of agricultural lands by forms of ownership.

The main goal of the "smart field" is to ensure stable growth of crop production through the introduction of innovative technologies for the collection, processing and use of information databases on the state of soils, plants and the environment. Successful implementation of the smart field can only be recognized if:

- an accuracy of 95% was achieved in assessing and predicting the yield of agricultural crops;
- an increase by 30% in yield has been achieved through the optimization of crop rotation, as well as the use of means to minimize the influence of factors that have a negative impact on the yield of agricultural crops;
- achieved minimization of direct production costs for the cultivation of agricultural crops in the amount of up to 20% by taking into account the technological properties and geographical location of agricultural land;

- achieved 50% introduction of a system of anti-erosion and environmental protection measures within the boundaries of fields among the total number of agricultural producers focused on the cultivation of agricultural crops;
- development and implementation of robotic technologies in field cultivation.

The main goal of the "smart garden" is the development and application of an intelligent technological system that automatically analyzes information about the state of the garden's agrobiocenosis, as well as effective management decisions and their implementation by robotic technical means. Successful implementation of a smart garden can only be recognized if:

- a complete digitization of the areas of garden perennial plantings and their introduction into a unified geographic information system has been achieved;
- the provision of more than 70% of garden organizations with information on the state of soils in areas of garden perennial plantations and the environment has been achieved;
- achieved 20% of field work robotization in a completely autonomous mode without human intervention;
- a 30% increase in yield and an increase in the quality of fruit crops has been achieved through the implementation of innovative technological processes;
- achieved 40% minimization of production costs due to the optimal use of materials.

The main goal of the "smart farm" is the development and application of innovative technical and technological solutions for the creation of new generation farms based on intelligent digital technologies. According to forecasts, the development of the market for agricultural robots, the volume of the robotization market for dairy farms in 2023 will reach 504 billion rubles, currently - 120 billion rubles. Successful implementation of a smart farm can only be recognized if:

- a 70% reduction in the incidence of mastitis in animals has been achieved;
- 40% profitability of production was achieved through robotization of the farm.

Note that at the present time, there is an active development of precision farming. Thus, such world leaders as the USA, Germany, Japan and China have reached a high level of development in this direction.

In the Russian Federation, in most cases, when implementing precision farming, innovative techniques and technologies of foreign manufacturers, for example, John Deere, are used. The reason for this is the low level of patent activity in the field of precision farming technologies in Russia relative to the world leaders.

Let us emphasize that for modern conditions, in the field of precision farming, rather active use of unmanned aerial vehicles is characteristic. Investigating the commodity structure of the robotechnics market for precision farming, it is worth noting that the production and sale of unmanned aerial vehicles makes up 20% of the total assortment. In Figure 1, we present the structure of the countries producing unmanned aerial vehicles.

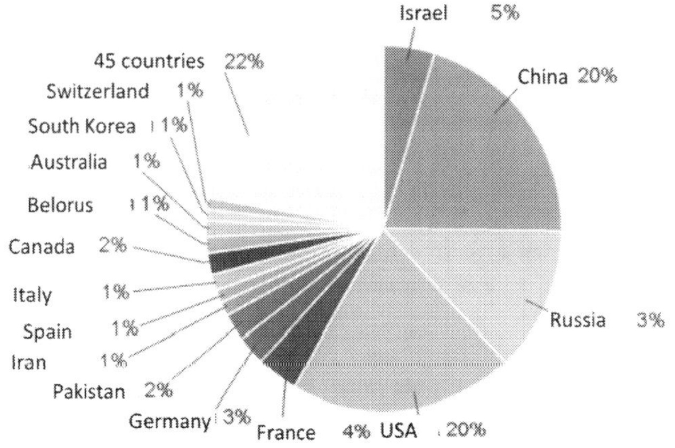

Fig. 1. Structure of countries producing unmanned aerial vehicles

So, according to Figure 1, the world leaders in the production of unmanned aerial vehicles are the United States and China with a total share of 40% of the total number of manufacturers, which produce more than 280 models annually. The Russian Federation is in third place with a 13% share and an annual production of 177 unmanned aerial vehicles.

At the present time, Russian manufacturers of innovative techniques and technologies for the crop production industry are:

- Cognitive Technologies, the main direction is the development and production of unmanned aerial vehicles. The key buyer is Special Economic Zone "Innopolis" ;

- Strizh-Telematica, the main direction is the development and production of detectors and sensors. The key buyer is "Valuikisakhar", California;

- RoboProb, the main direction is the development and production of land survey equipment. The key buyer is "White Bird";

- LLC "KSM-Intekh", the main direction is the development and production of precision farming systems. Key buyers are located in more than 50 regions of Russia.

Foreign manufacturers of innovative techniques and technologies for the Russian crop production industry are:

- Agroko, the main direction is the development and production of precision farming systems. The key buyer is Agroko (foreign software);

- Agrofirm "Progress", the main direction is the development and production of precision farming systems. The key buyer is Agrofirm "Progress" (foreign software).

In 2019, such innovative projects were announced:

- Project of a system for diagnosing the state of an animal during pasture keeping of cows. Initiator of the project is ZAO "Plant of electrical equipment";

- Project of a knowledge base in agriculture based on scientific publications and patents of Russia, USA,

Canada, Europe, China and other countries. Initiator of the project is Dataocean LLC;

- Project of a system for monitoring the health and reproduction of cattle. The initiator of the project is LLC "Perspektiva Group" with the support of the Altai State Agrarian University and the Moscow State Academy of Veterinary Medicine and Biotechnology named for K. I. Scriabin;

- Project of agrometeorological support for the activities of agricultural enterprises. Initiators of the project are LLC "CenterProgramSystem", LLC "Ilizium", Individual Undertaker Kurakov;

- Project of a system for remote monitoring of agricultural machinery with obtaining economically important parameters and their use in management and economic accounting. The project initiator is LLC "Farvater".

At the present time, among domestic agricultural commodity producers, the innovative potential is used only by 4–5%, while among economic entities in the US agricultural industry is by more than 50%. Today, only 1–2% of the total number of innovative developments intended for agricultural production is used in Russia, while in the USA is about 70%

Further, it is necessary to identify the factors that impede the active implementation of innovations in the activities of agricultural entities in Russia. They include:

- a shortage of highly qualified specialists with knowledge, skills, competence and experience that would allow the effective operation of innovative technologies in the production of agricultural products. The main reason for this problem is the high level of outflow of the able-bodied population from the village. The majority of active labor resources who have received education in the field of information and innovative technologies do not strive to implement their labor activity in the countryside;
- lack of effective training programs for training specialists in the use of modern innovative technologies for collecting and processing information on the state and use of agricultural lands;
- the high cost of foreign innovative developments, due to the close relationship with exchange rate changes in world currencies, as well as sanctions imposed by world leaders in relation to Russia and other trade restrictions within the framework of the low level of development of the Russian market for innovative technologies;
- lack of a regulatory framework and practice of interdepartmental interaction at the regional level;
- insufficient level of financing for the introduction of innovative techniques and technologies in most agricultural entities. Most of the agricultural producers, especially those in the "rural outback", have low solvency and for this reason they use outdated techniques and technologies in their production activities. In the current circumstances, due to the mechanisms of support for the subjects of the agricultural sector from the state in the conditions of available financing, in recent years, the volume of funds allocated for the purchase of modernized agricultural equipment at preferential prices has been significantly increased (in 2019, up to 5.2 billion rubles). However, the low economic potential of agricultural organizations leads to a minimization of the investment flow in fixed assets, savings on the purchase of equipment, which negatively affects the level of innovative activity of such economic entities.

To solve this problem, it is necessary to identify the relationship between the scientific potential and the direct application of innovative technology in production (Figure 2).

This diagram represents the connection between the 4 components of the process of creating and applying innovative technologies.

The scientific and educational structure is a supplier of ideas and results of scientific research, provides support and growth of the scientific level of production. The information structure is a database that contains all the information about the scientific achievements of the industry, as well as the cost of innovative technologies and how to use it.

State and municipal administration of the agro-industrial complex is a structure, the process of creating and applying innovative technologies, which performs the functions of financing both scientific research and their introduction into production, as well as monitoring the use of allocated funds.

The production structure is a unit of the process of creating and applying innovative technologies, which implements the request for innovation, as well as its subsequent implementation.

For the effective management of innovative technologies application, it is necessary to determine the role and place of innovation in the process of creation and its application. According to the data presented in Figure 3, the interaction of the scientific, industrial structure with universities within the framework of the functioning of the process of creating and applying innovative technologies will provide a significant increase in the quality of scientific achievements.

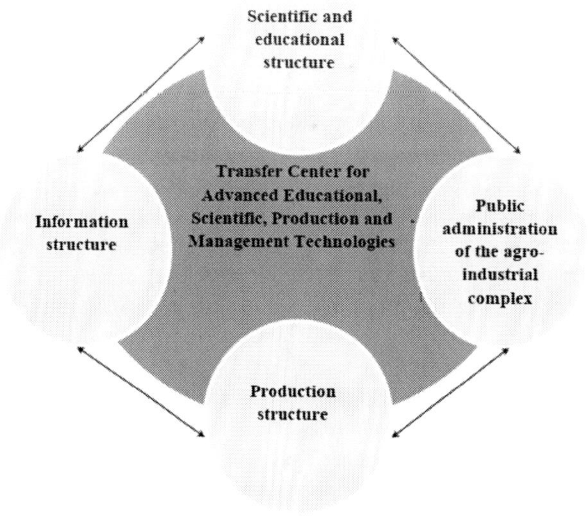

Fig. 2. Diagram of the relationship between the components of the process of creating and applying innovative technologies

It will also contribute to the development of infrastructure for the implementation of innovative projects and create preconditions for the optimal development of the system of existing innovations.

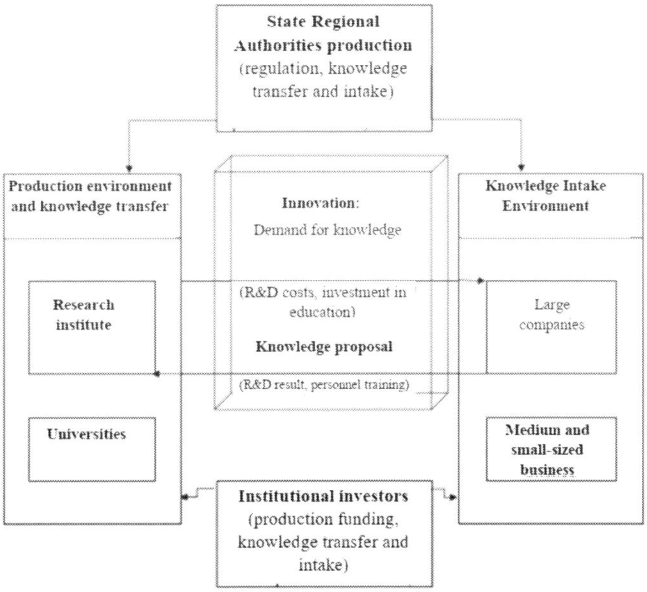

Fig. 3. The role and place of innovation in the process of creating and applying innovative technologies

Investors at the first stage are the state and specialized development funds. Further, at the second stage, the financing of proven technologies is carried out by customers on the basis of contracts concluded with universities and research institutes. The presented scheme is beneficial for agricultural enterprises seeking to modernize production.

4 Conclusions

As a result of the conducted research, we came to the conclusion that the use of innovative techniques and technologies, in modern conditions, has a high level of importance for the effective development of agriculture, as well as for increasing the competitiveness of agricultural products.

The identified problems, in our opinion, can be solved through public private partnerships in the creation and application of innovations in agriculture. It is under such conditions the effective intersection of science, government funding and agricultural production enterprises can be achieved.

References

1. V.I. Gaiduk, M.D. Kovalchuk, A. A. Ermakov, S.Y. Kamysheva. Revista Inclusiones, T. 7. No. S2-3, **188** (2020)
2. A.N. Bugara, http://apej.ru (2016)
3. I.O. Vlasova, Exploration Survey Company, **1(57)**, 207 (2016)
4. E.A. Egorov, Zh.A. Shadrina, G.A. Kochian Gardening and viticulture, **3**, 5 (2016)
5. M.D. Kovalchuk, *"Green Economy" in the Agro-Industrial Complex: Challenges and Development Prospects*. Materials of the All-Russian Scientific Conference, 243 (2018)

6. M.D. Kovalchuk, E.A. Shibanikhin, Materials of the international scientific conference, 152 (2017)
7. A.I. Trubilin, *Justification of directions and assessment of the effectiveness of innovative development of import-substituting industries in the Krasnodar Territory*, 150 (2019)
8. A.I. Trubilin., V.P. Chaika, A.I. Papakhchyan, A.V. Tolmachev, International Agricultural Journal, **2 (368)**, 22 (2019)

New perspectives of biogas use

Lyubov Zazykina*, and *Valentin* Gusev

Federal State Budget Scientific Institution Federal Scientific Center "All-Russian Research and Technological Poultry Institute" of Russian Academy of Sciences, Pticegradskaya str., 10, 141311 Sergiev Posad, Russia

Abstract. The article is a comparison of existing facilities for processing poultry waste into bio energy. Currently, poultry farms have difficulties selling their secondary production, because only some of them have land plots. Thus, according to the statistics, the use of organic fertilizers in Russia decreased by seven times over the past 20 years, and now only 0.9–1.3 tons of waste are used per 1 ha of sowing. Meanwhile, poultry population has not decreased, so this waste is not disposed properly. The most crucial task for modern managers is to neutralize and rational use waste, making recycling at the poultry farms. This problem can be solved only by taking into account social and economical consequences, when making economic decisions. Consequently, such a method of decision-making could improve the farms' competitiveness will be enhanced by using secondary production (particularly poultry waste) as biogas.

1 Introduction

The latter tend to decrease despite the achievement of several indicators declared in the Food Security Doctrine and positive dynamics of egg and poultry meat production growth. However, many enterprises experience the shortage of working assets for realizing their potential despite the presence of steady demand. As a rule, enterprises' policy is focused on satisfying customer requests for the range and quality of products.

The main factors contributing to this situation are the disparity of prices for resources and products, solvency, inflation, fluctuations in ruble exchange rate, insufficient government support under fierce competition. A manufacturer who produces goods with lower feed and energy costs always wins in the agricultural market. In this case, competitiveness can be enhanced by using secondary production, particularly poultry waste, as biogas.

2 Materials and Methods

Currently, there are such types of waste in the industry as feathers, poultry mortality, eggs and others. Therewith, poultry waste being improperly disposed pollutes the environment and increases business expenditure.

Its main uses can be:

* Corresponding author: l.zazykina@ya.ru

- Composting.
- Vermicomposting (option: processing poultry waste by insects and worms).
- Microbiological conversion of manure and poultry waste.
- Biogas production technology (option: methane digestion).
- Pyrolysis (option: free-oxygen thermal waste decomposition).
- Combustion technology for thermal energy production.
- Combustion technology for alternative electrical energy production.

Production of biogas is one of the promising areas.

Monographic, statistical, economic and abstract-logical methods as well as comparative analysis were used while studying the issue of poultry waste processing

3 Results and Discussion

Let us analyze the biofuel markets to identify trends and demand for various types of resources.

There are two world markets for biofuels, specifically, the liquid market and the solid biofuel market. Both are rapidly and dynamically developing. However, they have fundamentally different geography of production and consumption, pricing mechanisms as well as the structure of supply and demand. Such differences are primarily associated with production technologies, biofuels application range. According to analysts, oil consumption will increase until 2020 and reach 88 million barrels per day, then it will steadily fall and drop to almost 83–84 million barrels per day by the year 2030 and to 81 million barrels per day by the year 2035 [1]. However, the decline is significant in the last decade 2040-2050 where oil in transport is replaced by biofuels and electric vehicles. The oil share drops to around 15% in 2050 when following any of the examined main directions towards decarbonisation [3]. The United States, China and the European Union will play the main role in the oil demand dynamics, and the Middle East countries will critically affect the supply dynamics.

Table 1. Share of fuels in primary energy consumption in % [1]

	Reference scenario		**Current policy initiatives**		**Decarbonisation scenarios**	
	2030	2050	2030	2050	2030	2050
Renewable	18.4	19.9	19.3	23.3	21.9-25.6	40.8-59.6
Solids	12.4	11.4	12.0	9.4	7.2-9.1	2.1-10.2
Oil	37.1	32.8	34.1	32	33.4-34.4	14.1-15.5
Gas	22.2	20.4	22.7	21.9	23.4-25.2	18.6-25.9
Nuclear	14.3	16.7	12.1	13.5	8.4-13.2	2.6-17.5

The largest increase is observed in solar energy. It is from 2 % of global electricity production in 2019 to 22 % in 2050. It is expected that about 330 kW in each 1 MW of this photovoltaic energy will be used by households and businesses, and will be 5 % of global electricity in 2050 [9]. Wind energy will produce 26 % of the world's electricity in 2050, which is 5 % more than it is now.

Renewable energy production accounts for approximately a quarter of the world's total energy production, which is a new record [2]. New record values were also recorded for plants operating on solar and wind energy, which showed an increase of 94 GW for photovoltaic solar energy and 47 GW for wind energy, including 4 GW for wind energy in coastal waters [6]. Renewable energy production costs are declining. There is sufficient evidence that energy systems dominated by renewable energy sources can become a reality. Thus, the scale and pace of introducing renewable energy sources can be increased rather safely.

Biofuel is very rare in our country in comparison with the European and American markets. However, the market for solid biofuels is actively developing. Volumes of pellets and briquettes production are steadily increasing. In 2010, about 1 million wood and husk pellets were produced. The pellet market in our country is export-oriented. Specifically, up to 70–80 % of production is exported to European countries. Russia is the third largest exporter in the world after Canada and the USA [7] with regards to European countries. At the same time, wood pellets are mainly bought by the Scandinavian countries, Central and Northern Europe, and husk pellets are bought by Great Britain and Poland.

As for the granules of chicken manure pellets, they are high-quality organic fertilizers. They contain all the elements (in a favorable amount and combination) necessary for plant nutrition. Also in recent years, they are increasingly practicing them as fuel for local use in gas boiler plants [4].

All pellets, regardless of what raw materials are used for their production, are divided into 2 large segments:

1. Consumer segment (first class granules). Such granules are supplied mainly to private consumers for use in domestic boilers and fireplaces. In such granules there are no impurities (bark, branches), binders.
2. Commercial (industrial) segment. Granules of lower quality are used mainly in medium and large thermal installations.

4 Conclusion

Despite the fact that Russia mainly focuses on the development of biofuels derived from wood waste, there remains a niche in the market, which is gradually being filled by enterprises that produce pellets from manure. As it was already mentioned, one of the promising areas is the production of biogas and bioenergy from poultry waste in special generating units. The introduction of these facilities will allow enterprises, in particular poultry farms, to achieve the following:

- Independence from rising rates and limits concerning electricity, heat and gas;
- Personal source of domestic gas, electricity and heat;
- Exemption of fertile arable land from lagoons for poultry waste storage;
- Freedom from environmental and waste management payments;
- An opportunity to increase the competitiveness of products and receive additional income.

Cost reduction through the use of personal sources obtained from production waste, such as slaughterhouse waste, fat and grease traps helps the company to be a leader in the competition. In addition, the strategy of the enterprise should be aimed at generating additional income through their processing [10] in biogas generation plants. Biogas generation plants not only utilize all organic waste from agriculture, food and wood processing industries but also produce biogas (Table 2). Biogas is produced under anaerobic conditions and is analogous to natural gas in composition.

Table 2. Waste suitable for producing biogas and its output.

Source raw material (Humidity 86 %)	Biogas output (m^3/t)	Methane content in the resulting biogas (%)
Slaughter waste	250–300	60–70

Table 2. Continued

Poultry waste	90–130	60
Catch basin skimmings	230–250	60

In addition, the cost of this installation is at least 3 times cheaper than foreign analogues. When using a semi-fixed cost option, an enterprise can benefit in the form of free heat, cheap electricity and gas, thereby reducing the cost of manufacturing basic products.

There is also a mini combined heat and power unit (CHPU) on a poultry waste made for the "Voronezh Turkey" agroholding. The CHPU is designed to burn up to 110 thousand tons of poultry waste per year (300 tons/day) with all the needs of a poultry processing complex (including a feed factory; a slaughter waste processing workshop for a feed additive; treatment facilities) in electricity (up to 5 MW), process steam (2 t/h, 1.0 MPa, 190 °C), hot water supply (0.5 Gcal/h) and heating (up to 2.5 Gcal/h). An integral part of the CHPU is the workshop for processing ash into mineral fertilizer-soil conditioner with the capacity of up to 12 thousand tons per year [5]. As a result, the only waste is boiler flue gases. The level of harmful substances emissions does not exceed the level of burning fuel oil, and the most significant component is nitric oxide. However, the cost of this CHPU is quite higher than that of UPR. The installation of UPR allowed to reduce the cost of production by 25–30 % in JSC "Sharkanskoye RTP" of the Udmurt Republic, LLC "Complex utilization systems" of the Orenburg region, peasant farm Rashin Perm Territory. Bio-energy plants will automatically solve the problem of poultry waste disposal and get heat and electricity for their own needs, which can bring additional profit to the enterprise and increase its competitive advantage. Such plants cost recovery is 1.5–2 years.

References

1. *Biofuels and food security,* Report of the High Level Panel of Experts on Food Security and Nutrition of the Committee on World Food Security, Rome (2013)
2. J. B. Dahmus et al., IEEE Int. Symp. on Sustainable Syst. and Technol. (ISSST '09) (2009).
3. G. Eastwood, *The Biofuels Market Outlook* (2007)
4. Russia's energy strategy until 2020 (2003), http://www.energystrategy.ru
5. V. I. Fisinin, et al., *Environmental protection technologies for the prevention of local environmental pollution by the waste of poultry farms* (2018)
6. E. Hartman, *A Promising Oil Alternative: Algae Energy* (2008)
7. O. S. Sausheva, Internet j. Naukoved., **5**, 59 (2016)
8. D. P. Smith, Principles and Applications, 549–566 (2014)
9. *Weekly and monthly supply and demand of ethanol,* https://ethanolrfa.org
10. L. A. Zazykina, L. M. Roiter, V. A. Gusev, IOP Conf. Ser. Earth and Environmental Sci., 315(7) (2019)

The design of the working equipment for laying a drip tape based on the Grimme GF-75/4 ridge former

N.B. Martynova* and *V.I.* Balabanov

Russian State Agricultural University - MSHA named after K.A. Timiryazev, 127550 Moscow, Russian Federation

Abstract. The article examines statistical data on the amount of precipitation in the spring - summer period and substantiates the need for irrigation measures when growing potatoes in the Moscow region. The rationality of the use of drip irrigation as the most economical and effective method of watering has been proved. With drip irrigation, there is practically no danger of excessive watering, causing the leaching of nutrients from the root layer, an increase in the level of groundwater, and an additional load on the drainage network. The use of drip irrigation will make it possible to maintain an optimal water and air balance during the growing season. The possibility of combining technological operations for laying a drip tape and forming potato ridges is considered. It is found that the resistance of the drip tape laying will not create significant traction resistance, so the working equipment can be mounted on the ridge former. The design of a drip tape stacker based on the Grimme GF-75/4 ridge former was developed and a mechanized drip tape was laid in a potato ridge. It has been proven that the implementation of the drip tape laying and the formation of ridges with one machine will increase the productivity of the work and reduce the energy consumption.

1 Introduction

The amount of precipitation that falls during the growing season in the Moscow region varies widely, high-water years alternate with low-water ones. The difference in the amount of precipitation in the spring - summer period in different years varies 2 - 4 times [1]. Often precipitation is of a stormy nature, the precipitated moisture can not be absorbed into the soil for the lack of time and goes into the runoff. Precipitation during this period is extremely uneven. Days with heavy rains, when their daily fallout can amount to tens of millimeters, alternate with days and even weeks when there is no precipitation [2, 3].

When growing potatoes throughout the growing season, water consumption also changes. In the initial phase of development, the plant has sufficient moisture reserves of the parent root tuber and there is practically no need for additional moisture [4]. However, during this period, the soil is filled with moisture formed in the process of snow melting. In subsequent phases of development, the volume of daily water consumption increases. The maximum

* Corresponding author: nmartinova@rgau-msha.ru

values are observed during the period of flowering and the end of the growth of tops [5, 6]. Further, the daily water consumption decreases slightly (Fig. 1). As the graph shows, in order to obtain a sustainable harvest of a number of agricultural crops, in particular potatoes, additional irrigation is required.

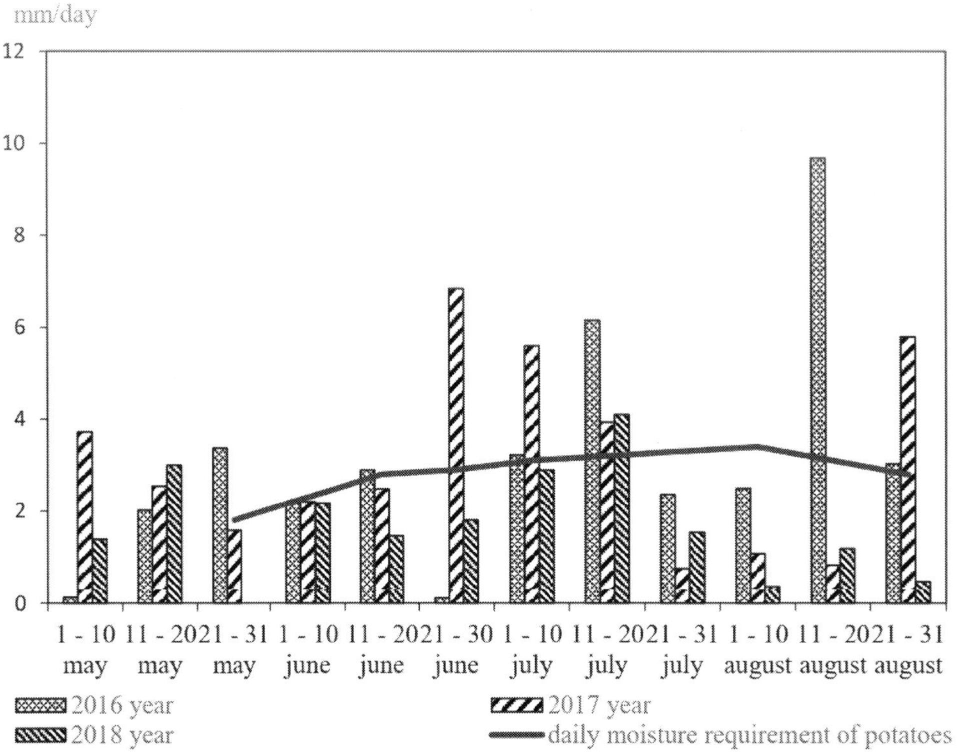

Fig. 1. Dynamics of precipitation in the spring-summer period and the need for moisture in potatoes

Among the irrigation agrarian-meliorative measures, drip irrigation is the most economical. The undoubted advantage of this irrigation method is the delivery of water directly to the root zone [7, 8]. This, in addition to saving water and electricity, practically eliminates excessive watering, and, consequently, leaching of nutrients from the root zone, raising the level of groundwater, and additional load on the drainage network. In addition, in contrast to surface irrigation, during drying, a dense crust does not form, which makes it difficult to aerate the soil. With drip irrigation, the spacing is practically not moistened, this leads to a slowdown in the growth of weeds [9, 10].

The area occupied by drip irrigation in the Russian Federation is about 51 thousand hectares, which is an insignificant percentage of 4633 thousand hectares of the area of all irrigated land. This index is below the world average and is significantly inferior to countries such as the United States or Israel. Pace of drip irrigation implementation is significantly lower than in a number of Asian countries [11].

One of the limiting factors is the low degree of mechanization when laying the drip tape. The equipment designed for mechanized laying of drip tape to the depth of 0.45 m is serially produced in the USA by Andros, Rain Flo, Kennco Manufacturing [12]. However, these tools are expensive and are not designed to put drip tape into a potato ridge. The use of primitive devices for laying the drip tape causes the uneven tension and breaks or bends of the tape.

2 Materials and methods

It is rational to lay the drip tape when growing potatoes in conjunction with the technological operation for the ridges formation, 10 - 14 days after planting. During this period, an increase in daily water consumption of the plant begins, and the combination of operations for ridges formation and laying of a drip tape will increase the work performance, reduce energy consumption [13, 14].

To lay the drip tape into the potato ridge, the equipment was developed for installation on the ridge former (Fig. 2), consisting of a frame on which a guide tube with a colter is fixed to place the drip tape in the ridge at the specified depth. The laying depth is adjusted by vertical movement of the pipe using the holder. On the vertical column of the frame there is an axle with clamping discs for fixing the coil. To prevent the tape from unwinding by inertia in the event of a sudden stop of the machine, a brake device with a tension roller and a spring-loaded pressure shoe is used. During the movement of the machine, the tape is unwound from the coil and enters the guide tube, along which it moves into the potato ridge.

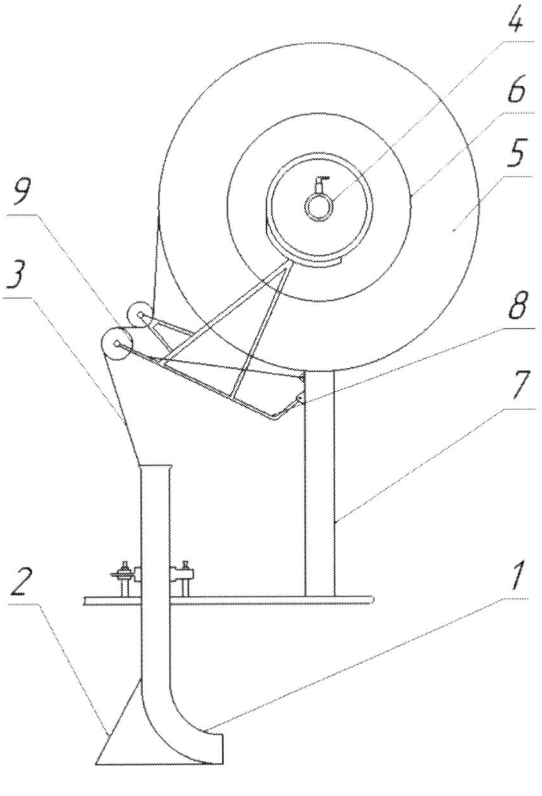

Fig. 2. Construction of drip tape stacker: 1 - guide tube, 2 - opener, 3 - drip tape, 4- shaft, 5 - coil of drip tape, 6 - protective disc with latch, 7 - vertical support post, 8 - braking device, 9 - tensioning roller.

To determine the possibility of installing additional equipment for laying the drip tape on the ridge former, you should calculate the resistance of the drip tape laying, which will be the sum of the soil resistance to development, the friction of the tape in the stacker, overcoming the unbalance of the coil and its friction on the axis [15, 16]:

$$F_{po} = R_{нг} + F_{лвр} + F_{лдр} + F_{лтр} =$$

$$= \mu \cdot G_p + k_p \cdot h \cdot b + \varepsilon \cdot h \cdot b \cdot v^2 +$$

$$+\left(f_o \cdot G_л \cdot d_o \cdot d_p^{-1} + 2 \cdot G_л \cdot l_э \cdot d_p^{-1}\right) \cdot \left(1 + e^{f_o \cdot \beta}\right), \quad (1)$$

Where F_{lv} is the tape pulling resistance, F_{lvr} - friction resistance of the tape in the stacker, kN; F_{ldr} - force to overcome the unbalance of the coil, kN; F_{ltr} - resistance to friction of the coil on the axis, kN; R_{ng} - soil resistance to the development, kN; f_o coefficient of friction of the axle in the bearings; G_l - weight of the tape coil, kN; d_o - axis diameter, m; d_p is the diameter of the coil, m; l_e - displacement of the spool center of gravity from the axis of rotation, m; β - tape wrap angle; μ coefficient of friction of the working tool on the ground; G_p - stacker weight, kN; k_p — resistivity to cutting, kN/m^2; ε - coefficient that takes into account the effect of cutting speed on the resistance to digging; h - development depth, m; b - development width, m.

As calculations have shown, when installing additional equipment on the ridge former, the pulling force will increase by 6 - 8%, which will not affect the performance of the machine, therefore, it is possible to carry out operations to form ridges and lay a drip tape by means of one machine in the same agrarian technical terms (Fig. 3).

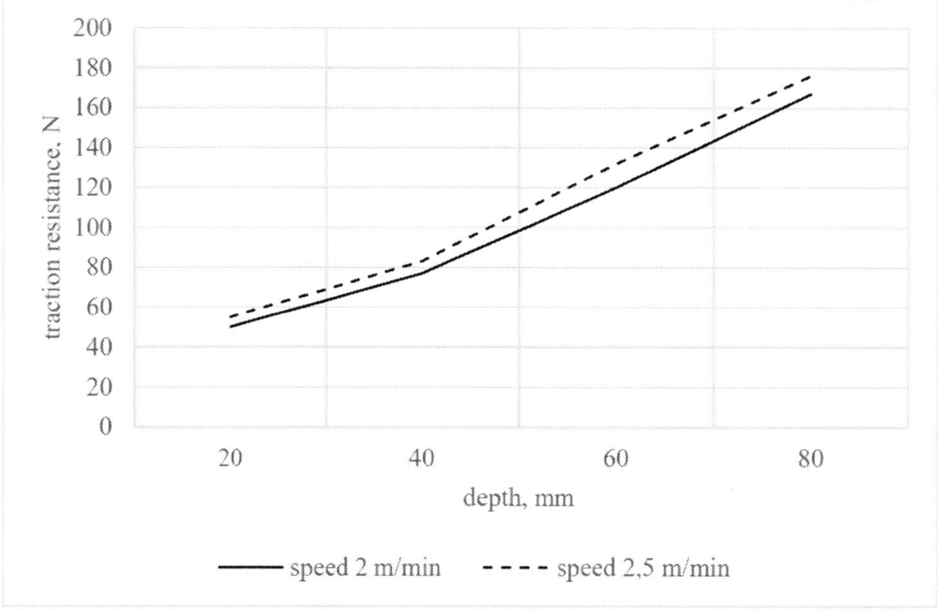

Fig. 3. Depth dependence of the drip tape laying resistance.

3 Results and discussion

According to the research results, the working equipment was mounted on the Grimme GF-75/4 ridge former and the drip tape was laid at the Field Experimental Station of the RGAU-MSHA (Russian State Agricultural University) named after K.A. Timiryazev. During the field research, the conditions for performing the technological operation for laying the drip tape were tried out, the optimal speed mode and the time for performing the operation were determined (Fig. 4).

Fig. 4. Ridge-based drip tape stacker Grimme GF-75/4.

Based on the potato planting scheme, a drip tape with a diameter of 16 mm with a distance between emitters of 300 mm was chosen. The parameters of the developed machine are presented in Table 1.

Table 1. Technical characteristics of the developed machine.

Working body weight	2150 kg
Working body width	3,0 m
Working speed	0,9 – 1,4 m/sec
Performance	0,51 – 0,97 ha/h
Drip tape diameter	16 mm
Dripper water consumption	1,6 L/h
Dripper spacing	300 mm

Watering was carried out in all subsequent phases of the growing season until the beginning of the wilting of the tops. Watering was carried out at a soil moisture content of 70% of the PPV at a constant rate, with the adjustment of the time between waterings (Fig. 5).

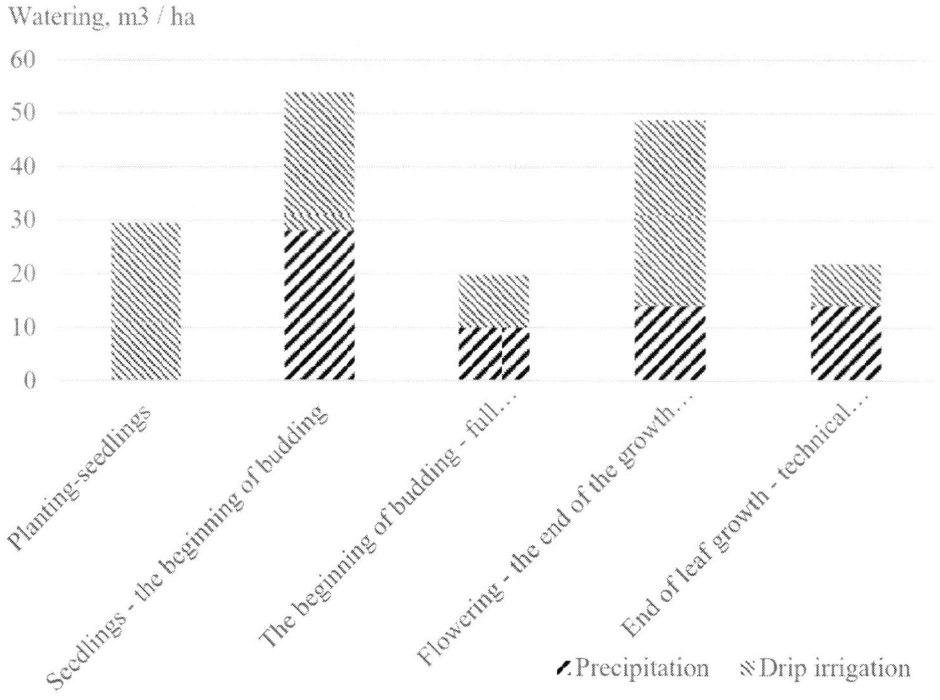

Fig. 5. Water consumption of potatoes at different stages of development.

4 Conclusion

When growing potatoes in the Moscow region, irrigation is required to obtain sustainable yields. Drip irrigation will make it possible to moisturize the root zone without over-watering. Drip irrigation will make it possible to maintain optimal water and air balance during the growing season. It is advisable to lay the drip tape together with the operation for the formation of ridges, as in the early phase of potato development there is no need for additional watering due to the moisture reserves of the parent root tuber. The resistance of the drip tape laying did not exceed 180 N, therefore, the installation of the working equipment for the drip tape laying on the ridge former will not cause a significant increase in traction resistance and will not affect the operational characteristics of the machine. The developed design of the drip tape stacker based on the Grimme GF-75/4 ridge former will increase work productivity and reduce energy costs due to the joint implementation of technological operations.

Acknowledgements

The article was prepared with the support of the Ministry of Education and Science of Russia within the framework of agreement No. 075-15-2021-032 dated March 23, 2021 on the provision of a grant in the form of a subsidy for the creation and development of an engineering center on the basis of an educational institution of higher education and (or) a scientific organization in within the framework of the federal project "Development of infrastructure for research and training" of the national project "Science and Universities".

References

1. N.N. Dubenok, Bolotin D.A., Novikov A.A., Bolotin A.G. Bulletin of the Nizhnevolzhsky agro-university complex: science and higher professional education, **51 (3)**, 81 (2018)
2. O.A. Starovojtova, N.E. Shabanov, The Herald of the Moscow State Agricultural Engineering University named after V.P. Goryachkin, **74 (4)**, 34 (2016)
3. J. Reyes-Cabrera, L. Zotarelli, M.D. Dukes, D.L. Rowland, S.A. Sargent, Agricultural Water Management, **169**, 183 (2016)
4. N.B. Martynova, A.Yu. Korneev, The Herald of the Moscow State Agricultural Engineering University named after V.P. Goryachkin, **84 (2)**, 18 (2018)
5. A.S. Shtahko, V.N. Shkura, Scientific journal of the Russian Research Institute of Melioration Problems, **31 (3)**, 39 (2018)
6. S.M. Vasiljev, V.N. Shkura, A.S. Shtahko, Bulletin of the Nizhnevolzhsky agro-university complex: science and higher professional education, **52 (4)**, 316 (2018)
7. Kh.A. Abdulmazhidov, A.S. Matveyev, The Herald of the Moscow State Agricultural Engineering University named after V.P. Goryachkin, **72(2)**, 40 (2016)
8. A.S. Ovchinnikov, A.A. Buber, Yu.P. Dobrachev, V.V. Borodychev, Bulletin of the Nizhnevolzhsky agro-university complex: science and higher professional education, **54 (4)**, 65 (2018)
9. A.S. Ovchinnikov, V.S. Bocharnikov, M.P. Mesherjakov, O.V. Bocharnikova, Bulletin of the Nizhnevolzhsky agro-university complex: science and higher professional education, **48 (4)**, 10 (2017)
10. N.N. Dubenok, D.A. Bolotin, S.D. Fomin, A.G. Bolotin, Bulletin of the Nizhnevolzhsky agro-university complex: science and higher professional education, **48 (4)**, 22 (2017)
11. M.N. Makani, S.A. Sargent, L. Zotarelli, D.J. Huber, C.A. Sims, Scientia Horticulturae, **197**, 428 (2015)
12. J. Reyes-Cabrera, L. Zotarelli, D.L. Rowland, M.D. Dukes, S.A. Sargent, American Journal of Potato Research, **91 (5)**, 504 (2015)
13. V.N. Krasnoshchekov, D.G. Olgarenko, Environmental Management, **4**, 51 (2016)
14. B.M. Kizjaev, N.B. Martinova, Irrigation and Reclamation, **13(3)**, 62 (2018)
15. E.V. Zhalnin, The Herald of the Moscow State Agricultural Engineering University named after V.P. Goryachkin, **82 (6)**, 10 (2017)
16. V.I. Balabanov, A. Li, The Herald of the Moscow State Agricultural Engineering University named after V.P. Goryachkin, **95(1)**, 10 (2020)

Mathematical model for calculating the parameters of machines for processing near-trunk strips in a terrace

Aslan Apazhev, *Artur* Egozhev*, *Magomed* Misirov, *Evgeny* Polishchuk, and *Asker* Egozhev

Kabardino-Balkarian State Agricultural University named after V.M.Kokov, Lenin Avenue, 1V, 360030 Nalchik, Russia

Abstract. Horticulture is one of the most dynamically developing branches of agricultural production. In recent years, in the Russian Federation, there has been an annual increase in the area for gardens, including on sloping lands, gross harvest and good yield. In the conditions of the foothill zone of the North Caucasus, the most promising direction is the development of slope lands, with favorable soil and climatic conditions for the cultivation of fruit crops. The design features of the terraces impose restrictions on the operating conditions of agricultural units, among the many it is necessary to emphasize the possibility of approaching the row line for processing the near-trunk circle from only one side. Most of the horticultural businesses have mainly equipment designed for work in flat gardening conditions, which for complete processing requires the passage of the unit along each side of the row line that cannot be provided in terraces. A design technology scheme of the machine working tool is proposed, the technical result of which lies in the performance of a high-quality processing in the zone of the near-trunk circle by ensuring the bypass of rotors with cutting segments around the tree trunk, without its damaging, in one passage of the unit along the row line.

1 Introduction

Terracing in the conditions of limited areas for land use makes it possible to develop the slopes of the North Caucasus for gardens [1, 2]. The air temperature is on average 0.5-0.6°C higher on terraces than on ordinary slopes, and the thermal regime is also significantly improved [2 4].

Among the variety of different methods of soil maintenance in gardens, in conditions of slope gardening, in order to minimize soil erosion processes, as well as provide plants with moisture and the required nutrients, a sod humus system is used, which provides for periodic mowing of vegetation growing in the spacing and near-trunk strip for mulch [5, 6].

Modern industrial horticulture at this stage is a capital-intensive branch of agriculture, the effectiveness of which is achieved through the use of new production means and

* Corresponding author: artyr-egozhev@yandex.ru

technologies. The main problem that producers face is the lack of modern technology for the care of the spacing and near-trunk strips of intensive fruit plantations.

The design features of the terraces impose restrictions on the operating conditions of agricultural units, among the many it is necessary to emphasize the possibility of approaching the row line for processing the near-trunk circle from only one side. Most of the horticultural businesses have mainly equipment designed for work in flat gardening conditions, which for complete processing requires the passage of the unit along each side of the row line that cannot be provided in terraces [7, 8].

Based on the above mentioned, it can be concluded that the introduction of new machines and units for processing the near-trunk circle (mowing and milling) in one passage of the unit, ensuring an increase in soil fertility on sloping lands, is relevant.

2 Materials and methods

A mathematical model is proposed for determining the basic kinematic and dynamic parameters of the machine working tool for processing the zone of the near-trunk circle of fruit trees of an intensive garden in one passage of the unit [8-12].

Let's consider the operation of the working tool during the technological process in the zone of the near-trunk circle. A conventionally fixed coordinate system centered at a fixed point is O denoted by Ox_1y_1. A movable coordinate system is also centered at a point O_1, the axis O_1y is directed along the line connecting the centers O and O_1, the axis O_1x is directed perpendicular to the axis O_1y. The coordinate system O_1xy is rigidly connected to the disk and, therefore, moves with it along the guide circle of the radius r_2 at a constant angular velocity ω_1. The movement of all points of the working tool in this case will be complex, and consists of a translational rotation around the axis with an O_1 angular velocity and ω relative movement around the O axis with an angular velocity ω_1.

We will preliminarily accept the condition that since the rotor speed to ensure the working process is much higher than the rotational speed of the rotary section around the tree trunk, we assume that the length of the arc is a circle $L = O_1O_1'$. In this case, the calculation of the areas run through by the mower knife or the working part of the cutter can be carried out as for a circle. Then, in accordance with these assumptions, the working body will ab process the area S_1, the next segment will cd process the area S_2. The cross-hatched area, bounded by the circumferential lines of both segments S_x, defines the idle area of the second segment, since this part is processed by the segment ab.

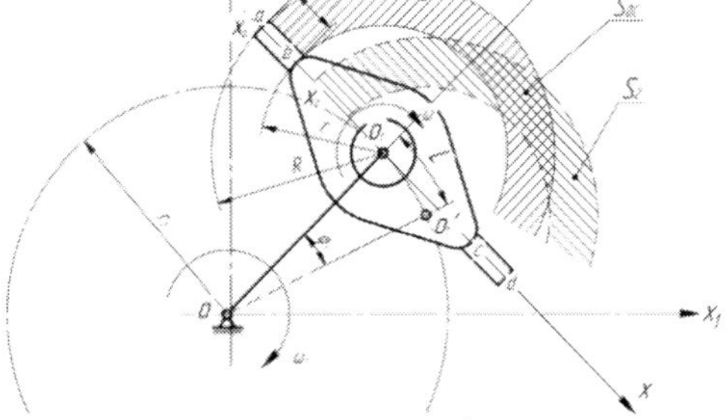

Fig. 1. Scheme for investigation of the working tool operation.

Let the t center of the disk O_1 move to the point O_1' at a moment of time, which will turn to the corner when the OO_1 crank turns $\varphi_1 = \omega_1 t$. At the same time, the center of the disk with working segments will turn to the corner $\varphi = \omega t$.

There will be no untreated site if

$$x_d - x_a = h, \tag{1}$$

where h - the length of the cutting segment, m.

3 Results and discussion

In order to perform the process of mowing vegetation or milling around the tree trunk with proper quality (no knives or cutters) the following condition shall be complied:

$$L = h, \tag{2}$$

where L is the size of the path of the rotor center passing with cutting segments along the arc of the guide circle when it turns to the corner φ, m.

The working part will cd pass through the axis x at the angle of the turn:

$$\varphi = \omega t_p = \pi, \tag{3}$$

where t_p there is a time that determines the rotation of the rotor with cutting segments at the desired angle, s.

The time of turning the disk to an φ angle (when installed on the rotor of two cutting segments) shall be:

$$t_p = \pi / \omega, \tag{4}$$

where there is an ω angular rotation velocity of the rotor with cutting segments, at c^{-1}.

The value of the path passed by the center of the rotor when turning relative to the tree trunk:

$$L = \pi r_2 \varphi_1 / 180 = \pi r_2 \omega_1 t_c / 180, \tag{5}$$

where φ_1 is the value of the rotation angle of the rotary section around the tree trunk, degrees; ω_1 is the angular velocity of rotation of the rotary section around the tree trunk, at $^{-1}$; t_c; is the time of rotation of the rotor with cutting segments at an angle φ_1, s.

The time in which the point O_1 will move into the position O_1':

$$t_c = 180 L / (\pi r_1 \omega_1). \tag{6}$$

The condition of the quality functioning of the working tool will be in compliance with the requirement:

$$t_p \le t_c. \tag{7}$$

Substituting values (4) and (6) to the expression (7), taking into account (2), we shall obtain:

$$\pi/\omega \leq 180h/(\pi r_2\omega_1). \tag{8}$$

The expression (8) links the mode (ω, ω_1) and the design parameters (h) of the unit.

Let's determine the angular velocity of ω_1 rotation by expressing it through the speed and movement of the energy vehicle. During the technical process, in general, the working body can be presented as a crank-slid mechanism with a guide axis that does not pass through the axis of the curve and the separated from it $d = const$ (Figure 2), and a takeaway rotary section with contacting, all over the technical process with the tree trunk jack wheels 1 and 2, in particular, as an epicyclical mechanism, in which moving wheels 1 and 2 (jack wheels) rolled over the stationary wheel 3 (tree trunk).

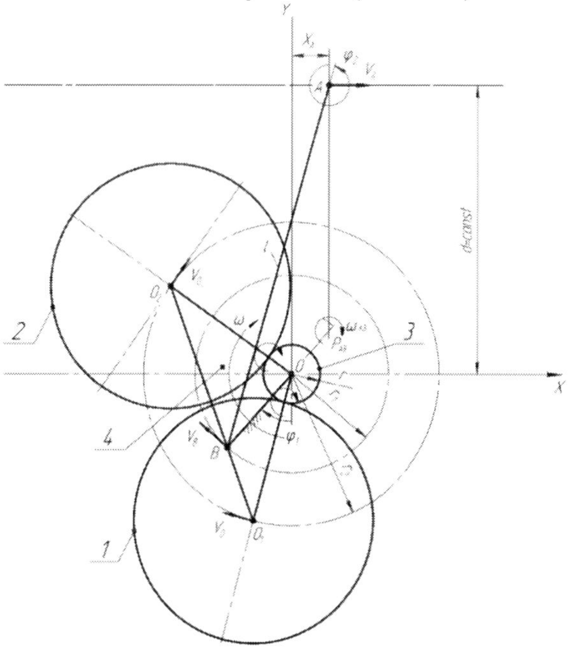

Fig. 2. Scheme to determine the main kinematic parameters of the takeaway rotary section.

The guide link in this mechanism is the A slider (energy vehicle), the driven link is the link 4 rigidly connected with the crank of the OV. The translational movement of the slider A at a constant speed V_A in the direction of the x axis causes the rotation of the OV crank, as well as the link 4 rigidly connected to it, at a constant angular velocity ω_1, forcing at the same time two wheels 1 and 2 to roll, without slipping, along a stationary wheel 3. The elastic elements included in the design provide continuous contact of jack wheels with a tree trunk, so that the rotor centers with cutting segments move around the tree trunk with constant angular velocity ω_1 along the guide circle.

The velocity of the point V, which belongs to the OV crank, which rotates around the point O in conjunction with the link 4, will be defined as:

$$V_B = r_1\omega_1, \tag{9}$$

where $r_1 = OB$ - the radius of the circle on which the point V, m; ω_1 moves; - the angular velocity of rotation of the crank of the OV, rad/s.

Point V also belongs to the link AV, which makes a plane motion.

The immediate velocity center of the *AV* crank rod is located at the intersection of perpendiculars restored at points *A* and *V* to their velocity directions.

At the same time, the velocity of point *V* is determined by the expression:

$$V_B = V_A \cdot AP_{AB} / BP_{AB}, \tag{10}$$

where AP_{AB} and BP_{AB} - distances from points *A* and *V* to the instant center of velocities, correspondingly.

Taking into account the expression (10) the angular velocity of the *OV* crank:

$$\omega_1 = V_B / r_1 = V_A / r_1 \cdot AP_{AB} / BP_{AB}. \tag{11}$$

Distances BP_{AB} and AP_{AB} depending on movement x_A can be defined as:

$$AP_{AB} = d - x_A / \tan\varphi_1 \,;\ BP_{AB} = r_1 - x_A / \sin\varphi_1 , \tag{12}$$

where φ_1 - the angle of the crank of the *OV* when moving the vehicle.

Then, taking into account expressions (12), the expression (11) takes the form:

$$\omega_1 = \frac{V_A}{r_1} \cdot \frac{d - \dfrac{x_A}{\tan\varphi_1}}{r_1 - \dfrac{x_A}{\sin\varphi_1}}. \tag{14}$$

To determine the value of the angle φ_1, depending on the movement x_A, we project a broken line *OVA* on the *0Y* and *0X* axis:

$$\begin{cases} r_1 \cos\varphi_1 + d - \ell\cos\varphi_2 = 0 \\ r_2 \sin\varphi_1 + x_A - \ell\sin\varphi_2 = 0 \end{cases}, \tag{15}$$

where x_A is a shift of the point *A* relative to the beginning of the coordinate system (the center of the tree trunk), m; d – distance between the row line and the vehicle movement line, m; φ_1 – the angle of the crank of the *OV*; φ_2 - the angle of the *AV* crank rod.

From the second equation of the system (15) after the transformation we shall obtain:

$$\cos\varphi_2 = \sqrt{1 - \left(\frac{r_1 \sin\varphi_1 + x_A}{\ell}\right)^2} = \frac{1}{\ell}\sqrt{\ell^2 - (r_1 \sin\varphi_1 + x_A)^2}. \tag{17}$$

Substituting the expression (17) in the first equation of the system (15), we shall obtain:

$$r_1 \cos\varphi_1 + d = \sqrt{\ell^2 - (r_1 \sin\varphi_1 + x_A)^2}. \tag{19}$$

From here, finally the expression for the φ_1 turn angle of the *OV* crank in the function of variable distance x_A will take the form:

$$\varphi_1 = \arctan\frac{x_A}{d} + \arccos\frac{\ell^2 - r_1^2 - d + x_A^2}{2r_1^2}. \tag{20}$$

4 Conclusions

When the unit moves along the line of the row, after the contact of the jack wheels with the tree trunk, through kinematic connections, the position of the design of the side-mounted turning section is changed relative to the tree trunk that allows, when turning the turning section around the tree trunk at the angle necessary for the complete mowing of vegetation in the zone of the near-trunk circle, which has common rotation axis with the jack wheels, rotors with cutting segments to move around the guide circle around the tree trunk.

References

1. A. K Apazhev, E. A. Polishchuk, Journal of Physics: Conference Series (JPCS), **1679** 042086 (2020)
2. A. M. Atlaskirov, Y. A. Shekihachev, L. A. Shomakhov, Agrotechnical and economic efficiency of a rotary garden mower, **2,** 164 (2020)
3. A. K Apazhev, Y. A. Shekikhachev, A. G. Fiapshev, L. M. Hazhmetov, International Scientific and Technical Conference Smart Energy Systems 2019 (SES-2019), **124** 05054 (2020)
4. V. S. Bzheumykhov, Y. A. Shekikhachev, Z. V. Bzheumykhova, AgroEcoInfo, **4(30)** (2017)
5. A. L. Khazhmetova, A. K. Apazhev, Yu. A. Shekikhachev, L. M. Khazhmetova, A. G. Fiapshev, Techniques and Equipment for countryside, **6(264),** 23 (2019)
6. A. L. Khazhmetova, A. K. Apazhev, Yu. A. Shekikhachev, L. M. Khazhmetova, A. G. Fiapshev, AgroEcoInfo, **2(36),** 29 (2019)
7. L. A. Shomakhov, Yu. A. Shekikhachev, R. A. Balkarov, Horticulture and Vine growing, **1**, 7 (1999)
8. A. M. Egozhev, E. A. Polishchuk, A. A. Egozhev, Rural mechanizer, **12,** 8 (2019)
9. A. K. Apazhev, Yu. A. Shekikhachev, L. M. Khazhmetov, The Gorsky State Agricultural University news, **53(2),** 138 (2016)
10. A. K. Apazhev, Yu. A. Shekikhachev, L. M. Khazhmetov, The Gorsky State Agricultural University news, **53(2),** 146 (2016)
11. A. Apazhev, V. Smelik, Y. Shekikhachev, L. Hazhmetov, Engineering for Rural Development, **18,** 192 (2019)
12. A. K Apazhev, A. G. Fiaphev, Y. A. Shekikhachev, L. M. Hazhmetov, L. Z. Shekikhacheva, Earth and Environmental Science, **315(5),** 052023 (2019)

Investigation of coking diesel injector spray nozzles in operation

Aslan Apazhev, *Yuri* Shekikhachev*, *Vladimir* Batyrov, *Lyudmila* Shekikhacheva, and *Anzor* Bolotokov

Kabardino-Balkarian State Agricultural University named after V.M.Kokov, 360030, Lenin Avenue, 1V, 360030 Nalchik, Russia

Abstract. One of the significant problems in ensuring the reliable operation of the nozzles is the intensive coking of the injector spray nozzle ports. Based on the assumption that all fuel left by injection under the needle burns, some researchers believe that the reason for coking is insufficiently emptying of this volume. There is also a well-known opinion about the impact of atomization quality at the final phase of injection. The lack of consensus and conflicting recommendations on the issue make the research relevant. A set of investigations has been carried out at OJSC "TsNITA" to study the influence of various factors, including design factors of the fuel system, on the coking of injector spray nozzle ports. This article describes the investigation results carried out on the basis of test materials for 24 variants of fuel systems on the D-240 engine, the analysis of varying parameters for the injection final phase is carried out depending on the combination of design factors and the relationship of coking of the injector spray nozzle ports with the final phase parameters is shown.

1 Introduction

During the injection process, coking occurs in the injector spray nozzle ports, even in the absence of visible violations of the fuel supply process (leaks, afterinjection, etc.). In scientific circles, there is no consensus on the causes of coking, recommendations are contradictory.

Some researchers suggest that the cause of coking is partial or complete emptying of the nozzle ports and the volume under the needle, causing the penetration of cylinder gases and oxidation of the remaining fuel film [1 3].

Reducing coking of sprayers at the final phase of injection is possible due to [4-7]: increasing the stroke of the needle and the diameter of its shut-off cone; reducing the density in the precision sprayer couple; reducing the area of the spray ports; reducing the stroke of the needle and its shut-off cone diameter. As a criterion for coking of the spray ports, the duration of the needle fit is often taken [8-15]. There is no experimental confirmation for the dependence of coking on the fuel equipment structural elements and

* Corresponding author: shek-fmep@mail.ru

the parameters of the fuel supply process, as well as the amount of fuel in the volume under the needle after injection.

2 Materials and methods

Tests on engines and a non-motorized bench were carried out only at the mode corresponding to the maximum torque. During oscillography of the fuel supply process, the injection was carried out into the bomb with an air back pressure of 5.0 MPa. All experiments were carried out on the same fuel delivery. The influence of the sprayer sizes was excluded by using one set of sprayers in the entire series of experiments.

3 Results and discussion

The coking degree of nozzle ports was determined by spilling nozzles out of a needle on a constant feed bench and was calculated by the formula:

$$k = \frac{\mu f_1 - \mu f_2}{\mu f_1}, \quad (1)$$

where μf_1 is the effective flow passage of the nozzle ports before the experiment, mm^2; μf_2 is the effective flow passage of the nozzle ports after the experiment, mm^2.

The dependence of the coking of the injector spray nozzle ports on the needle maximum stroke and the discharge volume of the pressure valve (Fig. 1) shows that the influence of the investigated design factors on coking is differentiated depending on their value.

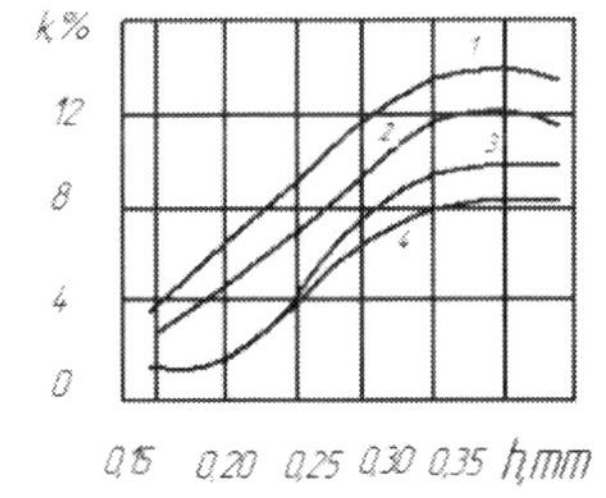

Fig. 1. Dependence of coking of the injector spray nozzle ports (k) on the needle stroke (h) and the discharge volume of the pressure valve (V_r) (1 – V_r = 35 mm^3; 2 – V_r = 50 mm^3; 3 – V_r = 65 mm^3; 4 – V_r = 80 mm^3.

In principle, an increase in the needle travel up to 0.35 mm contributes to the intensification of coking with all the investigated valves. With a decrease in the needle stroke, the influence of the valve weakens. It also follows from the results received that the same amount of coking can be obtained by a different combination of the needle stroke size and the discharge volume of the pressure valve. This indicates that the structural elements do not affect the coking of the nozzle ports directly through the structure, but due to the parameters of the fuel supply working process. Various combinations of the fuel system structural elements change the process of fuel supply, regulate the "time - section" value in the sprayer, and control the state of the fuel at the final phase.

Of the list of parameters (time, pressure) that determine the final phase of the fuel supply process, the parameter is directly related to the intensity of coking of the nozzle ports:

$$\lambda = \frac{P_a}{P_\varphi - P_a}, \tag{2}$$

where P_a is the pressure in the "appendix", MPa; P_φ is the pressure above the cone, MPa.

The ratio λ characterizes the dynamics of the fuel flow in the section of the flowing part of the sprayer inlet to the cone - "appendix".

The curve $k = f(\lambda)$ shows the dependence of the intensity of coking the sprayers on the parameter λ for all the variants of the fuel system of the D-240 engine under investigation, an increase λ facilitates mitigation of coking (Fig. 2).

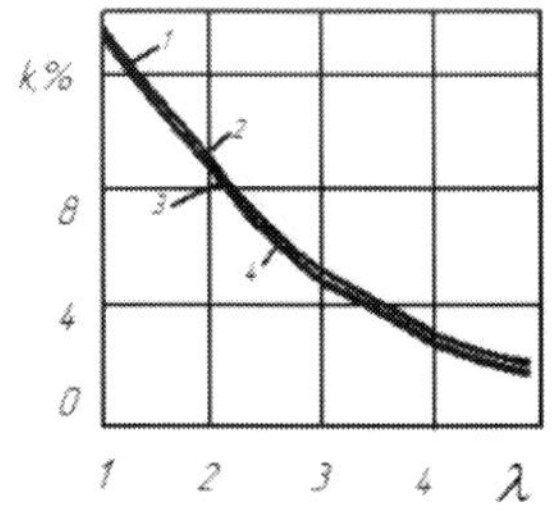

Fig. 2. Dependence of coking the injector spray nozzle ports (k) on the dynamics of the fuel flow λ at (h = 0.2-0.35 mm): 1 – V_r = 35 mm³; 2 – V_r = 50 mm³; 3 – V_r = 65 mm³; 4 – V_r = 80 mm³.

An exception is the fuel system variants with a discharge volume of the pressure valve of 80 mm³ and a needle stroke of more than 0.3 mm, for which coking increases (in comparison with V_r = 65 mm³ and h = 0.3 mm) due to the gases penetration into the sprayer due to the "overturning" of the fuel flow.

Thus, the influence of the investigated structural elements of the fuel equipment concerning coking the injector spray nozzle ports is carried out through the process of fuel supply by changing the operating mode of the sprayer at the final phase of injection.

The strongest effect on the parameters of the fuel supply process, and, consequently, the coking the injector spray nozzle ports is occurred by limiting the needle stroke to the value of h = 0.15-0.2 mm which helps to stabilize the fuel pressure drop across the cone during its fitting (Fig. 3)

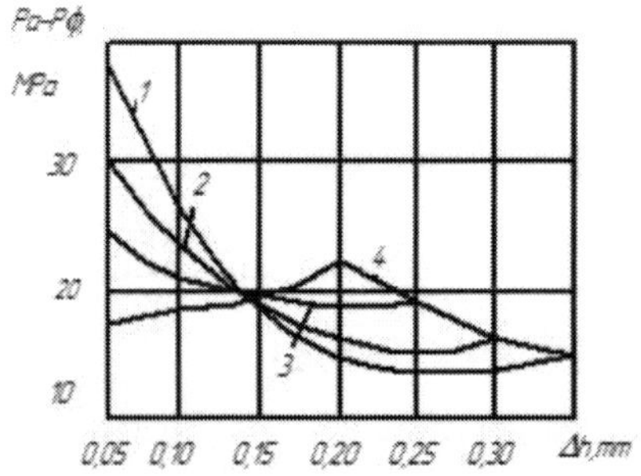

Fig. 3. Dependence of pressure drops across the nozzle cone at the final injection phase of the needle position (V_r = 35 mm³): 1 – h = 0.2 mm; 2 – h = 0.3 mm; 3 – h = 0.25 mm; 4 – h = 0.35 mm.

The specified stroke can be achieved by increasing the outlet section of the cone by increasing the diameter of the central channel under the needle.

Investigations carried out at JSC "TsNITA" on engines D-60, D-240, D-37E and AM-01 showed that an increase in the central channel under the needle to 1.6 mm with a simultaneous decrease of the needle stroke to h = 0.18-0.25 mm contributes to reducing the level of coking to a safe value (k =1-2%) [1-3].

Analysis of the test results (Fig. 4) indicates that the general regularity between the intensity of coking the injector spray nozzle ports and the time of the needle fitting at the final phase is not observed for all variants of diesel fuel systems.

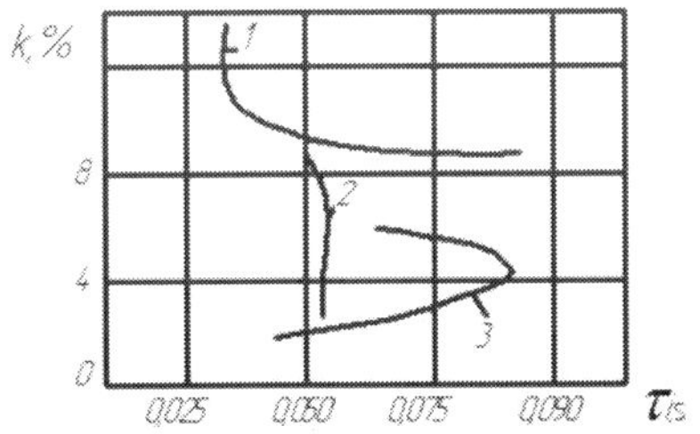

Fig. 4. Dependence of coking the nozzle ports from the time of needle fitting in position τ_i = 0.05 mm: 1 – h = 0.35 mm; 2 – h = 0.25 mm; 3 – h = 0.2 mm.

Depending on the system packaging, the influence of the fitting speed is differentiated. The duration of the needle fitting at the final phase, together with the discharge volume of the pressure valve and other structural elements of the fuel system, determines the rate of decrease in fuel pressure above and below the cone, i.e., it changes the value of the fuel pressure drop across the cone and, therefore, regulates the fuel flow rate.

4 Conclusions

The parameters (V_r , h) of the diesel fuel system regulate the ratio of the operating parameters for the final phase of fuel injection, and also significantly affect the intensity of coking the diesel injector spray nozzle ports. Limiting the needle stroke to h = 0.15-0.2 mm in the examined variants of the system makes it possible to stabilize the pressure drops on the cone at a minimum level and to reduce coking the nozzle ports to an acceptable value.

References

1. V. S. Koichev, A. K Kobozev, I. I. Shvetsov, D. I. Gritsai, E.V. Gerasimov, J. of Pharmac., Biolog. and Chemic. Sci., **8(5),** 642 (2017)
2. V. Coichev, K. Mosikyan, News of the National Agr. Univ. of Armenia **3,** 44 (2016)
3. V. S. Kurasov, V. V. Dragulenko, S. M. Sidorenko, *Theory of internal combustion engines* (2013)
4. A. K Apazhev, Y. A. Shekikhachev, V. I. Batyrov, L. Z. Shekikhacheva, J. of Phys.: Conf. Ser. (JPCS), **1679,** 042063 (2020)
5. A. K Apazhev, Y. A. Shekikhachev, V. I. Batyrov, Kh. L. Gubzhokov, A. L. Bolotokov, IOP Conf. Ser.: Mater. Sci. and Engin., **663(1),** 012049 (2019)
6. Y. A. Shekikhachev, R. A. Balkarov, M. M. Chechenov, H.B. Kardanov, L. Z. Shekikhacheva, J. of Phys.: Conf. Ser., **1515(4),** 042029 (2020)

7. Yu. A. Shekikhachev, V. I. Batyrov, R. A. Balkarov, L. Z. Shekikhacheva, Kh. L. Gubzhokov, Tech. and equip. for the vill., **4(262),** 14 (2019)
8. F. Kh. Khaliullin, A. V. Matyashin, R. R. Akhmetzyanov, V. M. Medvedev, M. A. Lushnov, IOP Conf. Ser.: Mater.s Sci. and Engin. Electr. Coll., 012016 (2019)
9. S. A. Sinitsky, V. M. Medvedev, R. R. Lukmanov, G. V. Pikmullin, O. I. Makarova, BIO Web of Conf. Internat. Sci.-Pract. Conf., 00025 (2020)
10. V. S. Koichev, K. A. Mosikyan, M. S. Barseghyan, *Problems of scientific and technological progress in the agro-industrial complex* (2017)
11. V. S. Kurasov, V. N. Pleshakov, E. E. Samurganov, A. V. Ponomarev, Proceed. of the Kuban St. Agr. Univ., **58**, 315
12. A. K Kobozev, I. I. Shvetsov, V. S. Koichev, I. I. Gazizov, N. V. Bakholdin, *Improvement of scientific and methodological work at the university* (2018)
13. A. K Kobozev, I. I. Shvetsov, V. S. Koichev, *Topical issues of engineering education* (2016)
14. A. K Kobozev, I. I. Shvetsov, V. S. Koichev, *Topical issues of engineering education* (2016)
15. B. L. Ivanov, B. G. Ziganshin, R. F. Sharafeev, I. R. Sagbiev, Bull. of Kazan St. Agr. Univ., **14,** 95 (2019)

Digitalization as a key factor in the development of the regional agricultural sector

Fatima Karaeva[1,*], *Rameta* Shokumova[1], *Pavel* Zhufan[2], and *Asiyat* Temrokova[1]

[1]Kabardino-Balkarian State Agricultural University named after V.M. Kokova, Lenin Avenue, 1V, 360030 Nalchik, Russia
[2]Mendel University of Agriculture and Forestry, Zemědělská 1, 61300, the Czech Republic

Abstract. Modern realities of sustainable development of the agro-industrial complex require new trends and tools. The article considers digital technologies as one of the key parameters for the development of the agricultural sector of the economy in increasing production efficiency. The greatest emphasis is placed on the formation of territorial - innovative models of the agrarian sector of the republic's economy, considering the growth of the gross domestic product, but at this time, the agrarian sector is not technologically modernized enough for the development of these platforms. First of all, it is necessary to resolve the issues of staffing with highly qualified specialists, which will allow to deploy digital technologies and reduce costs by 23%, since the average cost savings in land use using GPS navigation technologies is 11-14%, with differentiated fertilization - 8-12%. In addition to the growth of economic components, this development path ensures the environmental safety of agricultural production.

1 Introduction

Currently, digitalization is one of the most promising trends and tools of the state to achieve balanced sustainable economic development. Digitalization as a key factor is increasingly penetrating all segments of the region's agricultural sector and is gaining momentum by introducing digital innovations.

The role of digital technology in the development of the agricultural sector is great and consists in providing the population with safe, vital important products, in reducing costs, increasing the environmental, economic and production efficiency of production, which will attract workers of new professions to agriculture.

In this regard, a clear vector for the development of digitalization should be developed in the agricultural sector, aimed at the effectiveness of the implementation of technological operations based on digital platforms, allowing to achieve high results.

[*]Corresponding author: fatima64@mail.ru

2 Materials and methods

The new realities of strategic successful digitalization require a variety of industries in the application of the development of digital approaches aimed at the needs of market-oriented tasks, closely related to the practice of their implementation. To implement such approaches, standardized raw data is needed and demanded, allowing for the analysis of indicators, considering the emerging problems, and at the same time the development of recommendations aimed at action. Topical scientific publications within the framework of the digitalization of the economy, operational information arrays of the relevant ministries and departments of the republic have been examined. Following methods were applied in the process of working on the article: observation, logical, structural analysis, synthesis, as well as techniques of economic and statistical analysis.

3 Results and discussion

The digital economy can be defined as the integrity of the system of social-and-economic relations based on the use of digital information and communication technologies. It should be noted that the digital economy is basically aimed to simplify the interaction of participants in the production process both within the organization and in interaction with various external actors within the production process management framework.

This infrastructural superstructure should be considered conjointly with real production and consider its impact on the efficiency of technological processes. To obtain a general economic effect, the digital technologies should be implemented in parallel with the development of "traditional" production.

Digitalization may be considered at various levels (Fig. 1), and various research methods have been proposed by Russian and foreign economists [1].

The application of most of the methods is based on the collection of statistical information, which sometimes is not entirely accurate for assessing the real situation and, as a rule, the problem consists in the actual data collection [2].

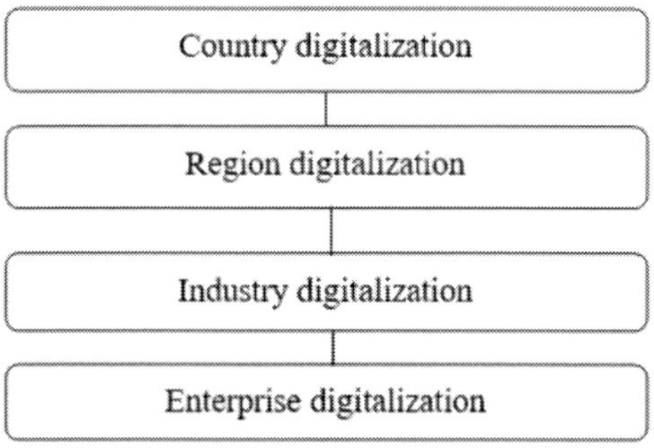

Fig. 1. Economic system digitalization levels.

It seems impossible to manage the digital transformation of the economy without an appropriate legislative framework, therefore, regulatory and legal acts have been adopted to regulate activities at various levels.

Decree of the President of Russia dated May 7, 2018 No. 204 "On national goals and strategic objectives of the development of the Russian Federation for the period up to 2024" set the task of transforming priority sectors of the economy and social sphere, including agriculture, through the introduction of digital technologies and platform solutions [3].

There are seven main directions of digital transformation of agriculture and scientific and technological development in the field of "Digital agriculture", which implies the introduction in the constituent entities of the Russian Federation of at least six projects of a complete innovative complex scientific and technical cycle of end-to-end digital systems: "Digital technologies in agro-industrial complex management", "Digital land use", "Smart field", "Smart garden", "Smart greenhouse", "Smart farm", based on modern competitive domestic technologies, methods, algorithms.

Currently, economic resilience is recognized as a vital factor in the survival of the rural population [4].

As part of the digitalization of the agricultural sector in 2021, a wider list of areas for the use of targeted concessional loans for the introduction of digital technologies and the automation of agricultural enterprises has been identified. In particular, the concessional short-term lending packages for up to 1 year and concessional investment loans from 2 to 5 years were provided. The first direction of lending implies the informatization and digitalization of agricultural production and the processing of own products, including the maintenance of machinery and equipment in this area.

In particular, the dairy sector of the industry gets the opportunity to use the loan funds for technical support of hardware and software and labeling of certain dairy product types.

Concessional investment lending may be aimed at expanding the fleet of agricultural machinery with automatic driving systems, spot application of materials, yield mapping, precise farming, or the data collecting and transmitting equipment [5].

Table 1. Costs of deployment and use of digital technologies in the North Caucasian Federal District in 2019.

Region	Total RUR mln	Including			
		Internal expenses		External expenses	
		RUR mln	%	RUR mln	%
North Caucasian federal district	13802.3	12405.51	89.9	1396.80	10.1
The Republic of Dagestan	1404.3	1368.03	97.4	36.30	2.6
The Republic of Ingushetia	306.9	304.50	99.2	2.41	0.8
Kabardino-Balkar Republic	572.0	514.82	90.0	57.21	10.0
Karachay-Cherkess Republic	651.5	590.01	90.6	61.50	9.4
Republic Northern Ossetia-Alania	1088.0	1004.05	92.3	84.01	7.7
The Chechen Republic	2298.9	2222.90	96.7	76.01	3.3
Stavropol Territory	7480.6	6401.23	85.6	1079.40	14.4

*Source: Russia region data [9].

In the North Caucasian Federal District in 2019, the cost of introducing and using digital technologies amounted to 13,802.3 RUR mln, of which 89.9% are internal costs and 10.1% are external. Analyzing by region, it can be seen that the largest expenditures on digital technologies are observed in the Stavropol Territory at 7480.6%, which is 54.2% in the share of the amount spent in the region.

In the Kabardino-Balkarian Republic, 572 mln RUR was invested in the deployment and use of digital technologies, where internal costs account for 90.0% in the amount of 514.8 mln RUR and for external costs - 10%, which amounted to 57.2 mln RUR.

The Kabardino-Balkarian Republic lags behind in information and communication technologies in general in Russia. Medium and small businesses have limited funds to pay for these services, everything is only at the nascent stage, and the level of IT services provided by external organizations is low as well. Therefore, the transformational transformation of the agro-industrial complex in the context of digitalization is the main prerogative of its modern development. The most important is the formation of territorial-innovative models of the agrarian sector of the republic's economy and increasing the volume of the gross domestic product. There is some improvement in the investment climate with a simultaneous improvement in its legal regulation. In particular, in the field of the agro-industrial complex, production capacities for the production of canned products have been expanded with the simultaneous creation of additional jobs.

If we assess the economic indicators of the region's agro-industrial complex, then, first of all, we will single out the volume of production by the end of 2020 of all agricultural producers in the amount of 61.4 RUR bln, the result is higher than the level of 2019 by 10%. Land reclamation measures are being intensively implemented in the republic within the framework of the regional project "Export of agricultural products" and the departmental program "Development of the land reclamation complex in Russia". Within this program, 7.9 th ha of irrigated land were commissioned in 2020, which is 22.2% higher than the program parameters.

Table 2. Dynamics of the main indicators of export of agricultural products of the Kabardino-Balkarian Republic.

By product types	Export as of 30.06, USD mln. USA			Average price, USD/ton		
	2019	2020	%	2019	2020	%
Oil and fat industry products	0.0	-	-	1069.3	-	-
Grain	0.4	2.0	by 4.5 times	847.1	3046.3	by 3.6 times
Fish and seafood	-	-	-	-	-	-
Meat and dairy products	0.2	0.1	-62.2	858.9	1026.2	+19.5
Food and processing industry products	2.5	2.7	+8.5	1036.3	1035.9	-
Other agricultural products	2.9	13.5	by 4.6 times	715.3	1015.3	+41.9
Total	6.0	18.3	by 3 times	835.6	1099.7	+31.6

*Source: [6].

The increase in the export of products by more than 3 times is due to grain farming exported to the near and far abroad (Azerbaijan, Kazakhstan, Belarus, Iran, etc.). Dairy and sour milk products, poultry meat, confectionery, mineral water, fresh fruits, etc. are exported as well. [7]. Pursuant to forecasts of the Ministry of Economy of the KBR, it is assumed that in 2020 the volume of exports of the Kabardino-Balkarian Republic will amount to 30 USD mln, or 140.8% in relation to the level of 2019.

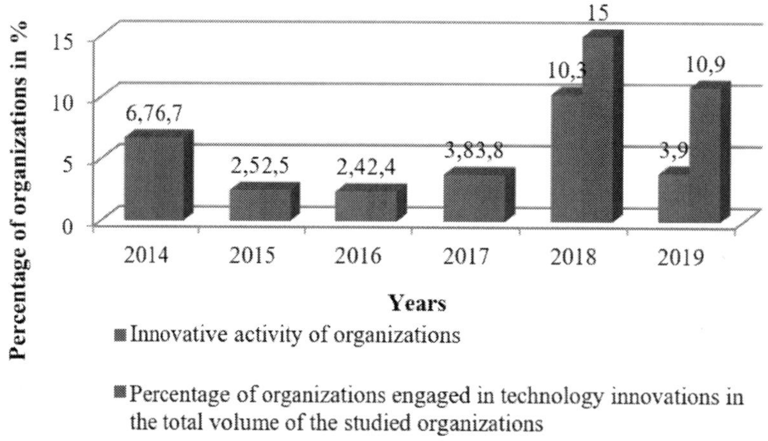

Fig. 2. Innovative activity of organizations in the Kabardino-Balkarian Republic.

In 2019, the innovative activity of the organization in the Kabardino-Balkarian Republic amounted to 3.9% against 10.3% in 2018. The decline in innovation activity was almost twofold (by 6.4 percentage points). The implementation of technological innovations in organizations in the reporting year in comparison with the previous period decreased by 4.1% and amounted to 10.9%.

For agro-industrial formations, the digitalization of agriculture will reduce costs by 23%, since the average cost savings for land use using GPS navigation technologies is 11-14%, with differential fertilization - 8-12%, and thanks to parallel driving systems - 8 -13% [8].

The volume of innovation implementation reflects an indicator that characterizes in monetary terms the costs of various types of technological innovation in the activities of the organization.

Based on the foregoing, let us consider the dynamics of technological costs within the NCFD.

Table 3. Dynamics of costs for technological innovation in the NCFD.

Region	2016		2017		2018	
	RUR mln	in % of the total volume of goods shipped, works and services performed	RUR mln	in % of the total volume of goods shipped, works and services performed	RUR mln	in % of the total volume of goods shipped, works and services performed
North Caucasian federal district	7896.5	1.4	8956.8	1.54	7142.4	0.8
The Republic of Dagestan	55.9	0.1	20.1	0.0	672.4	0.2
Republic of Ingushetia	-	-	0.7	0.0	-	-
Kabardino-Balkar Republic	86.2	0.3	296.5	0.9	326.0	1.0
Karachay-Cherkess Republic	8.1	0.0	26.7	0.1	12.9	0.0
Republic Northern Ossetia-Alania	19.4	0.1	49.5	0.2	36.1	0.1
The Chechen Republic	39.4	0.1	77.5	0.4	13.2	0.0
Stavropol Territory	7687.4	2.0	8485.8	2.1	6081.8	1.4

* Source: Russia region data [9].

In general, in the NCFD, costs are at the level of 7142.4 RUR mln and the indicator is lower than the previous periods, respectively, by 1,814.4 and 751.1 RUR mln. The cost of technological innovation in the Kabardino-Balkarian Republic increased compared to 2016 by 239.8 RUR mln and amounted to 326.0 RUR mln, which indicates the intensification of innovation processes over the past two years. In relation to the total volume of goods shipped, works and services performed in 2018, innovation costs amounted to 1.0%, which is more than 3 times more than in 2016. In comparison with the rest of the subjects of the North Caucasian Federal District in 2018, the largest share of costs falls on innovation processes in the Stavropol Territory (i.e. 1.4% or 6081.8 RUR mln). [7].

4 Conclusion

The process of globalization of the digitalization market is essentially a way of life, the basis for state management of sectors of the economy, society, and business. The digital transformation of the agricultural sector is gaining momentum, allowing to increase profitability by 2-3 times, the efficiency of the agro-industrial complex, contributing to leveling risks, reducing costs and production losses, introducing modern agricultural technologies such as remote sensing of land, digital twins of fields, precision farming technologies based on digital platforms and artificial intelligence.

The pace of implementation of technical and technological modernization and innovation is not yet sufficient for the introduction and implementation of these platforms, farmers need digitalization tools, solve the problem of a shortage of personnel with skills in working with digital products and technologies, and the key role of the state here should be reduced to a clear development of the development vector digitalization in all areas, which would ensure the competitiveness of production.

References

1. Skolkovo Digital Russia Methodology (2019) https://finance.skolkovo.ru
2. I. Yu. Merzlov, E. V. Shilova, E. A. Sannikova, M. A. Sedinin, Economics, Entrepreneurship and Law, **10(9)**, 2379 (2020)
3. Decree of the President of the Russian Federation of May 7, 2018 «On national goals and strategic objectives of the development of the Russian Federation for the period up to 2024»
4. A. K Dikinov, S. A. Chernyavskaya, F. E. Karaeva, K. N. Adaeva, S. A. Lyanova, Journal of Interdisc Iplinary Research, **10.2/13**, 86 (2020)
5. Materials of the Ministry of Agriculture of the Russian Federation, https://pravitelstvo.kbr.ru
6. Main export indicators of agro-industrial complex of Kabardino-Balkarian Republic, https://aemcx.ru
7. N. H. Kairova, R. E. Shokumova, Collection of Scientific Works of the III International Scientific and Practical Forum, 325 (2020)
8. Digital agriculture - opportunities and potential risks, https://ecfs.msu.ru
9. Regions of Russia. Socio-economic indicators. 2018 (Rosstat, 2018) http://www.gks.ru
10. A. M. Magomedov, Economics and management: problems, solutions, **1**, **12(96)**, 88 (2019)

Investigation of the travel depth of the disc colter and the conditions for its balance

Vladislav Mishkhozhev, *Aliy* Gabaev*, *Anatoly* Nam, *Vyacheslav* Dzuganov, and *Taymira* Pazova

Kabardino-Balkarian State Agricultural University named after V.M. Kokov, Lenin Avenue, 1V, 360030 Nalchik, Russia

Abstract. The seeding machine operation procedure consists of preparing the place where the seed should be placed, in placing the grain at a certain level below the soil surface in a certain order of arrangement and, finally, in the embedding – this work in large part falls on the share of the colter. The presented operation procedure of the seeder unit provides two main phases: reproduction of the maximum possible uniform feed of seed and its transporting to the colter; preparation of the groove, the required agrotechnical dimensions and distribution of seed material within it, followed by embedding them with soil. At the first phase of the seeding procedure, the issues of the outflow of seed material and its travel through the "seed tube - colter" system are relevant, and for the second phase, it is necessary to investigate the deformations of the soil layer reproduced by the colters, which is extremely necessary to select conditions that ensure uniform distribution of seed along the length and width of the row and at the desired depth. The article analyzes the investigation results by various authors devoted to the study of the stability of the depth of the opener stroke, its balance, and on their basis, analytical dependences are obtained.

1 Introduction

The vital task of agriculture is to guarantee the provision of our country with food, provided that soil fertility is preserved and increased, energy consumption is reduced, and the environment is protected. Its solution, especially at the stage of the formation of multi-structured forms of management, is facilitated by the introduction of the latest technologies and machines, including the comprehensive mechanization of crop and livestock production on the basis of a scientifically grounded system of machines [1-6].

Each agricultural machine performs a certain technological process, which includes one or several technological operations, the implementation of which is accompanied by qualitative changes in the processed material.

Agricultural machines, unlike industrial ones, make direct contact with living forms: seeds, plants, soil with its various microorganisms, etc. Therefore, their successful application is due to the introduction of zoned varieties of agricultural crops, adapted to machine technologies.

* Corresponding author: alii_gabaev@bk.ru

Agrotechnical requirements for the quality of technological operations have been developed for each agricultural machine.

It should be especially noted that each mechanized technological operation in the cultivation of agricultural crops should be optimized in terms of changeable application rates [7-11]. To do this, it is necessary for practical use to have a complete and suitable model of the relationship between such parameters as yield, amount of nutrients, moisture, density, soil fertility, field relief, and the like. Moreover, what is required is not a statistical fact, in what proportion are the mentioned parameters among themselves, namely, dynamic models of the relationship within a specific field, in other words we are talking about the need to develop a global model of agricultural production based on the laws of joint development of production and nature, modern systems theory, effective methods of substantiating decisions [12, 13].

Thus, as the main indicators of the work quality of the agricultural machines, in particular, grain seeders, in other words the output evaluation criteria of the technological process carried out by them, we can take such indicators as the uniformity of the distribution of seed and the uniformity of their embedding depth.

2 Materials and methods

Agrotechnical requirements for seeding of grain crops for embedding depths of 2-4, 4-5, 6-8 cm provide for permissible deviations: ±0,5; ±0,7; ±1,0 cm. Therefore, the colter shall provide the specified parameters of uniformity of seed embedding. Considering that soil shedding from the furrow walls has been reduced to a minimum, the fluctuation in the depth of the colters, which occurs as a result of external influences [14, 15], can be taken as the main factor in the unevenness of the seed embedding.

The disc colter system is represented by a drawbar, which is hinged at one end on a horizontal bar, and at the other end is rigidly mounted on the bolted casing with discs (Fig. 1). Thus, the colter system is a body that freely rotates about a horizontal axis parallel to the drawbar (with one degree of variance).

During the forward movement of the entire system, the following forces influence the colter:

- the weight G that is applied to the center of gravity of the colter with a drawbar;
- pulling force P, the line of influence of which passes through the suspension point of the drawbar to the horizontal bar;
- soil drag R.

Since the colter does not have a base and rests on one point, the reaction of the support and the soil drag at this point are inseparable from each other. Taking into account the symmetrical shape of the colter relative to the longitudinal vertical plane (Fig. 2), it should be noted that the influence of resistance forces on the right and left causes one resultant R, located in the plane of symmetry of the disc colter [5].

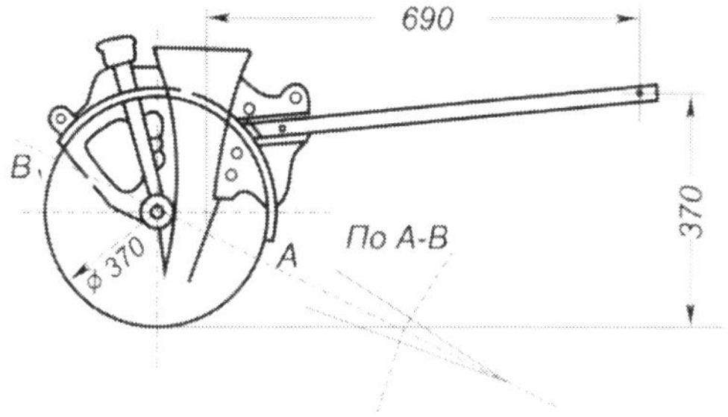

Fig. 1. Colter system with a drawbar.

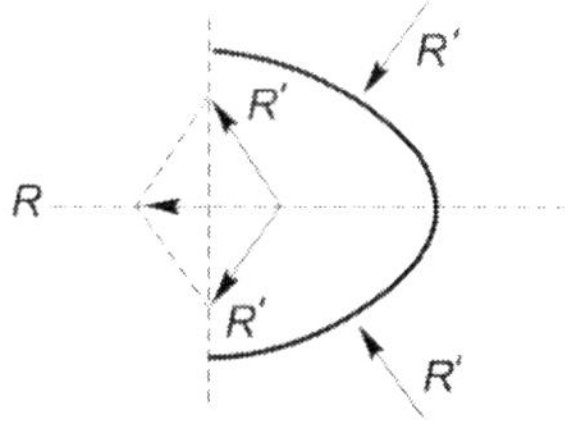

Fig. 2. Cross-section of the colter in the longitudinal vertical plane.

This conclusion is valid only when the colter operates in soil that is homogeneous in composition and condition. Otherwise, the emergence of one-sided impact is inevitable that is unbalanced on the other side.

Thus, when operating in a soil homogeneous in composition, due to the symmetrical shape of the disc colter, all three of the above forces G, P, and R can be considered located in the same plane, and, therefore, reduced to one resultant.

With this indicator, the disc colter differs significantly from the working bodies of other tillage implements, for example, the plow blade, the shape of which is asymmetric and always determines the combined action of forces causing one resultant pair of forces.

3 Results and discussion

Considering the steady motion of the disc colter, in which the above forces should be in equilibrium, it is possible for this case to determine R according to the given forces P and G. Thus, in order to comply with the equilibrium condition, the following condition shall be met:

$$R + P + G = 0, \tag{1}$$

and, therefore,

$$R = -(P + G), \tag{2}$$

in other words, the resultant of the given forces $(P + G)$ shall be equal and oppositely directed to the force R (Fig. 3).

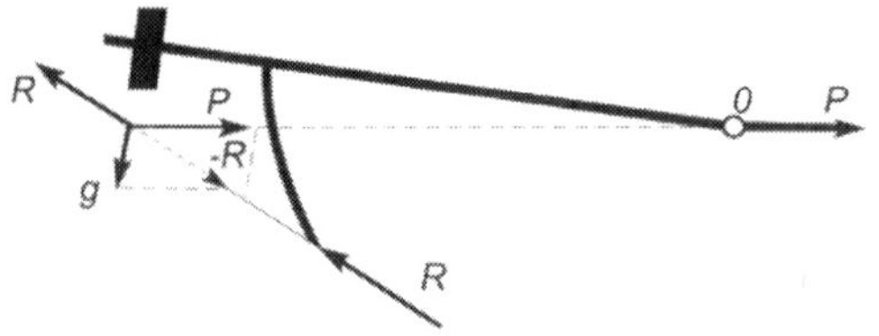

Fig. 3. Colter balance condition.

On the other hand, equation (2) can be represented in the following form:

$$R = (-P) + (-G), \tag{3}$$

that is, the force R, if it were known, can be decomposed into two components $(-P)$ and $(-G)$ so that the direction P passes through the hinge O of the suspension point. By transferring the force $(-P)$ to the hinge and decomposing it into two components: vertical V and horizontal H, it can be noted that the force V is transmitted through the hinge to the frame, and the force H shall resist to forward movement [6].

Components V and H are associated with forces R and G the following ratios:

- for vertical projections:

$$V = G - R\sin\alpha, \tag{4}$$

- for horizontal projections:

$$H = R\cos\alpha, \tag{5}$$

- for a moment relative to the point O:

$$Rr = G\ell, \tag{6}$$

where α – is a tilt of force R to the horizontal plane; r and ℓ – the shoulders of forces R and G.

Of the equality (4) and (5) after the exception α the R is determined:

$$R = \sqrt{H^2 + (G - V)^2}. \tag{7}$$

After the exception R, the angle α of the force R to the horizontal plane is determined:

$$tg\alpha = \frac{G - V}{H}. \tag{8}$$

The location of the force R, taking into account equality (6), is determined by the shoulder:

$$r = \frac{G\ell}{\sqrt{H^2 + (G - V)^2}}. \tag{9}$$

Thus, for the steady course of the disk colter, the force location value R is determined by the values G, H and V. Values H and V could be obtained using a special dynamometer, inserted into the hinge, where the drawbar of the colter is fastened.

In actual environment, in the during the operation, the force R does not remain constant, as a result of which the colter either rises or falls, rotating relative to the suspension point at the fixing point to the cross bar.

The position of the suspension point of the colter has a significant effect on the values of the colter weight G and shaft P when balanced. It is easy to make sure that, changing the vertical position of the drawbar fixing point, different forces G, P are required to balance the same resistance.

If, for example, the force R is balanced with the forces P and G, when the suspension point is O (Fig. 4). In this case $-R = P + G$, in other words resistance $-R$, equals in value R, but oppositely directed, is the resultant of forces P and G, moreover, the force P passes through the point O.

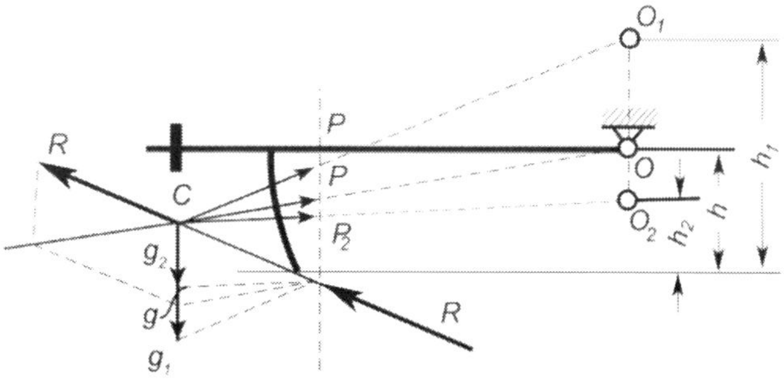

Fig. 4. Regulating of colter running depth

If we move the suspension point up the vertical O_1, then, obviously, the force R shall be decomposed into two forces, a force P_1, which direction will be determined by a straight line CO_1, and a force G_1, having the same direction as the force G. Accordingly, with the rise of the suspension point in order to maintain balance, it is necessary to increase P and G, conversely, decrease if the suspension point drops [7-8].

Leaving the suspension point of the colter at the same height, but moving it horizontally closer or further from the center of gravity of the colter, or, in other words, using short or long drawbars, different values G and P should be taken into account to balance the same resistance.

4 Conclusions

A short drawbar requires more colter weight and pulling force for balance, while a long drawbar requires less weight and less pulling force. The balance can be maintained with the same forces P, G and R, if the suspension point is raised and at the same time moved further from the center of gravity of the disc colter, lengthening the drawbar accordingly. However, do not use very long drawbars to reduce the load on the colter. Equilibrium can be achieved with a relatively short drawbar out of unnecessary load, but for this you should take the suspension point of the colter below.

References

1. E. V. Kyul, A. K Apazhev, A. B. Kudzaev, N. A. Borisova, Indian Journal of Ecology, **44(2)**, 239-243 (2017)
2. A. Apazhev, V. Smelik, Y. Shekikhachev, L. Hazhmetov, Engineering for Rural Development, **18**, 192-198 (2019)
3. A. K Apazhev, A. G. Fiaphev, Y. A. Shekikhachev, L. M. Hazhmetov, L. Z. Shekikhacheva, IOP Conference Series: Earth and Environmental Science, 315(5), 052023 (2019)
4. A. K Apazhev, Y. A. Shekikhachev, A. G. Fiapshev, L. M. Hazhmetov, E3S Web of Conferences / International Scientific and Technical Conference Smart Energy Systems (SES-2019), 124, 05054 (2019)

5. V. V. Dzuganov, Y. A. Shekikhachev, A. Sh Teshev., M. M. Chechenov, V. Kh. Mishkhozhev, IOP Conf. Series: Materials Science and Engineering, 919(3), 032015 (2020)
6. A. K Apazhev, E. A. Polishchuk, Journal of Physics: Conference Series (JPCS), 1679, 042086 (2020)
7. A. L. Khazhmetova, A. K. Apazhev, Yu. A. Shekikhachev, L. M. Khazhmetov, A. G. Fiapshev, V. S. Kurasov, Polythematic network electronic scientific journal of the Kuban State Agrarian University, **153**, 159-169 (2019)
8. H. H. Ashabokov, A. K. Apazhev, Yu. A. Shekikhachev, L. M. Khazhmetov, A. G. Fiapshev, AgroEcoInfo, **2(36)**, 32 (2019)
9. A. L. Khazhmetova, A. K. Apazhev, Yu. A. Shekikhachev, L. M. Khazhmetov, A. G. Fiapshev, AgroEcoInfo, **3(37),** 37 (2019)
10. A. K. Apazhev, Yu. A. Shekikhachev, L. M. Khazhmetov, The Gorsky State Agricultural University news, **53(2)**, 138-143 (2016)
11. A. K. Apazhev, Yu. A. Shekikhachev, L. M. Khazhmetov, The Gorsky State Agricultural University news, **53(2)**, 146-151 (2016)
12. A. L. Khazhmetova, A. K. Apazhev, Yu. A. Shekikhachev, L. M. Khazhmetov, A. G. Fiapshev, Techniques and Equipment for countryside, **6(264)**, 23-28 (2019)
13. A. L. Khazhmetova, A. K. Apazhev, Yu. A. Shekikhachev, L. M. Khazhmetov, A. G. Fiapshev, AgroEcoInfo, **2(36),** 29 (2019)
14. Y. A. Shekikhachev, V. H. Mishkhozhev, L. Z. Shekikhacheva, R. H. Zhigunov, Kan. V. Mishhozhev, Kaz. V. Mishhozhev, IOP Conference Series: Earth and Environmental Science, 548(2), 022004 (2020)
15. A. K. Apazhev, Yu. A. Shekikhachev, L. M. Khazhmetov, News of the Nizhnevolzhsky agrarian university complex: Science and higher professional education, **3(43)**, 238-245 (2016)

Optimization of production technology for encapsulated functional detoxicants

Maya Tamova[1,*], Elena *Barashkina*[1], *Rostislav* Zhuravlev[1], *Sergei* Usatikov[1], and Amina Dzhaboeva[2]

[1]KubSTU, st. Moskvovskaya, 2, 350072 Krasnodar, Russia
[2]Kabardino - Balkarian State Agricultural University named after V.M. Kokov, Lenin Avenue, 1V, 360030 Nalchik, Russia

Abstract. This study purpose was to optimize the production technology of encapsulated functional detoxicants based on combined dietary fibers (DF) obtained from beet pulp according to the developed technology. Production technology optimization for functional encapsulated food products was performed using mathematical programming methods based on Statistica v.10 and MathCAD v.15 software environments. An optimal technological production mode for encapsulated food products with high quality indicators, having an increased detoxification ability in relation to heavy metal ions, including combined DF, obtained from beet pulp.

1 Introduction

To protect human biochemical systems from technogenic factors, dietary fibers (DF), as physiologically active components of the human diet, should be included in the diet [1-3].

Secondary products of plant raw materials [4] processing, in particular, beet pulp, the most common by-product of sugar beet production seem to be perspective - about 20 million tons of this resource is produced in Europe every year [5, 6]. At present, highly efficient technologies have been developed to produce dietary fiber from beet pulp [7], they are distinguished by a high content of pectin substances and fiber, they may be used as a food additive in functional food production [8, 9].

At present, an urgent sector in food engineering, food industry and public catering industry is supported by innovative technologies, including natural structure-forming agents in the recipe formulation [10]. Alginates, a group of marine and bacterial structural exopolysaccharides, such as alginic acid, as well as its salts and derivatives, act as the most preferred and widespread materials for the preparation of bulk, macro- and microgels [11]. The use of sodium alginate is especially important in the encapsulation technology implementation, allowing for increased delivery stability and efficiency of encapsulated biologically active ingredients, including dietary fiber, with their more complete release in the human gastrointestinal tract [12].

[*] Corresponding author: tamova_maya@mail.ru

The study purpose was to develop an optimal technological production mode for encapsulated food products with high quality indicators, having an increased detoxification ability relative to heavy metal ions, including combined beet pulp DF.

2 Materials and Methods

2.1 Study objects

Sodium alginate (viscosity of 1 wt% dissolved in water is 350-550 mPa s) according to GOST 33310-2015, anhydrous calcium chloride according to GOST R 55973, distilled and deionized water, potassium dichromate according to GOST 2652, black chromogen according to TU TU6- 09-1760-72, lead (II) acetate according to GOST 1027. Combined DF were produced according to a previously developed technology from beet pulp using electrophysical and chemical exposure methods [13].

2.2 Preparation of model mixture samples to be encapsulated

The mixtures to be encapsulated were prepared as follows. A weighed portion of sodium alginate powder in an amount of 1% was dissolved in deionized water with constant stirring in Intllab MS-500 magnetic stirrer for 3-4 minutes [14]. To remove dissolved in the liquid phase, the solution was put in a VacModul vacuum module with VP1 vacuum pump for VD 53.Combined DF were added to the resulting structure former solution and stirred for 3-4 minutes until a homogeneous system was obtained.

The solution, forming a capsule shell containing polyvalent metal ions (required for the structure formation of a sodium alginate solution) was produced by dispersing calcium chloride in deionized water in an amount of 0.5%.

The mixture of biopolymers was extruded by axial feeding into the forming medium at a ratio of 1:10 through a device for encapsulated products with a given size of outlet nozzles, 2 mm in diameter. The resulting rounded-oval capsules were exposed in the interval from 1 to 3 minutes, washed with deionized water and dried by blowing air through a fine filter.

2.3 Determination of capsule syneresis

The syneresis value was calculated by gravimetric method to determine the amount separated from a 100 g sample of moisture over a regulated time interval.

2.4 Determination of capsule density

The capsule density was determined by hydrostatic weighing using VIBRA AFDK kit and SHINKO A.F.225 DRCE. analytical balance.

2.5 Determination of binding capacity

The binding capacity of the biopolymer mixture to be encapsulated was determined by titration. For this, 25 ml of a model solution of polysaccharides was mixed with 30 ml of a 0.1M solution of lead acetate ($Pb\ (CH^3COO)\ _2$). The mixture was kept with continuous shaking in SKO-180 Labtex laboratory rotation shaker and separated into fractions in Sartorius laboratory centrifuge, Sigma 2-6 10223. The precipitate was washed with distilled water until a negative qualitative reaction for lead ions with potassium dichromate ($K_2Cr_2O_7$). The resulting fraction and washings were combined and made up to the mark with distilled water in a 250 ml flask. A 10 ml aliquot of the resulting solution was titrated with 0.01 mol/l solution of ethylenediamine tetraacetate (EDTA) at 9 to 10 pH in the presence of black chromogen until the color changed from violet to blue.

The binding capacity (BC) was estimated by the value of metal ions binding and calculated by the formula:

$$BC = \frac{A_1 - A_2}{A_1} \cdot 100\%, \tag{1}$$

where A_1 is the total mass of metal, g

A_2 is mass of metal remaining in solution, g

In equation (1), values A_1 and A_2 were calculated from the results of titrometric tests using the following formulas:

$$A_1 = \frac{C_{EDTA} \cdot V_1 \cdot M}{1000}, \tag{2}$$

$$A_2 = \frac{C \cdot 250 \cdot V_{cp} \cdot M}{1000 \cdot V_2}, \tag{3}$$

where C is the concentration of initial metal solution, mol/l;

V_1 is initial volume of metal solution, ml;

V_2 is aliquot solution of the centrifugate taken for titration, ml;

V_{CP} is average volume of EDTA used for titration, ml;

M is metal molecular weight, Da.

2.6 Multicriteria optimization of study results

The multicriteria technology optimization for production of encapsulated food products with combined DF was performed using mathematical programming methods based on Statistica v.10 and MathCAD v.15 software.

3 Results and Discussion

To determine optimal technological parameters for encapsulated products production, a mathematical model of a two-factor experiment was developed.

Taking into account the previously obtained results [8, 14], the factors with the highest effect on target functions Y_1 (density of resulting capsules) $\leq$ 1.2 g/cm^3, Y_2 (capsule binding capacity relative to Pb^{2+} ions) $\rightarrow$ max and Y_3 (capsule syneresis after 72 hours of storage) $\rightarrow$ min:

X_1 is the amount of combined DF injected into a solution with a mass fraction of 1% sodium alginate,%;

X_2 is the residence time of capsules in the forming solution with a mass fraction of 0.5% calcium chloride, s.

For two-factor optimization, the solution can be obtained graphically and coincides with experimentally calculated solution.

A graphical display of the capsule syneresis dependence after 72 h of storage is presented (FIG. 1), the binding capacity of capsules relative to lead ions Pb^{2+} (FIG. 2) and density of resulting capsules (FIG. 3) depending on the amount of combined DF introduced into the solution with a mass fraction of 1% sodium alginate,% and the residence time of capsules in the forming solution with a mass fraction of 0.5% calcium chloride, s.

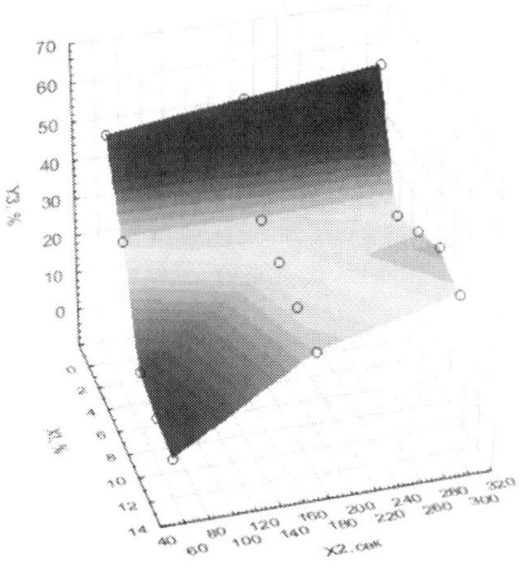

Fig. 1. Capsule syneresis after 72 hours of storage; $Y_3 \rightarrow min$.

Fig. 1 shows that when the value is limited, target function $Y_3 \rightarrow min$ is achieved with the following possible values $X_1 = 9\%$; $X_2 = 60$ s.

Fig. 2 shows that the max. value of Y_2 is set with the following possible valuesX_1 = 12%; $X_2 = 60$ s.

Fig. 3 shows that when target function value $Y_1 \leq 1.2$ g/cm^3 is limited at $X_1 = 9\%$; X_2 = 60 s.

To determine optimal values X_1,% and X_2, using the mode by means of MathCAD v.15.0 package, an attainability set and an optimal solution for Y_3 and Y_2 are constructed (Fig. 4).

According to Fig. 4, the optimal technological parameters to determine encapsulated products $X_1 = 9\%$; $X_2 = 60$ s, when the values of target functions $Y_3 \rightarrow min$ and $V_2 \rightarrow max$ and $Y_1 \leq 1.2$ g/cm^3 are achieved.

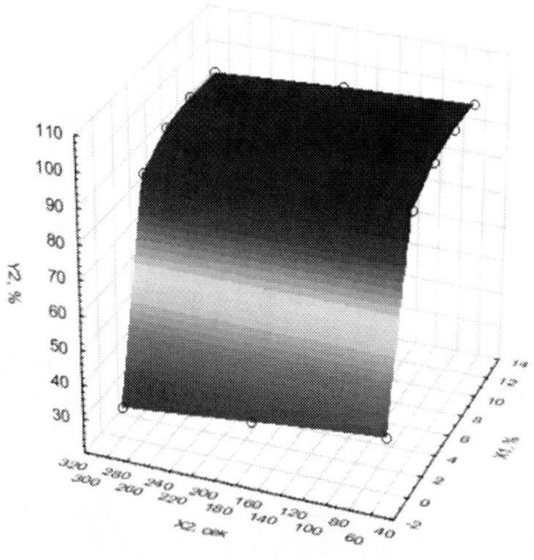

Fig. 2. Capsule binding capacity relative to ions Pb^{2+}; $Y_2 \rightarrow$ max.

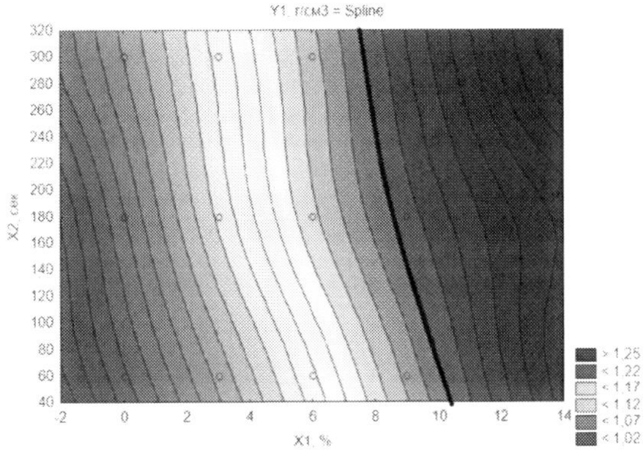

Fig. 3. Density of resulting capsules; $Y_1 \leq 1.2$ g/cm^3.

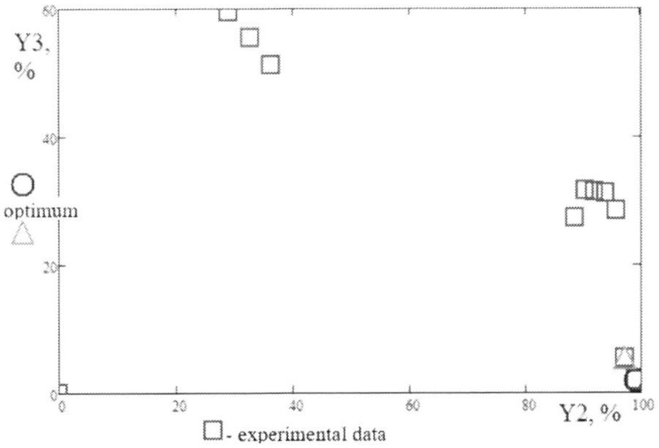

Fig. 4. Attainability set and an optimal solution for Y_3 and Y_2.

Thus, the optimal technological parameters for encapsulated products with combined DF were determined: the amount of combined DF introduced into the sodium alginate solution is 9%; exposure time of capsules in the forming solution of calcium chloride is 60 s. In this case, the capsule density is 1.18 g/cm^3, the binding capacity relative to lead ions Pb^{2+} is 99.1%, the capsule syneresis is 1.95%.

4 Conclusions

The multicriteria optimization of technological modes of combined food products was performed using mathematical programming methods implemented by means of Statistica v.10 and MathCAD v.15 software environments.

The optimal technological modes for the production of functional encapsulated food products were determined, with increased sorption capacity relative to heavy metal ions, with combined DF obtained from beet pulp.

References

1. N. M. Delzenne, M. Olivares, A. M. Neyrinck, M. Beaumont, L. Kjølbæk, T. M. Larsen, J.-W. van der Kamp, Clinical Nutrition. **39**, 414-424 (2019)

2. M. Y. Tamova, E. V. Barashkina, R. A. Zhuravlev, N. R. Tretyakova, E. S. Franchenko, Food Technology, **5-6**, 9-13 (2018)
3. Dietary Fiber: Properties, Recovery and Applications, 79-123 (2019)
4. T. A. Comuniana, M. P. Silva, C. J. F. Souza, Trends in Food Science & Technology, **108**, 269-280 (2021)
5. B. Joanna, B. Michal, D. Piotr, W. Agnieszka, K. Dorota, W. Izabela, *Advances in Biotechnology for Food Industry*, 359-392 (2018)
6. V. L. Finkenstadt, Sugar Tech, **16(4)**, 339-346 (2013)
7. M. Y. Tamova, E. V. Barashkina, R. A. Zhuravlev, N. R. Tretyakova, A. K. Shantyz, I. S. Koba, Food Technology, **5-6**, 107-110 (2019)
8. M. Y. Tamova, E. V. Barashkina, N. R. Tretyakova, R. A. Zhuravlev, N. D. Penov, Foods and Raw Materials, **1**, 2-9 (2021)
9. J. Berlowska, M. Binczarski, P. Dziugan, A. Wilkowska, D. Kregiel, I. Witonska, *Advances in Biotechnology for Food Industry*, 359-392 (Academic Press, 2018)
10. R. A. Zhuravlev, M. Y. Tamova, N. A. Bugaets, V. M. Poznyakovskiy, N. D. Penov, Foods and Raw Materials, **1**, 81-89 (2017)
11. I. G. O. Črnivec, N. P. Ulrih, *Biopolymer Nanostructures for Food Encapsulation Purposes*, 335-380 (Academic Press, 2019)
12. B. Lupo, A. Maestroa, J. M. Gutiérreza, C. Gonzáleza, Food Hydrocolloids, **49**, 25-34 (2015)
13. M. Y. Tamova, E. V. Barashkina, R. A. Zhuravlev, N. R. Tretyakova, S. V. Usatikov, A. Z. Todorova, Food Technology, **5-6**, 71-74 (2019)
14. M. Y. Tamova, E. V. Barashkina, R. A. Zhuravlev, N. R. Tretyakova, M. R. Musaeva, Food Technology, **1**, 62-65 (2020)

Development control system robotic platform for horticulture

Dmitriy Khort, *Alexey* Kutyrev*, *Rostislav* Filippov and *Stepan* Semichev

Federal Scientific Agroengineering Center VIM, 1-st Institutsky proezd, 5, Moscow, 109428, Russia.

Abstract. The article presents a control system for a robotic platform for horticulture. The electronic control system consists of a running engine control unit, a stepper motor steering unit, an electronic differential control unit, a power plant automatic on / off control unit, and battery charging balancing. The developed control system of the robotic vehicle contains a central computer that collects information from sensors and sensors, processes it and transmits control signals to the drives of the machine movement. The movement of the robotic platform is carried out both by a radio signal with a remote control, and in offline mode on a pre-set map of the area according to data from the GLONASS/GPS differential receiver of the satellite navigation system. It is also possible to independently control the movement of a robotic platform using a vision system. The autonomy of the robotic platform provides 10 hours of continuous operation in low-light conditions in various weather conditions.

1 Introduction

With the development of artificial intelligence, scientists, engineers and designers are increasingly turning to the topic of self-driving cars, tractors and robots [1,2]. Currently, the world is actively working on the development of robotic devices for agriculture that can solve a wide range of tasks: spraying, harvesting and caring for plantings [3,4]. Unmanned tractors and agricultural aggregates with intelligent motion control and vision systems that are able to recognize images, distinguish a cultivated plant from weeds, and perform a high-quality technological operation find practical application [5-7]. The use of agricultural robots is aimed at improving the environmental safety of products, minimizing the harmful effects of chemicals on humans and increasing the yield of products [8-10]. At the same time, it is worth noting that most of these machines have not yet been brought to mass production and the final consumer. Therefore, this direction is extremely relevant and in demand for both science and production. In modern conditions of agriculture, the use of robotic machines will allow creating highly intelligent, automated production of agricultural products, which will completely replace manual labor, minimize the negative impact of chemicals on humans and reduce the loss of working time associated with the human factor.

* Corresponding author: alexeykutyrev@gmail.com

 https://doi.org/10.1051/e3sconf/202126201024

2 Materials and Methods

A prototype of a universal robotic platform structurally consists of a frame with struts, a power plant, an electric transmission, and electronic control system blocks. Two ECMG600-033U geared motors mounted on the rear wheels drive the universal robotic platform. A LIFAN S-PRO 5500 gas generator and a LiFePo4 24V 105Ah power battery, which allows the platform to operate at full load for at least 10 hours of continuous operation and ensures that the battery is recharged in no more than two hours, power the geared motors. The steering system is a worm-type steering gear with a stepper motor and a stepper motor power controller. A schematic diagram of the control units of the robotic platform is shown in figure 1.

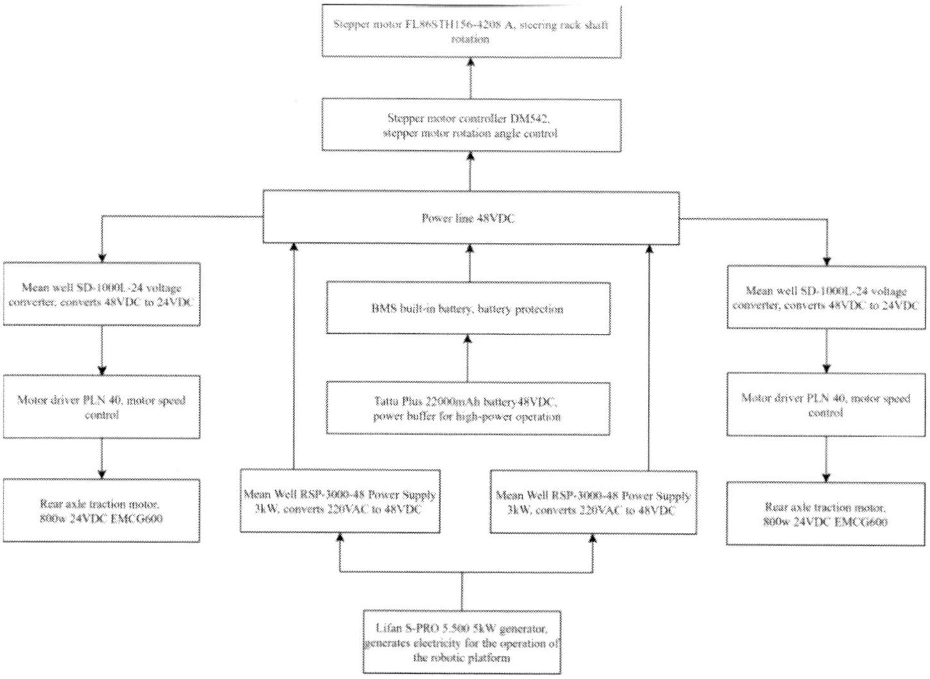

Fig. 1. Schematic diagram of the control of a robotic platform

The electronic control system consists of a control unit for running engines, a steering unit with a stepper motor, an electronic differential control unit, a control unit for automatic switching on and off of the power plant (gas generator) and balancing of battery charging.

The proposed motion control system, power plant and electric transmission, will allow the platform to operate in three modes: remote control, autonomous mode using electronic maps of the area using GPS signals, as well as in unmanned mode using machine vision modules. The control system is located inside the platform body. The technical characteristics of the platform are presented in Table 1.

Table 1. Main tailing dumps of the Sverdlovsk region

Basic parameters	Meaning
Curb weight, kg	850
Load capacity, kg	500
Characteristics of the running system: Wheel size, mm: - diameter, no more than, mm - width, no more than, mm Torque on each of the drive wheels, Nm Wheel formula	wheel type 650 200 330 4*2

Table 1. Continued

Machine dimensions, not more than, mm: - length - width, adjustable - height, no more than	 2800.0 1840.0;1960.0;2080.0 1600
Track width, base, mm: - front wheels - rear wheels - base	 1800 1800 1900
Ground clearance, mm	1200
The smallest turning radius, not more than, mm	3500
Forward speed, km/h - working - transport	2-6 8
Power plant: - type - generator power, W - power supply voltage, V	 Gasoline generator 5500 48
Drive: - type and quantity - power, kW - motor torque, Nm - motor shaft speed, rpm	Electric motor with two-stage planetary gearbox, 2 pcs. 0.6-0,8 kW 334 3000
Remote control operation	Remote control Fly Sky-T6 FS, transmitter frequency 2.4 GHz, channel width: 500kHz, modulation type: GFSK, antenna length 26 mm, sensitivity 105dBm
Management with technical vision	The autonomous control system contains an optical range finder (Lidar), 4 video cameras, laser distance sensors, a GPS navigation station, and an on-board computer
Steering, (type, design) Stepper motor:	Electric worm-type steering mechanism ST57-100 DM556E LEADSHINE Engine DriverStep-up power converter QS-488CCCV-1800W
Brake system (type, design)	Electric motor braking

3 Results and Discussion

The electrical connection diagram of the components of the robotic platform (Fig. 2) provides for the launch and power supply of the platform from two sources: the Lifans-pro 5500 generator and the Delta GX12-60 battery pack.

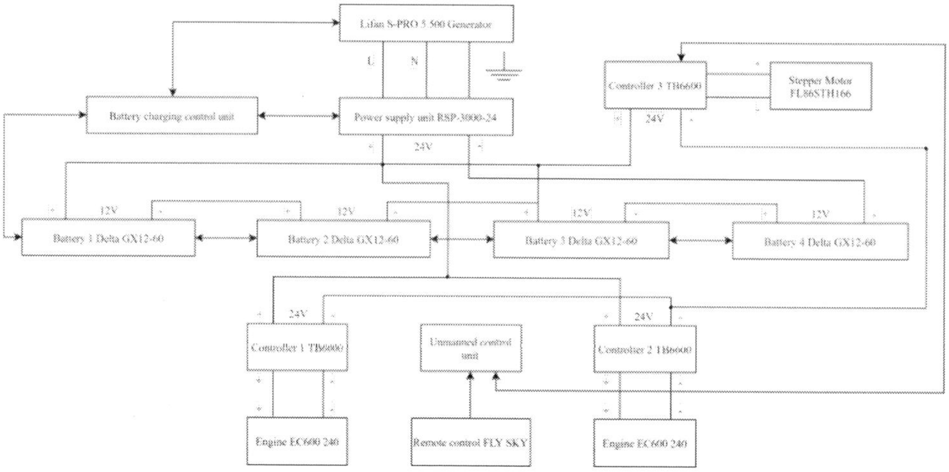

Fig. 2. Schematic electrical diagram of the control of a robotic platform

The main power source is batteries, which, thanks to a certain connection (in series with each other and in pairs in parallel), supply a voltage of 24 V to the controllers, which are controlled remotely by the control panel. The operator uses the remote control to send signals to the TB6600 and PLN40 controllers, and they in turn control the engines. When the battery is discharged less than 20%, the generator start system is activated, which includes: air speed control, automatic adjustment of engine speed when the load on the battery changes, autorun of the generator using the starter, monitoring the fuel level. If it is necessary to start the generator, the system that monitors the fuel level is activated. When the fuel level is sufficient, the generator turns on the ignition circuit, and the position of the air flap is checked. The air flap must be closed before starting the starter. If the flap is open, the air flap is moved to the closed position by means of a servo drive.

In the PLN 40 motor controller, the speed is regulated using a 5-10 kOhm potentiometer, this method is recommended by the manufacturer of the PLN 40 controllers. This method is adapted to the developed control system of the robotic device (Fig. 3).

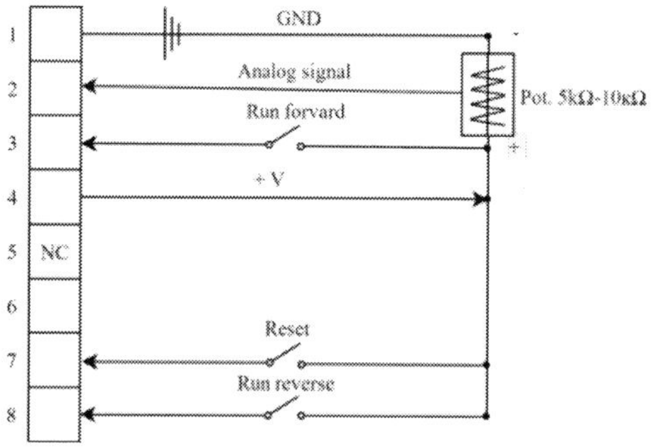

Fig. 3. Control scheme PLN 40

The diagram shows that the speed is controlled by changing the voltage at input 2. To change the voltage, a device has been developed that converts data from the central control unit to a voltage of 0-10 to control the speed of the motors using the PLN 40 controller (Fig.4).

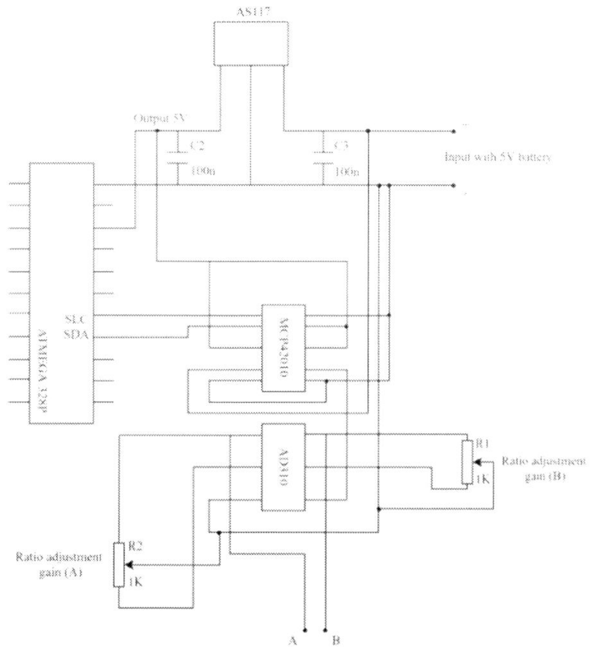

Fig. 4. Motor speed control scheme

To convert the digital signal to analog, the MCP42010-E/SL digital dual-channel potentiometer is used, which can split the voltage 0-5V into 128 points. The potentiometer is controlled via a serial peripheral interface (SPI). To form the reference voltage, the AS117 converter is used, which stabilizes the input voltages from 5-20V to a stabilized 5V. The PLN 40 controller reads the voltage from 0 to 10 V. To increase the voltage range, a two-channel AD310 operational amplifier with a gain of 2 is used. To adjust the gain, a subscript resistor is used for each channel. The voltage of 12V from the battery is supplied to the voltage stabilizer AS117, a stable voltage of 5V is formed at the output, a microcontroller and a digital potentiometer are powered from it. The ATMEGA 328P microcontroller generates a digital signal (the degree of motor speed) and is fed to a digital potentiometer, which converts the digital signal into a resistance. After the control voltage is formed, it is fed to the operational amplifier in which it is amplified to the values of the controller PLN 40.

The batteries must be recharged during their operation. For autonomous operation of the robotic platform and electric power management, a system has been developed that automatically supports the rated charging modes of the battery. To recharge the battery, an AC / DC converter is used, which converts an alternating voltage of 220V to a constant voltage of 24V and with a maximum current of 125 A. The robotic platform uses four 12 V batteries, two batteries for each engine. All batteries are connected to a single power supply system. All four batteries are charged simultaneously. The maximum battery charge current recommended by the manufacturer is 10 amps. When connecting the AC / DC converter directly to the batteries, the charging current will be equal to 125 amps, which will lead to premature failure of the battery. For this purpose, a current stabilizer is used (Fig.5).

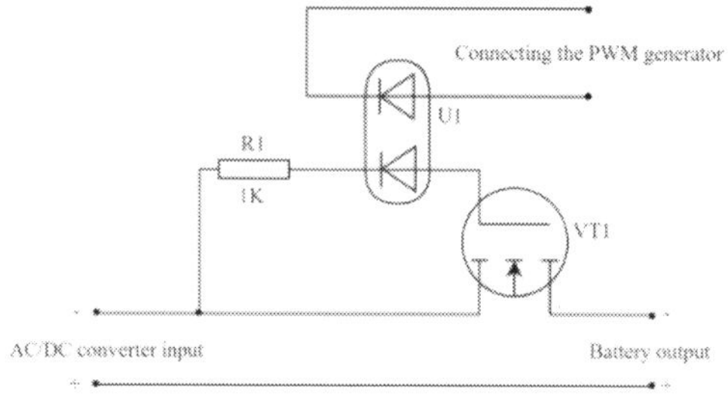

Fig. 5. Current stabilizer circuit

The use of a current stabilizer will limit the charging current of the battery, the current value is regulated by changing the duty cycle of the PWM generator, which is connected to the field-effect transistor VT1. The ICE2QR2280Z controller with a conversion frequency of 25 kHz is used as a PWM generator.

To turn the wheels of the robotic platform, an electric drive, a stepper motor is used. For precise positioning of the turning of the wheels, a control method has been developed that can be connected to a single control system of a robotic platform. A bipolar stepper motor ST57-100 is used as an actuator for the steering of the robotic platform, which is connected through a coupling to the steering gear. To control the stepper motor, the DM556E LEADSHINE driver is used, which is powered by a battery voltage of 24 V. 4 wires are connected to the microcontroller, VCC power supply of the microcontroller is 5 V, PUL is used to generate the number of steps using a PWM signal, DIR is the direction of rotation of the stepper motor, ENABLE permission to enable the driver.

As a result of the research, a robotic platform control system was developed (Fig. 6).

Fig. 6. Robotic platform with a developed control system

After installing the stepper motor on the robotic platform, the positioning accuracy was calibrated, the microstep, the torque on the shaft, and the driver supply current were adjusted. A sensor is connected to the steering gear output shaft to monitor the steering angle, as well as to monitor the skip of engine steps. A CNIKESIN 400 encoder is used as a rotation angle sensor. The encoder is connected to the microcontroller via a digital port, power is supplied via a 5V microcontroller circuit. When you turn the encoder, pulses are issued, which the MC reads. 1 impulse equals 0.90 rotation angle. If the stepper motor skips the steps, the encoder will fix this moment, then the microcontroller will command the driver to increase the current on the windings when the next step is applied. The current will increase until the encoder detects the rotation of the stepper motor shaft.

4 Conclusions

The developed control system of the robotic vehicle contains a central computer that collects information from sensors and sensors, processes it and transmits control signals to the drives of the machine movement. The movement of the robot is carried out both by a radio signal with a remote control, and in offline mode on a pre-set map of the area according to data from the GLONASS / GPS differential receiver of the satellite navigation system. A base correction station is used to provide a reliable satellite signal. Signals are transmitted from the robot to the control tablet via a WI-FI channel at a distance of up to 200 m. It is also possible to independently control the movement of a robotic platform using a vision system.

An analysis of technological application schemes on various horticultural crops showed that a robotic platform with various attachments can be used for harvesting horticultural crops, spraying, cultivating the soil on low-growing berry fields with a row spacing of 750-1000 mm, in intensive-type gardens with a row spacing of 2.5- 5 m and shrub berry fields with plantings between rows of 2.5-3.5 m (Fig.7).

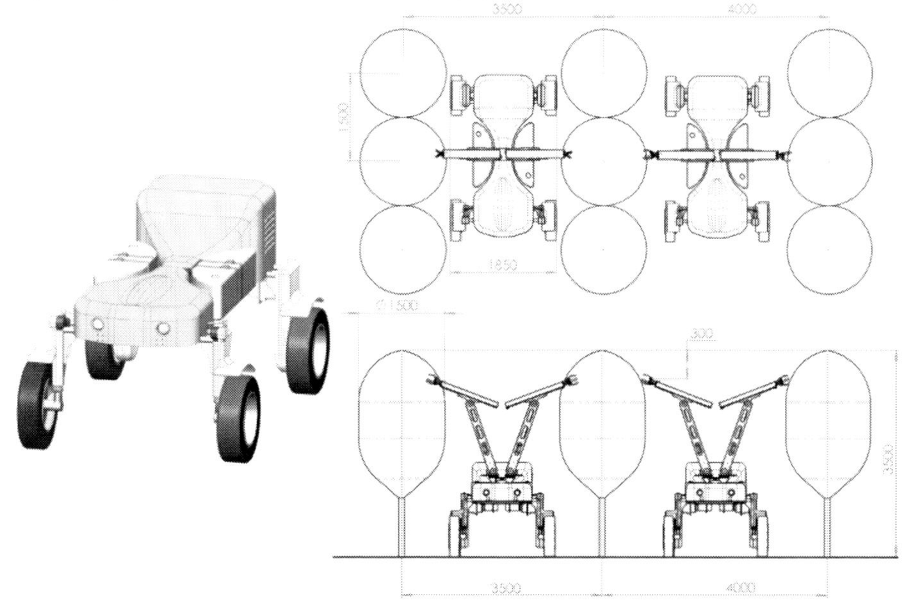

Fig. 7. Scheme the use of a robotic platform during the harvesting operation of horticulture crops

The autonomy of the robotic platform provides 10 hours of continuous operation in low light conditions in various weather conditions.

References

1. Zhenghao Fei, Stavros G. Vougioukas, Computers and Electronics in Agriculture, **180**, 105894 (2021)
2. M.E. De Kleine, M. Karkee, Trans. ASABE, **58(6)**, 1461 (2015)
3. D. Khort, A. Kutyrev, I. Smirnov, V. Osypenko, N. Kiktev, Communications in Computer and Information Science, Springer, Cham, 1158 (2020)
4. D. Khort, A. Kutyrev, R. Filippov, N. Kiktev, D. Komarchuk, Science and Technology (PIC S&T), Kyiv, Ukraine, 869 (2019)
5. Rui Suo, Fangfang Gao, Zhongxian Zhou, Longsheng Fu, Zhenzhen Song, Jaspreet Dhupia, Rui Li, Yongjie Cui, Computers and Electronics in Agriculture, **182**, 106052 (2021)
6. P.A. Dias, A. Tabb, H. Medeiros, Comput. Ind., **99**, 17 (2018)
7. F. Gao, L. Fu, X. Zhang, Y. Majeed, R. Li, M. Karkee, Q. Zhang, *Multi-class fruit-on-plant detection for apple in SNAP system using Faster R-CNNComput*, Electron. Agric., 105634, 176 (2020)
8. L. Fu, Y. Majeed, X. Zhang, M. Karkee, Q. Zhang, Biosyst. Eng., **197**, 245 (2020)
9. H. Kang, C. Chen, Comput. Electron. Agric., **168**, 105108 (2020)
10. Z. Zhang, C. Igathinathane, J. Li, H. Cen, Y. Lu, P. Flores, Comput. Electron. Agric., **175**, 105606 (2020)

Quality control of pork used in the recipe for kebabs in a technogenic zone

Atsamaz Biboyev[1,*], *Rustem* Temirayev[1], *Lada* Vitiuk[1], *Rita* Zhylova[2], and *Larisa* Shiritova[2]

[1]North Caucasus Mining Metallurgical Institute (State University of Technology), Nikolaeva 44, 362021 Vladikavkaz, Russia
[2]Kabardino-Balkarian State Agricultural University named after V.M. Kokov, Lenin Avenue, 1V, 360030 Nalchik, Russia

Abstract. The investigation purpose was to conduct a comparative control for the consumer qualities of the store pigs meat of the Landrace Danish breeding (LDB) and their cross-breeds of different pedigree on a large white breed (LWB) according to the scheme: control – well-bred Landrace (LDB) animals: 1-st check - 1/2 LDB 1/2 LWB yelts; 2-nd check - 3/4 LDB 1/4 LWB animals and 3-rd check - 7/8 LDB 1/8 LWB yelts, as well as dish samples No. 278 "Pork kebab", for which the animals meat of the specified pedigree was used in the recipe. The "pork kebab" dish samples from the pork loin of the 3-rd check group distinguished by the best chemical composition. The dish sample of the 3-rd generation pork meat (7/8 LDB × 1/8 LWB) distinguished by the best ecology nutritious features, so that the 3-rd check sample contained lower level of zinc, lead and cadmium. By the total score of the 3-rd check sample of this 3-rd generation Landrace pork dish surpassed the control sample.

1 Introduction

For cooking kebabs, meat from pigs of different breed types is used: meat, bacon, combined (porky) breeds, but to ensure high consumer properties of kebabs, first of all, the ratio of protein and fat in pork should be taken into account. In this respect, it is preferable to use the purely meat breed pigs with the more optimal ratio because of a lower proportion of fat, but protein is more [1, 2, 3, 4].

In the North Caucasus, including in North Ossetia - Alania, the most common breed of pigs is the large white breed (LWB), which belongs to the combined breeds. Animals of this breed are very well adapted to local conditions, including ecology. Due to the increased emission of high-density metals into the environment, the Republic of North Ossetia - Alania is one of the most polluted territories in the Russian Federation due to the large list of non-ferrous metallurgy enterprises in the city of Vladikavkaz, such as JSC "Electrozinc", JSC "Pobedit", etc. Recently, however, producers have begun to import high-yielding meat breeds, among which the animals of Landrace Danish breeding (LDB) distinguish for their excellent slaughter and meat qualities. But they are less adapted to local environmental

* Corresponding author: kapmar17@gmail.com

conditions that causes a decrease in their meat productivity and the quality of pork [5, 6, 7, 8].

Taking this into account, in order to increase the adaptive qualities, meat productivity and consumer qualities of pork produced in the technogenic zone of the Republic of North Ossetia - Alania, introductory mating of Landrace Danish breeding sows (LDB) with breeding boars producing large white breeds (LWB) is carried out, and subsequently, crossbreeds of the 1-st and 2-nd generations are mated with Landrace breeding boars [9, 10, 11, 12].

The investigation purpose was to conduct a comparative control for the consumer qualities of the store pigs meat of the Landrace Danish breeding (LDB) and their cross-breeds of different pedigree on a large white breed (LWB) according to the scheme: control – well-bred Landrace (LDB) animals: 1-st check - 1/2 LDB 1/2 LWB yelts; 2-nd check - 3/4 LDB 1/4 LWB animals and 3-rd check - 7/8 LDB 1/8 LWB yelts, fattened in the technogenic zone of RNO -Alania, as well as dish samples No. 278 "Pork kebab", for which the animals meat of the specified pedigree was used in the recipe.

2 Materials and methods

In the course of our investigations, the subjects of investigations were:

- when obtaining meat for processing - different pedigree hog carcasses on large white breeding (LWB) and Landrace Danish breeding (LDB), killed at the age of 7.5 months;
- dish samples, cooked according to the recipe No.278 "Pork kebab", from the pork loin of different pedigree.

To obtain different pedigree hog carcasses for large white breeding (LWB) and Landrace Danish breeding (LDB), we fed store pigs for meat production of different pedigree on large white breeding (combined or porky type), selected at the age of 2 months. At the same time, the introductory mating of Landrace Danish breeding sows (LDB) with breeding boars producing large white breeds (LWB) was carried out, and subsequently, crossbreeds of the 1-st and 2-nd generations were mated with Landrace breeding boars.

In accordance with the investigation scheme in the course of fattening store pigs at the age of two months formed, on the principle of analogue pairs (taking into account gender, age, breed, origin and live mass) 4 yelt groups of 15 heads each:

- control – well-bred Landrace (LDB) animals:
- 1-st check – 1/2 LDB x 1/2 LWB yelts;
- 2-nd check – 3/4 LDB x 1/4 LWB animals
- 3-rd check – 7/8 LDB x 1/8 LWB yelts.

3 Investigation results and discussion

In order to obtain different pedigree hog carcasses for large white breed (LWB) and Landrace Danish breeding (LDB) under the conditions of the Republic of North Ossetia - Alanya we fed Landrace store pigs for meat production of different pedigree on large white breeding. The slaughter indexes are shown in Figure 1 and the linear parameters of hog carcasses are shown in Figure 2.

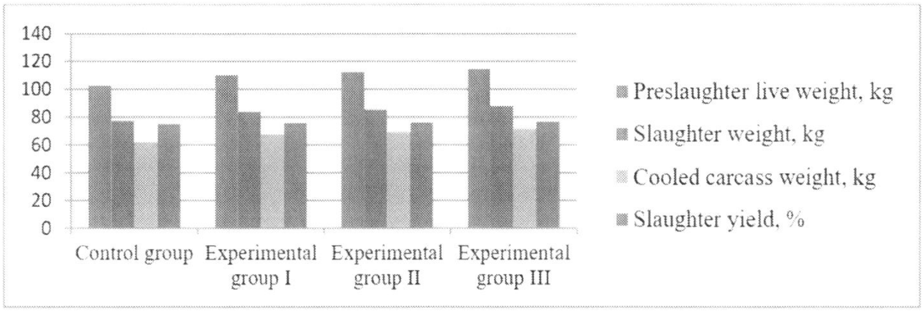

Fig. 1. Animal slaughter indexes of comparable groups.

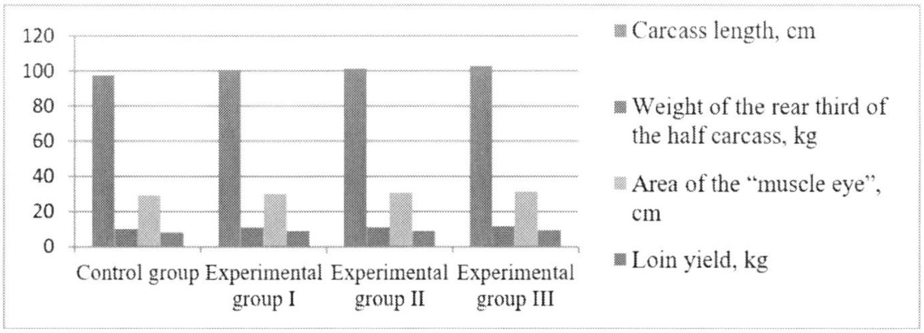

Fig. 2. Linear indexes of yelt carcasses.

Compared with the control group, there are significantly ($P<0.05$) better indexes of pre-slaughter live weight by 12.02 kg, slaughter weight - by 10.74 kg, slaughter yield - by 1.48%, chilled carcass weight - by 9.70 kg, the carcass yield - by 2.14% and the weight of the rear third of the half carcass - by 1.01 kg, the area of the "muscle eye" - l by 2.09 cm^2 and the loin yield - by 1.46 kg were yelts of the 3rd check group.

The chemical composition (Figure 3), the biological value of protein (Figure 4) and the availability of heavy elements (Figure 5) in samples of the animals longest muscle of the back are studied in the traditional way.

The best indexes of chemical composition and biological fullness were the yelt meat of the 3 check group, reliably ($P<0.05$) surpassing control on the content of dry matter by 0.72%, protein - by 0.8%, the size of protein quality index (PQI) - by 6.6%. Compared to the animals meat of the control group, the meat sample of the yelts of the 3-rd check group was inferior in terms of energy value by 2.03 kcal/100 g

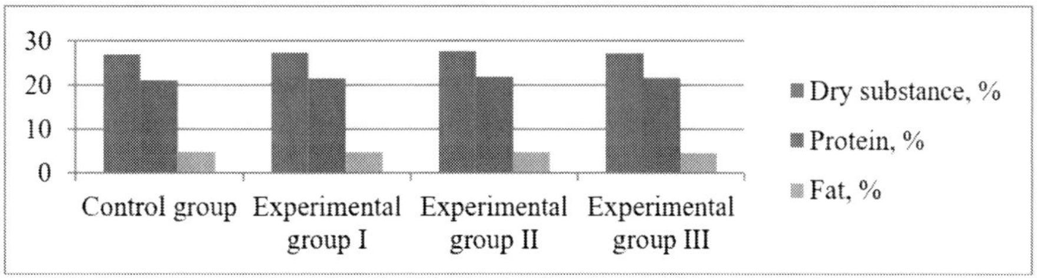

Fig. 3. The chemical composition of pork meat.

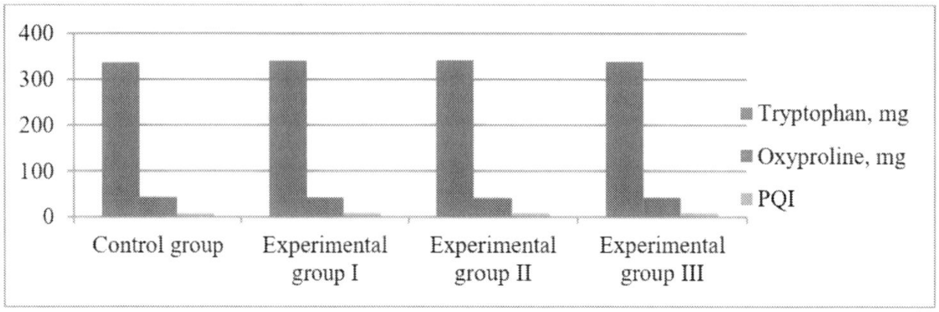

Fig. 4. The biological value of the yelt meat.

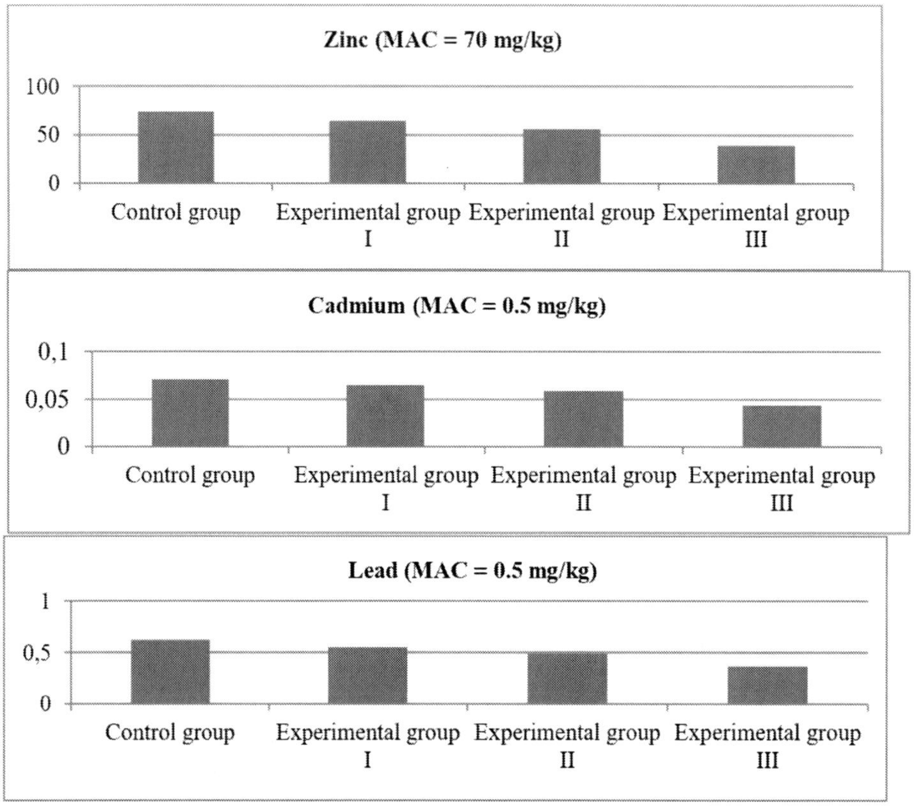

Fig. 5. Concentration of high-density metals in the meat of yelts.

The animals in the 3-year-old group (7/8 LDB x 1/8 LWB) had a reliably lower concentration of zinc by 1.90 times, lead by 1.68 and cadmium by 1.61 times, respectively. Moreover, the store pigs of the 3-rd check group have the concentration that did not exceed the maximum allowable concentrations (MAC) in any case.

In the compared samples of loin from the yelt carcasses of the compared groups studied indexes of water-holding capacity, diameter of muscle fiber (Figure 6) and pH of meat, determined immediately after slaughter, as well as after 24 hours of soaking.

According to the data obtained, the highest level of water-holding capacity, muscle fiber diameter was characteristic of the store pigs meat of the 3rd check group (7/8 LDB x 1/8 LWB), reliably ($P<0.05$) surpassing the control analogues in these indexes by 7.97% and 5.10 μm [55].

The animal meat pH value of the compared group immediately after slaughter, as well as after 24 hours of soaking, was within the normal range.

The following process chart and cooking scheme for dish No. 278 "Pork kebab" from the meat of animals of the compared groups (Table 1) is given below.

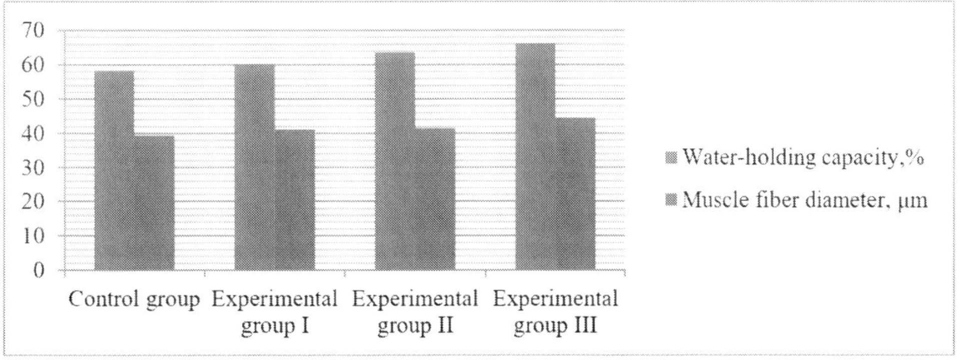

Fig. 6. Some physical indexes of yelt meat.

Table 1. Process chart Dish name: No. 278 "Pork kebab".

	Product name	**Weight of products per 1 portion, g**		**Weight of products per 10 portions, g**	
		Gross weight	**Net weight**	**Gross weight**	**Net weight**
1	Pork (loin)	177	127	1770	1270
2	Onion	26	21	260	210
3	Salt	4	4	40	40
4	Ground black pepper	4	4	40	40
5	Ground red pepper	4	4	40	40
6	Dried dill (khionthora)	4	4	40	40
7	Bay leaf	0.03	0.03	0.3	0.3
8	Mineral water	-	-	1500	1500
	Fried kebab weight	-	85	-	850
Side dish:					
9	Fresh tomatoes	126	107	1260	1070
10	Fresh cucumbers	120	100	1200	1000
11	Green onion	27	22	270	220
12	Onion	26	22	260	220
13	Salt	4	4	40	40
Spicy tomato sauce:					
14	Coriander	40	40	400	400
15	Spicy pepper	20	18	200	180
16	Onion	30	26	300	260
17	Ground black pepper	6	6	60	60
18	Tomato Paste	55	55	-	-
19	Water	10-20	10-20	-	-
	Yield of	-	250	-	2500

Meat, cut into 30-40 g cubes (3-4 pieces per portion), sprinkle with salt, pepper (black and red), dried dill, add raw chopped onions (lightly poured with vinegar), bay leaf and mineral water, mix, and put in cold place for 2-6 hours.

The marinated meat is put on a skewer and fried over hot coals or on the grill.

Sprinkle pickled onions (vinegar, pepper, salt) on the finished kebab and leave for a few minutes.

Garnish (fresh vegetable salad). Chop tomatoes, cucumbers, onions and green onions. Then add salt and pepper to taste, then mix.

Preparation of the sauce. Finely chop the onion (turnip-shaped), fresh pepper (spicy), coriander. Dilute the tomato with water and add the previously chopped onion, coriander and pepper. Then add pepper and salt to taste and mix well.

The meat shall be roasted on both sides and brown in color. The consistency of the meat cut is tender, the color is red-brown. The smell is specific, peculiar for marinated meat.

The chemical composition and nutritional value (Fig. 7) and the concentration of high-density metals (Fig. 8) of the "Pork kebab" compared samples are given below.

The best chemical composition was observed in the samples of the prepared dish from the pork loin of the 3-rd check group, which relative to the control sample was appeared in an increase in protein content by 0.87% ($P<0.05$) and a decrease in fat concentration by 0.84% and energy value by 3.02 kcal/100 g.

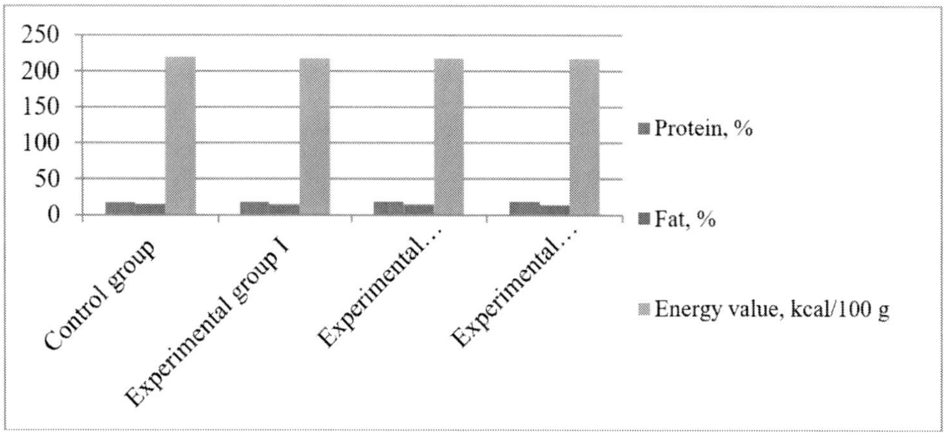

Fig. 7. Chemical composition of "Pork kebab" dish samples.

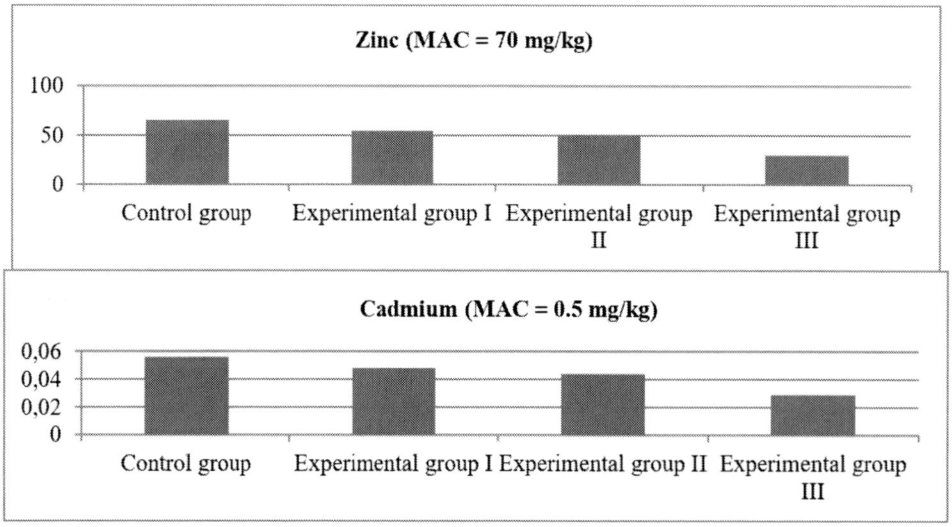

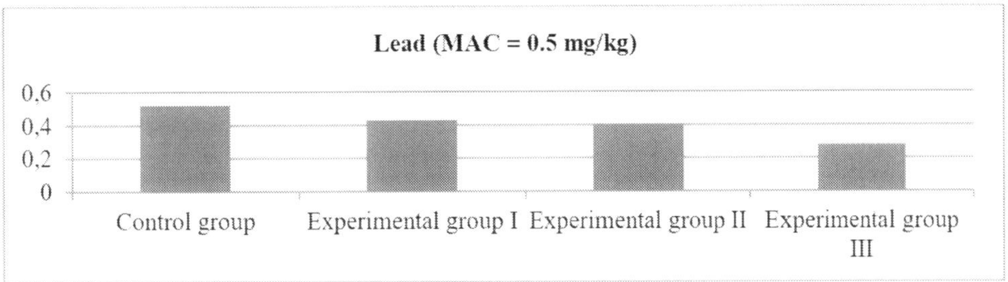

Fig. 8. Concentration of high-density metals in the "Pork kebab" dish samples.

The best ecological nutritional properties were distinguished by the No. 278 "Pork kebab" dish sample from the pork meat of the 3-rd generation, in which, relative to the control, there was a significant ($P<0.05$) lower content of zinc by 2.19 times, lead - by 1.86 and cadmium - 1.93 times. In the compared samples of the ready-made product, the content of high-density metals was lower than the MAC, with the exception of the availability of lead and cadmium in the control sample.

The taste panel consisting of 7 people carried out an organoleptic assessment of No. 278 "Pork kebab" dish samples (Table 2) according to the five-point scale.

Table 2. "Organoleptic assessment of No. 278 "Pork kebab" dish samples.

Indicator	sample			
	control	1 test	2 test	3 test
Appearance	4.73±0.04	4.83±0.05	4.86±0.05	4.89±0.02
Sectional color	4.67±0.04	4.70±0.05	4.71±0.05	4.87±0.05
Taste	4.71±0.04	4.77±0.05	4.82±0.05	4.90±0.04
Aroma	4.86±0.06	4.88±0.03	4.90±0.04	4.94±0.02
Juiciness	4.87±0.05	4.90±0.04	4.88±0.07	4.89±0.04
Tenderness	4.80±0.03	4.82±0.03	4.82±0.04	4.84±0.03
Overall grade	28.64±0.05	28.90±0.04	28.99±0.06	29.33±0.04

Statistical processing of the results for the organoleptic assessment of the compared samples showed that, according to the overall grades the 3-rd check sample of No. 278 "Pork kebab" dish from pork meat of the 3-rd generation (7/8 LDB ×1/8 LWB) for the Landrace breed is reliable ($P<0.05$) surpassed the control sample from well-bred Landrace pigs by 0.69 grade, primarily due to the assessments in taste, sectional color, appearance and aroma.

Consequently, under the conditions of the technogenic zone of North Ossetia - Alania, the best physicochemical properties and environmental safety are distinguished by the meat of the third generation Landrace Danish breeding for large white breed (7/8 LDS × 1/8 KBP), whose adaptive mechanisms are better developed, ensuring the level of accumulation of high-density metals in muscle tissue. This contributed to an increase in the nutritional value, sanitary and hygienic properties of meat and the organoleptic properties of the "Pork kebab" dish (recipe No. 278 of the Collection of recipes for dishes and culinary products for public catering enterprises).

4 Conclusions

1. It was found that with an excessive background of high-density metals in comparison with well-bred yelts of the Landrace Danish breeding for meat productivity (LDB), animals of the 3-rd generation (7/8 LDB x 1/8 LWB) have the best slaughter indexes for large white breed (LWB).

2. The results of the chemical analysis for the longest muscle of the back showed that, against the control sample, in the meat of animals of the 3-rd check group of calves (7/8 LDB 1/8 LWB), there was a significantly ($P<0.05$) higher dry matter content by 0.72%, protein - by 0.8%, the value of the protein-quality index (PQI) - by 6.6%, water-holding capacity by 7.97% and muscle fiber diameter - by 5.10 microns with a decrease in the level of zinc content by 1.90 times, lead - 1.68 times and cadmium - 1.61 times, respectively.

3. The No. 278 "Pork kebab" dish samples prepared from the pork loin of the 3-rd check group, (7/8 LDB × 1/8 LWB) distinguished by the best chemical composition, which relative to the control sample was appeared in an increase in protein content by 0.87% ($P<0.05$) and a decrease in fat concentration by 0.84% and energy value by 3.02 kcal/100 g.

4. The best ecological nutritional properties were distinguished by the No. 278 "Pork kebab" dish sample from the pork meat of the 3-rd generation (7/8 LDB × 1/8 LWB), thanks to it in the 3-rd check sample, relative to the control, there was a significant ($P<0.05$) lower content of zinc by 2.19 times, lead - by 1.86 and cadmium - 1.93 times. In addition, in the compared samples of the ready-made product, the content of high-density metals was lower than the MAC.

5. According to the overall grades the 3-rd check sample of No. 278 "Pork kebab" dish from pork meat of the 3-rd generation (7/8 LDB ×1/8 LWB) for the Landrace breed surpassed the control sample from well-bred Landrace pigs by 0.69 grade, primarily due to the assessments in taste, sectional color, appearance and aroma.

References

1. V. K Temiraev, V. R. Kairov, R. V. Kalagova, Biology and Medicine, **6, 3**, (2014)
2. V. A. Pogodaev, A. N. Arilov, A. F. Shevhuzhev, A. P. Marynich, R. H. Kochkarov, Research Journal of Pharmaceutical, Biological and Chemical Sciences, **8(1)**,1086-1090 (2017)
3. V. V. Tedtova, Z. T. Baeva, E. S. Dzodzieva, Z. Ya. Tsopanova, A. H. Pilov, Meat Industry, **3,** 60-62 (2013)
4. Y. A., Yuldashbaev, R. V. Temiraev, V. V. Tedtova, K. V. Temiraev, R. V. Osikina, M. S. Gazzaeva, L. H., Shugusheva, I. K Sattsaeva, M. M. Udychak, Journal of Livestock Science, **11**, 8-13 (2020)
5. S. F. Sukhanova, S. I. Kononenko, R. V. Temiraev, T. T. Tarchokov, Z. T. Baeva, L. A. Bobyleva, B. M. Shipshev, Journal of Pharmaceutical Sciences and Research, **10**, **11**, 2969-2971(2018)
6. S. F. Sukhanova, S. I. Kononenko, R. V. Temiraev, T. T. Tarchokov, Z. T. Baeva, L. A. Bobyleva, B. M. Shipshev, Journal of Pharmaceutical Sciences and Research, **10(11)**, 2969-2971(2018)
7. V. K Temiraev, V. R. Kairov, R. V. Temirayev, Z. A. Kubatieva and V. M. Gukezhev, Ecology, Environment and Conservation, **23(1)**, 554-561 (2017)
8. R. B. Temirayev, L. V. Tsaliyeva, I. G. Pliyeva, M. R. Dzutseva, The Gorsky State Agricultural University news, , **47(2)**, 85-87 (2010)
9. L. V. Tsaliyeva, R. B. Temirayev, F. R. Balikoyeva, N. A. Pyshmantseva, Meat Industry, **11,** 36-38 (2011)
10. R. Temirayev, V. Kairov, E. Khamitsayeva, T. Tuayeva, V. Gasiyeva, Compound feed, **4**, 70 (2008)
11. A. A. Sokolov, *Food Industry* (2011)

12. O. N. Krasulia, A. V. Smirnova, V. I. Bogush, Ye. A. Yushina, Meat Industry, **5,** 45-48 (2020)

The use of chickpea flour in the minced meat products formula

Amina Dzhaboeva[1,*], *Olesya* Byazrova[2], *Victoria* Tedtova[2], *Zarina* Baeva[2], and *Marina* Kokaeva[2]

[1]Kabardino-Balkarian State Agricultural University named after V.M. Kokov, Lenin Avenue, 1V, 360030 Nalchik, Russia
[2]North Caucasian Mining and Metallurgical Institute (State Technological University), Nikolaeva 44, 362021 Vladikavkaz, Russia

Abstract. The goal of the research was to substantiate the possibility of using various doses of chickpea flour in the recipe for the "Chopped Beefsteak" instead of beef to improve its consumer and protective properties. Taking into account the functional and technological properties and organoleptic characteristics of various minced meat models, the most optimal dose of including chopped chickpea flour in the formulation of semi-finished steak is 5.0% to the mass of minced meat. To assess the consumer qualities of the compared samples of the "Chopped beefsteak" dish with the addition of chickpea flour in a dosage of 5% to the mass of minced meat, technological charts were drawn up on the basis of recipe No. 604. When chickpea flour was included in the amount of 5.0% to the mass of minced meat, the consumer characteristics of the finished product were improved. Was found, that chickpea flour is a valuable dietary product, it is distinguished by high taste and health characteristics.

1 Introduction

It is known that in the modern domestic food market, the range of sold frozen semi-finished products from meat of different types of domestic and wild animals plays a very significant role in various regions of our country and is of great importance. One of the main value-based consumer benefits of this category of food and semi-finished products sold through the distribution network and public catering enterprises is a significant reduction in the preparation and duration of their preparation for consumption [1, 2, 3, 4].

At present, due to the presence of the facts of environmental pollution due to anthropogenic nature and an increase in the number of ecologically caused manifestations of various types of diseases, the prospect of creating plant groups of food plant additives that have antitoxic and protective actions on biological processes in the body is of particular importance and relevance. Extremely relevant interest in the domestic market is paid to the search for food additives that are of natural origin and have detoxifying and adsorbing properties. Moreover, they, unlike chemical preparations, should not cause numerous side effects on the physiological and biochemical state of consumers [5, 6, 7, 8].

* Corresponding author: kapmar17@gmail.com

Of all legumes, chickpeas are considered the leader in terms of nutrient content. It is a unique natural plant source of biogenic nutrients (proteins and amino acids, simple carbohydrates, easily and indigestible polysaccharides (including dietary fiber), lipids, vitamins of different groups, macro- and microelements), which are contained in optimal proportions. Due to the presence of a significant amount of fibrous polysaccharides (fiber) in chickpea flour, which are necessary for the consumer's body as natural enterosorbents for all types of xenobiotics, the functional activity of most body systems is optimized. The presence in the composition of food of coarse plant fibers contributes to the improvement of digestion and promotes the detoxification of various toxic substances, including heavy metals [9, 10, 11, 12].

The goal of the research was to substantiate the possibility of using various doses of chickpea flour in the recipe for the "Chopped Beefsteak" instead of beef to improve its consumer and protective properties.

2 Materials and Methods

To achieve this goal under conditions, an experiment was conducted to study the effectiveness of the inclusion of chickpea flour in the recipe for the "Chopped beefsteak" dish.

We took as the objects of our research:

- chopped semi-finished meat products (steaks) prepared according to traditional technology and recipes and with the addition of different amounts of chickpea flour.

- chickpea flour (besan) gluten free. We used chickpea flour "Vkusnoe Delo" produced by AltaiBio LLC (Barnaul) Altai Territory, Russian Federation

In this regard, according to the results of the research carried out, we proposed a method for optimizing the recipe for minced meat products by adding chickpea flour, which includes 4 stages of metroprocessing:

- at the first stage - the substantiation of the preparation of minced meat dishes, that is, the "Chopped beefsteak" dish, was carried out;

- at the second stage - formulations and the choice of technology for preparing beefsteak samples using chickpea flour were developed. At the same time, the optimal dose of introducing chickpea flour into the recipe of the studied dish was established;

at the third stage - a comparative merchandising and detoxification assessment of the compared steak samples was carried out;

- at the IV stage - a scientific, practical and economic substantiation of the inclusion of the best dose of chickpea flour in the recipe for the "Chopped beefsteak" dish was completed.

In the course of the research, steaks prepared according to the current recipe served as a control sample, and in the composition of 3 prototypes of finished products, 3, 5 and 7% of the mass of raw materials were replaced with chickpea flour.

3 Research results and discussion

Consumer properties of chickpea flour are primarily due to its chemical composition. Indicators characterizing the chemical composition and energy value of Desi chickpea flour are presented in Table 1.

Table 1. Indicators of the chemical composition and energy value of chickpea flour

Indicator	Indicator value
Mass fraction of dry substances, g/100 g	86.0
Mass fraction of proteins, g/100 g	22.0
Mass fraction of lipids, g/100 g	6.0
Nitrogen-free extractive substances, g/100 g	46.0
Mass fraction of fiber, g/100 g	11.0
Mass fraction of minerals, g/100 g	1.0
Energy value of 100 g of product, kcal	387
Vitamins, incl.	
fat soluble:	
A, µg/100 g	3.0
E, mg/100 g	0.8
water soluble:	
C, mg/100 g	6.0
PP, mg/100 g	1.6
B_1, mg/100 g	0.5
B_2, mg/100 g	0.2
B_5, mg/100 g	1.0
B_6, mg/100 g	0.5
B_9, µg/100 g	557.0

As can be seen from the data in Table 1, the water content in chickpea flour was 14.0% and dry matter - 86.0%, which indicates its good preservation for a long time, since with a similar amount of moisture in the product, mold mycoflora develops extremely slowly

A distinctive feature of chickpeas is the high content of structured carbohydrates - fiber (up to 11.0%). This factor contributes to the enhancement of intestinal motility. In addition, fiber in the gastrointestinal tract binds various toxic compounds and removes them from the body. Therefore, the addition of chickpea flour to the formulation of various products and dishes makes it possible to improve their protective qualities.

The energy value of 100 g of chickpea flour is 387 kcal, which makes it a high-value food additive.

Chickpea flour is rich in vitamins, including B vitamins, fat-soluble A and E. Particular attention should be paid to the high concentration of antioxidant and hypoxant vitamin C - up to 6.0 mg/100 g of chickpea flour, which indicates a high dexication capacity of this product.

When evaluating the best dosage for the inclusion of chickpea flour in the recipe for the "Chopped beefsteak" dish, we studied the chemical composition of semi-finished products (table. 2)

The control sample of the semi-finished product without the use of chickpea flour additives had the highest rates of moisture binding and water retention capacity. In addition, with an increase in the dose of the inclusion of flour from chickpea in the minced meat products formula due to the increased presence of fibre, an inverse relationship with the value of indicators characterizing the water-binding power and water-retaining power was observed. The consequence of this was a decrease in the structural and mechanical characteristics of prototypes of semi-finished products.

Table 2. The chemical composition of semi-finished steak

Indicator	Indicator value			
	control	experienced sample No. 1	experienced sample No. 2	experienced sample no. 3
Mass fraction of water, g/100 g	56.7	55.2	54.7	54.1
Mass fraction of proteins, g/100 g	27.2	27.7	27.9	28.2
Mass fraction of lipids, g/100 g	15.0	14.9	14.6	14.4
Mass fraction of carbohydrates, rg100 g	-	1.3	1.9	2.5
incl. fiber, g/100 g	-	0.45	0.65	0.85
Mass fraction of ash, g/100 g	0.96	0.94	0.93	0.92
Energy value, kcal	253	252	252	252

With an increase in the dose of chickpea flour additives from 3 to 7% in the composition of chopped steak in relation to the control sample, there was a tendency to a decrease in the mass fraction of water, fat, ash and calorie content with a simultaneous increase in the mass fraction of protein, carbohydrates, including fiber. This indicates an increase in the consumer and protective properties of the prototypes.

To establish the optimal dosage for the introduction of chickpea flour instead of a part of beef meat in the recipe for semi-finished steak products, it was necessary to study the effect of different dosages of the food additive on the water-binding power and water-retaining power (table. 3)

Table 3. Structural and mechanical characteristics of steaks

Name of sample	Indicator name	
	water-binding power, %	water-binding power, %
Control	90.7	67.8
1 test	89.4	65.5
2 test	87.6	63.9
3 test	85.8	62.3

Consequently, as a result of the studies carried out, taking into account the obtained functional and technological properties and sensory indicators of various minced model products, the optimal dosage of chickpea flour in the minced steak recipe is 5.0% to the mass of minced meat.

Technological maps of the control and experimental samples are presented in tables 4 and 5.

Table 4. Technological map number 1. Recipe No. 604 (II column) Chopped steak (control sample)

Raw materials	Net, g
Beef	80
Pork fat	12
Milk	6.76
Ground black pepper	0.04
Salt	1.2
Semi-finished product weight	100
Melted edible animal fat	7
Fried product weight	70

Table 5. Technological map number 2. Recipe No. 604 (II column) Chopped steak (prototype)

Raw materials	Net, g
Beef	75
Chickpea flour	5
Pork fat	12
Milk	6.76
Ground black pepper	0.04
Salt	1.2
Semi-finished product weight	100
Melted edible animal fat	7
Fried product weight	70

After the preparation of the compared samples of the chopped steak by the tasting commission of the Department of Technology of Public Food Products of the North Caucasus Mining and Metallurgical Institute (State Technological University) consisting of 7 people, their organoleptic assessment was carried out (table. 6)

Table 6. Organoleptic evaluation of the compared samples of the "Chopped beefsteak" dish, score

Indicator name	sample	
	control	experienced
Appearance	8,14±0,11	8,20±0,11
Sectional color	7,50±0,14	8,05±0,15
Taste	8,02±0,12	8,29±0,13
Smell	7,81±0,15	8,18±0,17
Texture	7,65±0,14	7,51±0,14
Juiciness	7,81±0,15	7,69±0,12
Total score	46,93±0,14	47,92±0,15

Products prepared with chickpea flour had an inversely proportional relationship with indicators of juiciness and consistency, therefore, according to these indicators, the experimental sample was inferior to the control one, which is associated with lower parameters of the moisture-binding and water-holding capacity of their minced meat due to an increase in the concentration of fiber.

The developed steak outperformed the control sample by 0.99 points in terms of the total quality score.

Therefore, to increase the consumer properties of the "Chopped beefsteak" dish, chickpea flour should be added to its recipe in an amount of 5.0% to the mass of minced meat instead of beef.

According to the methodology for assessing the detoxification properties of meat products on laboratory rats of the "Wistar" line, divided according to the principle of analogues into control and experimental groups, numbering 5 individuals, an experiment was carried out within 10 days.

Animals of the compared groups received lead sulfate ($PbSO_4$) in the diet at a dosage of 10 mg/kg feed. Along with this, rats from the control group consumed beefsteak prepared without chickpea flour. The animals of the experimental group were introduced into the diet of a steak with a 5.0% dosage of chickpea flour.

The results of the analysis of blood samples of laboratory animals in the compared groups are shown in Table 7.

Table7. Factor for determining the level of lead in the blood of experimental animals

Group of rats for and Analytical Research Department	MAC mg/kg	Lead content in rats in blood, mg/kg	
		at the beginning of the experiment	at the end of the experiment
Control	0.12	0,068 ± 0,003	0,136 ± 0,003
Test	0.12	0,069 ± 0,002	0,080 ± 0,003

By the end of the experiment, the level of lead in the blood of laboratory rats from the control group had doubled and exceeded the MAC. Against animals from the control group, by the end of the experiment, the content of lead in the blood of rat pups in the experimental group was significantly (P> 0.95) lower by 41.17%. Moreover, its concentration in their blood did not exceed the MAC.

Therefore, the addition of chickpea flour to the recipe of the "Chopped Beefsteak" dish in the amount of 5.0% to the mass of minced meat increases its detoxifying properties. In addition, the developed steak is enriched with fiber and differs from the traditional product in a pleasant "nutty" taste and aroma, which has a positive effect on the consumer characteristics of the finished product.

4 Conclusions

1. A distinctive feature of chickpea flour is a high content of structured carbohydrates - fiber (up to 11.0%), which binds various toxic compounds in the gastrointestinal tract and removes them from the body. Particular attention should be paid to the high concentration of antioxidant and hypoxant vitamin C in chickpea flour (up to 6.0 mg/100 g), which indicates a high detoxifying ability of the product. The presence of chickpea flour in minced meat recipes led to an increase in fiber content. It was found that with an increase in the dosage of flour from chickpea in the minced meat system, the values of the water-binding power and water-retaining power indexes decrease, which reduces the adhesion properties of the particles of the dispersed phase.

2. Considering the functional and technological properties and organoleptic characteristics of various minced meat models, the optimal dosage of chickpea flour, at which a quality-improving effect is achieved, is 5.0% to the mass of minced meat. To assess the consumer qualities of the compared samples of the "Chopped beefsteak" dish with chickpea flour in the amount of 5% to the mass of minced meat based on recipe No. 604 of the Collection of recipes for dishes and culinary products for public catering enterprises, technological documentation has been developed.

3. With the introduction of 5.0% chickpea flour into the minced beef steak recipe instead of a part of beef meat, an improvement in organoleptic parameters was observed. It was revealed that chickpea flour is a valuable dietary product, characterized by high taste and health benefits.

4. It was found that the addition of chickpea flour to chopped beefsteak significantly reduces the lead content in the blood of rat pups from the experimental group, whose diet was supplemented with lead sulfate ($PbSO_4$) at a dosage of 10 mg/kg of feed, which indicates high detoxification properties of the processed product.

References

1. S. F. Sukhanova, S. I. Kononenko, R. V. Temiraev, T. T. Tarchokov, Z. T. Baeva, L. A. Bobyleva, B. M. Shipshev, Journal of Pharmaceutical Sciences and Research, **10(11)**, 2969-2971 (2018)
2. V. V. Tedtova, Yu. A. Yuldashbaev, A. S. Dzhaboeva, V. R. Kairov, D. A. Yurin, Z. T. Baeva, A. A. Baeva, M. K Kozhokov, A. V. Khmelevskaya, Journal of Livestock Science, **10(1)**, 109-113 (2019)
3. I. R. Tlitseruk, F. N. Tsogoeva, S.V. Olisaev, T. A. Revazov, Bulletin of Maykop State Technological University, **3**, 34-37 (2011)
4. V.K. Temiraev, V.R. Kairov, R.V. Kalagova, Biology and Medicine, **6(3)**, BM-046-14. (2014)

5. V. A. Pogodaev, A. N. Arilov, A. F. Shevhuzhev, A. P. Marynich, R. H. Kochkarov, Research Journal of Pharmaceutical, Biological and Chemical Sciences, **8(1)**, 1086-1090 (2017)
6. V. V. Tedtova, Z. T. Baeva, E. S. Dzodzieva, Z. Ya. Tsopanova, A. H. Pilov, Meat Industry, **3**, 60-62 (2013)
7. Y. A. Yuldashbaev, R. V. Temiraev, V. V. Tedtova, K. V. Temiraev, R. V. Osikina, M. S. Gazzaeva, L. H. Shugusheva, I. K Sattsaeva, M. M. Udychak, Journal of Livestock Science, **11(8-13)** (2020)
8. D. S. Shkhalakhov, A. A. Nesterenko, Young Scientist, **23**, 109-111 (2016)
9. S.A. Gordynets, V. S. Vetrov, L. P. Shalumkova, Meat Industry, **11**, 44-46 (2004)
10. L. V. Antipova, A. A. Archipenko, A. L. Kulpina, Bulletin of the Russian Agricultural Academy, **4**, 73-74 (1998)
11. K. Yu. Shebela, N. Yu. Sarbatova, Innovative Science, **2-5(5)**, 149-152 (2015)
12. G. P. Maltsev, M. M. Sadulaev, N. M. Sivolobova A. L. Semenov, Food ingredients. Raw materials and additives, **8**, 73-74 (2003)

One the issue of the energy and resource saving

Zalimkhan Kudaev[1], *Aslan* Kumakhov[1,*], *Amur* Fiapshev[1], *Irina* Anikina[2], and *Sergey* Oskin[3]

[1]Kabardino-Balkarian State Agricultural University named after V.M. Kokova, Lenin Avenue, 1V, 360030 Nalchik, Russia
[2]Toraighyrov University, str. Lomova, 64, 140008 Pavlodar, Republic of Kazakhstan
[3]FSBEI HE "Kuban State Agrarian University named after I. T. Trubilin", str. Kalinina, 13, 350044 Krasnodar, Russia

Abstract. The issues of energy and resource saving are now especially relevant. In recent years, energy costs have skyrocketed, becoming a national issue. Considering all the factors of heat loss during design, it is possible to plan the required amount of energy. This will allow to achieve the so-called "heat balance". Currently, there are a huge number of energy audit companies in the country that work without budget funds, without implementing the tasks assigned to them. However, the law requires them to comply with the deadlines during which the owners of the objects must obtain passports. Under such conditions, auditor companies began to use approximate data, which identify the energy potential of enterprises. When identifying the potential, data are used that were given under different conditions and circumstances. At the regional level, auditors themselves find individuals capable of financing their activities, and implement the energy saving programs. All of the above omissions lead to the fact that the analysis is not accurate, which causes mistrust.

1 Introduction

The solution to energy problems is also at the state level, for example, the adopted Federal Law No. 261-FZ "On energy saving and increasing energy efficiency and on amending certain legislative acts of the Russian Federation" was supposed to clarify all controversial points [1- five]. However, since the day of the adoption of this law, there have been great disputes between specialists about its expediency. The law obliges all enterprises to conduct energy audits [6-10].

2 Materials and Methods

The list of objects that are subject to audit is described in the Law itself and is intended to follow them clearly. The clear prescriptions were supposed to make the energy survey clear

* Corresponding author: kymahov071@mail.ru

and simple, but they inspired even more questions. Russian energy auditing companies have begun to identify problems that need to be addressed at the state level.

3 Results and Discussion

In Russia, there are a huge number of energy audit companies that work without investing budget funds, without implementing the tasks assigned to them. However, the law requires them to comply with the deadlines during which the owners of the objects must obtain passports. Under such conditions, auditor companies began to use approximate data, which identify the energy potential of enterprises. When identifying the potential, data are used that were given under different conditions and circumstances. At the regional level, auditors themselves find individuals capable of financing their activities, and implement the energy saving programs. All of the above omissions lead to the fact that the analysis is not accurate, which causes mistrust.

One of the main problems is the huge amount of paperwork required when registering an object survey. Many energy companies still do not understand the true purpose of an energy passport. Statistics showed that the Ministry of Energy accepted only 25% of all passports prepared in 2017. This suggests that there are no clear requirements for the content of the passport, the procedure for its registration and approval. It is also unclear what significance this passport has for the facility. It turns out that there are norms and requirements for energy efficiency, but no one will comply therewith. It is technically impossible to describe this in the passport.

The constant use of energy resources causes the loss of energy resources and an increase in substances that can pollute the environment. Under these circumstances, any energy saving leads to lower emissions and solves some environmental problems [11-15].

Therefore, world experts suggest, when designing a building and structure, to consider the amount of energy required for their operation. When designing, it is necessary to take into account thermal insulation, according to climatic indicators, the production of its own energy.

The topic of saving energy and resources in the modern realities of the economy leads us to build various promising schemes to reduce resource and energy consumption at all stages of construction.

The transfer of the industry to energy and resource-saving direction is associated not only with changes in the regulatory and technical documentation, but also with the methods of its design. If compared with past years, namely the end of the last century, the issues of energy and resource conservation were solved by the construction of very massive buildings. Now the task is to lighten such structural elements at the expense of other, new design systems using modern, cheap materials for insulation.

Only a systematic approach to solving energy efficiency issues can bring the desired effect in solving this problem. A relatively small share is accounted for by losses in generation, transportation and heat energy metering. Usually, the largest percentage of energy losses occurs in the part of its consumption, which includes many components. It is in this direction that technological solutions are needed.

It is believed that the construction industry, when properly distributed, can consume a minimum amount of energy. This depends on both the insulation and the skillfully delivered heating systems. To solve environmental problems, it is necessary to reduce energy emissions, otherwise it will be unrealistic to achieve optimal performance. However, it is impossible to achieve this while the old buildings exist. This is explained by the fact that the last century buildings use 10 times more energy than modern ones. Therefore, a transition to a modern energy system is necessary, and this will take a huge amount of time.

Some countries even say that it is possible not to use energy at all. The development of science and technology leads to the emergence of modern materials in the field of construction, which will provide energy consumption from 300 kWh/m^2 to 20 kWh/m2^2. Moreover, these indicators already have their practical basis. Experts conducted an audit and identified these indicators in recently built buildings in accordance with modern energy saving requirements. Already now it can be argued that the chosen energy saving tactic is successful, subject to certain requirements:

— modern heat insulation of facilities;
— modern heating systems and installations;
— reuse of heat in ventilation systems;
— the presence of large windows allowing natural solar energy to enter the building.

A lot also depends on energy consumers, the room temperature selected by them determines the savings. When deploying an energy efficient building, it is necessary to consider the preferences of users, cooperate directly with them and act with their consent.

The thermal balance of a building reflects the energy savings in each building element. The difference between heat gain and heat loss indicates energy consumption.

There is an option of implementing the modern energy-saving technologies on old buildings. It is possible to achieve energy saving indicators with thermal insulation of a building, increase the infiltration capacity of a building, fill skylights.

The factor of the habitual vision of the thermal state of the building for its residents and the possibility of changing them is considered fundamental. When designing a building, it is necessary to consider the heat losses from additional energy sources, there must be a spare balance.

Taking serious, informed measures that can succeed in reducing energy costs for both new properties and old buildings is a major factor in the good performance of the national economy. Increasing the energy efficiency of buildings, it is necessary in parallel with the renewal of all structural elements. The task of a complete transition to zero energy losses in buildings is quite solvable, but this takes years, as well as a direct renovation of old buildings too. It is necessary to gradually develop goals and objectives and solve them.

The rise in the level of prices for consumed electricity made the issue of energy conservation socially important. The state is actively working to improve the environmental situation and the policy of the Ministry of Energy is aimed primarily at this. The tasks assigned to the ministry are quite solvable if you follow the methods of their implementation.

Today, almost half of the energy consumed is for construction, or rather for heating buildings. There is already a clear understanding of the necessity and obligation of energy efficient structures. Construction companies operate on energy and resource saving technologies. Architectural and construction decisions in new projects are made on the basis of energy performance. Only with the coordination of all links of the energy sector with the construction authorities will it be possible to improve the indicators of economic feasibility.

Considering that respect for the consumption of any type of energy is not at the required level in our society, it is still impossible to achieve the desired success. A systematic work with each subscriber is required. Of course, we cannot ignore the factor that replacing old systems with new ones is very difficult.

It is known that a strong state with a strong and developed economy is impossible without addressing the issues of energy efficiency and energy conservation. Here we must not forget the fact that it is the industry that is the first to face the inefficient use of energy resources and that ultimately this is reflected in the cost of industrial production. All types of energy and its correct use will be directly generated and reflected in income and production costs.

Here is a simple example: replacing lighting equipment (fluorescent lighting with LED) cuts costs in half.

The ongoing efforts to improve energy efficiency are becoming more and more relevant, given that the price of energy in our country, unfortunately, is not decreasing, but on the contrary is constantly growing.

4 Conclusions

The energy sector has been and remains one of the leading in maintaining the stability of the country's economy, therefore, all of the above recommendations should be implemented.

References

1. P. G. Grabovyi, *Textbook for universities*, 567 (1999)
2. A. V. Chapaev, Symbol of Science, **11**, 62 (2015)
3. A. K Apazhev, Y. A. Shekikhachev, A. G. Fiaphev, L. M. Hazhmetov, E3S Web of Conf., Int. Scientific and Technical Conf. Smart Energy Systems 2019 (SES-2019), **124,** 05054 (2019)
4. Z. R. Kudaev, A. A. Kumakhov, Izvestiya Kabardino-Balkar State Agrarian University named after V. M. Kokov, **3 (25)**, 83-85 (2019)
5. A. K. Apazhev, *Energy saving and energy efficiency: problems and solutions*, 8-11 (2020)
6. A. V. Baragunov, I. A. Savvateeva, S. H. Kushaev, A. A. Kumakhov, Z. R. Kudaev, Krasnoyarsk Science and Technology City Hall of the Russian Union of Scientific and Engineering Associations, 32012 (2020)
7. Z. R. Kudaev, Materials of the int. scientific and practical conf., 105-107 (2018)
8. T. V. Temukuev, A. G. Fiapshev, *Economic and technical mechanisms for stimulating energy saving* (2009)
9. A. G. Fiapshev, M. M. Khamokov, O. H. Kilchukova, Izvestiya of Kabardino-Balkar State Agrarian University named after V. M. Kokov, **1 (27)**, 63-68 (2020)
10. A. V. Chapaev, Kh. M. Karezhev, A. M. Sokhrokov, Vestnik Severo-Kavkazskogo federalnogo universiteta, **6 (63)**, 32-35 (2017)
11. A. K Apazhev, V. N. Berbekov, Y. A. Shekikhachev, L. M. Hazhmetov, G. V. Bystraya, L. Z. Shekikhacheva, IOP Conf. Series: Materials Science and Engineering, **919(6),** 062002 (2020)
12. A. K Apazhev, V. N. Berbekov, Y. A. Shekikhachev, L. M. Hazhmetov, G. V. Bystraya, L. Z. Shekikhacheva, IOP Conf. Series: Earth and Environ. Sci., **548(4),** 042022 (2020)
13. A. K. Apazhev, A. A. Gvaramiya, M. A. Marzhokhova, Siberian financial school, **5 (112)**, 22-26 (2015)
14. V. S. Bzheumykhov, Yu. A. Shekikhachev, AgroEkoInfo, **4 (30),** 2 (2017)
15. E. V. Kyul, A. K Apazhev, A. V. Kudzaev, N. A. Borisova, Indian J. of Ecology, **44 (2)**, 239-243 (2017)

The effectiveness of using biological preparations in the production technology for seed material of potatoes

Maya Orzaliyeva[1], *Yelena* Didanova[1], *Timur* Zherukov[1], *Khusen* Nazranov[1,*], and *Akrom* Ishnazarov[2]

[1]Kabardino-Balkarian State Agricultural University named after V.M. Kokov, Lenin Avenue, 1V, 360030 Nalchik, Russia
[2]Tashkent State University of Economics, Islam Karimov Str., 49, 100066 Tashkent, Republic of Uzbekistan

Abstract. An integrated plant protection system must focus on the biologization of this process. Used phytohormones and phytofungicides have a biostimulating and fungicidal effect on potato plants. The biological products used to one degree or another increase the adaptogenic capabilities of plants. Biological preparations have provided a high level of immunity of potato plants to major diseases. The fungicidal properties of the preparations were well appeared. No increase in infection or spread was observed in experiments using biofungicides. To a greater extent, the preparations acted as prophylactic ones, preventing the penetration of the phytopathogen into plants. At the same time, there is a long period of suppression of its development, the formation of resistance to late blight. The immunostimulating effect of Zircon and Epin extra in relation to late blight is found to a lesser extent in comparison with the Trichodermin, Rizoplan and Baktofit biofungicides. The fungicidal action of these preparations proved to be highly effective.

1 Introduction

An integrated plant protection system must focus on the biologization of this process. "It stands out from traditional methods, first of all, by the biocenotic approach, taking into account not individual species, but faunistic complexes of interconnected organs, the relationship between which can significantly affect the number of harmful organisms" [1].

As G. I. Bazdyrev puts it: *"The model of an integrated plant protection* system is: the use of methods for agrotechnical prevention and suppression of harmful objects; the use of techniques that preserve and promote the development of entomophages; sowing crops only with resistant varieties that are zoned in a given territory; application of methods for active suppression of harmfulness of harmful objects (autocidal method, biological and chemical plant protection products), taking into account the forecast of the development of harmful objects and economic thresholds of pests, diseases and weeds harmfulness "[1].

* Corresponding author: nazranov777@mail.ru

Taking into account the current policy trend for the use of chemicals and the greening of agricultural production, with the fact that the use of a chemical control method has significant consequences, it is necessary to increase the use of scientifically substantiated organizational, agronomic, biological methods. The complex use of various control methods allows the optimal way to protect crops from diseases, pests and weeds, and at the same time minimize the harmful effects on the environment [1,4,5,6,9].

You can eliminate unwanted results by searching for environmentally friendly methods of plant protection. Actually, the problem of environmental protection determines the development of non-chemical methods for plant protection, namely biological methods.

The problem of solving the task, providing the population of the southern region of the country with ecologically safe agricultural products is the most important at the present stage of development of the region and country agro-industrial complex.

The main task in the field of development, regulations for the use of growth regulators and their optimization is the problem of finding functional phytoregulators and inducers of plant resistance to diseases and pests, which could provide stimulation of potato growth and development. At the same time, it is important to note that great importance is attached to improving the quality indicators of products and their technological characteristics. [2,5,6,8].

Obtaining high-quality seed material is an *urgent issue* for every agricultural manufacturer. The yield and quality of products mainly depend on this.

The *purpose of this work* is to optimize the technological parameters of the use of biological preparations for the production of potato seed material in the mountainous zone of the KBR.

The *scientific novelty* lies in the identification of the fundamental possibility of using biological preparations to protect potatoes from pests and diseases, as well as for the general stimulation of crop growth.

The *practical significance* of the work lies in the greening of the production of potato planting material by using biological methods in an integrated crop protection system.

2 Materials and Methods

Research work to identify the effectiveness of plant growth regulators is carried out as part of the plan for a comprehensive scientific and technical project "Production of high-quality seed potatoes of competitive domestic varieties in a virus-free environment of the mountainous zone of the Kabardino-Balkarian Republic" in the fields of LLC "Zolsky Potato".

The cultivation technique in the potato seed multiplication nursery-garden is designed for using advanced technology with the application of high quality tools and machines. During investigations for planting, super elite seed tubers of a domestic variety with an average weight of 65 g were used. Winter barley was the predecessor in the crop rotation. The planting scheme provided for the distribution of 60 thousand tubers per hectare, the planting depth was 10-15 cm, when creating the ridge, its height was 18-20 cm. The working area of the plot is 72 m^2, the accounting area is 60 m^2.

Taking into account the large number of studied physiologically active substances (PAS) and various methods of their use on potatoes, depending on the tasks to be solved, specific schemes for the use of growth regulators and microbiological preparations were laid.

Field experiments, observations, laboratory analyzes were carried out according to generally accepted methods. Mathematical processing of the yield was carried out using computer programs for statistical data processing.

During the growing season of potatoes, observations, analyzes, and accountance were carried out in accordance with generally accepted methods. Phenological observations were carried out in accordance with the "Research Methods for the Potato Culture" [3].

Cleaning is performed manually.

Registration procedures for diseases are carried out without fail: late blight, ring rot, black leg, common scab, rhizoctonia and viral diseases. For all varieties of the same ripening period, the level of starch content in potato tubers is determined.

The tasting is carried out by groups of varieties ripeness twice.

The assimilation surface area was determined using a modified technique for accelerated calculation of the leaf surface area for agricultural crops using a scanner.

The potato crop structure analysis; the resistance of plants to diseases and pests was assessed before harvesting according to the "Standard scale for taking into account the damage to the surface of tubers by diseases" (Anisimov B.V., 2001) [4].

The quality indicators of the yield were determined by its content of dry matter - starch, nitrogen, ash content, and the content of vitamins. The dry matter yield in tubers was determined by drying at a temperature of 105 °C, starch - by washing and further drying the crushed mass of tubers, nitrate nitrogen - using the pH meter by an electrode method, protein - according to GOST 134964-84. ash content - found on the tubers burned in a muffle furnace, previously well washed, dried and weighed (according to GOST 26226-84), was determined on a Parov scales according to the method described in the tutorial on plant growing. Mathematical processing of the varieties yield was carried out by the variance analysis method on a computer [7].

Experimental design

The scheme of the use of biological products (table 1) consisted of processing tubers before planting and three times processing of tops in the phase of full germination, budding and after the flowering phase.

Table 1. Scheme of processing potato crops with biological preparations.

Option	Seed processing	Full germination phase	Budding phase	Flowering phase	Remarks
I. Control (water)	-	-	-		
II.Zircon	0.5мл/l per 8 h	0.35 ml/l	0.35 ml/l	0.5мл After flowering	350l/ha
III. Epin extra	1Ml/250 ml of water ,3 hours	60ml/ha	60ml/ha	60ml/ha, beginning of flowering	350l/ha
IV. Trichodermin	200ml/10 ml of water, 2-3 min	1.5kg/ha	2.8kg/ha	2.8kg/ha	350l/ha
V. Rizoplan	10l/1ton of tubers	-	10l/ha	10l/ha	350l/ha
VI. Baktofit	1ml/500ml of water	350ml/ha	700ml/ha	700ml/ha	350l/ha

3 Results and Discussion

The results obtained during the investigations to determine the effectiveness of the use of growth regulators on potatoes indicate an increase in the utilization rate of photosynthetically active radiation. In the control variant, this index was at the level of 1.29%. The use of phytoregulators of growth according to the recommended technology increased the utilization rate of FAR for phytohormones Zircon and Espin extra by 0.08 and 0.03%, respectively (table 2).

Table 2. Productivity index of potato early variety Zhukovsky.

Option	Leaf surface area, thous. m^2/ha	Dry biomass yield, t/ha	The productivity of the leaves (PRL) or the output of marketable products per 1 thousand units of FP, kg of tubers	Net productivity of photosynthesis (NPF), g/m^2 x day	FAR utilization factor,%
I. control	39.3	7.166	9.2	3.84	1.29
II.Zircon	44.5	7.988	10.3	4.18	1.37
III. Epin extra	43.7	7.884	10.1	4.08	1.32
IV. Trichodermin	43.1	7.845	10.1	4.01	1.35
V. Rizoplan	42.8	7.857	9.8	4.00	1.30
VI. Baktofit	43.7	7.864	9.9	4.07	1.35

The analysis of biometric indexes shows that the use of phytoregulators of growth and phytofungicides causes an increase in plant height by an average of 8.9 cm, and in terms of the number of stems, the first two options are higher by 0.7 pcs/bush. This suggests that the process of growth and development of the aboveground part of potato plants, which assimilates organic matter, is more productive on options with the use of Zircon and Epin extra phytohormones. Their net productivity of photosynthesis is higher, and this suggests that the yield on these options shall be higher, which was confirmed by the results of harvesting potatoes. The number of tubers increased in all variants of the use of biological products in comparison with the standard plant protection system in the technology of growing potatoes in this natural climatic zone.

Thus, the use of biological products in the technology of growing potatoes, in the system of integrated plant protection has a positive effect: the rate of growth processes, the duration of the period of photosithetically active radiation, stimulate the main metabolic processes, and increase the quantitative indexes of productivity parameters. Ultimately, all this causes the formation of a higher productivity of potato plants.

The analysis of the productivity results of Nart variety potatoes (table 3) showed that with the use of biological products, the percentage of standard products output significantly increases, this index was 10% higher than the control option. On the option with Epin extra, the number of tubers is greater than on all options, while the average weight of tubers from one bush is high - 631 g/bush. On this option, the average tuber weight is the lowest, which indicates the unevenness of the tuber laying. That is, in a bush of plants there is a large weight of medium tubers, the most suitable as seed material.

Table 3. Potato yield depending on the use of various biological preparations.

Experimental option	Replication, t/ha			Average biological yield by experimental options, t/ha	Standard products, %	Raise, %
	1	2	3			

Table 3. Continued.

I. control	18.2	19.3	19.2	18.9	85.7	-
II.Zircon	22.0	21.1	20.8	21.3	95.1	12.7
III. Epin extra	21.2	21.8	20.3	21.1	85.4	11.6
IV. Trichodermin	18.6	20.3	20.2	19.7	95.9	4.2
V. Rizoplan	20.6	19.8	20.2	20.2	94.6	6.8
VI. Baktofit	20.1	18.9	20.7	19.9	96.9	5.3
NSR0.5	0.4888	0.5444	0.4455	0.49		

Our investigation results were subjected to economic analysis, as a result of which it is possible to more accurately judge the effectiveness of a particular technique studied in the experiment.

For the analysis, we used the data of the main economic indexes of the economy in which the investigation was conducted. Economic analysis data are represented in Table 4.

Table 4. Economic efficiency of using biological preparations on potatoes.

Indicators	Crops					
	control	Zircon	Epin extra	Tricho dermi n	Rizo plan	Baktofit
Yield of standard tubers, t/ha	16.2	20.3	18.1	18.9	19.1	19.3
Direct costs of production, thousand rubles	105.3	118.5	119.2	118.9	118. 4	118.4
Price for 1 ton of young tuber, thousand rubles.	25	25	25	25	25	25
Gross cost, thousand rubles.	405	507.5	452.5	472.5	477. 5	482.5
Net income, thousand rubles	299.7	389.0	333.3	353.6	359. 1	364.1
Net cost, thousand rubles/ton	6.5	5.8	6.5	6.3	6.2	6.1
Profitability level, %	284.6	328.3	279.6	297.4	303. 3	307.5

They show that the Nart variety is responsive to the use of biological preparations in potato growing technology to obtain a young tuber. Biological preparations, both phytohormones and phytofungicides, increased productivity. This was especially clearly appeared when using Zircon, the yield of standard tubers exceeded the control option by 4.1 t/ha. The use of phytofungicides causes an increase in productivity by an average of 2.9 t/ha. The conditional net income for the Nart variety with the use of the biological preparation Zircon was 389 thousand rubles/ha, while the level of profitability was 328.3%.

The net cost of production for this option is 5.8 thousand rubles/ton, which is 0.7 thousand rubles/ton less than the control option.

4 Conclusions

Conclusions based on the results of investigations to optimize the regulations for the use of biological preparations in potato growing technology.

Processing of seed tubers and during the growing season with biological preparations gives a positive effect, while the effectiveness using the phytohormones Zircon and Epin extra has higher rates in the formation of an assimilation surface than when using phytofungicides. The use of the biological preparation Zircon increases the area to 44.5 thousand m^2/ha, the productivity of leaf work (PLW) or the yield of marketable products per 1 thousand units of FP 10.3 kg of tubers, with a net productivity of photosynthesis - 4.18 g/m^2 day.

The highest dry matter content in tubers was observed in the options with the use of the Epin extra and Trehodermin biological preparation, the maximum value for this index was 26.8%. A higher starch content was noted for the use options of phytohormone and Baktofit 16.1%, which is 5.9% higher than the control option. The use of biological preparations increases the quality indexes of commercial products in terms of dry matter content by 0.3-0.6%, and the starch content by 0.6-0.9%.

The use of the phytohormone Zircon in the cultivation technology provided 20.3-20.8 ton/ha of potato tubers of the Nart variety, with a marketable output of 95.1%.

When carrying out a general phytopathological assessment of the variety, only late blight was recorded among the diseases.

The low degree of infestation is primarily due to high-quality seed material and the choice of a site in the crop rotation. The use of biofungicides completely restrained the development of late blight, and the percentage of prevalence in other options is insignificant.

So, on the basis of the investigations carried out, the following conclusions can be drawn: the treatment of seed tubers of the Nart variety potatoes with biological products contributes to obtaining a larger assimilation surface and an increase in the duration of its work. The processes of development, an increase in the number of stems, bushiness, plant height, and the number of tubers in a bush are accelerated.

Used phytohormones and phytofungicides have a biostimulating and fungicidal effect on potato plants. The biological products used to one degree or another increase the adaptogenic capabilities of plants. Biological preparations have provided a high level of immunity of potato plants to major diseases. The fungicidal properties of the preparations were well appeared. No increase in infection or spread was observed in experiments using biofungicides. To a greater extent, the preparations acted as prophylactic ones, preventing the penetration of the phytopathogen into plants. At the same time, there is a long period of suppression of its development, the formation of resistance to late blight. The immunostimulating effect of Zircon and Epin extra in relation to late blight is found to a lesser extent in comparison with the Trichodermin, Rizoplan and Baktofit biofungicides. The fungicidal action of these preparations proved to be highly effective.

So, a systematic analysis of experience shows a high level of effectiveness of biological preparations.

The economic efficiency of using biological fertilizers, biostimulants and the biological method in the integrated plant protection system for growing potatoes is very high and amounts to more than 300%.

References

1. G. I. Bazdyrev, N. N. Tretyakov, O. O. Beloshapkina, *Integrated protection of plants against harmful organisms* (2014)
2. H. M. Nazranov, A. K. Yezaov, Ye. N. Didanova, et al., *Intensive technology for the production of organic vegetable products* (2019)
3. B. A. Dospekhov, *Field experiment technique* (1985)
4. A. I. Anisimov, O. V. Smirnov, S. A. Dobrokhotov, Agricultural news, **2**, 33-36 (2009)
5. I. V. Andreyeva, M. V. Shternshis, O. G. Tomilova, *Biological plant protection* (2018)
6. S. S. Litvinov, *Method of field experiment in vegetable growing* (2011)
7. A. K. Akhatov and others, *Diseases and pests of vegetables and potatoes* (2013)
8. S. A. Bekuzarova, I. M. Haniyeva , L. S. Gishkayeva, Ecological Consequences of Incre asing Crop Productivity: Plant Breeding and Biotic Diversity, 23-27 (2014)
9. M. M. Shakhmurzov, A. S. Gordeev, V. V. Kulintsev, Y. A. Yuldashbaev, N. V. Konik, R. A. Ulimbasheva, M. B. Ulimbashev, South of Russia: Ecology, development, **3**, 83-95 (2018)

Optimization of parameters and operating modes of a biomethane installation for maximum biomethane discharge

Amur Fiapshev[1,*], *Marat* Hamokov[1], *Olesya* Kilchukova[1], and *Irina* Anikina[2]

[1]Kabardino-Balkarian State Agricultural University named after V. M. Kokov, Lenin Avenue, 1V, 360030 Nalchik, Russia
[2]Toraighyrov University, str. Lomova, 64, 140008 Pavlodar, Republic of Kazakhstan

Abstract. The article covers the results of a theoretical investigation for dung and livestock manure anaerobic fermentation process, substantiated the structural technological scheme and operating modes of the biomethane installation. An investigation complex was carried out on the basis of substantiating the parameters of the technological process, equipment of a biomethane installation, where the goal is to increase the efficiency of substrate processing. Based on the analysis, it was found that the biomethane discharge is one of the main characteristics of processing the substrate, and this moment was taken as the main criterion for optimization work. The rational modes and parameters of the biomethane installation were determined using the mathematical method of planning a multifactorial experiment. The influence of the temperature for the fermentation process of the substrate, the time of mixing the substrate and the number of revolutions of the heat-exchanging mixer on the biogas discharge was established. The influence of design factors, including various factors, was carried out using mathematical statistics and mathematical modeling, where the dependence of the biogas discharge on the temperature of the substrate fermentation process, the time of mixing the substrate and the number of revolutions of the heat exchanging mixer were obtained. The proposed biomethane installation has passed production and laboratory tests in order to optimize the parameters of various capacities for their operation in various regions of the country, the creation of an experimental installation operating in the thermophilic mode, including the stage of preliminary preparation of livestock manure biomass in the technological chain, obtaining liquid and solid fertilizers and other waste-free technology products.

1 Introduction

The most important role of all human activities that contribute to the transformation of organic substances belongs to agriculture, which, due to the peculiarities of its production is engaged in all the processes accompanying the carbon cycle – the synthesis, destruction and mineralization of organic substances [1-5].

* Corresponding author: energo.kbr@rambler.ru

Under natural conditions, the rate of methanogenesis under the influence of the anaerobic biocenosis of microorganisms contained in livestock manure or other biomass is low. To intensify the process of methanogenesis, it is necessary to optimize the conditions under which the rate of enzymatic reactions involved in the methane formation would be maximum. An environment in which the concentration of dry matter is at the level of 10-12% is considered favorable for the life of the biocenosis of methane-forming microorganisms. In this case, the viscosity of the substrate allows the liquid to move freely with suspended solid particles of manure biomass and microbial cells in it, as well as gas bubbles.

Anaerobic fermentation of manure biomass is accompanied by an almost 50% decrease in the amount of dry organic matter in the sludge compared to its initial level in the reactor due to the inclusion of 10-15% of the substrate carbon in the microbial biomass, as well as components of biomethane such as methane and carbon dioxide.

The sludge contains less carbon than the original substrate; the presence of potassium and phosphorus gives the sludge the property of a high-quality organic fertilizer, in which nutrients are presented in a form more accessible to plants. The application of sludge into the soil, devoid of the specific odor inherent in processed livestock manure, does not encounter obstacles from an environmental point of view.

The fermented liquid, like the fermented substrate (sludge) does not have an unpleasant odor. This liquid contains 80% less organic matter, and its biological oxygen demand is 80% lower than before anaerobic fermentation.

2 Materials and Methods

The challenge is to design a high-performance, highly efficient, easy-to-operate and reliable installation.

The research carried out made it possible to design and manufacture an experimental biogas installation with a digester volume of 3.5 m^3, operating in the thermophilic mode of anaerobic fermentation, in which the mixing and heating devices are combined into one heat exchanging mixer unit. This technical solution makes it possible to heat the entire mass of the substrate due to the rotation of the heat-exchanging mixer.

3 Results and Discussion

To determine the optimal parameters of this installation, experimental investigations were carried out, a matrix of their planning was compiled, criteria of work efficiency and factors influencing them were determined.

To conduct investigations, planning an experiment and compiling a matrix, we set the number of experiments equal to 15, with the number of factors equal to 3, are given in table 1.

To analyze the data obtained from experimental investigations, a program for an electronic computing machine was used.

The criterion for evaluating the biomethane installation performance is accepted for biomethane discharge.

Table 1. Experiment planning matrix.

i	X_0	X_1	X_2	X_3	X_1^2	X_2^2	X_3^2	$X_1 X_2$	$X_1 X_3$	$X_2 X_3$	Y_i
1	1	1	1	0	1	1	0	1	0	0	Y_1

Table 1. Continued.

2	1	1	-1	0	1	1	0	-1	0	0	Y_2
3	1	-1	1	0	1	1	0	-1	0	0	Y_3
4	1	-1	-1	0	1	1	0	1	0	0	Y_4
5	1	0	0	0	0	0	0	0	0	0	Y_5
6	1	1	0	1	1	0	1	0	1	0	Y_6
7	1	1	0	-1	1	0	1	0	-1	0	Y_7
8	1	-1	0	1	1	0	1	0	-1	0	Y_8
9	1	-1	0	-1	1	0	1	0	1	0	Y_9
10	1	0	0	0	0	0	0	0	0	0	Y_{10}
11	1	0	1	1	0	1	1	0	0	1	Y_{11}
12	1	0	1	-1	0	1	1	0	0	-1	Y_{12}
13	1	0	-1	1	0	1	1	0	0	-1	Y_{13}
14	1	0	-1	-1	0	1	1	0	0	1	Y_{14}
15	1	0	0	0	0	0	0	0	0	0	Y_{15}

Theoretical investigations of the biomethane installation operation showed that the biomethane productivity is influenced by the following factors: temperature of the substrate fermentation process T_C; mixer revolutions n_T and substrate $t_П$ mixing time are presented in table 2.

Table 2. Levels and factors of their variation.

Varying factors step and levels	Coded value of factors (dimensionless)	Natural value of factors		
		X1 (T_c, degree)	X2 ($t_П$, min)	X3 (n_T, rpm)
Step	-	3	3	0.5
Low	-1	54	14	7
Zero	0	57	17	7.5
Upper	+1	60	20	8

In order to determine the optimal parameters and operating modes of the biogas installation, a regression equation was drawn up:

$$Y_{Q_{бг}} = 15,3967 - 0,6838X_1 + 0,0925X_2 - 0,2888X_3 - 0,27X_1X_2 - \\ -0,7725X_1X_3 + 0,775X_2X_3 - 1,2996X_1^2 - 1,0471X_2^2 - 1,1196X_3^2. \quad (1)$$

According to Fisher's criterion, the adequacy of the equation

(F_{design}= 0.765 < F_{table}=2.359).

The regression equation (1) in the decoded form is as follows:

$$Q_{БГ} = 15,3967 - 0,6838\left(\frac{T_П - 57}{3}\right) + 0,0925\left(\frac{t_П - 17}{3}\right) - 0,2888\left(\frac{n_T - 7,5}{0,5}\right) - 0,27\left(\frac{T_П - 57}{3}\right)\cdot$$
$$\cdot\left(\frac{t_П - 17}{3}\right) - 0,7725\cdot\left(\frac{T_П - 57}{3}\right)\left(\frac{n_T - 7,5}{0,5}\right) + 0,775\left(\frac{t_П - 17}{3}\right)\left(\frac{n_T - 7,5}{0,5}\right) - 1,2996\left(\frac{T_П - 57}{3}\right)^2 - \quad (2)$$
$$-1,0471\left(\frac{t_П - 17}{3}\right)^2 - 1,1196\left(\frac{n_T - 7,5}{0,5}\right)^2$$

Equation (2) after transformations shall take the following form:

$$Q_{БГ} = -905,8499 + 20,6062T_П + 1,8215t_П + 87,1701n_T - 0,03T_П t_П -$$
$$-0,515T_П n_T + 0,5167t_П n_T - 0,1444T_П^2 - 0,1163t_П^2 - 4,4784n_T^2. \quad (3)$$

Compiled system of differential equations (1), representing partial derivatives with respect to three factors, in order to establish the values of the factors for the maximum biomethane discharge.

$$\begin{cases} \dfrac{dY_{Q_{БГ}}}{dt} = -0,6838 - 0,27X_1 - 0,7725X_3 - 2,5992X_1 = 0 \\ \dfrac{dY_{Q_{БГ}}}{dt} = 0,0925 - 0,27X_1 + 0,775X_3 - 2,0942X_2 = 0 \\ \dfrac{dY_{Q_{БГ}}}{dt} = -0,2888 - 0,7725X_1 + 0,775X_2 - 2,2392X_3 \end{cases} \quad (4)$$

The optimal values of the factors in the coded form are determined as a result of solving the system of equations (4):

$$X_1 = -0,2519;\ X_2 = 0,0014;\ X_3 = -0,0381.$$

The maximum methane discharge is 15.5 m^3/day is achieved with decoded values of the factors: the number of revolutions of the heat-exchanging mixer - 7.5 rpm; mixing time of the substrate – 17 minutes; the temperature of the substrate fermentation process – 56.2°C.

The experiment reproducibility was checked according to the Cochran criterion:

$$G_{расч} = S^2(y_i)_{MAX} \Big/ \sum_{i=1}^{N} S^2(y_i) = 0,3003$$

The table value of the Cochran test is G_{table}= 0.335 at a 5% significance level f_1=2, f_2=15. The hypothesis about the homogeneity of variances is confirmed, since the value of the calculated Cochran criterion is less than the tabular one.

At zero temperature of the process of substrate fermentation ($T_П$ =57° C), the regression equation has the form:

$$Q_{БГ} = -200,4522 + 0,1115t_П + 57,8151n_T + 0,5167t_П n_T - 0,1163t_П^2 - 4,4784n_T^2. \quad (5)$$

Figure 1 shows the response surface for these characteristics.

At zero time of the process of substrate fermentation ($t_П$=17 min), the regression equation has the form:

$$Q_{БГ} = -908,4952 + 20,0962T_П + 95,954n_T - 0,515T_П n_T - 0,1444T_П^2 - 4,4784n_T^2. \quad (6)$$

Figure 2 shows the response surface for these characteristics.

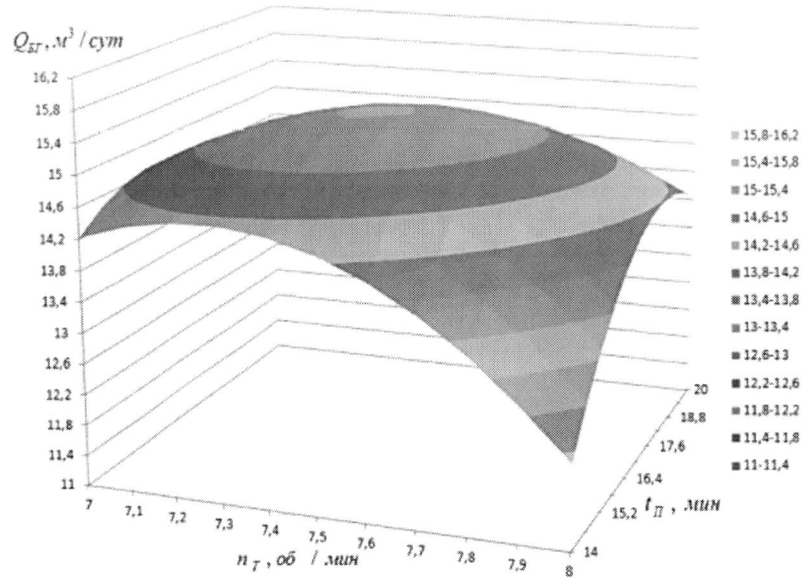

Fig. 1. Response surface $f(T_П, n_T)$ at zero level $T_П$ = 57°C.

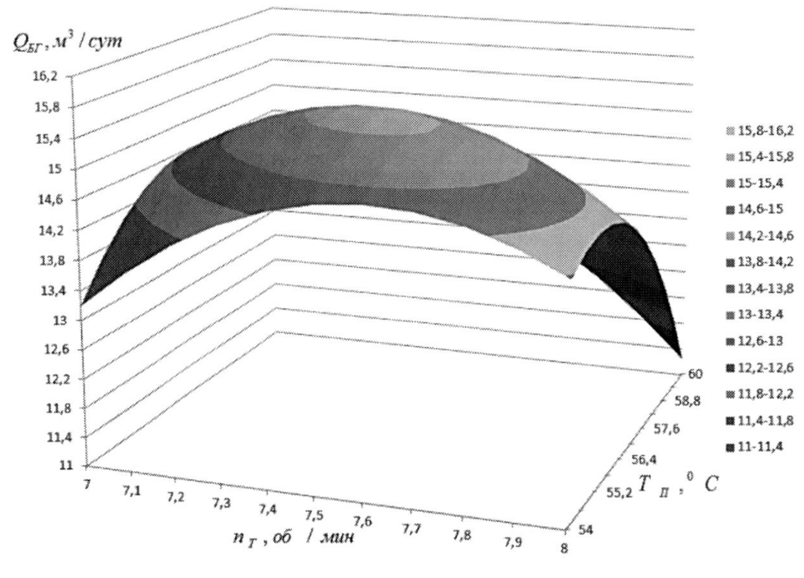

Fig. 2. Response surface at zero level = 17 min.

At zero revolutions of the heat-exchanging mixer (n_T=7.5 rpm), the regression equation has the form:

$$Q_{БГ} = -503,9842 + 16,7737T_{П} + 5,6967t_{П} - 0,03T_{П}t_{П} - 0,1444T_{П}^2 - 0,1163t_{П}^2. \quad (7)$$

Figure 3 shows the response when these characteristics change.

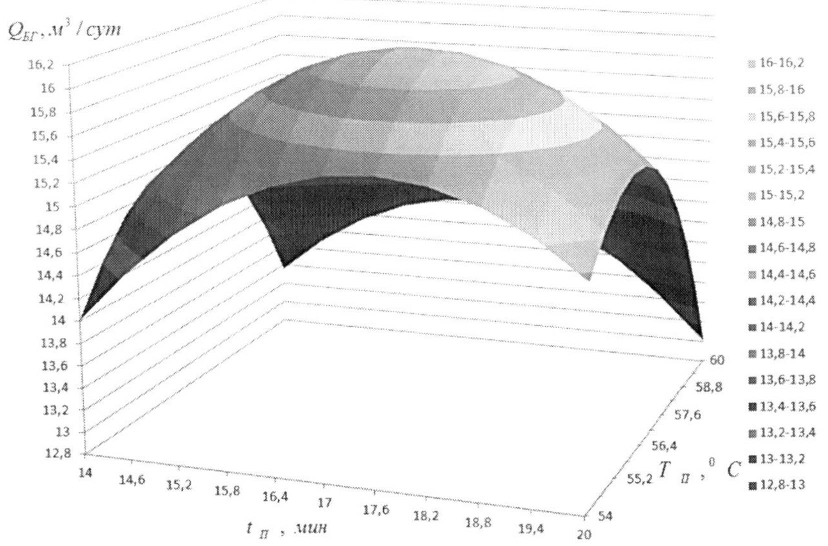

Fig. 3. Response surface $f(T_П, n_T)$ at zero level n_T = 7.5 rpm.

4 Conclusions

The conducted investigations allow us to conclude that the maximum value of the optimization criterion (biomethane discharge) 15.5 m³/day is achieved with the values of varying factors: temperature of the substrate fermentation process $T_П = 56,2^0 C$; mixing time of the substrate, $t_П = 17$ min; a number of revolutions of the heat-exchanging mixer $n_T = 7,5$ rpm.

References

1. A. K. Apazhev, Y. A. Shekikhachev, A. G. Fiapshev, L. M. Hazhmetov, E3S Web of Conferences, International Scientific and Technical Conference Smart Energy Systems 2019 (SES-2019), 124, 05054 (2019)
2. A. K. Apazhev, *Energy Saving and Energy Efficiency: Problems and Solutions*, 8-11 (2020)
3. A. K. Apazhev, Actual problems of agricultural science: applied and research aspects, **1**, 14-17 (2021)
4. A. K. Apazhev, Y. A. Shekikhachev, V. I. Batyrov, Kh. L. Gubzhokov, A. L. Bolotokov, IOP Conference Series: Materials Science and Engineering, 663(1), 012049 (2019)
5. A. K. Apazhev, Y. A. Shekikhachev, L. M. Hazhmetov, Z. L. Hazhmetova, D. T. Gabachiyev, J. of Physics: Conference Series, 1399(5), 055002 (2019)

6. I. M. Khaniyeva S.A. Bekuzarova, A. K. Apazhev, *Bioenergetic assessment of crop cultivation technologies and calculation of the economic efficiency of fertilization* (2019)
7. A. K. Apazhev, M. A. Marzhohova, L. Z. Halishkhova, *The Phenomenon of Sustainability of Economic and Environmental Development of Agrarian Territories* (2015)
8. A. K. Apazhev, Ts. B. Kagermazov, M. K. Kozhokov, A. S. Gordeyev, M. M. Kushkhova, Agrarian Russia, **1**, 39-42 (2015)
9. A. Fiapshev, O. Kilchukova, Y. Shekikhachev, M. Khamokov, L. Khazhmetov, MATEC Web of Conf., 212, 01032 (2018)
10. A. K. Apazhev, Y. A. Shekikhachev, A. G. Fiapshev, O. Kh. Kilchukova, Int. scientific and practical conference «AgroSMART – Smart solutions for agriculture», KnE Life Sciences, 40-50 (2019)
11. A. K. Apazhev, Yu. A. Shekikhachev, L. M. Khazhmetov, A. G. Fiapshev, A. B. Baragunov, M. M. Khamokov, O. H. Kilchukova, Bulletin of Agricultural Consulting, **4**, 42 (2015)
12. A. K. Apazhev, Yu. A. Shekikhachev, A. G. Khazhmetov, Bulletin of the Gorsky State Agricultural University, **11 (2)**, 60-63 (2016)
13. A. G. Fiapshev, M. M. Khamokov, O. Kh. Kilchukova, J. of Physics: Conference Series (JPCS), 1679, 052074 (2020)

Determination of critical stresses during prestressing along the entire length of a circular-cylindrical shell

Takhir Chapaev[1,*], and *Nurilla* Noraliev[2]

[1]Kabardino-Balkarian State Agricultural University named after V. M. Kokov, Lenin Avenue, 1V, 360030 Nalchik, Russia
[2]Tashkent State Agrarian University, Universitetskaya, 2, 100140 Kibray district, Tashkent region, Republic of Uzbekistan

Abstract. In the Russian Federation, there is an urgent issue of grain storage and processing to be resolved through the reconstruction of old and construction of new storage facilities. Currently, the most common construction is represented by steel granaries, erected by rolling using prestressing in shells. The relevance of the problem of stability of the wall of a granary, taking into account the main factors, affecting the strength, requires further theoretical and experimental study. One of the most frequently used in solving stability problems is the energy criterion in the form of Ritz-Timoshenko, which makes it possible to determine critical stresses in the shells. In this case, the problem solution becomes simpler, it is easier to consider the effect of such factors as initial perfection, friction forces between the shell and the winding, as well as other features of the stability problem of a prestressed shell. This article analyses the granary wall stability against lateral external pressure of prestressed winding or bandages. For large tanks, prestressing is generated along the entire length of the shell, and in vertical cylindrical granaries, prestressing is generated on a part of the length (height) of the shell.

1 Introduction

The actual ratios of tank diameters and granaries and their heights show that their bodies are shells of medium length.

The stability of cylindrical shells of medium length under the action of lateral uniform external pressure has been devoted to many works [1, 2, 3, 4, 5]. The analysis of these works shows that the effect of shell end grounding on its stability depends on the shell relative length.

Figure 1 shows the numerical analysis results of the effect of various conditions of the shell fastening on its stability from lateral uniform pressure [5]. The following designations are used here:

* Corresponding author: 7227229@mail.ru

$$Z_0 = \frac{L\sqrt[4]{1-\mu^2}}{\sqrt{Rt_1}}\ ;\ \varsigma_0 = \frac{q_{кр}}{q^*_{кр}}\ , \tag{1}$$

where:
L – shell length;
t_1 – shell thickness;
R – shell radius;
μ – Poisson's ratio;
$q^*_{кр}$ – determined by the formula of P. F. Papkovich [1].

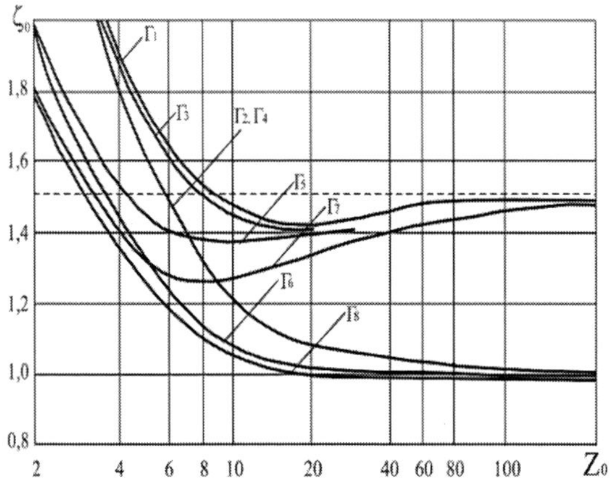

Fig. 1. Effect of boundary conditions on the cylindrical shell stability against lateral external pressure.

The shown curves correspond to the following variants of the boundary conditions at both shell ends:

$$\begin{array}{lllll} \Gamma_1 & w=0, & \frac{\partial w}{\partial x}=0, & U=0, & V=0; \\ \Gamma_2 & w=0, & \frac{\partial w}{\partial x}=0, & T_x=0, & V=0; \\ \Gamma_3 & w=0, & \frac{\partial w}{\partial x}=0, & U=0, & S_{xy}=0; \\ \Gamma_4 & w=0, & \frac{\partial w}{\partial x}=0, & T_x=0, & S_{xy}=0; \\ \Gamma_5 & w=0, & \frac{\partial w}{\partial x}=0, & U=0, & V=0; \\ \Gamma_6 & w=0, & M_x=0, & T_x=0, & V=0; \\ \Gamma_7 & w=0, & M_x=0, & U=0, & S_{xy}=0; \\ \Gamma_8 & w=0, & M_x=0, & T_x=0, & S_{xy}=0; \end{array} \tag{2}$$

where:
U and V – respectively, longitudinal and circumferential displacements of points of the shell middle surface;
T_x and S_{xy} – longitudinal and shear forces in the shell.

Parameter z_0 for vertical cylindrical tanks and granaries with a volume of 0.5 to 100 thous. m^3 is in the range from 20 to 80. The actual boundary conditions for fixing the body wall occupy an intermediate position between 2 and 8 variants of the boundary conditions given above. Thus, judging by the curves corresponding to variants 2 and 8 of the boundary conditions, we can say that for vertical cylindrical tanks and granaries with a volume of 5 to 100 thous. m^3, the presence of elastic grounding of the wall at the bottom practically does not affect the stability against lateral external pressure of the vacuum type. In other words, in calculating the stability of vertical cylindrical tanks or granaries of the indicated volume, it can be taken as a margin of stability that the body wall is hinged at the ends.

In the case of a spiral-wound silo, the stability problem of a separate shell is solved, $L \approx 40$ cm long, parameter z_0 has a value $z_0 = 2 \div 3$.

According to Fig. 1, the presence of wall clamping in the folds should increase its resistance to lateral pressure. However, in this case, we also assume that the shell is hinged at the ends as a margin of stability.

This article analyses the tank shell wall stability against lateral external pressure of a prestressed winding or bandages. For large tanks, prestressing is generated along the entire length of the shell, and in vertical cylindrical granaries, prestressing is generated on a part of the length (height) of the shell.

The shell stability is lost in one of the sections with the most significant initial dents. In the remaining sections, as the compression force increased, there was a slight increase in dents, however, with a loss of stability in one of shell sections, deflections stop growing, since forces in the winding or bandages decrease. Thus, for the limiting state of stability loss, we take the moment of the formation of a dent in one of shell sections.

2 Materials and Methods

The problem was solved using the Ritz-Timoshenko energy method. The energy balance equation for the prestressed shell is written in the form:

$$\Delta U = U_{обм} + U_{тр} + U_c + U_и + П - U_0 = 0, \tag{3}$$

where: $U_{обм}$ – potential energy of winding deformation during shell buckling;

$U_{тр}$ – potential energy of friction forces arising between the shell and the winding during their mutual displacement at the moment of buckling;

U_c – potential deformation energy of the shell middle surface;

$U_и$ – potential energy of shell bending;

$П$ – potential of winding pressure forces on the shell;

U_0 – potential energy of a prestressed shell before stability loss.

The critical force in the winding $N_{кр}$ is defined as the minimum tension force in the winding N_0, when the potential energy of the shell-winding system before buckling will be equal to the potential energy of the system in a deformed state after buckling.

To determine the parameters included in equation (3), it is necessary to set the shell deflection line.

To approximate the shell deflections, an expression is adopted that sufficiently corresponds to the real buckling shape shown in Figure 2:

$$w = f \cos \eta\theta \sin \frac{\pi x}{L}, \tag{4}$$

where: f – maximum amplitude of deflection at the dent center; $\eta = \dfrac{\pi}{2\theta_0}$;

θ_0 – the central angle defining the zone of winding separation from the shell during buckling.

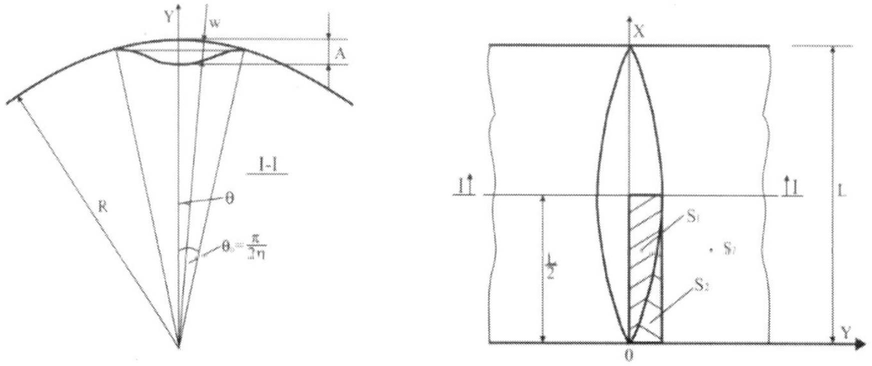

Fig. 2. Scheme of shell buckling along the entire winding length.

Formula (4) is valid for $0 < \theta \leq \theta_0$.

With the indicated indentation scheme, the value f can be determined through shell radius R and angle θ_0 by the formula:

$$f = 2{,}5R \sin\theta_0 tq\theta_0 , \tag{5}$$

where coefficient 2.5 is taken on the basis of experimentally obtained shape of the dent formed during stability loss and the analysis of the relationship between the value of shell deflections and delamination of the winding from the shell.

Thus, taking into account formula (5), the variable deflection parameter is angle θ_0. In other words, we assume that the shape of the initial wave formation is in "resonance" with the wave formation of the shell in the process of deformation, and that the only preset parameter is the maximum value of the initial dents A_0. This assumption somewhat enhances the effect of the initial inaccuracies, but it greatly simplifies the mathematical calculations. It is mentioned in works of L. G. Donell [5] and A. S. Volmir [6].

In real forms of initial dents, in contrast to those taken by formula (5), there is a buckling for the outer contour. The pressure from the prestressed winding in these areas with greater curvature is higher than in the areas with less curvature. Therefore, the real working conditions of the wall in the zone with stability loss can be considered similar to those assumed in this solution. That is, we assume that the shape of the initial indentations corresponds to formula (5) and that the pressure from the winding or bands is the same in all areas of their contact with the shell.

As a result of shell buckling and dent formation, the winding circumference is less than that of the shell. As, the force in the winding is reduced. The relative winding deformation ξ_1 is determined from the condition of compatibility of shell and winding deformation:

$$\xi_1 = 1 - \frac{S_1 + S_2 + S_3}{2\pi RL} , \tag{6}$$

where S_1, S_2, S_3 – sections of the reduced winding surface area after shell buckling (see Fig. 2), determined by the formulas:

$$S_1 = 2R\int_0^L \sin\theta_0 \sin\frac{\pi x}{L} dx = \frac{4RL}{\pi}\sin\frac{\pi}{2\eta} ;$$

$$S_2 = 4R\int_0^L \frac{\pi}{2\eta}\left(\frac{1}{2}-\frac{1}{\pi}\right)dx = \frac{2\pi RL}{\eta}\left(0,5-\frac{1}{\pi}\right);$$
$$S_3 = 2R\int_0^L (\pi-\theta_0)\,dx = 2\pi RL\left(1-\frac{1}{2\eta}\right).$$

After substituting them into formula (6), we have:

$$\xi_1 = \frac{1}{\eta\pi} - \frac{2}{\pi^2}\sin\frac{\pi}{2\eta}, \tag{7}$$

Taking into account the initial dent with an amplitude in the middle equal to f_0, the relative winding deformation with the shell stability loss will be less by an amount ξ_0 that can be determined as

$$\xi_0 = \frac{1}{\eta_0\pi} - \frac{2}{\pi^2}\sin\frac{\pi}{2\eta_0}, \tag{8}$$

where: $\eta_0 = \pi\sqrt{0,625\frac{R}{f_0}}$ characterizes the initial dent width.

Then the value of the winding relative deformation, taking into account formulas (7) and (8), will be determined by the formula:

$$\xi = \xi_1 - \xi_0 = \frac{1}{\pi\eta} - \frac{1}{\pi\eta_0} - \frac{2}{\pi^2}\left(\sin\frac{\pi}{2\eta} - \sin\frac{\pi}{2\eta_0}\right). \tag{9}$$

It makes it possible to proceed to the determination of the potential energy of deformation of the prestressed winding as a result of its length reduction during shell buckling. After loss of stability, the force in the winding will decrease by an amount $E_2 t_2 \xi = B_2 \xi$.

3 Results and Discussion

In the presence of significant dents $(2 \div 5t_1)$ and with a decreasing load, the process of stability loss is not so dynamic and is closer to the inertialless process. In this case, the behavior of the system is determined by the diagram of equilibrium states, and at any stage of deformation its resistance is equal to the load. The foregoing allows us to assume that the deformation of the winding occurs at an average value between the force before buckling and the force after buckling, i.e. with force. N_0 $N_0 - B_2\xi$ $N_0 - 0,5B_2\xi$ Then the potential energy of winding deformation can be determined by the formula:

$$U_{обм} = \frac{t_2}{E_2}\int_0^L R\int_0^\pi \frac{(N_0 - 0,5B_2\xi)^2}{t_2^2}\,d\theta dx = \frac{\pi RL}{B_2}(N_0 - 0,5B_2\xi)^2. \tag{10}$$

The energy of friction forces in this case is determined by the formula:

$$U_{тр} = 2\int_0^L R\int_{\frac{\theta_0}{2}}^{\pi} f_{тр}\frac{N_0}{R}\xi(\pi-\theta)Rd\theta dx = N_0 f_{тр}\xi RL\left(\pi - \frac{\pi}{4\eta}\right)^2. \tag{11}$$

The resulting formula makes it possible to approximately consider the effect of friction forces on the shell stability against the action of prestressing, since it does not take into account initial dents along the shell perimeter, they reduce shell contact zone with the winding, but also does not take into account the increase in radial pressure forces where the initial dents have a shape with slight bulging behind the outer contour, resulting in an increase in the shell curvature. Besides, a certain dynamism of the winding shift relative to the shell is not considered. Taking into account all these factors is a rather difficult task, difficult to accomplish within the framework of this article.

Shell buckling, as a rule, is accompanied by not only bending stresses, but also additional stresses in the middle surface. Part of the external load potential upon stability loss of the shell is spent on increasing the bending energy, and the other part - on changing the deformation energy of the middle surface. The ratio depends on the shell size and configuration taken on buckling. For medium length shells, the deformation energy of the middle surface becomes commensurate with shell bending energy. To determine the deformation energy, it is necessary to determine the stress distribution function in the middle surface Φ.

Since the shell buckling is accompanied by the dent formation with a large deflection amplitude, the problem should be solved in a geometrically nonlinear setting. To determine the stress distribution function in the middle surface, we will solve the fourth-order nonlinear differential equation of compatibility of deformations, used in the theory of flexible shells.

As there are initial decays in real shells, we use the equation given in [7, 8, 9]:

$$\frac{1}{E_1}\nabla^4\Phi = -\frac{1}{2}\left[L_{(w,w)} - L_{(w_0,w_0)}\right] - \frac{1}{R}\frac{\partial^2(w-w_0)}{\partial x^2}, \tag{11}$$

where operators $L_{(w,w)}$ and $L_{(w_0,w_0)}$, being nonlinear terms of this differential equation, are determined by the formulas [10]:

$$L_{(w,w)} = 2\left[\frac{\partial^2 w}{\partial x^2}\frac{\partial^2 w}{R^2\partial\theta^2} - \left(\frac{\partial^2 w}{\partial xR\partial\theta}\right)^2\right], \tag{12}$$

$$L_{(w_0,w_0)} = 2\left[\frac{\partial^2 w_0}{\partial x^2}\frac{\partial^2 w_0}{R^2\partial\theta^2} - \left(\frac{\partial^2 w_0}{\partial xR\partial\theta}\right)^2\right]. \tag{13}$$

To solve the differential equation (11), we use the following boundary conditions at shell ends (with hinged fastening): at $x = 0$ and $x = L$, $w = 0$ and $\frac{\partial^2 w}{\partial x^2} = 0$.

After integrating (11), taking into account formulas (12) and (13), we obtain the following expression for the stress distribution function in the shell middle surface Φ:

$$\Phi = E_1\left(f^2 - f_0^2\right)\beta\frac{\eta^2}{2R^2}\left(\cos\frac{2\pi x}{L} - \cos 2\eta\theta\right) + E_1\frac{f-f_0}{R}\beta\cos\eta\theta\sin\frac{\pi x}{L} - \frac{N_0}{t_1}\frac{x^2}{2}, \tag{14}$$

where: $\beta = \dfrac{\left(\dfrac{\pi}{L}\right)^2}{\left(\dfrac{\pi^2}{L^2} + \dfrac{\eta^2}{R^2}\right)^2}$.

Formula (14) is valid for $0 \le \theta \le \theta_0$.

The potential energy of deformation of the shell middle surface is determined taking into account a similar formula [4]:

$$U_c = \frac{t_1}{E_1} R \int_0^L \int_0^\theta \left[\left(\nabla^2 \Phi \right)^2 - (1+\mu) L_{(\Phi,\Phi)} \right] dx d\theta + \frac{t_1}{E_1} \int_0^L R \int_{\theta_0}^{\pi} \frac{N_0^2}{t_1^2} dx\theta , \qquad (15)$$

where: $L_{(\Phi,\Phi)} = 2 \left[\frac{\partial^2 \Phi}{\partial x^2} \frac{\partial^2 \Phi}{R^2 \partial \theta^2} - \left(\frac{\partial^2 \Phi}{\partial x R \partial \theta} \right)^2 \right]$.

As a result of integration and some transformations, we get:

$$U_c = B_1 C_1' + \frac{N_0^2 \pi R L}{B_1} , \qquad (16)$$

where: $B_1 = E_1 t_1$, $\beta_1 = \frac{\pi^2}{L^2} + \frac{\eta^2}{R^2}$, $C_1' = \left(f^2 - f_0^2 \right)^2 \beta^2 \left(\frac{\eta^3 \pi^5}{R^3 L^3} + \frac{\pi \eta^3 L}{R^3} \right) + \left(f - f_0 \right)^2 \frac{\beta^2 \beta_1^2}{8 \eta R} \pi L$.

The potential bending energy of the shell is determined taking into account the formula (12.95) [4]. However, in our problem, it is also necessary to take into account the bending stiffness of the winding in the annular direction. An approximate expression for the potential bending energy, taking into account the initial deflection, will have the form:

$$U_u = DR \int_0^L \int_0^{\theta_0} \left[\left(\nabla^2 w - w_0 \right)^2 - (1-\mu) L_{(w-w_0, w-w_0)} \right] \partial \theta dx =$$
$$= \frac{\pi R L (f - f_0)}{8\eta} \left[\frac{\pi^4}{L^4} D_1 + 2 \left(\frac{\pi \eta}{LR} \right)^2 D_1 + \frac{\eta^4}{R^4} \left(E_2 I_2 \frac{1}{2} + D_1 \right) \right] \qquad (17)$$

where: D – a parameter that takes into account the stiffness of the shell for bending in the annular and longitudinal directions, as well as the winding stiffness for bending in the annular direction;

$D_1 = \frac{E_1 t_1^3}{12(1-\mu^2)}$ – cylindrical shell stiffness;

$E_2 I_2 = \frac{E_2 t_2^3}{12}$ – reduced bending stiffness of the winding.

It should be noted that formula (17) takes into account possible mutual slippage between the winding and the shell, therefore, the flexural stiffness for the shell and the winding is determined separately as for two unconnected elements.

When the shell bulges in the dent zone, the radial pressure of the winding decreases, depending on the change in the shell curvature, to N_0/R zero. The larger the reduced winding thickness, the faster the tension force decreases, and, consequently, the radial pressure on the shell in this zone.

The potential of the radial pressure forces of the shell winding can be determined by analogy with the formula (13.91) [4]:

$$\Pi = -2R \int_0^L \int_0^{\theta_0} \frac{w}{2{,}5} \frac{1}{2} P d\theta dx = -\frac{N_0 - B_2 \xi}{2{,}5} \int_0^L \int_0^{\theta_0} \cos \eta \theta \sin \frac{\pi x}{L} d\theta dx = -\left(N_0 - B_2 \xi \right) C_2' , \qquad (18)$$

where: $P = \frac{N_0 - B_2 \xi}{R}$ – radial pressure of the winding on the shell after buckling;

$$C_2' = \frac{1}{1{,}25 \pi \eta} \left(f - f_0 \right) L .$$

It should be noted that the pressure P can be considered equal to zero when, in the section where the shell buckling occurs, its curvature becomes zero, that is, when (according to the diagram in Fig. 2 and formula (5)) the shell deflection becomes equal to $w/2,5$. Assuming that the winding pressure on the shell during buckling of the shell varies from P to zero according to a linear law, the average value of the pressure performing the work can be taken as $0,5P$. Accordingly, in formula (18), the deflection w is divided by a factor of 2.5, and the pressure by a factor of 2.

The potential system energy before stability loss is determined by the expression:

$$U_0 = \frac{t_1}{E_1} R \int_0^L \int_0^\pi \frac{N_0^2}{t_1^2} d\theta dx + \frac{t_2}{E_2} \int_0^L R \int_0^\pi \frac{N_0^2}{t_2^2} d\theta dx = \pi RLN_0^2 \left(\frac{1}{B_1} + \frac{1}{B_2} \right). \tag{19}$$

Substituting (10), (11), (16), (17), (18) and (19) into the energy balance equation (3), we solve it with respect to the winding tension force N_0 [10]:

$$N_0 = \frac{U_н + B_1 C_1^{'} + B_2 \xi \left(0/25 \pi RL\xi + C_2^{'} \right)}{\pi RL\xi + C_2^{'} - f_{mp} \xi RL\pi^2 \left(1 - \frac{1}{4\eta} \right)^2}, \tag{20}$$

In this formula, the first term of the numerator takes into account the effect of bending of the shell and winding, the second - the effect of deformations of the middle surface, and the third - the effect of the winding stiffness on tension. The third term included in the denominator characterizes the influence of friction forces between the shell and the winding during their mutual displacement at the moment of buckling.

4 Conclusions

To obtain the value of the critical force, it is necessary to minimize the values N_0 (20) by the parameters of the shell deflection, that is, to find the dimensions of the indentations that will correspond to the minimum force. Since in formula (5) the value f is expressed in terms of η, the minimization N_0 will be carried out by η, characterizing the dent width.

Minimization can be carried out by finding the derivative $\partial N_0 / \partial \eta$ and equating it to zero, however, such an equation is transcendental, and its solution is a very difficult problem. Therefore, it seems easier to carry out minimization using numerical analysis.

The value of the minimum critical effort according to the N_0 formula (20) is determined by varying the parameter η in the range from 1 to 60 with a step of 1. When determining the critical force, one can vary the flexibility of the shell, its relative length, the reduced thickness of the winding, the value of the initial deflections and the friction coefficient. Taking in the formula (20) $f_{mp} = 0$ we determine the value of critical efforts without taking into account the friction forces.

Summing up, we can draw the following conclusions: the given solution allows you to determine the value of critical efforts (or stresses $\sigma_{кр} = N_{кр}/t_1$) taking into account all the main factors [11]:

— shell length;
— shell flexibility;
— prestressing zone length;
— initial error value;
— reduced winding thickness;
— frictional forces between the shell and winding.

References

1. N. A. Alfutov, *Fundamentals of calculation for the stability of systems* (1978)
2. N. A. Alfutov, *On the influence of boundary conditions on the value of the upper critical pressure of a cylindrical shell* (1965)
3. V. L. Berezin, V. E. Shutov, *Strength and stability of tanks and* pipelines (1967)
4. A. V. Karmishin, V. N. Myachenkov, A. N. Frolov, *Statics and dynamics of thin-walled shell structures* (1975)
5. L. G. Donell, *Tanks, plates and shells* (1982)
6. A. S. Volmir, Reports of the USSR Academy of Sciences, **113 (2)** (1957)
7. N. A. Alumäe, Applied mathematics and mechanics, **1** (1949)
8. V. Z. Vlasov, *General theory of shells and its applications to technology* (1949)
9. V. Y. Vyalkov, V. A. Ivanov, V. N. Paimushin, J. of Applied Mathematics and Mechanics, **4** (2005)
10. V. V. Kozlov, Differential Equations, **4** (2009)
11. T. M. Chapaev, Scientific and Practical J. "Proceedings of Kabardino-Balkarian State Agrarian University named after V. M. Kokov", **1(15)** (2017)

Probability-theoretical approach to the accuracy of the component assembly of multilink mechanisms

Yuri Shekikhachev[1,*], *Vladimir* Batyrov[1], *Lyudmila* Shekikhacheva [1], *Ruslan* Balkarov[1], and *Nurilla* Noraliev[2]

[1]Kabardino-Balkarian State Agricultural University named after V. M. Kokov, Lenin Avenue, 1V, 360030 Nalchik, Russia
[2]Tashkent State Agrarian University, Universitetskaya, 2, 100140 Kibray district, Tashkent region, Republic of Uzbekistan

Abstract. The technological process, ensuring the accuracy of mechanism assembly is formally a sequential summation of primary manufacturing errors and an analysis of obtained, generalized values to develop and implement specific technological methods to meet process requirements for the accuracy of the relative position of parts. To ensure the accuracy of crank mechanisms for misalignment of piston and cylinder axes, assembled from specific, randomly received component parts, production personnel deal with strictly defined crankshafts, connecting rods, pistons, cylinder blocks, and, consequently, with one or another manufacturing errors. To ensure the accuracy of these parts assembly in a mathematical sense, with such a sequence, will represent nothing more than an algebraic summation of primary errors and compilation of their numerical values with the maximum permissible values. A different situation arises in the implementation of assembly technological processes with automatic achievement of accuracy parameters by varying individual parts from their total number. As in this case fairly representative aggregates are analyzed, the final result of combining parts into assemblies will occur with a greater or lesser probability, especially when the sum of the maximum deviations of constituent links (parts) is not equal and not less than the maximum permissible total values.

1 Introduction

The release probability from the assembly site of high-quality products with automatic support and accuracy will be determined by modules of individual deviations and by distribution laws of actual deviations within the tolerance range.

The process of manufacturing or repairing parts is accompanied by primary production errors in the relative position of base surfaces. Due to the influence of a large number of random factors on machining, these errors are also random variables, the distribution laws

* Corresponding author: shek-fmep@mail.ru

of which approach the Gaussian law the more fully the Lyapunov A.M. conditions are met. when summing up any primary errors.

Theoretical or basic patterns of distribution of primary errors are in good agreement with practical distributions, as a rule, only with strict identification of technological processes designed and adopted as standard (basic) ones. When solving the problem of assembly optimization, all these provisions are not significant, as in the case of constructing its general solution, the choice of a specific distribution law of actual deviations within the tolerance range is not so fundamental that one does not try to look for such a solution that would hold various initial arguments. In this case, any analogies in the distribution patterns will transform the general solution into particular variants with obvious simplifications of computational operations.

Thus, for a component assembly of crank mechanisms for the misalignment of piston and cylinder axes with automatic provision of the required accuracy by a sequential selection of deviations of parts on a computer with the preliminary formation of arrays of high-quality assembly units by calculation, the meaning of the main task of the mathematical model of the assembly operation in this case is reduced to theoretical-probabilistic analysis of real mechanisms to:

— construct a general solution for probability of high-quality assemblies with distortions of pistons within specified limits for arbitrary but known tolerance fields and distribution laws of actual deviations in the relative position of the base surfaces of the constituent parts within these tolerances of the probabilities of the sum of random variables falling into some previously specified scattering interval

— assessment of expansion possibilities and boundaries of considered maximum deviations without noticeable deterioration in the quality indicators of assembled mechanisms.

2 Materials and Methods

We believe that all considered errors in the relative position of base surfaces of the parts of crank-connecting rod mechanisms are the essence of values, independent and random in nature, conditionally distributed within the tolerance range according to patterns, analytically described by the following dependencies:

a) the error in the relative position of cylinder liner and crankshaft axes according to Gauss's law (FIG. 1, a):

$$F_1(\psi_1) = \frac{1}{\sigma_1\sqrt{2\pi}}\exp\left[-\frac{(\psi_1 - a_1)^2}{2\sigma_1^2}\right], -\infty \le \psi_1 \le \infty; \tag{1}$$

b) the error in the relative position of the main and connecting rod journal axes of the crankshaft according to the triangle law (FIG. 1, b):

$$F_2(\psi_2) = K|\psi_2|, -\xi \le \psi_2 \le \xi; \tag{2}$$

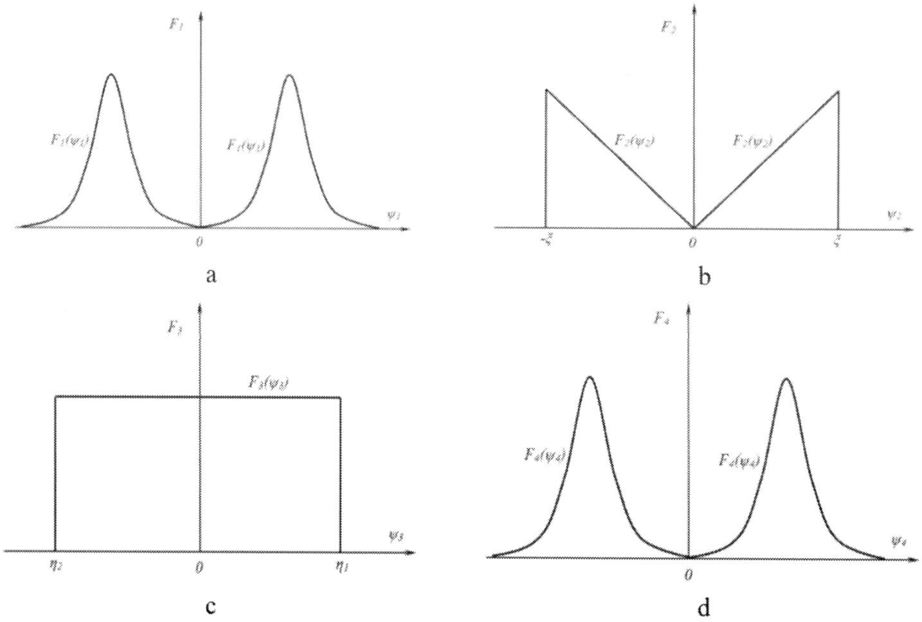

Fig. 1. Scheme of initial distributions of primary errors.

c) the error in the relative position of the connecting rod bearing axes according to the equal probability law (Fig. 1, c):

$$F_3(\psi_3) = \frac{1}{\eta_1 - \eta_2}, \eta_2 \le \psi_3 \le \eta_1; \tag{3}$$

d) the error in the relative position of piston generatrix piston and the piston pin hole according to the law (Fig. 1, d):

$$F_4(\psi_4) = \frac{1}{\sigma_4\sqrt{2\pi}} \exp\left[-\frac{(\psi_4 - a_4)^2}{2\sigma_4^2}\right], -\infty \le \psi_4 \le \infty. \tag{4}$$

In the future, to simplify calculations, we assume $a_1 = a_4$ and $\sigma_1 = \sigma_4$, accordingly, the mathematical expectations and standard deviations of normally distributed random variables.

Mathematically, misalignment of a piston in the cylinder for four-link crank mechanisms is calculated for each individual unit as the sum of random variables:

$$\psi_{\sum} = \sum_1^4 \psi_i = \sum_1^4 x_i = x_1 + x_2 + x_3 + x_4. \tag{5}$$

Then, if the probability distribution densities of these values $P_i(x)$ are known, then, for the pairwise representation, for example, $q_1 = x_1 + x_2$ and $q_2 = x_3 + x_4$, accordingly, we can write the following expressions for the probability distribution densities of such sums:

$$\overline{P}_1(y) = \int_{-\infty}^{\infty} P_1(y - x) P_2(x) dx; \tag{6}$$

$$\overline{P}_2(y) = \int_{-\infty}^{\infty} P_3(y - x) P_4(x) dx, \tag{7}$$

where $y_1 = f_i(x_i)$,

or finally:

$$P(y)=\int_{-\infty}^{\infty}\overline{P}_1(y-x)\overline{P}_2(x)dx. \tag{8}$$

Expression (9), under certain conditions, solves the problem. However, if there are more than two random variables, for example, as in this case, then its practical implementation becomes rather difficult even with simple laws for determining the probability densities of specified values x_i.

3 Results and Discussion

The applied aspect of the probabilistic-theoretical analysis of the accuracy of the assembly of four-link crank mechanisms for the misalignment of the piston in the cylinder always puts forward the problem of finding a solution more accessible for practical use, which for analyzed conditions can be obtained by solution approximation (8) by the method of moments or asymptotic method based on Lyapunov's limit theorem. In the latter case, for given distributions of the sums of interdependent random variables x_i with mathematical expectations $M(x)$ variances σ_i^2 and central moments of the third and fourth orders of the sum x_i of the values, the probability distribution density $P_i(x)$ can be approximated by a Gram-Charlier series, if $P_i(x)=0$ at $|x|\succ\alpha$ for certain α. Otherwise:

$$P_x(x)=\frac{1}{\sigma\sqrt{2\pi}}e^{-\frac{x_0^2}{2}}\left[1+\frac{1}{3!}\gamma_1\left(x_0^3-3x\right)+\frac{1}{4!}\gamma_2\left(x_0^4-6x_0^2+3\right)+\ldots\right], \tag{9}$$

where $x_0=\frac{x-M(x)}{\sigma};\gamma_1=\frac{M_3}{\sigma_3};\gamma_2=\frac{M_4}{\sigma_4}-3$ are the coefficients of asymmetry and kurtosis for x;

M_3, M_4 are the central moments, respectively, of the third and fourth orders for the probability density distribution.

With this problem formulation, the sought approximation is reduced to calculate the parameters of the total distribution using the well-known probability theory formulas:

$$M_{\sum}(x)=\sum_1^4 M(x); D(x)=\sum_1^4\sigma_i^2=M_2(x); \tag{10}$$

$$M_3(x)=\sum_1^4 M_3(x); M_4(x)=\sum_1^4 M_4(x). \tag{11}$$

For the initial distribution laws of considered primary production errors in the relative position of base surfaces of the parts of the crank mechanisms, we can write:

$$M(x_1)=M(x_4)=\frac{1}{\sigma\sqrt{2\pi}}\left[\int_{-\infty}^{\infty}xe^{-\frac{(x-\alpha_1)^2}{2\sigma^2}}dx\right]=0; \tag{12}$$

$$M(x_2)=K\int_{-\xi}^{\xi}x^2dx=0; \tag{13}$$

$$M(x_3)=\int_{-\eta_2}^{\eta_1}\frac{x}{\eta_1-\eta_2}dx=\frac{\eta_1-\eta_2}{2}. \tag{14}$$

Then:

$$M_{\sum}(x)=\sum_{1}^{4}M(x)=\frac{\eta_1-\eta_2}{2};M_{\sum}(x)=m. \tag{15}$$

By analogy, for variances or standard deviations, we have:

$$\sigma_1^2=\sigma_4^2=\left(\alpha_1^2+\sigma_1^2\right)\left[1+\Phi\left(\frac{\alpha_1}{\sigma_1\sqrt{2}}\right)\right]+\frac{\alpha_1\sigma_1\sqrt{2}}{\sqrt{\pi}}e^{-\frac{\alpha_1^2}{2\sigma_1^2}}; \tag{16}$$

$$\sigma_2^2=K\frac{\xi}{2};\sigma_3^2=\frac{(\eta_1-\eta_2)^2}{12}, \tag{17}$$

and consequently,

$$D(x)=\sigma^2=2\left(\alpha_1^2+\sigma_1^2\right)\left[1+\Phi\left(\frac{\alpha_1}{\sigma_1\sqrt{2}}\right)\right]+\frac{\alpha_1\sigma_1\sqrt{2}}{\sqrt{\pi}}e^{-\frac{\alpha_1^2}{2\sigma_1^2}}+ \tag{18}$$

$$+K\frac{\xi}{2}+\frac{(\eta_1-\eta_2)^2}{12}, \tag{19}$$

where $\Phi\left(\frac{\alpha_1}{\sigma_1\sqrt{2}}\right)$ is the probability integral, or error integral.

After calculation of corresponding integrals, the asymptotic representation can be rewritten as:

$$P_x(x)=\frac{1}{\sigma\sqrt{2\pi}}e^{-\frac{x_0^2}{2}}\left[1+\frac{1}{4!}\gamma_2\left(x_0^4-6x_0^2+3\right)+\ldots\right], \tag{20}$$

as $M_3=0$ and $\gamma_1=0$. Then the probability of hitting a random variable as a sum of random primary production errors in a certain interval will be:

$$P(\alpha_0\le x\le\alpha_1)=\frac{1}{\sigma\sqrt{2\pi}}e^{-\frac{x_0^2}{2}}\left[\left(1+\frac{3}{4!}\gamma_2\right)\int_{\alpha_0}^{\alpha_1}e^{-\frac{x_0^2}{2}}dx+\frac{\gamma_2}{4}\int_{\alpha_0}^{\alpha_1}e^{-\frac{x_0^2}{2}}\left(x_0^4-6x_0^2\right)dx+\ldots\right]. \tag{21}$$

4 Conclusions

Expression (21) was obtained, to solve the problem and, according to the initial laws of error distribution within certain tolerance fields, calculate the accuracy of the component assembly of crank mechanisms for the distortions of the piston in the cylinder, provided that the integrals included in this equality are represented in explicit form.

References

1. A. K. Apazhev, Y. A. Shekikhachev, A. M. Egozhev, L. Z. Shekikhacheva, A. A. Egozhev, IOP Conf. Series: Materials Science and Engineering, 862(3), 032005 (2020)
2. A. K. Apazhev, Y. A. Shekikhachev, V. I. Batyrov, J. of Physics: Conf. Series (JPCS), 1679, 042062 (2020)
3. Y. A. Shekikhachev, V. I. Batyrov, L. Z. Shekikhacheva, A. L. Bolotokov, H. L. Gubzhokov, IOP Conf. Series: Materials Science and Engineering, 862(3), 032001 (2020)
4. A. K Apazhev, E. A. Polishchuk, J. of Physics: Conf. Series (JPCS), 1679, 042086 (2020)

5. A. K. Apazhev, A. M. Egozhev, A. A. Egozhev, E. A. Polishchuk, National priorities and security: a collection of scientific papers based on the proceedings of the int. scientific and practical conf., 346-349 (2020)
6. L. A. Shomakhov, A. M. Egozhev, A. K. Apazhev, Farm Machinery Operator, **3**, 39-40 (2018)
7. Yu. A. Shekihachev, V. I. Batyrov, Kh. B. Kardanov, M. M. Chechenov, L. Z. Shekihacheva, Scientiic Life, **6(94)**, 929-937 (2019)
8. A. M. Egozhev, A. K. Apazhev, Tractors and Agricultural Machinery, **6**, 35-39 (2016)
9. A. K. Apazhev, A. M. Egozhev, Actual problems of scientific and technological progress in the agro-industrial complex, The 11th International Scientific and Practical Conference to the 65th anniversary of the Faculty of Agricultural Mechanization, 319-323 (2016)
10. A. K. Apazhev, A. M. Apazhev, Farm Machinery Operator, **10**, 36-37 (2015)
11. A. M. Egozhev, A. K. Apazhev, Proceedings of GOSNITI, **120**, 37-41 (2015)
12. A. K. Apazhev, L. A. Shomakhov, A. M. Egozhev, Tractors and Agricultural Machinery), **7**, 42-45 (2015)
13. A. M. Egozhev, A. K. Apazhev, Farm Machinery Operator, **10**, 38-39 (2014)
14. A. K. Apazhev, L. A. Shomakhov, A. M. Egozhev, Proceedings of GOSNITI, **117**, 57-61 (2014)
15. A. K. Apazhev, F. Kh. Kankulova, Reports of the Adyghe (Circassian) Int. Academy of Sciences, **8 (2)**, 119-122 (2006)

Tomato productivity depending on mineral nutrition and irrigation regimes in the conditions of film greenhouses in the mountain zone of the KBR

Zalim-Geri Shibzukhov[1,*], *Artur* Bagov[2], *Zalina* Shibzukhova[1], *Martin* Khantsev[1], and *Ilakhun* Akbar[3]

[1]Kabardino-Balkarian State Agricultural University named after V.M. Kokov, Lenin Avenue, 1V, 360030 Nalchik, Russia
[2]Kabardino-Balkarian State University named after H.M. Berbekov, st. Chernyshevskogo, 173, 360004 Nalchik, Russia
[3]Xinjiang University, Nongda East Road, 311, 830052 Urumqi, People's Republic of China

Abstract. The limited irrigation areas in the KBR, as well as the lack of a clear technology for the cultivation of this crop in film greenhouses using irrigation regimes and mineral fertilizers strongly affect the development and production of early tomatoes. In this regard, our research was devoted to the cultivation of tomatoes under irrigation conditions with the selection of the optimum doses of mineral fertilizers in spring film greenhouses. In the KBR settlements, vegetable crops are mainly grown in film greenhouses, which have numerous advantages over block greenhouses, mainly due to the cheapness of production several times over. Tomato seedlings were planted on April 10. Analysis of the data showed that the duration of the growing season increases on variants with optimum water and nutrient regimes of the soil, which led to higher quality products. This created favorable conditions for greater productivity of tomato fruits. Studies showed that the number of fruits and their weight on one plant were maximum according to the variants of the experiment on soil moisture content of 80% of normal moisture. The maximum value was obtained with a dose of mineral nutrition $N_{190}P_{160}K_{120}$.

1 Introduction

"In modern Russia, the share of vegetable production does not exceed 5% of the total volume of their production. The construction of film greenhouses is promising for the development of greenhouse vegetable growing in modern agro-economic conditions. These facilities are 2 times cheaper than stationary ones in terms of cost and in the conditions of spring crop rotation consume 4-5 times less energy and are able to ensure high yields of valuable, highly profitable vegetable crops" [2].

* Corresponding author: konf07@mail.ru

"The issues of development and application of new technologies for tomato cultivation in spring film greenhouses are very relevant and timely. It is necessary to qualitatively enhance the technology of growing crops, considering the fuller use of available natural and technical resources, the biological potential of productivity of varieties and hybrids in spring film greenhouses, as the most dynamic industry in modern Russia" [1].

Tomato is the KBR leading vegetable crop. "The value of growing early tomatoes is that they open up the supply of fortified outdoor products to the market. The production of early products, despite the additional material and labor costs, has a number of advantages and is economically feasible" [3].

"The restraining factors for the development of early tomato production using drip irrigation systems are the limited areas of irrigation by such systems and the lack of technology for growing this crop in film greenhouses, consistent with the peculiarities of water distribution over the irrigated area" [4].

In this regard, our research was devoted to the cultivation of tomatoes under irrigation conditions with the selection of the optimum doses of mineral fertilizers in spring film greenhouses.

Research purpose: optimization of the spring film greenhouse technology of growing tomatoes.

To achieve these goals, the following tasks were solved:

1. Establish the effect of drip irrigation on the growth, development and productivity of tomato crops, depending on irrigation rates and doses of mineral fertilizers;
2. Perform an economic analysis of the cultivation of early production of tomato fruits on various modes of irrigation and mineral nutrition.

The scientific novelty of the research consists in the optimization of the water balance of the soil, the introduction of a sufficient amount of mineral fertilizers to obtain the highest yield in the conditions of film greenhouses in the mountainous zone of the KBR.

Objects of research. Laboratory analyzes for the research were performed at the FSBI "SAS" Kabardino-Balkarskaya". The F1 Verlioka tomato cultivar was used. Drip irrigation is standard for plastic greenhouses. In the experiments, the following mineral fertilizers were used: nitrogen supply - ammonium sulfate, ammophos; phosphorus supply - superphosphate, ammophos; potassium supply - potassium sulfate.

2 Research methodology and conditions

The agricultural techniques in the experiments were developed on the basis of the current zonal recommendations with additions to their variants of the studied methods. The grown 40-day-old seedlings at the time of planting in the ground had a height of 25-30 cm with the flowers of the first cluster. Seeds were sown on March 5. Seedlings were planted in the greenhouse from 5 to 10 April.

The mathematical analysis of the obtained experimental data was performed according to B.A. Dospekhov's method.

The experiment scheme provided for the following options for the irrigation regime:

1. Maintaining soil moisture in the 0.4 m layer at 70% of normal moisture;
2. Maintaining soil moisture in the 0.4 m layer at 80% of normal moisture;
3. Maintaining soil moisture in the 0.4 m layer at 90% of normal moisture.

The calculation of the irrigation rate was performed in accordance with the research methodology.

Experiment scheme for doses of mineral fertilizers:

1. Application of mineral fertilizers in the application rate of $N_{90}P_{60}K_{60}$;
2. Application of mineral fertilizers in the application rate of $N_{140}P_{110}K_{90}$;
3. Application of mineral fertilizers in the application rate of $N_{190}P_{160}K_{120}$.

When calculating the doses of fertilizers, we relied on the recommendations for growing tomatoes in film greenhouses in literary sources, calculations for the removal of mineral elements by tomato plants and the background of the soil.

The experiments were repeated four times, randomized. The area of accounting plots for all options amounted to 100 m^2.

The experiments were performed in film greenhouses located in the Baksan district of the KBR in the village of Kishpek, at the LLC "Agro-K" enterprise (850 meters above sea level).

3 Results and Discussion

"The duration of the growing season, during which the tomato goes through different-quality phases of development, is one of the important criteria for assessing the conditions for the formation of the tomato fruit harvest. Their timing and the duration of interphase periods are usually significantly influenced by meteorological conditions, biological characteristics of the variety and cultivation technology" [4].

In our studies, the timing of the stages of development of tomato plants and the duration of interphase periods in different years are presented in Table 1.

Table 1. The influence of the timing of the onset of the phases of development of tomatoes from the introduction of mineral fertilizers and irrigation regimes.

Mineral fertilizer doses (kg of application rate/ha)	Humidity (% of normal moisture)	Seedling planting date	Flowering date	Start date of fruiting	Fruiting date	Date of last harvesting
$N_{90}P_{60}K_{60}$	70	10.04	19.04	01.05	08.06	15.09
	80	10.04	20.04	01.05	10.06	19.09
	90	10.04	20.04	02.05	09.06	18.09
$N_{140}P_{110}K_{90}$	70	10.04	19.04	03.05	10.06	22.09
	80	10.04	21.04	05.05	14.06	25.09
	90	10.04	21.04	09.05	12.06	21.09
$N_{190}P_{160}K_{120}$	70	10.04	20.04	08.05	13.06	27.09
	80	10.04	21.04	10.05	16.06	29.09
	90	10.04	21.04	10.05	13.06	24.09

Tomato seedlings were planted on April 10. The table shows that the variants with optimum water and nutrient regimes of the soil had the increased duration of the vegetation. This created favorable conditions for greater productivity of tomato fruits.

Studies showed that the number of fruits and their weight on one plant were maximal according to the variants of the experiment on soil moisture content of 80% of normal moisture. The maximum value was obtained with a dose of mineral nutrition $N_{190}P_{160}K_{120}$ (Tab. 2).

Table 2. Tomato yield structure depending on mineral nutrition and irrigation regime.

Mineral fertilizer doses (kg of application rate/ha)	Pre-irrigation humidity (% of normal moisture)	Number of fruits per plant (pcs.)	Fruit weight per plant (kg)	Weight of one fruit (kg)
$N_{90}P_{60}K_{60}$	70	24.2	3.45	0.151
	80	26.5	3.74	0.159
	90	25.4	3.65	0.154

Table 2. Continued.

$N_{140}P_{110}K_{90}$	70	30.7	4.22	0.162
	80	33.4	4.74	0.167
	90	31.8	4.56	0.164
$N_{190}P_{160}K_{120}$	70	36.5	5.34	0.166
	80	39.6	5.78	0.171
	90	37.6	5.59	0.169

In experiments, with an increase in the soil moisture threshold from 60 to 90% of normal moisture, the number of fruits increased by 2.2-3.5 pcs. The irrigation regime also influenced the weight of the fruits, with an increase in humidity, it increased.

Ensuring the quality products is an important condition for growing any agricultural crops. The special value of tomato fruits is given by their chemical composition, what chemicals are present in it, in what quantity, etc.

"The soil, climatic and weather conditions, varietal characteristics, growing conditions and other conditions affect the content of nutrients in tomato fruits. Irrigation and fertilization are among the most efficient factors contributing to improving the quality of the tomato crop" [5].

Table 3. Changes in the chemical composition of tomato fruits depending on the mineral nutrition and irrigation regime.

Doses of mineral fertilizers (kg of application rate/ha)	Humidity (IH %)	Chemical composition of tomato fruits				
		dry matter (%)	fiber (%)	sugar (%)	ash (%)	nitrates (mg/kg)
$N_{90}P_{60}K_{60}$	70	6.41	0.62	3.8	0.58	51.7
	80	6.35	0.66	3.7	0.59	49.3
	90	6.75	0.55	4.2	0.56	63.9
$N_{140}P_{110}K_{90}$	70	6.14	0.65	3.6	0.61	64.2
	80	6.07	0.71	3.4	0.62	61.4
	90	6.49	0.60	3.9	0.59	78.1
$N_{190}P_{160}K_{120}$	70	5.91	0.74	3.3	0.64	74.3
	80	5.82	0.79	3.2	0.66	71.6
	90	6.24	0.68	3.6	0.62	90.8

Data analysis (Table 3) showed that in variants with a dose of $N_{190}P_{160}K_{120}$, the nitrate content in fruits is maximum and ranges within 74.3 - 90.8 mg/kg, depending on the irrigation regime. An increase in the moisture content of the soil in some cases led to an increase in sugar in tomato fruits. A decrease in the content of nitrates was observed in the 80% of normal moisture variants. The content of nitrates for all variants of the experiments did not exceed the permissible limits.

"The most important parameter for assessing the efficiency of the technological process of sectoral agricultural production is the optimization criterion, that is, the parameter by which the object under study is evaluated and with which all the factors under study are linked into a single complex" [4].

"Indicators characterizing the economics of the functioning of production have a certain degree of importance in evaluating the efficiency of the technological process of cultivation of agricultural crops and the feasibility of using, reclamation agricultural practices" [3].

The feasibility of the technology was determined by the cost of gross production, items of expenditure for the full turnover of growing tomato plants. The average price of the product was taken for the entire vegetation (Table 4).

Table 4. Feasibility of growing tomato depending on mineral nutrition and irrigation regime.

Mineral fertilizer doses (kg of application rate/ha)	Soil moisture content (% of normal moisture)	Yield (t/ha)	Expenses, (RUR th/ha)	Prime cost (RUR th/ha)	Revenue (RUR th/ha)	Net income (RUR th/ha)
$N_{90}P_{60}K_{60}$	70	93.2	331.4	3.7	1025	693.6
	80	104.7	340.2	3.9	1151	810.8
	90	100.2	348.9	3.8	1102	753.1
$N_{140}P_{110}K_{90}$	70	111.5	399.8	3.4	1226	826.2
	80	131.4	410.5	3.5	1489	1078.5
	90	125.6	419.7	3.6	1381	961.3
$N_{190}P_{160}K_{120}$	70	128.7	476.5	4.1	1415	938.5
	80	140.2	487.4	4.2	1542	1054.6
	90	134.6	495.9	4.3	1480	984.1

With the increase in the doses of mineral fertilizers to $N_{190}P_{160}K_{120}$ and soil moisture to 90% of normal moisture, growing tomatoes required 495.9 RUR th/ha, this is the greatest value among other options, while the net income amounted to about 984.1 RUR th. Thus, at lower costs of 487.4 RUR th/ha on the same variant with fertilizers and moisture content of 80% of normal moisture, the net income level was about 1,054.6 RUR th, which is by 70.5 RUR th more.

With an optimum ration, with a humidity of 80% of normal moisture, expenses were about 410.5 RUR th/ha, and a net income amounted to 1,078.5 RUR th.

In real conditions, there is no ideal technology that would satisfy all the qualitative and quantitative indicators of production. We may only optimize some of the technological parameters to obtain the desired product in a certain quantity with the economic component of the manufactured product.

4 Conclusions

1. To maintain soil moisture in the 0.5 m layer at a level of 60-80% of normal moisture, 10 to 25 irrigations should be performed with an irrigation rate of 200- 300 m^3/ha.

2. The timing of the phases depended on the irrigation regimes and the application of doses of mineral fertilizers. The higher the irrigation rate and the higher the amount of applied fertilizer, the more the phase-to-phase periods increased.

3. Analysis of the yield structure showed that with an increase in the soil moisture threshold from 60 to 90% of normal moisture, the number of fruits increased by 2.2-3.5 pcs. The irrigation regime also influenced the weight of the fruits, with an increase in humidity it increased, and the indicators also increased in proportion to the application of mineral fertilizers.

4. The obtained chemical composition data showed that in variants with a maximum dose of $N_{190}P_{160}K_{120}$, the content of nitrates in fruits is correspondingly maximum and, depending on the irrigation regime, ranges within 74.3 - 90.8 mg/kg. The amount of nitrates for all items did not exceed the permissible environmental standards.

5. The greatest economic effect according to the variants of the experiments was obtained with a moisture supply mode of 80% of normal moisture with a dose of mineral nutrition $N_{140}P_{110}K_{90}$, where the net income per hectare was 1,078.5 RUR th.

References

1. A. K. Yezaov, Z. S. Shibzukhov, M. Kh. Nagoyev, Modern problems of science and education, **1(1)**, 1693 (2015)
2. A. K. Yezaov, Z. S. Shibzukhov, Current ecological state of the environment and scientific and practical aspects of rational environmental management: The II international scientific and practical Internet conf., 625-629 (2017)
3. Yu. B. Khushtov, Z. S. Shibzukhov, M. Kh. Indarokov, Current ecological state of the environment and scientific and practical aspects of rational environmental management: The II international scientific and practical Internet conf., FSBSI, 613-615 (2017)
4. Z. G. S. Shibzukhov, F. M. Kurzhiyeva, Innovative technologies for agrarian and industrial complex of the South of Russia: Materials of the All-Russian academic and research conf., 209-213 (2016)
5. Z. S. Shibzukhov, F. M. Kurzhiyeva, Current ecological state of the environment and scientific and practical aspects of rational environmental management: The I International scientific and practical Internet conf., 2130-2132 (2016)
6. Z. G. S. Shibzukhov, A. K. Yezaov, A. A. Shugushkhov, News of the Kabardino-Balkarian state agricultural university named after V. M. Kokov, **2(12)**, 27-32 (2016)
7. A. Ezov, Z.-G. Shibzukhov, B. Beslaneev, Z. Shibzukhova, M. Khantsev, Int. Scientific and Practical Conf. «Development of the Agro-Industrial Complex in the Context of Robotization and Digitalization of Production in Russia and Abroad» (2020)
8. K. Nazranov, E. Didanova, Z.-G. Shibzukhov, M. Orzalieva, B. Nazranov, Int. Scientific and Practical Conf. «Development of the Agro-Industrial Complex in the Context of Robotization and Digitalization of Production in Russia and Abroad» (2020)
9. M. S. Sidakova, E. N. Didanova, A. Y. Kishev, R. Z. Abdulkhalikov, Int. Scientific and Practical Conf. «Development of the Agro-Industrial Complex in the Context of Robotization and Digitalization of Production in Russia and Abroad» (2020)

Search, inspection and selection of new source material for millet breeding

Larisa Sokurova, and *Aishat* Yandieva

[1]Institute of agriculture – the branch of the RAS (Russian Academy of Sciences) Kabardino-Balkarian Research Center, Kirov str., 224, 360004 Nalchik

Abstract. Comprehensive inspection of millet original forms and their involvement in the breeding process to create new varieties and lines is an urgent problem and is of great theoretical and practical importance [3,4]. New varieties should be resistant to abiotic stresses and ecological plasticity, as global climate changes do not allow new varieties and cross-breeds to realize their potential possibilities, most of them are realized only by 20-40%. In this regard, it is important to take into account the specifics of global and local climate change, as well as the sharply increased frequency of extreme weather events over recent decades. [5] In general, in the genetic variability of cultivated plants, the most deficient category is resistance to the action of biotic and abiotic stressors.

1 Introduction

The success of crop breeding in Russia is facilitated by the plant genetic resources of VIR, which are among the four largest in the world, where there are more than 9 thousand samples of millet [1].

At the Institute of Agriculture of the Kabardino Balkarian Research Center of the Russian Academy of Sciences, there is a VIR control point. Annually, about 100-150 varieties of the millet unique world collection are kept alive, reproduced and inspected. The ones that distinguished from them on the basis of economically valuable traits are used in the further breeding of the institute.

The inspection of variability degree of genotypes in response to varying external conditions makes it possible to create widely adaptive varieties and cross-breeds, which, under favorable conditions, give the greatest and most stable yield [2].

The Institute of Agriculture of the Kabardino Balkarian Research Center is working on expanding, preserving and inspecting the characteristic collections of millet, isolating and creating new sources and donors of valuable traits for use in adaptive breeding. The ongoing investigations on millet are aimed at expanding the genetic diversity of the initial material, inspecting and improving the architectonics of plants and physiological adaptation mechanisms, and contributing to an increase in the adaptive potential of the created varieties [6,7].

Based on the inspection and assessment of samples from the collection of millet seeding VIR, promising initial material (donors and genetic sources) was identified, which will be involved in the breeding process to create new stress-resistant genotypes [8].

The most centralized, economically and environmentally effective means of increasing the size and quality of the crop are new varieties and cross-breeds due to resource-energy efficiency, environmental sustainability, environmental protection and profitability of agricultural production.

2 Materials and methods

The investigations were carried out in 2017-2019 on the experimental field of the Institute of Agriculture, located in the steppe zone of the KBR, which is characterized by insufficient moisture. The average annual precipitation according to multi-year data is 444 mm. This zone is characterized by a pronounced continentality. Winters with little snow, moderately cold, unstable, with frequent thaws.

A stable transition of air temperature over +10^{0}C is observed in the spring on April 15-20, in the autumn on November 5 - 10.

Soils in the steppe zone are represented by ordinary black soil (chernozem). The thickness of the humus layer reaches 70-90 cm, and the humus content in the plowing horizon ranges from 3 to 4.9%. The content of mobile phosphorus in the soil ranges from 15.6 to 28.7 mg/kg; exchangeable potassium 200-300 mg/kg (according to Machigin). The soil reaction is slightly alkaline.

The dynamics of plant growth was measured, the onset of phenophases, stages of organogenesis was marked and at the end of the growing season the yield was recorded. Resistance to lodging was determined visually according to a nine-point system in the phase of economic ripeness.

For laboratory analysis by quantitative traits before harvesting a bundle material in the amount of 25 plants of each sample was taken and analyzed for economically valuable traits, paying particular attention to plant height, length and type of panicle, its density, grain content, grain size, etc.

The inspection work on collection samples of millet was carried out in accordance with the VIR methodological instructions.

Over the years of investigation, we have inspected many different domestic and foreign samples of millet from the VIR world collection. They were sown annually on winter wheat using the RS-1 manual seeder at the optimum time for the zone. The seeding rate is 450 grains per 1 m^2, the sowing method is ordinary.

3 Results and discussion

Assessing the weather conditions of the growing season of plants, it should be noted that May - June 2017 was favorable for the growth and development of millet. July and August were drier, precipitation was 34.3-23.6 mm with an average daily air temperature of 26.8-26.0^{0}C.

Table 1. Mean 10-day precipitation data and air temperature in the investigation years.

Months	Ten-day periods	Amount of precipitations, mm					Air temperature, °C				
		2017	2018	2019	Average in three years	Average multi summer	2017	2018	2019	Average in three years	Average multi summer
May	1	11.2	11.2	39.1	20.5		18.0	18.0	21.0	19.0	
	2	34.4	16.9	17.2	22.8		16.3	19.0	23.5	19.6	
	3	33.9	59.0	16.0	36.3		16.6	20.2	27.0	21.2	

Table 1. Continued.

May	Amount monthly	79.5	87.1	72.3	79.6	65.3	16.9	19.0	23.8	19.9	16.2
June	1	30.4	14.1	1.2	15.2		21.0	21.3	36.0	26.1	
	2	30.8	26.9	7.7	21.8		20.4	21.5	36.0	25.9	
	3	8.4	1.8	20.0	10.0		23.8	24.3	38.0	28.7	
	Amount monthly	69.6	42.8	18.9	47.0	84.2	21.7	22.3	36.6	26.9	20.6
July	1	14.0	0.3	6.1	6.8		26.3	24.7	37.0	29.3	
	2	1.8	45.9	34.4	27.3		27.4	26.6	33.0	29.0	
	3	18.5	40.6	16.2	25.1		26.7	26.9	39.5	31.0	
	Amount monthly	34.3	86.8	56.7	59.2	59.4	26.8	26.0	36.5	29.7	23.0
August	1	0.7	42.1	32.2	25.0		29.1	23.7	36.0	29.6	
	2	11.7	1.5	3.3	5.5		24.3	22.6	40.0	28.9	
	3	11.2	0	4.2	5.1		24.8	22.8	39.5	29.0	
	Amount monthly	23.6	43.6	39.8	35.6	44.5	26.0	23.0	38.5	29.1	
Total:		207.0	260.3	197.7	221.4	253.4	22.8	22.5	33.8	26.4	22.5

In 2018, 267.5 mm of precipitation fell in May-August and they were distributed more evenly, while the air temperature averaged 22.6 °C for four months. During the months of May and June, 87.1-42.8 mm fell, with an average daily air temperature of 17.8-22.0^{0}C (tab. 1).

A concept of the degree of stability and the variation value of this trait in individual samples can be formed on the basis of their inspection.

There are three phases of the plant's response to stress factors - irritation, damage and adaptation. The first phase is characterized by a rapid deviation of many parameters from the standard and subsequent return to it. In the phase of damage, there is a discoordination of different metabolic links. The discoordination degree is positively correlated with the level of stress effort and negatively with plant resistance. Due to the interaction of opposite metabolic processes, the ratio of metabolites is standardized at a certain level, which is the final stage of adaptation.

The influence of unfavorable environmental factors (soil drought and high air temperature) made it possible to distribute millet samples into highly resistant, above average resistant, medium resistant and low resistant ones.

The inspected samples were distributed as follows: the highest resistance to overheating in the initial stages of development was shown by 6.2% of the samples to the total amount of material (81-100% of germinated seeds). The assessment for resistance to overheating was 8.8 grades (according to the nine-point system in accordance with the CMEA international classifier).

The second group - resistance above the average included 14.3% (44 samples) – from 61 to 80% of germinated seeds. The assessment for resistance to overheating was 8.1 grades.

The third group - medium resistant occupies 52.2% (160 samples) – 41-60% of germinated seeds. The assessment for resistance to overheating was 7.5-8.0 grades.

The fourth group occupies 27.3% (84 samples), 19-40% of seeds germinated – weak resistance to overheating in the initial stages of development, the assessment was 5.3 grades (Fig. 1).

The high productivity of plants and ecological plasticity in all the years of investigation were consistently demonstrated by the samples shown in Table 2.

The height of millet plants varies with an increase in the agricultural culture and, especially in humid years, accordingly, other morphological characteristics also change.

The inspected samples of millet were distributed in height as follows: very undersized – 60 cm high, low – 61-80 cm, medium-sized – 80 - 100, high – 100-12 cm, respectively (Fig. 2).

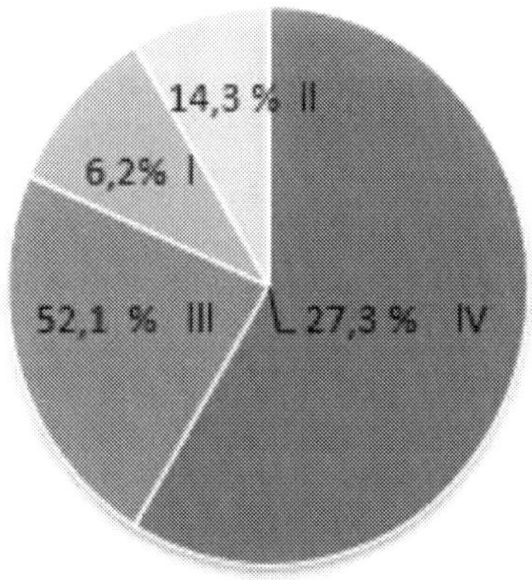

Fig. 1. Distribution of the inspected samples of seed millet by resistance groups (2017-2019).

Table 2. Donors and sources of millet isolated from the collection nursery-garden.

No.	VIR catalog number	Yield (ton/ha)	Deviation from standards, ton/ha	Mass of 1,000 grains, g	Grain weight from panicle, g	Number of grains from panicle, pcs	Plant height, cm	Panicle length, cm	Lodging, grade	Falling capacity, grade	Resistance to abiofactors	
											Above average	Highly resistant
1	Caucasian dawns	1.3	-	8.8	2.2	250	88.8	19.6	9	9	IV	
2	K – 6038	1.7	0.4	7.4	2.5	338	68.8	16.4	9	9	IV	
3	K – 6061	1.8	0.5	8.0	2.7	337	74.8	18.0	9	9	IV	
4	K – 6063	1.6	0.3	7.2	2.5	348	68.5	11.6	9	9	IV	
5	K – 6070	1.7	0.4	7.7	2.7	350	74.6	19.6	9	9	IV	
6	K – 6081	1.8	0.5	8.0	2.8	350	67.0	16.4	9	9	IV	
7	K – 6130	2.4	1.1	8.2	3.5	427	88.4	20.6	9	9		V
8	K – 6192	1.9	0.6	8.8	3.1	352	80.2	18.2	9	9		V

Table 2. Continued.

9	K – 6195	1.8	0.5	7.8	2.7	346	67.4	16.0	9	9		V
10	K – 6202	1.9	0.6	7.5	2.9	386	78.8	21.0	9	9	IV	
11	K – 6214	1.7	0.4	7.4	2.7	365	78.4	16.6	9	9	IV	
12	K – 6225	1.8	0.5	8.0	2.7	337	80.0	12.6	9	9		V
13	K – 6295	1.9	0.6	7.2	3.0	416	76.5	16.0	9	9		V
14	K – 6365	1.7	0.4	8.4	2.6	309	73.6	15.7	9	9	IV	
15	K – 6367	2.1	0.8	7.5	3.2	426	85.6	21.0	9	9		V
16	K – 6368	2.3	1.0	7.9	3.4	430	82.0	22.6	9	9		V
17	K – 6417	1.7	0.4	7.2	2.7	375	73.5	16.0	9	9	IV	
18	K – 10034	2.4	1.1	8.3	3.6	434	84.4	23.0	9	9		V
19	K – 10086	2.5	1.2	8.4	3.6	430	80.6	21.0	9	9		V
20	K – 6223	1.7	0.4	8.0	2.6	325	68.0	16.2	9	9	IV	
21	K – 1522	2.8	1.5	8.2	3.0	366	84.4	27.2	9	9		V
22	K – 1876	2.7	1.4	8.0	3.8	475	83.4	26.1	9	9		V
23	K – 5451	2.8	1.5	8.3	3.7	455	99.6	30.0	9	9		V
24	K – 5464	3.0	1.7	8.0	2.9	362	102. 1	34.0	9	9		V
25	K – 5466	2.9	1.6	8.8	3.5	397	88.0	28.2	9	9		V
26	K - 6054	2.9	1.6	7.8	4.4	500	84.2	28.8	9	9		V
	LCD_{05}	1.8										

In terms of plant height and panicle length, the distinguished genotypes are within the optimum zone, at which the highest yield is achieved.

One of the mechanisms of millet resistance to drought is the provision of water to plant bodies and tissues due to a well-developed water-conducting system of the stem and leaves. The more resistant samples to overheating and dehydration

turned out to be: K-6130, K-6-63, K-6070, K-6192, K-6214, K-6225, K-6295, K-6367, K-6368, K-10034, K-10086 withstood overheating and dehydration better than others.

Most of the grains were formed by plants of the sample K-6130, K-6295, K-6367, K-6368, K-10034, K-10086.

The excess over the Caucasian dawns standard variety is 60-185 pieces per plant. The index of the number of grains in a panicle is positively associated with the duration of the growing season and negatively with the grain size, but their various combinations are also possible. The selection of individual plants for this trait is difficult because of its great variability under the influence of environmental conditions.

Early ripeness was shown by 9 samples of the collection, which ripen 15-20 days earlier than the Caucasian dawns standard variety. Yield data for the investigation years are shown in Table 2. The yield data analysis shows that in the steppe zone of Kabardino Balkaria, the following samples were significantly distinguished by grain yield over the investigation years: K-6330, K-6367, K-6368, K-10034, K-10086, K-1522, K-1876, K-5451, K-5455, K-5464, K-5466, K-6054, K- 6130, K-6063, K-6070, K-6192, K-6214, K-6225, K-6295, K-2131, K-2306, K-5455, K-5456, K-5457. The excess over the standard variety is 0.5-1.7 ton/ha.

By the weight of 1,000 grains the samples were characterized in the range of 7.2-9.2 g. High weight of 1,000 grains was shown by K-1522, K-1876, K-2131, K-2306, K-5451, K-5433, K-5456, K-5464, K-5466, K-6061, K-6081, K-6130, K-6225, K-6365, K-10034, K-10086, K-6223 samples. The highest weight of 1,000 grains was shown by K-6192, K-5456, and K-5466 samples (table. 2).

4 Conclusions

The inspection of plant productivity relationship with structural elements in years contrasting with weather conditions enables the breeder to more effectively select the desired forms, taking into account the specific growing conditions.

The investigations carried out specify that it is promising to conduct selection according to grain size, panicle grain content, and an increase in their size.

References

1. N. P. Agofonov, *World Millet Collection and Its Use in Practical Breeding*, 69 (1970)
2. E. N. Zolotukhin, N. L. Tikhonov, L. N. Lizneva, et al., *Actual problems of breeding and seed production of grain crops in the southeastern region,* 46 (1999)
3. A. F. Kurtseva, *Genetic resources of the millet collection and their use in the breeding process,* 62 (1994)
4. N. I. Vavilov, *Theoretical Foundations of Breeding* (1987)
5. N. P. Agofonov, *Improvement of Breeding and Seed Production and Millet Cultivation Technology,* 9 (1985)
6. L.Kh. Sokurova, *Sustainable Development: Concepts,* 306 (2017)
7. L.Kh. Sokurova, Legumes and cereals, 3(67) (2018)
8. V. A. Ilyin, Yu. Ya. Mikhailova, O. S. Rasskazova, *Ways of creating high-quality varieties of millet. Breeding and Seed Production of Millet*, 61 (1973)

Breeding of early ripening potato varieties under the conditions of the Kabardino-Balkarian foothill zone

Aniuar Abazov[1], *Khaset* Abidov[1], *Sultan* Basiev[2], and *Khusein* Nazranov[3,*]

[1]Institute of agriculture – the branch of the RAS (Russian Academy of Sciences) Kabardino-Balkarian Research Center, Kirov str., 224, 360004 Nalchik, Russia,
[2]Federal State Budgetary Educational Institution of Higher Education Gorsky State Agrarian University, Kirov str., 37, 362040 Vladikavkaz
[3]Kabardino-Balkarian State Agrarian University named after V.M. Kokov, Lenin str., 1V, 360004 Nalchik, Russia

Abstract. Evolution and thousands of years of artificial selection have contributed to the formation of the widest variety of cultivated potatoes. Thanks to this, more than 4 thousand varieties of potatoes have been created in the world today. All of them are known to differ not only in habit, but also in the level of yield, ripening period, resistance degree to numerous diseases and pests, plasticity to certain climatic and soil conditions. The results of research and evaluation of 4 thousand hybrids of the 1st tuber generation (single-tuberous plants) of joint breeding of the Institute of Agriculture ISKh KBNTs RAN and FSBSI (Federal state budgetary scientific institution) "Federal Research Center (FRC) of Potato name after A.G. Lorkh" that passed the full scheme of breeding tests, 9 promising cross-breed populations with economically useful traits were identified: consistently high yield and resistance to a complex of viral, bacterial and fungal diseases. The selected cross-breeds are distinguished by early and medium early ripening periods with a growing season of 55-70 days and a maximum yield of 38.6-66.2 ton/ha, depending on the prevailing weather conditions during the growing season of potato plants in 2019-2020. The genotypes obtained are involved in the implementation of the federal scientific and technical program for the development of potato growing for 2017-2030. The main goal of the program is to create new varieties of early ripeness group of potatoes. The co-executors of this program are a number of research institutes of the Russian Federation, including the Institute of Agriculture ISKh KBNTs RAN. [1] As a result of many years of investigation, the promising cross-breed 2652-15 that emerged from breeding of the Malinovka variety with the backcross 93.20-12 under the name Zamir was transferred in 2020 for the State test of the Russian Federation in the North Caucasian region. Medium early variety for table use and mashed potatoes processing. Potential yield is 49.5 ton/ha. Tubers of an attractive rounded oval shape with shallow eyes weighing 85.5-108.4 g with a yellow color of the tuber and creamy flesh. The tubers are resistant to mechanical damage during combine harvesting, storability is good. Medium resistant to late

* Corresponding author: nazranov777@mail.ru

blight in tops and highly resistant to tubers. Relatively resistant to rhizoctonia and common scab, banded and wrinkled mosaic of leaves.

1 Introduction

Early "young" potatoes are distinguished by excellent taste and culinary qualities. At the same time, the tubers contain small amounts of vitamins B_1 (aneurin), PP (niacin), B_2 (riboflavin), A (carotene), as well as mineral salts of calcium, iron, potassium, sulfur and other substances that are extremely necessary for the normal functioning of the human body. [2, 3] However, the nutritional value of potatoes is significantly reduced during long-term storage in the summer and autumn period. So, if in new harvested tubers the content of vitamin "C" is 20-25 mg per 100 g of raw weight, after 2-3 months it decreases to 12-13 mg, and after 5-6 months to 7-9 mg [3]. Therefore, the creation of early ripening varieties of new generation potatoes is one of the most important areas of breeding, especially for the southern regions of Russia. Among more than 40 economically useful traits inherent in the new potato variety, resistance to the most common pathogens is one of the most significant. In particular, late blight remains one of the most harmful pathogens, which is very problematic to prevent [4, 5]. At the same time, most potato varieties created in the 90s of the XX century do not withstand the infectious load due to the emergence of new strains and phytopathogens, which leads to a significant shortage of harvest [6,7,8]. In this regard, modern breeding is experiencing great difficulties in the process of new genetic sources and effective donors in the creation of genotypes with complex resistance to the most harmful fungal diseases (late blight, rhizoctonia, alternaria) and severe forms of viruses (X, Y), combining early ripeness with high productivity and consumer qualities of tubers. The investigation purpose is to conduct breeding trials and selection of a new potato genotypes group and to identify the most promising early and mid-early varieties to ensure efficient production under the conditions of the Southern Federal District of Russia.

2 Materials and methods

To determine and clarify the potential of the isolated genotypes, a field experiment was laid with a randomized placement of plots in 4 several-fold repetition on the experimental fields of the laboratory for potato breeding and seed production of the Institute of Agriculture ISKh KBNTs RAN in the Zolsky district of Kabardino-Balkaria.

Field nursery-gardens for cross-breeds testing were located at an altitude of 1,500 m above sea level under the conditions of leached black soil for medium loamy granulometric composition with the following agrochemical characteristics: humus - 6.1-6.8%; the soil solution reaction - (KCl) - 5.3-6.6; (pH) - 6.4-6.7%. P_2O_5 content – 3.3-4.7; K_2O – 35-44 mg/100g. The predecessor is grain crops, bean cultures.

The meteorological conditions during the years of the investigation were distinguished in 2019 by the abundance of precipitation and their even distribution during the growing season. In 2020, on the contrary, it was characterized by high soil and air temperatures. During the period of budding and flowering, a lack of moisture was marked (second half of June, early July), which negatively affected the growth and development, first of all, of varieties of early ripening.

3 Research Results

The breeding process of highly productive new generation potato varieties included the solution of the following problems [9]:

— the initial material investigation for the effective selection of parental forms with high combinational ability;

— analysis of cross-breed offspring with the selection of recombinants with high productivity, early ripeness, good consumer qualities, resistant to the most common diseases;

— testing promising genotypes under various soil and climatic conditions in order to determine their adaptive potential.

Rational use of available genetic resources focused on achieving complex resistance to various pathogens of high productivity, early ripeness and quality of tubers made it possible to select ecologically adaptive genotypes in a relatively short time. In particular, when conducting trial digging in 2019, in 65 days after planting to clarify the early ripeness of genotypes, experimental data were obtained on yield, harvesting of grocery-ware tubers, average weight of grocery-ware tuber, starch content, dry matter, resistance to viral and fungal diseases. (Table 1)

Table 1. Yield structure of promising cross-breeds of breeding site (2019-2020).

Breeding number	Origin	Yield (t/ha)		Grocery-ware tubers, %	Average weight of 1 tuber, g	Starch content, %	Dry matter content, %	Disease resistance, grade		
		2019	2020					Viral	Late blight	Alternaria disease
St-d	Udacha	38.6	11.1	63.6	63.0	14.0	19.7	5	7	7
1755-55	Santana x Gala	45.4	15.1	85.5	64.4	14.3	20.1	7	7	5
2588-122	Udacha x 88.16/20	39.2	12.8	77.5	70.7	19.4	25.1	7	7	9
St-d	Nevskiy	42.5	14.1	68.4	63.2	16.8	22.5	5	5	9
4704-15	-	49.1	20.6	82.6	90.3	13.5	19.3	5	7	7
1871-2	Krasavitsa x Tiras	42.9	15.2	84.8	77.4	17.7	23.4	5	9	7
2652 -15	Malinovka x 93.20-12	49.5	29.3	82.3	71.1	15.7	21.5	7	5	5
2520-152	Udacha x 88.16/20	52.6	30.4	81.4	87.4	17.2	22.9	7	7	9
1869-1	Tabor x Tiras	50.7	12.4	74.4	50.7	13.3	19.0	7	9	7
2591 -1	Lira x Ausonia	66.2	14.8	67.2	59.9	21.0	26.8	7	7	5
2721-172	Kondor X 97.4-4	53.2	25.7	99.1	90.8	14.9	20.7	5	7	9

It should be marked that during the first trial digging (2020), there were no grocery-ware tubers in the samples.

During the 2019 harvesting period there was a high potato grocery-ware tubers in five promising cross-breeds: 1755-55 (Santana × Gala) - 85.5%; 4704-15 – 90.3%; 1871-2 (Krasavitsa × Tiras) – 84.8%; 2652-15 (Malinovka × 93.20-12) – 82.3%; 2520-152 (Udacha × 88.16/20 – 81%).

The largest average weight of 1 tuber was recorded in four investigated genotypes: 4704-15 – 90.3g; 2520-152 – 87.4g; 1871-2-77.4g and 2652-15 – 71.1g.

Increased starchiness of tubers was found in breeding cross-breeds: 2588-122 - 19.4%; 2520-152 - 17%; 2591-1 - 21.0%; 1871 - 2-17.7%, significantly exceeds the standard, dry matter - 25.1%; 22.9%; 26.8% and 23.4%, respectively.

Meteorological conditions in 2019 had the most significant influence on the productivity of cross-breeds, since this year the maximum potato yield was formed. At the same time, among the promising cross-breeds, the highest yield level was established in early ripeness cross-breeds: 1755-55 - 45.4 ton/ha; 2652-15 - 49.5 ton/ha; 2520-152 - 52.6 ton/ha and mid-early - 1871-2 - 42.9 ton/ha; 4704-15 - 49.1 ton/ha, i.e. the excess relative to the standard was 6.8; 10.6; 10.9; 14.0; 12.0 ton/ha.

The yield level of promising potato varieties in 2020 is formed significantly lower due to a lack of moisture, early death of tops and a shift in phenophases during the growing season of plants. Moreover, the negative reaction of genotypes of early and mid-early ripening periods was marked to the greater ripeness. It should be noted that under such extreme weather conditions, the selected cross-breeds turned out to be, nevertheless, the best in terms of productivity and grocery-ware tuber indicators that indirectly indicates a high resistance to bio- and abiostressors of the environment, heat and drought hardness. Among the best indices in terms of productivity there were the same cross-breeds as in 2019: 2652-15; 2520-152; 4704-15; 1755-55. Their yield level was 15.1-30.4 ton/ha.

Table 2. Results of table quality evaluation and culinary type for promising cross-breeds (2020).

Breeding number of the cross-breed	Origin	Indices of tuber table qualities, grade							
		Shell completeness	Flesh density	Friability of mealy	Watery	Flesh darkening after cooking	Smell	Taste	Culinary type
2520-152	Udacha x 88.16/20	7	5	5	7	9	7	7	VS
2652 - 15	Malinovka x 93.20-12	7	7	7	7	7	7	7	C
2588-122	Udacha x 88.16/20	7	5	7	7	7	9	7	C
2591 - 1	Lira x Ausonia	5	3	5	5	3	5	5	V.
2721-172	97.4 -4 x Kondor	3	7	1	3	9	7	5	A.
1755-55	Santana x Gala	3	7	3	5	9	7	7	AV
1871-2	Krasavitsa x Tiras	5	5	5	5	9	7	9	VS
1869-1	Tabor x Tiras	1	7	1	3	9	9	9	A.
4704-15	-	3	5	3	3	7	7	7	A.

High indices of taste and culinary properties of tubers were appeared in two promising cross-breeds: 2520-152 (Udacha x 88.16/20) and 2652-15 (Malinovka x 93.20-12), the second of which was submitted for the State Variety Test in 2020 (Table 2). In addition, promising cross-breeds are distinguished: 1755-55 (Santana x Gala), 1871-2 (Krasavitsa x Tiras) and genotype 4704-15, which after the "rehabilitation" will also be prepared for submission to the State Variety Test of the Russian Federation.

4 Conclusion

As a result of many years of investigation, the distinguished cross-breed 2652-15 that obtained as a result of introgressive breeding of the Malinovka variety with the backcross 93.20-12 under the name Zamir was transferred in 2020 for the State test of the Russian Federation in the North Caucasian region. The Zamir variety, in terms of ripening time, is one of the most demanded in the mid-early group. Suitable for table use and treatment into mashed potatoes. Potential yield is 49.5 ton/ha. A tuber of an attractive rounded oval shape with shallow eyes weighing 85.5-108.4 g has a yellow tuber color and creamy flesh. The tasting assessment of boiled tubers is good (6.0 to 7.0 grades), moderately boiled, resistant to darkening of the flesh in raw type and after boiling. The tubers are resistant to mechanical damage during combine harvesting, good storability is the main feature (90.0-95%).

According to the conclusion of the All-Russian point for testing potato cultivars for resistance to cancer and nematode at the Federal State Budgetary Scientific Institution of the Federal Research Center of potatoes named after A.G. Lorkh, it is resistant to the Dahlem cancer pathotype and potato golden cyst nematode (pathotype $R_0 1$). It is characterized by an average level of late blight resistance in tops and height in tubers. Relatively resistant to rhizoctonia and common scab, banded and wrinkled mosaic of leaves.

References

1. A. Kh. Abazov, H. K. Abidov, A. I. Sarbasheva, G. Kh. Abidova, R. A. Gazheva, Izvestia KBNTs RAN, **6(97)** (2020)
2. K. Z. Budin, A. I. Kuznetsov, I. M. Fomin, N.V. Shaburov, *Production of early potatoes in the Non-Black Earth Region* (1984)
3. E. E. Brown, *Early Potatoes* (1983)
4. S. D. Kiru, L.I. Kostina, E.V. Rogozin, I. A. Chalaya, *Global potato genetic resources and their use in modern breeding trends,* 44 (2012)
5. A.V. Filipov, *Supplement to the j. Plant Protection and Quarantine*, 5(27) (2012)
6. S. D. Kiru, *Scientific support and innovative development of potato growing,* 49 (2008)
7. E. A. Simakov, N. P. Sklyarova, I. M. Yashina, *Guidelines for the technology of the potato breeding process* (2006)
8. E. I. Koshkin, *Crop Pathophysiology* (2016)
9. A. N. Berezkin, A. M. Malko, E. L. Minin, V. N. Lapochkin, M. Yu. Cherednichenko, *Regulatory framework for breeding and seed production* (2019)

E3S Web of Conferences **262**, 01035 (2021) https://doi.org/10.1051/e3sconf/202126201035
ITEEA 2021

The calculation basis for a four-component jet mixer for fertilizer and water

Sergey Andreyevich Tarasyants[1,*], *Vladimir Alexandrovich* Rudakov[1], *Yulia Sergeyevna* Urzhumova[1], and *Karina Alexandrovna* Degtiareva[2]

[1]Novocherkassk Engineering and Meliorative Institute named after A.K. Kortunov – Don State University Branch, Novocherkassk, Russian Federation
[2]South Russian State Polytechnic University (NPI) named after M.I. Platonov, Novocherkassk, Russian Federation

Abstract. This article covers calculation basis for a four-component jet mixer for fertilizer and water. The calculation for a four-component jet mixer for fertilizer consists in determining the basic geometric and hydraulic parameters of a four-component jet mixer for fertilizer and water and supplying the mixture to irrigated areas, with fertilizing irrigation, which will allow supplying all types of fertilizers and changing the concentration of mixture in the required proportions. All geometric relative parameters are calculated on the basis of the need to scale the mixer dimensions and are taken as the ratio of the diameters of the mixer elements to the cylindrical part of the mixing chamber. The above calculation bases determine the main parameter of the mixer necessary for determining all the elements – the cross-sectional area of the annular nozzle. It was founded that when calculating the geometric characteristic, you should be guided by the desire to accept the value of the geometric characteristic m=2.5 ÷ 8.0, corresponding to the maximum value of the efficiency up to 40% and higher.

1 Introduction

The most common methods of mixing fertilizers and water, during fertilizing irrigation, are mixing tanks, injectors for applying fertilizers, devices mounted on a hydrant of an irrigated area, tanks for preparing mother solution, trailed spreaders, etc.

The analysis of devices in mixing systems shows that the above devices are complicated in design and inconvenient in operation. This publication covers calculation basis for a four-component jet mixer for fertilizer and water (figure 1, 2) [1].

Calculation of a four-component jet mixer for fertilizer consists of determining the optimal relative geometric and hydraulic parameters of the mixer for the possibility of designing and manufacturing its elements when applying fertilizers to a given area in case of fertirrigation.

* Corresponding author: starasyancz@mail.ru

https://doi.org/10.1051/e3sconf/202126201035

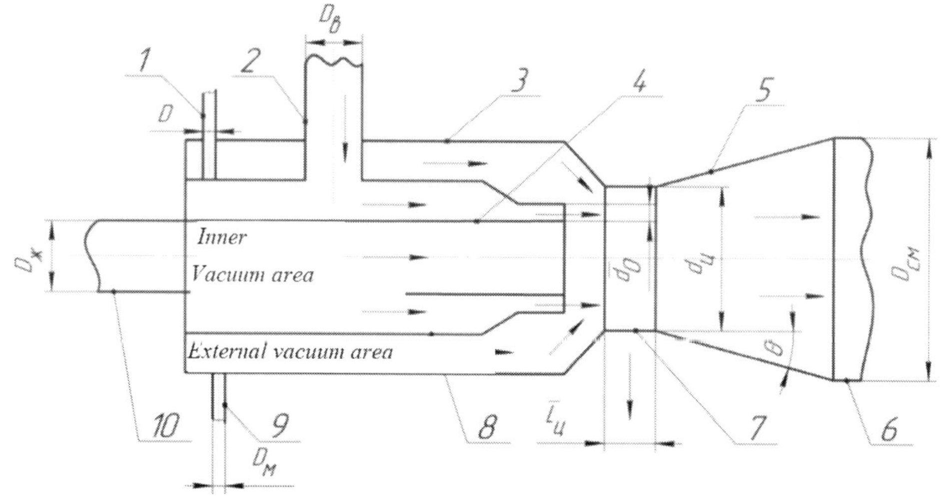

Fig. 1. Structural diagram of a four-component jet mixer for fertilizer.

Were: 1 - mineral fertilizers supply pipeline; 2 - working water supply pipeline; 3 - outside case of the mixer; 4 - inner adapter of the annular nozzle; 5 - diffuser; 6 - pipeline for discharging a mixture of fertilizers and water; 7 - mixing chamber; 8 - outer adapter of the annular nozzle; 9 - microelements supply pipeline; 10 - pipeline for supplying livestock waste.

Fig. 2. Four-component jet mixer Overall view.

2 Materials and Methods

The geometric parameters of the mixer include:

– the relative outer $\overline{d}_0^{/}$ and inner diameters $\overline{d}_0^{//}$ of the annular nozzle adapters;

– the relative distance from the nozzle edge to the beginning of the cylindrical part of the mixing chamber $\overline{Z}$;

– the actual diameter of the mixing chamber d_{ts};

– the relative length of the mixing chamber $\overline{L}_u$;

– the actual opening angle of the diffuser θ;

– diameters of supplying water pipelines D_w, mineral fertilizers D_m, livestock sewage D_l, a mixture of fertilizers and water D_{fw};

– *m* is the geometric characteristic of the mixer (the ratio of the cross-sectional area of the mixing chamber to the cross-sectional area of the annular nozzle end).

All geometric relative parameters are calculated on the basis of the need to scale the mixer dimensions and are taken as the ratio of the diameters of the calculated mixer elements to the cylindrical part of the mixing chamber.

Hydraulic parameters include:

– pump head of the water blower $H_{w.bl}$;

– flow rate in the nozzle V_0;

– kinetic energy in the nozzle $V_0^2/2g$;

– Flow rate in the mixing chamber V_c;

– kinetic energy in the mixing chamber $K_c = \frac{V_c^2}{2g}$;

– total energy in the mixing chamber $\frac{p_c}{g\rho_c}$;

– the amount of vacuum (pressure) in the mixing chamber $\frac{p_в}{g\rho_c}\left(\frac{p_c}{g\rho_c}\right)$;

– water flow consumption Q_0 in the nozzle;

– sucked fertilizer consumption in the nozzle Q_1;

– mixing ratio $\alpha_0 = \frac{Q_1}{Q_0}$.

3 Results and Discussion

The main parameter affecting the mixer operation is the cross-sectional area of the annular nozzle $\omega_0 = 0{,}785\left[d_0^{/2} - d_0^{//2}\right]$, which has the main effect on the outgoing flow rate of the working jet (water) (Figure 3).

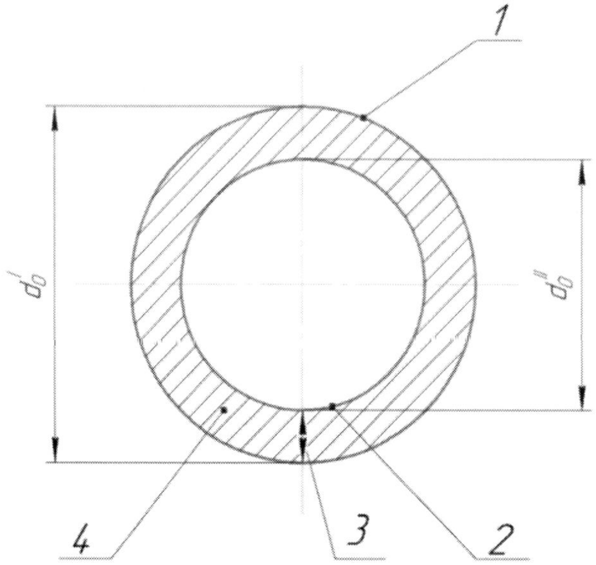

Fig. 3. Cross-sectional diagram of the nozzle end.

Were: 1 - inner diameter of the outer nozzle; 2 - outer diameter of the inner nozzle; 3 - the width of the slot h in the nozzle; 4 - outlet area of the working stream (water)

The main parameter, the kinetic energy in the mixing chamber, is determined according to the annular nozzle outlet flow rate V_0. The main principle for calculating the diameters $d_0^/$ and $d_0^{//}$ is the assignment of the slot width in the nozzle and the possibility of the presence of such a working flow rate that can create the vacuum necessary for fertilizers suction. The diameter of the outer adapter $d_0^/$ is assigned in the amount of 0.9–0.95 the diameter of the mixing chamber d_{ts} [2]. The outer diameter of the inner adapter $d_0^{//}$ (Figure 4) influences a lot – the width of the slot h, the flow rate in the nozzle V_0, the working flow supply Q_0, the flow rate in the mixing chamber V_c, the mixing ratio α_0. The inner adapter outer diameter of the annular nozzle $d_0^{//}$ plays a key role in the operation of the entire jet mixer. The required calculation is given below.

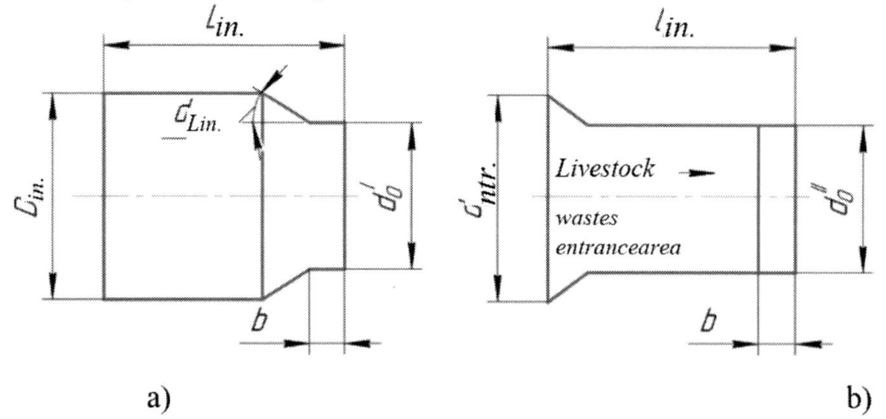

Fig. 4. Diagram of the external and internal adapters of the annular nozzle.

Were: a - external annular nozzle; b - internal annular nozzle

– the width of the slot h, if possible, is taken as the maximum and is determined by hydraulic calculations;

– the end length of the cylindrical sharpening is taken in the amount of 0.1–0.15 diameter and plays a passive role in the mixer operation;

– the length l_{int} is determined taking into account the possibility of inserting the fertilizer supply pipelines into the mixer case; it is taken as short as possible to reduce the pressure loss at the inlet ς_{in} of livestock sewage and mineral fertilizers;

– The inlet diameter d_{in} to the inner surface of the adapter is taken as the maximum possible based on the diameter of the external case of the mixer.

The relative distance from the edge of the annular nozzle to the beginning of the cylindrical part of the mixing chamber $\overline{Z}$ (Figure 5) is taken from the literature data as 0.1 ÷ 0.2d_{ts} [3].

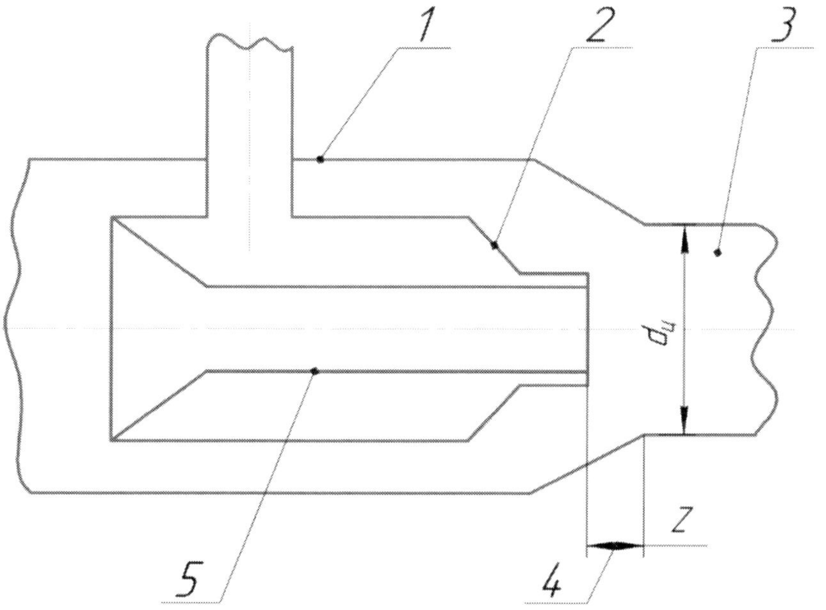

Fig. 5. Scheme of binding the mixing chamber to the edge of the annular nozzle.

Were: 1 - case; 2 - outer adapter of the annular nozzle; 3 - mixing chamber; 4 - distance from the nozzle edge to the beginning of the cylindrical part of the mixing chamber; 5 - inner adapter of the annular nozzle

In fact, the diameter of the mixing chamber d_{ts} also depends on many factors, and basically all other diameters of the mixing chamber in their final form are bound to it. The flows are mixed in the mixing chamber, the velocity diagram of the working and suction jet is leveled, according to the energy losses in the mixing chamber, the efficiency of the entire mixer is determined and the amount of vacuum required for fertilizers suction is determined. The calculation of the mixing chamber diameter is mainly based on the condition of the presence of such a value of kinetic energy in it, which is capable of creating the maximum value of negative pressure. The length of the mixing chamber determines the main energy losses in the mixer due to the presence of the maximum speed of the two flows. The design of the proposed annular two-surface four-component mixer [4] offers the shortest possible length of the mixing chamber due to the presence of a concave velocity curve diagram [5] (Figure 6).

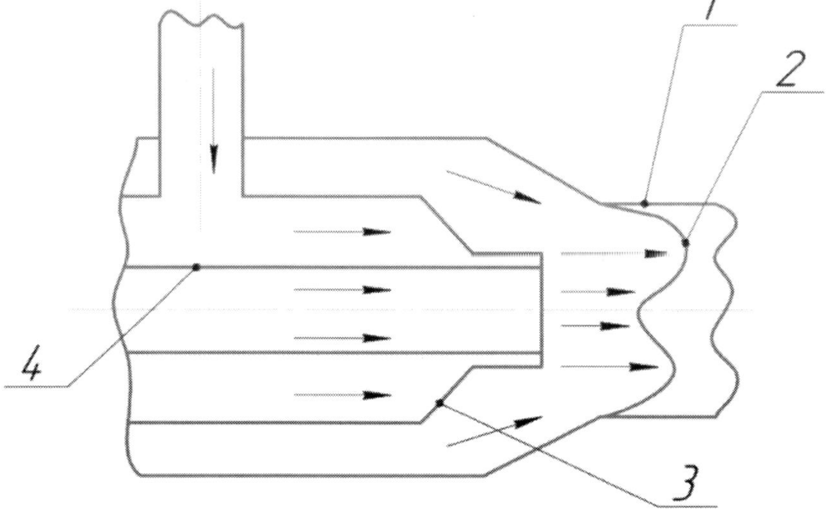

Fig. 6. Schematic diagram of the velocity curve diagram in the mixing chamber of the mixer.

Were: 1 - mixing chamber; 2 - concave velocity curve diagram in the mixing chamber; 3 - outer adapter; 4 - inner adapter

When determining the expansion angle value of the diffuser, you should proceed from two conditions:

– the requirement of reducing the diffuser length (reduce the length of the entire installation);

– reducing the coefficient of hydraulic resistance, reducing losses and, accordingly, increasing the efficiency of the mixer.

The above mentioned two conditions depend on one factor - the expansion angle, which, based on the literature data [6], corresponds to 6–10° for the smallest pressure losses.

When determining the diameters of pipelines, you should be guided by the known value of the relative flow rate in pressure pipelines, corresponding to 2–2.5 m/s [6].

When calculating the geometric characteristics $m = \frac{\omega_ч}{\omega_0}$ (the ratio of the cross-sectional area of the mixing chamber to the cross-sectional area of the annular nozzle), you should be guided by the desire to accept such a value that contributes to an increase in the efficiency of the mixer. Optimal values of the geometric characteristic *m* for jet mixers are taken in the range of 2.5 ÷ 8 and depends on the jet unit pattern of use. For fertilizer mixers, the optimal value "*m*" corresponds to values of 4.0-5.0 [7].

The above mentioned hydraulic parameters of the mixer are directly related to its geometric parameters. The main defining condition for the mixer operation is the value of the created reduced header pressure $H_{g.pr}$, the maximum value of which for jet mixers was determined earlier and corresponds to the value of 20–25 m. (This header pressure is created by the device of this type and was previously determined with all optimal geometric and hydraulic parameters).

When calculating the mixer header pressure, it is necessary to keep in mind the fact that the header pressure of the mixer shall exceed the header pressure in the network by at least 3-5 m. Before the theoretical calculation of the required header pressure in the mixer, it is necessary to calculate the required header pressure of the network $H_{g.pr}$, determined by the dependence

$$H_{г.пр} = H_г + \sum h_{wc} \quad (1),$$

where H_g is the geometric height of the mixture supply by the mixer;

$\sum h_{wc}$ – pressure losses in the system.

It is required to have in mind that the value of $H_{g.pr}$ depends on the technology used for irrigation.

The above-described dependence is suitable for use in case of using drip irrigation under the condition that the value of the header pressure loss Σh_w, the required pressure for the operation of drip lines is ≈20–25 m.

When using other types of irrigation equipment, it is necessary to increase the design header pressure by the amount of free header pressure, providing for the operation of the machine used. For all hydraulic parameters of the mixer, an important role is played by the working flow rate in the nozzle V_0, which depends on the pressure of the working water flow generated by different types of sources (centrifugal nozzles). The header pressure flow rate in the nozzle is determined by the known dependence [8]

$$V_0 = \varphi\sqrt{2gH_{н.пр}} \quad (2),$$

where φ is the rate coefficient (taken in this case 0.8);

$H_{h.pr}$ is given header pressure of a centrifugal pump (working header pressure).

The V_0 value plays the main role in the creation of kinetic energy in the nozzle $\frac{V_0^2}{2g}$.

When the rate V_0 is increased to the maximum, the kinetic energy also increases to the maximum (limited by cavitation phenomena in the nozzle) and the total energy $\frac{V_0^2}{2g} + \frac{p_0}{g\rho}$, having passed the zero threshold, the cross section 0-0 (Figure 7) becomes negative, which contributes to the creation of vacuum either in the nozzle or in the mixing chamber.

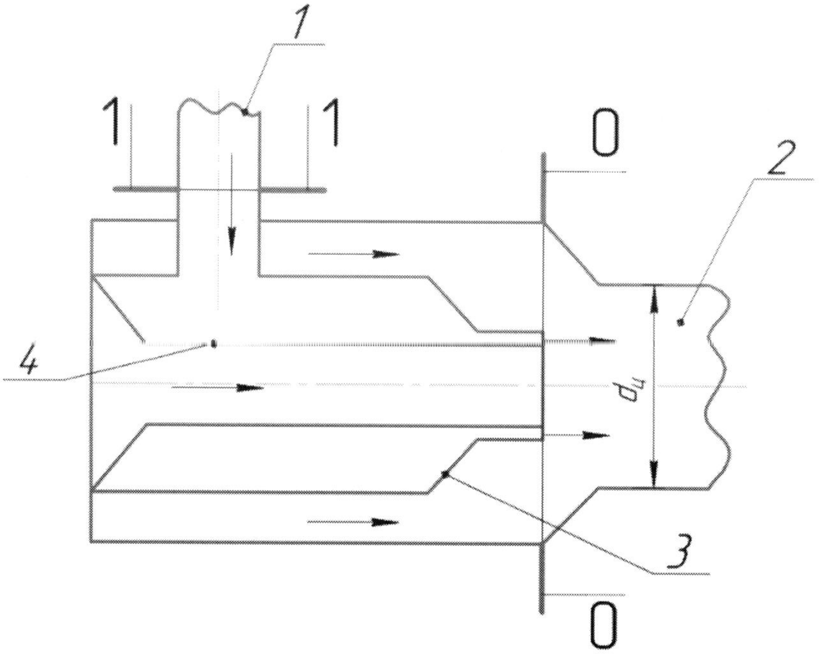

Fig. 7. Scheme of the working flow in the annular mixer.

Were: 1 - working water supply pipeline; 2 - mixing chamber; 3 - outer adapter; 4 - inner adapter.

The total energy in the cross-section 0-0 is determined by drawing up the D. Bernoulli equation for the cross-sections "1-1" and "0-0". For cross-section 1-1, the total energy is calculated according to the dependence $\frac{p_1}{g\rho_0}+\frac{V_1^2}{2g}$. As a rule, this energy is positive, because kinetic energy is $\frac{V_1^2}{2g}$ positive and insignificant (0.2–0.3 m) due to low speed (2–2.5 m/s) and high pressure $\frac{p_1}{g\rho_0}$ (30–40–50 m), depending on the parameters of the centrifugal pump. When the jet enters the nozzle, the rate is low V and turns into the higher rate V_0 (up to 30–40 m/s). In this case, the kinetic energy increases to values of 45.8–82 m and changes in the "0-0" $\frac{p_0}{g\rho_0}+\frac{V_0^2}{2g}$ cross-section: the value will $\frac{p_0}{g\rho_0}$ sharply fall, and the value $\frac{V_0^2}{2g}$ sharply increases, but its total value will remain the same, equal $\frac{p_1}{g\rho_0}+\frac{V_1^2}{2g}$, minus the header pressure losses in the portion from the cross-section 1-1 to cross-section 0-0, determined by the dependence $\varsigma_0\frac{V_0^2}{2g}$, where ς_0 is the coefficient of hydraulic resistance of the nozzle (as a rule, this value for nozzles of this kind is equal to 0.06–0.10 [9]). When considering these cross-sections according to the D. Bernoulli equation

$$\frac{p_1}{g\rho_0}+\frac{V_1^2}{2g}=\frac{p_0}{g\rho_0}+\frac{V_0^2}{2g}+\varsigma_0\frac{V_0^2}{2g} \tag{3}$$

the potential energy in the nozzle is determined $\frac{p_0}{g\rho_0}$

$$\frac{p_0}{g\rho_0}=\frac{p_1}{g\rho_0}+\frac{V_1^2}{2g}-\frac{V_0^2}{2g}-\varsigma_0\frac{V_0^2}{2g} \tag{4}$$

And as a result of the increased value $\frac{V_0^2}{2g}$, the potential energy $\frac{p_0}{g\rho_0}$ will obviously become negative, the maximum value of which reaches 9.5 m.

The value of such a vacuum is considered deep, which differs from the value of the vacuum in the suction pipelines of centrifugal pumps by almost 2–2.5 times. This calculation proves the superiority of the jet mixer suction capacity in comparison with the suction capacity of all types of pumping equipment.

When the flow enters the mixing chamber, the rate and kinetic energy decrease due to a change in the header pressure rate in the mixed flow. But it is required to have in mind that complete mixing of the working and suction flows occurs not at the beginning, but at the end of the mixing chamber. In jet mixers with a central supply, the velocity curve is elongated (Figure 8) and for complete mixing of flows, in such mixers the length of the mixing chamber, in contrast to the length of the mixing chamber of annular mixers (1.5–2 chamber diameters), is assumed to be 6–7 diameters, which when increased velocities play a significant role in both the total header pressure loss and the efficiency of the mixer.

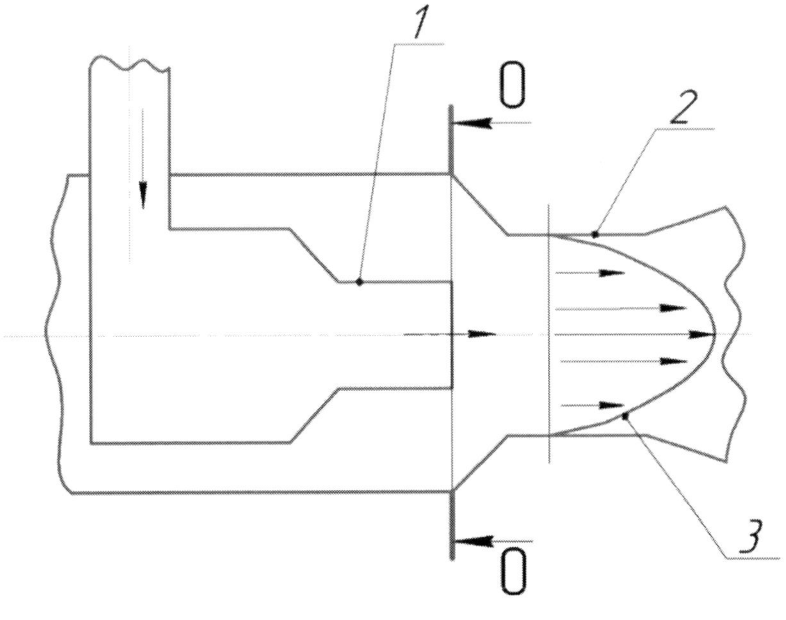

Fig. 8. Velocity curve in the mixing chamber of a mixer with a central inlet.

Were: 1 - nozzle; 2 - mixing chamber; 3 - velocity curve in the mixing chamber.

The mixing chamber in the existing designs of mixers begins at the nozzle end, in the cross-section 0-0, in which two flows are actually transported: calculated and sucked in, with different values of kinetic and potential energy. The kinetic energy of the working flow, as mentioned above, has a value that exceeds the kinetic energy of the sucked flow by many times due to the high difference in velocities and to determine the boundary of the transition of high vacuum to medium (vacuum in a mixed flow) in the mixing chamber, it is necessary to carry out special investigations, not carried out so far. According to the literature data, experiments [10, 11] show that the kinetic energy from the zero cross-section to the average length of the mixing chamber remains high, and its value differs from the energy in the 0-0 cross-section by 1-1.5%, which indicates a high suction capacity mixer and high vacuum in the suction lines. The rate $V_{m.c}$ and the rate header pressure in the mixing chamber $\frac{V_{\kappa.c}^2}{2g}$ play an important role in determining the value of the mixing coefficient of the two flows α_0. At a low value of α_0 (0.1, 0.2), there is practically no suction flow, and the velocity in the mixer is the ratio of the working flow rate to the cross-sectional area of the mixing chamber

$$V_c = \frac{Q_0}{\omega_c} \qquad (5).$$

With an increase in the coefficient α_0 to the value of 1.0; 2.0; 3.0; 4.0, etc. the sucked flow rate increases and the velocity in the mixing chamber is determined not by the ratio of the working flow rates Q_0, but by the ratio of the total flow rate Q_1+Q_0 to the area ω_s.

$$V_c = \frac{Q_1 + Q_0}{\omega_c} \qquad (6).$$

Accordingly, with an increase in the flow rate in the mixing chamber, the amount of kinetic energy will also change $\frac{V_c^2}{2g}$. At some values, α_0 can become equal to the value of the kinetic energy in the nozzle $\frac{V_0^2}{2g}$, in this case the values of the potential energy and vacuum in the mixing chamber $\frac{p_c}{g\rho_0}$ can become identical with the value of the potential energy in the nozzle $\frac{p_0}{g\rho_0}$.

When calculating the vacuum values in the suction pipelines, it is required to have in mind that if it is necessary to have a high suction header pressure in the mixer (suction vacuum), the mixer header pressure $H_{g.pr}$ should be calculated in such a way that the value of α_0 does not decrease below 1.0.

On the basis of the above mentioned recommendations, the possibility of determining the basic geometric and hydraulic parameters of a four-component jet mixer of fertilizers and water and supplying the mixture to irrigated areas during fertilizing irrigation has been developed, which will allow supplying all types of fertilizers and changing the mixture concentration in the required proportions.

4 Conclusions

The developed calculation bases are recommended for the theoretical substantiation of the geometric and hydraulic parameters of a four-component jet mixer of fertilizers and water during fertilizing irrigation. Recommendations for calculation are given for relative dimensions in case it is necessary to scale mixers for different options of their use. As the main parameter, the geometric characteristic of the mixer "m" shall be taken, in the range of 2.5 ÷ 8.0 corresponding to the maximum efficiency of the mixer up to 40%.

Reference

1. Pat. No. 193355. Russian Federation, MPK A01 C23/04, Four-component jet pump mixer / V.A. Rudakov, K.A. Degtiareva, Yu.S. Urzhumova, S.A. Tarasyants; applicant and patentee State Federal-Funded Educational Institution of Higher Professional Training "Don State Agrarian University" - No. 2019111839; declared 18/04/2019; published on 25/10/2019, Bull. No. 12
2. V.E. Fridman, *Hydroelevators*, 142 (1990)
3. Kh.Sh. Mustafin, Scientific and technical communication, **10**, 28 (1963)
4. Kh.Sh. Mustafin, *Calculation of the ejector on water and slurry: a collection of works*, 124 (1968)
5. Yu.L. Kirillovsky, *Energy balance and calculation of water-jet devices*, 23 (1957)
6. A.D. Altshul, Hydraulic resistance, **2**, 224 (1982)
7. S.P. Ogorodnikov, *Suction dredger injection*, 48 (1962)
8. P.G. Kiselev and others, *Guidance of hydraulic calculations*, 352 (1972)
9. N.A. Rzhanitsyn, Water jet pumps, (1988)
10. N.A. Rzhanitsyn, Water jet pumps, 176 (1980)
11. S.A. Tarasyants, *Environmentally sound technology and technical means for the preparation of livestock sewage for irrigation of forage crops*, 44 (1995)

Forecasting the impact on the environment of generating sources in agricultural production

E. Timofeev, and *A.* Erk

Federal State Budgetary Scientific Institution 'Federal Scientific Agroengineering Center VIM', branch in Saint Petersburg, Filtrovskoje highway, 3 p.o. Tiarlevo, 196625 Saint Petersburg, Russia

Abstract. The development of rural electrification is aimed at distributed energy, i.e. availability of autonomous sources of electricity and heat generation. Generation sources can use coal, fuel oil, gas, local and alternative energy sources as fuel and energy resources. This causes additional emissions of pollutants. Prediction of negative impact on the environment depends on the quantity and quality of emissions during the operation of various types of installations that generate electric and thermal energy. The purpose of the study is to select and substantiate the most attractive method for predicting the impact on the environment of generating sources in agricultural production. The widespread introduction of distributed energy using local and renewable energy sources will significantly reduce emissions of pollutants due to a decrease in energy intensity, matching the required load with the capacity of generation sources, using the most energy efficient sources, and widespread introduction of renewable energy sources. In this regard, it is advisable to forecast the impacts by the scenario method considering the implementation of the proposed distributed power supply system. The implementation of measures developed considering the analysis of the forecast of the negative impact on the environment of generating plants in agricultural enterprises will reduce the negative impact on the environment by introducing energy-efficient technologies into the energy balance of the enterprise, as well as increase production by up to 20% and increase sustainability. rural areas.

1 Introduction

The structure of consumption of fuel and energy resources (FER) in agricultural production of the Leningrad region of the Russian Federation is shown in Figures 1, 2, 3 [1,2].

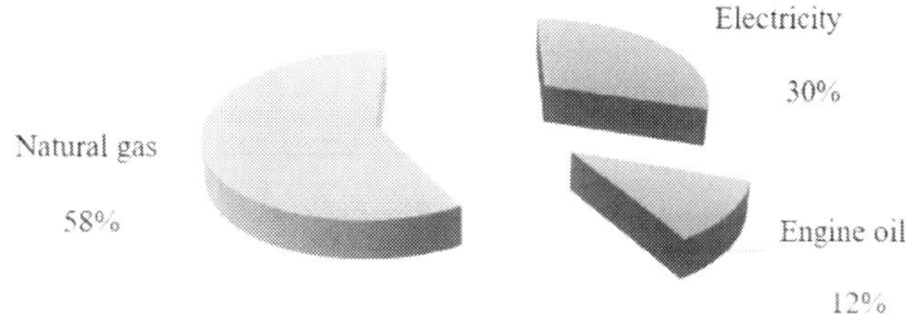

Fig. 1. Energy consumption by poultry enterprises.

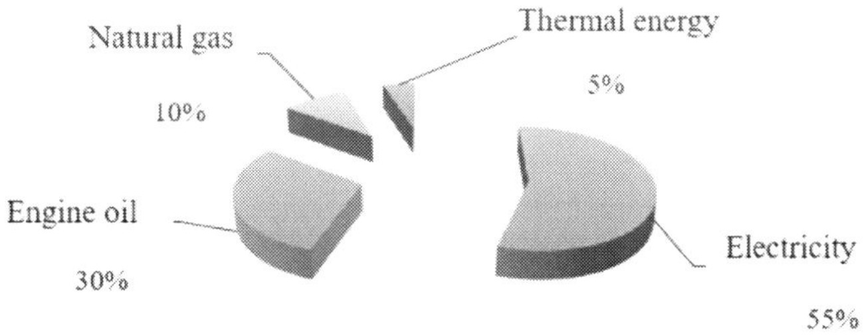

Fig. 2. Energy consumption by pig farms.

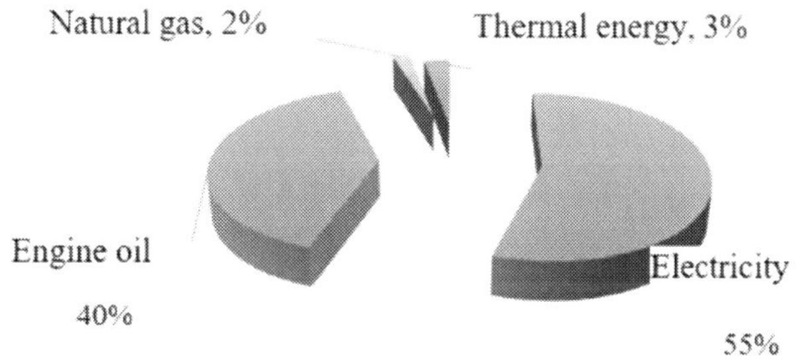

Fig. 3. Energy consumption of dairy livestock enterprises.

Electricity consumption ranges from 30 to 55%. Currently, the centralized power supply system is mainly used. However, the development of rural electrification is aimed at distributed energy [3,4,5], i.e. availability of autonomous sources of electricity and heat generation. Generation sources can use coal, fuel oil, gas, local and alternative energy sources as fuel and energy resources. This causes additional emissions of pollutants.

Prediction of negative impact on the environment depends on the quantity and quality of emissions during the operation of installations of various types that generate electrical and thermal energy [6,7].

The aim of the study is: selection and justification of the most attractive method for predicting the negative impact on the environment of generating sources in agricultural production.

2 Materials and methods

The analysis is based on the results of energy audits of farms in the Leningrad region, conducted by the Institute from 2011 to the present. Also, express surveys of five farms, conducted by the survey method in 1Q 2020.

Of particular interest in the express survey was the question of the availability of autonomous sources of generation using both traditional and renewable energy sources. Table 1 presents the initial data for calculating the negative impact on the environment.

Table 1. Initial data of indicators for farms for predicting the impact

Name of indicators	Years			
	2014	2015	2017	2018
Household 1				
Production value, RUR	101594	139903	133615	182388
FER consumption, c.f.t.	1679	1711	1652	1819
Household 2				
Production value, RUR	241462	250974	315364	336694
FER consumption, c.f.t.	2830.8	2666.7	2607.1	2544.6
Household 3				
Production value, RUR	93811	98195	114325	147337
FER consumption, c.f.t.	1,289.4	1187.7	1284.3	1194.5
Household 4				
Production value, RUR	52531	64984	86663	106370
FER consumption, c.f.t.	1147.7	1246	1503.8	1472
Household 5				
Production value, RUR	52531	64984	86663	106370
FER consumption, c.f.t.	1147.7	1246	1503.8	1472

3 Results and discussion

Forecasting the negative impact on the environment from the operation of sources of electricity and heat generation in agricultural production.

Autonomous generation sources: boilers running on gas, wood, solid or liquid fuels, etc., are sources of environmental pollution due to emissions of combustion products. In addition, the operation of heat generators in drying plants is also accompanied by emissions.

Table 2. Levels of emissions of pollutants into the atmosphere when burning various types of fuel

Fuel type	Emissions of pollutants into the air, t per 1,000 t of nat. fuel				
	CO_2	NO_2	SO_2	Solid particles (inorganic dust)	TOTAL
Natural gas	1.18	3.52	0.00	0.00	4.70
Bituminous coal	9.58	63.56	9.20	65.32	147.66
Fuel oil	5.20	5.20	35.30	0.30	45.90
Peat briquette	8.04	26.81	3.00	13.02	50.87
Wood briquettes, pellets	4.68	9.31	0.28	4.11	17.69
Firewood	4.9	9.4	0.3	4.3	18.9
Sawdust	5.0	9.6	0.5	5.0	20.0
Waste wood, scraps	5.2	9.9	0.4	5.2	20.7
Fast-growing wood	4.8	9.5	0.0	8.4	22.7
Spill, boughs, bark	5.6	11.4	0.8	13.4	31.3

Table 2 shows that wood fuel (primarily pellets and briquettes) after natural gas is more preferable, from the point of view of air pollution, in comparison with fuel oil (especially with coal), since it has practically "zero effect" on greenhouse gas emissions. gases, primarily CO2.

The results of energy surveys showed that at present, boiler houses for the generation of heat and electricity play an insignificant role in energy supply systems. Thus, the surveyed farms had and constantly applied: one boiler house on coal, two on fuel oil, eight on natural gas, four boilers on wood. On the other hand, almost all farms had 1-4 diesel generator sets to provide the first category of power supply to livestock farms during the period of a centralized power outage. The frequency of their activation was from 1 to ten times a year for a period of up to two days. In addition, the heat generators of the dryers operated during the harvesting period (two, three weeks) on liquid heating oil - fuel oil [8,9,10].

In this regard, at present, forecasting the negative impact on the environment from the operation of sources of electricity and heat generation in agricultural production is difficult due to insufficient information.

The direction of development of rural energy is the widespread introduction of distributed energy using local and renewable energy sources. This will significantly reduce emissions of pollutants due to a decrease in energy intensity, matching the required load with the capacity of generation sources, using the most energy efficient sources, and widespread introduction of renewable energy sources [11,12,13].

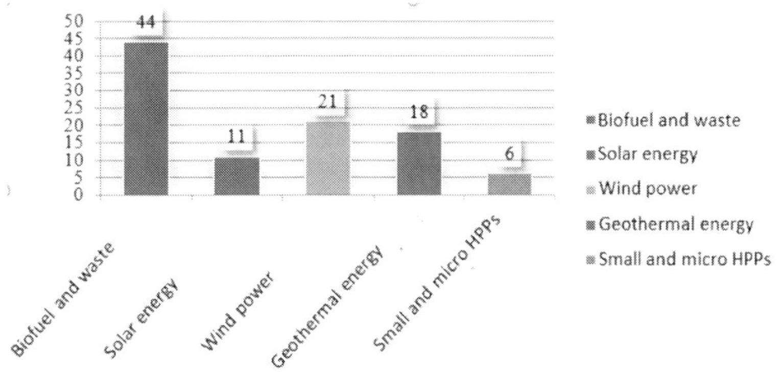

Fig. 4. The share of types of renewable energy sources in the distributed energy system.

In this regard, it is advisable to forecast the impacts by the scenario method considering the implementation of the proposed distributed power supply system.

The share of renewable energy sources in the agro-industrial complex is increasing every year and, pursuant to the forecast, by 2030, it will reach 25% in distributed energy networks [11]. Accordingly, emissions characterized by CO2 equivalent will be reduced by 21%.

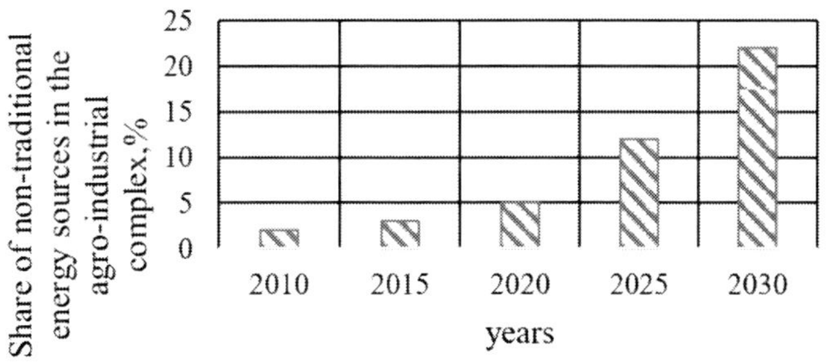

Fig. 5. Forecast of the development of the share of unconventional sources in the agro-industrial complex until 2030.

As can be seen from the figure, the share of unconventional energy sources in the agro-industrial complex is growing rapidly and may reach 22% by 2030.

4 Conclusion

1. The widespread introduction of distributed energy using local and renewable energy sources will significantly reduce emissions of pollutants due to a decrease in energy intensity, matching the required load with the capacity of generation sources, using the most energy efficient sources, and widespread introduction of renewable energy sources. In this regard, it is advisable to forecast the impacts by the scenario method considering the implementation of the proposed distributed power supply system.

2. The implementation of measures developed considering the analysis of the forecast of the negative impact on the environment of generating plants in agricultural enterprises will reduce the negative impact on the environment by introducing energy-efficient technologies into the energy balance of the enterprise, as well as increase production by up to 20% and increase sustainability. rural areas.

Testimonial.

A part of this work was performed within the framework of KS1675 project "Russian-Finnish Bioeconomy Competence Centre – BioCom" of the "South-East Finland – Russia Cross-Border Cooperation (CBC) Programme 2014–2020".

References

1. V. N. Brovtsin, A. F. Erk, O. V. Bychkova, Theoretical and scientific-practical journal "Mechanization and electrification of agriculture", **5,** 22 (2014)
2. V. N. Sudachenko, A. F. Erk, E. V. Timofeev, Selection of energy supply option for agricultural production facilities pursuant to economic criteria, technologies and technical means of mechanized production of crop and livestock products, **92,** 43 (2017)
3. A. V. Tikhomirov, M.: VIESH Bulletin., **1(22),** 11 (2016)
4. A. V. Tikhomirov, I. I. Sventitsky, E. K. Markelova, V. Yu. Ukhanova, *Energy strategy of agriculture in Russia for the period up to 2030,* 75 (2015)
5. D. A. Tikhomirov, A. V. Tikhomirov, Agrotechnics and energy supply, **1(18),** 47 (2018)
6. A. Shahsavari, M. Akbari, Renewable & Sustainable Energy Reviews, **90,** 275 (2018)
7. P. MacNaughton, X. Cao, J. Buonocore, J. Cedeno–Laurent, J. Spengler, A. Bernstein, J. Allen, Journal Of Exposure Science And Environmental Epidemiology, **28(4),** 307 (2018)
8. A. F. Erk, V. N. Sudachenko, E. V. Timofeev, V. A. Razmuk, Technologies and technical means of mechanized production of crop and livestock products, **89,** 19 (2016)
9. A. F. Erk, V. N. Sudachenko, E. V. Timofeev, V. A. Razmuk, Technologies and technical means of mechanized production of crop and livestock products, **1(94),** 63 (2018)
10. A. F. Erk, V. N. Sudachenko, E. V. Timofeev, V. A. Razmuk, Technologies and technical means of mechanized production of crop and livestock products, **89,** 23 (2016)
11. E.V. Shalavina, E.V. Vasilyev, Bulletin of APK Stavropol, **1(17),** 366 (2015)

12. C. Carnevale, F. Ferrari, G. Guariso, G. Maffeis, E. Turrini, M. Volta, Assessing the Economic and Environmental Sustainability of a Regional Air Quality Plan. Sustainability, **10(10)** (2018)
13. M. N. Naz, N. Naeem, M. Iqbal, Journal of Renewable and Sustainable Energy, **9(1),** (2017)
14. D. A. Tikhomirov, A.V. Tikhomirov, Electrotechnology and electrical equipment in the agro-industrial complex, **3(36),** 3 (2019)
15. S. Ardebili, A. Khademalrasoul, Journal of Cleaner Production, **204,** 819 (2018)

Study of the efficiency of drying grass seeds for forage crops on a rotary dryer

V.A. Smelik*, *A.N.* Perekopskiy, *A.V.* Dobrinov, and *S.V.* Chugunov

Federal Scientific Agroengineering Center VIM, branch in Saint Petersburg, Saint Petersburg, Russia

Abstract. In many countries of the world with high climate humidity, grass seeds after their harvesting cannot be preserved without artificial drying. The artificial drying process is associated with high energy costs, and its regimes should not reduce the sowing quality of seeds. Therefore, it is of particular importance to determine the most suitable drying regimes for seeds. For this purpose, for the conditions of seed farms in the North-West of the Russian Federation, experimental studies were performed to obtain information on the parameters and modes of drying seeds of perennial grasses for their various initial states and conditions. Experimental studies were performed on a rotary dryer during the drying of meadow fescue seeds. The experimental data were analyzed in Microsoft Excel 2010 using the methods of mathematical statistics. The obtained mathematical dependencies made it possible to establish that the effective modes of drying the seeds of the studied culture are provided during their layer-by-layer unloading from the drying chamber. Moreover, the lower layer of seeds should be unloaded every 2.5 hours of drying, while adding to the upper layer of the drying chamber a new volume of seeds equal to the unloaded one.

1 Introduction

Perennial grasses form the basis of animal feed in many countries around the world. The quality and condition of perennial herbage directly depend on the quality of the seed. To maintain the productivity of perennial herbage, agricultural enterprises should be provided with their own seed material, adapted in terms of biological characteristics to local cultivation conditions. The use of zoned selection seeds for sowing makes it possible to increase the yield of grasses by up to 10 kg/ha [1, 2, 3].

The experience of importing grass seeds from various regions of the country, including the EU, has shown their inefficiency for seed production. The sown areas in the North-West zone of the Russian Federation, sown with grasses that were brought from a number of European countries, as well as from the south of our country, did not provide the required productivity, while the death of certain types of grasses was quite often observed.

To meet the needs of livestock enterprises in forage from perennial grasses, it is necessary to organize their seed production, considering the biological characteristics of the cultivation conditions and in a sufficient quantity for consumers. The issues of post-harvest

*Corresponding author: smelik_va@mail.ru

processing of seeds before storing them for storage are also important, especially for regions of increased moisture.

In North-West Russia, seeds of high-moisture herbs cannot be preserved without artificial drying. Therefore, it is of particular importance to determine the most suitable drying regimes for seeds.

Drying is an integral part of the process of producing grass seeds in various countries of the world [4, 5, 6]. Due to bad weather conditions, almost all grass seeds in the Nordic countries should be dried. Therefore, many research projects of leading research centers and manufacturing companies are devoted to solving various problems associated with the drying processes of various agricultural crops, including grass seeds [7, 8, 9].

In the whole technology of grass seed production, the most important and energy-consuming, is the process of drying them, associated with the process of removing excess moisture at the optimum permissible temperature conditions. Sufficiently studied modes of drying grain crops cannot be mechanically used for drying grass seeds because of the significant difference in their physical and mechanical properties. Small grass seeds during convective drying, due to the lower duty cycle, create great resistance to the movement of the coolant. And the aerodynamic properties of grass seeds do not allow the coolant to be supplied at a higher speed in order to avoid losses. These and other features complicate and complicate the drying process [3, 10-12].

Today, dryers for seeds of perennial grasses of the following types are widely used: conveyor, rotary, floor and drum ones. When using floor dryers, labor costs increase sharply, since they are not sufficiently mechanized. In turn, drum dryers are less demanding on the cleanliness of the initial heap, but at the same time, due to the complexity of the design, their production was closed.

In connection with these circumstances, for the most part, at agricultural seed-growing enterprises for drying grass seeds, dryers of two types are purchased: conveyor and carousel [3, 4]. They are versatile, relatively simple in design and operation, and allow mechanization of all basic processes. However, up to now there are still questions on the choice and substantiation of the most effective parameters and modes of drying grass seeds in these dryers.

In this regard, the purpose of this study was to conduct experimental studies and, on the basis of the information obtained, establish effective modes of drying seeds of perennial grasses on a rotary dryer in the conditions of seed farms in the North-West of Russia, characterized by unfavorable harvesting conditions in terms of humidity.

2 Materials and methods

The studies were performed in the seed-growing enterprises of the Leningrad region, JSC Volkhovskoe, Volkhovsky district, and LLC Novoladozhsky, on a heap of increased initial moisture content without preliminary purification. For drying grass seeds, an SKM-0.5 dryer was used, designed for low-flowing materials [4].

The drying process in a rotary dryer is as follows. The raw material delivered for drying from the vehicle, trailer or car body is unloaded into the receiving hopper and fed into the drying chamber by the loading conveyor. A preheated coolant is blown under the perforated platform of the drying chamber of the rotary dryer by a fan. Which, passing the layers of grass seeds, heats them and takes away excess moisture. As a result of heat and mass transfer, the seeds are dried, and the heat carrier is saturated with moisture and removed from the drying chamber. Due to the fact that the heat carrier will have the greatest moisture-absorbing capacity at the entrance to the drying chamber, the seeds of the lower layers dry out faster and are unloaded by a belt conveyor for further processing.

During the experiment, a number of basic indicators were analyzed: the temperature of the coolant on various layers of the drying chamber, the initial and final temperatures of grass seeds, as well as their moisture content.

Herewith, samples of grass seeds were taken, their moisture content was determined pursuant to the requirements of GOST 13586.3-83 and GOST 13586.5-93.

Samples of samples in order to determine the values of temperature and moisture content of seeds before and after loading the dryer were taken from the flow of seeds falling from the loading device and the unloading screw of the dryer. A 0.5 liter plastic container with a lid was used for this. To measure the temperature of the seeds in the layer, the samples were taken with a sampling probe to a depth of 0.5 m and poured into plastic containers with a volume of at least 0.5 l. The seed temperature was measured with a T7 thermometer (Russia); the moisture content of control seed samples was measured using a Wile 35 moisture meter (Finland).

During the experiment, the temperature of the coolant was controlled by a digital thermostatic element installed in the air duct, which automatically regulated the operating temperature and the temperature of the seeds by automatically turning off the burner in case of overheating. The temperature of the coolant and, accordingly, its humidity were measured with a TKA-PKM hot-wire anemometer (model 60, RF) at the entrance to the drying room, in the supply air duct, and also in different layers of seeds.

The performance of the drying unit (actual) was measured based on the mass of sampling of the dried heap of seeds for a certain period. The experiment included at least five such samples. At each such cutoff, the seeds were loaded into a special container, and then weighing took place (AB-875 scales). To determine the cut-off time, a SOSpr-26-2-000 stopwatch was used.

As a result, the basic Q_d productivity (dry) in kg h^{-1} was determined as follows (1) [4]:

$$Q_d = 60\frac{\sum_{i=1}^{n} M_i}{\sum_{i=1}^{n} T_i}, \tag{1}$$

where M_i is the mass of seeds of the ith sample, kg;

T_i is the flow time of the i-th sample, s.

The moisture content of the seeds in the dryer was measured in real time with a FAUNA-P moisture meter (Russia) pursuant to the capacitive principle of moisture measurement. This is a microprocessor that displays the percentage of moisture, temperature and the name of the studied culture on a liquid crystal display. The device is equipped with two humidity and temperature sensors.

The moisture meter (Fig. 1) was a monoblock with a front panel with a liquid crystal display and a command button, as well as a battery compartment at the bottom of the device.

Fig. 1. General view of the Fauna-P flow moisture meter with humidity and temperature sensors.

Experimental studies of the drying process in a rotary dryer were performed when drying meadow fescue seeds with an initial (initial) moisture content of 22.2%. The coolant temperature was 55 °C. Drying time of seeds to final moisture was 2 hours (120 minutes). In connection with the design features of the rotary dryer, in order to study the process of changing the parameters of seeds in different parts of the drying chamber, the seed layer was divided into zones pursuant to its height, each 0.14 m thick. The first zone - *I* from 0 to 0.14 m, the second - zone *II* from 0.14 to 0.28 m and the third zone *III* from 0.28 to 0.42 m. In addition, the parameters of the seeds were measured in the upper and lower layers of the drying chamber.

In the course of the experiment using a rotary-type dryer, the influence of three important and significant factors was determined, such as the temperature of the coolant in the duct, the flow rate of the coolant, and the thickness of the seed layer on the drying process. The initial values of the levels of factors are determined on the basis of previously conducted experiments and analytical studies of a number of literary sources [3, 13, 14].

3 Results and discussion

Due to the fact that initially a pile of seeds has a high initial moisture content, it is necessary that the temperature of the coolant during the drying process be within the range of up to 50-55 °C in order to avoid the formation of cracks on the seeds. The presence of cracks on the surface negatively affects the period of storage and germination, which ultimately does not allow farms to have highly reproductive seeds at the stages of selection and seed production.

During warm weather with the presence of the sun, it is possible to dry grass seeds quite successfully without heating the coolant. In one of the farms in the North-West region of the Russian Federation, in JSC "Volkhovskoe", the drying of seeds was successfully performed without using a heat generator, only ambient air was used, while a heap of fescue seeds with meadow moisture of 21.4% was qualitatively dried within 7 hours to a standard humidity (13.8%) (Fig. 2).

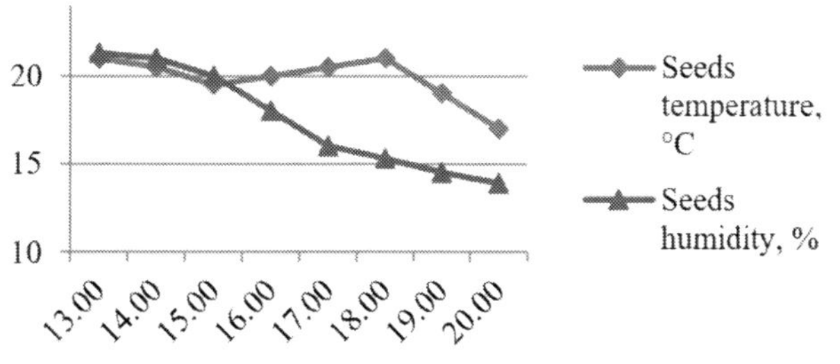

Fig. 2. The process of changing temperature (°C), humidity (%) depending on the daily drying time (h) of meadow fescue seeds.

If the combine harvesting of seeds is performed in the daytime up to 1 p.m., then during drying there is a good opportunity to save fuel, which is spent on heating the coolant. In this case, only electricity is consumed to drive the fan.

In the course of experimental studies, it was established that as the heated coolant passes through the layers of seeds, the temperature of heating the seeds decreases. As a result, there is a decrease in the drying speed from the lower layers to the upper ones (table).

Table 1. Results of experimental studies of the process of drying grass seeds to a rotary dryer.

Drying zones by layers			Results of measurements	
Designation	Thickness of layer, m		Temperature, t, C^{O}	Humidity
	from	to		
Low	0	0.01	49.2	14.1
I	0	0.14	44.6	15.9
II	0.14	0.28	36.2	21.2
III	0.28	0.42	24.1	22.7
Upper	0.42	-	20.4	23.3

As a result of the studies performed, it was found that the temperature of the seeds in the layer of the I-th zone at a steady state of the temperature of the coolant, already 30 minutes after the start of drying, becomes 12 °C lower. The temperature of the seeds in the layer of zone II reached 36.2 °C. In the layer III-th zone the temperature of the seeds after 1 hour of drying was 22 °C.

The results of the experiments made it possible to conclude that, depending on the speed, the drying process of seeds at different layers in height is not the same, while the drying time increases from the lower layer to the upper one (Fig. 3).

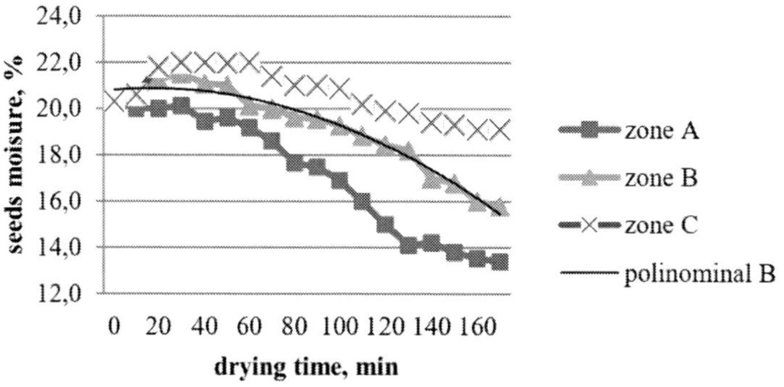

Fig. 3. The process of changing the moisture content of meadow fescue seeds in zones (I, II, III) at an initial moisture content of 20.2%

As a result of processing the experimental information, mathematical dependences of the change in the moisture content of seeds W, (%) on the drying time t, (min) were obtained. So, for a layer of seeds 0.14-0.28 (zone *II*), the indicated dependence has the form:

$$W = 20.69 + 0.1264 \cdot t - 0.0231t^2 \quad (2)$$

The results obtained made it possible to optimize the drying process (Fig. 3) and to propose to remove dried seeds from the lower layer from the drying chamber every 2.5 hours by means of an unloading conveyor. After that, the upper layers move to the lower ones, and a new batch of wet seeds is loaded into the freed upper layer.

The process of drying grass seeds in the drying chamber of the rotary dryer takes place in two stages. Initially, moisture from the lower layers of the seeds moves to the upper ones, moistening them, and then it is removed as it heats up and reaches an equilibrium moisture content with a moisture-saturated heat carrier. Therefore, it is necessary to maintain the optimal layer thickness in the drying chamber, since at high values in the upper layers, the seeds are moistened, and not dried. This occurs only when the temperature of the seeds is below the temperature of adiabatic saturation of air, corresponding to the completion of the drying process of seeds [3, 6].

4 Conclusions

In the North-West region of Russia, the seed heap arriving at the post-harvest processing point, as a rule, has a high humidity. To ensure the preservation of the conditioned properties of seeds, it is necessary to organize a high-quality reception and their immediate drying.

It is necessary to ensure the optimum thickness of the seed layer in the drying chamber, depending on the parameters of the heat carrier and the moisture content of the seeds.

The optimal thickness (height) of the seed layer in the drying chamber allows full use of the heat carrier (saturate it with moisture), which ensures the most efficient drying of seeds in terms of energy consumption

References

1. E. Kubeev, V.A. Smelik, E3S Web of Conferences. VIII International Scientific and Practical Conference of "Innovative technologies in science and education" (ITSE 2020) 03006 (2020)
2. Forage Seed Processing Line, http://www.petkus.com
3. A. N. Perekopskii, S. V. Chugunov, Technologies, machines and equipment for mechanized crop and livestock production, **91,** 126 (2017)
4. A. N. Perekopskiy, Rural Machine Operator, **5,** 6 (2015)
5. R. Stanisavljević, D. Djokić, J. Milenković, D. Terzić, V. Stevović, D. Tomić, D. Dodig, Zemdirbyste-Agriculture, **101(2),** 196 (2014)
6. O. I. Gaponyuk, M. V. Ostapchuk, G. M. Stankevich, I. I. Gaponyuk, Active ventilation and drying of grain (Odesa: VMV, 2014)
7. R. Stanisavljević, J. Milenković, D. Djokić, D. Terzić, D. Petrović, L. Djukanović, D. Dodig, Plant Soil and Environment, February, **59(1),** 37 (2018)
8. P. Savinyh, Y. Sychugov, V. Kazakov and S. Ivanovs, 17th Int Sci Conf "Engineering for Rural Development", **17,** 124 (2018)
9. J. R. Rasi, M. Neto, R. Bernardo, Int. J. Innov. Educ. Res., **9,** 410 (2021)
10. C. Li, B. Li, J. Huang, C. Li, Developing an online measurement device based on resistance sensor for measurement of single grain moisture content in drying process, **20,** 4102 (2020)
11. S. R. Parde, D. S. Jayas, N. D. G. White, Sciences Des Aliments, **23(5),** 589 (2003)
12. Development of Moisture Removal Process, Energy Consumption and Control of a Grain Dryer (2016–2019) https://ec.europa.eu
13. T. Jokiniemi, M. Hautala, T. Oksanen, J. Ahokas, Parallel plate heat exchanger for heat energy recovery in a farm grain dryer Drying Technology, **34(5),** 547 (2016)
14. Safe and Secure Supply of Grain with Innovative Resource-Efficient Drying, Handling and Storage 2019–2020 (Research project of RISE Research Institutes of Sweden AB - Livsmedelsacceleratorn, Agrifood and Bioscience, Göteborg), https://www.vinnova.se

Influence of crops of the cover crop when forming the ridge surface of landings of potatoes to dynamics of parameters of the soil state

A. A. Ustroyev, and *E. A.* Murzaev

Institute of Agroengineering and Environmental Problems of Agricultural Production (IAEP), branch of FSBRI FNATS VIM, St. Petersburg, Russia

Abstract. The main factors limiting the yield and quality of potatoes are the geometric parameters of the ridge, as well as the parameters of the soil state (moisture, hardness and temperature). In this regard, the task of maintaining the optimal values of these parameters during the potato growing season is urgent. For this, it is proposed, after planting potatoes, simultaneously with the formation of ridges, to sow a cover crop (yellow mustard) on them, followed by its destruction. The purpose of this research is to study the effect of sowing cover crops during the formation of the ridge surface of potato plantings on the dynamics of soil state parameters and potato yield. As a result of the studies, it was found that the use of a new technological method ensures a decrease in the effect of erosion processes on the ridge and the preservation of its geometric parameters, which makes it possible to exclude two inter-row cultivation of potatoes, as well as stabilize the daily soil temperature in the ridge by 2-5 ° C and retain moisture by 5 - 10%, which allows to increase the yield of potatoes by 11%.

1 Introduction

The technology of potato cultivation provides, after planting, the formation of full-volume ridges with a fine crumbly soil structure to ensure optimal conditions for the growth and development of plants [1-4]. Such a structure is subject to intense influence of erosion processes and requires hardening of their surface with the help of special technological methods to preserve the volume of the ridges. A patent search and analysis of literary sources [5-7] showed that the most promising method of strengthening the ridge surface can be a technological method of sowing cover crops on it and further destruction of seedlings during the growing season.

The following can be used as cover crops: mustard, oil radish, etc. The short growing season of cover crops allows them to develop a significant amount of green mass and root system within two or three weeks. The created vegetation cover is able to protect the soil surface with its leaf surface, reducing the speed of falling and excluding the ingress of large water droplets on the soil, and also significantly contributes to a decrease in wind speed near the earth's surface. All this can lead to a significant reduction or complete elimination of the development of erosion processes in potato plantings [8-10].

Due to the fact that when implementing the proposed technological method, it is necessary to sow cover crop seeds along the ridge surface, seed drills are not suitable for this operation. Fertilizer spreaders can be used as an alternative seeding medium, but such means have a high degree of uneven distribution of the spreading material over the surface of the field.

For sowing cover crops on the surface of the ridges, in our opinion, it is rational to use modern pneumatic seeding modules, which are installed on various agricultural units. Such modules are designed for sowing seeds of various crops and can be installed on various technical means. The module allows to make seeds of agricultural crops both separately and simultaneously with various technological operations performed by basic units [11].

When using standard baffle plates on such systems, the degree of uneven distribution of seeds on the ridge increases under the influence of external factors, for example, wind.

To eliminate this drawback, the existing system for distributing seeds of the pneumatic sowing device APV PS 300 M1 was enhanced by developing special diffusers that were installed instead of baffle plates. The diffuser design is divided into three parts. The first part is the fitting into which the seed tube from the pneumatic sowing device is connected. The second part is the diffuser body itself, inside which a divider is installed. The divider is designed to divide the air flow with the seeds, in order to better distribute it inside the diffuser. The third part is a confuser designed to combine two separated air streams and their orientation to the ridge surface of the field [12].

The purpose of this research is to study the effect of sowing cover crops during the formation of the ridge surface of potato plantings on the dynamics of soil state parameters and potato yield.

2 Materials and methods

Experimental studies were performed from June to early July 2020 at the IAEP experimental station in experimental and control versions.

The experiment cultivated potatoes of the "Udacha" variety, zoned for the Leningrad region. The potatoes were cultivated using organic technology without the use of mineral fertilizers and chemical plant protection products. As a cover crop, yellow mustard was used, the sowing of which was performed along the ridge surface, formed simultaneously with deep loosening of the row spacings after planting potatoes on May 28, 2020, using a developed combined unit consisting of a cultivator-deep cultivator for inter-row cultivation of potato plantings of the IAEP design and installed on its APV 300 M1 pneumatic seeder frame with an enhanced distribution system (Figure 1). The cover crop was destroyed on 20 July. Further treatments of potato plantings were not performed in the experiment. In the control variant, pre-emergence harrowing and post-emergence hilling with harrowing were performed.

Fig. 1. Multipurpose row-crop cultivator for inter-row cultivation of potato plantings with deep loosening of row spacings and simultaneous sowing of cover crops

A multipurpose row-crop cultivator works as follows. When operating in the field, the deep loosening share on a rigid stand performs deep loosening of the soil in order to eliminate compacted areas in the row spacings formed by the previous passes of the tractor wheels and the planting unit, and also forms favorable conditions for the development of the root system of cultivated crops. Adjustment of the depth of travel of the deep loosening share is achieved by changing the position of the column in height on the frame and can be up to 40 cm. The hilling body behind the hilling body performs the formation of a volumetric loose ridge, directing the soil from the row-spacing to the center of the row with feather dumps. In the course of work, the soil layer is subjected to crumbling, which provides a loose, finely lumpy structure inside the ridge. Seeds are sown into the prepared soil by a pneumatic seeding module through a distributor. The bracket for attaching the diffuser to the machine frame allows to adjust its distance to the profiled surface of the field, the distance from the machine frame to the embedding bodies and the angle of inclination relative to the ridge.

During the experiment, the parameters of the soil state were monitored in the composition: moisture (%), temperature (C °), hardness (MPa), geometric parameters of ridges (cm); and potato yield (t/ha). Soil moisture was measured using a soil moisture meter manufactured by TR di Turuni & c.Snc, and soil hardness was measured with a penetration logger from Eijkelkamp (Figure 2).

A B

Fig. 2. Soil moisture meter manufactured by "TR di Turuni & c.Snc" (a), penetration logger from Eijkelkamp (b)

With the aim of a deeper study of the effect of sowing a cover crop on the dynamics of temperature regimes inside the ridges, we have developed an information and measurement system that provides continuous recording of soil temperature in the root layer and the environment over time with a given degree of discreteness (Figure 3).

Fig. 3. Information-measuring system of temperature

An information-measuring system consists of several components:

- Debug board based on AVR microcontroller;
- RTC module on DS3231 microcircuit;
- A module for working with a micro SD card;
- Digital temperature sensors 18b20;
- Batteries (3 AAA batteries).

The information-measuring system was made in duplicate and installed at the experimental and control sections. It works pursuant to the software developed by us and accumulates temperature data on a micro SD card in .txt format. For further data processing, Microsoft Excel is used.

3 Results and discussion

Table 1 and Figure 4 show the dynamics of the geometric parameters of the ridges of potatoes in the experimental and control versions.

Table 1. Dynamics of the geometric parameters of the ridge, cm

Crest, cm					
	15 June	23 June	06 July	21 July	01 September
Experiment	18.3	18	18.2	17.1	16.9
Control	14.2	14.3	13.4	18.4	16.8

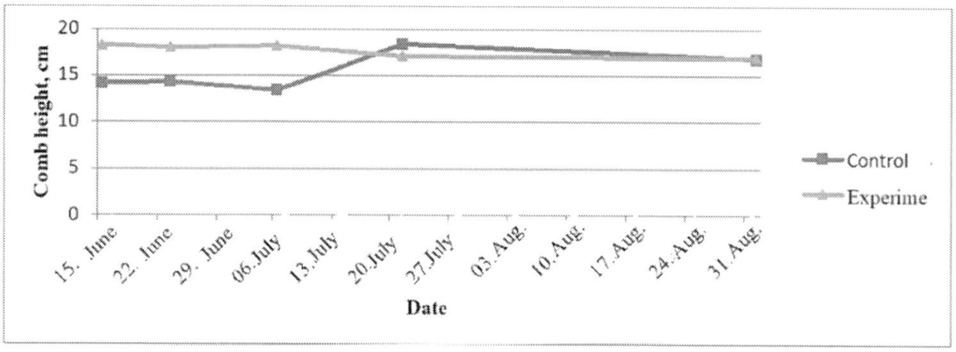

Fig. 4. Dynamics of geometric parameters of potato ridges

Analysis of the data obtained shows that sowing a cover crop on the ridge surface reduced the effect of erosion processes on the ridge, contributing to the preservation of its geometric parameters by 4-7 cm compared to the control during the growing season of the cover crop. Thanks to this, it was possible to exclude two inter-row cultivation of potatoes.

Figure 5 shows the dynamics of the ambient temperature, the temperature inside the tuberization zone in the control and experimental plots as of June 15.

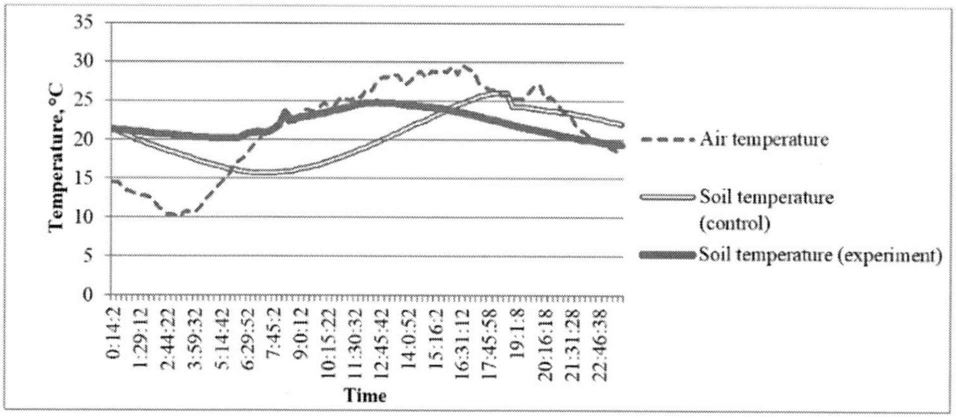

Fig. 5. Daily graph of changes in the temperature of the environment and soil

Analysis of the data obtained shows that with daily fluctuations in air temperature from 10 to 30 ° C, the soil temperature in the control ranged from 16 to 26 ° C, and in the experiment from 19 to 25 ° C. In this way, the cover crop protects the ridge from both overheating and sudden cooling. This trend was observed throughout the entire growing season of the cover crop on the profiled surface. After the destruction of the cover crop, its crop residues continued to restrain the effect of temperature on the ridge until its complete withering away (Figure 6).

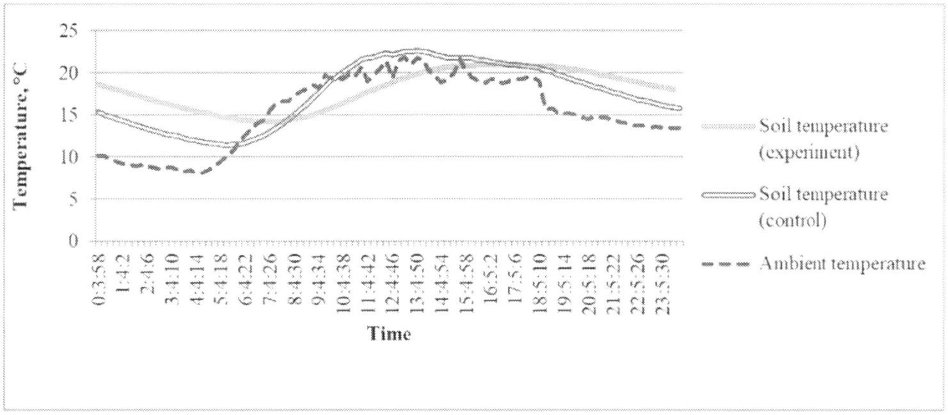

Fig. 6. Temperature dynamics after removing the cover crop

Table 2 and Figure 7 show the dynamics of soil moisture in potato ridges in the tuberization zone in the experimental and control variants.

Table 2. Dynamics of moisture in the ridge in the tuberization zone,%

	Humidity				
	15.06.2020	23.06.2020	06.07.2020	21.07.2020	01.09.2020
Experiment	29.18	22.9	11.56	16.98	34.3
Control	22.9	29.1	22.54	16.22	36.18

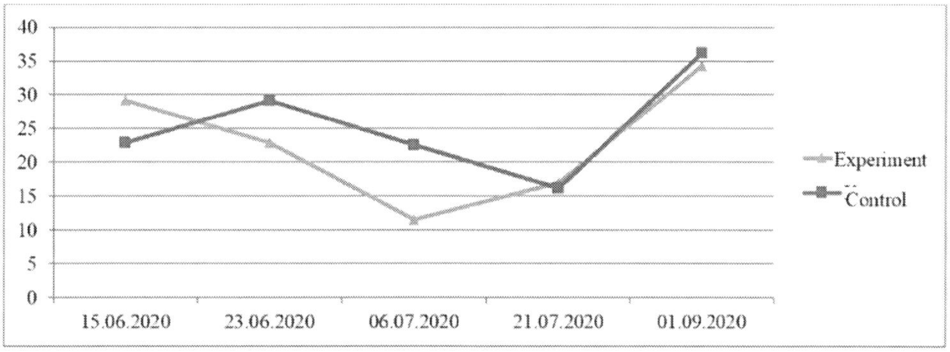

Fig. 7. Dynamics of soil moisture by layers

The analysis of the data obtained shows that the soil moisture as of 06/15/2020 in the experiment with the sowing of the cover crop was 5--10% higher than in the control (without sowing). This is due to the fact that on June 15, the cover crop was in the phase of two leaves, while a "rough soil surface" was created on the ridge and in the aisle, which prevented the free flow of precipitation along the ridge surface, as a result of which a more complete absorption of moisture by the soil took place.

In the course of further growth of the cover crop, the moisture content inside the ridge decreased due to the fact that it, while developing, consumed a large amount of moisture. After the destruction of the cover culture, the humidity was equal in the experiments.

Analysis of the obtained data on soil hardness inside the ridge by layers showed that no significant changes in hardness occurred due to sowing a cover crop on the ridge.

The average value of the yield of potatoes during experimental research is 18.03 t/ha in the experiment, which exceeds the value in the control 16.2 t/ha by 11%. This confirms the efficiency of the developed technological method.

4 Conclusion

A combined technological method has been developed for processing potato plantings on the ridges in organic cultivation technology using the operations of deep loosening of row spacings and sowing cover crops on the ridge surface with subsequent destruction of seedlings during the growing season.

The use of a new technological method ensures a decrease in the effect of erosion processes on the ridge and the preservation of its geometric parameters, which makes it possible to exclude two inter-row cultivation of potatoes, as well as stabilization of the daily soil temperature in the ridge by 2 - 5 ° C and moisture retention by 5 - 10%, which makes it possible to increase potato yield by 11%.

References

1. V. B. Minin, V. D. Popov, D. A. Maksimov, A. A. Ustroev, S. P. Melnikov, Agronomy Research, **2(18)**, 1359-1367 (2020)
2. A. B. Kalinin, I. Z. Teplinsky, A. A. Ustroev, P. P. Kudryavtsev, IOP Conference Series: Materials Science and Engineering, **516(1)**, 1-6 (2019)
3. D. Shpaar, A. Bykin, D. Draeger, *Potatoes*, 466 (2004)
4. A. B. Kalinin, V. A. Smelik, I. Z. Teplinsky, O. N. Pervukhina, Bulletin of the St. Petersburg State Agrarian University, **39**, 315-319 (2015)
5. E. A. Murzaev, Bulletin of the student scientific society of St. Petersburg State Agrarian University, **9(2)**, 65-68 (2018)
6. A. B. Kalinin, I. Z. Teplinsky, V. A. Ruzhev, *World trends and modern technical systems for potato cultivation*, 160 (2016)
7. A. B. Kalinin, I. Z. Teplinsky, Rural mechanic, **5**, 8-9 (2015)
8. V. D. Abashev, L. M. Kozlova, Agrarian science of the Euro - North – East, **6**, 169-178 (2015)
9. M. N. Novikov, A. M. Tamonov, L. D. Frolova, L. I. Ermakova, Agrochemical Bulletin, **4**, 20-26 (2013)
10. I. S. Kochetov, Agriculture, 3, 16 (2000)
11. Pneumatic seeding devices, https://www.apvrussia.ru/
12. A. B. Kalinin, A. A. Ustroev, I. Z. Teplinsky, E. A. Murzaev, Technologies and hardware components of the mechanized production of crop production and livestock production, **2(99)**, 148-157 (2012)

Impact of Organomineral Fertilization Systems on the Yield of Field Crops and Changes in the Productivity of Crop Rotation on Irrigated Ordinary Carbonate Chernozems of the Central Ciscaucasia

Ruslan Bizhoev, *Sarina* Konova, *Asiyat* Sarbasheva, *Olga* Batyrova, and *Rada* Gazheva

Institute of Agriculture - Branch of the Kabardino-Balkarian Scientific Center of the Russian Academy of Sciences, Kirova Street, 224, 360004 Nalchik, Kabardino-Balkarian Republic

Abstract. The paper presents the research results of the impact of different fertilization systems - mineral and organic, using intercropping of green manure crops, biological resources (straw of grain crops, foliar of maize) on crop yields of grain rotating crops - winter wheat, maize, peas and the productivity of hectare of arable black land of ordinary carbonate chernozem in dryland conditions of the Central Caucasus region with different indicators of growing seasons. The scientific novelty of the paper is the identification of optimum-rational fertilization systems and the justification of the influence of the studied fertilization systems and the use of by-products of the crop rotation, which make it possible to obtain stable crop yields and maintain the humus content in the soil. The work defines the efficiency of the use of different mineral fertilizer doses in combination with organic fertilizers, which form the highest crop yield and crop rotation productivity.

1 Introduction

The use of organic and mineral fertilizers is the most important tool in the system of agronomic measures to ensure stable crop yields. It is most effective in a certain system with existing crop rotation (agrocenosis), since only such application takes into account, as much as possible, not only the action but also the consequence of both organic and mineral forms. Only a science-based fertilization system in each crop rotation can provide planned crop yields while maintaining soil fertility [1,2,3].

The combined impact of these agricultural inputs on crop productivity and soil fertility can only be assessed through long-term stationary field studies.

It is known that, in the current economic climate, agricultural producers are forced to use fewer mineral fertilizers than the recommended norms for their specific farming conditions. This is primarily due to the cost and expenses of application, as well as the incomplete assembly of the market with the necessary fertilizers [4]. In this regard, research has studied the impact of different mineral fertilizer norms (full norm, half and one third of the recommended norm) when combined with organic fertilizers such as manure, green manure or straw, as well as their integrated application to winter wheat, maize, peas and crop rotation productivity.

2 Materials and Methods

The work was performed in a long-term stationary field experiment territorially located in Krasnoarmeysk village, Tersky district, Kabardino-Balkarian Republic on the experimental field of the Institute of Agriculture of the Kabardino-Balkarian Scientific Center of the Russian Academy of Sciences [5,6]. The experiment founded in 1979 and included in the Register of Long-term Experiments, registration No. 082, under the name "To develop technology of biological reproduction of the fertility of irrigated soil on the basis of the use of different types of organic fertilizers and rational use of mineral fertilizers in the agricultural landscape system of Kabardino-Balkarian Republic" [7,8]. Research is conducted according to classical methods. Annual reports on R&D are submitted to the Geographic Network, where the federal database is updated [9].

The experiments are based on a 9-grain field rotation with the following crop rotation: 1 - peas, 2 - winter wheat, 3-4 - dried maize, 5 - winter wheat, 6 - sunflower, 7 - winter wheat, 8 - dried maize, 9 - winter wheat. The experiment is based in the space of three fields, the alternation of crops in the fields of crop rotation occurs in time with the successive entry of fields into the crop rotation. Agricultural techniques for crop cultivation followed zonal guidelines. There are 60 options in the experiment, 240 plots, the plot area is 189 m^2 with a plot width of 6.3 m and a length of 30.0m. Fertilizers were applied according to a scheme of experiment.

Ammonium nitrate, simple and double superphosphate, potassium salt, semi-rotted manure were used as fertilizers. Variants and norms for different types of organic fertilizers are presented in Table 1. NPK norms are established on the basis of actual experimental material and are recommended as optimal in the area of cultivation.

In recent years, ammonium nitrate, ammophos and nitroammophos have been used due to the lack and cost of the required fertilizers, in particular superphosphate and potassium fertilizer.

3 Results and discussion

In 2018, in grain-and-grass crop rotations, according to the rotation Camilla maize was cultivated in 1 field, Cheget winter wheat in field 2, and Rassvet peas in field 3.

In this experiment, peas had an undoubted advantage as a predecessor (Table 1). The yield of Cheget winter wheat in control option1 without fertilizers for peas in 2018 was 46.2 dt/ha. The formation of such a yield was facilitated not only by weather conditions, but also by better timely soil preparation, planting crops at optimal times, the advantage of a good predecessor, and adherence to other agricultural practices.

In field 2 of the crop rotation according to the predecessor peas in option 2 with the use of the recommended mineral fertilization system - $N_{90}P_{120}K_{60}$ - the grain yield of Cheget winter wheat of 65.8 dt/ha was achieved. The effectiveness of the recommended mineral fertilizer system in options 6,10,14,18 when combined with organic resources (aftereffect of manure, green manure, grain straw, their complex application) by 11.5, 12.3, 13.2, 18, 8 kg/ha more than in variant 2, without organic fertilizers, which contributed to the grain yield of Cheget winter wheat after peas in 2018, respectively: 77.3; 78.1; 79.0; 84.6 dt/ha.

Table 1. Impact of organo-mineral fertilization systems on yields of maize, winter wheat and peas in the variants of field experience of 1979, dt/ha, 2018.

Variants	Fertilizers		Field 1 Camilla Maize		Field 2 Cheget winter wheat		Field 3 Rassvet peas	
	organic	mineral	crop yield	raise, %	crop yield	raise, %	crop yield	raise, %
1	0	O	11.1	0	46.2	0	24.9	0
2		NPK	25.4	14.3	65.8	19.6	33.8	8.9
3		1/2 NPK	21.4	10.3	60.1	13.9	32.0	7.1
4 5		1/3 NPK	15.3	4.2	53.9	7.7	30.8	5.9
6	Manure 50 t/ha once every 5 years	O	15.0	3.9	45.8	-0.4	26.9	2.0
7		NPK	27.7	16.6	77.3	31.1	35.8	10.9
8		1/2 NPK	23.2	12.1	67.6	21.4	34.8	9.9
		1/3 NPK	16.5	5.4	57.7	11.5	33.1	8.2
9	Green manure (rape)	O	15.3	4.2	46.6	0.4	26.6	1.7
10		NPK	27.9	16.8	78.1	31.9	35.6	10.7
11		1/2 NPK	22.9	11.8	73.4	27.2	34.4	9.5
12 13		1/3 NPK	16.7	5.6	68.2	22.0	32.2	7.3
14	Straw	O	14.9	3.8	57.4	11.2	26.8	1.9
15		NPK	28.9	17.8	79.0	32.8	38.0	13.1
16		1/2 NPK	24.3	13.2	74.9	28.7	35.2	10.3
17 18		1/3 NPK	16.0	4.9	69.7	23.5	33.4	8.5
19	Manure + straw + green manure	O	17.6	6.5	60.3	14.1	28.1	3.2
20		NPK	36.8	25.7	84.6	38.4	40.7	15.8
		1/2 NPK	29.3	18.2	76.8	30.6	38.6	13.7
		1/3 NPK	20.9	9.8	70.9	24.7	34.5	9.6
LSD_{05} for particular differences			1.09		1.22		0.78	
LSD_{05} for organic fertilizers			0.53		0.62		0.42	
LSD_{05} for mineral fertilizers and interactions			0.45		0.55		0.35	
Experiment accuracy, Sx%			2.54		1.13		1.74	

The reduction of recommended mineral fertilizer doses by half (1/2 RD - options 7.11.15.19) reduced the yield of irrigated crop rotation by 9-10% compared to the yield at the full dose of fertilizers. The decrease in winter wheat and other crops yields at 1/3 of the recommended dose was 17-18%.

The productivity of 1 ha of grain-and-grass crop rotation without fertilizer under irrigation (option 1) of the experiment in 2018 was 30.0 dt/ha (Table 2).

Analyzing the productivity of 1 crop rotation hectare with long-term use of different organic fertilization systems, we can conclude that it increases in the experimental options compared to control.

Table 2. Change in the productivity of crop rotation fields by crops and fertilization systems in the variants of the 1979 field experiment in 2018, dt.g.u./ha.

Variants	Fertilizers organic	Fertilizers mineral	Field 1 - maize Camilla SV	Field 2 - winter wheat Cheget	Field 3- peas Rassvet	Average
1	0	0	8.9	46.2	34.9	30.0
2		NPK	20.3	65.8	47.3	44.5
3		1/2 NPK	17.1	60.1	44.8	40.7
4		1/3 NPK	12.2	53.9	43.1	36.4
5	Manure, 50t/ha (once every 5 years) aftereffect	0	12.0	45.8	37.7	31.8
6		NPK	22.2	77.3	50.1	49.9
7		1/2 NPK	18.6	67.6	48.7	45.0
8		1/3 NPK	13.2	57.7	46.3	39.1
9	Green manure	0	12.2	46.6	37.2	32.0
10		NPK	22.3	78.1	49.8	50.1
11		1/2 NPK	18.3	73.4	48.2	46.6
12		1/3 NPK	13.4	68.2	45.1	42.2
13	Straw	0	11.9	57.4	37.5	35.6
14		NPK	23.1	79.0	53.2	51.8
15		1/2 NPK	19.4	74.9	49.3	47.9
16		1/3 NPK	12.8	69.7	46.8	43.1
17	Manure + straw + green manure	0	14.0	60.3	39.3	37.9
18		NPK	29.4	84.6	57.0	57.0
19		1/2 NPK	23.4	76.8	54.0	51.4
20		1/3 NPK	16.7	70.9	48.3	45.3

The climatic conditions of the 2019 growing season were very unfavourable for the growth, development and production of all field crops. During the autumn period of 2018 - from September to December inclusive - 51.2 mm of precipitation fell at a rate of 110.7 mm, which did not even allow winter wheat to grow in the autumn, despite the optimal planting time.

The relative air humidity in all months of growing season was less than the annual average by 2-14%. The climatic indicators were particularly unfavourable in June 2019.

In the first field of the experiment, after maize, Camilla SV maize was cultivated, the hybridization plot, in the second field after winter wheat, also Camilla SV maize, in the third field after peas - Cheget winter wheat.

Research has shown that the yield of Cheget winter wheat in field 3 under the predecessor of peas in 2019 was 2.0-2.4 times lower than that of wheat in previous years of research. In variant 2, the recommended mineral fertilization system achieved a grain yield of Cheget winter wheat of only 27.7 dt/ha (Table 3).

The efficiency of the recommended mineral fertilization system, when combined with organic resources, in options 6,10,14,18 (after manure, green manure, straw, integrated application) is 2.3, 1.5, 2.0, 5.3 dt/ha higher, than in variant 2, without organic fertilizers, which contributed to Cheget winter wheat grain yields under peas in 2019, respectively: 30.0; 29.2; 29.7; 33.0 dt/ha.

The reduction of the recommended mineral fertilizer rate by half (1/2 RD - options 7,11,15,19) reduced crop yields with irrigation in 2019 by 16-19% compared to total fertilizer yield. The crop yield decrease at 1/3 of the recommended dose was 30-39%.

The highest yields in Cheget winter wheat of 33.0 dt/ha and Camille maize of 70.7 dt/ha were achieved in 2019 in option 18 with the recommended mineral fertilizer rate combined with the integrated application of all studied bioresources.

Table 3. Impact of organo-mineral fertilization systems on the yield of corn and winter wheat in the variants of the 1979 field experiment, kg/ha, 2019.

Variants	Fertilizers		Field 1 Camilla SV maize, 1st generation		Field 2 Camilla SV maize, 1st generation		Field 3 Cheget winter wheat	
	organic	mineral	crop yield	raise, %	crop yield	raise, %	crop yield	raise, %
1	0	O	24.9	0	21.6	0	17.6	0
2		NPK	53.4	28.5	36.0	14.4	27.7	10.1
3		1/2 NPK	43.3	18.4	31.9	10.3	23.0	5.4
4		1/3 NPK	32.1	7.2	27.4	5.8	19.9	2.3
5	Manure 50 t/ha once every 5 years	O	32.8	7.9	23.9	2.3	17.9	0.3
6		NPK	55.8	30.9	37.9	16.3	30.0	12.4
7		1/2 NPK	47.7	22.8	33.8	12.2	24.1	6.5
8		1/3 NPK	38.4	13.5	28.0	6.4	20.6	3.0
9	Green manure (rape)	O	27.1	2.2	26.1	4.5	18.2	0.6
10		NPK	52.7	27.8	41.2	19.6	29.2	11.6
11		1/2 NPK	43.8	18.9	35.7	14.1	25.1	7.5
12		1/3 NPK	36.1	11.2	32.8	11.2	20.1	2.5
13	Straw	O	30.6	5.7	23.8	2.2	18.7	1.1
14		NPK	56.3	31.4	41.0	19.4	29.7	12.1
15		1/2 NPK	48.2	23.3	33.4	11.8	26.3	8.7
16		1/3 NPK	36.6	11.7	27.8	6.2	19.7	2.1
17	Manure + straw + green manure	O	35.7	10.8	28.7	7.1	20.6	3.0
18		NPK	70.7	45.8	43.4	21.8	33.0	15.4
19		1/2 NPK	60.7	35.8	37.0	15.4	29.9	12.3
20		1/3 NPK	40.1	15.2	35.0	13.4	22.8	5.2
LSD_{05} for particular differences			1.93		0.82		0.67	
LSD_{05} for organic fertilizers			0.96		0.41		0.15	
LSD_{05} for mineral fertilizers and interactions			0.84		0.37		0.22	
Experiment accuracy, Sx%			4.23		1.02		1.00	

Low crop yields (Table 4) compared to previous research years have affected the productivity of 1 ha of crop rotation. Analyzing the productivity of 1 crop rotation hectare, we can conclude that it differs in different options of experiments in comparison with the control - the cultivation of crops without fertilizers. In 2019, the production of 1 hectare of crop area without fertilizer in control was at the same level of 17.3-19.9 dt.g.u. for both winter wheat and maize.

Table 4. Change in the productivity of crop rotation fields by crops and fertilization systems in the options of the 1979 field experience in 2018 and 2019, dt.g.u/ha.

Variants	Fertilizers organic	Fertilizers mineral	Field 1 - maize Camilla SV	Field 2 maize Camilla SV	Field 3 - Cheret winter wheat	Medium productivity 2018	Medium productivity 2019
1	0	0	19.9	17.3	17.6	30.0	18.3
2		NPK	42.7	28.8	27.7	44.5	33.1
3		1/2 NPK	34.6	25.5	23.0	40.7	27.7
4		1/3 NPK	25.7	21.9	19.9	36.4	22.5
5	Manure, 50t/ha (once every 5 years) aftereffect	0	26.2	19.1	17.9	31.8	21.1
6		NPK	44.6	30.3	30.0	49.9	35.0
7		1/2 NPK	38.2	27.0	24.1	45.0	29.8
8		1/3 NPK	30.7	22.4	20.6	39.1	24.6
9	Green manure	0	21.7	20.9	18.2	32.0	20.3
10		NPK	42.2	33.0	29.2	50.1	34.8
11		1/2 NPK	35.0	28.6	25.1	46.6	29.6
12		1/3 NPK	28.9	26.2	20.1	42.2	25.1
13	Straw	0	24.5	19.0	18.7	35.6	20.7
14		NPK	45.0	32.8	29.7	51.8	35.8
15		1/2 NPK	38.6	26.7	26.3	47.9	30.5
16		1/3 NPK	29.3	22.2	19.7	43.1	23.7
17	Manure + straw + green manure	0	28.6	23.0	20.6	37.9	24.1
18		NPK	56.6	34.7	33.0	57.0	41.4
19		1/2 NPK	48.6	29.6	29.9	51.4	36.0
20		1/3 NPK	32.1	28.0	22.8	45.3	27.6

In 2020, in the experiment, Yuzhanka winter wheat was cultivated in field 1, after the Camilla SV maize, in field 2, after the Cheget winter wheat, Karat maize, in field 3, after the Cheget winter wheat, Karat maize.

In 2020, peas also showed an advantage as a predecessor in the experiment. The yield of Yuzhanka winter wheat in control variant 1, without fertilizers, according to the predecessor corn was 17.3 dt/ha, which is almost the same as the yield of winter wheat in 2019, the difference in yield did not exceed the smallest significant difference for this factor (Table 5).

Table 5. Impact of organo-mineral fertilization systems on the yield of corn and winter wheat in the variants of the 1979 field experiment, kg/ha, 2020.

Variants	Fertilizers		Field 1 Yuzhanka winter wheat		Field 2 Karat maize		Field 3 Karat maize	
	organic	mineral	crop yield	raise, %	crop yield	raise, %	crop yield	raise, %
1	0	O	17.3	0	30.7	0	28.5	0
2		NPK	29.0	11.7	51.6	20.9	53.7	25.2
3		1/2 NPK	23.6	6.3	46.7	16.0	44.7	16.2
4		1/3 NPK	20.0	2.7	45.2	14.5	32.6	4.1
5	Manure 50 t/ha once every 5 years	O	18.7	1.4	30.9	0.2	35.3	6.8
6		NPK	29.8	12.5	55.1	24.4	57.0	28.5
7		1/2 NPK	23.4	6.1	50.1	19.4	48.8	20.3
8		1/3 NPK	20.5	3.5	47.2	16.5	38.6	8.2
9	Green manure (rape)	O	18.4	1.1	31.0	0.3	28.5	0
10		NPK	29.0	11.7	52.1	21.4	54.0	25.5
11		1/2 NPK	24.8	7.5	49.2	18.5	44.3	15.8
12		1/3 NPK	20.2	2.9	48.2	16.5	36.3	7.8
13	Straw	O	18.8	1.5	31.3	0.6	30.7	2.2
14		NPK	29.7	12.4	54.4	23.7	58.3	29.8
15		1/2 NPK	26.0	8.7	51.3	20.6	49.1	20.6
16		1/3 NPK	19.7	2.4	48.1	17.4	38.3	9.8
17	Manure + straw + green manure	O	20.7	3.4	34.5	3.8	37.3	8.8
18		NPK	33.0	15.7	61.9	31.2	71.4	42.9
19		1/2 NPK	30.8	13.5	57.7	27.0	60.8	32.3
20		1/3 NPK	22.8	5.5	53.1	22.4	41.4	12.9
LSD_{05} for particular differences			0.23		1.22		1.46	
LSD_{05} for organic fertilizers			0.34		0.62		0.74	
LSD_{05} for mineral fertilizers and interactions			0.30		0.55		0.66	
Experiment accuracy, Sx%			1.03		1.13		2.84	

The effectiveness of the recommended mineral fertilization system when combined with organic resources in options 6, 10, 14, 18 (manure aftereffect, green manure, straw, their complex application) by 0.8; 0.0; 0.7; 4.3 dt/ha more than in variant 2, without organic fertilizers and amounted to: 29.8; 29.0; 29.7; 33.0 dt/ha.

The application of the recommended mineral fertilization systems, combined with the aftereffect and action of organic fertilizers (option 6,10,14,18), increased crop yields. Reducing the recommended mineral fertilizer doses by half (1/2 RD - options 7,11,15,19) reduced the yield of irrigated crop rotation by 13-15% compared to the yield at the full dose of fertilizers. The crop yield decrease at 1/3 of the recommended dose was 27-30%. The obtained yield indicators are processed mathematically, they are reliable.

The productivity of 1 crop rotation hectare also differs significantly in the direction of increase in the options of experiments in comparison with the control.

Studies have shown that it is recommended to use all possible techniques to replenish organic matter in the soil, especially in irrigation: manure application, straw dressing, maize foliar and other plant residues, as well as the cultivation of green manure crops.

4 Conclusion

Based on the research results conducted in 2018-2020, the following conclusions can be drawn:

1. The use of the recommended mineral fertilization system ($N_{90}P_{120}K_{60}$) in irrigation resulted in an increase in winter wheat yield of 10.0-19.6 dt/ha and in maize of 14.3-28.5 dt/ha compared to unprofitable cultivation.
2. The reduction of recommended mineral fertilizer doses by half (1/2 RD), reduced crop yields from 9 to 19% per year in crop rotation compared to crop yields with a full dose of fertilizers. With the use of 1/3 of the recommended dose of mineral fertilizers, the decrease in the yield of winter wheat and other crops of crop rotation was 17-39%.
3. The highest crop yield was produced in variants with the application of the fully recommended mineral fertilization system.
4. The highest crop yield for the period of research was formed in the 18 variant of the experiment, where the recommended full rate was applied and in different years according to the prevailing climatic conditions it was 83-88 for winter wheat and 101-232% higher for maize than without the use of fertilizers.
5. The productivity of 1 crop rotation hectare in different variants of experiments was 1.1-2.3 times, or 0.5-29.9 dt/ha higher than that of controlled crops without fertilizer.

References

1. A. S. Shpakov, *Adaptive Feed Production: Problems and Solutions* (2002)
2. E. D. Adinyaev *Cultivation of corn under irrigation* (1988)
3. V.I. Kiryushin, Agriculture, **5 (2)** (1987)
4. T. P. Lifanenkova, R.V. Bizhoev, *Alternative methods of reproduction of fertility of irrigated chernozem of ordinary steppe zone of the Central Ciscaucasia* (2011)
5. B. A. Armor, *Field experiment technique* (1985)
6. F. A. Yudin, *Methods of Agrochemical Research,* **11, 16, 100** (1971)
7. *Guidelines for conducting research in long-term experiments with fertilizers*, 1 (1975)
8. *Guidelines for research in long-term experiments with fertilizers,* 2 (1983)
9. *Register of long-term experiments with fertilizers and other agrochemicals of the Russian Federation* (2005)

Management of changes in companies of the agro-industrial complex: implementation of high technologies

Ekaterina Vinogradova, and *Anna* Galimova

Ural State University of Economics, Department of Information technology and statistics, 8 Marta St., 62, 620219 Ekaterinburg, Russia

Abstract. Digitalization is the most powerful tool for increasing the efficiency of agricultural enterprises. A computerized microclimate control system, the use of chip tags for animals and scanners to read information from these tags and enter data into the system, the use of flow meters, equipping machines with GPS sensors, the implementation of precision farming, space survey can improve the quality of production and monitoring, as well as neutralize the negative influence of the external environment on production efficiency. The ability to digitize agro-technological processes based on the creation of expert systems carrying out multiple factor analysis based on complex mathematical models. Monitoring and accounting of certain factors influence is performed on these platforms based on machine learning and a neural network. The introduction of high technologies should be carried out in close connection with the business processes reengineering, management of ongoing changes to achieve the maximum effect from the high technologies implementation.

1 Introduction

Currently, the main factors influencing the direction and intensity of organizational changes are the development of the market, technical progress and changes in the requirements for working conditions. The market makes new demands on the manufacturer due to the saturation of demand with consumer goods and the reduction in the life cycle of high-tech products. In this regard, companies are forced to expand the product range, strive to develop an innovative market and an increase production flexibility. Organizations can no longer cope with the increased needs of consumers without implementing high technologies into the production process.

Manufacturing of high quality products that can satisfy customer requirements is a complex process depended on many factors. In this regard, it is important for the manager to know in advance about all possible problems and be able to mitigate their negative impact in time. Economic planning is one of the most significant elements of the organization and management of production, which makes it possible to cover all areas of activity in their relationship with each other, as well as with external and internal factors.

A modern enterprise is a complex of interrelated factors of production, united by a technological process. In accordance with the marginalist theory, the classification of production factors includes labor (labor resources), land (natural resources), capital (financial resources), entrepreneurial abilities. Within the framework of the economic theory of post-industrial society, information resources are added to the factors of production, and then innovations, as an integral part of effective production. Under the influence of scientific and technological progress, the development of innovations plays a significant role in improving the product quality and the enterprise competitiveness.

This approach will enable to control over the dynamics of all activities of the enterprise in their relationship, which will ensure the coordinated and targeted work of individual subsystems. The main tasks of implementing high technologies in the agro-industrial complex are the maximum efficient use of resources, primarily land, and control over their use, large farms' management, and difficult climatic conditions.

2 Materials and methods

The specificity of the high-tech measures implementation in the agro-industrial complex is due to a complex assessment of its cost in the formation of planning documents. This is due to the following factors:

- a wide range of products created at different stages of the life cycle during the activities implementation;
- the use of initial data with various degrees of reliability and details to estimate the activity cost;
- a large number of factors affecting the cost of product creation [5];
- an actual duration of implementation may vary significantly from the planned one;
- the probability of the desired result's non-availability from the implementation.

The exclusive introduction of high technologies into the practice of agro-industrial enterprises does not ensure high efficiency of the company even in case of significant implementation costs [6]. This is due to recurring problems in many organizations, which are characteristic of the insufficient preparation of the management process for innovation implementation. Potential problems include:

1. Absence or formal implementation of the preparatory stage. Some owners or managers rely on intuitive desires and advertisements in the selection of implemented innovations. At the same time, manufacturers of high technologies are not sufficiently aware of the current problems of the agro-industrial complex [8]. Evidently, they are based on the experience of Western countries and do not correspond to the modern realities of Russian enterprises. As a result, the selected high technologies are not effective enough. For example, at present, the priority is the problem of inaccurate data from the fields, lack of technology, rather than heterogeneity of soil fertility.
2. Limitation of the implementation period. Based on the results of the implementation, agro-industrial enterprises pay tribute to innovations, avoiding verification and serious analysis based of the implementation results. The main achievement for them was the fact of implementation without further monitoring and improvement of changes [4]. Cosequently, significant errors are unnoticed, leading to a decrease in the implementation efficiency.
3. Inconsistency between the desired high-tech impact and the company's strategic objectives can cause irreparable damage to the enterprise due the following reasons:

- this difference causes inconsistency in actions of the structural divisions' heads;
- there may be a lack of ability to complete orders, in particular those of strategic importance;

- in connection with the resource allocation, not in accordance with strategic objectives, but according to the subjective opinions of the enterprise management, the main task of the employees is to fight for financial resources, and not to achieve the goals set [7].

Many companies do not have an approved list of strategic goals. In this regard, in such enterprises there is no certain system of resource allocation, subjective assessments of managers are the only factor influencing the formation of priority [10]. Nevertheless, a carefully formulated strategy in many cases does not justify the allocation of financial resources.

This factor in the future may be viewed as a reason for losing the company's competitive advantage, as well as its market share.

4. The main goal of the implementation is to reduce costs at the expense of creating value for the company's owners and customers. Enterprises aim at reducing costs across business units by leveraging resources efficiently [3]. This approach does not lead to an increase in the product quality and the development of the customer base and, as a result, it is not concerned with the creation of additional value for the customers.

5. Employees are not ready for changes. Managers conduct lectures formally on the need to introduce high technologies, and it does not lead to the staff's interest in the upcoming changes [1].

3 Results and discussion

The introduction of high technology refers to organizational changes and cannot be a one-time action. It should be carried out gradually, and then constantly support the changes made, and improve them. The authors suggest to use the PDCA cycle to improve the activities of the agro-industrial complex by implementing high technologies.

1. A thorough analysis of the actual state of the company's internal processes and changes in the external environment (political changes, instability of the situation in the national market, the emergence of new proposals in the field of enterprise activities and others) before choosing the technologies necessary for implementation will enable us to assess the improvement potential and analyze the implementation efficiency. The preliminary analysis will create a basis for assessing the improvement potential and conducting the implementation efficiency analysis.

An agro-industrial enterprise at the stage of concluding a contract should have the opportunity to study comprehensively the cost and time indicators of high technologies in order to maximize the leveling of information asymmetry and determine the correct price for the introduced type of high-tech products.

The cost of a high-tech product is influenced by several factors. For a high-tech enterprise, an important role is played by the amount of costs incurred and the expected profitability [2]. But due to the product uniqueness for the agro-industrial complex, the consumer value, which is the benefit to the buyer, is crucial in determining the value. Within the framework of this study, the use of the following groups of indicators is proposed for assessing consumer value:

– a set of basic characteristics of high-tech products planned for the use in the agro-industrial complex;

– cost indicators;

– product implementation effect.

The cost indicators, which are the basis for determining the consumer value of high-tech products, include the cost of implementing individual works and the entire life cycle of a prototype, the unit cost of the main characteristics of the sample, the effect of use [9]. The definition of customer value is based on a comparison of the listed indicators of the activity under consideration and the counterpart activity.

2. Approbation of a high-tech sample with a decomposition of the project into manageable elements of work, for which it is easy to determine costs and build performance schedules, study the results at the intermediate stages.
3. Control and rigorous analysis of the intermediate results obtained for large-scale improvement transfers. Monitoring of exceptional cases related to external causes is carried out. In the event of errors connected with process imperfection, the second stage is returned and corrected until a satisfactory result is obtained.
4. Large-scale introduction of high technologies, documentation and regular control. Repeating this cycle to debug and improve the work process.

To prepare employees for high-tech implementations, it is necessary to go through the stages of the K. Levin change model, based on the formula "The higher the social value of the group standard, the greater will be the resistance of an individual group member to move away from the level of this group standard." To introduce high technologies, managers of the agro-industrial complex companies need to implement consistently the following steps with the choice of some proposed methods (depending on the situation, type of high-tech products, policy of the company owner and managers):

1. defrosting (loosening of group standards if necessary):
- method of bridge burning (artificial crisis creation);
- command (order) method;
- MBO (management by objectives, enabling selection);
- argumentation;
- destabilization and search for ways of development;
- learning to accept changes;
- company's reorganization;

2. change (transition):
- method "boil a frog" (invisible implementation of changes);
- "take on weakly" (challenge the employee in order to motivate him);
- training and coaching;
- "first step" method (positive emotional perception of the first stage of changes);
- enrichment of job responsibilities (to enable the employee to feel his importance and significance in the process of change);
- "move - freeze" (changes in small parts with intermediate stabilization).

3. freezing (fixing at a new level):
- method of bridge burning (no way back),
- flow of facts (constant reminder of the success of change),
- "gold handcuffs" (favorable offers to key employees),
- money motivation
- socialization (attracting informal leaders to the side of change).

Involving employees in the change management process will lead to additional motivation and an increased desire to implement high technology, which will have a positive impact on the result.

4 Conclusions

Thus, there are some problems that enterprises face in the process of implementing high technologies, as well as controlling the plans implementation. But, nevertheless, the authors show possible ways of solving or minimizing these problems. The high-quality implementation of high technologies is currently necessary for the agro-industrial complex. Managers must pass the company through a series of the successive stages, without diminishing the significance of any of these stages. Then it will be possible to avoid the thoughtless implementation of foreign developments and the digitalization discrediting as a tool for solving production problems. Having been adapted to modern realities through the high technologies implementation, the Russian agricultural complex will be actively involved in developing precision farming technologies and a number of other technologies for translating data into a digital reality.

References

1. V. Botrić, L. Bozić, Int. J. Innov. Technol. Manage. **4**, 1850033 (2018)
2. H. Dzwigol, S. Shcherbak, M. Semikina, O. Vinichenko, V. Vasiuta, Acad. Strateg. Manag. J. **18 (1)**, 6 (2019)
3. O. V.Glinkina, S.A.Ganina, A. V. Maslennikova, T. A. Solostina, M. Viktorovna Soloveva, IJM. **3(19)**, 461 (2020)
4. N. I. Kuznetsov, N. V. Ukolova, S. V. Monakhov, J. A. Shikhanova, J. Environ. Manag. Tour. **3**, 608 (2017)
5. E. A. Skvortsov, E. G. Skvortsova, I. S. Sandu, G. A. Iovlev, Econ. Reg. **3**, 1021 (2018)
6. K. S. Ternovykh , V. V. Kurennaya, A. V. Agibalov, Vestnik VSAU. **2**, 98 (2020)
7. I. Trunina, O. Onyshchenko, O. Vartanova, O. Sushchenko, International J. of engineering and technology, **7(4.3)**, 530 (2018)
8. M. L. Vartanova, E.V. Drobot. Econ. relations, **1**, 12 (2018)
9. E. Yu. Vinogradova, A. V. Babkin, A. I. Galimova, S. L. Andreeva, Proceedings of 2017 SPUE, 63 (2017)
10. E. Vinogradova, O. Nikoliuk, A. Galimova. E3S Web of Conferences, 208 (2020)

E3S Web of Conferences **262**, 02001 (2021) https://doi.org/10.1051/e3sconf/202126202001
ITEEA 2021

New generation protein supplement in combined feeds for broiler chickens

E.N. Andrianova*, and *I. A.* Egorov

Federal Scientific Center All-Russian Research and Technological Poultry Institute RAS, 10 Ptitsegradskaya Str., 141311 Sergiev Posad, Moscow region, Russia

Abstract. In experiments on broiler chickens of the SGC "Smena" selection, it was found that the partial replacement of fish meal with an additive based on a protein of microbial origin, containing 68.3% of crude protein, allows to provide a live weight of chickens comparable to the control with high poultry safety, due to good protein digestibility and availability of amino acids. A comprehensive assessment of the productivity of broilers according to the productivity index showed that the replacement of 50% of fish meal with microbial protein contributed to an increase in the productivity index of chickens of the second group by 4.78 points in comparison with the control. The complete replacement of fish meal in the third experimental group was less effective, as evidenced by a decrease in the productivity index to 303.91 points compared to 308.73 in the control. It is shown that the introduction of a protein supplement into the feed for broiler chickens in Analysis of the chemical composition of the liver of 35-day-old broilers showed that the protein content in the liver of chickens of the second and third groups was higher than the control by 1.99 and 0.58%, while there was no increase in the fat content in the groups with a protein supplement. The crude fat content was at the level of 11.38 and 12.20% versus 13.0% in the control, which indirectly indicates the absence of a cytotoxic effect on liver cells from the use of microbial protein in the studied dosages.

1 Introduction

Use in feed production of protein sources based on non-traditional feed means, such as low-alkaloid lupine, rapeseed, peas, etc. [1,2,3,4] allows for reducing the use of soybeans and products of its processing in poultry rations, to reduce the inclusion of animal feed in compound feeds [5,6]. Protein products obtained with the use of improved technologies of microbiological synthesis [6,7,8,9], which replaced traditional technologies that allowed the production of microbial protein from hydrocarbon raw materials, are also worth noticing. Despite the high production volumes of paprine, eprin and haprin for the needs of agriculture in the 90s, the industrial production of microbial protein obtained from hydrocarbon raw materials at Russian biochemical plants was stopped due to the negative impact on the environment due to imperfect purification technologies, which led to

*Corresponding author: andrianova@vnitip.ru

pollution of the environment and atmosphere with protein dust and cells of producer microorganisms. Currently, the production in Russia of haprin, meprin and eprin using traditional technologies is practically absent, and from the products of microbiological synthesis, the feed industry mainly uses fodder yeast, which is obtained by the microbiological method on various nutrient media from the waste of timber processing sulfite-cellulose and alcohol production [9].

The purpose of these studies was to study the possibility of using a functional protein supplement obtained from plant resources based on microbiological synthesis technologies to replace animal feed in compound feed for broiler chickens.

2 Materials and methods

To accomplish this task, an experiment was carried out on broiler chickens of the SGC "Smena" selection from one day old to 35-day-old in the vivarium of the SGC "Zagorskoye", Federal Scientific Center All-Russian Research and Technological Poultry Institute RAS. Broilers were kept in a R-15 cage battery. The groups were formed by the method of analogs by live weight without division by sex, 35 animals per group. Chickens received dry full-feed loose mixed feed with nutritional value according to the standards of Federal Scientific Center All-Russian Research and Technological Poultry Institute 2019, ad libitum, according to the scheme shown in Table 1.

Table 1. Scheme of the broilers experiment.

Group	Feed characteristic
1 (c)	Nutrient-balanced compound feed containing fish meal
2	Compound feed, balanced in all nutrients, replacing 50% of fish meal with protein supplement during the entire growing period
3	Compound feed, balanced in all nutrients, with replacement of 50% of fish meal with protein supplement from 3-21 days and 100% replacement of fish meal from 22 to 35 days of growing

Table 2. Composition and nutritional value of compound feed for broiler chickens, %.

Constituent	Group					
	3-21 days			22-34 days		
	1 c	2	3	1 c	2	3
wheat	53.1	52.7	52.7	55.7	53.4	53.0
full-fat soya beans	10.0	10.0	10.0	25.0	27.0	27.0
soybean meal SP 42%	18.0	18.0	18.0	-	-	-
sunflower meal SP 36%	5.0	5.0	5.0	8.0	8.0	8.0
corn gluten	2.0	2.0	2.0	-	-	-
fish meal SP 67%	4.0	2.0	2.0	4.0	2.0	-
Protein supplement	**-**	**2.0**	**2.0**	**-**	**2.0**	**4.0**
soybean oil	4.7	4.7	4.7	4.0	4.0	4.0
lysine monochlorohydrate	0.27	0.31	0.31	0.30	0.30	0.35
DL-methionine	0.26	0.26	0.26	0.25	0.25	0.27
fine salt	0.30	0.33	0.33	0.30	0.33	0.36
monocalcium phosphate	0.7	0.8	0.8	0.8	0.90	1.0
limestone flour	1.3	1.5	1.5	1.3	1.5	1.70
premix	0.35	0.35	0.35	0.35	0.35	0.35
100 g of compound feed contains:						
Available energy, kcal MJ/kg	308 12.89	307 12.84	307 12.84	318 13.31	318 13.31	317 13.26
Crude protein	23.49	23.51	23.51	21.64	22.14	22.18

Table 2. Continued.

Crude fat	8.09	8.01	8.01	9.94	10.20	10.11
Linoleic acid	3.97	3.96	3.96	4.94	5.09	5.08
Crude fiber	4.32	4.42	4.42	4.24	4.39	4.49
Lysine	1.36	1.36	1.36	1.25	1.25	1.25
Methionine	0.64	0.63	0.63	0.60	0.59	0.59
Methionine + cystine	0.98	0.97	0.97	0.90	0.90	0.90
Assimilable lysine	1.21	1.19	1.19	1.10	1.09	1.08
Assimilable methionine	0.59	0.57	0.57	0.54	0.53	0.53
Methionine + assimilable cystine	0.87	0.85	0.85	0.77	0.77	0.77
Calcium	0.91	0.92	0.92	0.90	0.91	0.92
Phosphorus	0.67	0.66	0.66	0.69	0.69	0.69
Available phosphorus	0.40	0.40	0.40	0.41	0.41	0.41
Potassium	0.82	0.82	0.82	0.73	0.75	0.74
Sodium	0.17	0.17	0.17	0.17	0.17	0.16
Cl	0.30	0.31	0.31	0.3	0.31	0.32

During the first three days, broilers of all groups received pre-starter compound feed in the form of granules with a particle diameter of 0.9 to 1.2 mm of the same nutritional value.

Balance experiments to determine the digestibility and availability of nutrients in the diet were also carried out on males (n = 3 from each group) at 30-33 days.

The mean values (*M*) and mean errors (± m) were calculated for the indicators of counting and measurements. The significance of differences was assessed by Student's *t*-test, the differences were considered significant at $P \leq 0.05$. The analysis of the chemical and amino acid composition of the protein product, feed and litter, calcium and phosphorus in the tibia of broilers at the end of the experiment was carried out by specialists of the Department of Physiology and Biochemical Analysis of the Federal Research Center "Federal Scientific Center All-Russian Research and Technological Poultry Institute" RAS.

3 Results and discussion

The investigated additive is a fodder protein mixture of fermentolysate of plant raw materials and biomass of lactic acid and propionic acid bacteria, obtained as a result of bioconversion using non-pathogenic strains of a consortium of microorganisms. Its chemical and amino acid composition is shown in Table 3.

Due to the high content of non-protein nitrogen in the product - 3.41%, the complete replacement of fish meal for broilers of the third group was carried out in the final growing period.

Table 3. Chemical and amino acid composition of the protein supplement (% on air-dry basis).

Indicator	Protein supplement	Fish meal (tabular data)
Crude protein	**68.3**	**67.0**
Non-protein nitrogen	**3.41**	-
Crude ash	5.90	14.0
Fat	3.35	7.4
Calcium	0.33	4.26
Phosphorus	1.44	2.5
aspartic acid	6.52	6.0
threonine	3.1	2.81
Serine	2.19	2.86
glutamic acid	6.41	9.1
Proline	3.21	3.17
glycine	3.64	4.54
Alanine	4.82	3.93
Valin	4.45	3.67

Table 3. Continued.

Isoleucine	3.23	2.83
Leucine	5.11	4.65
Tyrosine	2.64	2.0
Phenylalanine	3.47	2.80
Histidine	1.08	1.44
Lysine	3.49	5.11
Arginine	3.31	3.89
Cystine	0.41	0.62
Methionine	1.63	1.88

The results of the experiment (Table 4) showed that at 14 days of age the live weight of the chickens of the second and third experimental groups, which received microbial protein instead of 50% fish meal, was higher than the control by 4.16 and 3.98% (the difference is significant at $P \leq 0.05$), at 21 days of age the advantage of the experimental groups in terms of live weight in comparison with the control was 1.06 and 0.71% (the difference is unreliable).

In the final period of rearing, the chickens of the second experimental group, which received 50% (2% by weight of the compound feed) protein supplement instead of fish meal, retained a higher growth rate in comparison with the control poultry. The live weight of the chickens in this group was 1.3% higher than the control (the difference is not statistically significant). Furthermore, the live weight of the males was 2.4% higher than that of the control, and the females - by 0.5%. The use of a protein supplement contributed to a decrease in feed costs per 1 kg of live weight gain by 0.17% in comparison with the control.

Complete replacement of fish meal with an additive in mixed feed for chickens of the third experimental group in the final period of growing provided a live weight of chickens comparable to the control, however, the backlog in live weight of males of the third group from the control was 1.2%, and that of females - 1.4% (the difference is not reliable). The feed conversion also deteriorated by 1.36%.

There were no significant differences in the viability of poultry in the studied groups. The safety of the flock in all groups at the end of cultivation was high and amounted to 100%.

A comprehensive assessment of the productivity of broilers according to the productivity index showed that replacing 50% of fish meal with an additive made it possible to increase the productivity index of chickens of the second group by 4.78 points in comparison with the control. The complete replacement of fish meal in the third experimental group was less effective, as evidenced by a decrease in the productivity index to 303.91 points compared to 308.73 in the control.

Table 4. Productivity of broiler chickens at different levels of feed additive inclusion ±.

Indicator	Group		
	1 (c)	2	3
Flock safety, %	100	100	100
Live weight, g at the age, days: 7	164.7 ± 1.1	164.8 ± 1.2	165.1 ± 1.0
14	430.1 ± 4.4	448.0±6.2* + 4.16%	447.2±6.1* + +3.98%
21	871.30 ± 12.0	880.5 ± 10.8 + +1.06%	877.5 ± 11.4 + +0.71%
35	1894.80 ± 33.62	1920.31 ± 34.23 + +1.3%	1890.72 ± 30.39 + -0.2%

Table 4. Continued.

Including male chicks female chicks	2062.20 ± 35.03 1769.25 ± 30.43	2110.93 ± 31.21 + +2.4% 1777.35 ± 24.92 + +0.5%	2036.89 ± 29.17 + -1.2% 1744.56 ± 20.92 + -1.4%
Feed costs per 1 bird, kg	3.27	3.31	3.308
Feed costs per 1 kg of live weight gain, kg	1.765	1.762	1.789
Average daily gain in live weight, g	54.49	55.24	54.37
EPI, point	308.73	313.51	303.91

*$P \leq 0.05$.

The data of the balance experiment presented in Table 3 are consistent with the obtained zootechnical results. Thus, the digestibility of protein, dry matter of feed, fat, the use of nitrogen, calcium, the availability of lysine and methionine were better in broilers of the second experimental group by 0.26; 1.72; 1.53; 1.1 and 0.7% compared to control. Fiber digestibility and methionine availability were at the control level.

Table 5. Digestibility and nutrient utilization (%), broilers aged 30-33 days.

Indicator	Group		
	1 (c)	2	3
Protein digestibility,	92.04 ± 0.47	92.3 ± 0.51	91.67 ± 0.42
dry basis of feed.	83.76 ± 0.39	85.48 ± 0.40	83.37 ± 0.41
fat,	81.8 ± 0.35	82.9 ± 0.33	82.0 ± 0.34
fiber	34.3 ± 0.27	34.5 ± 0.24	34.1 ± 0.22
Use of nitrogen	55.8 ± 0.30	56.5 ± 0.29	54.4 ± 0.25
calcium	60.57 ± 0.34	62.01 ± 0.37	60.2 ± 0.35
phosphorus	48.5 ± 0.30	48.9 ± 0.29	49.5 ± 0.25
Availability of lysine	91.1 ± 0.49	90.8 ± 0.54	87.5 ± 0.50
methionine	91.4 ± 0.50	91.7 ± 0.52	91.3 ± 051

Complete replacement of fish meal in the final growing period in broilers of the third experimental group led to a slight decrease in protein digestibility, feed nitrogen use and lysine availability - by 0.37; 1.4 and 3.6%. The digestibility of fat, dry matter of feed, fiber, availability of methionine in chickens of the third group was comparable to the control.

There were no significant differences between the groups in terms of the content of crude ash, the deposition of calcium and phosphorus in the skeleton of broilers from the control and experimental groups. The indicators corresponded to the physiological norm and the direction of poultry productivity.

Table 6. Chemical composition (%) and vitamin content (μg/g) in liver of 35-day-old broilers.

Indicator	Group		
	1 (c)	2	3
Moisture	78.01	74.01	74.94
Crude ash	4.85	4.92	4.90
Crude protein	71.24	73.23	71.82
Crude fat	13.00	11.38	12.20
Vitamin A	417.16	398.50	351.15
Vitamin E	17.22	19.84	20.68
Vitamin B_2	12.98	13.25	12.92

Analysis of the chemical composition of the liver of 35-day-old broilers showed (Table 4) that the protein content in the liver of chickens of the second and third groups was higher than the control by 1.99 and 0.58%, while there was no increase in the fat content in the groups with a protein supplement. The crude fat content was at the level of 11.38 and

12.20% versus 13.0% in the control, which indirectly indicates the absence of a cytotoxic effect on liver cells from the use of microbial protein in the studied dosages.

In terms of the storage of vitamin E in the liver, the chickens of the second and third experimental groups exceeded the control by 15.21 and 20.09% (the difference is significant), the content of vitamin B2 was at the control level. In the liver of chickens of the third experimental group, a significant deterioration in the deposition of vitamin A by 18.8% was noted.

Poultry meat is mainly a protein food and one of the important sources of fat in the body.

Analysis of the chemical and amino acid composition of the meat of experimental broilers showed (Table 5) that partial or complete replacement of fish meal with an additive based on microbial protein in the feed of chickens of the second and third experimental groups could not fully compensate for the protein of fish meal. The meat of the experienced broilers in terms of protein content was inferior to the control by 4.57 and 5.07%. The content of non-essential and essential amino acids also decreased by 4.22 - 4.26% and 1.19 - 1.28%.

It should be noted that the fat content in broiler meat was low, and was in the range of 2.04 - 1.84%, versus 1.82% in the control. There were no significant differences from the control in terms of the content of crude ash. The conducted tasting evaluation did not reveal differences in the taste of meat and broth between the control and experimental groups.

Table 7. Chemical composition and amino acid content in the breast muscles of broiler chickens at the age of 35 days, (%) on air-dry basis.

Indicator	Group		
	1 (c)	2	3
Moisture	73.69	72.48	71.16
crude protein	84.79	80.22	79.77
crude fat	1.82	2.04	1.84
crude ash	4.24	4.00	4.13
Total amino acids, %	**79.69**	**74.28**	**74.10**
Non-essential amino acids, %	**48.24**	**44.02**	**43.98**
Essential amino acids	**31.45**	**30.26**	**30.12**

4 Conclusion

Thus, based on the data obtained, it can be assumed that in terms of bioavailability and productive effect on poultry, a new generation protein supplement can be used to partially replace feed of animal origin in compound feed for broilers. The rational level of its inclusion in compound feed for broiler chickens up to 21 days of age is 2% by weight of compound feed, and from 22nd day of age it is possible to increase the input level up to 4% by weight of compound feed.

References

1. N. T. Bingol, S. I.Dede, M. A. Karsli, Y. Değer, D. Kiliçalp Kılınç, S. Çetin, Brazilian Journal of Poultry Science, **18(4),** 639 (2016)
2. G. Ciurescu, C. O. Pana, Romanian Biotechnological Letters, **22(1),** 12204 (2017)
3. E. N. Andrianova, I. A. Egorov, E. N. Grigorieva, A. N. Shevyakov, V. V. Pronin, Agricultural Biology, **54(2),** 326 (2019)
4. E. N. Andrianova, I. A. Egorov, V. V. Pronin, Agricultural Biology, **55(6),** 1245 (2020)

https://doi.org/10.1051/e3sconf/202126202001

5. V. I. Fisinin, I. A. Egorov, *Poultry feeding guide,* 215 (2019)
6. I. A. Egorov, *Unconventional feeds and additives,* 391 (1992)
7. Diana N, *Biotechnology and food of the future* (2016) https://www.agroxxi.ru
8. I. A. Egorov, V. A. Manukyan, T. M. Okolelova, T. N. Lenkova, E. N. Andrianova, A. N. Shevyakov, T. V. Egorova, T. A. Egorova, E. Yu. Baykovskaya, N. N. Gogina, et al, *Poultry feeding guide*, 226 (2018)
9. E. S. Volobueva, *Biotechnology for the production and use of functional additives based on microbial conversion of plant raw materials,* 21 (2020)

Influence of polymorphism of the kappa-casein gene of cows on the development of calves in the early postnatal period

Olga Polozyuk[1,*], *Margarita* Zabelina[2], *Tatiana* Preobrazhenskaya[2], *Igor* Tyurin[2], and *Elena* Lakota[3]

[1]FSBEI HE Don State Agrarian University, st. Michurina, 39/1, Persianovsky settlement, 346493 Oktyabrsky District, Rostov Region,
[2]FSBEI HE Saratov State Agrarian University named after N. I. Vavilov, Teatralnaya sq. 1, 410012 Saratov
[3]FSBRI "Federal Agrarian Scientific Center of the South-East", st. Tulaykova, 7, 410010 Saratov

Abstract. The authors found that the density of colostrum in cows with a homozygous genotype for the B allele of the kappa-casein gene was higher compared to the homozygous genotype for the A allele and had a greater amount of immunoglobulins by 29.5 mg/l, respectively. The formation of high colostral immunity in calves obtained from cows of mothers of genotypes AB and BB prevented the occurrence of diseases of the gastrointestinal tract, and in calves obtained from mothers of genotype AA genotype kappa-casein with a reduced amount of immunoglobulins in colostrum, it led to the development of diarrhea in 12% of newborn calves. The dynamics of the live weight of calves obtained from cows of mothers with the AA genotype of the kappa-casein gene, during the first 30 days of life, was 6.34 and 4.78 kg lower than in young cows with the BB and AB genotypes of the kappa-casein gene. At the age of one month, the leukocyte content in calves obtained from cows from mothers with a homozygous genotype for allele A of the kappa-casein gene was less by 10.8 and 16.2%, and hemoglobin by 10.5 and 10.7%.

1 Introduction

Assessment of the accuracy of the breeding value of an animal largely depends on the efficiency of selection and breeding work in animal husbandry. The introduction of molecular genetics into applied science made it possible to assess the genetic potential of animal productivity not only on the basis of the phenotype, but also at the DNA level. When using traditional breeding methods, there is a great danger of a decrease in the breed, herds, lines, families of genetic diversity, which significantly affects not only productivity and reproduction, but also sharply limits the adaptability of an individual to the conditions of keeping and viability of the resulting offspring.

*Corresponding author: polozyuk7@mail.ru

The use of DNA markers makes it possible to assess animals at the level of the hereditary inclinations of the organism, since the use of selection for genetic markers brings selection to a new level, allowing direct assessment of genotypes [1-6].

The creation of genomic databases allows for a quick search for information on localized genetic markers in a particular animal species, displays the general picture of the study of genomes, including information on mutant phenotypes. Knowing the nucleotide sequence and localization of a specific marker, it is possible to identify polymorphic regions, which subsequently makes it possible to establish a connection between the polymorphic variant and one or another economically useful trait, intrauterine pathology, neonatal disease, etc. [7-11]. One of the genes that control the quality of milk is the gene for kappa-casein, since the amount of immunoglobulins depends on its density. In this regard, it is an important selection criterion for dairy cattle breeds in terms of dairy production and enhancing the safety of newborn young animals [12].

Therefore, obtaining healthy young animals, resistant to diseases, unfavorable conditions of keeping, capable of realizing their genetic potential in the breeding process, is urgent.

The aim of the research was to examine Holsteinized black-and-white cows for the presence of a mutation in the kappa-casein gene, followed by the effect of immunoglobulins of the first portion of colostrum on the growth and development of newborn calves.

2 Material and methods

The experimental part of the work was performed in the conditions of the Matveevo-Kurgan district of the Rostov region on the basis of the MTF in LLC "Leninskoe Znamya" on Holsteinized black-and-white cows, as well as calves obtained from them. The cows were similar in calving season, age, live weight. The conditions of detention were the same. The frequency of occurrence of allelic variants of the kappa-casein gene was performed using PCR-RFLP methods.

For this study, 60 cows were tested during the dry period. For DNA genotyping, blood was taken from the tail vein into special vacuum tubes for PCR studies, which were placed in a thermos - refrigerator and delivered for research to the laboratory of molecular diagnostics and biotechnology of the regional laboratory. Amplification of the kappa-casein gene fragment was performed by PCR [13]. Based on the results of DNA genotyping, the influence of the cow's genotype for the kappa-casein gene on the immunoglobulin content of the first portion of colostrum and its effect on the development of newborn calves was assessed.

Throughout the experiment, as calves were born, the concentration of immunoglobulins in colostrum was determined by the kappa-casein gene in fresh cows of different genotypes in the first hours after calving. For this, before the first feeding of colostrum to the calf, colostrum was milked from fresh cows, and its density was determined using a lacto densimeter. The determination was performed by immersing the lacto densimeter in colostrum, previously poured into a cylinder with divisions from 1.020 to 1.080. The immersion mark was the density of the milk.

Determination of immunoglobulins in the blood serum of newborn calves was performed by the end of the first day after birth. In newborns, the amount of serum immunoglobulins was determined using a manual milk refractometer (MMR). For this, the blood of their jugular vein was taken into a test tube and placed in warm water, while the blood was standing. After that, a drop of serum was placed on a refractometer and the result was determined using a scale.

The first eight days after calving, newborn calves received three times colostrum from mothers cows, and then, up to 30 days of age, were fed with bulk whole milk pursuant to the drinking scheme approved by the farm. Newborn calves up to 8 days of age were kept in separate boxes for no more than 4 heads, the age difference in birth in each box was 2 days. Boxes for keeping newborn calves were filled on the principle "everything is empty everything is busy". After the transfer of the calves to the calf shed of the dispensary, mechanical cleaning, disinfection and bactericidal ultraviolet irradiation of the box were performed.

Morphological parameters: the concentration of hemoglobin was determined on a photoelectric calorimeter, the number of erythrocytes and leukocytes in the Goryaev chamber. Blood for research was taken from calves from the jugular vein in the morning at the same time, before feeding the animals on the 2nd and 30th days of life.

3 Results and discussion

According to the results of DNA testing performed by the researchers, three variants of the genotype were identified - AA, AB, BB, which have different frequencies of occurrence [14].

After DNA genotyping in 60 cows during the dry period, it was found (Table 1) that the frequency of individuals with the AB genotype was the most common, which was 63.3. In contrast to the AB genotype, cows with the BB genotype were encountered much less frequently, only 6.7%.

Table 1. Frequency of genotypes in cows by polymorphism of kappa-casein genes.

Breed	Genotypes		
	AA	AB	BB
Holstein-Frisian	Genotype frequency, animals		
	18	38	4
Holstein-Frisian	Genotype frequency, %		
	30	63.3	6.7

After birth, the only food for calves is colostrum. In this case, colostrum, during the transition of the calf from placental nutrition to nutrition in the environment, is the main link. Only in the first 24 hours after birth, the calf receives antibodies with the mother's colostrum, which are freely absorbed into the blood through the integumentary epithelium of the small intestine, and the amount of immunoglobulins that enter the body will determine its ability to respond to pathogenic agents. This is due to the fact that colostrum has a bactericidal effect, since it contains lysozyme, a substance capable of dissolving the membranes of microorganisms, as well as functionally active leukocytes and lymphocytes. Herewith, colostrum creates an acidic environment in the abomasum, as it has an increased acidity, which has a detrimental effect on conditionally pathogenic and pathogenic microflora, which prevents the development of putrefactive processes. Colostral route with colostrum is transmitted to newborns with autoantibodies and sensitized lymphocytes against certain cellular, tissue and organ antigens in case of autoimmune organ damage in mothers, which leads to upset gastrointestinal tract, indigestion, impaired enzyme formation and slowing down of enzymatic processes, intoxication and intoxication, the appearance of diarrhea [15].

All newborn calves were fed the first portion of freshly milked colostrum from mother cows when the sucking reflex appeared within two hours after birth. We found that cows with a homozygous genotype for allele B had the highest colostrum density (tab. 2), which averaged 1.067, which corresponded to the content of immunoglobulins of 106.7 mg/l. After the first feeding of colostrum, after 24 hours, blood was taken from the calves and the amount of immunoglobulins was determined. In calves obtained from cows of genotype BB of the kappa-casein gene, the amount of immunoglobulins in the blood serum of day-old calves averaged 5.8 g/dl.

Table 2. The amount of immunoglobulins in cow colostrum and calf blood serum.

Genotypes	The amount of immunoglobulins in colostrum, mg/l	The amount of immunoglobulins in the blood serum of calves, g/dl
BB	106.7±0.1***	5.8±0.1**
AB	97.8±0.1**	5.6±0.1*
AA	77.2±0.1	5.2±0.1

P>0.95*; P>0.99**; P>0.999***

In fresh cows with the most common frequency of the AB genotype of the kappa-casein gene, the relative density of colostrum was within 1.064, which corresponded to the content of immunoglobulins in colostrum of 97.8 mg/l. The amount of immunoglobulins in the blood serum of these newborn calves by the end of the first day of life averaged 5.6 g/dl.

The motor activity of newborn calves obtained from mothers with genotypes BB and AB of the kappa-casein gene was better expressed in the first hours after birth. They, on average, 18 minutes earlier than their peers received from mothers with the AA genotype of the kappa-casein gene, exhibited a sucking reflex, rose to their feet more actively, and reacted faster to external stimuli. As they grew and developed, these calves began to eat the proposed feeding earlier, no problems with disturbance of the gastrointestinal tract were observed throughout the experiment, and the safety of the calves was 100%.

In cows with homozygous genotype A of the kappa-casein gene, the relative density of colostrum was slightly lower and averaged 1.057, and the amount of immunoglobulins was 29.5 (P>0.999) and 20.6 mg/l (P>0.99) less than in cows of mothers with genotypes BB and AB of the kappa-casein gene. A decrease in the amount of immunoglobulins in colostrum resulted in a decrease in the amount of immunoglobulins in the blood of calves by 0.6 and 0.4 g/dL in comparison with analogs of young animals obtained from cows with BB and AB genotypes. A decrease in immunoglobulins in the blood serum of calves resulted in gastrointestinal upset in 12% of the calves born. The disease manifested itself in a decrease in appetite: the calves fiddled with the teat drinker, while milk consumption either did not occur or was not consumed in full. In such calves, an increase in the temperature in the oral cavity was noted, although the overall temperature was within the physiological norm. In the future, they noted an increase in the frequency and liquefaction of fecal masses. The calves lay for a long time with their eyes half-closed, with their heads stretched out or laid on their stomachs, they reacted poorly to external stimuli, pain was noted on palpation of the abdominal wall. The high incidence of morbidity in calves obtained from cows of mothers with a homozygous genotype for the allele A of the kappa-casein gene confirms the low efficiency of colostral immunity formation due to the low density of colostrum, and, consequently, the insufficient content of immunoglobulins in colostrum.

Thus, a direct relationship has been established between the content of immunoglobulins in the blood serum of calves and its presence in the colostrum of mothers cows. It follows from this that the higher the level of immunoglobulins in colostrum, the higher the amount of immunoglobulins in the blood of newborn calves, and therefore the higher the body's protective response to the action of opportunistic and pathogenic microflora.

The live weight of calves obtained from cows with the BB genotype of the kappa-casein gene at birth was 3.1 and 0.34 kg more than their peers obtained from cows with the AA and AB genotypes of the kappa-casein gene.

The growth and development of young animals in the postnatal period depends on the dynamics of growth, development and nutritional status. The results show that the growth rate of calves obtained from cows of different genotypes was significantly different. Thus, the dynamics of the live weight of calves (tab. 3), obtained from cows of mothers with the AA genotype of the kappa-casein gene, during the first 30 days of life was 6.34 ($P>0.99$) and 4.78 kg ($P>0.95$) lower than in young cows with genotype BB and AB of the kappa-casein gene. A higher growth rate was characteristic in the first month of life for calves obtained from cows of mothers with a homozygous genotype for the B allele of the kappa-casein gene, as a result of which the absolute weight gain was 3.1 kg higher than in young animals obtained from cows with a homozygous genotype for allele A of the kappa-casein gene.

Table 3. Growth rate of calves in the early postnatal period.

Genotype	AA	AB	BB
Live weight dynamics, kg			
At birth	28.46±1.18	31.28±1.14	31.62±1.10
At 30 days	48.64±1.24	53.76±1.36*	54.98±1.48**
Absolute weight gain, kg			
1-30	20.18±1.21	22.48±1.25	23.36±1.76
Daily gain, g			
1-30	672±24	749±30	778±26

$P>0.95^*$; $P>0.99^{**}$; $P>0.999^{***}$

Between young animals obtained from cows of mothers with genotypes AB and BB of the kappa-casein gene, the difference in absolute growth was insignificant and amounted to 0.88 kg in favor of calves obtained from cows of mothers with genotype BB of the kappa-casein gene.

In the study of morphological parameters of blood in all newborn calves after calving on the 3rd day of the study, physiological leukocytosis and erythrocytosis of newborns were noted.

Table 4. Morphological parameters of the blood of calves.

Indicators	Calves obtained from mothers of different genotypes		
	AA	AB	BB
	3 days after birth		
RBC, 10^{12}/l	9.1±0.2	9.3±0.1	9.2±0.2
Hb, g/l	102.4±0.6	112.4±0.6 **	113.6±0.7 **
WBC, 10^9/l	12.6±0.1	12.2±0.1	12.8±0.2
	After 30 days		
RBC, 10^{12}/l	5.4±0.1	6.9±0.1 **	7.1±0.1 ***
Hb, g/l	114.6±0.8	120.4±1.1 **	122.2±0.9 **
WBC, 10^9/l	7.4±0.2	8.2±0.2 **	8.6±0.1 ***

$P>0.95^*$; $P>0.99^{**}$; $P>0.999^{***}$

At the age of one month, these indicators were within the normal range. However, it should be noted that the number of leukocytes in calves obtained from mothers of the AA genotype of the kappa-casein gene was 10.8 and 16.2% less than the indices of calves obtained from cows of the AB and BB genotypes of the kappa-casein gene.

Hemoglobin, being a respiratory blood pigment involved in the transport of oxygen and carbon dioxide and performing its buffer functions, was in calves obtained from cows with a homozygous genotype for allele A of the kappa-casein gene by 10.5 and 10.7% lower than in peers obtained from cows AB and BB genotype. The hemoglobin content in the blood of calves obtained from cows of genotype AB and BB did not have significant differences, however, it was higher by 1.8 g/l in favor of calves obtained from cows of mothers with homozygous genotype for allele B of the kappa-casein gene.

4 Conclusion

1. The results of DNA genotyping indicated a low percentage of animals with the BB kappa-casein genotype, although this genotype has a high content of immunoglobulins in colostrum.
2. The density of colostrum in cows with a homozygous genotype for the B allele of the kappa-casein gene was higher compared to the homozygous genotype for the A allele and had a higher amount of immunoglobulins by 29.5 mg/l, respectively.
3. The absence of morbidity in young animals throughout the experiment was noted in calves obtained from cows with genotypes AB and BB of the kappa-casein gene, which is associated with the formation of high colostral immunity of calves, as a result of their consumption of colostrum, which has a high density and, as a result, contains a greater amount of immunoglobulins.
4. For the prevention of diseases of the gastrointestinal tract in LLC "Leninskoe Znamya" it is necessary to form a herd of dairy cows with the genotype BB of the kappa-casein gene.
5. The dynamics of the live weight of calves obtained from cows of mothers with the AA genotype of the kappa-casein gene, during the first 30 days of life, was 6.34 and 4.78 kg lower than in young cows with the BB and AB genotypes of the kappa-casein gene.
6. At the age of one month, the leukocyte content in calves obtained from cows from mothers with a homozygous genotype for allele A of the kappa-casein gene was less by 10.8 and 16.2%, and hemoglobin by 10.5 and 10.7%.

Acknowledgment

We express our deep gratitude to the organizers of the international scientific-practical conference "Innovative technologies in engineering and agroecosystems".

References

1. V. A. Bagirov, Sh. N. Nasibov, P. M. Klenovitsky, S. A. Lesin, et al, Reports of the Russian Academy of Agricultural Sciences, **37** (2010)
2. G. M. Dzhaparidze, Ya. A. Khabibrakhmanova, L. A. Kalashnikova, Actual problems of genetics and molecular biology in the framework of the Festival of Science, **125** (2012)
3. N. A. Zinovieva, O. V. Kostyunina, E. A. Gladyr, A.D. Bannikova, et al, Zootechniya, **8** (2010)
4. N. A. Zinovieva, S. V. Pozyabin, R. Yu. Chinarov, Agricultural Biology, **209** (2016)
5. A. A. Sermyagin, E.A. Gladyr, S. N. Kharitonov, A. N. Ermilov et al, Agricultural Biology, **182** (2016)
6. L. N. Chizhova, E. S., Surzhikova, M. V., Zabelina, E. D. Lutsiva, N. I. Efimova, Agrarian Scientific Journal, 75 (2020)

7. I. V. Lazebnaya, O. E. Lazebnyi, V. F. Maximenko, G. E. Sulimova, Agricultural Biology, **39** (2012)
8. M. I. Selionova, Bulletin AIC, **21** (2011)
9. S. V. Tyulkin et al, Scientific notes of the Kazan State Academy of Veterinary Medicine named after N. E. Bauman, **221** (2015)
10. G. E. Sulimova, Electronic Journal of the Laboratory of Comparative Animal Genetics (2004) http://www. lab-cga.ru
11. M. Ernst, J. Kuciel, T. Urban, Czech J. Anim. Sci, **533** (2003)
12. M. I. Selionova, Bulletin AIC Stavropol, **21** (2011)
13. E. M. Prinzenberg, H. Jianlin, G. Erhardt, J. Dairy Sci., **1198** (2008)
14. E. G. Medvedeva, V. I. Tsys, Bulletin of the Upper AIC, **18** (2008)
15. O. N. Polozyuk, V. A. Kavalerist, E. S. Polozyuk, T. M. Ovcharenko, Proceedings of the Kuban Agrarian University, **74** (2014)

New maternal line of the Cornish breed of the "Smena" selection and genetic center

Zh. V. Emanuylova[1], *A.V.* Egorova[1,2,*], *D. N.* Efimov[2], and *A.A.* Komarov[1]

[1]SGC "Smena" - a branch of the Federal State Budgetary Scientific Institution Federal Scientific Center "All-Russian Research and Technological Institute of Poultry" of the Russian Academy of Sciences (FSC "VNITIP" RAS), Russia
[2]FSBSI Federal Scientific Center "All-Russian Scientific Research and Technological Institute of Poultry" of the Russian Academy of Sciences (FSC "VNITIP" RAS), Russia

Abstract. The article is devoted to the assessment of the selection effect of the new line CM6. It was found that the five-year selection of meat hens of this line resulted in their egg production increase by 15.5 and 1.25% in 30 and 60 weeks of life (2020). The increase in the yield of hatching eggs, their fertilization, hatching of chickens and the yield of chickens per 1 hen was 1.4; 1.1; 1.7; 5.1%, respectively. The live weight of 5-week-old young birds is increased both for cockerels and hens by 10.2-8.7%; enhanced muscularity of the chest and legs at this age by 2.7-2.6% and 5.6-5.7%; increased feed conversion by 3.3-2.5% (2020) pursuant to gender. The integrated indicator - the productivity index of broilers of the combination CM5679 in 2020 was 389 units, which is 25.5% higher than in 2016. The productivity of meat hens of the mother line of the Cornish CM6 breed is high, the bird is competitive.

1 Introduction

Poultry farming is one of the most dynamically developing sectors of livestock farming on a global scale, playing an important role in the production of animal protein, which is the most important component of human nutrition.

In recent years, broiler production has been developing at a rapid pace, highly productive crosses of meat hens have been created, breeding programs are being enhanced [1-4, 6-10].

The similarity between individuals of the same species, as well as the greater similarity of descendants with their parents than with other individuals of a given species, indicates the existence of a hereditary relationship between generations. In general, this relationship is ensured through the transfer of genetic information, which represents, as it were, a plan for the development of the structure and functions of the future organism, received from the reproductive cells of the parents [1].

* Corresponding author: egorova@vnitip.ru

These programs envisage not only selection for increasing the genetic potential for live weight gain, feed conversion, fleshing of the chest, legs, meat qualities of broilers, as well as reproductive qualities of poultry [1,2].

A comprehensive assessment of broiler crosses depends on the number of broilers obtained from one layer of the breeder flock. Therefore, the selection of maternal parental forms to enhance reproductive qualities is necessary [2,4].

Intensive breeding of meat hens, especially in Cornish lines, for growth rate and breast flesh has led to a significant increase in the efficiency of broiler production worldwide. However, this progress was accompanied by negative trends: a decrease in the level of signs associated with reproduction, as well as the appearance and increase in the frequency of metabolic abnormalities of growth and development - ascites, sudden death syndrome.

In recent years, in breeding work, much attention is paid to methods of genomic selection, in particular, the use of color - and feathersex birds [2,5,22,23].

The research is aimed to evaluate the effect of breeding a new maternal line of the Cornish cross "Smena 9" in the production.

2 Materials and methods

The work was performed on the poultry of the original line CM6 and combinations: CM56 and CM5679 in production conditions of SGC "Smena".

The enterprise uses a combined selection method. The breeding nest was formed on the basis of 13 hens and 1 cockerel, their number was 60. 105-198 birds of offspring were taken from one cockerel, 8-15 birds from one laying hen. Poultry line CM6 (breeding group) was completed from broodstock homozygous for the rapid feathering gene, considering the main economically significant traits (live weight, fleshing of the chest, legs, payment for feed, egg production, egg weight, hatchability).

The accounting of the main indicators (live weight, egg production, egg weight, sexual maturity, fleshing of the chest, legs, feed conversion for selection poultry is individual [6].

Feathering of day-old chicks was determined by the phenotype: cover feathers are longer than flight feathers or equal to them (slow feathering), cover feathers are shorter than flight feathers and well developed (fast feathering).

At SGC Smena, 35-day-old cockerels, selected with the best indicators of live weight and meat body shape, are placed in cages for individual registration of feed conversion. From 36 to 42 days for each cockerel, feed consumption is considered, then they are weighed and feed costs kg/kg are calculated. In further breeding work at the enterprise, producers with a high live weight and feed conversion are used.

The bird was kept on a deep bed of coniferous sawdust, natural mating was used, and feeding was performed considering the recommendations of VNITIP [24].

To determine the meat qualities of the final hybrid, the carcasses were cut pursuant to the method [25].

3 Results and discussion

The egg production of chickens for 30 and 60 weeks of life of the CM6 line in 2020 is higher than in 2016 by 15.5 and 1.2%, respectively, by age (tab. 1)

Table 1. Poultry productivity of the CM6 line.

Indicator	Gender	Year		2020 by 2016, %
		2016	2020	
Egg production by laying hen, pcs.:				
in 30 weeks of life		18.1±0.573	20.9±0.415	+15.5
in 60 weeks of life		121.4±1.07	122.9±1.03	+1.2
Egg mass of chickens (g) at the age:				
30 weeks		58.7±0.13	59.4±0.10	+1.2
52 weeks		69.49±0.320	69.95±0.240	+0.7
Hatchery egg yield,%		93.3	94.7	+1.4
Fertility of eggs,%		91.7	92.8	+1.1
Chick hatching (individual),%		75.7	77.4	+1.7
Output of chickens on 1 laying hen, birds		85.7	90.1	+5.1
Hen preservation,%		98.0	98.7	+0.7
Live weight:				
1 week, g	♂	236.0±0.440	240.0±0.420	+1.7
	♀	229.0±0.390	237.0±0.405	+3.5
5 weeks, kg	♂	2.360±0.007	2.600±0.004	+10.2
	♀	2.070±0.006	2.250±0.003	+8.7
Muscularity, score:				
chest	♂	4.53±0.007	4.65±0.005	+2.7
	♀	4.58±0.006	4.70±0.004	+2.6
feet	♂	2.15±0.002	2.27±0.001	+5.6
	♀	2.10±0.001	2.22±0.002	+5.7
Feed conversion, kg/kg, for a period of 1-35 days	♂	1.55±0.012	1.50±0.010	- 3.3
	♀	1.59±0.015	1.55±0.017	- 2.5
Preservation of young birds,%		97.1	97.6	+0.5

The weight of eggs at 30 and 52 weeks increased in 2020 along this line by 1.2 and 0.7%, respectively, for age. The hatching egg yield was high and ranged within 93.3 - 94.7%. The chick yield per layer in 2020 was 5.1% higher than in 2016. The survival rate of adult birds during the productive period was high and amounted to 98.0% (2016) and 98.7% (2020).

As a result of five-year selection, the live weight of young birds in the first week increased by 1.7% for cockerels and 3.5% for chickens; and at 5 weeks of age - by 10.2-8.7%, respectively, gender. As for the fleshing of the chest and legs, this indicator enhanced in 2020 by 2.7-2.6% and 5.6-5.7%, respectively, for the indicator and gender. The feed conversion is increased by 3.3 - 2.5%. The survival rate of the young is high and ranged from 97.1 to 97.6%.

To obtain a two-line form of CM56, the cockerels of the paternal line CM5 are crossed with hens of the maternal line CM6.

The scheme for obtaining broilers is as follows:

♂♂ CM56 X ♀♀ CM79 → ♂♂ CM5679
♀♀ CM5679

Chickens CM79 - maternal parental form.

The bird of the paternal parental form CM56 is of the meat type, the Cornish breed, in which the high live weight of the CM5 line bird and the higher reproductive qualities of the CM6 line are combined.

Table 2 shows the data characterizing the cockerels of the cockerel parental form CM56 in comparison with linear cockerels CM5 and CM6.

Table 2. Economically useful qualities of two-line CM56 cockerels.

Sign	CM5	CM6	CM56	CM56 to CM5, %	CM56 to CM6, %
Fertility of eggs, %	91.4	9.4	94.3	+2.9	+1.9
Chick hatching, %	74.9	76.8	83.5	+8.6	+6.7
Preservation, % - young birds - adult cockerels	 97.0 97.5	 97.5 9.7	 9.7 97.8	 +0.7 +0.3	 +0.2 +0.1

Two-line cockerels CM56 have an advantage over single-line cockerels in terms of the considered characteristics (egg fertilization, hatching of chickens, poultry safety). When compared with the parameters of the CM5 line, the difference was 2.9 and 8.6% and that of the CM6 line - 1.9 and 6.7%, respectively. The survival rate of young and adult SM56 cockerels was 97.7 and 97.8%, which is slightly higher than for the same parameters of the CM5 line by 0.7 and 0.3% and the CM6 line - by 0.2 and 0.1%.

The new paternal form CM56 (cross "Smena 9") surpasses the bird B56 (cross "Smena 8") in survival rate by 0.7 and 0.5% (97.6% versus 96.9% - young birds and 97.9% versus 97.4% - adult cockerels), egg fertilization - by 3.3% (94.5 and 91.2%) and chick hatching by 5.3% (83.5 and 78.0%).

Table 3. The results of testing broilers with the combination of CM5679 in the SGC "Smena".

Indicator	Year		2020 by 2016, %
	2016	2020	
Chick hatching,%	81.0	84.3	+3.3
Live weight at 5 weeks, g	2009	2281	+13.5
Preservation, %	97.4	98.6	+1.2
Feed costs per 1 kg of live weight gain, kg	1.80	1.65	+8.3
The output of the pectoral muscles in% of live weight	21.0	22.5	+1.3
Broiler productivity index, units	310	389	+25.5

Broilers of the combination CM5679, when tested in production conditions of the SGC "Smena" in 2020, had the best performance in terms of chick hatching, live weight at 5 weeks of age, safety, feed costs, breast muscle yield in% to live weight, productivity index by 3.3; 13.5; 1.2; 8.3; 1.3; 25.5% compared to 2016 (Table 3).

4 Conclusions

It was found that the egg production of chickens of the CM6 maternal line at 30 and 60 weeks of life in 2020 was higher by 15.5 and 1.25%, respectively, by age compared to 2016.

The increase in the yield of hatching eggs, their fertilization, hatching of chickens, and the yield of chickens per 1 laying hen in 2020 was 1.4; 1.1; 1.7; 5.1% compared to these indicators in 2016, respectively.

The live weight of 5-week-old young birds in 2020 increased both for cockerels and chickens by 10.2-8.7%; enhanced muscularity of the chest and legs at this age by 2.7-2.6% and 5.6-5.7%; increased feed conversion by 3.3-2.5% compared to 2016, pursuant to gender.

The complex indicator - the productivity index of broilers of the combination CM5679 in 2020 amounted to 389 units, which is 25.5% higher than in 2016.

The productivity of meat hens of the mother line of the Cornish CM6 breed is high, the bird is competitive.

References

1. I. L. Galpern, A.V. Sinichkin, O. I. Stashevskaya, et al., *Breeding and genetic methods and programs for breeding new lines and creating competitive crosses of egg and meat hens,* 163 (2010)
2. L. I. Tuchemsky, K.V. Zlochevskaya, V. I. Fisinin, G. V. Gladkova, *The Selection of Broiler Chicken at State breeding Farm" Smena"*, 308 (2008)
3. A.V. Egorova., L. I. Tuchemsky, Zh. V. Emanuylova, D. N. Efimov, Zootechny, **6(2)**, 4 (2015)
4. L. S. Karpenko, The advancement of Plymouth Rock chicken carrying the gene of slow feathering, Proceedings of III International Conference "Poultry Production-World and Domestic Experience", 99 (2015)
5. T. Petrukovich, Animal Husbandry of Russia, **12(11)**, 12 (2017)
6. Ya. S. Roiter, A. V. Egorova, A. P. Konopleva, et al., *Selection and breeding work in poultry*, 287 (2016)
7. B. Carre, B. Meda, H. Juin, Progress in broiler selection: benefits, limitations as assessed by the didestive function, and consequence of dietary lysine concentration, Proc. XIV Eur Poultry Conf., 189-209(2014)
8. S. Cherepanov, O. Stanishevskaya, Preservation and usage of genetic resources in farm poultry, The Proc. XXV World's Poultry Cong., 343 (2016)
9. L. Lu. Dnutriev, O. Stanishevskaya, N. Dementyeva, Y. Turlova, M. Lapa, The Proc. XXV World's Poultry Cong., 345 (2016)
10. O. W. Willems, S. P. Miller, B. J. Wood, World's Poultry Science J., **69(1)**, 77-78 (2013)
11. D. N. Efimov, Zh. V. Emanuylova, E. V. Zhuravleva, A. V. Egorova, V. I. Fisinin, Agricultural Biology, **53**, 1162-1168 (2018)
12. P. K. Bramwell, *Sexing chicks in the backyard flock, The Poultry Site*, http://www.the poultry site.com/
13. Y. H. Cheng, T. E. Kuo, D. N. Lee, C. F. Weng, Zoological Studies, **45**, 104-113 (2006)
14. J. Zhao, J. Yao, F. Li, et al., Poultry Sci., **95(7)**, 1498-1503 (2016)
15. G. Bu, G. Huang, H. Fu, J. Li, S. Huang, Y. Wang, J. Mol. Endocrinol., **51**, 261-276 (2013)

16. L. Korshunova, Y. Royter, A. Egorova, The usage of gene modifiers in selection of new forms of color- and feather-sex poultry, Proc. XIV European Poultry Conf., 512 (2016)
17. Y. Royter, A. Egorova, A. Sevastianova, L. Korshunova, The selection of autosex interlinear forms of poultry (chicken, geese, Guinea fowl), Proc. XXV World's Poultry Congress, 257 (2016)
18. D. A. Dawson, S. Darby, F. M. Hunter, A. P. Krupa, I. L. Jones, T. A. Burke, Molecular Ecology Notes, **1(13)**, 201-204 (2001)
19. H. Ellegren Hens, EMBO Reports, **2(3)**, 192-196(2001)
20. H. Cerit, K. Avanus, Turk. J. Vet. Anim. Sci., **31(6)**, 371-374(2007)
21. H. Cerit, K. Avanus, World's Poultry Science J., **63(1)**, 91-99 (2007)
22. A.V. Egorova, L.V. Shakhnova, Poultry and poultry products, **3**, 41-43 (2013)
23. T. Petrukovich, Animal Husbandry of Russia, **12**, 11-12 (2017)
24. I. A. Egorov, V. A. Manukyan, T. M. Okolelova, et.al., Manual on Poultry Nutrition, **226** (2018)
25. V. S. Lukashenko, M. A. Lysenko, T. A. Stollyar, et al., *Manual on the Anatomical Dressing, Organoleptic Evaluation of Poultry meat and Eggs, Egg Morphology*, 35 (2013)

Evaluation of selenium status and effectiveness of selenolin in the treatment of selenium-dependent hepatopathies

Vladimir A. Grin, *Eugenia V.* Rogaleva, *Marina P.* Semenenko, *Ksenia A.* Semenenko, and *Elena V.* Kuzminova

Krasnodar Research Center for Animal Husbandry and Veterinary Medicine, Krasnodar, 350055, Russian Federation

Abstract. The article is devoted to the study of the assessment of selenium status in the pathogenesis of liver diseases and the effectiveness of the use of a selenium-containing drug. In the course of studies in animals with clinical signs and high rates of hepatopathies markers – transaminases, in 100% of cases a low level of selenium was revealed – 0.026 ±0.01 µg / ml, because of the increased consumption of this microelement on the background of increased free radical oxidation processes in the liver tissue. It was found that intramuscular administration of the drug selenolin at a dose of 20 mg / 100 kg four times with an interval of 15 days significantly improved the clinical status of the experimental animals; the effectiveness of therapy was 93.3%. Parenteral administration of selenolin favorably corrected the processes of lipid peroxidation and contributed to a significant ($p < 0.01$) decrease in the level of liver indicator enzymes in comparison with the control by 34.5% (aspartate aminotransferase) and by 45.8% (alanine aminotransferase), and also contributed to an increase in the level of selenium to the level of the physiological norm (0.068 ± 0.03 µg / ml).

1 Introduction

Intensification of animal husbandry, low adaptive potential of highly productive livestock, violations of the technology of feeding, keeping and exploitation of livestock very quickly lead to a decrease in the level of compensatory-restorative processes of the body, and, as a rule, the implementation of the pathological process in such animals is accompanied by a wide range of homeostatic deviations that take the form of a certain metabolic syndrome because of the functional liver weakness [1-5].

Since hepatopathies have multifactorial nature, they are quite widespread on farms and cause serious economic damage resulting in the decrease in livestock productivity, the birth of weak, often unviable young animals and early retirement of cows from the herd [1, 4].

Liver, being the center for maintaining homeostasis, plays an important role in the processes of adaptation and compensation for impaired functions. This is often realized in the form of dysfunction of the antioxidant system, one of the main enzymes of which is glutathione peroxidase, the active center of which includes selenium. The conclusion about the antioxidant function of selenium was first made by the English biochemist A. Diplock in

1970. The antioxidant role of selenium is mainly associated with the expression of selenoproteins containing selenium as a selenocysteine residue, fully ionized at physiological pH and acting as an effective redox catalyzer [6, 7].

The main biological role of glutathione peroxidase is to detoxify the reactive oxidant, hydrogen peroxide [8-11]. Classical glutathione peroxidase is a protein with a molecular weight of about 85 kDa, formed by four identical subunits with a molecular weight of 21 kDa, which are organized into square-shaped complexes. Each subunit consists of one selenium atom in the composition of selenocysteine [6, 7, 12-14].

According to the scientific studies of Aboul Soud et al (2011), as a result of selenium deficiency, damage of the cell membranes of hepatocytes is caused by the initiation of lipid peroxidation processes because of the decrease in the activity of the enzyme glutathione peroxidase, while an increase in the concentration of hydroperoxides, malondialdehyde and other lipid peroxidation products leads to the hepatocytes damage and contributes to the progression of the disease [15, 16].

Studies have determined the influence of organic forms of selenium on the regulation of antioxidant processes in the liver tissue of rats with experimental toxic hepatitis, and the administration of selenium drugs (sodium selenite, selenophene) promotes a slight acceleration of regeneration processes in liver after its *in vivo* partial extirpation [12, 16-18].

Thus, on the one hand, the development of hepatopathy leads to a reduced content of selenium in the body, on the other hand, selenium deficiency provokes a decrease in the processes of lipid peroxidation, which is one of the main triggers for the progression of liver diseases, since now there are already numerous data on the role and significance of oxidative stress in the development and progression of a number of pathological processes, primarily in the liver, as an organ most sensitive to the toxic effects of peroxide radicals.

However, the role of selenium status and the degree of its participation in the pathogenesis of the development of hepatopathies as one of the etiological factors in veterinary medicine has been insufficiently studied and requires further research.

The aim of the work was to determine the correlation between selenium deficiency and the occurrence of liver dysfunction, as well as the effectiveness of using the injectable drug selenolin in the treatment of hepatopathies in cows.

2 Materials and methods

The studies were carried out on Holstein cows with clinical and biochemical syndrome of hepatopathies. For the experiment, animals were selected with similar physiological and age parameters. The criterion for inclusion in the study was the clinical picture of metabolic disorders of the liver function such as depression, tachycardia, decreased appetite, rare and sluggish chewing, hypotension, dyspeptic manifestations, liver pain on percussion and palpation, increase in the percussion borders of liver, signs of bone tissue demineralization, which was accompanied by softening of the last caudal vertebrae, dullness and fragility of the covering.

To confirm the diagnosis and to clarify the degree of metabolic disorders in the surveyed livestock, blood was taken with its subsequent biochemical study to determine the main markers of liver pathology, indicating an increased load on this organ.

The effectiveness of the use of the injectable drug selenolin in the treatment of hepatopathies was determined in comparison with the positive (analogue drug) and negative control on three groups of cows (n = 30) 6-7 months of the dry period, selected taking into account body weight and productivity.

Blood biochemical parameters were studied on an automatic Vitalab Selectra Junior analyzer with software version 1.0. (open system for photometric tests, manufactured by

Vital Scientific N. V. Netherlands) using reagents from ELITech Clinical Systems (France) and Analyticon biotechnologies AG (Germany).

The correlation dependence between low selenium levels and manifestation of hepatopathy was determined by the level of selenium in the blood of sick animals by the atomic adsorption spectrometry using a Kvant-2 device (Russia).

The influence of selenolin on the processes of peroxidation of unsaturated fatty acids in the blood of cows was assessed by the level of primary products – diene conjugates (DC) and ketodienes (KD) as well as secondary products – malondialdehyde (MDA) and final products – fluorescent Schiff bases. The analysis of the dynamics of indicators of the peroxidation system was carried out in accordance with the methodological recommendations of FSBRI "All-Russian Research Veterinary Institute of Pathology, Pharmacology and Therapy" (2010) for the study of the processes of lipid peroxidation and the antioxidant defense system of the body in animals.

Statistical processing of the obtained results was carried out by the methods of mathematical statistics adopted in biology and medicine [19] using the software of the Mikrosoft® company and the Carl Zeiss® company. The criterion of reliability was determined according to the Student's table, while the average indicator (M) was calculated; to process the research data, descriptive statistics were used, followed by multiple pairwise comparisons using the Newman-Keuls test at a 5% significance level of differences ($p < 0.05$).

The object of the study is an injectable form of the drug selenolin, which is a 2% sterile solution of diacetophenonyl selenide (DAFS-25) in vegetable oil. It contains organically bound selenium, which forms, during destruction, selenium-containing fragments similar to natural metabolites of the animal body, which provides a relatively low toxicity and high assimilation of the drug in the animal body.

The animals of the experimental group were injected with selenolin intramuscularly at a dose of 5 ml (20 mg / 100 kg) 4 times with an interval of 15 days. The cows of the second experimental group in the same mode of administration received sodium selenite at a dose of 0.2 mg per 1 kg of animal weight. Animals of the control group were injected with soybean oil in a dose of 5 ml / 100 kg according to the similar scheme.

During the experiment, the animals were daily monitored, the severity of the pathological disorder of the main functions of the liver was noted, and the level of antioxidant activity of the body and the main biochemical parameters were studied. Blood samples for biochemical studies were taken at the beginning and at the end of the experiment two hours before morning feeding in accordance with generally accepted methods.

Background studies of blood serum found that the indicators of protein metabolism were characterized by the development of hypoproteinemia in 33.3% of animals and hyperproteinemia in 66.7%, which was accompanied by an increase in the gamma-globulin fraction and a positive thymol test. The indicators of pigment metabolism testified to the functional weakness of the liver: in 83.3% of cows, hyperbilirubinemia was determined with the appearance of a conjugated fraction in the blood.

Disorders of lipid-carbohydrate metabolism were characterized by hypocholesterolemia (in 93.3%) and hypoglycemia (in 53.3%), which can be regarded from the point of view of a deficit of metabolic energy in diets and functional disorders of liver.

With regard to the concentration of carotene in the blood, a decrease in this indicator was noted in 100% of animals because of the insufficient provision of diets and general metabolic disorders.

None of the animals with the detection of symptoms of hepatopathy have an optimal supply of selenium in the blood serum (0.06-0.12 μg / ml). The suboptimal selenium concentration, characterized by the level of micronutrient in the blood serum in the values of 0.06-0.056 μg/ml, was recorded only in 3.3% of cows, in the range of 0.057-0.038 μg / ml – in two cows. In the majority of animals (90%), the selenium concentration in the blood serum was significantly lower than the reference values, and averaged 0.026 ± 0.01 μg / ml, which can be classified as a deficient state for the microelement. Thus, it was found that in animals with a biochemical profile of hepatopathies, selenium deficiency was detected in 100% of cases.

The low concentration of selenium in the blood serum of animals can be explained by the depletion of endogenous reserves as a result of the activation of free radical oxidation processes in the liver tissue, for which a microelement is attracted, which is a cofactor of the antioxidant system. According to the literature data, a low level of selenium in the body of animals correlates with the incidence of hepatopathies, i.e., the higher the severity of the disease, the lower the selenium concentration [4, 11, 13, 16, 19-22].

At the same time, there is a clear parallel between an increased level of liver pathology markers – transaminases (alanine aminotransferase, aspartate aminotransferase, thymol test) with a decreased level of selenium. In sick animals in comparison with the reference values, dysfermentemia was noted, which in 84% of cases was caused by an increase in the activity of ALT and AST and in 90% – ALT and ALP, which can be considered as a differential diagnostic criterion for the violation of the integrity and lesions of hepatocytes, as well as an increase in thymol samples characteristic of chronic mesenchymal inflammatory liver syndrome, accompanied by cholestasis.

On the background of selenium deficiency and an increased level of transaminases in the blood of sick cows, a high status of peroxidation of unsaturated fatty acids (diene conjugates (DC $_{(232)}$) – 0.30 ± 0.01 AU / ml of blood, ketodienes (KD $_{(273)}$) – 0.22 ± 0.02 AU /ml of blood, secondary products – malondialdehyde (MDA $_{(537)}$) – 1.76 ± 0.04 μMl / l and final products – fluorescent Schiff bases $_{(365\text{-}370)}$) – 0.22 ± 0.02 AU / ml of blood.

3 Results and discussion

In a comparative aspect of the conducted studies, it was determined that the therapy with selenolin and sodium selenite improved the clinical condition of the experimental animals. However, it should be noted that the disappearance of the symptoms of hepatopathy in cows of the first experimental group occurred by 18-20 days of treatment, the therapeutic efficiency was 93.3% (28 animals), while the recovery of animals of the second group was noted only by 23-25 days, the therapeutic efficiency was at the level of 76.7% (23 animals). In the control group, the clinical signs of hepatopathy progressed and the disease turned into a recurrent form.

The normalization of the processes of peroxidation of unsaturated fatty acids with the administration of selenolin contributed to the optimization of metabolic processes in the liver by activating the synthesis of selenium-dependent glutathione peroxidase, which was accompanied by the restoration of the protein-synthetic function of liver and the decrease in the level of transaminases (alanine aminotransferase, aspartate aminotransferase) and alkaline phosphate transferase (Table 1).

In the experimental cows, there were positive changes in the biochemical status with the priority of indicators in the first experimental group.

In the experimental groups, the level of total protein returned to normal and was within the physiological norm (84.1 ± 6.12-82.7 ± 5.36 g / l). Whereas in the control group, on the background of progression of hepatopathy and increased load on the liver, higher values of total protein were recorded. However, this was due to an increase in the number of animals with hyperproteinemia – 25 heads (83.3%).

The restoration of the functional activity of hepatocytes of the hepatic tissue with the administration of selenolin inhibited the development of mesenchymal-inflammatory syndrome (negative thymol test) and contributed to a decrease in the synthesis of the gamma-globulin fraction. The level of gamma globulins in the first experimental group was lower than the indicators of cows of the second group and control by 5.1% and 21.5%, respectively.

Table 1. Influence of selenolin on the biochemical parameters of the blood of cows after intramuscular administration (M±m; n=30)

Indicators	Groups of animals			
	Background	1st experimental, selenolin	2nd experimental, sodium selenite	Control
Total protein, g / l	78.3±4.81	84.1±6.12*	82.7±5.36*	93.7±5.94
Albumins, %	38.8±5.53	44.6±7.04	46.3±6.82	36.4±5.17
α- globulins, %	11.75±1.88	13.5±2.13	10.24±1.62	12.6±1.59
β- globulins, %	7.05±0.62	6.49±0.70	6.17±0.57	5.85±0.46
γ- globulins, %	42.4±6.25	35.4±4.82	37.3±5.13	45.1±6.61
AST, AU / l	116.2±7.46	80.8±6.54**	108.5±6.02	123.3±8.34
ALT, AU / l	43.1±4.18	32.2±3.79**	37.6±5.41	59.6±7.22
ALP, AU / l	128.4±9.12	117.2±12.6	161.4±8.59	193.2±11.5
Thymol test	+	–	–	++
Glucose, mmol / l	3.1±0.13	3.7±0.11	3.3±0.14	2.9±0.12
Cholesterol, mmol / l	3.7±0.31	4.4±0.42**	3.9±0.36	3.5±0.27
Triglycerides, mmol / l	0.23±0.01	0.30±0.02**	0.27±0.02**	0.21±0.01
Carotene, mg%	0.42±0.05	0.57±0.03	0.44±0.06	0.39±0.04
Selenium, μg / ml	0.026±0.02	0.068±0.03**	0.054±0.02**	0.025±0.01
DC $_{(232)}$, AU./ ml of blood	0.30±0.01	0.21±0.03**	0.25±0.02**	0.31±0.03
KD $_{(273)}$, AU./ ml of blood	0.22±0.02	0.17±0.01**	0.20±0.02**	0.23±0.03
MDA $_{(537)}$, μMl / l	1.76±0.04	1.48±0.03*	1.57±0.07*	1.74±0.05
Schiff bases $_{(365\text{-}370)}$, AU / ml of blood	0.22±0.02	0.19±0.01**	0.20±0.02**	0.23±0.03

Note: * – $p \leq 0.05$, ** – $p \leq 0.01$, *** – $p \leq 0.001$

Analysis of the activity of aspartate aminotransferase and alanine aminotransferase in experimental animals showed that the level of these indicator enzymes was significantly ($p < 0.01$) lower than the control values by the end of the experimental period, the difference was 34.5% and 45.8%, respectively.

The introduction of sodium selenite did not allow to have an equally significant effect on the activity of transaminases in the blood of cows of the second experimental group: the concentration of aspartate aminotransferase was at the level of the upper limit values – 108.5±6.02 AU / l, while the level of alanine aminotransferase at the end of the experiment exceeded the reference values for this indicator and amounted to 37.6 ± 5.41 AU / l at a norm of 6.9-35 AU / l.

Under conditions of increased generation of active oxygen forms in the body, when the self-induced chain mechanism of free radical reactions goes beyond the stationary level, selenolin showed a significant ($p < 0.05$; $p < 0.01$) inhibitory effect on the synthesis of active oxygen forms in the blood of cows at different levels of their generation.

Therefore, in the dynamics in the experimental group, the decrease in DC was at the level of 30%, KD – 22.7%, MDA – 15.9% and Schiff bases – 13.6%. In a comparative aspect with the indicators of the cows of the second experimental group, the level of lipid peroxidation products was lower by 16.0% (diene conjugates), 15% (ketodienes), 5.7% (malondialdehyde) and 5% (Schiff bases) in relation to the control values it was lower by 32.3%, 26.1%, 14.9% and 17.4%, respectively.

The beneficial effect of selenolin and sodium selenite on the course of metabolic processes in liver was accompanied by the correction of lipid and carbohydrate metabolism.

The level of glucose and lipids (cholesterol and triglycerides) in both groups at the end of the experimental period had higher values in comparison with control analogs with a tendency of higher values in the first experimental group. In the first experimental group, the excess in terms of glucose in comparison with the control was 27.6%, in the group with an analogue drug it was 13.8%. The level of cholesterol and triglycerides was higher than the indicators of the second experimental group by 12.8% and 11.1% and higher than the control indicators by 25.7% and 42.9%, respectively.

The reparative effect of selenolin on the metabolic functions of liver was also manifested by higher values of carotene – 0.57 ± 0.03 mg /%, which exceeded the indications of cows that were injected with an analogue drug by 29.5% and control by 46.2%.

The conducted biochemical studies of blood serum made it possible to reveal the positive dynamics of the selenium level in the experimental animals. Parenteral administration of selenolin contributed to a significant ($p \leq 0.01$) increase in selenium levels within the physiological norm (0.068±0.03 μg / ml) and exceeded the indicators of the second experimental group by 25.93%, and the control indicators in 2.62 times. Higher selenium indices in animals of the first experimental group were due to the fact that selenolin contains selenium in its composition in the form of an organic compound, which has high bioavailability and less toxicity for animals compared to inorganic preparations. According to N.I. Konova (1992), when sodium selenite is used, the absorption of selenium is much lower: in monogastric animals, it is about 74%, and in ruminants only 30%, which is because sodium selenite is rapidly reduced to elemental selenium and forms insoluble forms with other components of mixtures [22].

By the end of the experimental period, the selenium level in the blood of cows from the control group was still at a physiologically low level – 0.025±0.01 μg / ml, which is apparently associated with an increase in pathological processes in liver and an increase in the functional activity of the antioxidant system. These processes are accompanied by an increased use of selenium for the construction of enzymes specific for a given system and lead to a further depletion of the reserves of this element.

The hepatoprotective mechanism of action of selenolin lays in organic selenium, which is part of the drug, as a cofactor, participates in the detoxification of hydrogen peroxide and hydroperoxides of unsaturated fatty acids, thereby inhibiting free radical oxidation processes leading to the formation of extremely reactive hydroxyl radicals, the excessive accumulation of which contributes to the development metabolic syndrome because of the functional liver weakness.

4 Conclusion

Thus, intramuscular administration of the drug selenolin into the body of cows helped leveling the manifestation of clinical symptoms of hepatopathy, reducing high levels of hepatic transaminases, thereby ensuring the restoration of metabolic processes in liver. Selenolin in the treatment of hepatopathies weakened the negative effect of xenobiotics and other factors by maintaining the antiradical activity of liver at a sufficient level, activating reduced glutathione and reducing the formation of toxic lipid peroxidation products, thereby providing a pronounced hepatoprotective effect.

References

1. A. S. Erokhin, Selenium and animal reproductive function, 142 (Moscow, 2008)
2. V. Yu. Kozlovsky, *Forest Glades,* 274 (2010)

3. E. V. Rogaleva, V. A. Grin, M. P. Semenenko, Collection of scientific papers of the Krasnodar Research Center for Animal Husbandry and Veterinary Medicine, **9(1)**, 262 (2020)
4. A. Kalkan, V. Bulut, S. Avci et al, J. Trace Elem. Med. Biol., **16(4)**, 227 (2002)
5. M. P. Semenenko, D. V. Osepchuk, N. A. Yurina, E. V. Kuzminova, K.A. Semenenko, IOP Conference Series: Earth and Environmental Science, 42048 (2020)
6. H Zeng, Molecules, **14(3)**, 1263 (2009)
7. V. A. Tutelyan, V. A. Knyazhev, S. A. Khotimchenko et al, Selenium in the human body, 224 (M: Publishing house RAMN, 2002)
8. V.G. Skopichev, L.V. Zhichkina, O.M. Popova et al, Microelementosis of animals: tutorial, 288 (SPb.: Prospectus of Science, 2015)
9. S.T. Kokhan, E.V. Fefelova, M.V. Maksimenya et al., Fundamental research, **11-4**, 837 (2012)
10. V.A. Grin, A.A. Abramov, M.P. Semenenko, E.V. Kuzminova, E.V. Rogaleva, E.N. Rud, Veterinary medicine of the Kuban, **2,** 6 (2020)
11. S.K. Jain, P.W. Pemberton, A. Smith et al, J. Hepatol., **36(6)**, 805 (2002)
12. L.L. Zubkova, Kazan science, **3**, 241 (2010)
13. G.N. Bliznetsova, PhD thesis abstract, 46 (Voronezh, 2010)
14. A.S. Gasanov, A.A. Gatina, Z.M. Zukhrabova et al, Kazan: Information Technology Center KGAVM, 51 (2020)
15. Yu.M. Stepanov, V.V. Belitsky, S.V. Kosinskaya, Frequent gastroenterology, **3**, 65 (2012)
16. M.A. Abou Soud, A.M. Othman, G.E. Desoky et al, J. Toxicol. Sci., **36(3)**, 285 (2011)
17. A.I. Berdnikov, Scientific potential - modern agro-industrial complex, Izhev State Agr. Acad., **2**, 76 (2009)
18. A.V. Boroda, PhD thesis, 101 (Kursk, 2005)
19. A.A. Zverev, T.L. Zefirov, Biometrics ,42 (Kazan: Kazan Federal University, 2013)
20. V.A. Chelnokov, PhD thesis, 169 (Kursk, 2013)
21. T. Himoto, H. Yoneyama, K. Kurokohchi et al., Nutr. Res., **31(11)**, 829 (2011)
22. A.V. Zholnin, I.A. Myakishev, P.N. Popkov, Bulletin of SUSU, **19**, 126 (2010)

Application of intelligent methods for diagnosing animal diseases based on determining the structures of blood facies

Marina V. Yanaeva[1], *Elena V.* Kuzminova[2], *Nikita A.* Akindinov[1], *Denis V.* Osepchuk[2], and *Marina P.* Semenenko[2]

[1] Kuban State Technological University, Krasnodar, 350072, Russian Federation
[2] Krasnodar Research Center for Animal Husbandry and Veterinary Medicine, Krasnodar, 350055, Russian Federation

Abstract. The intensive development of computer technologies, the growing importance of the problems of intellectualization and decision-making determine the need for software tools that make it possible to increase the efficiency of laboratory diagnostics in veterinary medicine. An analysis of the structure of the solid phase of biological fluids (facies) is proposed as a universal method for assessing the state of homeostasis of an organism. The paper highlights the problem of diagnosing animal diseases based on the determination of indicators of the structures of blood facies, substantiates the relevance of using neural networks to recognize the obtained images of blood facies, presents the main methods of an intelligent information system for diagnosing animal diseases.

1 Introduction

The intensive development of computer technologies, the growing importance of the problems of intellectualization and decision-making determine the need for software tools that make it possible to increase the efficiency of laboratory diagnostics in veterinary medicine.

It has been scientifically proven that pathological processes already at the early stages of their development are accompanied by changes in the physicochemical properties of the internal environment of the body. An analysis of the structure of the solid phase of biological fluids (facies) is proposed as a universal method for assessing the state of homeostasis of an organism. This method is based on the theory of self-organization, according to which, during dehydration, the elements dissolved in biofluids form stable functional bonds that form specific structures characteristic of certain pathological processes. Crystallographic methods of laboratory diagnostics are economical, available for wide practice, capable of providing integral information about the state of the body with a high degree of reliability at the earliest stages of the development of pathological processes, which opens up broad prospects for this direction in clinical medicine and other areas of study of living systems [1, 2].

However, the application of crystallographic methods in veterinary medicine (obtaining and describing crystallograms in pathological conditions in animals) is at an early stage. A limiting factor in the development of this promising direction is a certain difficulty in interpreting the crystal pattern in normal and pathological conditions, due to a significant variation in the types of crystal growth in a dehydrated drop of biological fluid.

The aim of this work is to substantiate the possibility of using neural networks in veterinary medicine for recognizing images of crystallograms of animal biofluids [3].

2 Materials and Methods

At the Krasnodar Research Centre for Animal Husbandry and Veterinary Medicine, a crystallographic study of the blood of healthy calves and those with colibacillosis was carried out. Whole blood sample was taken from animals, which, then, in the form of a drop was applied to a previously defatted glass slide and dried at t – 20-25 °C, relative humidity – 65-70% and minimum mobility of the surrounding air, strictly in a horizontal position. The duration of the drying period (before the analysis of the structure) ranged from 18 to 24 hours. The study of the dehydrated drop – "facies" was carried out using an optical microscope "Mikmed-3" with a built-in digital camera "Canon" (5.0 Mpx). The resulting image was transmitted to the monitor screen.

3 Results and Discussion

A visual comparison of the crystalloscopy pictures showed (Fig. 1) that the features of the facies of sick animals relative to healthy ones include the following: the peripheral zone is characterized by a strip of dried plasma of a darker color; in the center there is a reduced destruction of the crystal pattern; reduction in the number of fractures; the number of cases of asymmetric, chaotic arrangement of fractures dominates over symmetric ones.

A –healthy **B –sick**

Fig. 1. Crystalloscopic facies of calf blood (healthy and sick with colibacillosis).

When a biological fluid is dehydrated, the structure formation of the solid phase proceeds taking into account these relationships, which make it possible to analyze them.

Proceeding from this, conducting research to establish the relationship between the state of the body and the morphotexture of crystals of the corresponding biofluids, the formation of databases and knowledge bases on their basis, the construction and training of a convolutional neural computer network, makes it possible to develop a new information technology for recognizing the pathology of the body [4].

Currently, neural network technologies are used in humane medicine. Some of the most requested are listed below:

"Ideal anesthesiologist based on artificial intelligence" (2020) – a neural network developed by a group of scientists from the Massachusetts Institute of Technology and Massachusetts Hospital, which is able to monitor the patient's level of unconsciousness during surgery and determine / administer the optimal dose of anesthetic based on pharmacokinetic and pharmacodynamic models with random parameters. Testing has shown that the new algorithm outperforms the proportional-integral-derivative regulator, which is often used to determine the ideal dose of anesthetic. The main advantages of the new approach are the ability to scale clinical variables, as well as the relationship of the deep network with the input variables and the recommended dosage [5].

"Artificial Intelligence for Tuberculosis Diagnosis" (2019) is an AI system BacillAi, developed by the English company of Cambridge Consultants, which allows to partially solve the problem of the spread of tuberculosis in developing countries, where this disease is one of the most frequent causes of premature death. To diagnose tuberculosis using BacillAi artificial intelligence, a doctor only needs a microscope and a smartphone with Internet access. The doctor simply takes a photo of the patient's stained sputum sample with the application using a conventional laboratory microscope. Further, the image is automatically processed by a neural network on a remote server, and soon the doctor receives information about the number and type of tuberculosis cells in the sample.

"Botkin. AI - Russian AI system for diagnosing cancer" (2019) - a startup that develops an artificial intelligence system for automatic analysis of medical images and diagnostics of diseases. Botkin. AI already knows how to analyze CT scans of the chest. It provides the specialist not only with information about the detection of a tumor, but also highlights the areas that require its attention. The accuracy of the algorithm when analyzing a computed tomogram of the chest is up to 95%. The system is improved not only by processing a large amount of images, but also in the process of practical work with doctors (when the doctor confirms or refutes the diagnosis). In the near future, the developers promise to add modules for the analysis of mammography and digital x-rays, which will increase the number of pathologies recognized.

In 2018, Google created AI to diagnose eye diseases. In medical centres, optical coherence tomography is used to diagnose eye diseases - a modern non-invasive non-contact method that allows visualizing various structures of the eye with a higher resolution (from 1 to 15 microns) than ultrasound. However, this technology has a significant drawback - it takes a very long time to process and interpret the images (in some cases, about a week), and in some conditions it can be fatal for the patient. The Google DeepMind team has created a neural network that can process these images very quickly and automatically detect the presence of fifty different eye diseases. The neural network was trained on a huge database of images from the British Eye Center Moorfields Eye Hospital, verified by qualified specialists.

In 2018, Google also created an augmented reality microscope for diagnosing cancer, as examining blood samples under a microscope is still a necessary procedure for diagnosing cancer. Moreover, the accuracy of the diagnosis depends on the doctor's eyes. For this procedure, Google employees developed a neural network trained on millions of already processed samples. Their augmented reality microscope system automatically draws the doctor's attention to suspicious areas, so the diagnostic accuracy is significantly increased. The development allows real-time analysis of the image and presentation of the results of recognition of the cells with which the study is carried out. The image refresh rate is 10 frames per second, which is quite enough for diagnostics. The algorithm itself is based on machine vision and machine learning technologies.

An intelligent information system for the diagnosis of animal diseases will be implemented on the basis of a convolutional neural network. The following are the planned ways to implement an intelligent information system for the diagnosis of animal diseases [6, 7].

Wavelet large-scale analysis, texture analysis of images, segmentation method, spatial methods of image enhancement, digital image processing algorithms, one-dimensional signal processing, statistical data processing, system analysis, and pattern recognition are used as digital image processing methods. Wavelets are widely used for filtering and preprocessing data and analyzing state, pattern recognition, for solving problems of image compression and processing, for training neural networks. Our vision is designed in such a way that we focus our attention on the essential details of the image, cutting out the unnecessary. Using the wavelet transform, you can smooth or highlight some of the details of an image, enlarge or reduce it, highlight important details, and even improve its quality.

Segmentation methods can be divided into two classes: interactive - using user input directly in the process of work and automatic - not requiring any interaction with the user. For morphological processing of images, methods of biomedical morphometry, morphological filtering, thinning and truncation will be applied.

A neural network is a mathematical model, as well as its software or hardware implementation, built on the principle of the organization and functioning of biological neural networks - networks of nerve cells of a living organism. An artificial neural network is a system of interconnected and interacting processors (artificial neurons). Such processors are usually quite simple (especially when compared to processors used in personal computers). Each processor on such a network deals only with signals that it periodically receives and signals that it periodically sends to other processors. And, nevertheless, being connected in a sufficiently large network with controlled interaction, such individually simple processors together are capable of performing rather complex tasks [4,6].

From the point of view of machine learning, a neural network is a special case of pattern recognition, discriminant analysis and clustering methods. In terms of mathematics, training neural networks is a multi-parameter nonlinear optimization problem. Neural networks are not programmed in the usual sense of the word, they are trained. Learning is one of the main advantages of neural networks over traditional algorithms. Technically, training consists in finding the coefficients of connections between neurons. During training, the neural network is able to identify complex dependencies between input and output data, as well as perform generalization. This means that in case of successful training, the network will be able to return the correct result based on data that were absent in the training set, as well as incomplete and / or "noisy", partially distorted data.

The main element in image processing using neural networks is analysis, the first phase of which will be image recognition. After that, artificial intelligence using machine learning recognizes the actions and classifies them.

In order to recognize an image, the neural network must first be trained on the data. This is very similar to the neural connections in the human brain - we have certain knowledge, see the object, analyze it and identify it [8, 9].

Neural networks are demanding on the size and quality of the dataset on which it will be trained. The dataset can be downloaded from open sources or collected on your own.

A neural network for image recognition is perhaps the most popular way to use neural networks. At the same time, regardless of the features of the tasks being solved, it works in stages (Fig. 2).

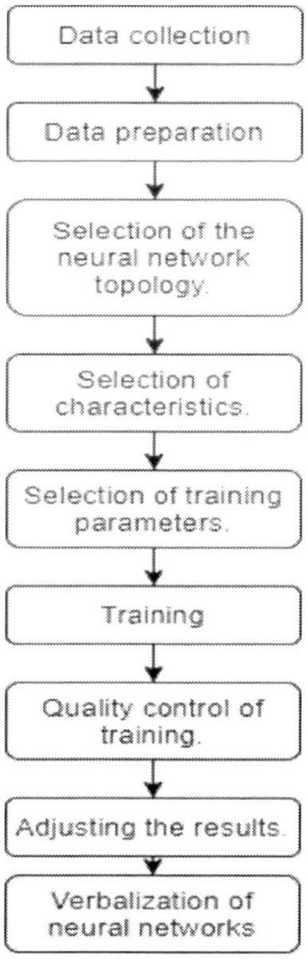

Fig. 2. Scheme of creating a neural network for image recognition.

The images of the crystallographic facies of the blood of animals will be the recognized images. When training a network, various samples with a label of what type they can be attributed to are offered. A vector of feature values is used as a sample, and the set of features under these conditions should allow to unambiguously determine which class of images neural networks deal with [10, 11].

When training, it is important to teach the network to determine not only a sufficient number and values of features to give good accuracy in new images, but also not to overfit, that is, not to "adjust" unnecessarily to the training set from images. After completing the correct training, the neural network should be able to identify images that it did not deal with during the training process.

It is important to take into account that the initial data for the neural network must be unambiguous and consistent, so that situations do not arise when the neural network will produce high probabilities of belonging of one object to several classes.

A convolutional neural network (CNN) contains one or more concatenated or connected convolutional layers. CNN uses a variation of the multilayer perceptron. Convolutional layers use convolution on the input and pass the result to the next layer. This operation allows the network to go deeper with fewer parameters. Convolutional networks show outstanding results in applications requiring image recognition. It is the convolutional neural network that will be built and trained in the intelligent information system for the diagnosis of animal diseases [12].

4 Conclusion

The main expected scientific results are:

— development of methods for diagnosing diseases based on the determination of indicators of the structures of blood facies in animals;
— building and training a convolutional neural network;
— development of methods for recognition of medical images (facies of biofluids);
— building an intelligent information system for the diagnosis of animal diseases.

References

1. A.I. Andyushkin, Bulletin of the Chuvash University, **3**, 355-9 (2013)
2. V.N. Kidalov, A.A. Khadartsev, *Tula polygraphist*, 244 (2009)
3. S.A. Kraevoy, N.A. Koltovoy, Electronic Mathematical and Biomedical J. "Mathematical morphology", 320 (2016)
4. G. Aurelien, *OOO Dalektika*, 688 (2019)
5. S. Kraevoy, N. Koltovoy, Electronic Mathematical and Biomedical J. "Mathematical Morphology", **1**, 321 (2016)
6. V.O. Petrov, V.A. Kamaev, S.V. Poroisky, Modern problems of science and education, **6**, 106-10 (2009)
7. G. Sreenu, M. A. Saleem Durai, J. of Big Data, **6 (1)** (2019)
8. S. Francois, *Peter,* 400 (2018)
9. A. Trask, *Peter*, 352 (2020)
10. A.S. Bolkhovitinov, A.G. Verhovcev, O.V. Gradov, Electronic Mathematical and Biomedical J. "Mathematical morphology", **9 (1)** (2010)
11. R-C. Mihailescu, P. Davidsson, U. Eklund, J.A. Persson, Knowl. Eng. Rev., **33**, (2018)
12. C. Szegedy, et al., *MA*, 1-9 (2015)

The efficiency of growing large meat chickens of various crosses in cage batteries

Rustam Abdulkhalikov[1,*], and *Victor* Malorodov[2]

[1]Kabardino-Balkarian State Agricultural University named after V.M.Kokov, Lenin Avenue, 1V, 360030 Nalchik, Russia
[2]FSBEI RSAU - Moscow Timiryazev Agricultural Academy, str. Timiryazevskaya 49, 127550 Moscow, Russia

Abstract. The article presents the results of a comprehensive assessment of the possibility of producing large meat chickens of domestic and foreign crosses with the cage method of growing, considering the separation by gender. The highest broiler productivity index was 213 units, which is by 27 units higher than when growing broilers of a foreign cross. The content of histidine is significantly higher in the meat of broilers of the domestic cross (by 0.50 g in the muscles of cockerels and by 0.54 g in the muscles of hens). It was found that it is advisable to produce large meat chickens using the domestic SK-Rus 613 cross for deep processing of individual parts of carcasses. The results are confirmed by the zootechnical efficiency of broiler production, the biological usefulness of the muscle composition and meat qualities, as well as the taste and aroma properties of meat products.

1 Introduction

The intensification of broiler meat production contributes to the expansion of the range of poultry products and determines the need to enhance technology, considering the increasing pre-slaughter poultry live weight. Growing large meat chickens causes the need to implement deep processing of individual parts of carcasses. Broiler meat industrial production is aimed at increasing the uniformity of the poultry live weight, which is ensured by technological standards for growing [4, 8, 10, 18]. The stocking density of poultry, calculated depending on the final average live weight, is one of the important indicators.

Currently, the technological standards providing for the rearing of large meat chickens on deep litter have been developed and tested [22, 25]. However, the possibility of producing broiler meat for deep processing with the cage method of growing poultry, depending on the cross of poultry at a standard stocking density, has not been studied in detail [1, 2, 12]. In most cases, birds were observed to have breast folds, in connection with which studies on the assessment of the use of crosses of domestic and foreign selection for the production of large meat chickens in cages are relevant. It is important to take consider the latest advances in

* Corresponding author: rustam742008@mail.ru

breeding, feeding and broiler growing technology [5-7, 9, 11, 13-17, 19-21, 23, 24]. The optimum period for raising the number of gender-segregated broilers for the production of large meat chickens has not been identified as well.

The purpose of the research is to determine the efficiency of the cage production of large meat chickens of various crosses, considering the division by gender, and to assess the biochemical composition and quality of broiler meat.

2 Materials and methods

The study was performed at the poultry farm of OOO Veles-Agro (Kabardino-Balkarian Republic, Prokhladnensky district), which is the epizootic free state. Broilers of crosses SK-Rus 61 3 and Ross-308 were grown up to 49 days old (cockerels) and up to 56 days old (hens) in a poultry house using the cage method pursuant to the scheme presented in Table 1. Chickens were divided by gender at day old age. The number of birds in groups 1 and 2 was 110 birds each, in groups 3 and 4 - 120 birds each. The total initial livestock in the experiment is 460 chickens. The stocking density per cockerel was 410 cm^2; per chicken - 370 cm^2; load on 1 nipple - 9 animals, feeding area - 2.3 cm per 1 animal. Air exchange in the halls was provided by a supply and exhaust ventilation system operating on the principle of negative pressure with end exhaust ventilation.

Table 1. Experiment scheme.

Indicator	Group			
	1	2	3	4
Broiler cross	SK-Rus 61 3	Ross-308	SK-Rus 61 3	Ross-308
Chicken gender	♂	♂	♀	♀

The samples of breast muscle tissue were taken from three clinically healthy broilers with an equal average pre-slaughter body weight (separately for cockerels and hens) in order to study the amino acid composition of broiler meat. Slaughter and anatomical cutting were performed to assess the meat qualities of broilers. The experiment conditions comply with the requirements of the recommendations "On measures on further enhancement of the organizational forms of work using experimental animals " (Order of the USSR Ministry of Health No. 755 dated August 12, 1977) and "The Guide for Care and Use of Laboratory Animals" (National Academy Press, Washington, DC, 1996). All efforts to minimize bird suffering and reduce the number of birds under study were made. The juiciness of the meat was assessed using the Grau-Hamm method; tenderness or roughness of meat was assessed using a consistometer; weight loss of carcasses before and after cooking was assessed as well. The tasting assessment was performed in accordance with the methodological recommendations of VNITIP.

3 Results and discussion

The average pre-slaughter live weight of meat chickens significantly differs in all groups, regardless of gender. In addition, broilers of domestic cross-country reliably exceed broilers of foreign cross-breed in average live weight by 166 g (cockerels) and by 118 g (hens). The broilers of group 1 exceeded the average daily gain of broilers in group 2 by 3.4 g, at the same time, among the hens, the difference was 2.1 g in favor of the domestic cross. The preservation in groups was in the range of 95.4-96.9% with the highest indicator in group 3. The feed consumption per 1 kg of live weight gain was lower in the cross of domestic selection among cockerels by 0.21 kg; among hens it was 0.28 kg in comparison with foreign cross of broilers.

There was a tendency for hens to consume more feed than cockerels, which is associated with a lower intensity of weight gain. In group 1, the highest broiler productivity index was obtained - 213 units, which is 27-102 units higher than in groups 2-4 (Table 2).

Table 2. Broiler farming results.

Indicator	Group			
	1	2	3	4
Average pre-slaughter live weight, g	2852.0 ± 50.4 a	2686.0 ± 49.4 b	2342.0 ± 45.9 c	2224.0 ± 36.7 d
Daily gain, g	57.4	54.0	41.1	39.0
Preservation, %	95.4	95.8	96.9	96.8
Feed costs per 1 kg gain, kg	2.61	2.82	3.18	3.46
Broiler productivity index, units	213	186	127	111

Note: the difference between the mean values in the groups within the indicator, indicated by different letters, is significant at P≥0.95.

As a result of the analysis of the amino acid composition of broiler muscles, no regular differences were found in the quantitative content of nonessential, semi-nonessential and essential amino acids, depending on the cross (Table 3). Nevertheless, it was noted that the level of histidine, which is an essential amino acid for children, was significantly higher in the meat of domestic broilers (by 0.50 g in the muscles of cockerels and by 0.54 g in the muscles of hens). Histidine is a precursor in the biosynthesis of histamine that promotes the growth and repair of human tissues. In the case of severe deficiency of this amino acid, speech impairments and decreased mental activity are possible.

Additionally, the fatty acid composition of broiler meat has been studied. It was revealed that in terms of the total content of fatty acids, the meat of chickens of the Ross-308 cross was superior to that of the SK-Rus 61 3 chicken. The level of fatty acids in the meat of broilers of foreign breeding was noted in the range of 27.7-29.9 g per 100 g, while the same indicator in the meat of domestic chickens was 24.3-25.3 g per 100 g. The advantage can be explained by the increased content of oleic acid, which is not indispensable for humans and is not considered when determining the biological value of the fatty acid composition of broiler meat [3]. The high biological value of lipid compounds in broilers is provided by polyunsaturated fatty acids (linoleic and linolenic), which are easily assimilated by the human body and are irreplaceable. In terms of the PUFA amount, the meat of poultry of domestic cross is 15.1-19.2% higher than the meat of broilers of foreign cross.

Table 3. Amino acid composition of broiler muscles (g) per 100 g.

Amino acid	Group			
	1	2	3	4
Valin	1.02	0.98	1.04	0.98
Isoleucine	0.96	1.00	0.98	0.98
Leucine	1.65	1.76	1.57	1.75
Lysine	1.86	1.86	1.94	1.87
Methionine	0.64	0.63	0.75	0.63
Trenin	1.02	0.95	1.02	1.03
Phenylalanine	1.03	0.94	1.04	0.86
Alanine	1.12	1.08	1.16	1.19
Arginine	1.62	1.49	1.55	1.38
Aspartic acid	2.03	1.20	2.08	1.86
Histidine	1.48	0.98	1.53	0.99
Glutamic	3.31	3.20	3.48	3.41
Proline	0.99	0.99	0.99	0.94
Tyrosine	0.59	0.69	0.61	0.56
Serine	0.85	0.75	0.87	0.85
Cystine	0.20	0.18	0.22	0.18

The meat qualities of broilers are presented in table 4. Domestic cross broilers outnumber the foreign cross broilers in terms of the gutted carcass weight by 144.5 g (cockerels) and by 112.5 g (chicken). A similar trend can be seen in terms of the muscle mass ration to the gutted carcass mass. There were no significant differences between cockerels and hens of different crosses in terms of the weight of bones from the weight of the gutted carcass and the weight of inedible parts from the pre-slaughter live weight. Significant superiority was revealed in terms of the weight of edible parts relatively to the pre-slaughter live weight of the bird. Thus, the cockerels of the domestic cross had an advantage of 145.6 g, the hens - by 117.9 g, in comparison with the foreign cross of broilers. The ratio of edible parts to inedible parts in the domestic poultry is generally higher.

Table 4. Broiler meat quality.

Indicator	Group			
	1	2	3	4
Gutted carcass weight, g	2067.7	1923.2	1658.1	1545.6
Muscle mass relatively to the gutted carcass mass, g	1497.0	1386.6	1193.8	1105.1
Bone weight relatively to the gutted carcass weight	322.6	321.2	253.7	255.0
Weight of edible parts relatively to pre-slaughter live weight, g	1726.5	1580.9	1396.1	1278.2
Mass of inedible parts relatively to pre-slaughter live weight, g	1125.5	1105.1	945.9	945.8
The ratio of edible parts to inedible parts	1.53	1.43	1.48	1.35

The main physical indicators of the meat qualities of broilers are shown in table 5. Judging by the juiciness of the meat of the pectoral and leg muscles, there is a tendency of superiority in the studied indicator of hens over cockerels, regardless of the cross. In general, it should be noted that large meat chickens of domestic cross had a slight advantage in terms of meat juiciness. The tenderness of the meat of the pectoral and leg muscles is significantly lower in broilers of foreign cross, both among cockerels and chickens.

Table 5. Physical indicators of the meat qualities of broilers.

Indicator	Group			
	1	2	3	4
Juiciness of meat, %:				
pectoral muscles	59.41	58.75	60.04	59.42
leg muscles	59.90	58.98	60.21	59.93
Tenderness of meat, mm:				
pectoral muscles	4.1	33.6	4.31	34.2
leg muscles	53.87	6.40	4.02	0.92

The results of meat moisture loss after cooking and organoleptic evaluation of broiler meat are shown in Table 6. It was found that when cooking and roasting meat of domestic cross-breed chickens, the moisture loss is lower in comparison with the foreign cross of broilers. Judging by all indicators of the organoleptic assessment, the meat of chickens of groups 1 and 3 is slightly superior to those of groups 2 and 4, which allows us to conclude about the greatest taste value of meat obtained from poultry of domestic selection.

Table 6. Moisture loss and organoleptic evaluation of broiler meat.

Indicator	*Group*			
	1	*2*	*3*	*4*
Loss of moisture in meat, %:				
when cooking	24.2	25.7	24.5	25.1
when roasting	33.1	35.1	33.7	34.8
Organoleptic assessment, score:				
bouillon	5.00	4.94	5.00	4.95
cooked meat	4.97	4.87	4.98	4.90
grilled meat	5.00	4.92	5.00	4.93

4 Conclusions

The studies performed indicate the feasibility of production using the domestic cross SK-Rus 61 3 for cage growing of broilers in the form of large meat chickens for deep processing of individual parts of carcasses. The results are confirmed by the zootechnical efficiency of broiler production, the biological usefulness of the muscle composition and meat qualities, as well as the taste and aroma properties of meat products.

References

1. V. S. Buyarov, I. P. Saleeva, Ye. A. Buyarova, Bulletin of the Oryol State Agrarian University, **2(17),** 54 (2009)
2. V. S. Buyarov, Ye. A. Buyarova, V. A. Borodin, Animal Science, **9**, 24 (2003)
3. I. V. Zhuravlev, T. G. Aydinyan, V. I. Fisinin, D. A. Davtyan, Agricultural Biology, **6,** 66 (1999)
4. A. Osmanyan, L. Bakaeva, Yu. Plaksin et al Growing large broilers in cages, Poultry farming, **4(11)** (1993)
5. A. Osmanyan, I. Saleeva, V. Malorodov, R. Gaifullin, Chief zootechnician, **7**, 52 (2019)
6. A. K. Osmanyan, V. V. Malorodov, Poultry and poultry products, **1**, 13 (2021)
7. A. Osmanyan, R. Mahdavi, V. Malorodov, Chief zootechnician, **3**, 50 (2018)
8. A. K. Osmanyan, *Abstract of thesis ... of Doctor of Agricultural Sciences* (1998)
9. A. K. Osmanyan, V. V. Malorodov, N. G. Cherepanova, I. P. Saleyeva, Poultry farming, **12,** 42 **(**2020)
10. *Technology of growing large meat chickens for deep processing of meat* (1994)
11. V. I. Fisinin, Bulletin of the Russian Academy of Sciences, **85(9),** 785 (2015)
12. V. I. Fisinin, T. A. Stollyar, V. S. Buyarov, Bulletin of the Oryol State Agrarian University, **1(4)**, 6 (2007)
13. V. I. Fisinin, *World and Russian poultry farming: realities and challenges of the future* (2019)
14. V. I. Fisinin, L. I. Tuchemsky, J. V. Yemanuylova, Agricultural Biology, **40(2),** 39 (2005)
15. V. I. Fisinin, Bulletin of the Russian Academy of Sciences, **87(4),** 333 (2017)
16. V. I. Fisinin, K. V. Zlochevskaya, Agricultural Biology, **5**, 3 (1984)
17. V. Z. Khamitova, A. K. Osmanyan, R. A. Yerigina et al. Zootechnics, **9,** 15 (2019)

18. V. Z. Khamitova, A. A. Gerasimov, I. V. Cheredov, A. K. Osmanyan, Poultry farming, **12,** 13 (2012)
19. V. Z. Khamitova, A. K. Osmanyan, V. V. Malorodov, Poultry farming **1,** 22 (2021)
20. G. A. Bobyleva, Poultry & Chikenproducts, **3**, 12 (2010)
21. V. I. Ficinine, A. K Ocmanyan, R. Mahdavi, I. A. Egorov, Animal Nutrition, **3(3)**, 258 (2017)
22. Poultry, **5,**16 (1994)
23. R. Mahdavi, A. K Osmanyan, V. I. Fisinin et al, Journal of Animal Physiology and Animal Nutrition, **102(5),** 1266 (2018)
24. A. K Osmanyan, R. Mahdavi, S. Ghazi Harsini et al., Amino Acids, **50(2),** 353 (2018)
25. T. A. Stollar, V. S. Lukashenko, A. N. Tishenko, Pat. 2057459 Russian Federation, IPC and 23 K 1/00. /; applicant and patentee VNITIP. №. 94005721/15, Bul. 10.3

The use of gene technologies in the genomic-index assessment of producers of Canadian breeding

Nurmagomed Bukarov[1], *Vasily* Zakharov[2], *Shamsutdin* Hashegulgov[3], *Ibragim* Taov[4], and *Tsray* Kagermazov[4,*]

[1]JSC "Moskovskoye" breeding enterprise, str. Soyedinitelnaya, 7, 142403 Noginsk, Moscow region, Russia
[2]Federal State Budgetary Scientific Institution "All-Russian Research Institute of Breeding", str. Lenina, 13, Moscow Region, 141212 Pushkino, Lesnye Polyany, Russia
[3]FSBEI HE "Ingush State University", I.B. Zyazikova av., 7, 366700 Magas, Russia
[4]Kabardino-Balkarian State Agricultural University named after V.M.Kokov, Lenin Avenue, 1V, 360030 Nalchik, Russia

Abstract. The article provides an assessment of the use of gene technologies in the genomic-index assessment of producers in the conditions of farms in the Moscow region, which allows to provide dairy farming with high-quality breeding material. Genomic index assessment accelerates the breeding goal achievement process. It contributed to an increase in the productivity of cattle, an increase in the profitability of cattle breeding, which allows to reduce the time for creating breeding herds of dairy cattle by 2 times. Of the 15 bulls, 10 (67%) were evaluated by the A1-A3 breeding categories. Two bulls were categorized as neutral. The bull Ecuador received the A2B1 breeding categories. The daughters of this bull are fat and milk enhancers, which is 4.46%, with a milk yield of 12432 kg of milk.

1 Introduction

Genomic scanning methods allow to determine the origin of animals, the genetic potential of productivity, as well as to accelerate the creation of breeding herds. Genomic assessment is a powerful tool for use in breeding programs. Marker breeding increases the importance of all breeding procedures, thereby accelerating the genetic improvement of herds in terms of productive qualities. For instance, this method is already being used to assess the fertility of bulls (DPR, the pregnancy index of the daughters of a bull) and the resistance of animals to diseases.

Genomic technologies may significantly improve breeding efficiency and eliminate the disadvantages of the livestock breeding traditional methods.

Pursuant to the research in the Russian Federation and abroad, it should be noted that the biotechnological innovations allow to create new genotypes of livestock exceeding the original animals by 30-40% [1, 2, 5,6-10].

*Corresponding author: laura07@yandex.ru

Pursuant to the regional concept (Program for the improvement of dairy cattle of the Moscow region for the period until 2022), by 2023, the average milk yield per forage cow will reach 7300-7500 kg in the farms of the Moscow region, with a fat content of 4.10% and a protein content of 3.30-3.35 %. More than 40% of animals should be breeding, with a cow milk yield of 10,000 kg or more.

The accelerated formation of highly productive breeding herds will be ensured using the breeding genetic resources of the countries of the leading market suppliers.

In this regard, we are interested in the adaptive potential of cattle of Canadian origin. In the Russian Federation, there is still insufficient data on the suitability for use of livestock of Canadian origin in milk production. This determined the need to import Canadian cattle into our country and assess the prospects for this decision.

The paper is aimed is to examine the genome-valued seed bulls to study their productive qualities in the conditions of farms in the Moscow Region and to provide dairy farming with high-quality breeding material, which allows to reduce the time for creating pedigree herds of dairy cattle by 2 times, while increasing the profitability of farmed livestock.

2 Material and methods

The research material is data on the daughters of 15 genome evaluated bulls of Canadian origin. Genomic assessment was performed for 2 breeding indices Pro$ (economic assessment) and GLPI (indicator of genomic assessment of lifelong productivity). The value of LPI - consists of 4 important components: indicators of milk production, durability of cows, health and fertility. Herewith, the value of the economic index Pro$ undergoes an annual update. It is associated with changes in milk prices in Canada, as well as an increase in the cost of livestock production. It is noted that the values of the quantities obtained by the Pro$ index have significantly expanded the interval of their distribution. In highly productive animals, an increase in the value of the index is observed, while the indices with low productivity indicators drop an order of magnitude lower. In the Holstein breed bred in Canada, the average growth of the Pro$ index at the beginning of 2019 amounted to USD 371 and reached USD 670. Pro$ index has been introduced into practice since August 2015 [3, 4]. The second index was introduced due to the need to improve the selection accuracy and satisfy the desire of breeders to have a reliable breeding tool capable of improving the entire set of specified breeding indicators. Both indices have found practical use.

The research was performed in the breeding farms of the Moscow region. The information database on the assessment of dairy cattle in the Moscow region (Regional Breeding Information Center "Mospleminform") was used as of 1.05.2019. We also used information from the catalogs of breeding bulls of JSC "Moskovskoye" for breeding, published in 2018-2020. In JSC "Moskovskoe", high requirements are imposed on Holstein bulls purchased for the breeding enterprise.

The selection of bulls born in 2011-2012 at the age of 7 months and older was performed pursuant to the genomic assessment. We selected animals with good health, development, conformation, milky body type, strong limbs, developed genitals.

In genetic terms, the elimination of bulls-carriers of genes, which in a homozygous state lead to the death of embryos and fetuses, was performed. These are primarily BLAD, CVM, BY, HCD mutations, HH1-HH5 haplotypes. The work performed in this direction to date has made it possible to identify more than 10 haplotypes that negatively affect fertility and remove them from breeding use.

3 Results and discussion

Pursuant to the results of the analysis of the data in Table 1, it follows that representatives of two lines were selected as fathers (n = 15). Table 1 presents data on the prediction of the potential of young bulls.

Table 1. Genomic forecast of productivity indicators for young Holstein bulls of Canadian origin.

Bull, original no.	Pro$	GLPI	Milk, kg	Fat, %	Protein, %	Father	Line
HOCANM106739661	846	2400	1880	0.15	0.06	C. GarrettET52805710	UI
HOCANM107359022	742	2416	1777	0.37	0.13	F. Snowman 388965513	UI
HOCANM107522503	891	2524	1143	0.24	0.14	V-E-D Sudan ET62768990	RS
HOCANM107870105	1016	2637	1814	-0.12	0.12	D. Mayfield ET69473980	RS
HOCANM11218165	674	2345	1686	0.15	0.01	E. Bakon ET136800233	UI
HOCANM107359094	762	2403	2218	0.00	-0.01	F.G. Snowman 388965513	UI
HOCANM107870103	1407	2679	1897	0.09	0.11	M.DSI Mogul ET840M306972816	UI
HOCANM11230463	547	2341	1771	0.11	0.16	L. Oman-OmanET135746776	UI
HOCANM11492044	562	2303	1473	0.24	0.19	D. Phoenix 588ET66636724	RS
HOCANM107870111	1176	2554	1698	0.16	0.21	D. Phoenix 588ET66636724	RS
HOCANM11230547	1020	2533	879	0.36	0.23	D. BukemET66636657	RS
HOCANM107522499	1019	2571	1586	0.35	0.14	V-E-D Sudan ET62768990	RS
HOCANM11203408	563	2476	1570	-0.08	0.10	F.G. Snowman 388965513	UI
HOCANM107359084	762	2403	2311	0.00	0.00	F.G. Snowman 388965513	UI
HOCANM11492002	79	2119	2050	-0.08	-0.04	V.Explode138905680	UI

Bulls are mostly represented by the Wes Ideal line 933122 (UI; n = 9), to a lesser extent the Reflection Sovering 198998 (RS, n = 6). Bulls-producers were evaluated pursuant to 8 indicators of productive qualities. The characteristics of the bull-producer begins with the origin, for instance, based on the designation of the breed-HO-Holstein-Holstein, CAN-Canada, M-bull belonging to JSC "Moskovskoe" for breeding work and its number. This is followed by the projected estimates of economic indicators based on the genomic forecast of lifetime productivity, followed by the forecast for milk yield, as well as indicators of the content of fat and protein in milk.

For those wishing to learn the details of the genomic forecast for each of the young bulls delivered to the Russian Federation, there are detailed information on the Canadian Dairy Network website. Bull numbers are mostly 9-digit, but there are also 8-digit numbers. The latter are included in the list of genome-valued young bulls. By typing 8-digit numbers in the Query search engine, one can access all the information about the parameters of interest to the manager-livestock breeder.

Table 1 shows mainly the data on the productivity potential of bulls. Herewith, other parameters were not used, for example, the productivity of mothers of bulls, indicators of conformation and reproduction. All these and some other parameters were obtained by genomic scanning and are used in breeding work. It should be noted that 9 out of 15 bulls were obtained by embryo transfer.

Table 2 shows the performance indicators of the daughters of bulls. In it, in 8 columns, the results of evaluating bulls by the quality of offspring are presented.

Table 2. Performance indicators of the daughters of 15 genome-rated sire sires.

Bull	Productivity indicators					Daughters of the same age		
	Milk yield of daughters n	kg	Fat, %	Protein, %	Breeding category	Milk, kg	Fat, %	Protein, %
K. Lotong M106739661	295	12139	4.28	3.23	B1	+52	+0.05	+0.01
Colin M107359022	189	12122	4.27	3.36	A3B3	+83	+0.02	+0.02
Crete M107522503	81	12089	4.05	3.29	A3	+123	0.00	+0.04
Lightup M107870105	30	9499	4.22	3.37	H	+33	-0.24	+0.01
Lifey M11218165	100	11761	4.34	3.31	A3	+136	-0.04	-0.02
Loni M 107359094	80	12131	4.17	3.31	A1	+534	-0.01	+0.06
Movi M107870103	33	12416	4.22	3.24	A1	+311	-0.04	0.00
O-Lexus M11230463	62	12026	4.25	3.26	A3	+103	0.00	+0.05
Poncho M11492044	53	131364	4.10	3.28	A1	+580	-0.01	+0.01
Poirot M107870111	21	12001	4.24	3.40	H.	-62	0.00	+0.06
Real M11230547	74	11038	4.05	3.26	B2	+2	+0.03	0.04
Suban M107522499	111	11595	4.09	3.38	B1	+29	+0.05	+0.07
Flevo M11203408	87	12084	4.15	3.36	A1	+260	-0.06	+0.01
Ecuador M107359084	32	12432	4.46	3.17	A2B1	+262	+0.06	-006
Exo M11492002	37	12281	3.99	3.40	A1	+521	-0.04	-0.02

Herewith, 10 out of 15 bulls (67%) were assessed by the breeding categories A1-A3. Two bulls were categorized as neutral. Note also that the Ecuador bull received breeding categories A2B1. The daughters of this bull improve the fat-milk content, which is 4.46%, with a milk yield of 12432 kg of milk. These parameters almost reach the final standards: a milk yield of more than 10,000 kg, with a fat content of 4.5% and a protein-milk content of 3.45%. Thus, the assessments of the daughters of Canadian bulls showed that a significant increase in productivity and profitability of production was obtained in farms raising dairy cattle from cows. Milk yield exceeded the level of 10,000 kg of milk.

4 Conclusions

Bulls of the Holstein cattle breed of Canadian origin significantly increase the productive qualities of the bred cattle in the conditions of our country, the cattle has a good adaptive potential and is recommended for use in breeding work.

References

1. N. Yanchukov, *Thesis ... of doctor in agricultural sciences,* 345 (2011)
2. Dairy and beef cattle breeding, **8,** 6 (2013)
3. L. Beavers, B. V. Doormaal, Canadian Dairy Network-Information (2015)
4. B. V. Doormaal, Canadian Dairy Network-Information (2019)
5. Janchukov, Elite Genetics. Catalogue of Dairy sires, **24** (2020)

6. T. T. Tarchokov, IOP Conference Series: Earth and Environmental Science, **341(1),** 012047 (2019)
7. A. A. Mishkhozhev, Animal breeding, **11,** 24 (2017)
8. A. A. Zhilyaev, Animal breeding, **9,** 8 (2020)
9. T. T. Tarchokov, Animal breeding, **5,** 27 (1996)
10. M. O. Baytaev, Agrarian Russia, **1,** 28 (2013)

E3S Web of Conferences **262**, 02008 (2021) https://doi.org/10.1051/e3sconf/202126202008
ITEEA 2021

Experimental results on pancreatic juice parenteral application in broiler growing

Vladimir Vertiprakhov[1,*], *Alyona* Grozina[1], *Vladimir* Fisinin[1], *Natalya* Ovchinnikova[1], and *Muhammed* Kozhokov[2]

[1]FSBSI Federal state budget scientific institution, Federal scientific center "All-Russian research and technological Institute of poultry" of the Russian Academy of Sciences, 141315 Sergiyev Posad, Russia
[2]Kabardino-Balkarian State Agricultural University named after V.M.Kokov, Lenin Avenue, 1V, 360030 Nalchik, Russia

Abstract. The paper presents the results of the use of pancreatic enzymes in the form of intramuscular injection of pure pancreatic juice obtained from fistulated chickens to broiler chickens. At 7 days of age during the formation of groups, the differences between the control and experimental chickens were 17.9% ($p<0.05$) in favor of the former. It was found that after the injection of the drug 1 time in 7 days, there was a stimulation of metabolism in the experimental group broilers. By 14 days of age, the differences between the groups were 17.4% ($p<0.05$), to 21 days of age - 9.1% ($p<0.05$), to 28 days of age - 5.3%, to 35 - day-old age - 4.9%. Consequently, as the chickens were raised by 28 days of age, the chickens of the experimental group caught up with the live weight of the control ones, and, considering the increase in live weight, even exceeded this indicator (50.1 g/day versus 42.7). The use of parenteral pancreatic juice of chickens at a dose of 0.1ml/kg of live weight of broiler chickens allows to conclude that the drug has a stimulating effect on poultry metabolism, manifested by an increase in live weight gain, an increase in the content of lymphocytes, as an indicator of immunity, activation of protein and mineral metabolism in the body when using an enzyme preparation.

1 Introduction

Pure pancreatic juice of chickens is a clear or yellowish alkaline liquid. The specific weight of juice is 1.0064-1.0108, the content of solids is 1.8-2.2% and the pH level is 7.2-8.1. The juice has proteolytic, amylolytic and lipolytic activity, and, in terms of the concentration of proteases and especially amylase, it exceeds the pancreatic juice of mammals, while being inferior to it in terms of the concentration of lipase [1]. The pancreatic proteases are trypsin, chymotrypsin, carboxypeptidases, and elastase. In the pancreas, they are in an inactive state and are activated when they enter the intestine by the enzyme enterokinase, discovered by N.P. Shepovalnikov (1899) in the laboratory of I.P. Pavlov. Among carbohydrases, α-amylase and invertase were found in the pancreatic juice of birds. Pancreatic lipase

*Corresponding author: vertiprakhov63@mail.ru

disintegrates fats into monoglycerides and unsaturated fatty acids (with the release of glycerol), hydrolyzing the ester bonds of triglycerides in positions 1 and 3.

The scientific literature contains the data on the effect of intrinsic enzymes on the function of the pancreas in rats [2] and dogs [3]. Experimental data allowed the authors to come to the conclusion on the presence of a regulatory role of pancreatic enzymes in the secretory process pursuant to the feedback principle [4-6]. Such experiments have not been performed on birds that have the specific digestive features. Therefore, we decided to fill this gap and, for the first time, to study the development in ontogenesis and morphological-and-biochemical parameters of blood in broiler chickens against the background of parenteral use of pancreatic enzymes.

2 Material and methods

The experiments were performed on 20 broiler chickens of cross-selection SGC Smena from one day to 42 days of age in accordance with the requirements of the European Convention for the Protection of Vertebrate Animals used for Experiments or for other Scientific Purposes (ETS No. 123, Strasbourg, 1986) [7]. Broiler chickens were kept in the VNITIP vivarium in compliance with optimum conditions: room temperature, duration of lighting, rations (start, growth and finish) were in accordance with the recommendations of VNITIP (2014) [8]. The control group consisted of chickens with a higher live weight within the age (10 birds), the experimental group consisted of chickens with a live weight deficit (10 birds).

Pancreatic juice was obtained from laying hens with chronic pancreatic duct fistula [9] and stored in a sterile container at -20 °C. Before use, the juice was defrosted and applied being previously 10 times diluted with sterile saline solution. Chickens of the control group were injected with saline at a dose of 1.0 ml per 1 kg of live weight with the frequency of 1 time in 7 days, and the chickens of the experimental group were injected intramuscularly with 0.1 ml/kg of live weight of pancreatic juice, diluted to 1.0 ml with saline solution weekly, starting from 7 days of age.

Blood was taken from fasting chickens in the morning from the axillary vein 1 time in 7 days. For morphological studies, special vacuum tubes were used, where EDTA was used as an anticoagulant - a purple cap (sterile vacuum tubes, 4.0 ml, Jiangxi Hongda Medical Equipment Group Ltd, 39, China). For biochemical studies, tubes with lithium heparin (green cap) or with a coagulation activator (red cap) were used. Immediately after blood collection, the tube was slowly inverted 6-8 times to completely mix the blood with the reagents. A blood sample was centrifuged at 3000 rpm for 3 minutes to separate the formed elements from plasma or serum.

Biochemical methods. The activity of trypsin in blood plasma was studied using benzoyl DL-arginine nitroanilide (BAPNA) as a substrate on a BS3000P semi-automatic biochemical analyzer (China) by the kinetic method [10]. The activity of alkaline phosphatase, the content of glucose, cholesterol, total protein, calcium, phosphorus was performed on a BS3000P semi-automatic biochemical analyzer (China) using kits from the DIAKON-VET company (RF).

Statistical methods. For statistical processing of the results, the JMP Trial 14.1.0 software was used, with the help of which the mean value (M) and the standard deviation (± m) were calculated, the significance of differences was determined by the Student's t-test. Differences were considered statistically significant at $p<0.05$.

3 Results and their discussion

The results of the study showed that at the day-old age, broiler chickens in the experimental group had a live weight lower by 8.7% than chickens in the control group (Table 1).

Table 1. Live weight of broiler chickens of the cross-breeding SGC Smena in ontogenesis using parenteral pancreatic juice.

Age, days	Group	
	Control	Experimental
1	47.1±0.74	43.0±0.50*
7	158.3±4.36	130.0±0.50*
14	419.8±11.40	346.8±12.50*
21	943.9±23.60	857.8±26.20*
28	1525.1±35.30	1444.8±51.00
35	2239.2±82.10	2130.8±78.80

Note: * - the difference with the control group is significant at $p<0.05$

It can be seen from this table that during the first 7 days of growing, the differences between the groups increased: in live weight, the experimental chickens lagged behind the control ones by 17.9% ($p< 0.05$). On the seventh day, pancreatic juice was injected into the chickens of the experimental group in accordance with the test procedure. By 14 days of age, the differences between the groups were 17.4% ($p<0.05$), by 21 days of age - 9.1% ($p<0.05$), by 28 days of age - 5.3%, by 35 days of age - 4.9%. Consequently, as the chickens were raised by 28 days of age, the chickens in the experimental group did not differ significantly from the control ones, and, considering the increase in live weight, even overtook them (Fig. 1).

To study the mechanism of the stimulating action of pancreatic enzymes used parenterally, we performed the determination of the morphological and biochemical parameters of the blood of broiler chickens (Tables 2, 3).

From this table 2 it can be seen that in the experimental group at 21 and 28 days of age, there is a decrease in the number of leukocytes in the blood (respectively, by 10.7% and 19.6%, $p< 0.05$). At the same time, there is a change in the ratio of different forms of white blood cells in the leukocyte formula: the content of lymphocytes increases at 21 days of age (by 14.2%) and 28 days of age (29.3%), which indicates an increase in specific protection against bacteria and viruses. and a decrease in the overall resistance that neutrophils and eosinophils provide. A decrease in the number of erythrocytes and hemoglobin (with the exception of 35 days of age) shows a decrease in metabolism within the physiological norm. Indicators of RDW-CV and RDW-SD are connected with the erythrocyte value. The first index determines the ratio of the width of the erythrocyte to the volume (coefficient of variation). The RDW-CV index indicates the deviation of the erythrocyte volume from the mean value. Thus, pursuant to the morphological parameters of blood, broiler chickens receiving injection of pancreatic juice of chickens had lower values in the number of leukocytes and erythrocytes, which did not go beyond the physiological norm.

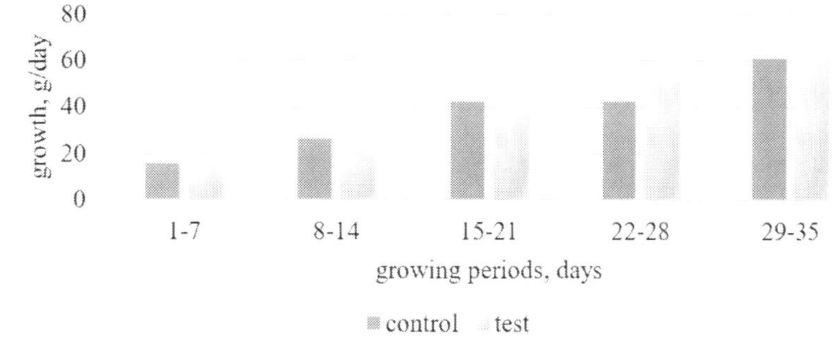

Fig. 1. Average daily gain in live weight of broiler chickens when using intramuscular injection of pancreatic juice of 0.1 ml/kg of live weight.

Table 2. Morphological parameters of the blood of broiler chickens of the SGC Smena selection using intramuscular pancreatic juice of chickens.

Haematological indicators	Group age - 21 days		Group age - 28 days		Group age - 35 days	
	Control	Experimental	Control	Experimental	Control	Experimental
WBC, $^{10\ 9}$/l	28.1±0.87	25.1±0.10*	32.1±2.47	25.8±1.25*	30.7±1.26	28.9±2.02
Neu, %	33.1±2.29	26.9±1.86	52.2±3.00	43.8±1.76*	38.2±2.29	39.9±2.6
Lym, %	62.8±2.87	71.7±2.07*	40.3±3.55	52.1±2.46*	59.3±2.22	57.3±2.75
Mon, %	0.3±0.05	0.3±0.07	0.6±0.14	0.5±0.11	0.2±0.04	0.2±0.01
Eos, %	3.7±0.7	1±0.24*	6.8±0.61	3.4±0.97*	2.1±0.26	2.4±0.42
Bas, %	0.1±0.01	0.1±0.01	0.1±0.01	0.1±0.01	0.2±0.02	0.2±0.02
RBC, 10^{12}/l	2.5±0.03	2.3±0.04*	3.0±0.09	2.6±0.07*	2.9±0.04	2.7±0.06*
HGB, g/l	134.9±1.67	123.5±0.89*	155.1±4.33	135.5±2.59*	153±1.56	145±3.76
HCT, %	34.3±0.31	31.4±0.34*	38.4±1.1	33.9±0.72*	36.9±0.36	34.7±0.84
MCV, fl	135.5±0.55	134.6±1.01	129.7±0.57	129.2±1.07	126±0.68	127.4±0.66
MCH, pg	53.4±0.27	53±0.55	52.48±0.26	51.8±0.66	52.3±0.29	53.1±0.43
MCHC, g/l	393.8±1.67	393.8±1.57	404.4±1.07	401±1.79	414.8±1.48	417±2.15
RDW-CV, %	11.2±0.21	12.4±0.38*	10.4±0.07	10.9±0.04*	10.0±0.05	10.7±0.20*
RDW-SD, fl	62.9±1.31	69±2.32*	55.2±0.54	57.6±0.38*	51.6±0.45	55.4±0.77*

Note: * - the difference with the control group is significant at $p<0.05$

Biochemical blood tests make it possible to monitor the functional state of the body, the work of the liver, kidneys, pancreas and other organs, as well as monitor the processes of protein, carbohydrate, fat and mineral metabolism. The data are presented in Table 3.

Table 3. Biochemical parameters of broiler chickens with intramuscular administration of chicken pancreatic juice.

Biochemical indicators	Group, age 21 days		Group, age 28 days		Group, age 35 days	
	Control	Experimental	Control	Experimental	Control	Experimental
Total protein, g/l	30.4 ± 0.33	31.7 ± 0.78	1)0.76	35.2 ± 0.51*	36.4 ± 0.76	38.2 ± 0.50
Trypsin, un/l	1)56.1	1)68.98	961 ± 49.4	969.9 ± 13.06*	1016 ± 52.9	1125.0 ± 69.8
Alkaline phosphatase, un/l	1)940.20	1)878.88	4567.0 ± 313.60	6492.7 ± 610.00*	2429 ± 183.40	3919.0 ± 189.30
Calcium, mmol/l	2.7 ± 0.04	2.6 ± 0.04	2.6 ± 0.09	2.4 ± 0.03	2.4 ± 0.13	2.1 ± 0.02
Phosphorus, mmol/l	3.9 ± 0.08	4.2 ± 0.29	2.7 ± 0.07	2.3 ± 0.01*	2.2 ± 0.04	2.4 ± 0.11
Cholesterol, mmol/l	3.1 ± 0.09	3.2 ± 0.12	3.1 ± 0.05	3.0 ± 0.10	3.4 ± 0.03	3.5 ± 0.07
Glucose, mmol/l	13.3 ±0.27	12.1 ±0.53	11.0 ±0.15	11.2 ±0.21	12.9 ±0.26	11.6 ±0.81

Note: * - the difference with the control group is significant at $p<0.05$

The results showed that parenteral administration of pancreatic juice affects protein and mineral metabolism. In the age range of 21-28 days, the content of total protein in the blood changes, increasing by 7.3% ($p<0.05$), and the amount of total phosphorus in the experimental group also decreases by 14.8% ($p<0.05$) and the activity of alkaline phosphatase increases by 42.2% ($p<0.05$).

Enzyme preparations are widely used in cattle breeding and poultry farming, usually as feed additives [11]. In medicine, there is a drug for parenteral administration - crystalline trypsin, but such drugs are not widely used due to the poor knowledge of the mechanism of action of pancreatic enzymes when administered intramuscularly to the body as a whole [12]. We studied the level of trypsin and nitric oxide metabolites in the postprandial phase. The regularity of the parallel increase in the content of physiological NO donors and the activity of trypsin in the blood after feeding the bird was established [13], which is associated with parasympathetic regulation, since the atropine administered preliminarily blocked an increase in the secretory function of the pancreas. In this work, for the first time, an experiment was performed on poultry, which allows to study the mechanism of action of native pancreatic juice on the organism of broiler chickens. As a result, it was found that the most significant changes in the metabolism of broiler chickens occur in the experimental group due to the stimulation of protein and mineral metabolism. This is consistent with an increase in live weight and morphological changes in the composition of the bird's blood. Questions of further use of the preparation in the form of an aerosol for mass processing of poultry are discussed.

4 Conclusions

The use of parenteral pancreatic juice of chickens at a dose of 0.1ml/kg of live weight of broiler chickens allows to conclude that the drug has a stimulating effect on poultry metabolism, which manifests itself in an increase in live weight gain, an increase in the content of lymphocytes, as an indicator of immunity, activation of protein and mineral metabolism. in the body when using an enzyme preparation.

References

1. V. I. Fisinin, V. G. Vertiprakhov, Ye. L. Kharitonov, A. A. Grozina, Agricultural Biology, **54(6)**, 1122 (2019)
2. J. S. Laporte, J. Tremolieres, Comptes rendus de l'Académie des Sciences, **273**, 1205 (1971)
3. P. K. Klimov, A. A. Fokina, *Physiology of the Pancreas. Regulation of exocrine function* (1987)
4. A. S. Loginov, N. Sh. Amirov, O. D. Chernoyarova, Physiological Journal of the USSR, **67**, 957 (1981)
5. G. F. Korotko, Journal of Fundamental Medicine and Biology, **2**, 4 (2016)
6. S. Rothman, C. Leibow and L. Isenman, Physiological Reviews, **82**, 1 (2002)
7. European Convention for the Protection of Vertebrate Animals used for Experimental or Other Scientific Purposes (ETS №. 123) (Strasbourg, 1986) http://www.conventions.ru
8. *Guide to Optimizing Poultry Compound Feed Recipes* (2014)
9. Ts. Zh. Batoev, *Physiology of Bird Digestion* (2001)
10. V. G. Vertiprakhov, A. A. Grozina, Veterinary, **6**, 51 (2018)
11. I. A. Yegorov, V. A. Manukyan, V. G. Vertiprakhov, T. N. Lenkova, T. A. Yegorova, I. N. Nikonov, A. A. Grozin, K. V. Borisenko, Poultry farming, **9-10**, 61 (2019)

12. V. G. Vertiprakhov, A. A. Grozina, Method for normalizing digestion in animals by parenteral administration of pancreatic enzymes. Patent for invention 2738930, 12.18.2020. Claim №. 2019139198 dated 03.12.2019
13. V. Y. Titov, A. V. Ivanova, A. N. Osipov, A. M. Dolgorukova, V. G. Vertiprakhov, N. A. Slesarenko, I. I. Kochish, Bulletin of Experimental Biology and Medicine, **168(3)**, 321 (2020)

Impact of various reagents during blood collection from poultry to analyze morpho-biochemical indicators

Vladimir Vertiprakhov[1,*], *Alena* Grozina[1], *Vladimir* Fisinin[1], *Natalya* Ovchinnikova[1], and *Anatoly* Bittirov[2]

[1]Federal State Budgetary Scientific Institution "Federal Scientific Center "All-Russian Research and Technological Poultry Institute" of the Russian Academy of Science, 141315 Sergiev Posad, Moscow region, Russian
[2]Kabardino-Balkarian State Agricultural University named after V.M. Kokov, Lenin Avenue, 1V, 360030 Nalchik, Russia

Abstract. The work provides experimental data on the impact of anticoagulants (K3 - EDTA, 3.8% sodium citrate and lithium heparin) and coagulation activator on morpho-biochemical indicators in the blood samples of broiler chickens of cross-selection of Agricultural Center "Smena" at the age of 35 days. The studies were carried out with the use of semi-automatic flow analyzer Sinnowa BS-3000P (SINNOWA Medical Science & Technology Co., Ltd, China) and veterinary automatic blood analyzer DF-50 by Dymind Biotech (PRC) with the use of original reagents. The results showed that the most optimal reagent for determination of morphological indicators of the poultry blood was EDTA. The average white blood count (WBC) is lower in blood samples with citrate and heparin in contrast to EDTA by 30.5% and 24.1%, respectively. The average red blood count (RBC) is lower in tubes with sodium citrate (by 37.9 %) and lithium heparin (by 13.8 %) in contrast to K3-EDTA. When the sodium citrate anticoagulant is used, the blood biochemistry results for all positions of the experiment, excluding calcium, are lower than in heparin, and its closest values are lower than in coagulation activator, excluding trypsin activity. That's why, the most optimal reagents for biochemistry, including for blood enzyme tests, should be considered the tubes with heparin or coagulation activator.

1 Introduction

The morphological and biochemical blood indicators are important for determining the physical status and health of the live-stock animals and poultry, including for mycotoxicosis diagnostics. Recently, the classical research methods have been replaced with automatic blood and biochemical analyzers with a set of reagents, i.e. there has been the significant upgrading of the instrumental base of scientific laboratories that has led to a high accuracy of research results requiring the determination of new reference values, as

*Corresponding author: vertiprakhov63@mail.ru

well as to increase in the speed of analysis. The scientific literature has accumulated wide and contradictory material on blood morphology and biochemistry in poultry [1], this is often related to the use of various reagents for blood plasma and serum production.

The tubes of different manufacturers that are currently used for blood sampling contain necessary concentrations of anticoagulants (sodium citrate, ethylenediaminotetraacetic acid (EDTA), lithium heparin and others). The existing anticoagulants ensure the preservation of the components without significant changes in the course of sample analysis and are characterized by the properties that inhibit the blood components coagulation process. This prevents blood coagulation due to the binding of calcium ions (EDTA, sodium citrate), inhibition of thrombin activity (heparin, hirudin) [2,3]. There is enough information in the literature on the effect of various types of anticoagulants on hematological indicators in human or animals, but there is limited information on the use of tubes for studying blood samples of poultry. In this regard, in order to obtain more correct analysis results, it is important to choose the correct anticoagulant for blood collection. Therefore, the purpose of this work was to compare the impact of anticoagulants (K3 - EDTA, 3.8% sodium citrate and lithium heparin) on morpho-biochemical indicators in the blood samples of broiler chickens of cross-selection of Agricultural Center "Smena" at the age of 35 days.

2 Materials and research methods

The blood collected from broiler chickens of cross-selection of Agricultural Center "Smena" at the age of 35 days in the morning hours, in fasting state, was tested during the experiment. The produced plasma was tested on the semi-automatic flow analyzer Sinnowa BS-3000P (SINNOWA Medical Science & Technology Co., Ltd, China) with the use of biochemical sets (DIAKON-VET, Russia) by determining total protein, alkaline phosphatase, glucose, cholesterol, calcium, phosphorus. The trypsin activity in blood plasma was tested by using the benzoyl-DL-arginine nitroanilide (BAPNA) as a substrate on the semi-automatic biochemical analyzer BS3000P (KHP) by means of kinetic method [4]. The hematological tests were performed on the veterinary automatic blood analyzer DF-50 by Dymind Biotech (PRC) using the original reagents. The tests were performed with simultaneous use of different tubes: with K3 EDTA, lithium heparin, sodium citrate and coagulation activator. The samples were centrifuged at 3000 rpm for 5 minutes to separate plasma from the formed elements.

Currently, the most preferable anticoagulant for hematological tests is ethylenediaminetetraacetate (EDTA) with the most effective concentration of 1.2 mg/ml of blood which is used for the blood plasma production and can form the soluble high-stable complexes due to formation of chelate compounds with calcium ions. The effectiveness of the analysis for blood plasma separation depends on a number of factors during preparation of the material, i.e. the quality of the analysis will increase when the tube is inverted 5-7 times immediately after blood collection to ensure better mixing of the blood and the anticoagulant and centrifugation at 2000-3000 rpm. During the hematological tests related to blood cells counting, determining the ESR, etc., during the PCR tests (qualitative and quantitative methods), it is advisable to use tubes containing 1.95 mg of EDTA/1 ml of blood. It is advisable to store tubes with collected blood samples for up to 6-10 hours at 4°C, further storage of tubes with blood samples is not recommended for more than 24 hours due to low red cell and white cell counts [5].

The anticoagulant heparin is used to test the electrolyte composition, gas composition, and blood alcohol content. However, heparin is acidic and is not recommended for using in the morphological studies, white cell and platelet counting, polymerase chain reaction (PCR) diagnostics.

The anticoagulant sodium citrate is used to collect venous blood and in testing the coagulation properties of blood. In the process of blood coagulation, the enzyme thrombin converts fibrinogen into fibrin, i.e. activation of initially inactive enzymes occurs due to successive complex reactions with the participation of active enzymes. The ratio of citrate to the amount of blood taken should be 1/9 that is achieved by using tubes with 3.8% or 3.2% sodium citrate solution (0.129 mol/l). When conducting coagulation tests, it is recommended to observe certain rules: immediately after venipuncture, the released tissue thromboplastin can affect the test results, therefore, it is necessary to mix the blood and the anticoagulant, and the amount of blood taken should be at the level as set forth in the requirements [5].

Serum can be produced by blood coagulation and clot retraction (thickening), but external activator (silicon dioxide) is required. In practice, centrifugation of tubes with blood accelerates the clot retraction. To produce the maximum pure serum, the following conditions should be observed: 1. Shake and mix the tube contents; 2 Wait until the blood coagulation process is completed with the tube in vertical position; 3. Perform centrifugation for at least 10 minutes at 3000 rpm.

It is important to note that with subsequent shaking or overturning, the blood serum contamination with clot components can occur, it is allowed to store the blood samples at room temperature for up to 6 hours, and in the cold-storage holding conditions (+ 4°C) - for up to 24 hours. The separated blood serum is used for biochemical and immunological tests in the study of protein composition, enzymes, hormones, tumor markers, HIV infection, hepatitis, etc. [5].

Statistical methods. For statistical processing of the results, the JMP Trial 14.1.0 software was used to calculate the mean value (M) and the standard deviation (±m) and determine the significance of differences by the Student's t-test. Differences were considered statistically significant at $p < 0.05$.

3 Study results and discussion

EDTA is recommended as anticoagulant for hematological tests, since it allows for the best preservation of cellular components and blood cell morphology. We considered this group as control when determining the blood cell morphology and compared the effect of other reagents on morphological indicators of blood (Table 1).

Table 1. Results of morphological blood tests of broiler chickens at the age of 35 days with the use of tubes with different anticoagulants (n=60, M±m).

Hematological indicators	Tubes with anticoagulants		
	K3-EDTA (control)	3.8 % sodium citrate	lithium heparin
WBC, 10^9/l	34.4 ± 0.77	23.9 ± 0.85*	26.1 ± 0.15*
RBC, 10^{12}/l	2.9 ± 0.06	1.8 ± 0.04*	2.5 ± 0.03*
HGB, g/l	146.0 ± 2.56	90.6 ± 1.82*	132.3 ± 2.18*
HCT, %	36.9 ± 0.58	22.2 ± 0.45*	31.7 ± 0.38*

Note: * the differences with the control group are reliable at $p<0.05$

The study results showed the reliable difference between the control and the other tubes. The average white blood count (WBC) is lower in blood samples with citrate and heparin in contrast to EDTA by 30.5% and 24.1%, respectively.

The average red blood count (RBC) is lower in tubes with sodium citrate (by 37.9%) and lithium heparin (by 13.8%) in contrast to K3-EDTA. Hemoglobin (HGB) is also lower in the citrated and heparinized blood samples in contrast to EDTA by 37.9% and 9.4%, respectively. Hematocrit (HCT) is lower with citrate by 14.7% and with heparin by 5.2% regarding K3-EDTA.

The use of K3-EDTA as anticoagulant caused a significant difference in concentrations of the measured blood parameters. This may be related to the buffer system of the blood. Blood plasma proteins, which have buffer properties, are able to bind both acids and alkalis, and are ampholytes. Hemoglobin - red blood cell protein - hemoglobin (Hb) and oxyhemoglobin (HbO_2) have considerable buffering effect. In the blood capillaries, HbO_2 dissociates into oxygen and Hb. Such dissociation creates favorable conditions for binding of carbonic and other acids due to the alkali reserve. EDTA is the tetrabasic carboxylic acid, soluble in alkali, that facilitates the addition of the Hb molecule to the free residue of this acid. That's why, the tubes with EDTA anticoagulant are better to be used for determining such blood parameters as red blood cells, hemoglobin, hematocrit and erythrocytic indices. In the leucogram indicators (Fig. 1), similar values were obtained when using EDTA and sodium citrate, and they differed significantly in any manner when using the lithium heparin anticoagulant.

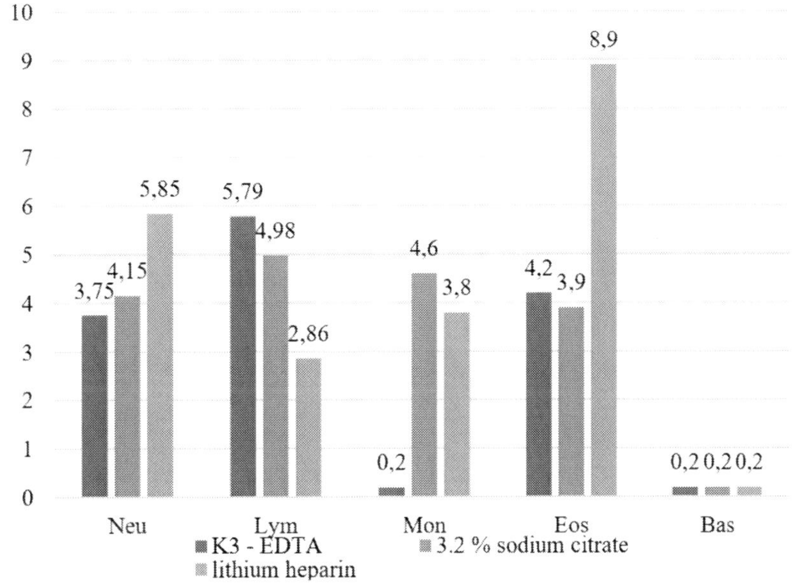

Fig. 1. Leucogram of 35-day old broiler chickens when the tubes with different anticoagulants are used, %.

The Neu and Lym indicators are decreased by 10 times

Figure 1 shows that the percentage of neutrophils (Neu) is the highest in tubes with heparin - 58.5% that is 17% lower than in tubes with citrate, and the lowest in tubes with EDT - 37.5%. The lymphotite content (Lym) is the highest in the samples with EDTA - 57.9%, with citrate it is lower by 8.1%, and the lowest with heparin - 28.6%. The citrated blood shows high monocyte content - 4.6%, the heparinized blood contains 3.8%, EDTA - 0.2%. The eosinophil content (Eos) is the highest with lithium - 8.9%, lower by 4.7% and 5.0% in EDTA and citrate, respectively. There was no difference in the percentage of basophils (Bas) regardless of the type of anticoagulant used.

Therefore, the experiment results showed that the most optimal reagent for determination of morphological indicators of the poultry blood was EDTA.

The blood biochemistry results of broilers are given in Table 2.

Table 2. Biochemical indicators of blood of broiler chickens at the age of 35 days with the use of different tubes (n=60, M±m).

Biochemical indicators	Lithium heparin (control)	K3 EDTA	3.8% sodium citrate	Coagulation activator
Total protein, g/l	36.3 ± 0.68	39.8 ± 0.56*	30.2 ± 0.80*	38.6 ± 0.58*
Trypsin, U/l	1086.0 ± 36.67	83.2 ± 3.20*	74.1 ± 2.22*	68.2 ± 2.43*
Glucose, mmol/l	11.3 ± 0.49	12.3 ± 0.16	6.9 ± 0.56*	12.3 ± 0.24
Alkaline phosphatase, U/l	3324.0 ± 195.50	5266.0 ± 395.20*	1826 ± 86.7*	3929 ± 358.7
Cholesterol, mmol/l	3.4 ± 0.05	3.2 ± 0.06*	1.8 ± 0.18*	3.3 ± 0.07
Calcium, mmol/l	2.5 ± 0.05	0.6 ± 0.01*	3.2 ± 0.12*	2.9 ± 0.18
Phosphorus, mmol/l	2.6 ± 0.06	2.7 ± 0.05	2.3 ± 0.06*	2.0 ± 0.07*

Note: * the differences are reliable in contrast to the control at $p<0.05$

These tables show that when using the lithium heparin anticoagulant for broiler blood, the total blood protein content differed from EDTA by 9.6% downward, from the coagulation activator - by 6.3%, the control sample exceeded the tube with sodium citrate by 16,8%. The trypsin activity with the lithium heparin anticoagulant exceeded the EDTA samples by 13.0 times, with sodium citrate by 14.6 times, with coagulation activator by 15.9 times. The glucose indicators differ from the lithium heparin specimen when using sodium citrate where the level is 38.9% lower ($p<0.05$). The alkaline phosphatase activity reaches its maximum value when using EDTA that is 58.4% higher than the tube with heparin, and in the tube with sodium citrate, the minimum value is observed which is 45.1% lower than the control. The blood cholesterol content below the control is observed in the tube with EDTA (by 5.9%) and sodium citrate (by 47.1%). The calcium content is significantly less than the control in the tube with EDTA (by 76.0%), and with sodium citrate, on the contrary, it is higher by 28.0%. The phosphorus content is noticeably lower with sodium citrate (by 11.5%) and coagulation activator (by 23.1%). Consequently, when the sodium citrate anticoagulant is used, the blood biochemistry results for all positions of the experiment, excluding calcium, are lower than in heparin, and its closest values are lower than in coagulation activator, excluding trypsin activity.

It is known [5-8] that the mechanism of action of sodium citrate is based on binding of calcium ions in the blood; therefore, it can be assumed that there is a relationship between the calcium ion and the trypsin activity in the blood plasma. A negative correlation was identified in the chicken blood between the trypsin activity and the calcium content [9]. This is consistent with the study results [10,11], which evidence that protease inhibitors are among the most widespread components of plasma protein, that is in significant excess of active proteases, except for cases when activation occurs on surfaces or other pathogens. Chelating agents such as EDTA and citrate do not directly inhibit proteases, but they limit the protease activation in the blood coagulation system by affecting calcium-mediated surface binding and allowing inhibitors to dominate. Our data on the difference in research results when using different blood anticoagulants are consistent with the literature [12].

4 Conclusions

The conducted experimental researches have shown that when taking blood samples, the morpho-biochemical test results differ significantly depending on the type of anticoagulants or coagulation activators. For morphological blood tests of poultry blood, it is recommended to use the tubes with EDTA, and for biochemistry, including for blood enzyme tests, the tubes with heparin or coagulation activator are recommended.

Acknowledgements

The work was performed based on the grant from the Russian Science Foundation 20-76-10003.

References

1. V. A. Ponomarev, V. V. Pronin, L. V. Kletikova, L. V. Malovichko, N. N. Yakimenko, *Clinical and biochemical parameters of the blood of poultry* (2014)
2. V. V. Dolgov, *Laboratory hematology* (2014)
3. L. A. Khorovskaya, S.N. Kovalevskaya, N. B. Vynornova, A. Kallner, Transactions of St. Petersburg State I.P. Pavlov Medical University, **XVII(4)**, 23 (2010)
4. V. G. Vertiprakhov, A. A. Grozina, Veterinary, **6**, 51(2018)
5. Supply Medical Institutions, http://medtehural.ru
6. Assembly and method to improve vacuum retention in evacuated specimen containers (2013) www.freshpatents.com
7. Procedures for the Collection of Diagnostic Blood Specimens by Venipuncture; Approved Standard - Fourth Edition. CLSI document H3-A4. (Wayne, PA: Clinical and Laboratory Standards Institute, 2005)
8. S. A. Burtis, E. R. Ashwood, D. E. Bruns, *Textbook of Clinical Chemistry and Molecular Diagnostics* (2006)
9. G. L. Hortin, D. Sviridov, N. L. Anderson, *High-abundance polypeptides of the human plasma proteome comprising the top 4 logs of polypeptide abundance* (2008)
10. R. A. R. Bowen, A. T. Remaley, Biochemia Medica, **24(1)**, 31 (2014)
11. M. Mohri, H. Rezapoor, Research in Veterinary Science, **86(1),** 111 (2009)
12. V. G. Vertiprakhov, A. A. Grozina, T. M. Rebrakova, I. V. Kislova, S. V. Lebedev, O. V. Kvan, I. A. Vershinina, B. V. Usha, S. V. Shabunin, Iranian Journal of Applied Animal Science, **10(2)**, 327 (2020)

Biotechnological aspects of transformation of heavy metal salts and their detoxication in milk of milk cows

Boris Dzagurov[1], *Oleg* Getokov[2,*], *Vladimir* Gukezhev[3], *Taladin* Kokov[2], and *Vitaly* Vorokov[5]

[1]Gorsky (Mountain) State Agrarian University, 37 Kirova str., 362040 Vladikavkaz, North Ossetia-Alania
[2]Kabardino-Balkarian State Agricultural University named after V.M.Kokov, Lenin Avenue, 1V, 360030 Nalchik, Russia
[3]Kabardino-Balkarian Scientific Center of the Russian Academy of Sciences
[4]Kabardino-Balkarian State University, str. Chernyshevskogo, 173, 360004 Nalchik, Russia
[5]SFEI of HPT "I.T. Trubilin Kuban State Agrarian University", 13 Kalinina St., 350044 Krasnodar, Russia

Abstract. Based on the results of previous studies on pigs and poultry, in which encouraging results of a significant decrease in the concentration of heavy metals in the body were obtained, respectively, in pork, poultry and eggs, the study provided for the use of the bentonite clay of the Zamankul deposit (RNO-Alania) as an enterosorbent in relation to heavy metals and detoxification of the organism of dairy cows. In order to study the feasibility of using bentonite as an enterosorbent, studies were performed on dairy cows of the Simmental breed in the Kaloyev farm located in st. Zmeyskaya, Kirovsky district of North Ossetia-Alania. To conduct research on the principle of analog pairs, two experimental groups of cows (control and experimental, five cows in each group) of the 3rd and 4th lactation were formed. The experiment lasted for 305 days of lactation from March 2018 to January 2019. Herewith, the control group of cows was fed with a basic diet balanced in all nutritional elements, the experimental livestock, together with the main diet, daily in the composition of the concentrates was injected with crushed bentonite with a particle diameter of 4-6 mm based on the dry matter of the feed (137 g/animal). The study of the concentration of heavy metals (cadmium, lead and zinc) in soil, feed, blood and milk was performed in two periods of lactation (on the 250th and 300th days of lactation). Herewith, an increased content of heavy metals in the soil was established, relative to the MPL: cadmium - 10.1; lead - 7.4; zinc - 9.7 times more, in drinking water: cadmium - 2.5 times; lead - 9.0 times and zinc - 9.7 times more, in feed: cadmium - from 1.7 to 5.3 times; lead - from 1.1 to 1.7; zinc - from 1.0 to 2.5 times the MPL. The inclusion of bentonite to the ration of the animals of the experimental group, in comparison with the control group, in both study periods (250 and 300 days of lactation) contributed to a significant decrease in the concentration of the studied heavy metals in the blood from 17 to 20%, in milk - from 16 to 18 %. The transformation ration values of

*Corresponding author: getokov777@mail.ru

heavy metals from feed into the body, respectively, into milk, were also lower in the animals of the experimental group, relative to the control.

1 Introduction

As known, heavy metals are considered one of the most dangerous toxicants that enter the biosphere along with the combustion products of diesel engines, industrial emissions, especially when processing the non-ferrous metal ore. Once in the human or animal body, these pollutants accumulate in muscle tissue, liver and kidneys, causing cytolysis of cellular structures and necrotic manifestations, a decrease in hematopoiesis, thereby causing a weakening of immunity, both cellular and humoral, increasing the risk of developing various pathologies, and in animals also decrease in productivity [1, 3, 5,7,8,9,10].

The biosphere, including the soil of the Republic of North Ossetia-Alania, is contaminated with heavy metals, due to emissions of pollutants by industrial enterprises of non-ferrous metallurgy, exhaust gases from vehicles, etc. In this case, following MPL values have been exceeded: - zinc by 10 times, - lead by 8 times, - cadmium by 9 times [5]. In connection with the above, we conducted studies to study the possibility of using as an enterosorbent, in relation to heavy metals, feeding dairy cows with bentonite clay from the local Zamankul deposit, located near the village of Zamankul and the transformation of heavy metals (cadmium, lead, zinc) along the food chain: soil - plants (feed - cows' organism (blood) - milk. The studies performed are *relevant* due to the fact that bentonite dressings from this deposit, considering the sorbing properties, were previously used by us for the first time, as an enterosorbent for removing heavy metals from the body of pigs and poultry of the meat production direction [2, 3, 4, 6], accordingly, from a product that has established encouraging research results.

The research was aimed to study the effect of bentonite supplements (as enterosorbents) of dairy cows and their ability to detoxify and remove heavy metals from the body, respectively, to reduce their concentration in milk.

2 Material and methods

The experiment was performed in the farm "Kaloyev", located in st. Zmeyskaya of Kirovsky district of North Ossetia-Alania from March 2018 to January 2019.

For the research, two experimental groups of dairy cows of the 3rd and 4th lactation of the Simmental breed (control and experimental, five heads in each group) were formed by the analog-pair method. The conditions of keeping the cows corresponded to the zoo-hygienic regulations. The system of keeping the experimental livestock in the spring-summer-autumn period was stall-pasture, in the winter - stall. The experiment lasted for 305 days of lactation, in which the control group of cows was fed a basic diet balanced in all nutrients, mineral elements and vitamins pursuant to the standards of Federal Cattle Breeding Research Center (2003) [1]. Following ingredients were the basis of the fodder ration in the winter period: meadow hay - 16%, corn silage - 44%, mixed fodder - 9%, grain chaff - 5%, grain stillage - 14%, fodder molasses - 12%. Herewith, dry matter contained in the winter ration of feeding amounted to 13.7%. The sugar-protein ratio was 0.8:1.1; the ratio of calcium to phosphorus is 1.6:1.0. The summer ration of feeding consisted of: artificial pasture grass - 88%, compound feed - 5%, grain stillage - 7%. The dry matter in the ration was 12.9%. The sugar-protein and calcium-phosphorus ratios in the summer feed ration were within the recommended limits. For the experimental group, in addition to the main ration, the concentrates included crushed bentonite, with a particle diameter of 4-6 mm, daily in an amount of 1% (137 g/animal), based on the dry weight of the feed. The specified amount of fed bentonite was established during a reconnaissance experiment in the same farm, in which out of three different tested levels of feeding dairy cows, the optimal dose was recognized as 1% of the dry mass of feed, at which the milk productivity of cows and the qualitative and technological characteristics of milk significantly increased [2, 11].

In accordance with the purpose of research and study of the transformation of heavy metals along the food chain (soil-feed-organism-milk), during the accounting periods (on the 250th and 300th days of lactation), 5 average samples of soil, feed, blood and milk were taken. pursuant to generally accepted methods. Ashing of these samples was performed in a muffle furnace at temperatures up to 500°C pursuant to GOST 26929-94. Ash samples were subjected to atomic adsorption analysis on an AAZ-115-M1 spectrophotometer at the Republican Veterinary Laboratory.

Herewith, the transformation ration values of heavy metals from feed and water into the body of cows, respectively, into milk, were calculated pursuant to the following formula:

$$\text{KT} = \frac{\text{A milk}}{\text{A ration} + \text{A water}} \times 100\%, \quad (1)$$

where TR is the transformation ratio;
A milk - the content of heavy metals in the daily milk yield, mg/kg;
A ration - the content of heavy metals in the daily diet, mg/kg;
A water - the content of heavy metals in drinking water, mg/l;

The transformation ratio of heavy metals into the body of cows (into the blood) was calculated by the formula:

$$\text{KT} = \frac{\text{A blood}}{\text{ACB}} \times 100\%, \quad (2)$$

where TR is the transformation ratio;
A blood - the content of heavy metals in the blood of cows, mg/kg;
ACB - the content of heavy metals in the dry matter of the ration, mg/kg;

Soil samples, taken to a depth of 10 centimeters, were taken in five places on the territory of the farm, where mowing and preparation of roughage, as well as mowing of green mass and grazing of cows in the summer grazing period, was performed.

3 Results and discussion

In the laboratory, spectral analyzes of average samples of soil, water and feed, the concentration of heavy metals, given in Table 1, was established.

From the results of spectral analyzes of the studied samples for the content of heavy metals, given in Table 1, it follows that in soil samples taken in different parts of the territory of the economy, an excess concentration of cadmium from the MPL was established - by 10.1; lead - by 7.4; zinc - by 9.7 times ($P \leq 0.01$). The content of heavy metals in drinking water was also higher than the MPL: cadmium - by 2.5; lead - by 9.0; zinc by - 9.7 times.

Table 1. Concentration of heavy metals in the test samples, mg/kg, n = 5.

Sample designation	Heavy metals		
	Cadmium	Lead	Zinc
HM mobile forms in soil	2.01±0.11	44.5±1.34	225±2.66
MPL	0.2	6	23
Drinking water mg/l	0.0025±0.0001	0.27 ± 0.03	0.02 ± 0.001
MPL	0.001	0.03	1.0

Table 1. Continued.

Meadow hay	1.61 ±0.13	8.6±0.23	77.2±8.54
Corn silage	1.53±0.34	7.8±0.53	61.3±4.52
Grain chopping	0.51 ±0.31	9.3± 0.98	58.4±8.53
Grain distillery refuse	0.55 ±0.42	5.84±0.42	52.1±3.57
Artificial pasture grass	0.77 ±0.55	7.32 ±0.43	63.4 ± 4.32
MPL	0.3	5.0	50.0

The content of heavy metals in the studied fodder was also higher than the MPL: cadmium in meadow hay - 5.3 times; corn silage - 5.1 times; grain cut - 1.7; in grain stillage - 1.8; grass from artificial pasture - by 2.5 times. Lead was contained in the feed used: meadow hay - 1.7 times more than the MPL; corn silage - by 1.5; grain cut - by 1.6 times; grain stillage by 1.1; grass from artificial pasture - by 1.4 times. Zinc was also contained more from the MPL: in meadow hay - by 1.5 times; corn silage - by 1.2; grain cut - by 2.5; grain stillage - by 1.0 times; in grass from artificial pastures - by 1.2 times.

The morphological and biochemical composition of the blood, respectively, the physiological activity of individual organs and tissues of the whole organism depends on the composition of the animal feed ration, the presence of certain nutrients in it. In this regard, in order to study the effect of feeding with bentonite, as an enterosorbent in relation to heavy metals, on the 250th and 300th days of lactation, blood was taken from 2 compared experimental groups of cows, from the jugular vein, in the morning before feeding and preserved her heparin. Herewith, the conducted hematological studies established that the content of the studied heavy metals in the blood (in the body) of cows in the control group was higher than the MPL, which corresponded to the concentration of heavy metals in the fed feed. The inclusion of bentonite as an enterosorbent in animals of the experimental group contributed to a decrease in the amount of studied heavy metals in the body of cows. The results of spectral analyzes of blood and indicators of the transfer of heavy metals from feed to the body (into the blood) are given in Table 2.

Table 2. Concentration of heavy metals in the blood of experimental cows and the transformation ration values from feed to the body, mg/l n=5.

Experimenta l groups	250th day of lactation					
	Cadmium		Lead		Zinc	
	MPL	0.01-0.015	0.05-0.25		3.0-5.0	
	content in blood	transf. values., %	content in blood	transf. values., %	content in blood	transf. values., %
Control	0.044±0.002	1.04	0.29±0.003	3.62	7.84±0.11	1.87
Experimental	0.036±0.005	0.85	0.24±0.005	3.11	6.52±0.16	1.56
% of the control group	-118.1	-0.19	-117.2	-0.51	-116.8	-0.31
	300th day of lactation					
Control	0.031±0.003	1.69	0.24±0.003	3.64	7.21±0.14	2.06
Experimental	0.025±0.006	1.36	0.19±0.004	3.12	5.92±0.15	1.69
% of the control group	-119.3	-0.33	-120.8	-0.52	-117.8	-0.37

The results of the studies given in Table 2 show that the inclusion of bentonite in the diet of cows in the experimental group, considering its detoxification (enterosorbent) properties, contributed to the elimination of the studied heavy metals from the body and a decrease in their concentration in the blood from 16 to 19% ($P \leq 0, 01$) in both study periods. A proportional decrease in the coefficients of the transfer of the studied heavy metals from the fed feed into the organism (into the blood) of animals was noted. It was found that in the first study (at 250 days of lactation), which fell on the winter period, a higher concentration of heavy metals was noted in the feed, respectively, in the blood of cows than in the second study (at 300 days of lactation) in the summer period of the study.

One of the factors that determine the degree of environmental safety of milk and dairy products and their use for human nutrition is the level of pollution with heavy metal salts. Partial removal of heavy metal salts from the body of animals of the experimental group in the zone with increased environmental pollution contributed to the production of relatively environmentally friendly products and the improvement of the commercial and technological properties of milk (B.A. Dzagurov, A.G. Karlov, 2020). In this regard, during the period of control milk yields on the 250th and 300th days of lactation of experimental cows, milk samples were taken from cows of the compared experimental groups and subjected to spectral analyzes for the presence of heavy metals (cadmium, lead and zinc). The analysis results are shown in Table 3.

Table 3. The content of heavy metals in milk and the coefficients of their transformation from feed into milk, n = 5.

Experimental groups	250th day of lactation					
	Cadmium		Lead		Zinc	
	MPL mg/kg	0.03	0.1		5.0	
	content in milk	transf. values., %	content in milk	transf. values., %	content in milk	transf. values., %
Control	0.041±0.003	1.05	0.13±0.006	3.61	6.98±0.14	1.85
Experimental	0.034±0.006	0.86	0.11±0.007	3.22	5.87 ±0.16	1.54
% of the control group	-117.0	-0.19	-115.4	1)	-115.9	-0.31
	300th day of lactation					
Control	0.037±0.003	1.51	0.16±0.004	3.62	6.53±0.15	1.82
Experimental	0.031±0.006	1.31	0.13±0.005	3.21	5.36±0.17	1.54
% of the control group	-116.3	-0.20	-118.8	-0.41	-117.9	-0.29

The digital material presented in Table 3 shows that the addition of bentonite to animal feed contributed to a decrease in the concentration of cadmium in the milk of the experimental group of cows by 17%, at the 250th day of lactation, and at the 300th day of lactation by 16.3% at $P \leq 0.01$, in relation to similar indicators of milk samples, milked from cows of the control group. The coefficients of the transition of cadmium into milk from the cows of the experimental group, in the first period of the study, are less by 0.19%, in the second period - by 0.20%, compared with the control.

The content of lead in the milk of the experimental group of cows decreased in relation to the control samples of milk in the first period of the study - by 15.4%, in the second period - by 18.8% ($P \leq 0.01$). In the first period of the study (250 days of lactation), the coefficient of transformation of lead from feed into milk of the experimental group of animals was less than the control - by 0.39%, in the second period of analyzes (300th day of lactation) - by 0.41%.

It should be noted that the conversion ratios of lead from feed to milk were significantly higher compared to the analogous indicators of the conversion of cadmium and zinc into the milk of experimental cows.

Zinc in the milk of cows from the control group in the first period of research (250 days of lactation) contained 6.98 mg/l, in milk samples from the experimental group - 5.87 mg/l, which is 15.9% less ($P \leq 0.01$). The coefficient of transformation of zinc from feed into milk of the experimental group of cows was 0.31% less than in the milk of cows from the control milk samples. In the second period of analyzes (300 days of lactation), the milk of cows in the control group contained 6.53 mg / l zinc, in the milk of experimental milk samples -

5.36 mg / l, i.e. by 17.9% ($P \leq 0,01$) less control milk analogues. Coefficients of zinc conversion from feed to milk in the first period of analyzes were 0.31% less in milk from the experimental group of cows, in the second period - by 0.29%, in relation to milk samples from the control group.

The results of the research have established that the inclusion of bentonite in the composition of the feed ration of cows contributed to a significant (reliable) decrease in the concentration of the studied heavy metals both in the blood of animals and in milk produced, in comparison with similar indicators of control blood and milk samples.

4 Conclusions

1. In the studied soil samples on the territory of the farm where the research was performed, an increased content of heavy metals relative to the MPL was noted: cadmium - 10.1; lead - 7.4; zinc - 9.7 times more.
2. The drinking water samples contained: cadmium - 2.5 times; lead - 9.0 times and zinc - 9.7 times more in comparison with the MPL.
3. In the fed fodder the concentration of the studied heavy metals exceeded: cadmium - from 1.7 to 5.3 times; lead - from 1.1 to 1.7; zinc - from 1.0 to 2.5 times the MPL.
4. In the blood (body) of cows of the control group on the 250th day of lactation in relation to the animals of the experimental group, more heavy metals were contained: cadmium by - 18.1%; lead - by 17.2%; zinc - by 16.8%. On the 300th day of lactation: cadmium - by 19.3%; lead - by 20.8%; zinc - by 17.8%. Pursuantly, the transformation ration values of heavy metals from feed into the body (blood) were lower in the animals of the experimental group compared to the control in both study periods from 0.19 to 0.43%.
5. Milk samples dairy from cows of the experimental group during the study periods (250 and 300 days of lactation) contained significantly less heavy metals in relation to the control from 16 to 17%. Coefficients of transformation of heavy metals from feed into milk were lower in animals of the experimental group in relation to the control by 0.19-0.41%.

5 Recommendation to production

In order to reduce the content of heavy metals in the milk produced and reduce the coefficients of the transfer of heavy metals from feed to the body of cows, respectively, into milk, feed dairy cows with crushed (bentonite particle diameter - 4-6 mm) bentonite in an amount of 1% of the dry weight of the feed, in kind 137 g/animal per day.

References

1. V. A. Alikaev, E. A. Petukhova, L. D. Haleneva, et al, Animal Feeding and Management Handbook, **320** (1982)
2. B. A. Dzagurov, R. H. Gadzaonov, A. G. Karlov, Bulletin of the Gorsky State Agrarian University, **57(1),** 54 (2020)
3. B. A. Dzagurov, A. G. Karlov, Bulletin of the Gorsky State Agrarian University, **57(2),** 97 (2020)
4. B. A. Dzagurov, Z. A. Ktsoeva, *Bentonite feeding of pigs* (2018)
5. V. R. Kairov, *Actual problems of scientific support for increasing production, improving the quality of feed and their effective use*, 165 (2001)
6. S. A. Lapshin, Chemicalization of agriculture, **8**, 62 (1988)

7. M. G. Chabaev, Journal of Pharmaceutical Sciences and Research, **10(5),** 1122 (2018)
8. A. A. Baeva, Indo American Journal of Pharmaceutical Sciences, **6(9),** 15992 (2019)
9. A. B. Baragunov, E. A. Kushaeva, M. M. Hamokov, A. V. Sagatelian, E3S Web of Conferences, **193** (2020)
10. A. B. Baragunov, I. A. Savvateeva , S. H. Kushaev, A. A. Kumakhov, Z. R. Kudaev, IOP: Earth and Environmental Science, **3** (2019)
11. T. T. Tarchokov, Z. M. Aisanov, S. F. Sukhanova, M. G. Tleinsheva, A. A. Mishhozhev, D. S. Balpanov, IOP: Earth and Environmental Science, **1** (2019)

Effect of probiotics and selenium-containing preparations on nonspecific resistance of calves

Victor Eremenko[1], *Oleg* Getokov[2,*], *Boris* Dzagurov[3], *Safarbi* Kazanchev[2], and *Murat* Ulimbashev[2]

[1]Kursk State Agricultural Academy named after I.I. Ivanov, 70 Marksa Str., 305021 Kursk, Russia
[2]Kabardino-Balkarian State Agricultural University named after V. M. Kokov, Lenin Avenue, 1V, 360030 Nalchik, Russia
[3]Gorsky State Agrarian University, 37 Kirova Str., 362040 Vladikavkaz, Russia

Abstract. Black-and-white Holstein calves were studied. 3 experimental groups were formed: they included calves of two weeks of age, 10 animals each. The calves of the first group received "Intestevit" probiotic with milk at the rate of 3 doses per day for 1 animal for 15 days. The second group of calves, in addition to "Intestevit" preparation, also received selenium-containing preparation "Selen Plex™" at a dose of 1 mg/kg. The third group was the control group. Blood for testing was taken from the tail vein before the morning feeding and the start of preparation feeding, 30 and 60 days after the use of these preparations. Feeding "Intestevit" preparation to calves from 15 days of age for 60 days increases the following indicators relative to the control group: total immunoglobulins by 8.4%; Leukocyte phagocytic activity (LPA) - 10.5%; serum lysozyme activity (SLA) - 20.1%; serum bactericidal activity (SBA) - 4.6%; T-lymphocytes (absolute content) - 14.1%; B-lymphocytes (absolute content) - 1.5%. Feeding "Intestevit + Selen Plex™" complex preparation to calves from 15 days of age for 60 days increases the following indicators relative to the control group: total immunoglobulins by 11.7%; LPA - 12.1%; SLA - 15.8%; SBA - 7.6%; T-lymphocytes (absolute content) - 17.8%; B-lymphocytes (absolute content) - 2.2%.

1 Introduction

It is not possible to consider all aspects to satisfy animal boy needs in industrial animal husbandry. It results in various diseases and a decrease in productivity [1-5]. In this regard, an increase in the level of productivity of animals, their health will be associated with satisfaction of their physiological need for nutrients to contribute to implementation of their genetic potential. To solve this problem, the most intense ontogenesis periods should be considered, when the future productivity of animals is formed. The most vulnerable is the immune system, highly affected by various factors [6-8]. Therefore, it is necessary to search for the most effective methods to enhance the body resistance, especially in the early

*Corresponding author: getokov777@mail.ru

period of ontogenesis. One of these methods is the use of probiotic and selenium-containing preparations in diets [9-14]. The main advantage of probiotic preparations is the absence of side effects and harmlessness to the animal body and resulting products. Probiotic preparations balance the intestinal microbiocenosis and provide a beneficial effect on the animal body as a whole. Selenium-containing preparations are also used in animal husbandry. Selenium is part of the enzyme glutathione peroxidase. Therefore, selenium-containing preparations help to reduce the activity of lipid peroxidation and increase antioxidant protection. One of these selenium-containing preparations is Selen Plex™: it is a selenium protein. According to its effect on the body, it is non-toxic, compared to inorganic selenium preparations.

Selenprotein has more pronounced antioxidant properties. Therefore, the study of selenium-containing and probiotic preparations effect on the specific resistance of growing calves is relevant.

Study Purpose. The study purpose was to analyze the state of nonspecific resistance in growing calves when using "Intestevit" probiotic and selenium-containing "Selen Plex ™" preparation in their diet.

2 Study methods

Black-and-white Holstein calves were studied. 3 experimental groups were formed: they included calves of two weeks of age, 10 animals each. The calves of the first group received "Intestevit" probiotic with milk at the rate of 3 doses per day for 1 animal for 15 days. The second group of calves, in addition to "Intestevit" preparation, also received selenium-containing preparation "Selen Plex™" at a dose of 1 mg/kg. The third group was the control group. The conditions for keeping and feeding the calves were identical. Blood for testing was taken from the tail vein before the morning feeding and the start of preparation feeding, 30 and 60 days after the use of these preparations. In blood samples, total immunoglobulins were determined by the zinc sulfate method using a turbidity reaction. T-lymphocytes, B-lymphocytes, phagocytic activity of leukocytes, phagocytic index, phagocytic number were studied according to V.S. Gostev. Lysozyme activity of blood serum was determined using Micrococcus lisodeikticus culture, and bactericidal activity of blood serum using Staphylococcus aureus culture.

The obtained study results were subjected to biometric processing.

3 Study results

3.1 Total immunoglobulins

The concentration of total immunoglobulins in the blood of experimental calves is shown in Figure 1.

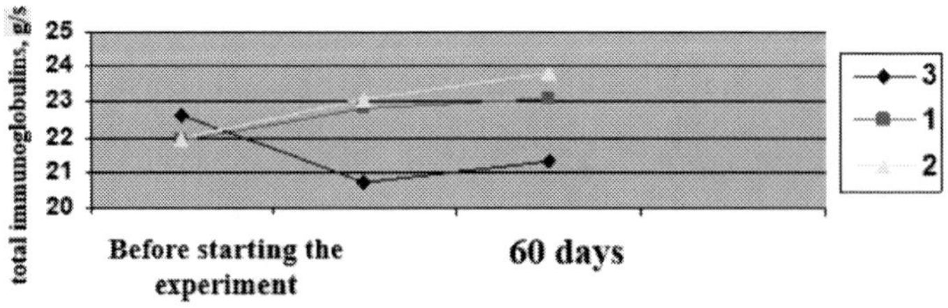

Fig. 1. Concentration of total immunoglobulins in the blood of experimental calves.

The presented data show that before using the preparations in all three groups, the level of this indicator was practically the same and amounted to 21.6 ± 0.4 g/l in the control group, 21.9 ± 0.4 l in the first experimental group, in the second experimental group 22.0 ± 0.5 l. 30 days after the start of "Intestevit" administration, the level of total immunoglobulins increased to 22.8±0.5 g/l, and in the group of calves that received Intestevit + Selen Plex™ preparations, 23.1±0, 3 g/l. Then, with increasing age of calves, 2 months after the start of administration, the level of total immunoglobulins continued to increase in both experimental groups. The calves that received "Intestevit", after 60 days in their blood, the concentration of total immunoglobulins in comparison with the control group was higher by 1.8 g/l and amounted to 23.1±0.4 g/l. In the group of calves that received "Intestevit" + "Selen Plex™", after 60 days the level of total immunoglobulins was 2.5 g/l higher than in the control group and amounted to 23.8±0.4 g/l. In the control group, the concentration of total immunoglobulins was at a lower level relative to similar data of experimental groups.

3.2 Leukocyte phagocytic activity

The performed studies of leukocyte phagocytic activity are shown in Figure 2.

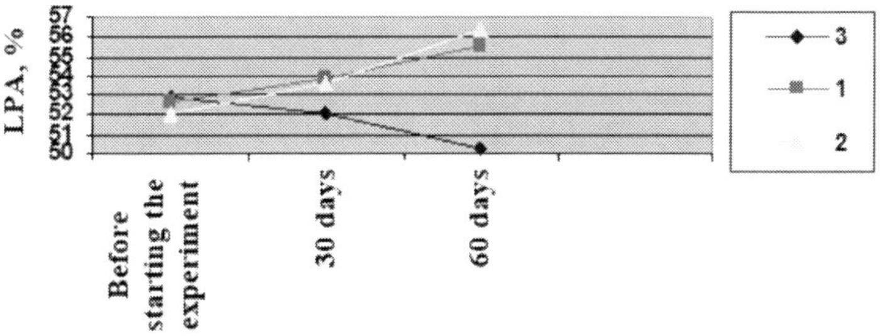

Fig. 2. Leukocyte phagocytic activity in the blood of experimental calves.

Analysis of the results shows that the cellular protection of blood in the calves of experimental groups was relatively more pronounced relative to the control group, where they did not receive preparations. Before using preparations in the control group, LPA was 52.9 ± 1.8%. In compared groups, this indicator was approximately the same and amounted to 52.6±1.2% and 52.0±1.7%. After 30 days from the beginning of the "Intensiveit" preparation, LPA was 53.9±1.6%, and in the group of calves that received "Intestevit + Selen Plex™" preparations - 53.6±1.1%. In the subsequent period of the experiment, LPA in both experimental groups increased. In the group of calves that received "Intestevit", after 60 days, LPA was at the level of 55.6±1.3%, and in the group of calves that received "Intestevit + Selen Plex™" - 56.4±1.2%. The increase relative to the data before administration was by 5.7% and 8.4%, respectively, by groups. It proves that the complex use of "Intestevit + Selen Plex™" preparations has a more significant effect on the blood cellular defense. During the experiment, the phagocytic index (PI) had the same direction of change as the leukocyte phagocytic activity, Figure 3.

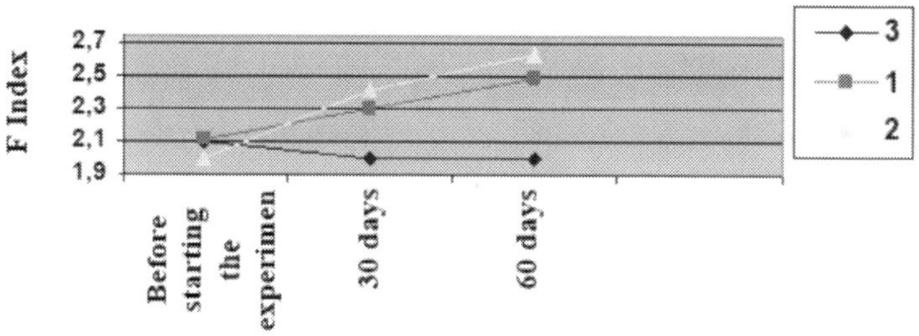

Fig. 3. Phagocytic blood index of experimental calves.

So, in the period before administration in the control group, the phagocytic blood index was 2.1±0.08 and in subsequent periods did not change and remained at the same level. In experimental groups, it was subject to changes in the direction of increase. So, in the group of calves that received "Intensive" after 30 and 60 days from the beginning of admission, it was 2.31±0.11 and 2.49±0.12, respectively. Regarding the data before administration, this increase by the 60th day was 18%. In the group of calves that received complex preparation "Intestevit + Selen Plex™", the dynamics of changes in this preparation was as follows: So, after 30 and 60 days, the phagocytic index was 2.42±2.10 and 2.63±0.14. Regarding the data before administration, this increase by the 60th day was 31.5%.

Lysozyme and bactericidal activity of blood serum play a significant role in the natural resistance formation.

3.3 Serum lysozyme activity (SLA)

The study results of serum lysozyme activity are shown in Figure 4.

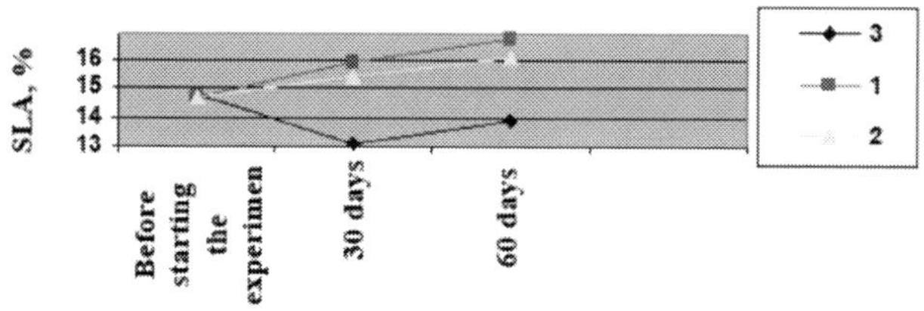

Fig. 4. Serum lysozyme activity of experimental calves

The above data demonstrate that before the start of administration in experimental and control groups, SLA was the same and within the range of 14.7-14.8%. In subsequent experimental periods in the control group, it decreased and was 13.1±0.20% and 13.9±0.20% on days 30 and 60, respectively. In experimental groups, on the contrary, this indicator increased after administration of preparations. So, after 30 days and 60 days following the start of "Intestevit" administration SLA was 15.9±0.18% and 16.7±0.19%, respectively. In the experimental group, where calves received "Intestevit + Selen Plex™" complex preparation, these values after 30 and 60 days were 15.4±0.18% and 16.1±0.19%.

3.4 Serum bactericidal activity (SBA)

Before the experiment, SBA was slightly higher in the control group of calves and amounted to 43.5±2.1%, Figure 5.

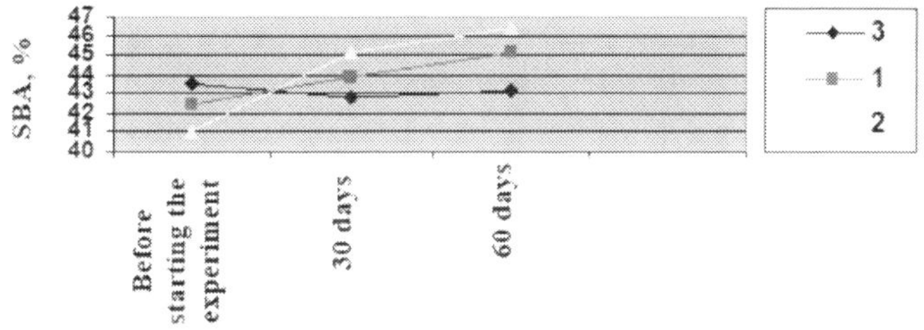

Fig. 5. Serum bactericidal activity of experimental calves

In compared groups, this indicator amounted to 42.4±2.4% and 41.0±1.3%. In the control group, in the subsequent periods of the experiment after 30 and 60 days, this indicator was 42.8±2.2% and 43.1±2.0%. In calves that received "Intestevit" preparation, SBA, 30 and 60 days after the start of administration, was 43.9±1.7% and 45.1±2.0%, respectively. In the ccompared experimental group, where calves received "Intestevit + Selen Plex™" complex preparation, these values after 30 and 60 days were 45.1±1.1% and 46.4±1.2%. These values were higher than in the group of calves that received "Intestevit".

3.5 T-lymphocytes

The results of the relative content of T-lymphocytes are shown in Figure 6.

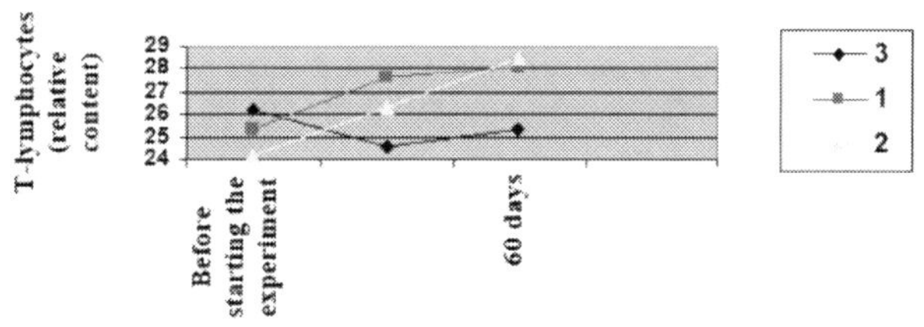

Fig. 6. The content (relative) of T-lymphocytes in the blood of experimental calves.

As the above data show, the absolute and relative content of T-lymphocytes in experimental calves that received "Intestevit" and "Intestevit + Selen Plex™" preparations was higher than in the control group. Before testing, the relative content of T-lymphocytes in all groups was approximately the same. So in the control group it was 25.2±1.1%, in the group that received "Intestevit + Selen Plex™" - 24.2±1.1%. The maximum level of T-lymphocytes in experimental groups reached 60 days after administration. In the group of calves that received "Intestevit", the relative content after this interval of "Intestevit" administration was 28.1±1.0%, and in the group that received "Intestevit + Selen Plex™" complex preparation - 28.5±1.1%. The absolute content of T-lymphocytes is shown in Figure 7.

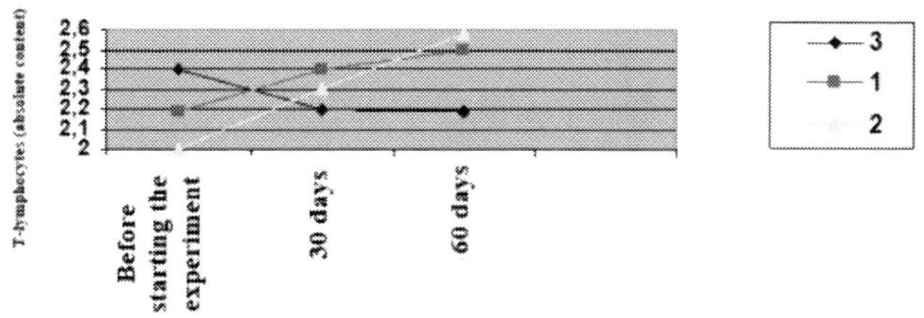

Fig. 7. The content (absolute) of T-lymphocytes in the blood of experimental calves.

 https://doi.org/10.1051/e3sconf/202126202011

From the above data it is obvious that in the control group with increasing age of calves, the absolute content of this indicator decreased from 2.4±0.11•10⁹/l to 2.19±0.08 • 10⁹/l. In the group of cow calves that received "Intestevit", the absolute content of T-lymphocytes increased from 2.19±0.12 x 10⁹/l to 2.50±0.09 x 10⁹/l, in the group that received "Intestevit + Selen Plex™" complex preparation it increased from 2.0±0.12 x 10⁹/l to 2.58±0.14 x 10⁹/l.

3.6 B-lymphocytes

Figure 8 shows the study results of the relative content of B-lymphocytes.

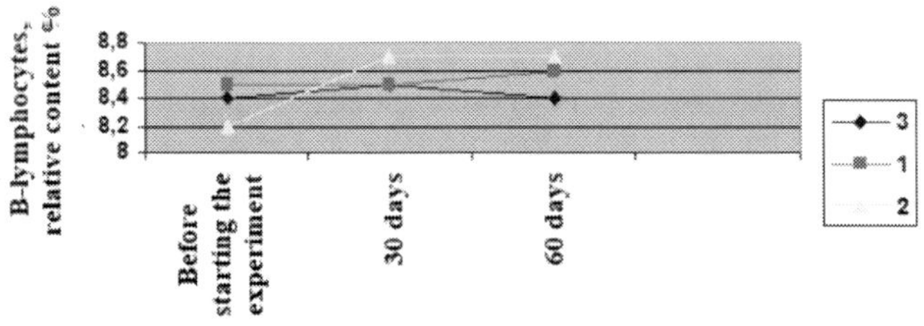

Fig. 8. The content (relative) of B-lymphocytes in the blood of experimental calves.

These data prove that in control group animals the relative content of B-lymphocytes did not change with increasing age and amounted to 8.4%, the absolute content changed from 1.35±0.08 to 1.33±0.09•10⁹/l. Figure 9. In the group of calves that received "Intestevit" preparation, there were practically no changes in the relative and absolute content of B-lymphocytes. So, the relative content was within the range of 8.5±0.13-8.6±0.15%, and the absolute 1.34 ± 0.09 - 1.35 ± 0.08•10⁹/l.

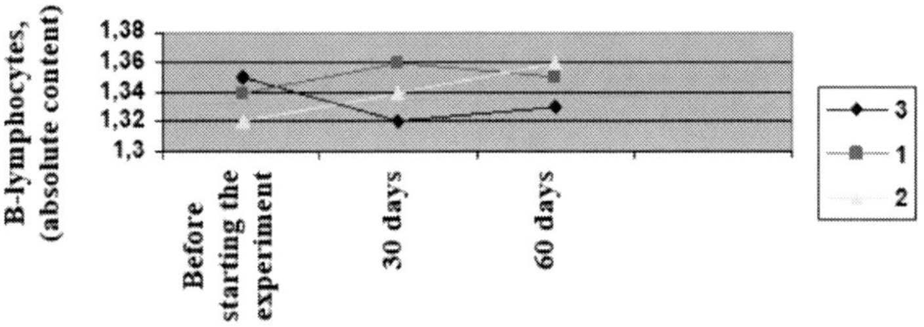

Fig. 9. The content of B-lymphocytes (absolute) in the blood of experimental calves.

The increase in the relative and absolute value of B-lymphocytes occurred in the group of calves that received "Intestevit + Selen Plex ™" preparation. Thus, the relative content of B-lymphocytes increased from 8.2±0.5 to 8.7±0.18%, the absolute content from 1.32±0.08 to 1.36±0.10 x 10⁹/l.

4 Conclusions

1. Feeding "Intestevit" preparation to calves from 15 days of age for 60 days increases the following indicators relative to the control group: total immunoglobulins by 8.4%; Leukocyte phagocytic activity (LPA) - 10.5%; serum lysozyme activity (SLA) - 20.1%; serum bactericidal activity (SBA) - 4.6%; T-lymphocytes (absolute content) - 14.1%; B-lymphocytes (absolute content) - 1.5%.

2. Feeding "Intestevit + Selen Plex™" complex preparation to calves from 15 days of age for 60 days increases the following indicators relative to the control group: total immunoglobulins by 11.7%; LPA - 12.1%; SLA - 15.8%; SBA - 7.6%; T-lymphocytes (absolute content) - 17.8%; B-lymphocytes (absolute content) - 2.2%.

References

1. M. Averenkova, Problems of Veterinary Sanitation, Hygiene and Ecology, **118**, 190 (2006)
2. T. D. Buzanova, Abstract of Thesis of Master of Vet. Sciences, 22 (2007)
3. N. V. Burnysheva, Abstract of a Thesis of Master of Agriculture, 22 (2007)
4. V. I. Eremenko, O. B. Sein, *Metabolic status. Nonspecific resistance and their correlation in cattle* (2011)
5. V. I., Eremenko, O. B. Sein, A. V. Titova, O. A., Prelikov, M. A. Blednov, Zootechnics, **7,** 27 (2009)
6. Ya. Slabitskyi, S. Voloshanskaya, Agricultural Biology, **1,** 21 (1987)
7. G. I. Boryaev, Abstract of a Thesis of Doctor of Biological Sciences, 43 (2000)
8. A. Bedenko, Milk and Feed. Management, **4,** 32 (2007)
9. V. V. Ermakov, V. V. Kovwalskiy, *Biological Significance of Selenium* (1974)
10. N. G. Ignatiev, N. K. Kirillov, Veterinarian, **4,** 55 (2001)
11. B. Seremak, Folia Univ. agr. Stetin. Zoctechn, **37,** 63 (1999)
12. L. Tucker, Jnt. Hatchery Pract, **21,** 7 (2007)
13. J. Kruze, H. Ceballos, A. Stryhn, A. Mella et al, Vet. Med., **9(54),** 478 (2007)
14. T. D. Buzanova, Abstract of a Thesis for Master of Vet. Sciences, 22 (2007)

Immunogenetic status and breeding value of Holstein cows

Vladimir Kaishev[1], *Anzor* Zhilyaev[2], *Tamirlan* Kudaev[2] , *Madina* Tleinsheva[2], and *Ibragim* Taov[2]

[1]Federal State Budgetary Educational Institution of Higher Education Southern Federal University, 105/42 B. Sadovaya Str., 344006 Rostov-on-Don, Russia
[2] Kabardino-Balkarian State Agricultural University named after V. M. Kokov, Lenin Avenue, 1V, 360030 Nalchik, Russia

Abstract. In the genealogical structure of the breeding stock, the highest specific weight is occupied by the Reflection Sovering and Vis bek Aydiala lines, accounting for 56.8 and 39.9% of animals, respectively. The study results of Holstein cattle blood-group factors can be used to monitor its gene pool, genetic intrabreed differentiation, assess selection parameters, and conduct targeted selection of parental pairs. It was determined that in EAA blood group system in Holstein cows of reproduction from the USA, the frequency of A1 antigen is 58.7%, and in domestic reproduction animals, this antigen is found with a lower frequency of 50.7%. At the same time, in the specified system of blood groups in both groups of experimental animals, antigen "a" is characterized by the highest frequency, the lowest occurrence is A2 antigen, and animals with A1 antigen took an intermediate position. In EAB system, the US Holstein cows have a higher frequency of such antigens as E_3', O_3 и I_1 compared to the reproductive animals of Agro-Soyuz LLC. The EAF blood group system is represented by two antigens, where F antigen frequency in both groups of experimental animals is 100%. Of all the blood group systems, the greatest differences between compared groups of animals were detected for the antigens of EAZ blood group system.

1 Introduction

Within the framework of “Agro-Industrial Complex Development" national project implementation in the republics of the North Caucasus region, Holstein breed animals are imported from various countries, distinguished by high indicators of productivity. Under new conditions the imported animals are characterized by high productivity rates, which is due, along with the genotype, to paratypical factors [1, 2, 3, 4, 5, 6, 7, 8, 9]. At the same time, the problem of a more complete implementation of the productivity potential of imported animals and its preservation in subsequent generations, considering adaptation to new conditions of feeding and keeping, determination of immunogenetic status of animals of the Holstein breed remains relevant.

2 Materials and methods

To assess the hereditary qualities and breeding value of Holstein breed animals, we analyzed the data on the productivity of the female ancestors of stud bulls. For this purpose, we used the method of N.A. Kravchenko (1973) on the assessment of the origin and determination of breeding indices [10]. To study the immunogenetic status of cows in the laboratory of immunogenetics, biochemistry and general chemistry of the State Scientific Institution "Stavropol Scientific Research Institute of livestock breeding and forage production", blood group systems and antigenic differences (antigens belonging to 7 blood group systems) were determined according to generally accepted method in 75 Holstein cows imported by heifers from the USA, as well as from daughters received under conditions of Agro-Soyuz LLC. The actual frequency of individual antigens was calculated by the ratio of the number of animals, carrying this antigen, to the total number of individuals in the group.

3 Results and discussion

To characterize animal hereditary qualities of the Holstein breed, we analyzed genealogical structure data of the breeding stock of Agro-Soyuz LLC. It was determined that In the genealogical structure of the breeding stock, the highest specific weight is occupied by the Reflection Sovering and Vis bek Aydiala lines, accounting for 56.8 and 39.9% of animals, respectively. The share of other lines in the herd structure is insignificant. The used stud bulls of these lines are characterized by a high genetic potential of milk productivity, high hereditary inclinations of the productivity of female ancestors, they improve such features as milk yield, body type and protein-milk content. Parental indices of Holstein bulls ranged in milk yield from 10806 to 13718 kg, in fat and milk content in the range of 3.65-4.47% and 3.17-3.5%. Differences between analyzed animals were revealed in terms of the parental index of the bull, due to linear affiliation. Thus, Holstein breed bulls of the Reflection Sovering line exceeded the bulls of the Vis Bek Aydiala line in milk yield by 0.86%, in terms of the mass fraction of fat and protein - by 0.04 and 0.04 abs.%, in terms of the total amount of milk fat and protein - by 11.0 and 8.0 kg, or 2.3 and 2.05%, respectively.

The study of blood-group factors in Holstein cattle is an urgent task, because the results obtained can be used to monitor its gene pool, genetic intrabreed differentiation and assessment of selection parameters, conduct targeted selection of parental pairs, and also supplement the available materials on breeding and improving this breed.

To characterize blood groups, as well as the immunogenetic status of Holstein cattle, the frequency of antigens was studied (Table 1).

Table 1. Frequency of antigens in Holstein cows (n = 75).

Blood group system	Antigen	Number of animals with antigen		Antigen frequency, %	
		Cows	Heifers	Cows	Heifers
EAA	A1	44	38	58.7	50.7
	A2	15	20	20.0	26.7
	a	57	56	76.0	74.7
EAB	E_3'	26	21	34.7	28.0
	I_2	29	30	38.7	40.0
	Y_2	47	45	62.7	60.0
	Q'	43	44	57.3	58.7
	I_1	32	25	42.7	33.3
	G_2	13	17	17.3	22.7

Table 1. Continued

EAB	$O^{/}$	53	53	70.7	70.7
	O_3	50	45	66.7	60.0
	$G^{//}$	1	3	1.3	4.0
	O_2	3	5	4.0	6.7
	P^1	10	19	13.3	25.3
EAC	X_2	27	28	36.0	37.3
	C_1	58	57	77.3	76.0
	C_2	38	32	50.7	42.7
	W	13	17	17.3	22.7
	R_2	36	36	48.0	48.0
EAF	V	5	4	6.7	5.3
	F	75	75	100	100
EAS	$H^{\|\|}$	11	13	14.7	17.3
	S_1	25	26	33.3	34.7
	$H^{\|}$	-	-		
	$U^{\|\|}$	-	-		
EAZ	Z	19	35	25.3	46.7
	z	53	108	70.7	144.0
EAT	T_2	3	4	4.0	5.3

It was found that EAA blood group system is represented by 3 antigens, they have different frequency, due to the genotype of the compared animals. Thus, in Holstein cows of reproduction from the USA, the frequency of A1 antigen is 58.7%, and in domestic reproduction animals, this antigen is found with a lower frequency of 50.7%. In this system of blood groups, A2 antigen occurs with a frequency of 20.0% in imported animals. In comparison, the proportion of animals with A2 antigen is 26.7%. Along with this, in both groups of experimental animals, the same occurrence of "a" antigen is observed, which is 76.0 and 74.7%, respectively. At the same time, in the specified system of blood groups in both groups of experimental animals, antigen "a" is characterized by the highest frequency, the lowest occurrence is A2 antigen, and animals with A1 antigen took an intermediate position.

The EAB blood group system is represented by 11 antigens. In the specified system of blood groups, O^1 antigen is distinguished by the highest frequency of occurrence, which is similar in both groups of experimental animals and is 70.7%. In both groups of animals, O_3 and Y_2 antigens also differ in high frequency, somewhat lower than the previous antigen, but amounting to 66.7 and 60.0 for O_3 antigen, and 62.7 and 60.0% for Y2 antigen. At the same time, G// and O2 antigens differ in the lowest frequency, its specific gravity was insignificant. Comparison of groups of experimental animals for antigenic differences showed that Holstein breed cows of reproduction from the USA have a higher frequency of such antigens as $E_3^{/}$, O_3 и I_1 compared to animals of reproduction of Agro-Soyuz LLC Along with this, the second group animals were distinguished by a higher frequency of antigens G_2, $G^{//}$, O_2, P1 compared to the first group animals.

The EAC blood group system is represented by five antigens, among which C_1 antigen was the most frequent, its proportion was 77.3% in US cows and 76.0% in Agro-Soyuz LLC reproductive animals. In this system of blood groups in both groups of experimental animals, some antigens have similar frequency values. These features are characteristic of antigens R_2 C_1, X_2. At the same time, differences in the frequency of C_2 antigen manifestation in favor of the first group of cows are observed between compared groups of experimental animals. The indicated superiority of the first group animals over the second group animals in terms of C_2 antigen frequency is 8.0%. However, the opposite tendency is observed in W antigen frequency, the second group animals in terms of W antigen frequency exceeded the first group animals by 5.4%.

The EAF blood group system is represented by two antigens, where F antigen frequency in both groups of experimental animals is 100%, i.e. all analyzed animals contain the specified antigen in their genotype. In the analyzed groups of experimental animals, V antigen frequency is insignificant and ranges within 5.3-6.7%.

The EAS blood group system is represented by two antigens, of which S_1 antigen has a higher and similar frequency in both groups, amounting to 33.3 and 34.7%, respectively. The second group animals were characterized by a higher frequency of H^{II} antigen compared with animals of the first group, the indicated difference is 2.6%.

Of all analyzed blood group systems, the greatest differences between compared groups of animals were detected for the antigens of EAZ blood group system. In the first group animals, Z and z antigens of this system are manifested with a much lower frequency compared with the second group animals. The indicated difference for the antigens of this blood group system is 21.4-73.3%. At the same time, in this system of blood groups in both groups of experimental animals, z antigen has a higher frequency, amounting to 70.7 and 144.0%, respectively, compared to Z antigen.

One T_2 antigen was detected in the EAT blood group system. It is worth mentioning that a feature of this herd and the analyzed groups of animals is the low frequency of T_2 antigen in the EAT blood group system, which is 4.0 and 5.3%, respectively.

Along with the study of frequency of antigen distribution in the Holstein breed animals, we analyzed the data on the milk productivity of cows selected in the group of cows-mothers o bulls, showing the genetic potential of productivity in breeding conditions (Table 2).

It was found that Holstein cows selected for the group of bull mothers were distinguished by high productivity indicators. In this group of cows, milk yield for 305 days of lactation is 9657 kg with sufficient variability indicators for breeding, i.e. a standard deviation of 1787 kg and a coefficient of variation of 18.5%. The presented data on the realization of the genetic potential of productivity suggests the idea of developing a new experimental high fat content milk line of Holstein cattle, as evidenced by rather high values of the fat content in milk at the level of 4.16% and high values of this sign variability.

Table 2. Performance indicators of the best cows in a herd (n = 20).

Indicator	Parameters		
	$X \pm m_x$	δ	C_v
Milk yield for 305 days of lactation, kg	9657±410	1787	18.5
Fat content of milk, %	4.16±0.08	0.37	8.8
Protein content of milk	3.26±0.01	0.06	1.96

At the present stage in scientific research, considerable attention is paid to the techniques and methods of accelerated selection of farm animals according to the leading economically useful features, analysis of international experience in formation of new animal types under the influence of genetic and paratypical factors. At the same time, the issues of increasing the fat-milk content of the Holstein breed base on promising lines and families are still to be developed. In this regard, the problems associated with increase in implementation of the model animal potential. There are no such scientific studies on the massif of the Holstein breed on the North Caucasian Federal District farms.

These tables also characterize high protein-milk content of the bovine-producing composition of this herd cows, which is 3.26%.

4 Conclusions

In general, Holstein cows of different reproduction revealed differences in antigen carriage, indicating the preservation of sufficient genetic heterogeneity for targeted selection and breeding to increase the milk productivity of animals.

References

1. T. T. Tarchokov, Z. M. Aisanov, S. F. Sukhanova, A. A. Mishhozhev, D. S. Balpanov, IOP Conference Series: Earth and Environmental Science, **341(1)** (2019)
2. A. A. Mishhozhev, M. G. Tleinsheva, Z. M. Aisanov, T. T. Tarchokov, Zootechnics, **11**, 24 (2017)
3. A. A. Zhilyaev, T. T. Tarchokov, I. G. Sudorgina, R. Z. Abdulkhalikov, M. G. Tleinsheva, D. S. Balpanov, A. A. Kogotyzhev, Zootechnics, **9**, 8 (2020)
4. T. T. Tarchokov, Zootechnics, **5**, 27 (1996)
5. M. O. Baytaev, Ts. B. Kagermazov, T. T. Tarchokov, Agrarian Russia, **1**, 28 (2013)
6. I. M. Dunin, S.E. Tyapugin, R. K. Meshcherov, V. P. Khodykov, Sh. R. Meshcherov, N. S. Nikulkin, Zootechnics, **2**, 5 (2020)
7. A. T. Mysik, Zootechnics, **1**, 2 (2010)
8. O. A. Basonov, N. P. Shkilev, A. O. Basonova, N. I. Ivanova, S.G. Arutyunyan, Zootechnics, **10**, 6 (2019)
9. O. A. Basonov, N. E. Pavlova, Zootechnics, **11**, 11 (2018)
10. N. A. Kravchenko, Farm Animal Breeding (1973)

Effect of antioxidant and adsorbent on broiler meat production and consumer qualities

Irina Ktsoeva[1,*], *Rustam* Abdulkhalikov[2], *Rustem* Temiraev[1], and *Svetlana* Savkhalova[2]

[1]Gorsky State Agrarian University, 37 Kirova Str., 362040 Vladikavkaz, Russia
[2]Kabardino-Balkarian State Agricultural University named after V. M. Kokov, Lenin Avenue, 1V, 360030 Nalchik, Russia

Abstract. To reduce the risk of aflatoxicosis, antioxidants have proven to be effective feed additives in combination with adsorbents as excellent inhibitors of intoxication processes in the poultry body. The study purpose was to determine the effect of Tox-O adsorbent in combination with Oxy-Nil antioxidant drug on the broiler meat production and quality and reduce the risk of aflatoxicosis. It was demonstrated that to optimize meat production in the diets of farmed poultry with a tolerant amount of B_1 aflatoxin, it is necessary to introduce Oxy-Nil antioxidant in the amount of 600 g/t and Tox-O adsorbent in the amount of 750 g/t of feed. Due to synergistic effect of these drugs, the nutritional value of poultry meat was optimized, which was expressed in control group comparison in the enrichment of pectoral muscle samples with dry matter and protein. When feeding tested drugs in pectoral muscle samples of chickens of experimental group III, the value of meat biological usefulness was improved.

1 Introduction

In modern years, one of the most acute problems considered by specialists in the field of feed production is a high level of compound feed quality during storage. Moreover, these ingredients, used in poultry feed formulation, first of all, become contaminated with molds. The latter secrete extremely dangerous toxic compounds - mycotoxins. These toxins are considered to be natural "pollutants" of compound feed and poultry products used for food [1, 2, 3].

It is also known that the main contamination sources of grain ingredients with mycotoxins during their ripening, transportation and storage of crops is the so-called "field" mold, that is, in natural conditions, these fungi are already present on the grain surface. But during long-term storage in violation of the requirements and operating mode of storage facilities, there is already a process of "secondary contamination of grain components with mycotoxins, which most often prevails over primary mold contamination in the field. First of all, it resulted in excessive contamination of poultry feed ingredients with mold poisons - myctoxins. At the same time, B_1 aflatoxin, a producer of Aspergillus fungi, poses a particular danger to the poultry organism and its products [4, 5, 6].

* Corresponding author: zarina_kt@mail.ru

To avoid the risk of aflatoxicosis, as well as reduce the growth energy and quality characteristics of broiler meat, specialists in the field of feed production are developing methods to detoxify this mold poison. According to some researchers [7, 8, 9], one of the most effective methods to reduce the risk of this mycotoxicosis in poultry is the intake of adsorbent drugs in the feed. There are new generation drugs that are also inhibitors for molds. Besides, to reduce the risk of aflatoxicosis, antioxidants have proven to be effective feed additives in combination with adsorbents as excellent inhibitors of intoxication processes in the poultry body. They are also involved in the elimination of mold poisons from poultry body, accompanied by the optimization of broiler meat production and quality [10, 11].

Considering the above, the study purpose was to determine the effect of Tox-O adsorbent in combination with Oxy-Nil antioxidant drug on the broiler meat production and quality and reduce the risk of aflatoxicosis.

2 Materials and methods

To solve this problem in the steppe zone of the Republic of North Ossetia - Alania on four groups of meat chickens of the Ross-308 cross, 100 chickens in each, an experiment was performed according to the scheme shown in Table 1.

Table 1. Substantiation of the research concept.

Group of broilers	Compound feeds and doses of drug additives
Control	Formulation of standard compound feed with a tolerant dose of B_1 aflatoxin (SCF)
Experimental group I	SCF + Oxy-Nil antioxidant in the amount of 600 g/t feed
Experimental group II	SCF + Tox-O adsorbent in the amount of 750 g/t feed
Experimental group III	SCF + Oxy-Nil antioxidant in the amount of 600 g/t + Tox-O adsorbent in the amount of 750 g/t feed

The duration of broilers farming for meat during this experiment was six weeks.

The standard complete poultry feed was based on cereal (barley and wheat) and protein (sunflower meal) components, infected with Aspergillus fungi. To comply with the tolerant dosage of B_1 aflatoxin - 0.25 mg/kg [12], these ingredients, contaminated with the Aspergillus fungus were mixed with other safe (for this mycotoxin) components of poultry compound feeds using standard dispensers.

The economically useful features of the experimental meat chickens were determined using generally accepted approaches.

Control slaughter after the experiment for broilers of compared groups was performed according to the standard method [13].

The obtained digital experimental material after the study was processed by the statistical method with the Student's test of validity.

3 Results and Discussion

The detoxification efficiency of mold toxins by introducing adsorbents, as well as biologically active compounds into the diet, is often accompanied by an increase in the viability of young poultry, that is why we present livability indicators of meat poultry in compared groups in Figure 1.

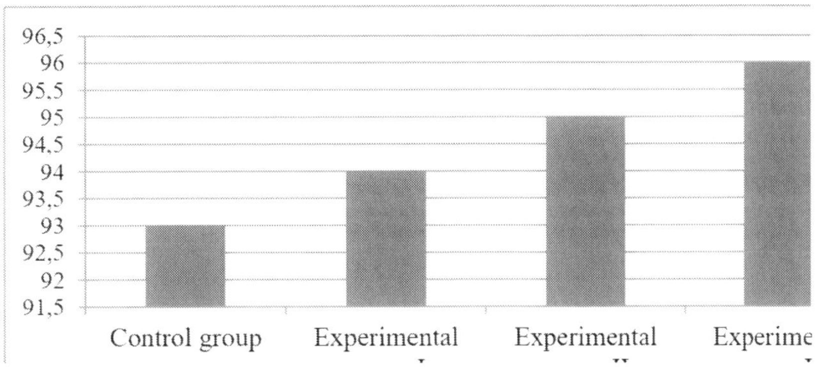

Fig. 1 - Poultry livability in compared groups, %.

It was determined that due to the better mycotoxin elimination when feeding in the composition of compound feed instead of an adsorbent and an antioxidant, broilers of experimental group III surpassed their alternatives in the control group by 3.0% in terms of livability.

The influence level on the growth rate of broilers, fed with the drug, special attention should be paid to changes in their average daily gains (Fig. 2).

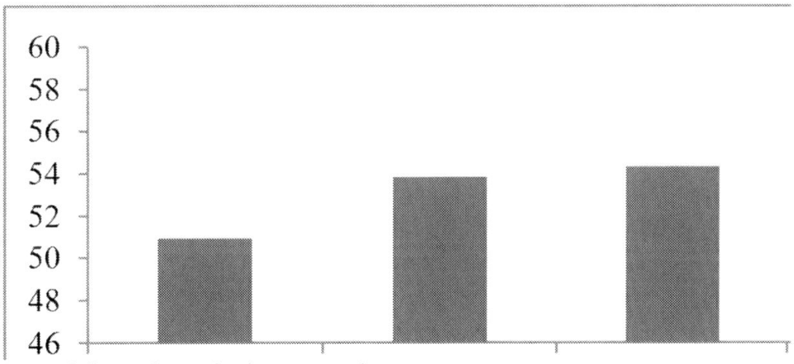

Fig. 2. Average daily poultry gains in compared groups, g.

During the experiment, with the joint introduction of these feed additives into the rations, meat chickens of experimental group III were distinguished by the highest level of growth energy, which outstripped the bird of the control group in terms of the average daily body mass gain by 12.63% ($P > 0.95$).

At the same time, it was important to evaluate the efficiency of Tox-O adsorbent effect in combination with Oxy-Nil antioxidant drug on feed conversion, studied by calculating the cost of compound feed per unit of absolute gain in poultry of compared groups (Fig. 3).

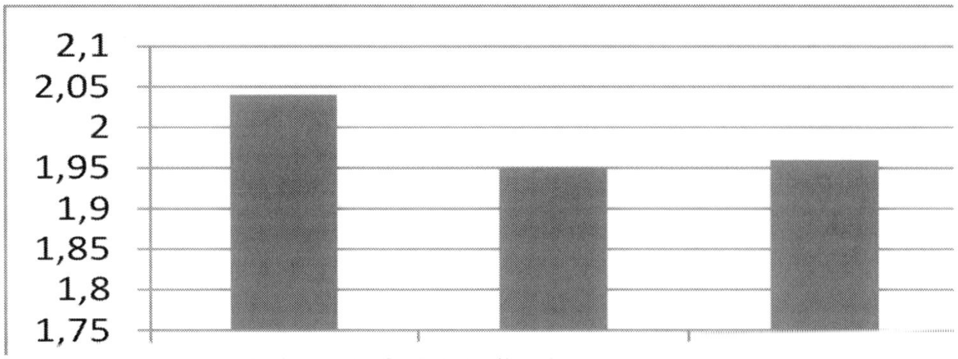

Fig. 3 - Cost of compound feed per unit of gain in broilers, kg.

Based on meat poultry production data, more efficient use of compound feed during these experiments was observed in poultry broilers of experimental group III, in comparison with control equivalents, consumed 8.82% less compound feed per kilogram of body mass gain.

The effect of feed drugs on poultry meat products depends on its slaughter indicators, estimated by the mass yield of half-dresses and dressed poultry (Fig. 4) and the slaughter yield indicator in broilers of compared groups (Fig. 5).

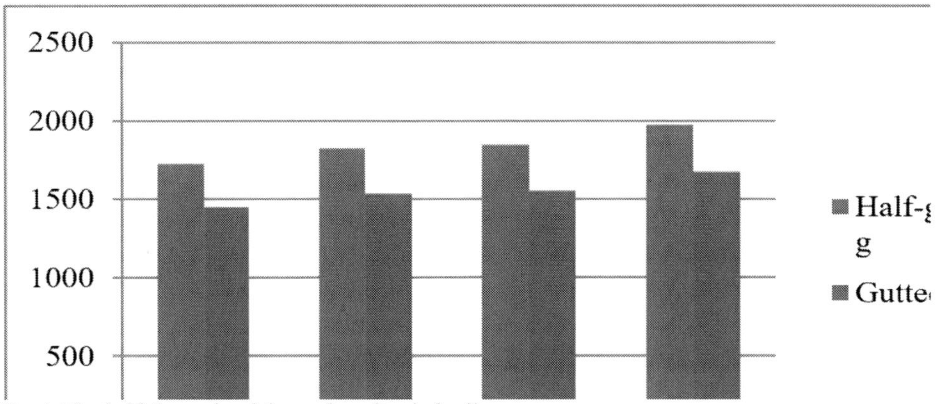

Fig. 4. The half-dressed and dressed poultry in broilers, g.

In the present experiment, due to synergistic effect between tested feed drugs, it was possible to achieve an increase in detoxification capabilities in the poultry body while reducing the risk of aflatoxicosis. Considering above mentioned, there was an improvement in the slaughter parameters in meat chickens of experimental group III, as evidenced by their statistically significant ($P > 0.95$) superiority in terms of the half-dressed poultry mass by 14.62%, dressed - by 15.45%, as well as slaughter yield - by 0.69%.

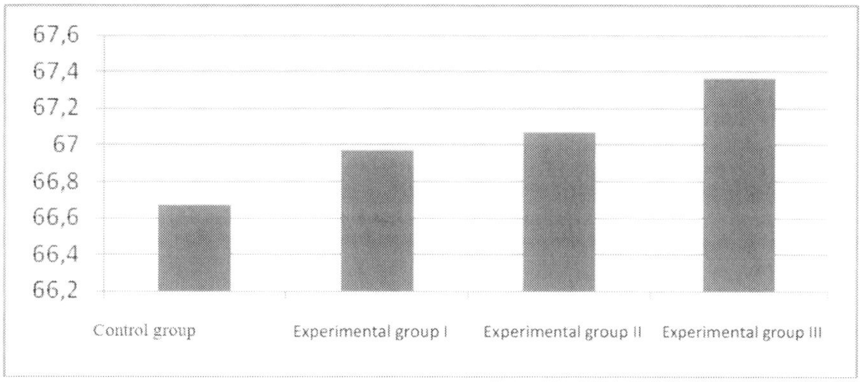

Fig. 5. Slaughter yield in experimental broilers,%.

The consumer qualities of poultry meat are primarily determined by its chemical composition parameters. These indicators are greatly affected by ecological characteristics of consumed feed, as well as efficient elimination of various toxins from the body of fattened broilers, including mold.

Considering the above, in the presence of a tolerant amount of B_1 aflatoxin in the consumed mixed feed, the effect of Tox-O adsorbent in combination with Oxy-Nil antioxidant drug on the main parameters of the chemical composition of pectoral muscle samples of broilers from compared groups was studied. The data is shown in Figure 6.

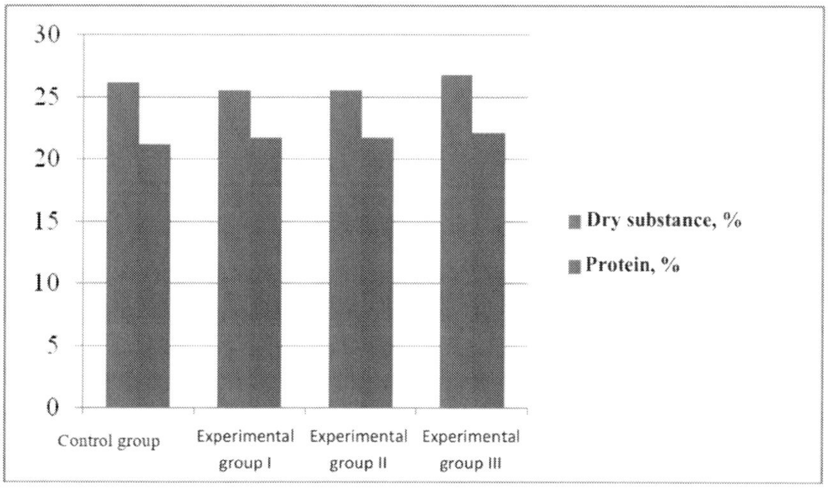

Fig. 6 - Chemical composition of poultry meat samples of compared groups,%.

As shown by study results, in the course of this experiment, with combined additions of these drugs to the compound feed, a positive effect on the chemical composition of broiler meat samples from experimental group III was established. It was demonstrated in comparison with control group equivalents in a statistically significant (P>0.95) enrichment of the pectoral muscle samples with the presence of a dry matter mass fraction by 0.61 and protein - by 0.95% with a parallel lipid reduction - by 0.38% (P>0.95), respectively.

At the same time, the nutritional value of poultry meat is largely due to its biological usefulness, studied by protein-quality indicator (PCI) in pectoral muscle sample, the ratio between the mass fraction of the amino acids tryptophan and hydroxyproline. The protein-quality indicator value of poultry meat in compared groups is shown in Figure 7.

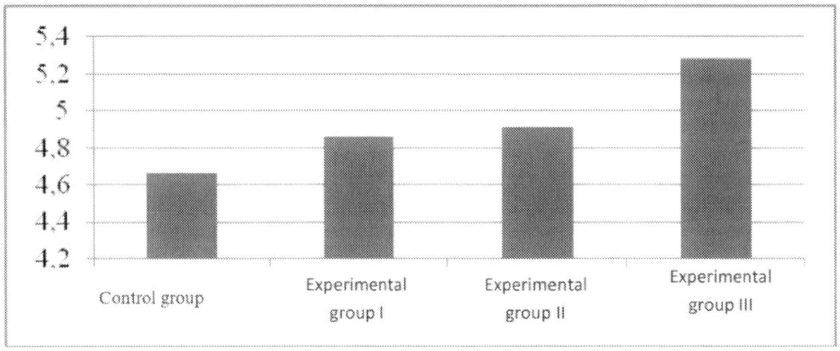

Fig. 7. The protein-quality indicator of poultry meat of the compared groups, units

According to study results, it was found that when feeding in the poultry feed composition with a tolerant amount of mold poison B_1 aflatoxin in these pectoral muscle samples of chickens of experimental group III, in comparison with their control equivalents, there was a significant ($P > 0.95$) increase in the protein-quality indicator value of poultry meat by 12.54%.

4 Conclusions

1. To optimize the productivity and nutritional value of poultry meat, it is necessary to introduce Oxy-Nil antioxidant in the amount of 600 g/t and Tox-O adsorbent in the amount of 750 g/t of feed into diets containing a tolerant amount of B_1 aflatoxin.

2. Due to the more effective detoxification of the specified mold poison due to synergistic effect of these feed drugs, the nutritional value of poultry meat was optimized, expressed in comparison with control group equivalents in a reliable ($P>0.95$) enrichment of pectoral muscle samples with dry matter by 0.61 and protein - by 0.95% with a parallel reduction in lipids - by 0.38% ($P>0.95$).

3. When feeding the tested drugs in the composition of compound feeds at a tolerant level of mycotoxin in the pectoral muscle samples of meat chickens of experimental group III in comparison with their control equivalents, an increase in the value of the protein-quality indicator of poultry meat by 12.54% was observed.

References

1. R. Kh. Gadzaonov, A. A. Stolbovskaya, A. A. Baeva, G. K. Kibizov, Poultry Breeding, **4,** 23-24 (2009)
2. R. B. Temiraev, Z. T. Baeva, N. G. Ter-Teryan, A. A. Gazdarov, L. R. Tebloeva, (Cheese and Butter Making, **5**, 56 (2009)
3. F. F. Kokaeva, R. B. Temiraev, A. A. Stolbovskaya, O. Yu. Leontyeva, Meat Industry, **2**, 59-61 (2012)
4. R. B. Temiraev, Z. R. Ibragimova, L. H. Albegova, M. Sh. Gadieva, A. T. Bagaeva, S.K. Abaeva, Storage and Processing of Agricultural Raw Materials, **11**, 74-76 (2007)
5. V. K Temiraev, V. R. Kairov, R. B. Temiraev, Z. A. Kubatieva, V. M. Gukezhev, Ecology, Environment and Conservation, **23(1),** 554-561 (2017)
6. S. Kononenko, R. Temiraev, Z. Baeva, A. Gazdarov, Compound Feed, **6**, 104-105 (2011)

7. A. A. Baeva, A. A. Stolbovskaya, F. F. Kokaeva, Z. G. Dzidzoeva, Yu. S. Tseboeva (Yu.S. Gusova), O. Yu. Leontyeva, G. K. Kibizov, Proceedings of the Kuban State Agrarian University, **4(13)**, 179-182 (2008)
8. R. Temiraev, S. Lokhova, I. Kokoeva, D. Tsarukaeva, Poultry Breeding, **10**, 35 (2006)
9. R. B. Temiraev, M. K Kozhokov, S. K Cherchesova, F. F. Kokaeva, I. R. Tletseruk, Journal of Environmental Management and Tourism, 8, **3(19)**, 567-573, (2017)
10. L.V. Tsalieva, R. B. Temiraev, S. I. Kononenko, B. A. Dzagurov, M. S. Gazzaeva and S. A. Grevtsova, Journal of Pharmaceutical Sciences and Research, **9(12)**, 2397-2400 (2017)
11. R. B. Temiraev, S. G. Kozyrev, S. I. Kononenko, A. A. Baeva, L. A. Bobyleva, A. L. Kalabekov, Journal of Pharmaceutical Sciences and Research, **9(6)**, 780-784 (2017)
12. Y. A. Yuldashbaev, R. B. Temiraev, V. V. Tedtova, K. B. Temiraev, R. V. Osikina, M. S. Gazzaeva, L. H. Shugusheva, I. K Sattsaeva, M. M. Udychak, Journal of Livestock Science, **11**, 8-13 (2020)
13. L. A. Vityuk, A. A. Baeva, I. V. Kochieva, A. A. Stolbovskaya, S. I. Kononenko, A. V. Yarmoc, I. R. Tletseruk, L. A. Bobyleva, B. G. Tsugkiev, I. K Sattsaeva, Pollution Research, **36(4)**, 748-754 (2017)

Determining the degree of kinship of sarlik yaks bred in the mountainous regions of the Kabardino-Balkarian Republic by blood groups

Lyudmila Sabanchieva[1], *Kristina* Kanukova[1], *Ismail* Gazaev [1], *Murat* Gubzhokov[2], and *Zalina* Bogotova[1,2]

[1]Federal State Budgetary Research Institution "Federal Research Center "Kabardino-Balkarian Research Center of the Russian Academy of Sciences", Balkarova Str., 2, 360002 Nalchik, Russia
[2]Institute of Agriculture - branch of the Kabardino-Balkarian Research Center of the Russian Academy of Sciences, Kirova Str., 224, 360004 Nalchik, Russia
[1,2]Federal State Budgetary Educational Institution of Higher Education "H.M. Berbekov Kabardino-Balkarian State University", Chernyshevsky Str., 173, 360004 Nalchik, Russia

Abstract. The formation of immunogenetics as a science was the result of numerous studies on the study of blood groups and other polymorphic systems, the results of which are successfully applied both in animal husbandry and in animal breeding. Modern biotechnological methods used to analyze blood groups enable us to determine the true origin of animals. To determine the correctness of the breeding processes occurring in the breed, the antigenic composition of blood groups, which are also used to study the reliability of origin, needs to be studied. The results obtained make it possible to analyze genetic similarity, determine the level of polymorphism and the nature of structural changes in the result of breeding work. Also, by analyzing the distribution of the frequency of occurrence of antigens, it is possible to determine the genetic distances and genetic similarity of breeds and lines among themselves [1].

1 Introduction

Currently, the control of the origin of individual lines, breeds and whole herds by means of immunogenetic analysis to determine animals for breeding purposes is very relevant and contributes to an increase in the efficiency of selection and breeding in animal husbandry [1,2,3].

To confirm the authenticity of the origin of an individual, this method is more optimal, due to its information content, development and accessibility. The principle of immunogenetic analysis is to determine blood groups in parental forms and offspring and to analyze their correspondence, which makes it possible to solve the following issues:

- establishing the origin of the offspring by sire and by dam (reliability of origin);
- identification of genetic abnormalities;
- control of genetic safety;
- assessment of the prospects for animal genetic resources.

A scientific article by Koltsov D.N., Romanov Yu.D., Gontov M.E. presents the results of the works of such scientists as Rendel J. et al. on the obtained interbreed discrepancies in the blood groups of cattle. The determining factors contributing to the establishment of the degree of kinship in livestock breeding are the results of studies of blood groups, establishing the origin of both breeds and individual groups of animals and their relationships. According to the laboratory research carried out by I. Matoushek, it was revealed that several breeds took part in the formation of the Czech red-and-white cattle breed [4].

Data from numerous studies show the relationship of blood groups with the productivity of certain traits of animals that play an important role in animal husbandry. Thus, for example, the scientific works of I.R. Giller, according to the definition of the blood group of a cow of the Trostyanets breeding plant named Vorotka, reflect this theory. Later it was found that some representatives of her offspring inherited high milk fat and alleles from the B system (O, I, T, G′, K′), while others, without alleles from this system, did not differ in significant indicators of fat milk. Thus, the establishment of genetic similarity between parents and offspring in terms of the frequency of occurrence of blood groups of parents with offspring has become the most important, and the results are very valuable in the management of breeding animals [5].

Recently, genetic examination has become very relevant. The expediency of this analysis lies in the possibility of eliminating errors when taking into account data on the origin of purebred animals arising from some features of the technology of livestock breeding, where the percentage of error is 15-20%. And sometimes the number of records with incorrect data can reach up to 50%.

On the basis of the order of the Ministry of Agriculture of Russia, immunogenetic laboratories are organized in animal breeding areas, the work of which consists in the correctness and accuracy of testing to breed significant purebred livestock. The factors that allow for controlling the reliability of the origin of animals are:

— codominance of inheritance of antigenic factors and their invariability during ontogenesis;

— numerous combinations of blood groups [6,7].

To obtain objective data on the origin, it is necessary to take blood from the alleged parents and offspring. The presence of identical antigens in animals indicates their relationship, and antigens found in the offspring, which are not included in the genetic structure of the parents, exclude the fact of their relationship. The reliability of the data can be seen in (Table 1) [8].

Table 1. Differences in determining the degree of kinship.

		Calf	Result
Bull	I_1 A'_2 P'_2 R_2 F Z	G_2 I_2 A'_2 P'_2 O′	Kinship is confirmed
Cow	G_2 I_2 Y_2 E'_3 O′ U′, H		
Bull	I_1 A'_2 P'_2 R_2 F Z	G_2 T B′ D′ Q′ H″ Z	Kinship is not confirmed
Cow	G_2 I_2 Y_2 E'_3 O′ U′ H		

Thus, an immunogenetic analysis of the blood of cattle in order to establish the kinship is relevant and expedient, since it eliminates errors in the study and accounting of the reliability of the origin of breeding animals caused by objective and subjective reasons.

2 Methods and Materials

The object of research were sarlik yaks bred in the "Selskie dary" pedigree breeding unit. Determining the analysis of blood cells for erythrocyte antigens and their relationship with economically useful traits allows you to effectively manage the genetic structure of the livestock.

Group erythrocyte antigens were determined according to the rules of genetic examination of cattle breeding material (Federal State Research Institution Rossinformagrotech, 2003. - 48 p.) in the laboratory of molecular selection and biotechnology of Kabardino-Balkarian Research Center of the Russian Academy of Sciences using 40 standard reagents 7 genetic systems produced by "Samarplem" JSC, which made it possible to determine the frequency of occurrence of blood group alleles in sarlik yaks and confirmation of the truth of the origin (Table 2, Figure 1,2,3) [8].

Table 2. List of standard reagents used.

Systems	Antigens
EAA	A_1 A_2
EAB	G_2, G_3, I_1, I_2, P_2, O_3, O_4, A'_2, T B′ D′ Q′ Y_2 Y'_2 E'_3 G′ O′ P'_2 T_1
EAC	C_1 C′ X_1 X_2 R_1 R_2 W L′
EAF	F V
EAS	S_1 S_2 U U′ U″ H H′ H″
EAZ	Z
EAL	L.

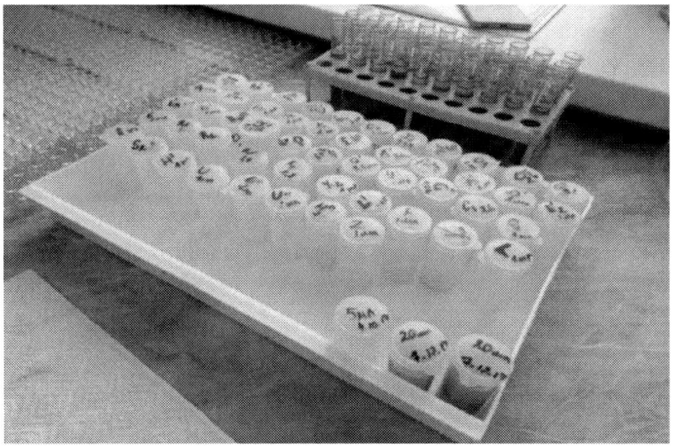

Fig. 1. Standard reagents.

3 Results and discussion

The very process of selection work on farms influences the formation of their gene pool and the determination of genotypes by blood groups, which will make it possible to divide the livestock according to their genetic differences. (Table 3).

The number of alleles, genetic systems and genotypes at each locus can be calculated by the characteristics of the blood groups of the livestock.

For statistical analysis, the computer program Microsoft Office Excel was used, and the allele coefficient was calculated using the formula:

$q = F/n$, where

F is the number of alleles in the studied livestock;

n is the total number of examined yaks.

Table 3. Coefficient of occurrence of blood group alleles in a herd "Selskie dary" pedigree breeding unit.

Systems	Antigens	Allele frequency, %	Systems	Antigens	Allele frequency, %
EAA	A_1	0.96	EAC	C_1	0.04
	A_2	0.38		S.	0.02
EAB	G_2	0.16		X_1	0.10
	G_3	0.32		X_2	0.23
	I_1	0.64		R_1	
	I_2	0.40		R_2	0.25
	P_2	0.05		W	0.05
	O_3	0.80		L.	0.08
	O_4	0.21	EAF	F	0.82
	A'_2	0.58		V	0.02
	T.	0.03	EAS	S_1	0.21
	V.	0.06		S_2	0.14
	D′	0.28		U.	0.21
	Q.	0.32		U.	0.54
	Y_2	0.35		U.	0.09
	Y'_2	0.41		H.	0.59
	E'_3	0.24		H.	0.13
	G′	0.12		H.	0.66
	O.	0.57	EAZ	Z	0.62
	P'_2	0.36	EAL	L.	0.06
	T_1	0.03			

Fig. 2. Adding reagents to plates.

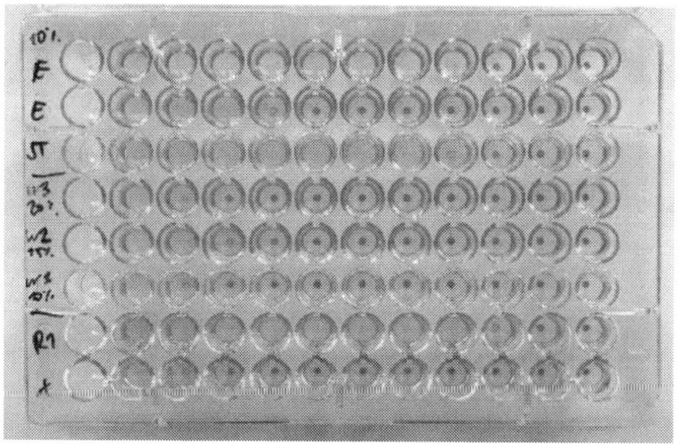

Fig. 3. Result of the hemolysis reaction.

The results of research on sarlik yaks at "Selskie dary" pedigree breeding unit show that in the systems:

— EAA - A_1 and A_2 antigens are detected, with the frequency of occurrence 0.38 and 0.96%, respectively.

— EAB - 19 antigens are detected, with antigens O_3 - 0.80, I_1- 0.64, A'_2- 0.58, O′- 0.57, I_1- 0.40 being most frequent.

— EAC - 7 antigenic factors are detected, with X_2, R_2, X_1 being most frequent - 0.23, 0.25, 0.10, respectively.

— EAS - H″- 0.66, H- 0.59, U′- 0.54 were detected with the maximum frequency.

— The antigens in the following systems were rare and not typical for sarlik yaks:

— EAB - B′, P_2, T, T_1, varied from 0.03 to 0.06%;

— EAC - C′, C1, W, L′, varied from 0.02 to 0.08 (R_1 was not detected at all);

— EAF - V antigen, 0.02%;

— EAS - U″ antigen, 0.09%;

— EAL - L antigen, 0.06%.

Thanks to the studies carried out, it was possible to establish the true origin of offspring by sires and dams (Table 4). From the presented biological material (blood) and family analysis, the reliability of the origin of replacement chicks was 99.0%, and the discrepancy for both parents was 1.0%.

Table 4. Overall reliability score of kinship.

Samples investigated					Blood groups systems, %: EAA, EAB, EAC, EAF, EAS, EAZ, EAL					
					detected inconsistency, %			detected consistency, %		
Total	Sires	Dams	Bulls	Heifers	By sire	By dam	By parents	By sire	By dam	By parents
231	37	97	36	61	0.0	0.0	1.0	99.0	99.0	99.0

4 Conclusion

The possibility of determining the blood groups of an animal immediately after its birth suggests that it is possible to predict the future productivity of the offspring from them. As a result of the successful solution of this issue in the breeding work and the work of breeders, significant changes would follow.

Identification of different forms of the same gene located at loci allows in some cases to determine its genotype. Regardless of the current situation, this will give us the opportunity to select high-quality animals in selection and breeding work, which in turn will expand genetic variability and optimize the structure of the gene pool. The work carried out has shown that it is possible to improve the productive qualities of sarlik yaks by crossing animals with a rare blood group, and its combination with heterogeneous genetic markers made it possible to increase genetic variability in the livestock.

The results of the conducted research make it possible to preserve the gene pool of unique breeds of farm animals, which in turn preserves high-quality breeds of animals for crossing, and further maintains the level of the population while solving the assigned tasks.

This technique using 40 standard reagents from 7 genetic systems made it possible to determine the frequency of occurrence of blood group alleles in the herd. Also, the immunogenetic analysis of the blood of the investigated herd of the farm of the "Selskie dary" pedigree breeding unit showed that the majority of the livestock constituting the breeding base coincides with the intended offspring. Therefore, according to the studies carried out, an assessment of the reliability control of 99.0% was given, which is practically significant for the economy.

References

1. I. R. Giller, Tribal business and breeding biology of agricultural animals, 35 (1973)
2. I. Johanson, J. Rendel, O. Gravert, *Genetics and breeding of farm animals* (1970)
3. J. Matoushek, *Blood group in cattle* (1964)
4. D. N Koltsov, Yu. D. Romanov, M. E. Gontov, Achievement of Science and Technology, **57** (2011)
5. I. R. Giller, L. A. Oleinik, Immunogenetic similarity in cattle inbreeding, **51** (1978)
6. S. L. Gridina, V. F. Gridin, F. A. Sagitdinov, O. I. Leshonok, G. A. Romanenko, Agroindustrial complex of Russia, **857** (2016)
7. R. V. Manoilov, V. F. Gridin, *Youth and science* (2016), http://min.usaca.ru/
8. *Rules of genetic examination of breeding material of cattle* (2003)

Modeling of feed additives for poultry, with the inclusion of probiotic strains

Galina Zelenkova[1,*], *Alexey* Zelenkov[1,2], *Antonina* Pahomova[3] and *Alexander* Pakhomov[4]

[1]Don State Technical University, sq. Gagarina, 1, 344003 Rostov-on-Don, Russia
[2]Ministry of Agriculture and Food of the Rostov region, Krasnoarmeyskaya st., 33, 344010 Rostov-on-Don, Russia
[3]Platov South-Russian State Polytechnic University (NPI), Novocherkassk, Russia
[4]Don State Agrarian University, o. Persianovsky, 346493, Russia

Abstract. The distribution of feed containing probiotics is one of the most attractive trends in modern animal husbandry for consumers. The introduction of a live component makes the feed not only more effective, but also reduces the content of toxic components, making animal products more environmentally friendly. However, the introduction of viable cells of beneficial microorganisms into the feed significantly increases the requirements for the level of feed design, which in turn forces the development of new, unusual algorithms for such design for zootechnics. First of all, the development of feed additives with the inclusion of probiotic strains should be carried out only taking into account the specific application (type of animal of agricultural significance) and the extent to which the studied microorganisms and the substrate for microbial fermentation are suitable for this animal. The most important issue in this regard is the interaction of the living component with the inanimate one. At a minimum, substances that are part of probiotic feeds and feed additives should not inhibit the growth of target microorganisms. The ideal option is the ability to stimulate their growth or prebiotic activity.
Keywords: model, feed additives, poultry, probiotic strains, laying hens, repair young chickens

1 Introduction

The search and testing of new cheap and environmentally safe feed additives that increase the productivity of poultry, have a positive impact on its health, increase the safety of livestock and increase productivity, is an urgent problem of modern poultry farming.

Today, a wide range of various feed additives, preparations, vitamin and mineral premixes has been developed for poultry, but the task of designing effective probiotic preparations for veterinary use, optimized for a living organism and specific application tasks, has not yet been solved.

The distribution of feed containing probiotics is one of the most popular trends in modern poultry farming. The introduction of a live component makes the feed not only

* Corresponding author: zelenkovalex@rambler.ru

more effective, but also reduces the content of toxic components, making poultry products more environmentally friendly.

First of all, the development of feed additives with the inclusion of probiotic strains should be carried out only taking into account the specific application (type of poultry and its purpose) and the extent to which the studied microorganisms and the substrate for microbial fermentation are suitable for this bird. The most important issue in this regard is the interaction of the living component with the inanimate one. At a minimum, substances that are part of probiotic feeds and feed additives should not inhibit the growth of target microorganisms. The ideal option is the ability to stimulate their growth or prebiotic activity.

2 Materials and Methods

Probiotics are attractive for use in animal husbandry due to their ability to stimulate growth and prevent animal infections. The developed feed additives use two strains of probiotic bacteria *Bacillus subtilis* and *Bacillus amyloliquefaciens*.

Bentonite was used as one of the components of the feed additive.

Industrial reserves of bentonite clays in the Rostov region exceed 200 million tons and are among the most powerful in the world. Bentonites are colloidal clays of volcanic origin, consisting mainly of minerals of the Montmorillic group (up to 90%), including sodium, magnesium, calcium and aluminum salts of silicic acid. The porous structure containing active exchange cations determines the unique adsorption, cation exchange, and catalytic properties of these minerals, which are both highly acid-resistant and thermally stable. The presence of a complex of biologically active substances in the bentonite clay is of great importance. It contains from 30 to 40 mobile forms of macro-and microelements. The largest specific mass among them is occupied by oxides of silicon, aluminum, iron, calcium, magnesium, sodium, potassium and phosphorus, as well as trace elements-copper, zinc, manganese, cobalt, iodine, etc. At the same time, the clay contains no arsenic, bismuth, mercury, antimony. As an active sorbent, bentonite adsorbs alkaloids, toxins, some microorganisms, harmful metabolic products and removes them from the body.

The principle of solid-phase fermentation, implemented in the development of feed additives, is a cost-effective method for converting food by-products or readily available natural raw materials into probiotic feed additives for poultry. This is a fermentation process carried out on an insoluble carrier material, a source of nutrients, in the absence of a liquid phase, and it may be preferable to liquid-phase fermentation due to higher yields and better characteristics of the resulting product.

When developing the composition of feed additives, it was taken into account that the needs of laying hens and repair young animals in the mineral components that come with the feed are different. In general, the following components were used:

- waste content from filters of the oil extraction industry, containing residues of vegetable oil, phosphates, fatty acids);
- ecobentokorm (similar to bentonite clay);
- probiotic preparations based on the *Bacillus subtilis* strain - content of viable spores 10^{10}-10^{11} CFU / g, based on the *Bacillus amyloliquefaciens* strain-content of viable spores 10^{9}-10^{10} CFU/g;
- synthetic preparations of fat-soluble vitamins-A, D_3, E.

The developed technological scheme involves combining the components of the supplement and probiotic preparations immediately before feeding the poultry, according to the developed algorithm, the effect of such an active component of the feed additive as bentonite on the growth and survival of the target strains was tested.

Two variants of bentonite were used: pure bentonite and bentonite after being used as a filter for sunflower oil purification.

To determine the number of microorganisms in bentonite, 1 g of the test powder was ground in 10 ml of sterile water until a homogeneous suspension was formed, from which a series of successive dilutions was prepared and then sown on a solid culture medium LB and on soy agar. Incubated at a temperature of 37°C for one day.

To determine the toxicity of bentonite to bacteria of the genus *Bacillus*, both types of bentonite were sterilized at a temperature of 120°C for 20 minutes. Then each of the bentonite types was ground in sterile water (2 ml of water per 1 g of bentonite) until a homogeneous suspension was obtained, which was subsequently mixed with soy agar at a concentration of 1, 2,5, 5, 10, 20, 50%. The resulting nutrient medium was poured into Petri dishes, the control was soy agar without the addition of bentonite.

Suspensions of *B. subtilis* bacteria (MacFarland turbidity 0.7, corresponding to about $2 \cdot 10^8$ CFU/ml) or *B. amyloliquefaciens* (MacFarland turbidity 1.9, corresponding to about $5 \cdot 10^8$ CFU/ml) were prepared. From the obtained suspensions, a number of successive dilutions were prepared and seeding was performed from the fifth and sixth dilutions. The cups were incubated for 1 day at a temperature of 36°C or 42°C. To continue to produce the counting of colonies from the fifth breeding.

3 Results

Both types of bentonite were seeded with different types of bacteria at concentrations of 5x102 and 3x102, respectively, for pure bentonite and bentonite after oil filtration. The bacteria had different colony forms, and belonged to different Gram-positive and Gram-negative species.

Data on the effect of bentonite on bacteria of the genus *Bacillus* are presented in Tables 1 and 2.

Table 1. Effect of bentonite in different concentrations on the growth of *B. subtilis*, number of colonies per cup

Concentration, %	Pure Bentonite		Bentonite after sunflower oil filtration	
	36°C	42°C	36°C	42°C
0	23±5	22±2	26±3	19±3
1	21±4	20±4	20±2	23±5
2.5	20±3	24±3	12±5*	13±6*
5	25±3	24±5	0*	0*
10	26±4	21±4	0*	0*
20	21±5	27±5	0*	0*
50	18±3	24±3	0*	0*

* significant difference from the control ($p<0.05$)

Table 2. Effect of bentonite in different concentrations on the growth of *B. amyloliquefaciens* number of colonies per cup

Concentration, %	Pure Bentonite		Bentonite after sunflower oil filtration	
	36°C	42°C	36°C	42°C
0	48±3	51±5	44±6	47±3
1	50±4	50±3	40±4	45±3
2.5	51±6	45±6	24±6*	27±5*
5	46±3	40±5	0*	0*
10	54±2	53±4	0*	0*
20	47±5	51±4	0*	0*
50	43±4	49±3	0*	0*

* significant difference from the control ($p<0.05$)

Analyzing the results obtained, it can be concluded that pure bentonite had no effect on the growth of the studied bacteria. Even when bentonite was added to the medium in high concentrations, no reduction in bacterial survival was observed.

Bentonite obtained from filters after cleaning sunflower oil had a depressing effect on the growth of bacteria, both *B. subtilis* and *B. amyloliquefaciens*. If at a one-percent concentration of such bentonite in the medium, no significant differences in bacterial growth from the control were observed, then already at an increase in the concentration to 2.5%, a decrease in survival was noted by 47-50% for *B. amyloliquefaciens* and 41-48% for *B. subtilis*. A further increase in the concentration of spent bentonite completely suppressed the growth of bacteria.

Obviously, this difference in the action of bentonite is associated with the processes occurring in the oil filtration process. It can be assumed that the growth of bacteria suppresses the remaining sunflower oil in the bentonite, but this is unlikely. Yes, it is known that vegetable oil is not a good enough substrate for the growth of bacteria of the genus *Bacillus*, however, it is not bactericidal for them.

On the other hand, bentonite is able to adsorb various substances on itself. It is logical to assume that in the process of oil purification, toxic substances are deposited on the bentonite and accumulate in it. This assumption is more likely, but requires further research.

Conclusion: pure bentonite is considered by us as a growth substrate for probiotics intended for use in industrial and farm poultry farming. It is proposed to study the possibility of using spent bentonite, after purification of sunflower oil, as a carrier for probiotic crops aimed at use in poultry farming, taking into account the use of spent bentonite in small concentrations (no more than 1%).

The developed algorithm for designing probiotic additives assumes, as part of the module for confirming the quality of the resulting product, in the event of a negative result, the replacement of toxic components for the target strains with non-toxic ones.

It may also be necessary to adjust the formulation of the supplement in the direction of increasing the content of probiotic components.

Parameters of feeding the chickens with the use of feed additives on the basis of the studied probiotic strains.

Method of application of a feed additive based on the studied probiotic strains for birds:

- a feed additive in the amount of 1.5% of the weight of mineral feed is introduced into the diet of repair young chickens;
- in the diet of laying hens in the amount of 3% by weight of mineral feed.

This amount of feed additive introduced into the diet of poultry of different age periods of cultivation was determined by the main limiting indicator of the test product – calcium.

Feeding feed additives is carried out at all stages of the period of the poultry egg production efficiency, and in a separate periods: the rearing chickens – 17 weeks (from 10 to 120 days), the contents of rearing in the shop laying hens (prekladovy period) until 21 weeks of age, the contents of laying hens: renoprotective period to 45 weeks of age, the period of operation – up to 72 weeks.

Feed mixtures of mixed feeds PK-3-51 and PK-1-0-88 developed and compiled on the basis of the applied feed products at LLC "Aksayskaya poultry farm" (Aksaysky district, Rostov region).

Composition of the feed mixture of mixed feed PK-3-51:

Young animals 6-17 weeks: corn – 30,000%; wheat – 24,199%; extruded soy – 17,304%; sunflower cake – 24,999%; salt – 0.279%; monocalcium phosphate – 0.725%; shell flour – 1,400%; megafos 5000 – 0.015%; agrocele – 0.050%; feed lysine 78% – 0.305%; feed methionine – 0.114%; mycofix select – 0.050%; premix young – 0.500%; vitamin B_4 – 0.060%.

In the experiment, they were introduced on the basis of the studied probiotic strains instead of the shell flour used in the conditions of the poultry farm.

The composition of the feed mixture feed PC-1-0-88:

Laying hens 17-57 weeks.: corn – 44,558%; wheat – 10,000%; sunflower cake – 31,000%; yeast hydrolysis – 3,219%; salt of 0.263%; monocalcium phosphate – 0,726%; Shelly flour – 9,099%; megafos 5000 – 0,012%; agrocele – 0,050%; lysine feed 78% – 0,333%; methionine feed – 0,110%; mycofix select – 0,050%; premix – 0,500%; vitamin B_4 – 0.080%.

In the experiment, a feed additive was introduced based on the probiotic strains studied, reducing the shell flour to 6.0%.

4 Conclusions

As a result of the research, an algorithm for obtaining new feed additives for poultry is presented, as well as the parameters of feeding chickens and the composition of feed additives with the inclusion of the studied probiotic strains are developed.

References

1. G.A. Zelenkova, I.F. Gorlov, Fodder bentonite for farm animals and a bird (ekobentokorm) TU 9283-199-10514645-13-2013 specifications
2. A. Pahomova, S. Halász, A. Pakhomov et al., E3S Web of Conferences, 217, 06002 (2020)
3. A. Zelenkov, G. Zelenkova, S. Tresnitskii et al., E3S Web of Conferences, 217, 09010 (2020)
4. A. Pahomova, S. Halász, V. Fedorchuk, G. Zelenkova, A. Pakhomov, E3S Web of Conferences, 210, 11003 (2020)
5. A. Zelenkov, A. Ermakov, G. Zelenkova et al., E3S Web of Conferences 135, 01088 (2019)
6. V. Berikashvili, K. Sokhadze, E. Kachlishvili, V. Elisashvili, ML. Chikindas, Probiotics Antimicrob Proteins, **10(4)**, 755 (2018)
7. A. Zelenkov, A. Ermakov, G. Zelenkova et al., IOP Conf. Series: Earth and Environmental Science, 403 (1), 012020 (2019)
8. A. Zelenkov, G. Zelenkova, A. Ermakov et al., E3S Web of Conferences 164, 06032 (2020)
9. A. Pahomova, A. Pakhomov et al. Vitamin and mineral adsorption food additives for poultry and the method of its application (A23K 50/70, A23K 20/158, A23K 20/174)

Results of using different breed studs in commercial fine wool sheep breeding

Vasily Aboneev[1,*], *Dmitry* Aboneev[2], *Ekaterina* Aboneeva[2], *Safarbi* Kazanchev[3], and *Dastanbek* Baimukanov[4]

[1]FSBSI Krasnodar Scientific Center for Animal Science and Veterinary Medicine, 4, Pervomaiskaya Str., 350055 Krasnodar, Russia
[2]North Caucasian Federal University, 1, Pushkina Str., 355017 Stavropol, Russia
[3]Kabardino-Balkarian State Agricultural University named after V.M. Kokov, Lenin Avenue, 1V, 360030 Nalchik, Russia
[4]Kazakh National Agrarian Research University, 8, Abai Avenue, 050010 Almaty, Republic of Kazakhstan

Abstract. This work presents the material of research and production experiments on the crossing of stud rams of the North Caucasian meat and wool breed and Australian meat merino with fine wool ewes of a commercial herd. It is identified that the use of North Caucasian rams on fine wool ewes serves to increase the fertilizing ability and milk production of the breeding stock, to increase viability and resistance of young animals, as confirmed by the best morphometrical differences of crossbred offspring placentas. Crossbred offspring from semi-fine wool rams had a large live weight at birth at 21 days and at 4 months of age. The use of Australian producers did not have a significant effect on the above indicators of economically useful traits in comparison with purebred breeding.

1 Introduction

One of the most important tasks in the development of the current industry is to increase the competitiveness of sheep breeding production. Changes in the contemporary social and economic conditions and, as a result, a decrease in the demand for most of the products of sheep breeding led to a sharp decrease in the number of animals, their productivity, a change in the breed composition and structure of breeds of different directions. Nevertheless, the main types of fine wool sheep products, wool and meat, have not lost their value, while their value has changed in the direction of mutton meat. This type of product was of the utmost importance at all stages of the industry's development. Specifically, in the period from 1985 to 2005, new types of sheep with improved indicators of meat production were bred in commercial fine wool herds. The development of techniques and methods for increasing the fertilizing ability, ewes fertility, resistance and viability of young animals, especially in the first periods of an animal's life, including

* Corresponding author: aboneev49@mail.ru

before separation from mothers is an important element contributing to an increase in the production of mutton meat. The study of these issues is the main objective of our research.

2 Materials and Methods

To achieve the objectives, reproductive qualities of Caucasian breed ewes were identified when they were crossed with North Caucasian meat and wool rams (group 1), Australian meat merino rams (group 2) and purebred Caucasian studs (group 3), as well as some indicators of productivity, resistance and viability of offspring of different origin. To do this, the flock of fine wool ewes of the senior shepherd of agricultural enterprise "Novomaryevskoe" M.M. Magomedov in the amount of 425 animals was inseminated by rams of the above breeds in approximately the same amount by the method of their daily sampling in the hunt. In the first group 140 animals were inseminated, in the second group - 130 animals and in the third group - 155 animals. At the same time, the reproductive qualities of rams and ewes, viability, resistance and live weight of the produced offspring, including some morphometrical differences of the placenta were studied. To achieve the objective, the volume and quality of sperm production of stud rams of different breeds, the fertility of mothers by the number of lambs per 100 lambed ewes, the mortality of lambs and their live weight from birth to 4 months of age, some morphometric differences of the placenta, blood serum bactericidal (SBA) and lysozyme (SLA) activity, according to existing zootechnical and biological methods. The feeding level of ewes, stud rams and the produced offspring corresponded to the set norms.

3 Results and Discussion

Two-year-old studs, which met the economically useful requirements of the elite class and had features typical to the breeds listed above were selected from the existing stud ram flock to carry out research and production experiments. Fresh semen of Australian meat merino was imported from the Stavropol Research Institute of Livestock and Fodder Production experimental station. The volume of ejaculate in the Caucasian breed was 1.42 ml, with a mobility of 9.5 points, and in the North Caucasian breed and Australian meat merino these indicators were 1.53 ml, 9.5 points and 1.12 ml, 9.0 points accordingly.

An important biological peculiarity of different breeds of sheep, which significantly affects the economic efficiency and the overall profitability of the industry, is the fertility of the ewes and the viability of the produced offspring. Scientists [2, 4, 5, 6-11] pointed out that the fertility of ewes and the viability of young animals can be increased with different methods, including targeted selection of multiple ewes and rams, as well as interbreeding of sheep of different productivity directions. They also pointed out that ewes with twins and triplets produced more meat in body weight and teg of finer assortments than ewes with single lamb, despite the lower average weight of each lamb.

Our experimental studies on the individual accounting of the results of insemination and lambing of fine wool ewes inseminated with the sperm of rams of different breeds showed significant differences between the compared animals, including the resulted offspring.

Table 1 shows information on fertilization rate, ewe reproduction capability, viability and live weight of young animals.

Table 1. Reproductive qualities of ewes, viability and live weight of lambs.

Indicators	Group		
	1	2	3
Inseminated ewes, animals	140	130	155
Lambed ewes, animals	134	119	142
%	95.9	91.2	91.5
Total lambs produced, animals	152	123	147
Reproduction capability, %	114.0	103.3	103.7
Lamb viability up to 4 months, animals	142	110	132
%	93.4	89.6	88.9
Live weight of young ewe, kg at birth on day 21 at 4 months	4.2±0.10 11.8±0.16 24.1±0.19	3.8±0.08 10.9±0.15 21.9±0.19	3.8±0.07 10.7±0.14 21.5±0.19

The analysis of the information from Table 1 indicates that the ewes inseminated by the North Caucasian rams had the best fertilizing ability, and somewhat worse was this indicator by the Caucasian and Australian meat merinos. At the same time, 30 animals or 19.4% of the ewes, inseminated in the first hunt by rams of the Australian meat merino, remained barren and were re-inseminated by the rams of the same group. 5 ewes from rams of the North Caucasian breed and purebred Caucasian studs, or 3.6% and 3.8%, respectively remained barren. The fertility of ewes of the compared groups was 114.0% in 1st group, and in 2nd and 3rd groups 103.3% and 103.7%, respectively, that is, the ewes of 1st group prevailed over the peers of 2nd and 3rd groups by 11.3% and 11.7%.

An important indicator that determines the efficiency of the breeding process and the profitability of the sheep breeding industry is the resistance of the produced offspring and its subsequent viability. The resistance is caused by complex protective reactions of the body and ensures the viability of young sheep of different genotypes in postnatal ontogenesis [1, 2, 10, 12, 13, 14].

In our studies, the estimation of the protective potential of the experimental groups of animals was carried out by monitoring the activity of humoral factors (serum bactericidal activity - SBA, serum lysozyme activity - SLA).

The analysis of obtained results, shown in Table 2, made it possible to indicate a number of patterns due not only to the age characteristics of the produced offspring, but also due to their breed.

Experimental lambs in the early postnatal period had the lowest rates of humoral immunity. At birth, the level of serum bactericidal and lysozyme activity (SBA, SLA) in young animals of different origins varied from 35.94% to 37.10% and from 22.26% to 23.71%. At the same time, SBA indicators were higher in the crossbred offspring of North Caucasian rams by 0.78 abs.% and 1.16 abs.% related to the peers of 2nd and 3rd groups. According to the level of blood serum lysozyme activity, the lambs of the 1st group prevailed over the animals of the 2nd and 3rd groups. The difference reached 0.23 abs.% and 1.45 abs.% in favor of the offspring of 1st group, respectively. By the age of two months, there was an increase in indicators on the protective potential of the experimental animals (Table 2). It can be assumed that a sharp increase in the level of blood serum bactericidal activity in lambs of the experimental groups in this age period relates to the gradual formation, development of the immune system, which provides the protective potential of the growing organism, as evidenced by the indicators of the growth rate of young animals. At the same time, an increase in bactericidal activity was higher in the offspring of North Caucasian rams related to the lambs of 1st and 3rd groups. Referring to the level of blood serum lysozyme activity, no significant differences were found between the experimental groups in this indicator of humoral immunity in the studied period of ontogenesis.

Table 2. Age dynamics of the natural resistance level of young ewes of different genotypes,%

Indicator	Age periods	Group		
		1	2	3
SBA	At birth	37.10±0.83	36.32±0.70	35.94±0.83
	2 months	55.57±0.78	55.01±0.84	54.37±0.87
	4 months	46.72±0.51	46.09±0.86	45.71±0.36
SLA	At birth	23.71±1.06	23.48±0.77	22.26±0.93
	2 months	24.75±0.61	24.62±0.59	24.06±0.62
	4 months	37.74±0.47	37.06±0.92	36.43±0.47

The reactivity of the organism gradually improves with an increase in the age of young animals in the experimental groups and by the age of 4 months a decrease in the serum bactericidal activity and an increase in lysozyme activity are observed.

For a specific tendency of age-related changes in the indices of natural resistance in the experimental young lambs, the superiority of the offspring of North Caucasian rams over the peers of young lambs of 2nd and 3rd groups in terms of the level of activity of humoral factors (SBA, SLA) was revealed. At the age of 4 months, this advantage in favor of the young lambs of 1st group was 1.4 abs.% and 2.2 abs.% in bactericidal activity and 1.8 abs.% and 3.6 abs.% in lysozyme activity.

Consequently, a comparative analysis of the results obtained revealed the advantage of the organism protective potential of the offspring of semi-fine wool rams of the North Caucasian meat and wool breed in all periods of postnatal ontogenesis over animals of other variants of selection according to the level of humoral factors of natural defense. At the same time, the amplitude of the revealed changes was within the physiological norm.

The best protective properties of the offspring of the compared groups had a certain effect on the viability of young animals from birth to the separation from ewes. It is one of the major indicators that determine the economic efficiency of the industry. In our studies, the offspring produced by North Caucasian rams turned out to be the most viable from birth to the separation. This indicator reached 93.4%, which is 3.8% and 4.5% more than in animals of 2nd and 3rd groups.

Sampled placentas, 8 from each ewe, incident with rams of different breeds, were subjected to laboratory studies to determine some morphometrical indicators. It was identified that the weight of the placenta from the ewes of the first group was equal on average to 276 g, while in animals of the second and third groups it was, respectively, 245 g and 234 g, or 12.7 and 17.9 percent less. The measurement of the cotyledon number in the placenta of the ewes of the compared groups made it possible to identified the following patterns. While the ewes inseminated by the North Caucasian rams had 76.7 cotyledons, the peers of the 2nd and 3rd groups had 5.5 and 6.5 less, or 7.7% and 9.3%. Significant differences were also found in the size of cotyledons in the placenta of the ewes of the compared groups. Thus, the size of cotyledons in the placenta of ewes inseminated by semi-fine wool rams of the North Caucasian breed was 2.2/3.1 cm, and for ewes of 2nd group, this indicator was 1.8/2.2 cm. According to this indicator the ewes of the 3rd group are in the last place. The size of their cotyledons was 1.6/2.0 cm. The distance between cotyledons in animals of 1st, 2nd, 3rd groups was 1.8/3.1 cm; 2.9/3.7 cm and 3.2/4.3 cm, respectively. These results indicate that the density of cotyledons in the placenta of crossbred young ewes from North Caucasian rams is much higher related to the peers of 2nd and 3rd groups, which means that the efficiency of fetal nutrition in this group of animals was higher. The observed patterns indicate that the insemination of ewes by rams of the North Caucasian meat and wool breed during the period of pregnancy contributes to the better development of the placenta, and as a result, better development of its offspring during the period of embryonic development. It is confirmed by our experimental data.

Individual weighing of lambs at birth (Table 1) showed that the lambs produced by the ewes of the 1st group in live weight exceeded the peers of the 2nd and 3rd groups by 0.4 kg or 10.5% ($P > 0.05$). The milk production of the ewes, calculated as the product of the average live weight of the lambs at the age of 21 days plus the coefficient 5, was the highest in the animals of the first group. This advantage over the ewes of 2nd and 3rd groups was 8.3% and 10.2%, respectively, ($P > 0.05$). Weighing of lambs when separating from ewes showed a similar pattern between the compared groups of animals. If the average live weight of the lambs of the first group was 24.1 kg, then in the offspring of the second and third groups it was 2.2 kg and 2.6 kg less, or 10.0% and 12.1% ($P > 0.05$)

The calculation of economic efficiency indicators based on the methodology of scientists (3) showed that, with the higher fertility of ewes, the viability and live weight of lambs in the first group of animals, the level of profitability was 20.3% and 21.7% higher, related to the offspring of 2nd and 3rd groups.

4 Conclusion

Consequently, the use of stud rams of the North Caucasian meat and wool breed in commercial fine wool sheep breeding allows to increase the fertility and milk production of ewes, the viability, live weight of offspring and the profitability of young-stock breeding in comparison with the use of Australian meat merino and purebred breeding.

References

1. V. V. Aboneev, L. N. Skorykh, Sheep, Goats, Woolen Business, **3**, 20-22 (2002)
2. V. V. Aboneev, A. I. Surov, K. G. Chukhno, Zootechniya, **8**, 26-27 (2008)
3. V. V. Aboneev, N.K. Tymoshenko, T.P. Rusanova et al., *Methodology for calculating the economic efficiency of the production of sheep products in order to more fully realize the economic potential of the industry* (2013)
4. V. V. Aboneev, V. V. Marchenko, R. P. Larionov, *Problems of Livestock and Fodder Production in Russia*, 14-17 (2015)
5. V. V. Aboneev, L. G. Gorkovenko, A. Ya. Kulikova, N. I. Tsapkina, Scientific basis for increasing the productivity of farm animals, **2**, 4-9 (2016)
6. V. V. Aboneev, V. V. Marchenko, L. A. Gnezdilova, N. I. Tsapkina, International bulletin of Veterinary Medicine, **2**, 35-42 (2017)
7. V. V. Aboneev, L. A. Gnezdilova, V. T. Ranyuk, D. V. Aboneev, Veterinary Medicine, Animal Science and Biotechnology, **1**, 120-125 (2019)
8. V. V. Aboneev, T. T. Tarchokov, S. F. Sukhanova, D. V. Aboneev, E. V. Aboneeva, V. V. Marchenko, IOP Conf. Series: Earth and Environmental Science, 341 (2019) 012045
9. L. A. Gnezdilova, V. V. Aboneev, D. V. Aboneev, *Current state, problems and prospects for the development of agricultural science*, 275-276 (2018)
10. A. A. Erokhin, V. V. Aboneev, E. A. Karasev, S. A. Erokhin, D. V. Aboneev, *Forecasting the productivity, reproduction and resistance of sheep* (2010)
11. Yu. A. Kolosov, I. S. Gubanov, V. V. Aboneev, Izvestiya Orenburskogo SAU, **4(72)**, 310-312 (2018)
12. Y. Kolosov, A. Klimenko, V. Vasilenko, N. Shirokova, L. Getmantseva, A. Kolosov, V. Aboneev, L. Chizhova, V. Marchenko, A. Mikhailenko, D. Aboneev, OnLine Journal of Biological Sciences, **17(4)**, 343-347 (2017)

13. L. N. Chizhova, V. V. Aboneev, S.N. Shumaenko, Sheep, Goats, Woolen Business, **4**, 13-15 (2014)
14. L. N. Chizhova, V. V. Aboneev, D. V. Aboneev, Proceedings of the international research conference, 143-146 (2014)

Exterior and constitutional features of first-calf cows of black-and-white cattle of different genotypes

Orest Basonov[1,*], *Dmitry* Petrov[1], *Amanbai* Karynbaev[2,3], *Zaurbek* Aisanov[4], and *Tsrai* Kagermazov[4]

[1]FSBEI HE Nizhny Novgorod State Agricultural Academy, 97, Gagarin Avenue, 603107 Nizhny Novgorod, Russia

[2]Non-commercial JSC "National Agrarian Science and Educational Centre", 4A, Kurgaldzhinskoe shosse, 010000 Nur-Sultan, Republic of Kazakhstan

[3]FSBEI HE Southern Federal University, 105/42 Bolshaya Sadovaya Str., 344006 Rostov-on-Don, Russia

[4]Kabardino-Balkarian State Agricultural University named after V.M. Kokov, Lenin Avenue, 1V, 360030 Nalchik, Russia

Abstract. The dairy cattle breeding development depends largely on the culture of raising young animals of the cattle. Production and maintaining healthy young animals is an extremely challenging task for livestock breeders [1,3,12,14]. Breeding of dairy replacement heifers is considered a major element of the breeding system and milk production technology, since the formation of milk production takes place just in the process of growth and development. Individual selection of parental pairs affects not only the cattle reproductive ability, but also the growth of offspring [2, 4,13], [5-11].

1 Introduction

The dairy cattle breeding development depends largely on the culture of raising young animals of the cattle. It is very difficult to produce and maintain healthy young animals. Correct dairy young-stock breeding is the reason for the optimal expression of genetically coded productive capabilities of animals in the first stage of their growth and development. The peculiarities of the growth and development of the animal, determined by heredity and environmental conditions, especially feeding are taken into account when organizing breeding and feeding.

This article provides a comparative analysis of the dynamics of live weight, body measurements and body build indices of first-calf cattle of several combinations of selection. First-calf heifers of the fifth experimental group, produced by fathers and mothers with the cross line affinity, exceeded in live weight and gain of animals included in the control group, whose parents had pure lines.

* Corresponding author: dekanat.zootexnia@yandex.ru

Under the same conditions of keeping and feeding, the heifers of the studied groups grew in a different way, and in the process of ontogenesis were revealed several differences in the indices. The first-calf heifers of the 5th group in different age periods had the highest live weight and exceeded the peers of the 1st, 3rd and 4th groups from 10 to 23 kg with a significant difference. The highest average daily gain up to 6 months of age were observed in the animals of the 5th experimental group. For the whole growing period (18 months), the animals in the control group had the highest absolute gain, which was significantly higher than that in the experimental groups.

The body build indices as leggy, stretchiness, pelvis-chest, chest, blockiness and boniness were calculated, as well as the coefficient of milk production, which was used to judge the direction of the animal productivity. First-calf cattle belongs to the dairy-meat direction of productivity according to the leggy and stretchiness index. All groups belong to the milk production direction according to the downwardness and chest index, and to the meat one according to the boniness index.

Study purpose and objectives. The purpose of our study was to determine the growth and development of first-calf heifers depending on the breeding pairs combinations.

2 Materials and Methods

Scientific and economic study was carried out at Plemzavod im. Lenin LLC in the Koverninsky district of the Nizhny Novgorod region. The animals were divided into groups depending on the combinations of selection with subsequent biometric processing of the primary research material. Five groups of black-and-white first-calf heifers were formed. The first control group consisted of animals produced from pure lines with both the paternal and maternal ancestors of the same line (n=30). The second group consisted of cows with the paternal ancestors of one line, and the maternal ancestors of another line (n=8). The third group consisted of first-calf heifers with the paternal ancestors of the same pure line, and the maternal ancestors with the cross line affinity (n=14). The fourth group consisted of first-calf heifers with the paternal ancestors with the cross line affinity, and the maternal ancestors of the same pure line (n=20). The fifth group consisted of animals with paternal and maternal ancestors with the cross line affinity (n=27).

3 Results and Discussion

The growth and development of an animal organism at different stages of life does not proceed equally quickly. The major factor affecting the growth and development of young animals is feeding. The correct development of a young animal related to the provision with fresh air, light and warmth. Well-organized loose housing creates especially favorable conditions for the development of young cattle. Growth and development are studied according to the individual weighing data from the moment of birth to 18 months of age (Table 1).

It was identified that under the same conditions of feeding and keeping, the animals of the studied groups grew in a different way, and in the process of ontogenesis, several differences in indicators were revealed. The first-calf heifers of the 5th group have the highest live weight at the age of 6 months related to the 3rd and 4th group. The difference is 14 kg with $P > 0.999$ and 8 kg with $P > 0.99$, as well as the control group prevails over the peers of the 3rd group, 7 kg with a significant difference $P>0.95$. The first-calf heifers of the 5th experimental group at 10 months of age are leading in terms of live weight related to the groups 1, 3 and 4 - 10, 23 kg ($P>0.99$) and 17 kg with $P>0.999$.

The control group animals are superior at the age of 12 months over their peers of the 3rd and 4th groups by 19 and 10 kg with a significant difference P>0.95, as well as the first-calf heifers of the 5th experimental group - 26 and 17 kg (P>0.99). The cattle of the 5th group are distinguished by a high live weight at the age of 18 months related to the 2nd and 3rd group, the difference is 31 kg and 21 kg with P>0.99, as well as animals of the 4th group - 25 kg (P>0.99) and 15 kg (P>0.95). The first-calf heifers of the control group prevail over the peers of the 2nd experimental group by 20 kg with P>0.95. The animals of the 3rd group are smaller in live weight at the first insemination to the 1st, 4th and 5th groups by 20 kg, 22 kg and 24 kg (P>0.95), and the indicator of the 2nd experimental group is smaller than the 4th by 20 kg with a significant difference P>0.95.

Table 1. Dynamics of change in live weight of black-and-white heifers, kg

Age, months	Group									
	1		2		3		4		5	
	X̄±m	C_v	X̄±m	C_v	X̄±m	C_v	X̄±m	C_v	X̄±m	C_v
At birth	36.2±0.7	12.5	37±1.9	14.7	37±1.2	12.5	38 ±0.9	11.5	37±0.9	13.8
6	182±2.5*	8.7	185±3.8	5.8	175±3.1	6.7	181±0.3	0.1	189±2.4 ***	6.5
10	280±4.9	9.3	277±6.8	6.9	267±5.8	8.2	273±0.4	0.7	290±4.4 ***	7.8
12	330±5.5*	10.6	323±9.2	8.1	311±6.8	8.2	320±1.5	2.1	337±5.4**	8.3
18	432±4.6*	6.8	412±8.6	5.9	422±5.9	5.2	437±1.9**	1.9	443±5.2**	6.2
At the first productive insemination	401±3.2*	5.1	385±9.0	6.7	381±9.3	9.1	405±2.2 *	2.4	403±3.6*	4.7

Table 2 presents the indicators characterizing the dynamics of the average daily gains of animals.

Table 2. Average daily gains, g

Age, months	Group				
	1	2	3	4	5
6	807.2±16.92	820.1±31.18	768.3±20.07	793.9±24.36	841.6±12.81**
10	816.9±13.83*, **	772.9±37.00	764.3±16.43	769.2±16.12	840.1±14.82**
12	845.0±12.33**, ***	766.7±22.82	732.1±11.26	788.3±14.22	788.9±11.72
18	563.2±26.01	492.4±51.37	616.7±28.33*	650.3±52.07*	589.7±±27.24

Table 2 shows that the greatest average daily gain at the age of 6 months have the animals of the 5th experimental group related to the 3rd group, the difference is 73.3 g (9.1%), P>0.99. According to the gain at the age of 10 months, first-calf heifers of the control group predominate over 3rd and 4th experimental groups - 52.6 g (6.4%) with P>0.99 and 47.7 g (5.8%) with P>0.95, as well as cows of the 5th group by 75.8 g (9.2%) and 70.9 g (8.6%) with P>0.99.

At 12 months, the control group has the highest gains related to the 2nd, 3rd, 4th and 5th experimental groups, the difference is 78.3 g (9.3%) with P>0.99, 112.9 g (13.4%) with

P>0.999, 56.7 g (6.7%) with P>0.99 and 56.1 g (6.6%) with P>0.999. According to the average daily gain at the age of 18 months, first-calf heifers of the 3rd and 4th experimental groups prevail over 2nd group - 124.3 g (22.1%) and 157.3 g (28.1%) with a significant difference P>0.95.

Generally, it should be noted that the animals had rather high gains, which indicates that the animals were well fed and maintained. The absolute gains are shown in Table 3.

The first-calf heifers of the 5th experimental group have the greatest absolute gains over the 3rd group in 6 months, the difference is 13.2 kg (9.1%) with P>0.99. The animals retained their superiority over the 3rd and 4th groups for the next 4 months, the difference was 9.1 kg (9.3%) and 8.8 kg (8.7%) with P>0.95. The first-calf heifers of the control group at the age of 10-12 months prevailed over the peers of the 3rd experimental group - 6.8 kg (13.4%) with a significant difference P>0.95. The cows of the 3rd, 4th and 5th experimental groups had the greatest gains at the age of 12-18 months, the superiority over first-calf heifers of the 2nd group is 22.4 kg (22.1%), 28.8 kg (28.1%) and 17.6 kg (17.3%) with P>0.95. The animals of the 5th experimental group have the highest absolute gain in 18 months, the difference with the peers of the 2nd and 3rd groups are 30.8 kg (7.8%) with a significant difference P>0.99 and 20.9 kg (5.3%) with P>0.95.

Table 3. Absolute gains, kg

Age, months	Group				
	1	2	3	4	5
0-6	145.3±3.05	147.6±3.74	138.3±3.61	142.9±4.38	151.5±2.31**
6-10	98.0±2.49	92.8±4.44	91.7±2.96	92.3±2.90	100.8±2.67*
10-12	50.7±2.22*	46.0±2.74	43.9±2.03	47.3±2.56	47.3±2.11
12-18	101.4±6.68	88.6±6.16	111.0±5.10*	117.1±9.37*	106.2±4.90*
0-18	395.4±5.96	375.0±9.41	384.9±6.12	399.6±9.64	405.8±5.30*, **

Table 4 shows that the control group has the highest relative gain in 6 months (133.4%), as well as over the next 4 months - 42.5%. The difference in the period from 0-6 months between the first group with the rest of the experimental groups is 0.1-2.9%, and from 10-12 months - 0.3-2.3%. According to the relative gains at the age of 10-12 months, groups 1, 2, 4 and 5 have superiority over their peers of the 3rd group - 4.7% (P>0.999), 3.4% (P>0.99), 4% (P>0.999) and 3.2% (P>0.99) accordingly.

Table 4. Relative gains, %

Age, months	Group				
	1	2	3	4	5
0-6	133.5±1.46	133.3±2.76	130.5±±2.39	131.1±2.01	133.3±1.47
6-10	42.5±0.73	40.2±1.47	41.5± ±0.63	40.7±±0.73	42.2±0.75
10-12	16.6±0.54***	15.3±0.59**	11.9±0.73	15.9±±0.74***	15.1±0.59**
12-18	26.6±1.39	24.1±1.85	30.3±1.54*	30.9±2.27*	27.2±1.29
0-18	181.0±0.68	167.1±1.86	167.9±.12	168.4±1.01	168.8±0.85

At the age of 12-18 months, animals of the 3rd and 4th groups have the highest gain related to the first-calf heifers of the 2nd group - 6.2 and 6.8%, with a significant difference P>0.95. For the whole growing period (18 months), the control group has superiority over the experimental group (181%).

Many scientists believe that the highest milk production has well-developed, but not the largest animals. Each breed has its own standard. Table 5 shows body measurements of the black-and-white first-calf heifers.

Table 5. First-calf heifer body measurements, cm

		Group				
		1	2	3	4	5
		X̄±m				
Body measurements	WH	139.9± 0.21**, ***	137.5± 0.24***	136.1± 0.21	139.0± 0.24***	137.7± 0.27***
	CW	46.1± 0.29***	44.0± 0.24**, ***	43.0± 0.18***	41.5± 0.24	47.9± 1.18**, ***
	CD	68.7± 1.09	69.9± 0.19*, ***	69.1± 0.21**	68.0± 0.24	70.3± 1.19
	HG	191.6±0.84	191.0±0.24	192.8±0.28 *, **, ***	192.0±0.24 **	191.0±0.47
	BL	160.8± 0.28*	163.5± 0.24***	160.1± 0.21	161.5± 0.24***	160.3±1.96
	RW	51.8± 0.62**	50.0± 0.24	50.6± 0.37	51.5± 0.24*, ***	50.0± 1.25
	CC	19.1± 0.04**, ***	18.2± 0.12	18.7± 0.23	18.4± 0.17	18.6± 0.19

Note: WH - withers height, CW - chest width, CD - chest depth, HG - heart girth, BL - body length, RW - rump width, CC - cannon bone circumference.

According to the WH among the studied groups, the first control group prevails over the peers of the second, fourth and fifth groups, the difference is 2.4 cm (1.7%) with P>0.999, 0.9 cm (0.6%) with P>0.99 and 2.2 cm (1.6%) with P>0.999; the second group prevails over the first-calf heifers of the third group - 1.4 cm (1.01%) with a significant difference P>0.999; the fourth group prevails over the animals of the second, third and fifth groups - 1.5 cm (1.1%), 2.9 cm (2.1%) and 1.3 cm (1%) with P>0.999; as well as the fifth group over the cows of the third group - 1.6 cm (1.1%) with a significant difference P>0.999.

The largest CW has the animals of the first and fifth groups related to the second, third and fourth groups, the difference is 2.1 cm (4.6%), 3.1 cm (6.7%), 4.6 cm (10%) with P>0.999 and 3.9 cm (8.5%) with P>0.99, 4.9 cm (10.6%), 6.4 cm (13.9%) with P>0.999; the second group related to the third and fourth groups - 1 cm (2.2%) with P>0.99 and 2.5 cm (5.5%) with P>0.999; the third group related to the fourth group - 1.5 cm (3.3%) with P>0.999.

According to the chest depth, the difference between the second group with peers of the third and fourth groups is 0.8 cm (1.1%) with a significant difference P>0.95 and 1.9 cm (2.7%) with P>0.999, and first-calf heifers of the third group prevail the animals of the fourth group by 1.1 cm (1.6%) with a significant difference P>0.99. The highest HG indicator has the first-calf heifers of the third group related to the second, fourth and fifth groups - 1.8 cm (0.9%) with P>0.999, 0.8 cm (0.4%) with P>0.95 and 1.8 cm (0.9%) with a significant difference P>0.99; as well as the fourth group related to the second group - 1 cm (0.5%) with P>0.99. According to the BL the difference between the second group with the first, third and fourth groups is 2.7 cm (1.7%), 3.4 cm (3.8%) and 2 cm (1.3%) with P>0.999; fourth and third groups - 1.4 cm (0.8%) with P>0.999; the first and third groups - 0.7 cm (0.4%) with a significant difference P>0.95. According to the RW the animals of the first group prevail over their peers of the second group –1.8 cm (3.5%) with P>0.99; the fourth group over the second and third groups - 1.5 cm (2.9) with P0.999 and 0.9 cm (1.1%) with P>0.95. The largest CC have the heifers of the control group, the difference with the second, fourth and fifth groups is 0.9 cm (4.7%), 0.7 cm (3.7%) with P>0.999, and 0.5 cm (2.6 %) with P>0.99.

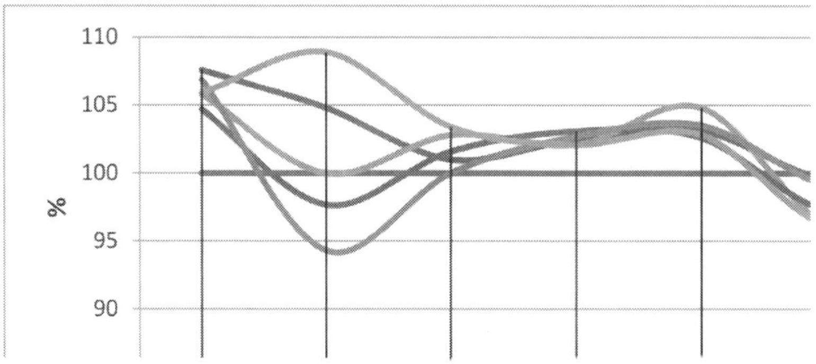

Fig. 1. Exterior profile of first-calf heifers

In addition to body build indices, body measurements are used to create exterior profiles. An exterior profile is a graphical representation of difference in measurements or body build indexes of an animal or a group of animals related to a standard or other model criteria. Fig. 1 shows the exterior profile of first-calf heifers with different combinations of selection in Plemzavod im. Lenin LLC related to the breed standard.

All cows are tall, the second group corresponds to the breed standard in chest width, while the first and fifth have wide chests, and the third and fourth groups have narrow ones. The animals have a well-developed chest, the fourth group corresponds to the standard, the bones are strong, they have a stretched constitution, the rump width does not correspond to the norm, the cannon bone circumference of the control group is close to the standard related to the to the experimental groups.

The absolute values of body measurements allow to compare the development of selected indicators of animals, but do not characterize the proportions of their body build. To judge about the type of animal constitution and the relative development of one or another, the body build indices are calculated, which characterize the ratio (in %) of anatomically related body measurements. When determining the body type of animals with various linear combinations, we calculated the following body build indices: leggy, stretchiness, pelvis-chest, chest, blockiness and boniness were calculated, as well as the coefficient of milk production, by which the direction of the animal productivity can be judged (Table 6).

Table 6. Body build indices and the coefficient of milk production of first-calf heifers, %

Indices	**Groups**				
	1	**2**	**3**	**4**	**5**
Leggy	50.9±0.82	49.1±±0.20	49.2±0.08	51.1±0.09	48.9±0.96
Stretchiness	114.9±0.28	118.9±0.03	117.6±0.03	116.2±0.03	116.5±1.65
Pelvis-chest	88.9±1.06	88.0±0.06	84.9±0.31	80.6±0.09	95.7±4.48
Chest	67.0±0.76	62.9±0.27	62.2±0.14	61.0±0.14	68.1±2.61
Blockiness	119.2±0.65	116.8±0.02	120.4±0.22	118.9±0.28	119.1±1.55
Boniness	13.7±0.02	13.3±0.08	13.7±0.15	13.2±0.10	13.5±0.16
Coefficient of milk production	1592±31	1610±52	1601±33	1648±29	1523±20

Table 6 shows that all groups of first-calf heifers, according to the leggy and stretchiness index, belong to the milk-meat direction of productivity, as well as according to the pelvic-chest index, however, 3rd group belongs to the meat direction of productivity, and 4th group - to the dairy direction. All groups belong to the milk production direction according to the downwardness and chest index, and to the meat one according to the boniness index.

4 Conclusions

Findings indicate that the highest daily average and absolute gains have the animals of the 5th experimental group (animals with the cross line affinity according to their paternal and maternal ancestors) (P>0.99), and according to the relative gains - the control group (pure lines).

It is proved that first-calf heifers of all studied groups are tall, the first and fifth group have wide chest, the second group corresponds to the breed standard, and the third and fourth groups have narrow chest. Each studied group of animals have well-developed chest depth, strong bones, and stretched constitution. The rump width and cannon bone circumference does not correspond to the norm.

It was identified that first-calf heifers in Plemzavod im. Lenin LLC have a high milk production coefficient, which indicates their milk type.

References

1. O. A. Basonov, V. V. Klipova, N. P. Shkilev, I. A. Elfimova, *The role of sire bulls in increasing the level of realization of the genetic potential of dairy herds* (2019)
2. O. A. Basonov, N. V. Vorobyova, M. E. Taigunov, S.S. Basonova, Zootechniya, **7**, 14-15 (2010)
3. I. V. Malyavko, V. A. Malyavko, Zootechniya, **5**, 15-17 (2016)
4. Yu. A. Svetova, T. A. Guseva, Zootechniya, **10**, 17-18 (2014)
5. T.T.Tarchokov, Z.M.Aisanov , S.F. Sukhanova, Mishhozhev, A.A., Balpanov, D.S., IOP Conference Series: Earth and Environmental Science, 341(1), 012047 (2019)
6. A.A.Mishkhozhev, M.G.Tleinsheva, Z.M.Aisanov, T.T.Tarchokov, Zootechniya, **11**, 24-27 (2017)
7. A.A. Zhilyaev, T.T.Tarchokov, I.G. Sudorgina, R.Z.Abdulkhalikov, M.G.Tleinsheva, D.S.Balpanov, A.A.Kogotyzhev, Zootechniya, **9**, 8-11 (2020)
8. T.T. Tarchokov, Zootechniya, **5**, 27-30 (1996)
9. M.O. Baytaev, Ts. B. Kagermazov, T. T.Tarchokov, Agrarnaya Russia, **1**, 28-30 (2013)
10. M. B. Ulimbashev, T. T. Tarchokov, Agrarian Science, **6**, 24-25 (2005)
11. T. T. Tarchokov. A. M. Tkhazeplov, J. Zootechniya, **12**, 5-7 (2004)
12. A. B. Baragunov, E. A. Kushaeva, M. M. Hamokov, A. V. Sagatelian, E3S Web of Conferences, International Conference on Modern Trends in Manufacturing Technologies and Equipment, ICMTMTE 2020, 193 (2020)
13. A. B. Baragunov, I. A. Savvateeva, S. H. Kushaev, A. A. Kumakhov, Z. R. Kudaev, 2nd International Scientific Conference on Agribusiness, Environmental Engineering and Biotechnologies, AGRITECH-II 2019, IOP Conference Series: Earth and Environmental Science, 3 (2019)

https://doi.org/10.1051/e3sconf/202126202017

14. T. T. Tarchokov, Z. M. Aisanov, S. F. Sukhanova, M. G. Tleinsheva, A. A. Mishhozhev, D. S. Balpanov, Russian Conference on Innovations in Agricultural and Rural Development 2019, AgroCON 2019, IOP Conference Series: Earth and Environmental Science, 1 (2019)

Intraspecific differentiation by exterior in horses of different breeds

Vladimir Demin[1,*], *Timur* Tarchokov[2], *Elena* Ryabova[1], *Inna* Tsyganok[1], and *Natalia* Kulmakova[1]

[1]Federal State Budgetary Educational Institution of Higher Vocational Education Russian State Agrarian University - Moscow Timiryazev Agricultural Academy 49 Timiryazevskaya str, 127550 Moscow, Russia
[2]Kabardino-Balkarian State Agricultural University named after V. M. Kokov, Lenin Avenue, 1V, 360030 Nalchik, Russia

Abstract. The analysis of the exterior features of horses of the roadster and heavy draft horse breeds by basic and supplementary measurements of body articles and conformation indexes was carried out in order to study the intraspecific differentiation. The results showed significant differences between the racing roadster breeds and heavy draft breeds. The racing roadster breeds were similar in most respects, but the Russian Saddle Horse had more differences compared to other roadster breeds. Soviet Heavy Draft mares were reliably superior to riding mares in most measurements and conformation indices, but inferior to them in terms of neck length. Our research has shown that the selection of highly specialized types of horses leads to a distinct differentiation between breeds of horses in the exterior structure. The resulting digital data should be used for the development of standards and regulations in the breeding of horses for different uses.

1 Introduction

The large number of horse breeds is due to the great variety of uses for them. As far back as antiquity, different horse types began to appear associated with the predominant use of horses in harness in chariots or for horseback riding. The treatises of the time described the selection of a horse suitable for particular tasks.

As the horse breeding and hippology developed, the science of exterior design, which became a separate discipline in the 18th century, underwent various formative stages, but at the same time included completely opposite views and currents that could be completely opposite: the search for the "ideal exterior" based on golden proportions or the proportions of famous horses.

The 19th century saw the intensification of agriculture and the emergence of highly specialized breeds of animals, including horses. New breeds and types of horses, while

* Corresponding author: demin@rgau-msha.ru

having high performance in certain areas, differ dramatically in size, physique, and points. The attempt to generalize the characteristics of different breeds and to describe ideal forms in connection with the specialization of breeds instead of finding just the ideal conformation was made in the 19th century, but the effort to reduce the variety of breeds to a few ideal types has been criticized and found to be untenable [13].

The next stage in the development of the doctrine of external conformation was the trend towards the exterior functionality, in which the unity of form and function is inextricably linked. For each type and use specific exterior requirements need to be developed. Nowadays, the breed shaping process in horse breeding continues, and breeds and types of horses that meet modern requirements are created and recreated. A large proportion of the combined use breeds – riding draft and working draft – which were previously used in agriculture and the army, have been reoriented towards sporting use and have acquired (transformed into) a riding horse type. There has been an increase in the specialization of horse breeds according to the use, so that the Russian Saddle breed has not just been re-bred as a sporting horse, but primarily as a horse for one type of equestrian sport – dressage [3, 13].

The specific requirements in the different uses of the horses lead to changes in the exterior and physique of the horses and reinforce the intra-breed differentiation in terms of exterior. In horses of different breeds, the biomechanics of movement, performance and other attributes are largely determined by conformation. This is characteristic of both cultured breeds and local horses [1, 2, 4, 5, 7-11, 14-18]. For example, a study of the measurements of the exterior of thoroughbred horses showed a mutual influence of both measurements and joint angles on each other. The use of linear measurements and angulation values will, according to the researchers, reveal the correctness of limb structure and performance potential of thoroughbred horses. This will contribute to the choice of horse for the sport with less risk of lameness [11]. In Iranian Turkoman horses, a medium to high heritability of measurements was found, which suggests the possibility of breeding for these traits [5]. Detailed measurements of horse skull have shown that more information is available with geometric morphometry compared to linear morphometry [12]. Measurements of joint angles, such as the hip joint, are an aid in preventing lameness in the horse, as it has been proven that there is a relationship between the angle of the hip joint and ligament damage [14]. It has been found that the treading and walking speed in Icelandic horses were most was the most influenced by height at the withers, height at hips, trunk length and forelimb length ($p<0.05$) [16]. A gene influence on the angular size of certain joints in Lippiciana and Franche-Montagne horses was found. To improve the assessment of horse exterior features, scientists have investigated the use of a spatial model of the horse shape. Significant relationships ($p<0.05$) with withers height, shoulder length and inclination, leg structure, stride length at a trot were found [6, 7]. A comparison of the aboriginal (primitive) horse breeds – Konik and Hutsul, based on 25 measurements and 24 body indices, showed that, despite the fact that both breeds are similar, Konyk and Hutsul horses have different measurement ratios. The authors attribute the differences between Hutsul and Konik horses to the breeding of Hutsul horses in the highlands. Konik horses were also compared and Polish half-bred and thoroughbred horses. It has been revealed that half-bred and thoroughbred breeds clearly differ in their exterior structure from Tarpan-like horses. The specific structure of the exterior of aboriginal Tuvinian horses makes it relatively easy to identify them visually. Conformation studies, as the authors point out, are important for establishing the characteristics of different horse populations [8-10].

Thus, in order to intensify the breeding work with the various horse breeds, it is necessary to have a clear and precise understanding of the requirements for exteriors and how these can be characterized in numerical terms.

2. Materials and Methods

The aim of our work is to study the exterior traits in horses of different breeds.

Objectives

1. Comparison of the basic and supplementary measurements of heavy draft and riding horses.
2. Analysis of primary and secondary conformation indexes in horses of different breeds.

The study is based on data on basic and additional measurements of Russian Saddle Horse mares of Starozhilov stud farm – 86 animals, 17 animals of Hanoverian breed of Elitar breeding farm, 13 animals of Budenny breed of Velikoknyazhessk stud farm, 54 animals of Soviet Heavy Draft breed of Perevozsk stud farm. The mares were measured between 2017 and 2020.

Four main and 18 additional body article measurements were taken, from which 31 conformation indexes were calculated.

The results were processed by conventional methods of statistical analysis

3 Results and Discussion

The selection of horses to increase height in different breeds occurs for different reasons. Horses used in classical equestrian sports tend to be higher at withers, because larger horses can jump over high obstacles more easily in show jumping and triathlon, and a larger horse looks better in dressage, which can give the judges some advantage when judging; with heavy draft breeds, a larger height is associated with greater live weight and therefore greater pulling power.

The mares of the studied breeds were very tall, with the largest of them being the Soviet Heavy Draft breed. Their average height at withers was 166.8 cm, which is reliably higher than the average height of the Russian Saddle breed at 165.3 cm. The average height at withers between mares of sporting riding breeds was not found to be a significant difference. As can be seen from Table 1, Soviet Heavy Draft mares were reliably superior to other mares in terms of basic measurements (with the exception of withers height compared to Budenny and Hanoverian mares). They differed considerably from riding mares in terms of physique, as shown by the indices calculated from the basic measurements: Soviet Heavy Draft mares had a more elongated body (by 7.5-9%), a higher value of chest girth index (by 15-16%), compactness (by 6%) and bone structure (by 2-3%) than riding horses; which had a format index close to 100% (square format), a lighter physique and less boniness.

Table 1. Basic measurements and body indices of producing mares of different breeds.

Breed	Q-ty	Body measurements (cm)				Conformation indexes (%)			
		height at the shoulder	Body length	Chest girth	metacarpal girth	Format	Chest girth	Blackness	Metacarpus girth
Russian Saddle Horse	86	165.3 ±0.34	166.0 ±0.49	194.9 ±0.68	20.2 ±0.07	100.4	117.91	117.41	12.22
Budyonny	13	165.3 ±1.3	167.1 ±1.6	196.2 ±2.0	21.0 ±0.4	101.1	118.69	117.41	12.70
Hanoverian	17	166.2 ±0.78	169.2 ±1.48	197.5 ±2.41	21.7 ±0.21	101.8	118.83	116.72	13.06
Soviet Heavy Draft	54	166.8 ± 0.51	182.3 ±1.0	224.2 ±1.7	25.5 ±0.1	109.3	134.4	122.98	15.3

On the average values of the main measurements between riding mares, there was no significant difference, except for the metacarpal girth, in which the Hanoverian mare was superior to the Russian Saddle mare (21.7 and 20.2 cm respectively). Overall the Hanoverian mares were somewhat larger than Budenny and Russian Saddle mares. Russian Saddle mares have a square format (100.4%), less massiveness and boniness compared with representatives of other sporting breeds; chest girth index reached 117.91% and metacarpus girth index was 12.22%.

The different breeds of riding horses differ not only in the physique, but also in the size of individual body articles. When describing the exterior of heavy draft horses, it is common to mention the large head. In our research, the average length of the heads of the Soviet Heavy Draft mares was greater than that of the representatives of riding breeds – 65.9 cm, but the reliability of the difference between the averages established only with Russian Saddle mares, being the smallest in the head length – 60.52 cm.

Russian Saddle breed mares have significantly shorter length of the head, as compared to Budenny and Hanoverian horses (60.52, 64.5 and 64.3 cm, respectively). This is indicated by the lower values of the big head index, which was 2-3% lower in Russian Saddle mares than in mares of other breeds. The ratio of the length of the head to the length of the body for Russian Saddle mares is also smaller than for Hanoverian and Budenny mares, which again shows that Russian Saddle horses are characterized by a small head. This is largely due to the fact that, in recreating the Russian Saddle breed, great care was taken to preserve the distinctive type of riding horse characteristic of the Orlov-Rostopchin horse (13). A small value of the ratio of head length to body length in Soviet Heavy Draft mares is due to their long body.

The length of the neck is important to the biomechanics of movement, as the neck and head act as a lever to change the position of the horse's centre of gravity, helping to change direction and pace. The long neck makes it easier to control the horse, especially during under-saddle work. Riding breed horses in our study had a neck length reliably 10 cm longer than the mares of the Soviet Heavy Draft breed. This is also evidenced by the larger ratio of neck length to withers height and body length in riding horses. Riding breed mares had no significant differences in neck length. However, the ratio of neck length to head length in Russian Saddle mares was the highest – 152.84%, which gives the impression of a very long neck, while the ratio of neck length to head length in Soviet Heavy Draft mares was lower – 120.49%, which emphasizes the short neck and makes the head visually larger.

Table 2. Measurements and indices that characterize the size of the head and neck in mares of different breeds.

Breed	Q-ty	Body measurements (cm)				Conformation indexes (%)				
		height at the shoulder	Body length	Head length	Neck length	Head length / height at the shoulder	Head length / Body length	Neck length / Height at withers	Neck length / Body length	Neck length / Head length
Russian Saddle Horse	86	165.3 ±0.34	166.0 ±0.49	60.52 ±0.27	92.5 ±1.04	36.61	36.46	55.96	55.72	152.84
Budyonny	13	165.3 ±1.3	167.1 ±1.6	64.5 ±1.12	93.3 ±1.22	39.02	38.6	56.44	55.83	144.65
Hanoverian	17	166.2 ±0.78	169.2 ±1.48	64.3 ±0.6	91.2 ±0.71	38.69	38.0	54.87	53.90	141.83
Soviet Heavy Draft	54	166.8 ± 0.51	182.3 ±1.0	65.9 ±0.6	79.4 ±1.01	39.51	36.15	47.6	43.55	120.49

The following table shows the average height at withers, back and hip and the indices that describe the topline of sport mares and Soviet Heavy Draft mares. Soviet Heavy Draft is not only superior to other breeds mares in height at the withers, but at the height at back and hip (significantly in comparison with the Russian Saddle mares).

Horses of sport riding breeds had greater height at withers than at hip, and heavy draft mares had a slight modification – their height at hip was slightly higher than the height at withers. The traditional view is that modification is not desirable for riding horses, as it increases the load on the forelegs, but modification is not a serious disadvantage for heavy draft horses working under saddle, so it is not a serious exterior disadvantage. The topline curve can be characterized by the ratio of height at withers and hip to height at the back, as can be seen from Table 3, in Soviet Heavy Draft horses compared to riding mares, this curve was more pronounced because of the greater height at hips.

Among the groups of horses examined, the highest withers were of the Russian Saddle breed, with a difference of more than 10 cm between the withers height and the back height, as evidenced by the size of the ratio between these measurements. The ratio of height at withers and hip to height at back shows the curve of the topline and indicates that the curve was more pronounced in Russian Saddle mares than in Hanoverian and Budenny mares. Russian Saddle mares had a lower hip height than other horses. In Budenny mares height at hip was approaching the height at withers, which is quite acceptable for the breed, used mainly in show jumping, but not desirable for dressage horses.

Table 3. Measurements and indices that characterize the topline in mares of different breeds.

Breed	Q-ty	Body measurements (cm)			Conformation indexes (%)		
		height at the shoulder	Height at back	height at hips	modification	height at withers / height at back	height at hip / height at back
Russian Saddle Horse	86	165.3 ± 0.34	154.94 ± 0.39	162.48 ± 0.33	98.29	106.68	104.87
Budyonny	13	165.3 ± 1.3	157.3 ± 0.92	164.3 ± 1.31	99.39	105.08	104.45
Hanoverian	17	166.2 ± 0.78	157.3 ± 0.61	164.0 ± 0.83	98.68	105.66	104.26
Soviet Heavy Draft	54	166.8 ± 0.51	158.5 ± 0.5	167.0 ± 0.5	100.12	105.24	105.36

The performance of heavy draft breeds correlates with the size and weight of the horse, which explains the greater massiveness and greater chest girth of heavy draft horses. In our study, Soviet Heavy Draft mares had not only a reliably larger chest girth compared to riding mares, but also a reliably larger chest width of 54.5 cm, which is 12.5-16.5 cm more than that of riding mares. The chest depth of Soviet Heavy Draft mares is significantly greater than that of ridden mares. The relatively greater mass and more developed rib cage of the Soviet Heavy Draft mare is indicated by the larger index values, which characterize this body point compared to riding mares. The chest index shows the development of the rib cage in width and depth, as shown in Table 4, the more rounded in cross section chest cage of the Soviet Heavy Draft mares – the index was 64.19%, and more flattened – 47.8% for the Budenny mares; Russian Saddle and Hanoverian mares occupy an intermediate position. More narrow and relatively flat chest was observed in Budenny mares, perhaps due to the fact that the horses of this breed are tested for performance in flat racing and such a structure creates less air resistance and provides advantage in the race.

Table 5 presents measurements and indices characterizing croup size in horses of different breeds. The highest croup length was observed in Soviet Heavy Draft mares, 61.2 cm, which is reliably longer than in mares of other breeds, as evidenced by the higher croup to withers ratio, compared to riding mares, but the croup length to body length ratio was about the same as in riding mares. Soviet Heavy Draft horses while generally more massive and broad-bodied, had a reliably larger croup width at the hook bone – 66.7 cm. In contrast to riding mares, heavy draft mares have a larger croup compared to the length – the ratio of these measurements is 108.99%, while riding mares' croup was relatively narrower – the value of this ratio was 89.68 - 96.53%.

Table 4. Measures and indices of chest development in mares of different breeds.

Breed	Q-ty	Body measurements (cm)			Conformation indexes (%)				
		Chest girth	Chest width	Chest depth	Massiveness	Blackness	Chest depth	Chest	Chest width / croup width At the hook bone
Russian Saddle Horse	86	194.9 ±0.68	41.9 ±0.26	80.8 ±0.29	117.91	117.41	48.88	51.85	76.74
Budyonny	13	196.2 ±2.0	38.0 ±0.74	79.5 ±0.89	118.69	117.41	48.09	47.80	71.97
Hanoverian	17	197.5 ±2.41	40.3 ±0.67	78.4 ±1.03	118.83	116.72	47.17	51.40	78.56
Soviet Heavy Draft	54	224.2 ±1.7	54.5 ±0.94	84.9 ±0.5	134.4	122.98	50.9	64.19	81.71

Russian Saddle breed mares have more developed croup in comparison with mares of other sporting riding breeds, their croup was reliably longer and wider (58 and 54.6 cm) than that of Budenny and reliably wider than that of Hanoverian mares. The relative croup size of Russian Saddler mares was also larger, at almost 35% of the body length, which suggests that Russian Saddle mares have a long croup. In addition, they have relatively wider croup than the Hanoverian breed mares. This should be considered a positive feature for the Russian Saddle breed, which specializes in dressage, a form of equestrian sport where the execution of most elements requires a significant transfer of the centre of gravity to the rear and load on the croup and hindquarters muscles.

Table 5. Measurements and indices of croup size for mares of different breeds.

Breed	Q-ty	Body measurements (cm)		Conformation indexes (%)		
		Croup length	Croup width at hook bones	Croup length / Height at withers	Croup length / Body length	Croup width at hook bones / Croup length
Russian Saddle Horse	86	58.0 ± 0.23	54.6 ± 0.29	35.09	34.94	94.14
Budyonny	13	54.7 ± 0.72	52.8 ± 0.62	33.09	32.73	96.53
Hanoverian	17	57.2 ± 0.51	51.3 ± 0.77	34.4	33.8	89.68
Soviet Heavy Draft	54	61.2 ± 0.62	66.7 ± 0.7	36.8	33.57	108.99

Structural features of the forelimbs of horses of different breeds are shown in Tables 6 and 7. Soviet Heavy Draft breed horses have shorter limbs, this is evidenced by the smaller leg height at the elbow compared with riding breed horses (significantly) and lower index of leg height. However, the extremities of the Soviet Heavy Draft mares are more bony, evidenced not only by the significantly larger metacarpus girth and the corresponding index (Table1), but also by significantly higher ratio of metacarpal girth to leg height at the elbow – 4-6% higher than in riding horses. The highest leg height at the elbow was found in Russian Saddle mares, but the difference between the averages was not significant.

Table 6. Measurements of the articles of the front legs of the mares of different breeds (cm).

Breed	Q-ty	Shoulder blade length	Shoulder length	Forearm length	Metacarpus length	Pastern length	Leg length at elbow
Russian Saddle Horse	86	56.5 ± 0.29	39.3 ± 0.24	48.7 ± 0.19	26.7 ± 0.16	13.2 ± 0.12	98.52 ± 0.19
Budyonny	13	58.1 ± 0.48	45.3 ± 0.52	48.1 ± 0.53	24.7 ± 0.21	13.7 ± 0.11	97.9 ± 0.38
Hanoverian	17	58.3 ± 0.72	41.3 ± 0.59	47.9 ± 0.76	23.9 ± 0.41	13.3 ± 0.19	96.5 ± 0.23
Soviet Heavy Draft	54	62.7 ± 0.64	48.1 ± 0.53	45.3 ± 0.45	30.3 ± 0.5	10.9 ± 0.24	94.8 ± 0.63

The longest shoulder blade and shoulder were observed in the Soviet Heavy Draft breed mares (significantly greater than in the riding horses), which is probably due not only to the larger size of the chest, but also to the fact that heavy draft horses have to develop more pulling power and need the large musculature in the shoulder blade and shoulder area.

The mares of the Russian Saddle breed had shorter shoulder blade and shoulder length, the shoulder blade length was reliably shorter than in the mares of other breeds, and the shoulder length was reliably shorter than in the mares of the Soviet Heavy Draft and Budenny breeds. Although the forelegs of dressage horses are not as loaded as those of show-jumping, triathlon or racing horses, a significantly shorter shoulder blade and shoulder length compared to other sports breeds is undesirable as it leads to reduced pushing force in the limbs.

Table 7. Indices characterizing the forelegs in mares of different breeds, %.

Breed	Q-ty	Leg height	Metacarpus girth / Leg length at elbow	Shoulder blade length / height at the shoulder	Shoulder blade length / Chest depth	Shoulder blade length / Leg length at elbow	Shoulder length / Shoulder blade length	Shoulder length / Forearm length	Metacarpus length / Forearm length	Pastern length / Metacarpus length	Pastern length / Leg length at elbow
Russian Saddle Horse	86	59.6	20.5	34.18	69.93	57.35	69.56	80.7	54.83	49.44	13.4
Budyonny	13	59.23	21.45	35.15	73.08	59.35	77.97	94.18	51.35	55.47	13.99
Hanoverian	17	58.06	22.49	35.08	74.36	60.41	70.84	86.22	49.9	55.6	13.78
Soviet Heavy Draft	54	56.83	26.9	37.59	73.85	66.14	76.7	106.18	66.89	35.97	11.5

Long forearm is considered a positive attribute for horses, as it provides a greater grasp of space when striding and trotting; this attribute is particularly important for sporting horses, whose quality of movement is assessed when competing [15]. Sporting horses have reliably longer forearms than that of Soviet Heavy Draft horses, and shorter metatarsals and pasterns. The indices showing the ratio of foreleg article lengths in horses also confirm that the forelimbs of heavy draft and riding horses have significant differences in the development of individual articles that are leveraged in the biomechanics of movement.

Table 8. Measurements and indices of the hind legs in mares of different breeds.

Breed	Q-ty	Body measurements (cm)				Conformation indexes (%)		
		Femoral length	Lower thigh length	Metatarsal length	Pastern length	Femoral length / Lower thigh length	Metatarsal length / Lower thigh length	Pastern length / Metatarsal length
Russian Saddle Horse	86	56.3 ± 0.32	56.2 ± 0.25	46.8 ± 0.21	13.8 ± 0.13	100.18	83.27	29.49
Budyonny	13	58.0 ± 0.27	56.5 ± 0.26	46.3 ± 0.32	13.8 ± 0.12	102.65	81.95	29.81
Hanoverian	17	57.6 ± 0.69	57.3 ± 0.78	44.1 ± 0.52	13.3 ± 0.17	100.52	76.96	30.16
Soviet Heavy Draft	43	59.2 ± 0.64	53.3 ± 0.71	38.7 ± 0.6	11.6 ± 0.2	111.07	72.61	29.97

When analyzing the structure of the hind legs of horses of different breeds, the same pattern as in the front legs is observed: a longer femur in Soviet Heavy Draft mares (the difference between the averages is only valid with Russian Saddle mares), shorter tibia and pastern. The ratios between the lengths of the hindquarters' articles confirm this trend; the exception is the ratio of pastern length to metatarsals length, which was close in value in mares of all breeds.

The results showed significant differences between the riding sporting breeds and heavy draft breeds. The riding sporting breeds were similar in most respects, but the Russian Saddle Horse had more differences compared to other riding breeds. Our research has shown that the selection of highly specialized types of horses leads to a distinct differentiation between breeds of horses in the exterior structure. The resulting digital data should be used for the development of standards and regulations in the breeding of horses for different uses. The significant differences between the breeds studied led to the following conclusions.

4 Conclusions

1. Soviet Heavy Draft mares were reliably superior to riding mares in most measurements and conformation indices, but inferior to them in terms of neck length.

2. Russian Saddle horses had significantly shorter head than the mares of other breeds, they were significantly lower in height at back, compared with other sporting breeds mares

3. Soviet Heavy Draft horses differed considerably from riding mares in forelimb structure, but had significantly longer shoulder blade, shoulder, metacarpus and hip lengths, but shorter leg heights at the elbow, forearm and pastern lengths.

4. The mares of the Russian Saddle breed had the shoulder blade and shoulder length reliably shorter than those of other breeds. They also had significantly smaller cannon metacarpus girth than Hanoverian breed horses.

5. In the structure of the hindquarters, there are fewer significant differences between Soviet Heavy Draft mares and riding mares: Soviet Heavy Draft mares have a longer femur, but shorter pastern.

References

1. V. A. Demin, I. B. Tsyganyuk, Zootechnics, **1**, 15-18 (2020)
2. V. A. Demin, I. B. Tsyganyuk, Agricultural Science, **11**, 19-20 (2014)
3. V. A. Demin, E. V. Ryabova, *Current Achievements and Current Issues in Horse Breeding*, 73-79 (2019)
4. V. A. Demin, E. V. Ryabova, Zootechnics, **12**, 16-17 (2014)
5. H. Ghezelsoflou, P. Hamidi, S. Gharahveysi, Journal of Equine Science, **29(4)**, 91-96
6. A. I. Gmel, T. Druml, R. von Niederhäusern, T. Leeb, M. Neuditschko, Genes, **10**, 370 (2019)
7. AI. Gmel, T. Druml, K. Portele; R. von Niederhäusern, M. Neuditschko, PLoS ONE, **13(8)**
8. M. Komosa, H. Frąckowiak, H. Purzyc, M. Wojnowska, A. Gramacki, J. Gramacki, Journal of Animal Science, **91**, 1660-1668 (2013)
9. M. Komosa, H. Purzyc, Journal of Animal Science, **87**, **7**, 2245–2254 (2009)
10. B.M. Mongush, A. M. Zaitsev, M. M. Atroschenko, Yu. A. Yuldashbaev, V. A. Demin, Agricultural Science, **4**, 40-43 (2020)
11. M. Mostafa, N. Senna, A. Abu-Seida, Y. Elemmawy, Journal of the Hellenic Veterinary Medical Society, **70(2)**, 1533-1540 (2019)
12. P. M. Parés-Casanova, A. Salamanca-Carreño, R. A. Crosby-Granados, J. A. Bentez-Molano, Animals, **10**, 118 (2020)
13. V. A. Parfenov, *Historical sketch of the formation and transformation of the Russian Saddle breed horses. State Studbook of Russian Saddle Breed Horses* (2000)
14. J. Routh, C. Strang, S. Gilligan, and Dyson, Equine Vet Educ, **32**, 183-192 (2020)
15. E. V. Ryabova, Science Alley, **5 (21)** (2018)
16. G. J. Stefánsdóttir, A. Jansson, S. Ragnarsson, V. Gunnarsson, Comparative Exercise Physiology, **17(2)**, 151-160 (2020)
17. I. B. Tsyganyuk E. A. Yatsenko, TAA Reports, 347-351 (2019)
18. I. B. Tsyganyuk, Hippology and Veterinary, **2(20)**, 36-40 (2016)

Exterior features and productivity of the Kazakh fine-wool breed of sheep and its crossbreeds with meat breeds

Narzhan Zhumadillayev[1], *Yusupzhan* Yuldashbaev[2,*], *Amanbai* Karynbaev[3], *Akmal* Khudaiberdiev[2], and *Beslan* Efendiev[4]

[1]LLP "Kazakh Research Institute of Animal Husbandry and Forage Production", branch of the "Research Institute of Sheep Farming named after K. U. Medeubekov", Zhibek Zholy 15, 040622 Mynbayevo, Kazakhstan
[2]Russian State Agrarian University of the Moscow Agricultural Academy named after K. A. Timiryazev, 49, Timiryazevskaya St., 127550 Moscow, Russia
[3]Non-profit Joint Stock Company "National Agrarian Scientific and Educational Center", Kurgaldzhinskoe highway 4a, 010000 Nur-Sultan, Republic of Kazakhstan
[4]Kabardino-Balkarian State Agricultural University named after V.I. V.M. Kokov, Lenin Avenue, 1V, 360030 Nalchik, Russia

Abstract: The article presents the dynamics of changes in live weight for a year, measurements, body build indices and productivity of one-year-old pure-bred Kazakh fine-wool female lambs and crossbreeds of different bloods with German meat merino, when creating a fine-wool meat sheep breed in the south-east of Kazakhstan - "Etti merino". The resulting crossbred offspring are characterized as large animals, with a high live weight and good meat forms.

1 Introduction

Sheep breeding is a traditional, economically important, but also socially significant industry for Kazakhstan. Earlier, the economy of sheep breeding in Kazakhstan was based on wool production, and now there is an increased demand for mutton, especially lamb and young mutton, as a result of which it has become in demand in many countries of the world. It is known that lamb contains 2.5-4.3 times less cholesterol than beef and pork, and due to the fact that animals are kept on pasture all year round, sheep meat corresponds to environmentally friendly products [1, 2, 3] ...

Currently, the most high-quality and competitive meat products can be provided by specialized meat and meat-wool breeds, which are distinguished by high meat productivity, are distinguished by such biological characteristics as early maturity, intensive growth, economical transformation of feed into products [4, 5, 6,7,8,9 ,10].

In this direction, an urgent problem is the creation of domestic meat breeds of sheep, by crossing local queens with rams-producers of specialized meat breeds. So in Kazakhstan in 2001 in the RESEARCH INSTITUTE OF SHEEP-BREEDINGnamed K.U. Medeubekov, a

* Corresponding author: zoo@rgau-msha.ru

branch of Kazakh Research Institute of Animal Husbandry and Fodder Production LLP, work has begun on the creation of a highly productive meat fine-wool breed "etti merino"(meat merino) for export. The first step on this path was the use of sheep-breeding merinofleischschaf (MFS) of German meat merino with an array of queens of the Kazakh fine-wool breed (KFW).

Research on crossing of Kazakh fine-wooled queens with rams of German meat merino in SPK "Plemzavod Almaty" Talgar region and PK named after Eskeldy (now KH "Sariev S.M.) Eskeldinsky district of Almaty region indicates that the live weight of the offspring increases by 10-15.0%, the fertility of queens increases by 15-20%. When slaughtering rams at 4^x months of age, the carcasses have a mass of 19.0-20.0 kg. The average daily gain of lambs during the suckling period is 290-300 grams.

Purpose of work. Creation in the southeast of Kazakhstan of a highly productive meat fine-wool breed of sheep - "Etti Merino", which combines high meat productivity and early maturity, as well as the ability to produce high-quality merino wool at the level of world analogues. Due to the high economic efficiency, the sheep of the new breed are in great demand in the farms of the Republic.

2 Methodology

The first stage of work is the creation in the 80s of the last century of the Sarybulak intra-breed type of Kazakh fine-wool sheep, large fine-wool sheep with a high live weight, wool of 60-64 quality and a high shear, obtained as a result of crossing of Kazakh fine-wool queens with rams of the Polvars breed. At the second stage, the selection work was aimed at increasing muscle mass, slaughter yield, the severity of meat forms and indicators of early maturity. For this purpose, in 2002, rams-producers of German meat merino were used on the uterus of the Sarybulak intra-breed type.

3 Results

As a result of the crosses, the live weight and its growth were studied in crossbred female lambs MFS of various types of blood and purebred KFW, from birth to one year of age (Table. 1)

From the data in the table it can be seen that from birth to beating the greatest average daily gains in live weight in female lamb ¾ bloods according to merinofleischschaf and from breeding "inter se" ½ MFS KFW (288 and 276 grams, respectively) or more by 23-51 and 11-39 grams in comparison with other peers. In the period from the beating from mothers to the age of one year, the difference in the increase in live weight is somewhat preserved in half-bred animals (the average daily gain is 7-10 grams more) and in general, from birth to one year of age, the difference is within 16-15 grams.

From the data in the table it can be seen that from birth to beating, the greatest average daily gains in live weight in female lamb ¾ bloods according to merinofleischschaf and from breeding "inter se" ½ MFS KFW (288 and 276 grams, respectively) or more by 23-51 and 11-39 grams in comparison with others. In the period from the beating from mothers to the age of one year, the difference in the increase in live weight is somewhat preserved in half-bred animals (the average daily gain is 7-10 grams more) and in general, from birth to one year of age, the difference is within 16-15 grams.

Table 1. Live weight and its growth in one-year-old female lambs of the Kazakh fine-wool breed and crossbreeds of different proportions of blood.

Share of blood	Qty (head)	Live weight (kg)			Live weight gain					
					suckling period		from 4 months up to a year		from birth up to a year	
		at birth	4 months	1 year	total (kg)	middle daily (g)	total (kg)	middle daily (g)	total (kg)	middle daily (g)
¾ MFSKFW	71	4.3 ± 0.08	34.6± 0.41	51.4± 0.45	30.3	288	16.8	64	47.1	129
½ MFSKFW "inter se"	86	4.1± 0.16	33.2± 0.47	50.9± 0.41	29.1	276	17.7	68	46.8	128
½ MFSKFW	104	4.0± 0.07	31.9± 0.45	50.6± 0.48	27.9	265	18.7	71	46.6	127
¼ MFSKFW	96	4.0 ± 0.11	29.9± 0.42	45.1± 0.52	25.9	247	15.2	58	41.1	113
KFW	93	3.8± 0.11	28.7± 0.43	44.8± 0.38	24.9	237	16.1	61	41.0	112

The main indicator of the growth and development of animals, which determines the meat qualities of young animals, is its physique. The exterior is one of the main indicators of the animal's adaptability to certain conditions.

In this regard, exterior features were studied in the experimental young animals, for which 9 body measurements were taken from 20 female lambs of each group (Table. 2)

Table 2. Measurements of one-year-old female lambs of the Kazakh fine-wool breed and hybrids of different proportions of blood.

Indicators	Groups				
	¾ MFSKFW	½ MFSKFW "inter se"	½ MFSKFW	¼ MFSKFW	KFW
Quantity, head	20	20	20	20	20
Height at the shoulder	65.2±0.35	64.4±0.20	60.9±0.34	59.9±0.31	56.3±0.26
Height at hips	66.7±0.40	64.5±0.33	61.3±0.47	60.9±0.32	56.3±0.19
Diagonal body length	67.6±0.40	66.2±0.41	62.1±0.36	60.8±0.51	56.2±0.41
Chest depth	36.3±0.24	36.2±0.28	35.8±0.28	34.8±0.21	31.7±0.31
Chest width	32.2±0.22	31.5±0.31	31.0±0.20	29.4±0.31	26.7±0.30
Chest girth	135.0±0.9	133.1±0.8	124.0±0.8	121.9±0.8	113.6±0.6
Width in bricks	20.0±0.35	19.5±0.15	19.0±0.37	18.7±0.16	17.6±0.23
Half grip of the rear	68.5±0.49	67.5±0.31	64.0±0.48	62.2±0.67	58.4±0.44
Metacarpal girth	10.1±0.05	10.1±0.06	10.0±0.07	9.7±0.03	9.5±0.08

In terms of height at the withers, the highest indicators were found in the female lambs of genotypes ¾, ½ "inter se", ½ MFSKFW. At the same time, higher indicators at the withers were observed in hybrids of ¾ blood, which exceeded their contemporaries - by 0.8; 4.3; 5.3; 8.9 cm or 1.3; 6.5; 8.1; 13.6%. The same pattern was found for the height in the sacrum.

The oblique body length is of great importance in the development of the meat content of animals. The measurements of the oblique body length were somewhat larger in crossbred female lambs ¾, ½ "inter se", ½, ¼ blood (67.6; 66.2; 62.1; 59.9) than in purebred fine-wool female lambs (56.3) ... Measurements that characterize the volumetric indicators of crossbred groups female lambs have been well developed: depth, width and chest girth. In particular, in terms of chest depth of female lamb ¾, ½ "inter se", ½, ¼ blood levels have higher absolute indices of this feature than in female lamb and exceed them, respectively, by 4.6; 4.5; 4.1; 3.1 or 14.5, 14.1, 12.9, 9.7%. In terms of measurements of the width of the chest, the superiority of crossbred female lamb is ¾, ½ "inter se", ½ - 31.9; 31.5; 31.0. Largest chest girth possessed female lambs ¾, ½ "inter se".

For evaluating animals for the development of meat content, measurements are of great importance - the width in the hips and the half-girth of the rear. So, in width in hips, the greatest indicators of this trait were noted in female lambs, ½ "inter se", ½, ¼ bloods, predominance over purebred peers, respectively, by 13.6; 10.8; 7.9; 6.3%. The superiority in terms of half-girth of the hind is also in crossbreed female lambs 68.0; 67.5; 65.5; 61.9 against 59.4 in female lambs of Kazakh fine-wool sheep.

Body measurements do not give a complete picture of the constitution of the animal as a whole. In this regard, to characterize the exterior, we used body build indices (Tab. 3), by which we judged the degree of development of the organism, the proportions of the physique and the general constitutional type of the animal.

Table 3. Body build indices in one-year-old female lambs of the Kazakh fine-wool breeds and crossbreeds of different proportions of blood.

Indicators	**Groups**				
	¾ MFSKFW	**½ MFSKFW "inter se"**	**½ MFSKFW**	**¼ MFS KFW**	**KFW**
Leggy	44.3	43.7	41.2	41.9	43.7
Stretchiness	103.7	102.7	101.9	101.5	99.8
Hip-chest	161.0	161.5	163.2	157.2	159.6
Chest	88.7	87.0	86.5	84.5	84.2
Blockiness	199.7	201.1	199.6	200.5	201.6
Overgrow	102.3	100.2	100.7	101.7	100.0
Boniness	15.4	15.7	16.4	16.1	16.8
Massiveness	207.1	206.6	203.3	203.5	201.8
Meat	105.1	104.8	105.1	103.8	103.7

The index of elongation characterized the crossbred female lambs as long-bodied animals with an elongated format - ¾ - 103.7, ½ "inter se" - 102.7, ½ - 101.9, ¼ - 101.5 versus KT-99.8.

The indices of the pelvic-thoracic and thoracic, characteristic the volumetric indicators, were higher in the female lambs received from German producers and their sons. The chest index in female lamb MFSKFW ¾, ½ "inter se", ½, ¼ bloods, exceeded their KFW contemporaries, respectively, by 4.5; 2.8; 2.3; 0.3 or 5.3; 3.3; 2.7; 0.4%. The indices of massiveness and meatiness characterize the development of the trunk in width According to these indices, the female lambs MFSKFW surpass the KFW purebred female lambs and characterize them as wider-bodied, with developed meat forms.

Along with the indicators of growth and development, in order to obtain a complete characterization of the offspring of purebred and crossbred female lambs breeds, we analyzed productivity data at one year of age (Table. 4)

In one-year-old female lamb, the highest indicators of live weight have three-quarter-blooded and half-blooded ones from breeding "inter se" (respectively, 51.38 and 50.93 kg), which exceeds pure-bred Kazakh fine-wool female lambs by 6.58-6.13 kg or 14, 7-13.7 per cent. However, for quarter-blooded MFSKFW, no noticeable differences are observed for this feature.

In terms of the length of the wool fiber, the most "long-haired" were the crossbreeds of ½ MFSKFW from breeding "inter se" (by 0.56 cm or 6.02%). There are no significant differences between the rest of the groups according to the indicated characteristics.

A higher cut of wool is characteristic of the female lambs obtained from breeding "inter se" half-blooded MFSKFW hybrids, which surpassed the rest of the groups in this indicator by 0.3-0.17 kg. In addition, among them there are more animals of the elite and class I (by 15-4%) in comparison with purebred Kazakh fine-wool sheep and other crosses.

Table 4 Productivity of one-year-old female lambs of different proportions of blood.

Breed bloodiness	n	Live weight, kg	Cut wool, kg	Wool length, cm	Output el + I class. %
¾ MFSKFW	56	51.38 ± 0.45	5.06± 0.07	9.36 ± 0.10	72
½ MFSKFW "inter se"	50	50.93 ±0.41	5.23 ±0.07	9.86 ±0.10	76
½ MFSKFW	75	50.65 ±0.48	4.98 ±0.05	9.52 ± 0.09	68
¼ MFSKFW	46	45.13 ±0.52	4.88 ±0.06	9.36± 0.10	64
KFW	120	44.80 ±0.33	4.96 ±0.03	9.30 ± 0.06	61

4 Conclusions

Thus, German meat merinos have a positive effect on the indicators of meat productivity of the Kazakh fine-wool breed: they increase muscle mass, slaughter yield, the severity of meat forms and indicators of early maturity, which indicates the possibility of keeping them in the conditions of deserts and semi-deserts of Kazakhstan and are distinguished by high meat productivity.

References

1. A. I. Erokhin, E. A. Karasev, Yu. A. Yuldashbaev, Izvestiya TSKHA, **2**, 126-135 (2012)
2. A. I. Erokhin, T. A. Magomadov, E. A. Karasev, *Formation of meat content in sheep in postnatal ontogenesis* (2010)
3. M. E. Karabaeva, Diss. Dr. Biol. Sciences (2016)
4. V. I. Kosilov, E. A. Nikonova, B. B. Traisov, Sheep, goats, woolen business, **3**, 25-26 (2018)
5. Aboneev V.V., Skorykh L.N., *Strategy and main directions of development of sheep and goat breeding in Russia*, 45-48 (2002)
6. A. V. Derevyankin, Author's abstract. Cand. diss. (2004)
7. O.K.Gogaev, K.E. Kessaev, B.S.Kaloev, M.E. Kebekov, T.T.Tarchokov, Asian Journal of Microbiology, Biotechnology and Environmental Sciences **18(4)**, 1029-1038 (2016)
8. V V Aboneev, T .T .Tarchokov, S. F. Sukhanova, D. V. Aboneev, E. V. Aboneev and V. V. Marchenko, Magazine "IOP: Earth and Environmental Science», 341(1) (2019)
9. I. F. Gorlov, N. V. Shirokova, E. Y. Anisimova, M. I. Slozhenkina, Y. A. Kolosov, A. K. Natyrov, A.Y. Kolosov, N. I. Mosolova, M. A.Kolosova, T. T. Tarchokov, Journal of Applied Animal Research, **49(1)** (2021)
10. M. B. Ulimbashev, V. V. Kulintsev, M. I. Selionova, R. A. Ulimbasheva, B. T. Abilov, Z. T. Alagirova , South of Russia: Ecology, Development, **13** (2018)

Frequency distribution of polymorphic variants of the beta-casein locus in Russian various breed cattle subpopulations

Natalia Kovalyuk[1,*], *Elena* Shiryaeva[1], *Lyudmila* Yakusheva[1], *Vladimir* Kaishev[2], and *Zaurbek* Aisanov[3]

[1]Federal State Funded Research Institution "Krasnodar Research Center for Animal Husbandry and Veterinary Medicine", Pervomayskaya Str., 4, 350055 Krasnodar, Russia
[2]Federal State Budgetary Educational Institution of Higher Education Southern Federal University, Sadovaya Str., 105/42 B., 344006 Rostov-on-Don, Russia
[3]Kabardino-Balkarian State Agricultural University named after V. M. Kokov, Lenin Avenue, 1V, 360030 Nalchik, Russia

Abstract. Using the author's test system, 2607 Holstein, Ayrshire and Jersey breeds were genotyped at CSN2 locus. It was determined that the frequency of preferred genotype A2A2 is in Holstein, Ayrshire and Jersey groups of animals - 0.21; 0.25 and 0.60 respectively. The frequency of A2 allele in Holstein breed is 0.48; in Ayrshire - 0.51; in Jersey - 0.78. Thus, the Russian subpopulations of Ayrshire, Jersey, and Holstein cattle subpopulations according to frequencies of A2A2 genotypes (CSN2 locus), the Jersey breed has the best indicators (0.60). Russian breeding enterprises selling semen production of Ayrshire and Holstein bulls should increase the share of breeders with CSN2 genotypes A2A2.

1 Introduction

For β-casein gene (*CSN2*), 12 polymorphic variants were described. The A1 and A2 alleles are known to be the most frequent genetic variants of the CSN2 casein protein [1,2]. It was concluded that the mutation causing differences in the casein protein is considered the result of a single nucleotide polymorphism (C/A) in codon 67 in exon 7 of the beta-casein gene, that is the CCT codon (A2, proline) is changed to CAT (A1, histidine)). Many researchers [3,4], [4-7] believe that the bioactive peptide β-casomorphin 7, derived from casein A1 variants, can cause coronary heart disease in humans, atherosclerosis, type I diabetes mellitus, sudden infant death and autism syndromes.

It is believed that this mutation arose about 5000 years ago and spread widely in various cattle breeds. It was detected [4] that in such dairy cattle breeds as Guernsey, Schwyz, Jersey, Holstein, Ayrshire and Red Danish the frequency of the A1 allele ranges within 4-2%, 34-30%, 50-37%, 56-47%, 60-51% and 77%, respectively.

* Corresponding author: skniig@yandex.ru

In light of the ever-increasing requirements for dairy products quality, it is important to determine the potential of Russian subpopulations of Holstein, Ayrshire and Jersey breeds in terms of beta-casein locus selection.

2 Materials and Methods

Studies on the frequency of genotypes and alleles, tested at the beta-casein locus of cows were based on Federal State Funded Research Institution "Krasnodar Research Center for Animal Husbandry and Veterinary Medicine" and NPO Yug-Plem LLC in the laboratories of biotechnology and molecular genetic expertise. 2607 animals of the following cattle breeds were analyzed: Holstein (n = 1081) (Agroobedinenie Kuban JSC), Ayrshire n = 917 (Ilchenko Yu.V. Peasant Farm, A.I. Ezubov Peasant Farm) and Jersey n = 609 (Agroalliance Invest LLC).

United States Patent 7094949 (McLachlan CN (2006) Breeding and milking cows for milk free of ß-casein A1) describes a technology for animal genotyping at beta-casein locus (CSN2) to identify cows, carrying the A2 genotype. Studies by M. Miluchova, M. Gabor, A. Trakovicka (2013) and others [8] used the PCR/RFLP technology described in the patent. The studies detected an unsuccessful structure of one of the primers, characterized by a length of 33 base pairs and forming hairpins with quite a high probability, which is the main feature of this test system. Therefore, we have proposed our own genotyping system.

Diatom™ DNA Prep 100 systems (Isogen Laboratory LLC, Moscow) were used for DNA isolation from blood. To set up a PCR reaction, we used Gene Pak PCR Core systems (Laboratory Izogen, Moscow), primer selection was performed using the Primer Premier program, and amplification was performed empirically.

To create a restriction site in the required SNP (T → G) area, it was necessary to introduce a nucleotide substitution into one of the primers. The following primers were used:

5´AGG GAT GTT TTG TGG GAG GCT CTT3´

5´ATA AAA TCC ACC CCT TTG CCC AGA 3´

As a result, the fragments, amplified from the region of the ß-casein variant A2 gene, were cleaved by BstDEI endonuclease (No. E227, NPO SibEnzyme) into 2 fragments: 64 and 22 bp. The fragment amplified from the A1 allele had no restriction site (its size was 86 bp). Restriction results were evaluated on a 2.5% agarose gel (Fig. 1).

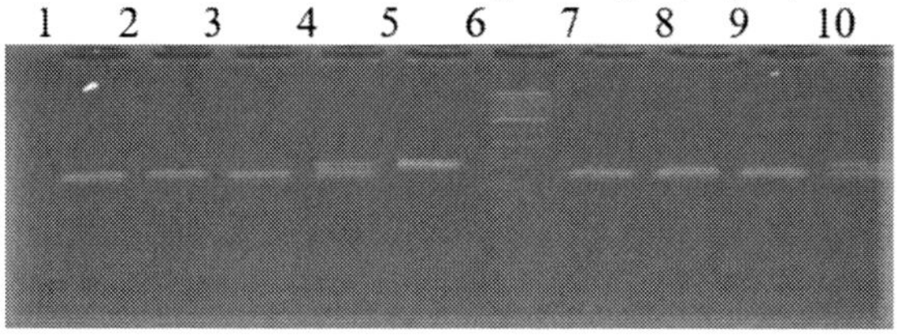

Fig. 1. Electropherogram of hydrolysis products of the CSN2 gene region amplificates by the BstDEI endonuclease restriction 1,2,3,7,8,9 - A2A2 genotype; 4, 10 - A1A2 genotype; 5 - A1A1 genotype; 6 - molecular weight marker;

3 Results and Discussion

The actual frequency of individual genotypes was determined by the ratio of the number of carriers of a particular genotype to the total number of individuals in the analyzed group. In this case, the allele frequencies were calculated using the following formula: $P(A) = (2N1 + N2)/2n$, where N1 is the number of homozygotes at the considered allele, N2 is the number of heterozygotes, n is the sample size.

The frequency distribution of genotypes and alleles, tested at the beta-casein locus of various breed cows is shown in Table 1.

Table 1. Frequency distribution of genotypes and alleles in dairy breed cows.

Genotype/allele	Frequency of CSN2 genotypes and alleles in		
	Holstein breed, (n = 1081)	Ayrshire breed, (n = 917)	Jersey breed, (n = 609)
A1A1	0.26	0.24	0.04
A1A2	0.53	0.51	0.36
A2A2	0.21	0.25	0.60
A1	0.52	0.49	0.22
A2	0.48	0.51	0.78

It was determined that the frequency of preferred genotype A2A2 is in Holstein, Ayrshire and Jersey groups of animals - 0.21; 0.25 and 0.60 respectively. The frequency of A2 allele in Holstein breed is 0.48; in Ayrshire - 0.51; in Jersey - 0.78. The obtained results are, in general, consistent with those obtained by other researchers.

3 Conclusions

Thus, the Russian subpopulations of Ayrshire, Jersey, and Holstein cattle subpopulations according to frequencies of A2A2 genotypes (CSN2 locus), the Jersey breed has the best indicators (0.60). We consider it expedient for domestic breeding enterprises selling semen production of Ayrshire and Holstein bulls to increase the share of breeders with CSN2 genotypes A2A2 (CSN2 locus).

References

1. H. M. Farrell, R. Jimenez-Flores, G. T. Bleck, E. M. Brown, J. E. Butler, L. K Creamer, H. E. Swaisgood, *Nomenclature of the proteins of cows' milk* (2004)
2. S. Kaminski, A. Cieslinska, E. Kostyra, Polymorphism of bovine -casein and its potential effect on human health, **48**, 189-198 (2007)
3. A. Cieslinska, E. Kostyra, H. Kostyra, K. Olenski, E. Fiedorowicz, S. Kaminski, Int. J. Food Sci. Nutr., **63**, 426-430 (2012)
4. A. Parashar, R. K Saini, Int. J. of Bioassays, **4(12)**, 4611-4619 (2015)
5. A. Cieslinska, E. Sienkiewicz-Szłapka, J. Wasilewska, E. Fiedorowicz, B. Chwała, M. Moszynska-Dumara, E. Kostyra, Influence of candidate polymorphisms on the dipeptidyl peptidase IV and opioid receptor genes expression in aspect of the ß - casomorphin-7 modulation functions in autism Peptides, 65, 6-11 (2015)
6. V. Petrat-Melin, P. Andersen, J. T. Rasmussen, N. A. Poulsen, L. V. Larsen, J. F. Young, J. Dairy Sci., **98**, 15-26 (2015)
7. K L. Reichelt, Bioengineer D. Tveiten, A. M. Knivsberg, G. Brønstad, Microb. Ecol. Health D, **23**, 18958 (2012)
8. M. Miluchova, M. Gabor, A. Trakovicka, ABP, **60(4)**, 799-801 (2013)

Use of natural bentonite clays in pond fish farming to create optimal concentrations of ammonium ions

Timur Tlupov[1,*], *Karina* Bambetova[2], *Kamaludin* Magomedov[1], *Ruslan* Kumykov[1], and *Nikolay* Khalko[3]

[1]Kabardino-Balkarian State Agricultural University named after V. M. Kokov, Lenin Avenue, 1V, 360030 Nalchik, Russia
[2]Kabardino-Balkarian State University named after Kh.M. Berbekov, Chernyshevskogo St., 173, 360004 Nalchik, Russia
[3]Grodno State Agrarian University, Tereshkova St., 28, 230008 Grodno, Republic of Belarus

Abstract. The composition and some sorption properties of bentonite clays, the efficiency of natural waters treatment in pond fish farming from excess ammonium ions using the natural bentonite clay from the Gerpegezhskoye deposit have been studied. The maximum sorption capacity of bentonite clay of this material has been determined. It is found out that when a smaller amount of the sample is added to each liter of the standard solution (0.1 and 0.01 g), higher values of the sorption capacity are identified. To achieve the optimal degree of cleaning of ponds from excess ammonium ions and expediency in terms of reducing the material and labor costs to a minimum, the optimal amount of bentonite was calculated, the use of which would be sufficient to maintain the ammonium ion concentration. The conditions were chosen so that to ensure a decrease in the ammonium nitrogen concentration from 2-3 maximum allowable concentration (MAC) (5-7.5 mg/l) to 0.2-0.8 mg/l. It is shown that the sorption capacity is realized more efficiently when the bentonite weigh decreases to 0.01 g/L, provided that the ammonium ion content in the aqueous medium exceeds the MAC by 2-3 times.

1 Introduction

Today, zeolites, glauconites and clay rocks are mainly used in pond fish farming to create an ecologically clean habitat [1-5]. Despite the growing market needs and the availability of wide resource base in Russia [6-8], the available range of natural compounds, the sorbents used in pond fish farming, is limited [6,7]. About 75% of the Russian pond fish farmers use natural sorbents developed over 40 years ago. The development of new effective formulations with the use of affordable and cheap natural bentonites remains relevant. In this regard, to study Gerpegezhskoye deposit of bentonite clays as sorbents (taking into account the sufficient cheapness and high sorption capacity) in pond fish farming for

* Corresponding author: timyrtlypov@mail.ru

creation of the ecologically clean habitat is, in our opinion, of significant practical and scientific interest.

Their use is due to sufficiently high sorption capacity [9], ion exchange properties [9-10], availability and low cost [11]. The most common sorbents are clay minerals that include bentonite clays. Bentonites are included in the group of layered band aluminosilicates with expanding structure, which determines their high sorption capacity and swelling capacity [9-11]. The Gerpegezhskoye deposit bentonite clay content SiO_2 - 62%, Al_2O_3 – 13.5%, CaO – 7.8%, MgO – 2.36%, Fe_2O_3 – 5.3%, K_2O – 2.5%, Na_2O – 1.3%. Bentonite clays include a large set of macro- and micronutrients being vital for the organism - Cu, Zn, Mn, Co, Ag, Ca, Mg, Cr, J, etc. The authors [12] found that bentonite clays showed increase in the sorption exchange capacity in the acid - salt - soda activation series. It is noted that soda activation increases the maximum value of the sorption exchange capacity by almost 3 times in comparison with the native (in the natural state) clay. The used samples were not subjected to acidic, saline and soda activation and were used in their natural state.

For the reared fish production, the high productivity is constrained by a number of important stress factors: the accumulation of gas - ammonia (NH_3) which is formed during biochemical decomposition of nitrogen-containing organic compounds when dissolved in water [13, 14]; oxygen deficiency, that has a detrimental effect on the vital parameters of fish, can cause developmental disorders and lead to death [15, 16]; pH of the medium - neutral pH is the most favorable for fish. With significant shifts in pH by acidic or alkaline parameter, the fish respiration rate decreases. The deterioration of the toxicological situation occurs due to the accumulation of heavy metals, pesticides, oil products in the surface waters and due to acid rains [17, 18]. We have considered the comprehensive effect of all these factors on the state of hydrobionts and fish capacity. However, in this work, we give only the results of studies on the pond water treatment for fish farming from ammonium ions with the use of bentonites with their sorption capacity, and we assess the effectiveness of their effect on the hydrochemical and toxicological situation.

2 Materials and Methods

To determine the maximum capacity, the bentonite weigh fraction 0.25-0.5 mm was put into the 25 ml conical flask with 5 ml of 0.25 M ammonium chloride solution. The sample was saturated with ammonium for 6, 8 and 10 hours and subject to mechanical stirring. Then the solid phase was separated by filtration and aliquot (0.5 ml) was taken to determine the ammonium ion content with the Nessler's reagent [18]. The 5 g sample weigh was put into the 250 ml flask with 50 ml of 0.25 M ammonium chloride solution. The saturation with ammonium was performed for 6, 8 and 10 hours and subject to mechanical stirring. Then the solid phase was separated by centrifugation. The solution was analyzed for ammonium, potassium, sodium, calcium and magnesium ion content. Simultaneously, a blank experiment was carried out to make a correction for possible ingress of the same ions into the solution due to the material dissolution. In the blank experiment, the same bentonite weigh (5 g) was poured into 50 ml of bidistilled water, mechanical stirring was performed during the corresponding experiment. After centrifugation, the potassium, sodium, calcium and magnesium contents were determined in the solution. Bentonite fraction 0.25-0.5 mm was added to the standard solution at the rate of 20; 10; 1; 0.1; 0.01 g of bentonite per 1 liter of the solution. The comparison of the treatment efficiency with different amounts of bentonites was carried out at a sorption time of 2, 6, 24, 72, 96, 264 hours.

The ammonium ion content was determined by the colorimetric method with the Nessler's reagent [18]. The iodometric method was used to determine the dissolved oxygen

 https://doi.org/10.1051/e3sconf/202126202021

[19, 20]. The hydrogen ion concentration was determined by the electrometric method [20, 21]. The calcium and magnesium exchangeable cation content was determined trilonometri cally [22], and the alkali metals, potassium and sodium cation content - by the flame photometry [23, 24]. To determine the hydrochemical parameters, the following devices were used: colorimeter photo-electric KFK-2, pH-ion meter EV-74, analytical balance, atomic absorption spectrophotometer AAS-30.

3 Results and Discussion

To determine the maximum sorption capacity, weighing of the tested montmorillonite clay was carried out according to the method [18]. It is found out that when a smaller amount of the sample is added to each liter of the standard solution (0.1 and 0.01 g), higher values of the sorption capacity are received (Table 1). The experiment was carried out in the fishery pond simulating conditions, the initial ammonium ion concentration in the solution corresponded to 6.3 mg·eq. N/l

Table 1. Bentonite capacity in the fishery pond simulating conditions, mg.eq. N/l.

Amount of bentonite per 1 liter of solution, g	Sorption time, h			
	2	4	24	72
0.1	1.02	1.23	1.88	1.43
0.01	11.98	14.76	15.97	12.94

As it can be seen from Table 1, the maximum amount of ammonium ions absorbed by bentonite clay made 13.9 mg eq. N/l per each gram of bentonite introduced. This bentonite capacity value exceeds the theoretically possible ion-exchange bentonite capacity calculated for the tetrahedral aluminum content in its structure. The sorption process seems to include not only the exchange process on the montmorillonite, but also processes not associated with ion exchange both on the montmorillonite and on other minerals that make up the rock.

Thus, the exchange capacity value obtained in the standard test conditions averaged 0.59 mg eq. N/l of bentonite clay. The maximum amount of absorbed ammonium ions in the conditions corresponding to fish-farming ponds when 0.01 g of the rock was added to each liter of ammonium ion solution made 13.9 mg eq. N/l per each gram of montmorillonite clay.

Since the use of bentonites in the aquatic environment is based on the cation exchange, we have established which ions and in what quantity are released from bentonite during the ammonium ion sorption. The ion exchange was carried out according to the standard method for the 0.25M ammonium chloride solution, both by stirring and in the steady-state conditions for 6, 10, 12 and 36 hours. Upon expiration of this period of time, the amount of calcium, magnesium, sodium and potassium cations and the amount of absorbed ammonium ions were determined by the difference in its concentration in the solution before and after sorption. The study results are given in Table 2.

Table 2. The composition of bentonite exchange cations released during ammonium ion exchange.

Sorption time, h	Amount of exchange cation, mg.eq N/g						Equivalent composition of exchange cations, %			
	Released					Absorbed				
	Ca^{2+}	Mg^{2+}	Na^+	K^+	∑ions	NH_4^+	Ca^{2+}	Mg^{2+}	Na^+	K^+
6	0.27	0.01	0.20	0.10	0.58	0.60	44.6	4.7	33.1	16.9

Table 2. Continued.

10	0.36	0.01	0.19	0.10	0.66	0.59	52.3	3.3	29.1	14.5
12	0.33	0.02	0.18	0.10	0.63	0.63	49.2	5.7	29.0	15.3
36	0.46	0.02	0.20	0.09	0.77	0.79	57.4	4.4	25.9	11.5

Table 2 shows that the sum of exchange cations is equivalent to the amount of absorbed ammonium ions during 36 hours of sorption. This evidences that, in the experimental conditions, the sorption process proceeds by the mechanism of ion exchange. The calcium ion is most actively exchanged for the ammonium ion. After 36 hours of sorption, each gram of bentonite released 0.46 mg·eq. of Ca^{2+} (or 9.2 mg) to the external solution that is 0.92% of the mass of the bentonite weigh and about 59.7% of the sum of all exchange cations. Thus, the research data allow us to state that in the process of ion exchange, the calcium ions mainly enter the aqueous medium in contrast to the magnesium, sodium and potassium ions. All these prevent the creation of favorable conditions for the development of hydrobionts.

To achieve the optimal degree of cleaning of ponds from excess ammonium ions and expediency in terms of reducing the material and labor costs to a minimum, the optimal amount of bentonite was calculated, the use of which would be sufficient to maintain the ammonium ion concentration. Taking into account that, on the one hand, exceeding a certain level of ammonium ion content in the pond water can cause toxic damage to hydrobionts, fish-farming standards establish the maximum allowable concentration (MAC) of ammonium ions up to 2.5 mg N/L. On the other hand, the presence of certain amount of ammonium ions in the water reservoirs is necessary for development of phytoplankton. The optimal concentrations of ammonium nitrogen are within the range of 0.2-0.8 mg/l. Based on the above, the conditions were selected so that to ensure a decrease in the ammonium nitrogen concentration from 2-3 MAC (5-7.5 mg/l) to 0.2-0.8 mg/l, and the concentrations less than 0.2 mg/l are undesirable. The bentonite weigh fraction 0.25-0.5 mm varied from 20 to 0.01 g per liter of ammonium ion solution. The experiment time ranged from 2 hours to 528 hours (22 days). The results of the experiment are given in Table 3 and Figure 1.

Table 3. Dependence of the bentonite sorption capacity on its amount and sorption time.

Amount of bentonite per 1 liter of solution, g	Sorption time, h						
	2	4	12	24	456	480	528
20.00	0.011	0.013	0.015	0.017	0.021	0.022	0.022
10.00	0.013	0.011	0.223	0.033	0.031	0.036	0.039
1.00	0.120	0.250	0.287	0.305	0.217	0.274	0.332
0.10	1.021	1.203	1.527	1.890	0.934	0.879	0.816
0.01	11.980	14.025	15.125	15.990	6.928	7.896	8.427

The analysis of the tested bentonite sorption capacity showed that the efficiency of the maximum capacity depends inversely on the amount of the introduced rock. When the reduced amount of bentonite 20 to 0.01 g/l is introduced into the solution, its capacity increases during the first day by about 1000 times and then by about 300 times.

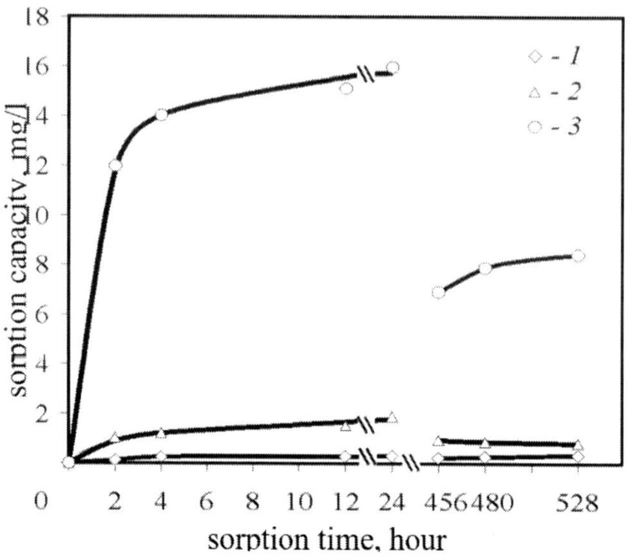

Fig. 1. Time dependence of the bentonite sorption capacity with bentonite content of, g/l: 1.00 (1); 0.10 (2) and 0.01 (3).

Along with the determination of the bentonite sorption capacity in the same conditions, their ability to sorb ammonium ions was studied. For this purpose, the degree of ammonium ion sorption by bentonites in the amount from 0.01 to 20 g/l was studied in the solutions with 3 MAC for a certain period of time. The study results are given in Table 4 and Figure 2.

Table 4. Dependence of the ammonium ion content in the solution on the amount of bentonite, mg N/l.

Amount of bentonite per 1 liter of solution, g	Sorption time, h							
	0	2	4	24	72	456	504	528
20.00	6.970	3.704	3.092	2.156	1.623	0.515	0.501	0.560
10.00	6.621	4.759	4.615	1.510	1.510	1.031	0.960	0.740
1.00	7.069	5.187	3.590	3.058	2.384	2.490	2.436	2.390
0.10	6.250	4.460	4.175	3.263	3.880	4.797	4.986	5.024
0.01	6.256	4.449	4.145	3.880	4.317	4.289	4.653	5.075

Table 4 shows that after 2 hours, the ammonium ion concentration in the solutions decreases to 2 MAC in the presence of 1 and 10 g/l of bentonite, to 1.8 MAC - 0.01 and 0.1 g/l and to 1.56 MPC 20 g/l.

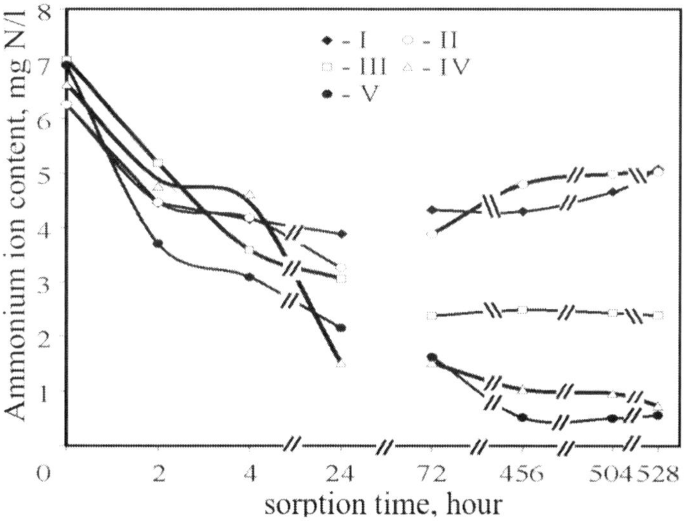

Fig. 2. Dependence of the ammonium ion concentration in the solution on the sorption time. Bentonite content, g/l: 20.00 (I); 10.00 (II); 1.00 (III); 0.10 (IV); 0.01 (V).

It follows that the 2000-times difference in the maximum and minimum amounts of the sorbent ensures the difference in the degree of ion concentration only by 0.24 MAC. It should be noted that the same trend is observed after 4 hours and after a day, and so on. This natural process creates the conditions for further decrease of the ammonium ion concentration to the optimal value. Some fluctuations in the process of sorption shown by curves 1-5 in Fig. 2 are explained by the complexity of sorption processes.

4 Conclusion

Taking into account such factors as time, degree of treatment of the aquatic medium, amount of sorbent, its cost and labor costs, the amount of bentonite of 0.01 g/l is the most appropriate for reducing the ammonium ion content from 3 MAC to the safe level.

References

1. M. W. Deem, R. Pophale, P. A. Cheeseman, D. J. Earl. J. Phys. Chem. C., **113(51)** (2009)
2. S. Baerlocher, L. V. McCusker, D. H. Olson, *Atlas of Zeolite Framework Types* (2007)
3. V. A. Drits, A. G. Kossovskaya, *Clay minerals, smectites, mixed - layer formations* (1990)
4. A. I. Shevelev, Exploration and protection of mineral resources, **9**, 15-19 (2005)
5. F. Ayari, E. Srasra, M. Trabelsi-Ayadi, Desalination, **185 (1-3)**, 391-397(2005)
6. A. Salem., L. Karimi, Korean J. of Chemical Engineering, **26 (4)**, 980-984 (2009)
7. L. V. Kurtukova, Success of the modern natural science, **12**, 27-34 (2011)
8. V. P. Finevich, Russian Chemical J., **4**, 69-76 (2007)
9. O. V. Chistik, *Ecology* (2000)
10. I. M. Potravny, *Environmental audit, Theory and practice: Training book* (2013)
11. L. M. Chukhlebova, Regional problems, **1**, 54-58 (2011)

12. A. S. Cherkasov, V. A. Somin, L.F. Komarova, L.V. Kurtukova, Success of the modern natural science, **12**, 34-38 (2011)
13. N. V. Shekhovtseva, *Ecology of aquatic microorganisms* (2008)
14. N. G. Nalivayko. *Water microbiology* (2006)
15. O. N. Bauer, V. A. Musselius, V. M. Nikolaeva, Yu. A. Strelkov, Pond fish - diseases, 320 (1981)
16. L. I. Grishchenko, M. Sh. Akbaev, G. V. Vasilkov, *Fish diseases and fundamentals of fish-farming* (1999)
17. D. A. Serov, E. V. Ryabukhina, *Influence of heavy metals on hydrobionts in the ecosystem of water bodies* (2013)
18. *Instruction of the Scientific Board on Analytical Methods No. 235 Thermochemical quantification of pegasin in rocks* (1987)
19. S. A. Alekhin, *Guideline for chemical analysis of inland waters* (1973)
20. *Guideline RD, Volume concentration of dissolved oxygen in seawaters, Measurement by iodometric method* (2010)
21. *Guideline RD, for methods of chemical analysis of seawaters* (2010)
22. *Guideline for chemical analysis of inland surface waters* (2010)
23. A. F. Zhukov, I. F. Kolosova, V.V. Kuznetsov, *Analytical chemistry, Physical and physicochemical methods of analysis* (2001)
24. E. M. Kats, N. K. Galkina, I. B. Serova, Geochemistry, **11**, 17-22, (2011)

Effect of adsorbent and phospholipide on meat production and nutritional value of fattened calf bull meat

Maksim Shabanov[1, *], *Victor* Temiraev[1], *Taladin* Kokov[2], *Beslan* Efendiev[2], and *Shamurat* Herremov[3]

[1]Federal State Budgetary Educational Institution of Higher Education Gorsky State Agrarian University, Kirova Str., 37, 362040 Vladikavkaz, Russia
[2]Kabardino-Balkarian State Agricultural University named after V.M. Kokov, Lenin Avenue, 1V, 360030 Nalchik, Russia
[3]Union of Industrialists and Entrepreneurs of Turkmenistan, Atamurad Niyazova Av., 174, 744000 Ashgabat, Turkmenistan

Abstract. One of the most effective ways of removing HM salts that have entered the digestive tract of fattened animals is the technological method of including adsorbent preparations in their diets, which, binding various toxins, excrete them from the body. The study purpose is to develop a method to increase the meat productivity and quality of meat of young cattle for fattening in a technogenic zone by inclusion of ToxySorb adsorbent in their diets with a high content of heavy metal salts in the amount of 1.25 kg/t of compound feed and lecithin in the amount of 10 g/100 kg of live weight. According to study results, it was shown that when these drugs were fed together in the indicated quantities, it contributed to better detoxification of xenobiotics. The calf bulls of experimental group 3 had the best slaughter indicators, as well as the best consumer, sanitary and hygienic properties of meat products.

1 Introduction

At present, beef production is increased to satisfy the Russian consumer needs with complete protein of animal origin. At the same time, a greater share in beef production in our country is occupied by meat as a result of young livestock fattening, primarily dairy and mixed breed calf bulls. At the same time, the duration of calf bull fattening is one year, starting from six months of age, until they reach a body weight of more than 400 kg. If these requirements are met, with good nutrition of fattened animals, an optimal ratio of proteins, lipids, mineral and biologically active substances is observed in beef [1, 2, 3].

Considering such requirements, with the rational organization of beef production by calf bull fattening, successful implementation of genetically determined potential of their body, first of all, helps to ensure physiological body needs for energy and all necessary nutrients. However, this problem cannot be solved only by a full-fledged balanced feeding

* Corresponding author: kapmar17@gmail.com

of young fattening cattle according to current rations. The effective fattening of calf bulls is supported by the use of high-quality feed in the diet. Moreover, in modern conditions, with this approach to feeding calf bulls for meat, close attention is paid to ecological safety of feed [4, 5].

In recent decades, anthropogenic loads on the environment have increased due to intensive development of industrial enterprises, transport, construction, chemicalization of agriculture, etc. It results in the contamination of soil, plants, plant and animal origin food with various chemical toxicants. One of the most dangerous toxicants for the human and animal body are heavy metal (HM) salts. They gradually accumulate in organs and tissues (they are not excreted) and subject the body to intoxication in growing dynamics [6, 7, 8].

It is known that the North Ossetia Republic territory, due to the presence of a number of large enterprises of non-ferrous metallurgy and vehicles, is characterized by a high level of soil pollution and plant feed crops with heavy metal salts (first of all, such elements as copper, lead, zinc, cadmium, etc.). Therefore, the development of methods to prevent contamination with these compounds of meat products produced in the technogenic zone of the Republic of North Ossetia-Alania is constantly underway [9, 10, 11]

One of the most effective ways of removing HM salts that have entered the digestive tract of fattened animals is the technological method of including adsorbent preparations in their diets, which, binding various toxins, excrete them from the body. There is good experience in the synergistic effect of adsorbents with other biologically active additives (BAA), including phospholipids, for elimination of these toxins [12, 13, 14].

As a result, our purpose was to develop a method to increase the meat productivity and quality of fattening calf bulls in a technogenic zone by including ToxySorb adsorbent and phospholipid lecithin in their diets with a high content of HM salts.

2 Materials and Methods

We studied Schwyz calf bulls in the Republic of North Ossetia-Alania. Of these, at the age of six months, according to the principle of analogous pairs, we formed four groups of ten animals each. They were fed during fattening (lasting 12 months) with balanced diets according to the scheme shown in Table 1.

The tested feed additives in the compound feed shop were introduced into the compound feed for experimental animals with the help of industrial dispensers and mixed.

Upon reaching the age of 18 months, 3 calf bulls, typical in body weight and fatness, were selected from each group, they were slaughtered according to the generally accepted procedure.

In selected average samples of the rib eye muscle of calf bulls from compared groups, the chemical composition, biological value and ecological safety of meat were studied using conventional methods.

Table 1. Substantiation of schematic diagrams for introduction of preparations into the diets of calf bulls, n = 10.

Group	Feeding features
Control	Basic diet (BD) with excessive levels of Zn, Cd, Pb salts
Experimental 1	BD + ToxySorb adsorbent in the amount of 1.25 g/t compound feed
Experimental 2	BD + lecithin preparation in the amount of 10 g/100 kg of live weight
Experimental 3	BD + ToxySorb adsorbent in the amount of 1.25 kg/t of compound feed + lecithin preparation in the amount of 10 g/100 kg of live weight

All the obtained digital material on a personal computer was biometrically processed with the establishment of the Student's test of validity.

3 Results and Discussion

During the experiment, the concentration of these toxicants in the diet for the experimental calf bulls was regularly studied. It was determined that fattening animals in winter diet had an excess of maximum permissible concentrations (MPC) for Pb (lead) by 63.1-64.3%, Zn (zinc) - by 67.0-67.3%, and Cd (cadmium) - by 62.6-63.0%, in summer diet - by 61.2-61.4%; 66.1-66.4 and 60.3-60.6%, respectively.

During the slaughter, we studied the effect of the feed preparations on the slaughter indicators of the fattened calf bulls, shown in Fig. 1.

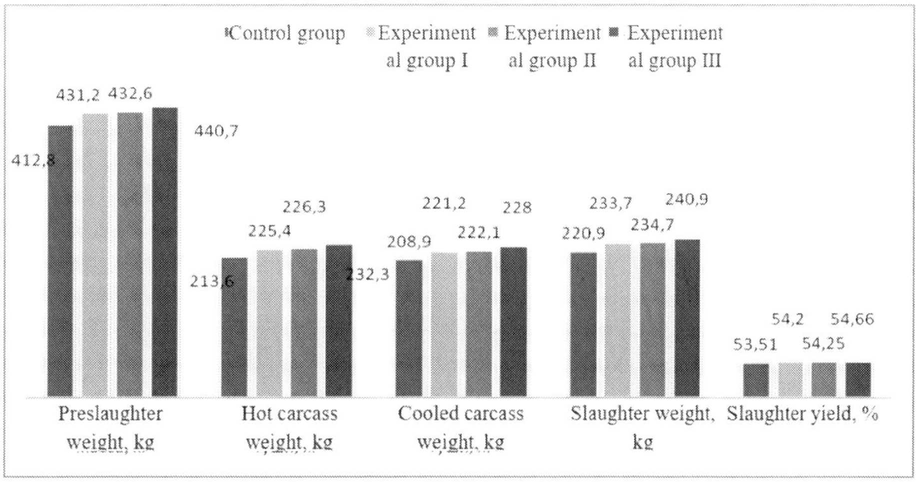

Fig. 1. Effect of tested preparations on slaughter indicators of calf bulls.

It was determined that joint additions of adsorbent and phospholipid complex to compound feed had a more favorable effect on slaughter indicators in experimental fattening of young cattle. As a result, in comparison with the control group, animals of experimental group 3 had higher parameters characterizing their pre-slaughter weight by 6.76% ($P < 0.05$), slaughter weight - by 9.05% ($P < 0.05$), hot carcass weight - by 8.75% ($P < 0.05$), chilled carcass weight - by 9.14% ($P < 0.05$) slaughter yield - by 1.15% ($P < 0.05$). It is considered as a consequence of the best detoxification effect, thanks to the rational intake of ToxySorb adsorbent for feeding, showing synergistic effect for removal of HM salts from the body with lecithin, helped to optimize the slaughter indicators of fattened animals.

The nutritional value of calf bull meat of compared groups is more clearly characterized by the chemical composition of eye rib samples, their indicators are shown in Fig. 2.

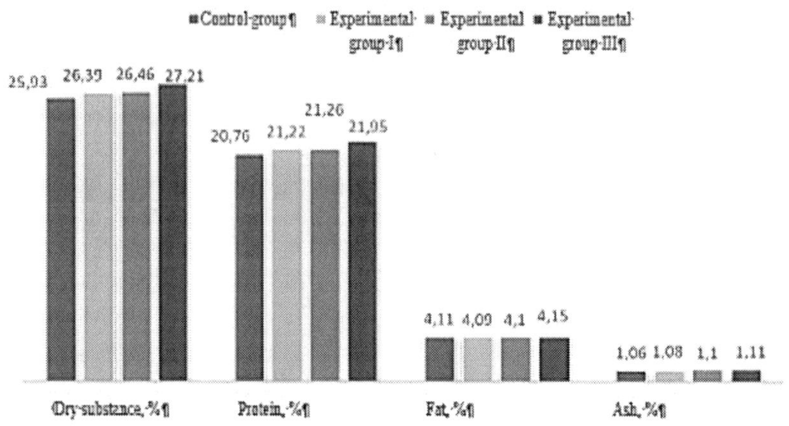

Fig. 2. Effect of tested preparations on the chemical composition of calf bull meat.

Analysis of the chemical composition of the eye rib samples showed that the joint feeding of the tested preparations contributed to the improvement of the nutritional value of the produced beef. This versus animals of the control group was manifested in an increase in the meat samples of analogs from experimental group 3 of indicators of dry matter concentration by 1.28% ($P < 0.05$) and protein - by 1.19% ($P < 0.05$).

An equally important indicator for evaluation of meat nutritional quality is the determination of its biological value, taking into account tryptophan and hydroxyproline levels (Fig. 3), determined for the animals of compared groups by calculating the protein-quality indicator (PQI) (Fig. 4).

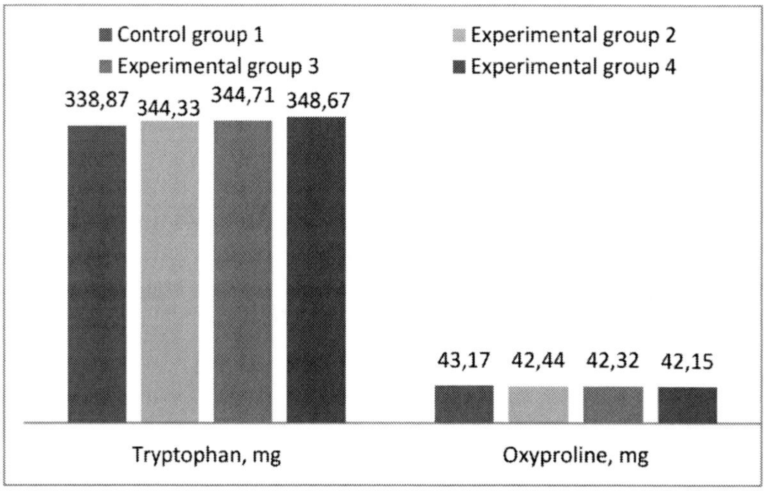

Fig. 3. Concentration of tryptophan and hydroxyproline in calf bull meat.

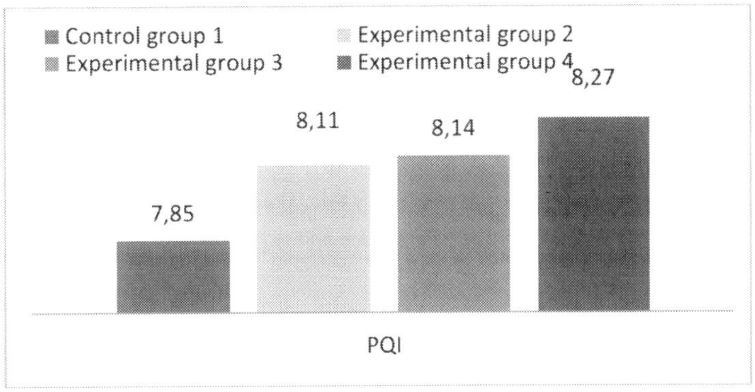

Fig. 4. Protein-quality indicator of calf bull meat.

It was determined that due to combined adsorbent and lecithin preparation feeding, it was possible to achieve a significant ($P < 0.05$) increase in the concentration of the essential amino acid tryptophan by 2.89% in the bulls of experimental group 3 in rib eye samples versus control group. Thanks to this, it became possible to increase the biological value of meat samples from animals of experimental group 3, which also surpassed their control equivalents in terms of PQI by 5.4%. It demonstrates protein metabolism improvement in animals of experimental group 3 due to better elimination of toxicants.

It is confirmed by the content of zinc salts (Fig. 5), lead (Fig. 6) and cadmium (Fig. 7) in calf bull meat samples from compared groups.

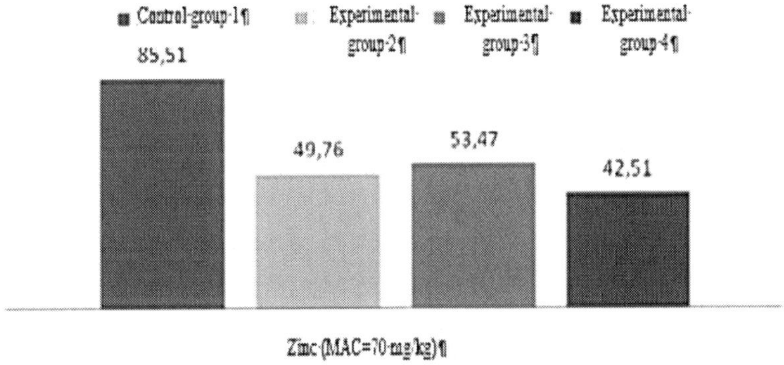

Fig. 5. Zinc content in meat samples of calf bulls from compared groups.

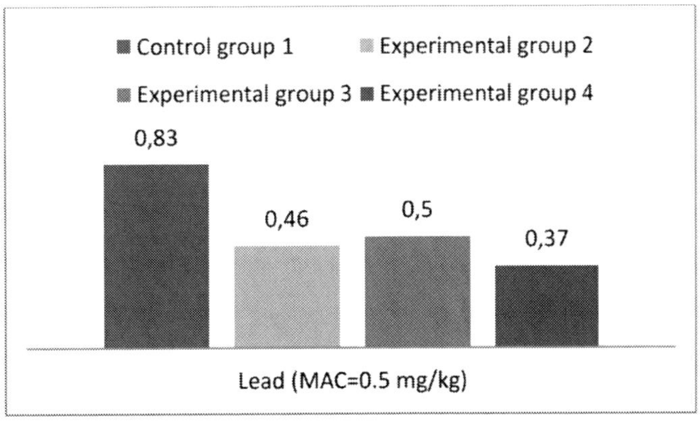

Fig. 6. Lead content in meat samples of calf bulls from compared groups.

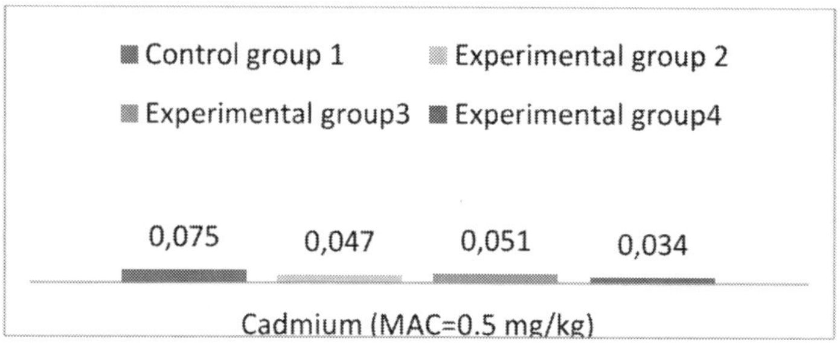

Fig. 7. Cadmium content in meat samples of calf bulls from compared groups.

It was determined that the joint feeding of adsorbent and lecithin preparations had a more favorable effect on the ecological safety of produced beef during the experiment. It was manifested relative to control group equivalents in a decrease in the eye rib samples of animals of experimental group 3 in the concentration of zinc by 2.01 times ($P < 0.05$), lead - by 2.24 ($P < 0.05$) and cadmium - 2.21 times ($P < 0.05$). It should be noted that in the calf bull meat samples of experimental group 3, there was no excess of MPC for any of these elements.

4 Conclusions

An effective method to increase the meat productivity and nutritional quality of calf bull meat in the technogenic zone of North Ossetia-Alania is the joint feeding of ToxySorb adsorbent in the diet with an increased level of heavy metal (HM) salts in the amount of 1.25 kg/t of compound feed and lecithin in the amount of 10 g/100 kg of live weight.

References

1. Y. A. Yuldashbaev, R. B. Temiraev, V. V. Tedtova, et al., J. of Livestock Science, **11**, 8-13 (2020)
2. S. I. Kononenko, R. B. Temiraev, A. A. Gazdarov, *Modern problems of dairy and beef cattle breeding, milk and beef production: Proceedings of International scientific and practical conference*, 181-183 (2012)
3. R. B. Temiraev, Z. T. Baeva, A. V. Muzaeva, N. G. Ter-Teryan, I. A. Arishina, Proceedings of the Kuban State Agrarian University, **21**, 140-143 (2009)
4. V. V. Tedtova, Z. T. Baeva, E. S. Dzodzieva, Z. Ya. Tsopanova, A. Kh. Pilov, Meat Industry, **3**, 60-62 (2013)
5. V. K. Temiraev, V. R. Kairov, R. B. Temiraev, Z. A. Kubatieva and V. M. Gukezhev, Ecology, Environment and Conservation, **23(1)**, 554-561 (2017)
6. B. Temiraev, L. V. Tsalieva, I. G. Plieva, M. R. Dzutseva, Bulletin of the Gorsky State Agrarian University, **47(2)**, 85-87 (2010)
7. R. B. Temiraev, Z. T. Baeva, U. I. Teziev, A. A. Gazdarov, Dairy Industry, **5**, 73-74 (2009)
8. E. S. Dzodzieva, M. G. Kokaeva, R. B. Temiraev, G. A. Abramova, D. O. Gurtsieva, Meat Industry, **2**, 46-48 (2015)
9. V. K. Temiraev, V. R. Kairov, R. V. Kalagova, Biology and Medicine, **6(3)**, BM-046-14 (2014)

10. L. V. Tsalieva, R. B. Temiraev, S. I. Kononenko, B. A. Dzagurov, M. S. Gazzaeva and S. A. Grevtsova, J. of Pharmaceutical Sciences and Research, **9(12)**, 2397-2400 (2017)
11. S. Kononenko, R. Temiraev, Z. Baeva, A. Gazdarov, Kombikorma (Compound Feed), **6**, 104-105 (2011)
12. V. A. Pogodaev, A. N. Arilov, A. F. Shevhuzhev, A. P. Marynich, R. Kh. Kochkarov, Research J. of Pharmaceutical, Biological and Chemical Sciences, RJPBCS, **8(1)**, 1086-1090 (2017)
13. V. A. Pogodaev, A. P. Marynich, S. V. Frolko, E. G. Mishvelov, and A. Y. Glushko, Research J. of Pharmaceutical, Biological and Chemical Sciences, RJPBC, **7(4)**, 1345-1348 (2016)
14. R. B. Temiraev, M. K. Kozhokov, S. K. Cherchesova, F. F. Kokaeva, I. R. Tletseruk, J. of Environmental Management and Tourism, **3(19)**, 567-573 (2017)

Body types of Aberdeen Angus bulls and their relationship with meat production

Mukhamed Shakhmurzov[1], *Anatoliy* Shevkhuzhev[2], *Vladimir* Pogodaev[2,*], *Yusupzhan* Yuldashbaev[3], and *Shamurat* Kherremov[4]

[1]Kabardino-Balkarian State Agricultural University named after V. M. Kokov, Lenin Avenue, 1V, 360030 Nalchik, Russia
[2]Federal State Budgetary Scientific Institution North Caucasian Federal Scientific Agrarian Center, Nikonova Str., 49, 546241 Mikhailovsk, Russia
[3]Russian State Agrarian University - Moscow Agricultural Academy named after K. A. Timiryazev, Timiryazevskaya St., 49, 127550 Moscow, Russia
[4]Union of Industrialists and Entrepreneurs of Turkmenistan, Atamurad Niyazova Av., 174, 744000 Ashgabat, Turkmenistan

Abstract. The study results of Aberdeen-Angus breed bull meat production, depending on the body type, are presented. The first group consisted of bulls descended from stud bulls of large tall type, the second group consisted of bulls descended from stud bulls of the small compact type. It was determined that at the age of 18 months, bulls from large-type parents reached a weight of 442 kg, and from small-type parents - 413 kg. The first group bulls had a higher pre-slaughter weight by 31.4 kg and heavy carcasses compared to the second group bulls by 28.3 kg or 13.5% ($P > 0.99$). The first group bulls, being higher-legged, produced carcasses with a higher bone content (by 4.08 kg) during slaughter compared to the second group. In their carcasses, there was 4.52 kg of boneless meat per 1 kg of bones, while in the second group carcasses - 5.03 kg. However, in general, when boning semi-carcasses of the first group, more boneless meat was obtained, by 9.9 kg. or by 11.72% ($P > 0.99$) compared with the second group.

1 Introduction

The main purpose of beef cattle breeding in the coming decades is to increase production volumes and improve the quality of the products obtained enhance the competitive capacity of the industry [1, 2].

Therefore, the development of regional programs for the rational use of breed and feed resources to increase the quantity and improve the quality of the produced beef is an urgent goal [2, 3, 4].

The growing demand for meat as a food product and the need to obtain it with the lowest cost of feed and funds put forward as the main task of breeding in beef cattle breeding - development of animals with high vigor, good feed pay, capable of reaching

* Corresponding author: pogodaev_1954@mail.ru

large livestock by the age of 15-18 months, with a high slaughter yield and good quality meat indicators [5-8].

In this regard, in beef cattle breeding there is a tendency towards a transition from the extreme compact (solid) type with a short and square body on low legs to a type that is larger in live weight and well-defined meat forms [9, 10].

The Aberdeen-Angus breed, widespread in Russia, including in Karachay-Cherkessia, belongs to one of the world's best meat breeds [11-14]. However, in the last two decades, this breed has also developed towards the formation of a relatively small, but very early maturing type. In this regard, there is a need for a certain breed restructuring, establishment of the most desirable and promising type, improvement of breeding and rearing pedigree young stock methods.

The study purpose was Aberdeen-Angus breed bull meat production, depending on the body type.

2 Materials and Methods

In studies on cattle meat production, much attention is paid to meat qualities, based on the assessment of its distribution in the carcass.

Hammer LLC of Karachay-Cherkessia conducted the experiment: from the offspring of 4 large bulls and 3 small bulls of the Aberdeen-Angus breed, 14 bulls were selected in each group.

The fathers of the first group bulls had a higher level of live weight compared to the fathers of the second group bulls (by 80 kg) and differed from them in high-leggedness, less broad body and massiveness. The mothers of the first group bulls also had a higher live weight (by 49 kg) in comparison with the mothers of the second group bulls, the body type differences were insignificant.

Immediately after weaning, they were placed on control rearing. The average age of the first group bulls at weaning was 8 months 22 days, and the second group bulls - 8 months 20 days.

Selection according to the principle of analogs in experimental groups was based on live weight, age, health status and origin.

The first group consisted of bulls descended from stud bulls of large tall type, the second group consisted of bulls descended from stud bulls of the small compact type.

The conditions of bull feeding and keeping in experimental groups were the same and corresponded to the farm rearing system.

The feeding level was designed to obtain an average daily gain of 650-750 g in the period from 8 to 12 months and 750-850 g from 12 to 18 months of age.

The study of the growth of experimental animals was performed by weighing, when setting up for experiment, at the age of 12, 15 and 18 months.

To get a more complete picture of the growth intensity in experimental animals, the relative growth rate was calculated.

To study the quantity and quality of meat products of experimental animals, a control slaughter of young animals was performed at the age of 18 months. For control slaughter, 3 of the most typical animals of the same age were selected from each group.

he slaughter was carried out at the Circassian meat-packing plant LLC Kavkaz-Myaso according to All-Russian Research Institute of Metrological Service and Federal Research Center for Animal Husbandry named after Academy Member L.K. Ernst procedures.

To study the morphological composition of the carcasses, left semi-carcasses were divided into five cuts with their subsequent boning.

To characterize full meatiness of the carcasses, the following measurements were taken:

1. Body length in cm - from the sacrum extreme anterior point on the cut to the middle of the first rib anterior edge.
2. Thigh length in cm - from the highest point of the heel hock joint to sacrum anterior point.
3. Carcass length in cm - body length + thigh length.
4. Thigh circumference in cm - in a plane 60% apart from the beginning of the thigh length measurement line and perpendicular to this length.

Coefficients "K1" of the carcass full meatiness and "K2" of the thigh were calculated using the following formulas:

$$\text{Coefficient K1} = \frac{Carcass\ weight*100}{Carcass\ length}$$

$$\text{Coefficient K2} = \frac{Thigh\ girth*100}{Thigh\ length}$$

The obtained experimental material was biometrically processed using the Microsoft Office 2007 software package.

3 Study Results

The live weight dynamics is one of the main indicators in the study of animal meat production.

It was found that the first group bulls exceeded the second group equivalents in live weight (FIG.1). The confidence criterion value, calculated for the difference in live weight is $t_d = 3.26$. It indicates the confidence of the obtained difference ($P > 0.99$).

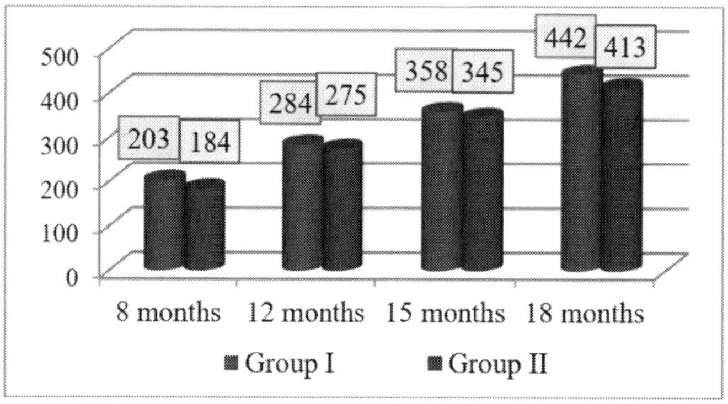

Fig. 1. The live weight dynamics of experimental bulls.

The presented data on changes in the average daily gain in live weight of experimental bulls indicate that, during the suckling period, the first group bulls demonstrated higher gains than the second group bulls (FIG. 2).

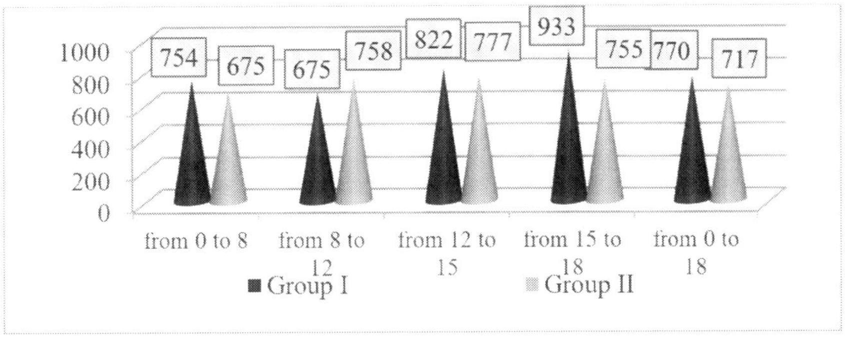

Fig. 2. Change in average daily gains of experimental bulls by growth periods.

The reason for this is the better milk yield of mothers (larger ones in the first group), because the conditions of feeding and keeping cows with suckling calves were approximately the same, the heritability of live weight at this age was rather low (h = 0.28-0.30).

After weaning and up to 12 months of age, the first group bulls grew slightly worse and demonstrated average daily gains 11% lower than the second group bulls. During this period, the difference in average body weight between groups decreased from 10.3% at weaning to 3.3% at 12 months of age.

Starting from the age of 13 months, the second group bulls grew somewhat more slowly than the first group bulls, which until the end of the experiment had higher average daily gains. Over the entire experimental period, the first group bulls demonstrated average daily gains by 4.4% higher than the second group bulls and at the end of the experiment had a large live weight of 29 kg, or 7.02%.

Data on slaughter yield, carcass quality and meat amount were determined based on control slaughter at the end of the experiment when the bulls reached the age of 18 months.

The first group bulls had a well-pronounced tall body type, the second group bulls were distinguished by their low legs and compactness. The average age of the first group bulls was 526 days, the removable weight was 436 kg, the second group bulls were 530 days and 406 kg, respectively.

Before being sent to the meat packing plant, the animals were weighed, upon arrival and put on a 24-hour fasting period, during which they were watered. Before being taken to the slaughterhouse, the animals were weighed again and their lifetime fatness was determined. All animals for slaughter, when accepted at the meat packing plant, were classified as having the highest nutritional status.

The slaughter results showed (Table 1) that the first group bulls had a higher pre-slaughter weight by 31.4 kg and heavy carcasses compared to the second group bulls by 28.3 kg or 13.5% (P> 0.99).

Table 1. Control slaughter results of experimental bulls (n = 3).

Indicators	Group 1	Group 2
Preslaughter weight, kg	408.4±3.74	377.0±4.20
Carcass weight, kg	238.0±2.95	209.7±3.02
Interior fat weight, kg	9.9±0.14	11.7±0.16
Internal fat yield, %	3.0±0.03	3.1±0.03
Slaughter weight, kg	247.9±3.45	221.4±3.52
Carcass yield, %	58.27±0.25	55.62±0.23
Slaughter yield, %	60.71±0.28	58.73±0.27

There were some differences in the internal fat yield. The first group bulls had a higher carcass yield (by 2.65 abs.%) and a lower content of internal fat by 1.8 kg.

By the slaughter weight, the first group bulls surpassed their equivalents by 26.5 kg (P > 0.99), and by the slaughter yield, respectively, by 1.98 abs.%.

Offals of the first group bulls had a larger weight (Table 2). However, in relation to the pre-slaughter weight, there were almost no differences between the groups or they were very insignificant (intestines, head, legs).

Table 2. Weight and yield of offal, fat and skins.

Name of	Group 1		Group 2	
	Weight (kg)	% to live weight	Weight (kg)	% to live weight
Head	16.0	3.81	14	3.71
Legs	8.16	1.94	7.1	1.88
Lung	2.58	0.60	2.36	0.62
Heart	1.97	0.47	1.67	1.44
Liver	5.3	1.26	4.86	1.29
Kidney	1.10	0.26	1.10	0.29
Spleen	1.2	0.28	0.88	0.22
Intestine without contents	3.96	0.94	4.50	1.19
Total fat	9.86	2.35	11.73	3.11
Incl. caul fat	2.63	0.62	2.70	0.72
suet fat	1.13	0.27	1.53	0.41
intestinal	4.00	0.95	4.70	1.25
mouse round	2.10	0.50	2.80	0.74
Hide	30.6	7.28	29.3	7.77

There is no objective method to evaluate carcass quality and production conditions. Carcasses are evaluated mainly by the presence of fatty deposits on their surface. Finally, such an important indicator of the meat carcass quality, as the development of muscle tissue, is not sufficiently taken into account as there are no objective characteristics.

To characterize the full meatiness of carcasses, their measurements have recently been used. As a result of a number of studies in our country and abroad, a relationship was established between bone weight and carcass length (r = 0.78). The meat layer thickness, measured at the first rib middle, is related to the total weight of the meat in the carcass (r = 0.91). A direct relationship was established between the fore shank girth and the weight of bones in the semi-carcass, between the thigh girth and boneless meat yield during carcass boning.

To determine the full meatiness of the carcass, the size and quality of fat coating during the control slaughter, their expert eye assessment was performed and the main measurements were taken.

All carcasses of killed bulls are distinguished by well-developed muscles, which was especially noticeable in the thigh, waist and back. Fat coating on carcasses of both the first and second groups was thin, uniform, with gaps in the area of shoulder blades and ribs. The fat layer thickness at the last rib level was 1–2 mm. The carcasses obtained from the slaughter of the second group bulls had a slightly better coating (3.13 points) in comparison with the first group (2.74 points).

To characterize the full meatiness of carcasses, they were measured (Table 3).

Table 3. Carcass measurements and assessment of their full meatiness.

Carcass measurements	Group 1	Group 2
Body length	103.0±1.56	98.7±1.48
Femoral length	90.7±0.69	83.3±0.74
Thigh girth	94.3±0.76	90.3±0.68
Carcass length	193.7±2.57	182.0±2.63
Coefficients: K_1	120±0.88	115±0.80
K_2	104±0.64	108±0.70

If the value of "K_1" is higher than 100, i.e. the carcass weight exceeds its length, it indicates very good meatiness. The correlation coefficient between "K_1" indicators and meat yield indicators is 0.84.

In our experience, the full meatiness of carcasses was quite high, both in the first and in the second group. In the first group, the full meatiness of carcasses was slightly higher, since they had a large eight (by 12%) and length (by 6%). The second group carcasses had a higher indicator in terms of the thigh, it was shorter and relatively wider. The differences in the measurements of the carcasses were a reflection of the body type of the killed animals.

The study of the morphological composition of the carcasses of the experimental animals was performed using by boning of semi-carcasses, resulted in the following: the weight of meat, fat, bones and tendons.

The boning of semi-carcasses showed that the first and second group carcasses had some differences in the content of bones and boneless meat (Table 4).

Table 4. Results of semi-carcass boning.

Indicators	Group 1		Group 2	
	kg	% to semi-carcass weight	kg	% to semi-carcass weight
Average chilled semi-carcass weight	119.26±1.24	100	104.89±1.09	100
Produced from boning: boneless meat	94.34±0.93	79.10	84.44±0.87	80.51
bones	20.87±0.59	17.50	16.79±0.49	16.00
tendons	4.05±0.21	3.40	3.66±0.19	3.49

Table 5. Weight and morphological composition of individual carcass cuts.

Cuts	Weight (kg)	To semi-carcass weight (%)	Cut content (%)		
			boneless meat	bones	tendons
Group 1					
Neck	10.69	8.9	82.1	13.7	4.2
Scapulohumeral part	20.42	17.2	78.1	18.4	3.5
Dorsal part	37.91	31.8	77.5	20.8	1.7
Saddle	11.32	9.5	82.9	11.6	5.5
Hip	38.92	32.6	79.3	16.6	4.1
Group 2					

Table 5. Continued.

Neck	9.54	9.2	85.3	12.3	2.4
Scapulohumeral part	18.28	17.4	79.5	16.9	3.6
Dorsal part	32.30	30.8	78.6	19.5	1.9
Saddle	10.94	10.4	83.1	9.9	7.0
Hip	33.83	32.2	80.7	15.3	4.0

The difference in the content of bones in the carcasses is consistent with the differences in the constitution of animals. The first group bulls, being higher-legged, produced carcasses with a higher bone content (by 4.08 kg) during slaughter compared to the second group. In the first group carcasses, there was 4.52 kg of boneless meat per 1 kg of bones, while in the second group carcasses - 5.03 kg. However, in general, when boning the first group semi-carcasses, more boneless meat was obtained, by 9.9 kg or 11.72% ($P > 0.99$), compared with the second group.

The study of the relative weight of the carcass individual parts did not reveal significant differences between the groups (Table 5).

Differences in the ratio of individual cuts to carcass weight were very small. Boning of each cut separately showed that the cut weight of the first group carcasses, as well as the carcases as a whole, contains more bones than the second group.

4 Conclusion

The bulls from stud bulls of a large type surpassed bulls from parents of a small type in meat production, which was reflected in a higher growth rate and better feed pay with an increase in live weight. At the age of 18 months, bulls from large-type parents reached a weight of 442 kg, and from small ones - 413 kg. These differences are statistically significant ($P > 0.99$).

Small-type animals were distinguished by a more pronounced meat type of constitution. They were relatively low-legged and broad-bodied, with more thigh completeness. In the first group carcasses, there were 4.52 kg of boneless meat per 1 kg of bones, while in the second group carcasses - 5.03 kg. However, in general, when boning semi-carcasses of the first group, more boneless meat was obtained, by 9.9 kg. or by 11.72% ($P > 0.99$) compared with the second group.

References

1. R. G. Iskhakov, V. I. Levakhin, M. G. Titov, Animal science, **3**, 22-25 (2007)
2. V. Kulintsev, M. Shakhmurzov, A. Shevkhuzhev, et al., J. of Engineer. and Advan. Technol. (IJEAT), **9(1)**, 2249-8958 (2019)
3. S. E. Andresen, C. L. Goad, L. Kriese-Anderson, et al., Animal Sci, **35(1)**, 57-65 (2019)
4. D. A. Randelin, V. V. Randelina, I. S. Bushueva, Bulletin of meat cattle breeding, **1(60)**, 244-247 (2007)
5. V. A. Pogodaev, V. V. Golembovsky, V. I. Komlatsky, L. F. Velichko and L. I. Konkov, IOP Conf. Series: Earth and Environ. Science, 624, 012134 (2021)
6. P. A. C. Luz, C. Andrighetto, et al., Effect of integrated croplivestock systems in carcass and meat quality of Nellore cattle Livestock Sci., 220, 83-92 (2019)

7. V. V. Onishchenko, N. S. Dega, F. H. Bostanova, Int. J. of Humanities and Natural Sciences, **1(1)**, 29-35 (2019)
8. V. A. Pogodaev, A. N. Arilov, A. I. Petenko, A. A. Soldatov, T. I. Pakhomova, Research J. of Pharmaceutical, Biological and Chemical Sciences, **9(4)**, 766-759 (2018)
9. M. V. Tarasov, V. M. Gabidulin, V. Yu. Shmakov, Vestnik myasnogo skotovodstva, **63 (3)**, 71-76 (2010)
10. M. Tofastrud, A. Hessle and Y. Z. B. Rekdal, Weight gain of free-ranging beef cattle grazing in the boreal forest of south-eastern Norway Livestock Sci., 2 (2020)
11. L. I. Kibkalo, V. V. Bychkov, Bulletin of the Kursk State Agricultural Academy, **1**, 70-71 (2010)
12. A. M. Pardo, M. A. Elzo, L. T. Gama and L. M. Melucci, Genetic parameters for growth and cow productivity traits in Angus, Hereford and crossbred cattle Livestock Sci., 233, 103952 (2020)
13. M. J. D'Occhio and P. S. Baruselli, Giuseppe Campanile, Influence of nutrition, body condition, and metabolic status on reproduction in female beef cattle, 125, 277–84 (2019)
14. A. Kucherenko, Animal breeding in Russia, **3**, 6-7 (2009)

Growth and development indices of Aberdeen-Angus bulls originating from sires of different body types

Mukhamed Shakhmurzov[1], *Anatoly* Shevkhuzhev[2], *Vladimir* Pogodaev[2,*], *Vladimir* Gukezhev[3], and *Vitaly* Vorokov[4]

[1]Kabardino-Balkarian State Agricultural University named after V.M. Kokov, Lenin Avenue, 1V, 360030 Nalchik, Russia
[2]Federal State Budgetary Scientific Institution North Caucasian Federal Scientific Agrarian Center, Nikonova str., 49, 546241 Mikhailovsk, Russia
[3]Institute of agriculture KBNTs RAN, Kirov str., 224, 360024 Nalchik, Russia
[4]Federal State Budgetary Educational Institution of Higher Education "Kuban State Agricultural University named after I.T. T. Trubilin", Kalinina str., 13, 350044 Krasnodar, Russia

Abstract. The purpose of this work was to establish the characteristics of growth, exterior, payment for feed by gain in live weight among young Aberdeen-Angus breed, originating from bulls of different body types. Hammer LLC of Karachay-Cherkessia conducted the experiment: from the offspring of 4 large bulls and 3 small bulls of the Aberdeen-Angus breed, 14 bulls were selected in each group. The investigation results on the intensive rearing of different young stock from weaning to 18 months of age showed that large-type bulls were superior to small-type bulls in terms of growth rate and better payment for feed by gain in live weight. At the age of 18 months, bulls from large-type parents reached a weight of 442 kg, and from small ones - 413 kg. These differences are statistically significant ($P > 0.99$) for the period from weaning to one and a half years old bulls, originating from bulls of a large type spent 8.1 EFU per 1 kg of gain, and bulls originating from bulls of a small type – 8.4 EFU.

1 Introduction

Beef cattle breeding is the most promising and economically efficient in districts where there are large areas of natural forage lands, which include the Karachay-Cherkessia Republic [1, 2]. Here you can keep large herds of beef cows and raise calves with full milk feeding up to 7-8 months of age, which, in combination with intensive fattening and post-weaning of young stock, makes it possible to organize beef production with minimal labour costs and expenses [3-5].

In this regard, in beef cattle breeding there is a tendency towards a transition from the extreme compact (solid) type with a short and square body on low legs to a type that is larger in live weight and well-defined meat forms [6-8].

* Corresponding author: pogodaev_1954@mail.ru

The Aberdeen-Angus breed, widespread in Russia, including in Karachay-Cherkessia, belongs to one of the world's best meat breeds [9-11]. However, in the last two decades, this breed has also developed towards the formation of a relatively small, but very early maturing type [12-14]. This raises the requirement of establishing the most desirable and promising type, improving the breeding methods and the methods for growing breeding young stock.

The purpose of this work was to establish the peculiarities of growth, exterior, payment for feed by gain in live weight among young Aberdeen-Angus breed, originating from stud bulls of different body types.

2 Materials and Methods

Hammer LLC of Karachay-Cherkessia conducted the experiment: from the offspring of 4 large bulls and 3 small bulls of the Aberdeen-Angus breed, 14 bulls were selected in each group.

The characteristics of bull parents in terms of live weight and the nature of body type are shown in Table 1.

The fathers of the first group bulls had a higher level of live weight compared to the fathers of the second group bulls (by 80 kg) and differed from them in high-leggedness, less broad body and massiveness. The mothers of the first group bulls also had a higher live weight (by 49 kg) in comparison with the mothers of the second group bulls, the body type differences were insignificant.

Immediately after weaning, they were placed on control rearing. The average age of the first group bulls at weaning was 8 months 22 days, and the second group bulls - 8 months 20 days.

Selection according to the principle of analogs in experimental groups was based on live weight, age, health status and origin.

The first group consisted of bulls descended from stud bulls of large tall type, the second group consisted of bulls descended from stud bulls of the small compact type.

The conditions of bull feeding and keeping in experimental groups were the same and corresponded to the farm rearing system.

During the experiment, according to the methodology, the following investigations were carried out: to inspect the growth and development of the experimental animals, they were weighed monthly, when setting up for the experiment at the age of 12, 15 and 18 months, and their main body measurements were taken.

In order to identify a more complete picture of the growth tension degree for experimental animals, the relative growth rate was calculated according to the formula:

$$К = \frac{Wt - W0 * 100}{Wt + W0 : 2},$$

where K is the relative increase in live weight (increase in percentage over a certain period of time); Wt is the weight of animals at the end of the reference period; Wo is the weight of the animal at the beginning of the reference period.

In order to investigate the degree of payment for feed by the experimental animals, a constant accounting of consumed feed was made by weighing feed on two adjacent days with the subsequent removal of residues;

After weaning, the experimental animals were housed in the barnyard. From December to March, the animals of the experimental groups were tethered at night, the rest of the time they were kept outdoors in the bases separately in groups. Feeding during this period was carried out three times a day. The animals received silage and concentrates indoors, hay in troughs outside.

Beginning from April until the end of the experiment, the animals were kept with no strings in the bases and only in bad weather they were housing indoors for the night. The animals were fed from troughs located in the bases 3 times a day.

The diet of the experimental animals included: alfalfa hay; corn silage, crushed barley, compound feed and sunflower meal. The corn silage and hay were of average quality.

The needs of animals for mineral feed were met both by the ration feed and by the additional supply of salt licks and chalk.

On a ration consisting of the above feed, the animals were kept until the end of June. At the end of June, last year's hay was replaced by a new crop of alfalfa hay, as well as a small amount of green wheat and Sudanese grass.

From August 18 to the end of the experiment, the experimental animals received chopped corn.

The feeding level was designed to obtain an average daily gain of 650-750 g in the period from 8 to 12 months and 750-850 g from 12 to 18 months of age. The protein nutritional value of the rations during the entire rearing of the animals was quite satisfactory. The content of phosphorus and especially calcium in the ration feed corresponded to the feeding norms of the young stock.

3 Results of the study

The weight growth of animals is one of the main indices in the inspection of meat productivity. Thanks to its accounting, it is possible, even during the life of the animal, to fairly accurately judge the amount of meat products that can be obtained from it, the development of the whole body, its requirements for nutrients, the cost of feed per unit of gain obtained.

In our experiment, the animals were weighed in the morning before feeding.

The change in the weight growth of bulls from weaning to 18 months return on average for the groups is shown in Table 1.

Table 1. The live weight dynamics of experimental animals.

Age, months	Group	n	M ± m	G	Cv	t_d
8	1	14	203 ± 4.8	17.55	8.6	2.50
	2	14	184±5.9	21.42	11.6	
12	1	14	284±5.2	19.08	6.7	1.34
	2	14	275±3,9	14.14	5.1	
15	1	14	358±7.9	28.57	7.9	1.32
	2	14	345±5,9	21.26	6.2	
18	1	14	442±7.4	26.7	6.0	3.26
	2	14	413±5,0	18.14	4.4	

Biometric treatment of the live weight of the experimental bulls showed that at the age of 8, 12 and 15 months, the difference between the groups was insignificant. At the end of the experiment, at the age of 18 months, the bulls of the first group exceeded in live weight the bulls of the second group somewhat less than at weaning (at 18 months by 7.0 and at 8 months by 10.3%), however, due to a decrease in the variability of the live weight of animals within the groups, the differences between them become statistically significant ($P > 0.99$).

The data given for changes in the amount of daily average increases (table 2) of the live weight of experimental bulls indicate that during sucking period the bulls of the first group gave higher increases than bulls of the second group.

Table 2. Change in average daily gains of experimental bulls by growth periods.

Group	Age periods, months				
	0-8	8-12	12-15	15-18	0-18
1	754	675	822	933	770
2	675	758	777	755	717

The reason for this is the better milk yield of mothers (larger ones in the first group), because the conditions of feeding and keeping cows with suckling calves were approximately the same, the heritability of live weight at this age was rather low (h = 0.28-0.30).

After weaning and up to 12 months of age, the first group bulls grew slightly worse and demonstrated average daily gains 11% lower than the second group bulls. During this period, the difference in average body weight between groups decreased from 10.3% at weaning to 3.3% at 12 months of age.

Starting from the age of 13 months, the second group bulls grew somewhat more slowly than the first group bulls, which until the end of the experiment had higher average daily gains. Over the entire experimental period, the first group bulls demonstrated average daily gains by 4.4% higher than the second group bulls and at the end of the experiment had a large live weight of 29 kg, or 7.0%.

Calculation of the relative growth rate once again confirms that up to 12 months of age the bulls of the second group had a higher growth energy compared to the bulls of the first group.

Starting from 13 months old, they lag behind in growth energy from the bulls of the first group. Both in the first and in the second group, the energy of growth decreases with a return, but in the second group this decrease occurs more sharply than in the first one (by 13.2% in group 1 and by 21.8% in group 2).

Feed payment is determined by the amount of nutrients expended per unit of gain. Throughout the experiment, the consumption of feed per one head, both in the first and in the second group, was practically the same. Therefore, the consumption of feed by animals in the experimental groups changed over the growing periods, corresponding to the changes in the average daily gains (Table 3).

In the period from 8 to 12 months of age, the bulls of the second group gave higher average daily gains and spent 6.6% less feed units on it than the bulls of the first group. Subsequently, a decrease in the average daily gain of bulls in the second group led to an increase in feed consumption by 1 kg of gain in comparison with the first group.

The difference in the consumption of feed per 1 kg of gain, both on average for the entire time of the experiment, and for individual periods, was small and amounted to 6-8%. The level of digestible protein consumption by groups almost completely corresponded to the consumption of feed units.

Table 3. Feed costs per 1 kg of live weight gain by experimental animals according to rearing periods.

Period (month.)	Spent per 1 kg of gain					
	Group 1				Group 2	
	Energetic feed unit	% to 2nd gr.	Protein, g	% to 2nd gr.	Energetic feed unit	Protein, g
0-8	6.0	93.7	1038	93.3	6.4	1113
8-12	9.6	106.6	1013	107.0	9.0	950

12-18	7.3	91.2	740	90.2	8.0	820
8-18	8.1	96.6	830	95.8	8.4	867
0-18	7.1	94.6	930	94.7	7.5	982

The differences in the efficiency of feed use, obtained in our experience, are very insignificant and statistically uncertain.

The second index of animal growth is linear growth. Its accounting through the systematic taking of linear measurements is additional evidence of the general level of the animal development.

The main purpose of introducing measurements of animals and assessing their exterior by measurements was to make it more accurate and get rid of subjectivity, which is possible with eye assessment.

In our experiment, in order to investigate the features of linear growth and differences in the body type of bulls of compared groups, 11 basic measurements were taken (Fig. 1, 2, 3, 4).

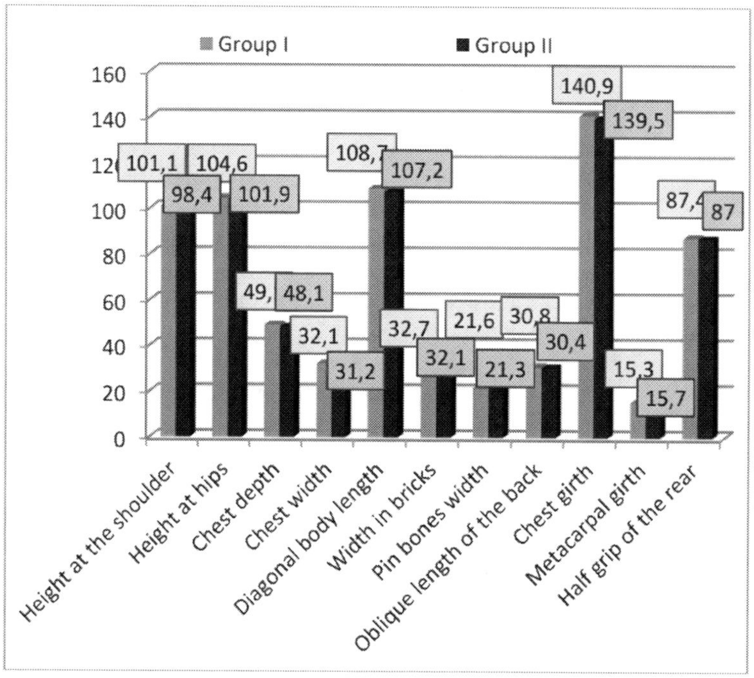

Fig. 1. Measurements of bulls body points at the age of 8 months (n=14), cm.

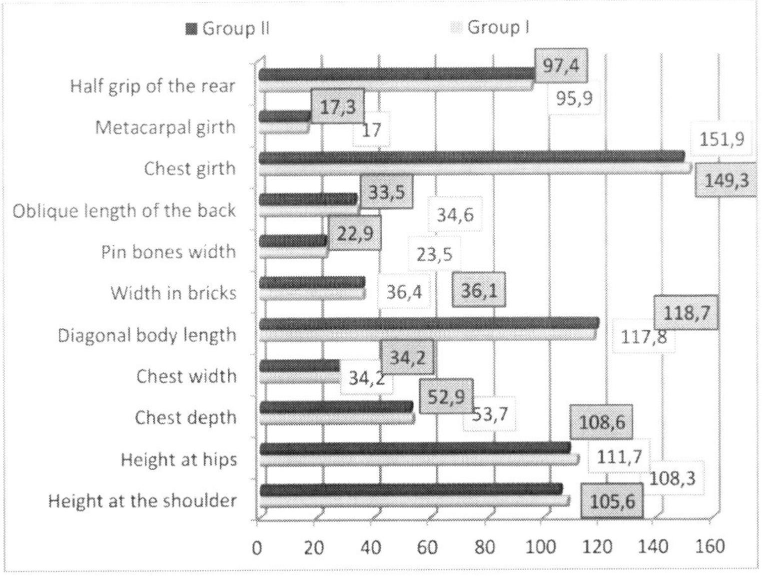

Fig. 2. Measurements of bulls body points at the age of 12 months (n=14), cm.

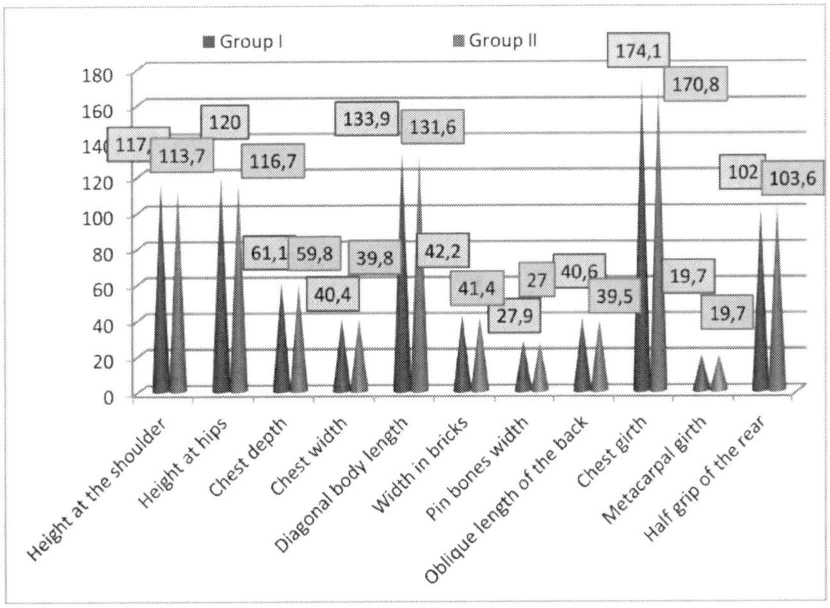

Fig. 3. Measurements of bulls body points at the age of 18 months (n=14), cm.

It was established that the animals of the first group, as larger, exceeded the animals of the second group on all measures (except semi-capture of the back) and, in particular, on such measures as height at the withers and sacrum, the depth of the chest, the oblique length of the back and the width in the pin bones. And these differences took place in all age periods. The external profile 3 shows that with age the differences in measurements changed differently. Differences in width measurements in the pin bones and oblique length of the back, slightly smaller in the depth and girth of the chest and least of all in height at the withers and sacrums, have increased the most.

The difference in the size of the chest width measurements decreased from 8 to 18 months of age. The first group bulls, who were eight months old and cede, were equal to the bulls of the second group in the metacarpal girth at 18 months. The double-capture pro sulfur rate, almost identical at 8 months of age, at the age of 18 months was higher for the second group.

At the end of the experiment, the most typical bulls of the first group looked relatively higher-legged and narrow-bodied.

The age changes obtained in our experience indicate that the animals of the first and second groups gave the smallest increase in height at the withers and in height at the sacrum. Measurements such as chest width, maclock width, width in the pin bones and oblique length of the back gave the greatest increase.

Age changes for animals of the first and second groups were similar. Insignificant differences were only for changing the measurements of the width in the pin bones, the oblique length of the back and the metacarpal girth. The size of these measurements, during the investigation period, increased slightly faster for bulls of the first group than the second one.

Currently, in the practice of zootechnical investigations a large number of indices is used, characterizing the body type of the animal. In our work, we used such indices which are able to most clearly show the differences in the body type nature of large tall and small compact animals. Indices of wideness, massiveness, blockiness, long-leggedness and stretch meet the task most of all.

Body indices show that the bulls of the first group in all age periods were more high-legged than the bulls of the second group, which were relatively more wide-bodied and low-legged (Figure 4, 5, 6)

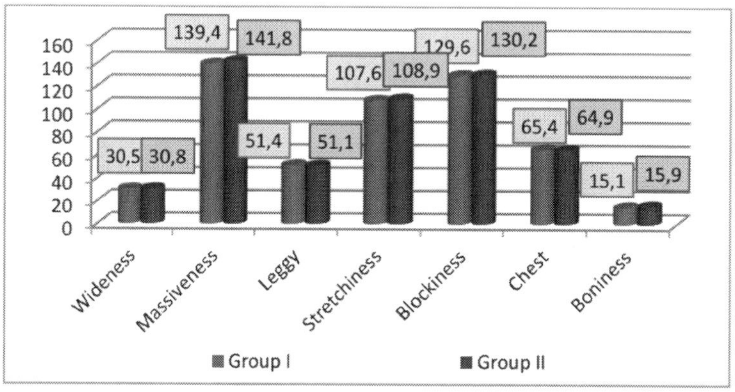

Fig. 4. Body indices of experimental bulls at 8 months, %.

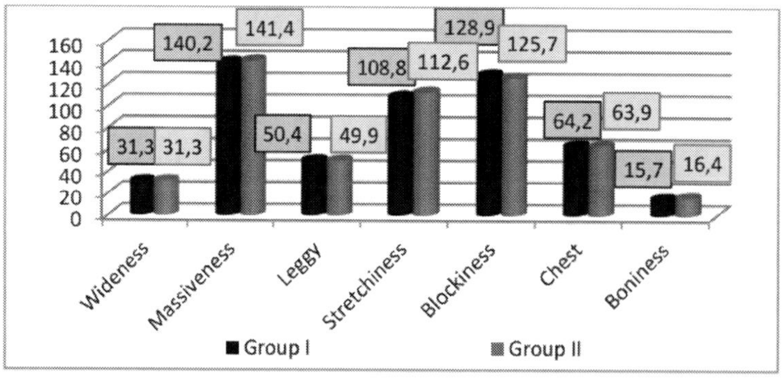

Fig. 5. Body indices of experimental bulls at 12 months, %.

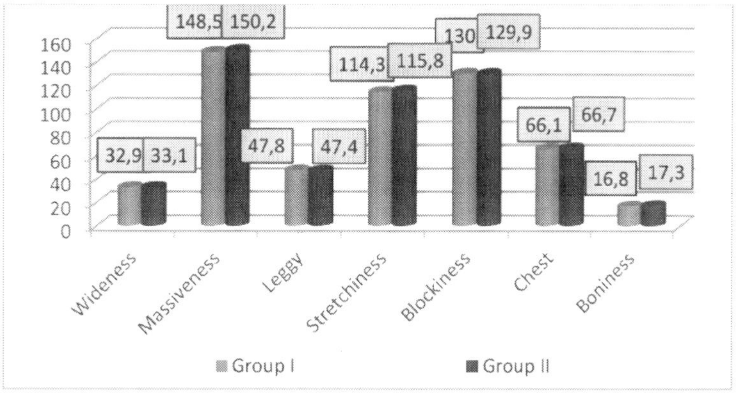

Fig. 6. Body indices of experimental bulls at 18 months, %.

The bone index of the bulls of the second group was higher, because they had a shorter but relatively wider index. We see a similar picture for stud bulls. Bulls of the small compact type have the bone index 2.8% higher than bulls of the large tall type.

With age, the long-legged index decreased (by 7%), while wideness, massiveness and stretch indices increased (by 7.9%; 6.5%; 6.2% for the first group and 7.5%; 7.2%; 6.3%; for the second group). The bone index (11.3% for the first group and 8.8% for the second group) increased significantly with age.

It should be noted that only carrying out the experiment under conditions of sufficient abundant feeding allowed large animals of a tall body type to show genetic capabilities at the age of 1.5 years, while when grown under normal economic conditions, differences in the live weight of the offspring of large and small parents become reliable only when they are older.

As shown by a number of investigations, small-scale animals finish their development earlier and at an earlier age begin to put off a significant amount of fat in the body. These animals, as more precocious are able to form a body type characteristic of adult animals much earlier. Increased fat deposition and its share in the resulting increase causes increased feed consumption per unit of growth.

As a result of the early completion of development and increased fat formation for small-scale animals, there is a faster decrease in the growth energy and when fattened they do not reach such a living mass as older animals. In the latter, the process of development is longer, fat in the body is deposited less, they last longer retain high growth energy and are able to achieve a higher live mass.

4 Conclusion

The results of the Aberdeen-Angus breeding bulls assessment made it possible to distinguish two main types of animals by the body type and live mass: a large tall and small compact.

The investigation results on the intensive rearing of different young stock from weaning to 18 months of age showed that bulls originating from the bulls of large-scale stud bulls exceeded bulls from small type parents for meat productivity which was expressed in a higher growth rate and better feed payment by the increase in live mass. At the age of 18 months, bulls from large-type parents reached a weight of 442 kg, and from small ones - 413 kg. These differences are statistically significant (P>0.99) for the period from weaning to one and a half age, large type bulls spent 8.1 EFU per 1 kg of gain, and small type bulls spent 8.4 EFU.

References

1. V. Kulintsev, M. Shakhmurzov, A. Shevkhuzhev, V. Pogodaev, D. Smakuev, Int. J. of Engineering and Advanced Technology, **9(1)**, 4561-4564 (2019)
2. V. V. Onishchenko, N. S. Dega, F. H. Bostanova, Int. J. of Humanities and Natural Sciences, **1(1)**, 29-35 (2019)
3. P. Aparecida C. Luz, C. Andrighetto, et al., Effect of integrated croplivestock systems in carcass and meat quality of Nellore cattle Livestock Sci., **220**, 83-92 (2019)
4. M. V. Tarasov, V. M. Gabidulin, V. Yu. Shmakov, Beef cattle bulletin, **63(3)**, 71-76 (2010)
5. O. Sheveleva, Production of beef on the basis of the development of specialized meat cattle breeding, **11**, 23-27 (2008)
6. B. Chugai, A. Betin, A. Frolov, Animal husbandry of Russia, **2**, 45-46 (2010)
7. A. F. Shevhuzhev, N. I. Belik, D. R. Smakuev, Research J. of Pharmaceutical, Biological and Chemical Sciences, **7**,**(4)**, 430-434 (2016)
8. Kucherenko, Animal breeding in Russia, **3**, 6-7 (2009)
9. V. A. Pogodaev, V. V. Golembovsky, V. I. Komlatsky, L. F. Velichko and L. I. Konkov, IOP Conf. Series: Earth and Environmental Science, 624, 012134
10. M. Tofastrud, A. Hessle and Y. Z. B. Rekdal, *Weight gain of free-ranging beef cattle grazing in the boreal forest of south-eastern Norway Livestock Sci.,* 2 (2020)
11. L. I. Kibkalo, V. V. Bychkov, Bulletin of the Kursk State Agricultural Academy, **1**, 70-71 (2010)
12. V. Pogodaev, A. Arilov, A. Petenko, A. Soldatov, T. Pakhomova, Research J. of Pharmaceutical, Biological and Chemical Sciences, RJPBCS, **9(4)**, 755-759 (2018)
13. A. M. Pardo, M. A. Elzo, L. T. Gama and L. M. Melucci, *Genetic parameters for growth and cow productivity traits in Angus, Hereford and crossbred cattle Livestock Sci.*, 233 (2020)
14. M. J. D'Occhio, P. S. Baruselli and G. Campanile, Influence of nutrition, body condition, and metabolic status on reproduction in female beef cattle, 125, 277-284 (2019)

Effect of breeding chickens (gallus gallus l.) of the plymouth rock domestic meat cross "smena 9"

Zh. V. Emanuylova[1], *A.V.* Egorova[2,*], *D. N.* Efimov[2], *I. A.* Egorov[2], and *A. A.* Komarov[1]

[1]SGC "Smena" - a branch of the Federal State Budgetary Scientific Institution Federal Scientific Center "All-Russian Research and Technological Institute of Poultry" of the Russian Academy of Sciences (FSC "VNITIP" RAS), 141357 Russia
[2]FSBSI Federal Scientific Center "All-Russian Scientific Research and Technological Institute of Poultry" of the Russian Academy of Sciences (FSC "VNITIP" RAS), 141311 Russia

Abstract. The article is devoted to the assessment of the efficiency of selection of the initial lines of the Plymouth Rock breed of the domestic meat cross "Smena 9". It was found that breeding of meat chickens of lines CM7 and CM9 (from 2016 to 2020) resulted in an increase in the live weight of the young, fleshing of the chest and legs at 35 days of age, the feed conversion for both cockerels and hens by 8.8-10.0%; 0.9-1.9% and 1.0-1.9%; 1.2-1.8%. In 2020, all juveniles of the paternal line CM7 at one day of age were the carriers of the marker gene *K* (100%), the maternal line Cm9 - the carriers of the marker gene *K* (100%). The accuracy of sexing of the maternal parental form CM79 is 99.7%. From one parental pair of the Smena 9 cross, 307.6 kg of meat was obtained, which is 14.2% higher than in the previous cross. The average daily gain of the final hybrid of the cross "Smen. 66 kg/kg; a 9" grown in the production conditions of the SGC "Smena" was 63.5 g; feed costs were at 1.66 kg/kg; preservation rate - 98.8%; the output of the pectoral muscles and the slaughter yield - 23.5 and 73.1% (against 56.3 and 71.4%). The bird of the cross "Smena 9" is competitive.

1 Introduction

Dynamic development of domestic poultry farming causes the necessity of constant creative search of ways and methods of enhancing the efficiency of poultry and the quality of its products [1-3].

Intensive selection of highly productive chickens of meat crosses inevitably leads to a decrease in genotypic diversity in the original lines, which reduces the effect of selection to further increase the genetic potential of the bird. Therefore, after 7-8 years of in-depth breeding work with industrial crosses of meat chickens, it becomes necessary to include new traits in breeding programs that will increase the yield of broiler meat from layers of the parent flock [1,4,5].

* Corresponding author: egorova@vnitip.ru

The efficiency of breeders in the line, nucleus stock and breeder flocks is determined by the number of quality eggs obtained for incubation and quality chicks obtained from these eggs. This process begins from SGC, breeding farms, where pure lines are bred, enhanced and crossed. The gene flow then travels through the breeder and breeder flocks to broilers. On this path, there are numerous factors affecting the manifestation of traits determined by genes. The success of the pedigree and industrial poultry farming depends on continuous genetic enhancement and poultry creation at all levels of appropriate feeding conditions, housing, biosafety.

To obtain the effect of heterosis, breeders use a differentiated approach to the issues of directed selection of hens of the parental and maternal forms, which is part of the scheme for the implementation of the cross breeding program [2,4].

The genetic value of the meat cross largely depends on the productive potential of the original lines, which are the basis for the construction of the cross [5,6].

To obtain the effect of heterosis, breeders use a differentiated approach to the issues of directed selection of hens of the parental and maternal forms, which is part of the scheme for the implementation of the cross breeding program [2,4].

In recent years, the poultry industry has increasingly begun to use marker genes in breeding work allowing to enhance the economic efficiency in the production [7-11].

Increasing the meat yield from laying hens is one of the most important tasks for breeders. In this regard, the breeding qualities of birds are being enhanced by using new techniques and breeding methods in combination with technological and veterinary programs. One of these techniques is the use of the slow feathering gene when creating initial lines, further crossing of which allows to obtain autosex offspring in the parental form or in the final product - broiler [2,7,10].

Further breeding progress should be based on the study of these markers and other DNA-related methods [12-16]. The use of marker genes greatly facilitates the separation of the day old cockerels and hens. There are genes that determine the autosexuality of chickens by the color of the plumage (colorsex) and by the growth rate of the feathers of the covering and primary plumage (feathersex) [17-23].

2 Material and methods

The research was performed in the production of the SGC "Smena" (Moscow region). The object of research was the poultry of the original lines of the Plymouth rock breed of the breeding of this enterprise (cm7 and CM9), maternal parental forms CM79, B79 and broilers (CM5679) of the Smena 9 cross.

The livestock of the evaluated young stock pursuant to the CM7 line in 2016 amounted to 1182 birds (cockerels) and 1175 birds (hens); on the line CM9 -1235- 1242 birds; in 2020 -5695 -5708 (lines CM7), 7400-7398 birds. (line CM9) for each maternal parental form (CM79 and B79) - 500 birds; for the final hybrid - 1000 birds.

Keeping chickens and raising young stock, broilers - on deep bedding. The adult birds were kept in natural mating.

The main technological parameters, light and temperature-humidity conditions, the poultry feeding program corresponded to the standards used for meat chickens, methodological guidelines for poultry feeding [24].

Linear birds (CM7 and CM9) were kept in breeding nests for 13 chickens and one cockerel, which were completed considering the following indicators: live weight, fleshiness, homozygosity for the rapid feathering gene *kk* and k^- (line CM7); egg production, hatching eggs yield, their weight, hatchability, live weight of poultry, fleshing of the chest, legs, feed costs, homozygosity for the gene for slow feathering *KK* and K^- (line CM9).

The autosex maternal parental form was obtained by crossing the lines of the Plymouth Rock breed of the paternal line CM7 and the maternal line CM9.

In the original lines (CM7 CM9) and the maternal parental form (CM79), feathering was determined by the development of wing feathers (slow-feathering individuals - coverts longer than flight feathers or equal to them; fast-feathering - coverts are shorter than flight feathers and well developed). In addition, with the use of PCR (PCR-RT), molecular genetic typing of feather pulp samples was performed, which was performed at JSC "Syntol" (g. Moscow), in order to maintain the homozygosity of the original lines for the *K* and *k* alleles.

Anatomical cutting of broiler carcasses was performed pursuant to the method [25].

For biochemical studies, blood was obtained from the jugular vein, sodium citrate solution was added to the test tube, then the blood was centrifuged for 5 minutes (5000 rpm). The obtained plasma was studied on a semi-automatic analyzer Sinnowa BS3000P (China), cholesterol, triglycerides, total protein, calcium, phosphorus were determined.

All experimental data obtained in the course of the research were processed by the method of variation statistics pursuant to Student. The data were also processed using the Statistica 10.0 software package (StatSoft Inc., USA) and Microsoft Excel. The reliability of the differences between the compared indicators was determined by the Student's *t*-criterion within the following levels of significance: *- $P \leq 0.05$; **- $P \leq 0.01$; ***- $P \leq 0.001$.

3 Results and discussion

A feature of the breeding program for the creation of a cross of meat chickens with an autosex maternal parental form based on marker genes of slow and fast feathering was the selection of birds of the paternal line CM7 for fast feathering, and the maternal line CM9 for slow feathering. Evaluation and selection of cockerels in breeding nests by the feathering rate of offspring at one day of age for a number of years was performed in order to consolidate the entire population of the line CM7 (2357-11408 birds - fast-fledging) and CM9 (2477-14797 birds - slow-fledging). Heterozygous cockerels and their offspring were excluded from the breeding process, leaving only homozygous for the *K* and *k alleles.* In addition, molecular genetic typing of chickens for these alleles was performed simultaneously using quantitative PCR (RT-PCR).

The results of assessment and selection for alleles *K* and *k* 1-day-old chickens are presented in Table 1.

Table 1. Feathering of day old young stock of the original lines CM7 and CM9 in the process of selection.

Line	Feathering type of day-old chicks	Gender	Year		
			2016	2018	2020
CM7	wing coverts longer than or equal to flight feathers (slow-feathering)	cockerels	89.1	60.4	0
		hens	99.2	58.9	0
	wing coverts are shorter than flight feathers and well developed (fast feathering)	cockerels	10.9	39.6	100
		hens	0.8	41.1	100

Table 1. Continued.

CM9	wing coverts longer than or equal to flight feathers (slow-feathering)	cock-erels	93.0	10	100
		hens	83.1	100	100
	wing coverts are shorter than flight feathers and well developed (fast feathering)	cock-erels	7.0	0	0
		hens	16.9	0	0

Table 2. Productive qualities of birds in the selection process in the original lines of the Plymouth Rock breed in 2016 – 2020.

Sign	Line			
	paternal (CM7)		maternal (CM9)	
	2016	2020	2016	2020
Live weight at 7 days of age, g: ♂ ♀	211.0±0.367 206.8±0.362	216.0***±0.404 213yu2 ***±0.415	199.8±0.31 0 194.0±0.31 1	207.5 ***±0.305 200.9 ***±0.306
Live weight of chickens in 35 days, kg ♂ ♀	2.231±0.004 1.880±0.004	2.428 ***±0.004 2.059 ***±0.003	1.970±0.00 3 1.700±0.00 4	2.167 ***±0.004 1.821 ***±0.003
Muscularity at 35 days, score ♂ - chest	4.58±0.006 4.58±0.005	4.63 ***±0.005 4.62 ***±0.004	4.36±0.006 4.40±0.005	4.41 ***±0.005 4.45 ***±0.004
♀ ♂ - legs ♀	2.20±0.002 2.10±0.002	2.24 ***±0.002 2.13 ***±0.002	2.09±0.003 2.08±0.002	2.13 ***±0.002 2.10 ***±0.002
Conversion of forage, kg/kg (for 1-35 days) ♂ ♀	1.67±0.014 1.69±0.016	1.64±0.011 1.67±0.013	1.68±0.012 1.73±0.012	1.66±0.010 1.70±0.012

Note. Significant differences compared to F1 *** - P≤0.001.

The above data indicate that in the process of breeding meat chickens of the Plymouth Rock breed of lines CM7 and CM9 from 2016 to 2020 there was an increase in live weight, fleshing of the chest and legs at 35 days of age, feed conversion for both cockerels and chickens by 8.8-10.0%; 0.9-1.9% and 1.0-1.9%; 1.2-1.8% (Table 2).

The breeding progress in one year for the main breeding traits of the lines CM7 and CM9 was: live weight at 35 days of age in cockerels - 39.4 g; chickens - 35.8-24.2 g; fleshing of the chest, legs - 0.22-0.38% (cockerels); 0.17-0.28% (chickens); feed conversion - 0.36-0.24% (cockerels); 0.24-0.35 (chickens).

During 60 weeks of life, 136 heads were obtained from layers of the parental CM79 line, which is 27.1% higher than for the CM7 line (107 birds), for the CM9 line (12 birds) - by 9.7%, and compared to the parental form of the cross "Smena 8" (B79 - 134 head) - by 1.5%, and in terms of meat yield from one parental pair - by 14.2% (CM79 -307.6; B79 - 209.5)

The average daily gain of the final hybrid of the Smena 9 cross was 63.5 g; feed costs were at 1.66 kg/kg; safety - 98.8%, the yield of pectoral muscles and slaughter yield - 23.5 and 73.1% (versus 56.3g; 1.78 kg/kg; 97.7%; 21.4% and 71.4%) corresponding to the indicators of broilers of the Smena 8 cross.

Table 3. Biochemical blood parameters in meat chickens of the Smena 9 cross at different age periods (n = 20, M ±SEM).

Line and form	Indicators				
	cholesterol, mmol/l	triglycerides, mmol/l	Total protein, g/l	calcium, mmol/l	phosphorus, mmol/l
1 day					
CM7 line	5.4±0.03	2.2±0.06	28.0±0.33	3.2±0.12	3.6±0.12
CM9 line	5.3±0.05	2.3±0.07	28.4±0.40	3.4±0.11	3.8±0.15
MPF SM79	5.3±0.07	2.3±0.06	29.7±0.51	3.4±0.10	4.2±0.16
1 week					
CM7 line	5.2±0.04	2.2±0.04	39.5±0.61	4.6±0.10	2.2±0.08
CM9 line	5.1±0.03	2.4±0.08	41.0±0.57	4.6±0.08	2.3±0.06
MPF SM79	5.1±0.04	2.2±0.06	39.3±0.74	4.7±0.07	2.5±0.13
2-week					
CM7 line	5.4±0.06	3.2±0.06	34.3±1.12	4.6±0.12	2.8±0.06
CM9 line	5.3±0.06	3.2±0.08	34.0±1.24	5.1±0.11	2.8±0.10
MPF SM79	5.7±0.12	3.5±0.12	35.2±0.61	4.4±0.08	2.4±0.08
3 week					
CM7 line	3.5±0.04	3.1±0.05	35.0±0.77	3.9±0.09	1.9±0.07
CM9 line	3.6±0.03	3.0±0.07	34.9±1.02	4.4±0.11	1.9±0.15
MPF SM79	3.6±0.04	3.1±0.06	33.5±0.51	4.5±0.11	1.8±0.11
4 week					
CM7 line	3.7±0.05	3.2±0.05	39.8±1.01	3.4±0.12	2.0±0.08
CM9 line	3.8±0.05	3.2±0.06	41.7±0.88	4.6±0.14	2.0±0.06
MPF SM79	4.1±0.11	3.4±0.08	40.9±1.07	4.5±0.07	2.5±0.11
5 week					
CM7 line	4.4±0.10	3.1±0.11	42.6±1.02	2.9±0.04	2.4±0.05
CM9 line	4.6±0.04	2.9±0.12	42.9±0.80	3.1±0.06	2.3±0.04
MPF SM79	4.4±0.06	3.3±0.16	41.9±1.07	3.2±0.12	2.4±0.06

Table 3 shows the biochemical parameters of blood in meat chickens at different age periods (1-day-old, 1-, 2-, 3-, 4-, 5-week old) along the lines CM7, CM9 and the maternal parental form CM79.

In day-old chickens, the level of cholesterol and triglycerides did not differ significantly from that of older chicks. A decrease in cholesterol indices was established at 3 weeks of age in chickens of the CM7 paternal line by 35.2%, in the CM9 maternal line and the maternal parental form (CM79-MPF) - by 30.3% compared with the daily age and practically remained at the same level until 5 weeks of age. At this age, an increase in cholesterol values was noted as compared with the 4-week age by 18.9%; 21.1%; 7.3% pursuant to lines and shape. Our data on changes in cholesterol with age are consistent with other researchers [26-28]. Lipid metabolism is influenced by limited feeding of young meat chickens [29]. The amount of triglycerides in the blood of young animals of the initial lines CM7, CM9 and two-line combination CM79 remained at the same level until 2 weeks of age. In chickens of the CM7 line at 2 weeks of age, an increase in this indicator was observed by 45.5%, in the CM9 line - by 39.1%, and in the MRF - by 52.2% compared to the one-day-old age. This level of triglycerides in the blood was maintained up to 5 weeks of age of the young. According to a number of authors, the level of triglycerides in broiler chickens changes with age [30,31].

In our studies, the concentration of total protein in the blood plasma of 1-day-old chickens of lines CM7, CM9 and MRF CM79 was lower than at 1-, 2-, 3-, 4-, 5- weeks of age, which is consistent with the data of a number of authors (27.28). B.I. Kuznik [33] associates this with a low function of protein biosynthesis [32,33]

The index of total blood protein at 1 week of age was higher than in 1 day old chickens by 32.3-44.4%. The juveniles of the CM9 line exceeded this indicator in the blood plasma of the CM7 poultry by 3.8% and the MRF - by 4.3%. At 2 weeks of age, the amount of total protein in the blood plasma of chickens of this line was 34.0 g/L and remained at this level until 3 weeks of age. Then this indicator increased to 41.7-42.9 g/L ($P < 0.05$). In broilers, the authors of [34] note the growth of total protein from 14 to 42 days of age.

The indicator of calcium in the blood plasma of chickens at the age of 1 day had a low level, then in the first week it increased in the CM7 line by 43.8, in the CM9 line - by 35.3%; in the MPF - by 38.2%.

In our studies, it was found that in chickens MPF CM79 and chickens of the CM9 line at 2 weeks of age there were differences in the content of calcium in the blood plasma, in the CM9 line the calcium content was 15.9% higher than in the MPF CM79 ($P < 0.001$) and chickens. CM7 - by 10.9% ($P < 0.01$). At 4 weeks of age, an increase in the calcium content in the blood of chickens of the CM9 line was noted by 35.3% and the MPF - by 32.4% in comparison with the CM7 line. By the age of 5 weeks, the plasma calcium level reached 2.9 - 3.2 mmol/l.

Day-old chicks showed a high level of phosphorus, then it decreased 1.6-1.7 times (1 week), then (4 and 5 weeks) there was an increase in this indicator. At 2 weeks of age in chickens of lines CM7 and CM9, it was higher by 16.7% compared to the combination of CM79, and after 2 weeks, the phosphorus content in the blood plasma of MPF was 25.0% higher than in chickens of lines CM7 and CM9.

The ratio between calcium and phosphorus at the day-old age of the bird was within 1:0.8 - 1:0.9, which is associated with the embryonic period [35]. At 1 week of age, the ratio changed in the direction of increasing calcium up to 1.9: 1 - 2.1: 1; in the subsequent age period up to 2.5:1 (MPF); 2.3:1 (line CM9); 2.1:1 (line CM7). At 4 weeks, the calcium-phosphorus ratio changed in chickens of the CM9 line (2.3:1), and at 5 weeks of age it decreased to 1.2: 1 - 1.4:1.

The level of total protein in blood plasma increases with the age of the experimental bird, which was associated with the enhancement of the protein-educational function (27). Triglyceride levels rise by 33.3-59.1 by 2 weeks of age. There are differences in the level of calcium, phosphorus, and cholesterol at 2- and 4 weeks of age, and triglycerides - at 1-week of age between the MPF chickens and the parental lines.

For the first time, we obtained data on the biochemical parameters of blood plasma in birds of the original lines CM7, CM9 and the maternal parental form CM79 of the new cross "Smena 9", which are consistent with the data of other authors [30-32,34,35].

4 Conclusions

In the course of breeding meat chickens of lines CM7 and CM9 (from 2016 to 2020), there was an increase in live weight of young birds, fleshing of the chest and legs at 35 days of age, feed conversion for both cockerels and chickens by 8.8-10.0%; 0.9-1.9% and 1.0-1.9%; 1.2-1.8%. To maintain cockerel homozygosity for the *K* and *k* alleles, the offspring were evaluated annually by phenotype and using PCR. In 2020, all juveniles of the paternal line CM7 at one day of age were the carriers of the marker gene *K* (100%), the maternal line Cm9 - the carriers of the marker gene *K* (100%). The accuracy of sexing of the maternal parental form CM79 is 99.7%. From one parental pair of the Smena 9 cross, 307.6 kg of meat was obtained, which is 14.2% higher than in the previous cross.

The average daily gain of the final hybrid of the cross "Smen. 66 kg/kg; a 9" grown in the production conditions of the SGC "Smena" was 63.5 g; feed costs were at 1.66 kg/kg; preservation rate - 98.8%; the output of the pectoral muscles and the slaughter yield - 23.5 and 73.1% (against 56.3 and 71.4%).

The bird of the cross "Smena 9" is competitive.

References

1. S.V. Cherepanov, Topical issues of breeding work in the poultry industry in Russia, Poultry farming, **9**, 2-4 (2018)
2. A.V. Egorova, The main directions of work with meat chickens of the parent herd, Poultry farming, **3**, 16-21 (2017)
3. S. Cherepanov, O. Stanishevskaya, Preservation and usage of genetic resources in farm poultry, The Proc. XXV World's Poultry Cong., **343** (2016)
4. Ya. S. Roiter, A.V. Egorova, A.P. Konopleva, et al., *Selection and breeding work in poultry farming* (2016)
5. I. L. Galpern, V.V. Sinichkin, O. I. Stanishevskaya, A.G. Bychaev, E.S. Fedorova, *Acceleration of the rate of genetic progress of the productive traits of egg and meat chickens* (2009)
6. A. V. Egorova, E. Zh. V. Manuylova, D. N. Efimov, L. I. Tuchemskiy, Evaluation of meat hens of the original lines of a selection herd by growth rate, Poultry farming, **6**, 8-13 (2018)
7. A.V. Egorova, L.V. Shakhnova, Separation of autosex meat chickens by sex, Poultry and Poultry Products, **3**, 43 (2013)
8. T. Petrukovich, Separate growing of broilers, Livestock Russia, **12**, 11-12 (2017)
9. D. N. Efimov, Zh.V. Emanuylova, E. V. Zhuravleva, A.V. Egorova, V. I. Fisinin, Agricultural Biology, **53**,1162-1168 (2018)
10. O. W. Willems, S. P. Miller, B. J. Wood, World's Poultry Science J., **69(1)**, 77-78 (2013)
11. G. Bu, G. Huang, H. Fu, J. Li, S. Huang, Y. Wang, J. Mol. Endocrinol, **51**, 261-276 (2013)
12. H. Cerit, K. Avanus, Turk.J. Vet. Anim. Sci., **31(6)**, 371-374 (2007)
13. H. Cerit, K. Avanus, World's Poultry Science J., **63(1)**, 91-99 (2007)
14. L.G. Korshunova, *Transgenesis and gene expression in poultry: summary of thesis* (2012)
15. C. Lessells, C. Mateman, Molecular Ecology, **7(2)**,187-195 (2002)

16. J. Zhao, J. Yao F. Li, Z. Yang, et al., Poultry Sci., **95(7)**, 1498-1503 (2016)
17. Ya. I. Alekseev, A. M. Borodin, A. V. Nikulin, Zh. V. Emanuilov, D. N. Efimov, V. I. Fisinin, Agricultural Biology, **52(2)**, 367-373 (2017)
18. L. Korshunova, Y. Royter, A. Egorova, The usage of gene modifiers in selection of new forms of color- and feather-sex poultry, Proc. XIV European Poultry Conf., 512 (2014)
19. Y. Roiter, A. Egorova, A. Sevastianova, L. Korshunova, The selection of autosex interlinear forms of poultry (chicken, geese, Guinea fow), Proc. XXV World's Poultry Congress, 257 (2016)
20. P. K. Bramwell, *Sexing chicks in the backyard flock, The Poultry Site*, http://www.thepoultry site.com/
21. Y. H. Cheng, T. E. Kuo, D. N. Lee, C. F. Weng, Zoological Studies, **45**, 104-113 (2006)
22. D. A. Dawson, S. Darby, F. M. Hunter, A. P. Krupa, I. L. Jones, T. A. Burke, Molecular Ecology Notes, **1(13)**, 201-204 (2001)
23. H. Ellegren Hens, EMBO Reports, **2(3)**, 192-196 (2001)
24. I. A. Egorov, V. A. Manukyan, T. M. Okolelova, et al., *Guide to feeding agricultural bird / Under* (2018)
25. V. S. Lukashenko, M. A. Lysenko, T. A. Stollyar, A. Sh. Kavtarashvili, V. V.Dyachko vskaya, A. I. Kalashnikov, *Methods of anatomical cutting of carcasses, organoleptic assessment of the quality of meat and eggs of poultry and morphology of eggs* (2013)
26. O. S. Kotlyarova, *Characteristics of immunomorphological and biochemical parameters of broilers in ontogenesis in conditions of industrial poultry farming in Western Siberia* (2013)
27. S. A. Ermolina, K. V. Buldakova, V. A. Sozinov, Successes of modern natural science, **9**, 34-37 (2014)
28. A. G. Koschaev, *Biotechnology of production and use of functional feed additives for poultry* (2008)
29. X. Yang, J. Zhuang, K. Rao, X. Li, R. Zhao, Res. Vet. Sci. Dec., **89(3)**, 438-444 (2010)
30. E. A. Nazarova, *Physiological and biochemical status and productive qualities of broiler chickens with the combined use of lactoamilovorin and sodium selenite* (2012)
31. T. O. Azarnova, M. Naydensky, A. Bobylkova, Hypothesis of early development of embryos, Cattle breeding in Russia, **7(13)**, 1-5 (2012)
32. I. A. Egorov, A. A. Grozina, V.G. Vertiprakhov, T.N. Lenkova, V. A. Manukyan, T. V. Egorova, M. V. Koshcheeva, Agricultural biology, **53(4)**, 820-830 (2018)
33. B.I. Kuznik, *Physiology and pathology of the blood system* (2002)
34. A. Piotrowsska, K. Burlikowska, R. Szymeczko, Folia biologica Krakow, **59(3-4)**, 18 (2011)
35. I. A. Egorov, T. N. Lenkova, V. A. Manukyan, et al., *Physiological and microbiological features of digestion chickens of meat breeds in the embryonic and postembryonic periods to create new feeding technologies that ensure the fullest possible realization of the genetic potential of the bird* (2019)

Polymorphism of CAST gene in five sheep breeds in Bulgaria

Ivona Dimitrova[1,*], *Milena* Bozhilova-Sakova[2], *Atanaska* Teneva[1], *Tanya* Ivanova[2], and *Ina* Stoycheva[3]

[1]Agronomy Faculty, University of Forestry, Sofia, Bulgaria
[2]Institute of Animal Science, Kostinbrod, Bulgaria
[3]Institute of Forage Crops, Pleven, Bulgaria

Abstract. CAST gene is a candidate marker that influences the intensity of growth and meat quality. The aim of this study was to investigate the genetic variability of CAST gene in five Bulgarian sheep breeds – two merino, two local and one for milk. A total of 150 ewes, belonging to these breeds were investigated for polymorphisms of CAST gene by using PCR-RFLP method. A 622 bp fragment of *Ovis aries* CAST gene was amplified using PCR. After restriction with endonuclease *MspI* two alleles were observed in Ascanian, Caucasian, Breznik and Pleven Blackhead breeds. One allele (M) and one genotype (MM) were detected in Cooper-Red Shumen breed. The highest frequency of the allele N was established in the Ascanian merino breed (0.27) followed by Caucasian merino (0.13), Breznik (0.12) and Pleven Blackhead (0.07). Two genotypes - MM and MN, were observed in Caucasian, Breznik and Pleven Blackhead breeds. The three possible genotypes were found only in Ascanian merino ewes - MM, MN and NN. The lowest frequency of the homozygous genotype MM (0.50) and the highest frequency of the heterozygous genotype MN was established in the Ascanian merino breed (0.47). The obtained results confirm that the PCR-RFLP method can be used to identify different genotypic variation of CAST gene in Bulgarian sheep breeds. The established genetic diversity in the calpastatin gene indicates that, after further associative studies, this gene may be included in the breeding programs of certain sheep breeds.

1 Introduction

Meat quality and carcass weight are the traits, which are formed of a complex of genes and the environmental conditions. Molecular genetic studies have identified several genes that affect some quantitative characteristics, such as meat production [1, 2]. Marker assisted selection allows creation of individuals with specific DNA variations and hence improved productive traits which is much faster than conventional selection. The calpastatin gene (CAST) is considered as one of the promising markers for the intensity of growth and meat quality in sheep [3, 4, 5, 6]. The calpain-calpastatin system includes m-calpain (CAPN1),

* Corresponding author: ivonna_dimitrova@yahoo.co.uk

μ-calpain (CAPN2) and calpastatin and plays a key role in skeletal muscle development, participates in myoblast migration, protein metabolism, apoptosis, muscle growth and development, as well as fat deposition [7]. The calpastatin gene inhibits calpain activity and is important for regulating meat post mortem tenderness, birth weight, and growth rate to weaning [8]. In sheep, there is an inversely correlation between the level of muscle calpastatin after slaughter with the meat tenderness. CAST gene shows major and significant effects on birth weight and fat thickness [9].

The CAST gene is located at chromosome 5 locus 5q15 (NC_040256.1) of the sheep genome (Ovis aries L.) consisting of 29 exons. The first identification of the calpastatin gene in sheep was carried out by Palmer et al. [10] using the PCR-RFLP method. Polymorphism was detected in the amplified fragment of 622 bp from exon and intron 1 of CAST gene using MspI endonuclease, which digests the nucleotide sequence at the following specific site: 5 '...C↓CGG...3'. In this way have been identified two polymorphic variants, as a result from presence or absence of the point mutation (SNP) and changing of CC**G**G to CC**A**G [11, 12].

Establishing genetic polymorphism in existing sheep breeds is a major task, on the one hand for local breeds with the aim of conserving genetic resources, and on the other hand, for the efficient exploitation of existing breeds. Different sheep breeds there are in Bulgaria - including local breeds and others intended for specific purposes such as the production of milk, meat and wool. Seven breeds of sheep raised in Bulgaria have been studied so far [13 - 18] and four of them showed calpastatin gene polymorphism.

The purpose of the present study was to investigate the genetic polymorphism of the calpastatin gene in ewes of five Bulgarian breeds (two merino, two local and one for milk production) using PCR-RFLP.

2 Materials and Methods

The investigation was carried out in the Laboratory of Genetics of Agronomy Faculty, the University of Forestry.

2.1 Animals and blood collection

In present study were tested 150 ewes of five Bulgarian breeds -two merino (Ascanian and Caucasian), two local (Cooper-Red Shumen and Breznik,) and for milk production - Pleven Blackhead. Blood samples were collected from *v. jugularis* in vacuum tubes containing EDTA.

2.2 DNA extraction

Genomic DNA was extracted from whole blood using a commercial kit for DNA purification according to the manufacturer's instruction (Illustra Blood Genomic Prep DNA Purification Kit, GE Healthcare). DNA concentration and purity were determined using spectrophotometer Biodrop and agarose electrophoresis on 1% agarose gel (Bioline) and 1x TBE buffer (Jena Bioscience).

2.3 PCR amplifications

PCR amplification reactions were conducted in total volume 10 μl containing 40 ng genomic DNA, 0.2 μl dd H2O, 20 pM of each primer and 5 μl of ready-to-use 2×(1.5 mM MgCl2) MyTaqTM HS Red Mix (Bioline). For the amplification of CAST gene was used the primers' pair suggested by Palmer et al., (1997) - forward primer: 5'-TGG GGC CCA ATG ACG CCA TCG ATG-3' and reversed primer: 5'-GGT GGA GCA GCA CTT CTG ATC ACC-3'. PCR reactions were carried out by using the QB-96 thermocycler (Quanta Biotech) under the following conditions: primary denaturation at 94°C for 5 min, followed by 30 cycles of denaturation at 94°C for 30 s, annealing at 62°C for 45 s, elongation at 72°C for 1 min. The reaction was finished by final extension at 72°C for 10 min and was stored at 10°C forever.

2.4 Restriction Fragment Length Polymorphism

The genotypes of investigated animals were established using RFLP method. The restriction reactions were carried out in 10 µl final volume containing 6 µl PCR product, 10 U/µl endonuclease *MspI* (Bioneer), buffer and ddH_2O. The incubation process was performed in heat-block for 15h at 37°C. The fragment sizes were determined by agarose gel electrophoresis using 50 bp DNA Ladder (Thermo) on 2% agarose gel (Bioline) stained by 10000x RedGel™ NucleicAcid Stain (Biotuim) and 1x TBE buffer (Jena Bioscience). The results were visualized under UV light.

3 Results and Discussion

The PCR results showed a 622 bp fragment of the CAST gene. Enzyme MspI produces two fragments of 336 bp and 286 bp at allele M, whereas at allele N, the PCR product remains uncut. After restriction with endonuclease MspI two alleles were observed in Ascanian, Caucasian, Breznik and Pleven Blackhead breeds and only one (M) in Cooper-Red Shumen. Only one genotype was detected in Cooper-Red Shumen – genotype MM (with two fragments - 286 bp and 336 bp). Two genotypes were observed in Caucasian, Breznik and Pleven Blackhead breeds - genotype MM (286 bp and 336 bp) and genotype MN (with three fragments - 286 bp, 336 bp and 622 bp). Three genotypes were found in Ascanian merino ewes MM, genotype MN and genotype NN (with one fragment – 622 bp) (Figure 1).

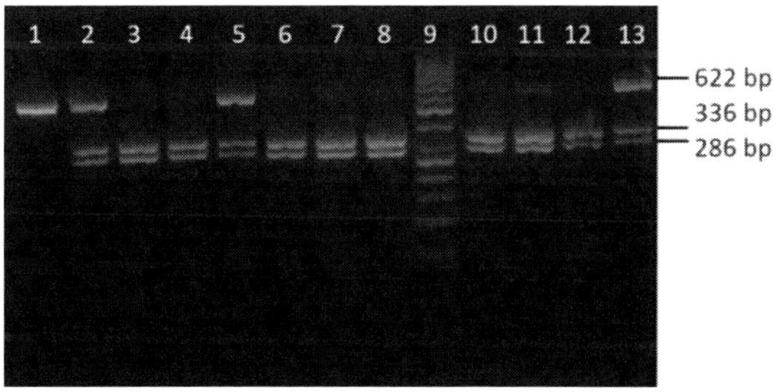

Fig. 1. DNA electrophoresis of CAST amplicons after digestion with *MspI* restriction enzyme: lane 9 - 50 bp DNA ladder, lane 1 – NN genotype, lanes 2, 5, 13 - MN genotype and lanes 3, 4, 6, 7, 8, 10, 11 and 12 – genotype MM.

The allele N was most common in the Ascanian merino breed (0.27) followed by Caucasian merino (0.13), Breznik (0.12) and Pleven Blackhead (0.07) (Table 1).

The highest frequency of the heterozygous genotype MN was at the Ascanian merino breed (0.47) followed by Caucasian merino (0.27), Brezhnik (0.23) and Pleven Blackhead

(0.13). The observed and the expected heterozygosity were calculated on the base of chi-square test using Hardy-Weinberg equilibrium. Populations Ascanian, Caucasian, Breznik and Pleven Blackhead were discovered to be in Hardy-Weinberg equilibrium for CAST locus. The coefficient of inbreeding varied from -0.182 in Ascanian breed, -0.177 in Caucasian and -0.111 in Pleven Blackhead sheep breeds to 0,104 in Breznik. It appears that the level of coefficient of inbreeding is very high in local Cooper-Red Shumen (Table 1).

Table 1. Number of animals, values of allele and genotype frequencies, average heterozygosity (observed Ho, expected He), coefficient of inbreeding and degree of probability for CAST gene in investigated breeds.

Breeds	n	Allele frequency		Genotype frequency			Ho	He	Inbreeding coefficient	p
		M	N	MM	MN	NN				
Ascanian	30	0.73	0.27	0.50	0.47	0.03	0.466	0.394	-0.182	P>0.01
Caucasian	30	0.87	0.13	0.73	0.27	0.00	0.266	0.226	-0.177	
Cooper-Red Shumen	30	1.00	0.00	1.00	0.00	0.00	0.000	0.000	1.000	
Pleven Blackhead	30	0.93	0.07	0.87	0.13	0.00	0.200	0.180	-0.111	
Breznik	30	0.88	0.12	0.77	0.23	0.00	0.233	0.211	-0.104	

The obtained results are in agreement with these reported earlier in other Bulgarian sheep breeds (Figures 2 and 3).

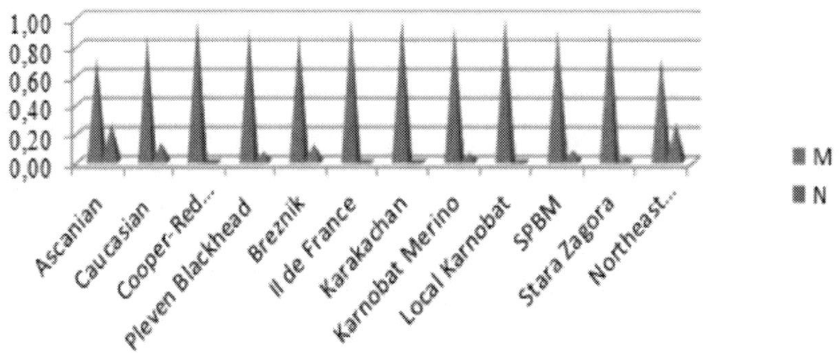

Fig. 2. Allele frequencies of CAST gene in investigated Bulgarian breeds.

The lack of diversity found in the local Cooper-Red Shumen breed coincides with that found for the other two aboriginal breeds - Karakachan [15] and local Karnobat [14], but similar results were also obtained for animals of the breed Il de France [16] (Figures 2 and 3).

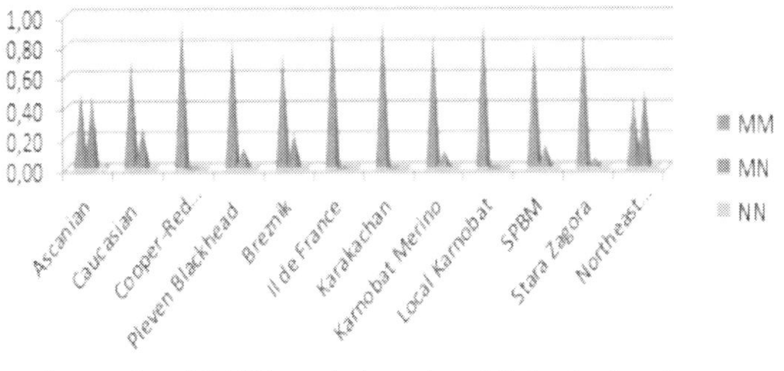

Fig. 3. Genotype frequencies of CAST gene in investigated Bulgarian breeds.

Animals of the Ascanian breed show the presence of the three possible genotypes MM, MN and NN at frequencies 0.50, 0.47 and 0.03, respectively. Three genotypes were identified in the study of the Synthetic population Bulgarian Milk sheep breed with frequencies 0.84, 0.15 and 0.01 for MM, MN and NN, respectively [13]. Similarly, three genotypes with frequencies of 0.82, 0.12 and 0.06 for MM, MN and NN, respectively, were identified at the Soviet merino breed [19], and at Turkish Kıvırcık crossbred ewes with frequencies of 0.82, 0.16 and 0.02, respectively [21]. At Tuvan short fat-tailed sheep breed three genotypes - MM, MN and NN, with similar frequency distribution have been identified (80.39%, 17.65% and 1.96% in steppe type and 77.00%, 22.00% and 1.00% in the mountain type, respectively) [5]. At the Prydniprovska meat sheep also have been reported the three genotypes, for MM the frequency is 0.77, for MN - 0.13 and for NN - 0.10 [22]. These three genotypes were also found in the CAST locus at the colombian creole sheep, the MM genotype was the most frequent (83.9%), followed by the other genotypes - MN (15.5%) and NN (6.0%) [23].

Many authors have conducted associative studies of the calpastatin gene with economically important traits in different sheep breeds. Pakistani sheep breeds Balkhi and Kajli with heterozygous (MN) genotype of this gene exhibited higher weight gain from birth to four months and eight months of age, respectively [24]. At Prydniprovska breed there is a tendency to increase the live weight of lambs carriers of N allele at 90 days of age (NN or MN) [23]. At sheep from Jordan breed Awasi with MN genotype have been reported higher final body weight, higher daily gain with a low conversion ratio and heavier longissimus muscle weight compared to MM animals, but the meat tenderness of the animals with MN genotype was lower than the ones with MM genotype [25]. West Siberian mutton breed sheep with the CAST MN genotype were also distinguished by the best average daily weight gain from birth to weaning by 6.9% and live weight by 20%, in relation to the sheep with genotype NN [26].

In this study, we identified satisfactory genetic diversity in the studied herds except in Cooper-Red Shumen. The highest genetic diversity was observed in both Merino breeds. The obtained results in genotyping allow tracing possible relationship with some characteristics of growth and meat quality in the future studies.

4 Conclusions

Improving the quantity and quality of sheep meat through traditional methods of selection is a costly and time-consuming process that can be accelerated by applying candidate genes with established genetic diversity. A suitable gene for this purpose in sheep is the calpastatin (CAST) gene. The obtained results confirm that the PCR-RFLP method can be used to identify different genotypic variation in Bulgarian sheep breeds. Genetic diversity in CAST gene was identified in four of the studied breeds - Ascanian, Caucasian, Breznik and Pleven Blackhead. Monomorphism in this gene was observed only in Cooper-Red Shumen breed. Further studies are needed to establish associations between the different genotypes and their impact on the meat productivity characteristics of distinct sheep breeds and the possibilities for their inclusion in breeding programs.

Acknowledgements

This research was part of the project КП-06-Н 26-9/18.12.2018 (2018–2021) "Investigation of DNA markers associated with production in sheep breeds reared in Bulgaria" financed by NSF - the Ministry of Education and Science, Republic of Bulgaria.

References

1. A.V. Deykin, M.I. Selionova, A.Y. Krivoruchko, D.V. Kovalenko, V.I. Truhachev, Vavilov J. of Genetics and Breeding, **20 (5)**, 576 (2016)
2. J. Knapik, K. Ropka-Molik, M. Pieszka, Ann. Anim. Sci., **17 (1)**, 23 (2017)
3. H. Chung, M. Davis, AJAVA, **7**, 641 (2012)
4. A. Ibrahim, I. Ismail, M.F. Shehata, A.R.El-Beltagy, J. Am. Sci., **11 (3)**, 106 (2015)
5. K. Kulikova, Y. Yuldashbaev, M. Petrovic, V. Caro Petrovic, D. RuzicMuslic, et al., Genetika, **50 (3)**, 885 (2018)
6. G. Gebreselassie, H. Berihulay, L. Jiang, Y. Ma, Animals, **10 (1)**, 33 (2020)
7. M. Greguła-Kania, T.M. Gruszecki, A. Junkuszew, E. Juszczuk-Kubiak, M. Florek, Meat Science, **154**, 69 (2019)
8. S. Byun, H. Zhou, R. Forrest, C. Frampton, J. Hickford,, Anim. Genet, **39**, 572 (2008)
9. E. Armstrong, G. Ciappesoni, W. Iriarte, C. Da Silva, F. Macedo, et al., Meat Science, **145**, 202 (2018)
10. B. Palmer, N. Roberts, J. Hickford, R. Bickerstaffe, J. Anim. Sci., **67**, 1499 (1998)
11. M. Greguła-Kania, Afr. J. Biotechnol., **10(61)**, 13082 (2011)
12. M. AhaniAzari, E. Dehnavi, S. Yousefi, L. Shahmohamadi, Slovak J. Anim. Sci., **45(1)**, 1 (2012)
13. S. Georgieva, D. Hristova, I. Dimitrova, N. Stancheva, M. Bozhilova-Sakova, J. BioSci. Biotech. **4(1)**, 95 (2015)
14. D. Hristova, S. Georgieva, S. Tanchev, AST **7(1)**, 36 (2015)
15. M. Bozhilova-Sakova, I. Dimitrova, Athens: ATINER'S Conference Paper Series (2016)
16. M. Bozhilova-Sakova, I. Dimitrova, In: Proceedings of Scientific Conference with International Participation “Animal Science - Challenges and Innovations”, 1(3), 332 (2017)
17. I. Dimitrova, M. Bozhilova-Sakova, M. Iliev, IOSR-JAVS, **10 (8)**, 61 (2017)
18. M. Bozhilova-Sakova, I. Dimitrova, T. Tzonev, N. Petrov, Bulg. J. Agric. Sci., **26(2)**, 475 (2020)
19. I. Gorlov, N. Shirokova, A. Randelin, V. Voronkova, N. Mosolova, et al., Turkish J. Vet. Anim. Sci. **40(4)**, 399 (2016)
20. C. Sumantri, R. Diyono, A. Farajallah, I. Inounu, JITV, **13 (2)**, 117 (2008)

21. S. Kaplan, S. Atalay, Selcuk J. Agric. Food Sci., **31(3)**, 147 (2017)
22. I. Pomitun, V. Rossokha, Y. Boyko, O. Guzevatyi, M. Shpilka, et al., Agric. Sci. Pract., **6(2)**, 58 (2019)
23. D. Montes Vergara, C. Lenis Valencia, D. Hernandez Herrera, J. MVZ Cordoba, **24 (1)**, 7113 (2018)
24. S. Khan, M. Riaz, A. Ghaffar, M.Ullah, Pak J. Zool., **44(2)**, 377 (2012)
25. K. Jawasreh, R. Jadallah, A. Al-Amareen, A. Abdullah, A. Al-Qaisi, et al., Indian J. Anim. Sci., **87(5)**, 635 (2017)
26. A. Afanasyeva, V. Sarychev, G. Goncharenko, In: Proceedings of the International Scientific and Practical Conference "Digital agriculture - development strategy", Advances in Intelligent Systems Research, 167, 116 (2019)

Improving the efficiency of corporate management of agricultural enterprises

Sergey Yekimov[1,*], *Vladimir* Purtov[2], *Ievgen* Buriak[3] , *Dmytro* Kabachenko[4], and *Anastasiia* Poltorak[5]

[1]Publishing House "Education and Science", Olstynska 607/1, Praha 8, 18100, Czech Republic
[2]V.N. Karazin Kharkiv National University, Svobody Square, 4, Kharkiv, Ukraine
[3]Kremenchuk Mykhailo Ostrohradskyi National University, Kremenchuk, Ukraine
[4]Dnipro University of Technology, Dnipro, Ukraine
[5]Mykolayiv National Agrarian University, Mykolayiv,Ukraine

Abstract. Financial indicators play an important role in the management of an agricultural enterprise. the tasks of the management of the agricultural enterprise are to control the efficiency of employees and maintain the financial stability of the enterprise. In this paper, the problem of ensuring effective management of an agricultural enterprise was investigated. The activity of an agricultural enterprise depends on the influence of external factors: political and legal, technical and economic, ecological and economic, demographic, socio-cultural. We studied the question of how to take into account the influence of variable external factors in the management of an agricultural enterprise. Traditional methods based on financial analysis data, in our opinion, do not fully reflect the situation in the agricultural enterprise. International management accounting standards are optional, but they allow for the use of more flexible indicators in management accounting that take into account the activities of a particular agricultural enterprise. This will facilitate short - and medium-term planning at the agricultural enterprise and facilitate the achievement of the enterprise's tactical and strategic goals.

1 Introduction

Due to the high cost of debt capital for countries with economies in transition, including the CIS countries, the possibility of borrowing through the issuance of securities is very limited. as a rule, this is due to corruption, shortcomings in legislation, structural deformation of the economy, imperfection of the judicial system, and flaws in corporate governance. Therefore, the search for methods to improve the efficiency of corporate governance can have a positive effect without increasing the cost of capital of the enterprise and without increasing the level of its capitalization.

According to [1] corporate governance is a system of enterprise management provided for by law in order to improve the efficiency of its functioning, taking into account the interests of its owners and other counterparties.

* Corresponding author: 3701313@mail.ru

According to [2] corporate governance should combine the practice of economic activity and legislative norms, which in turn allows you to profitably attract human and material resources.

According to [3],[4] the functions of corporate regulation can be divided into general and special. The general ones are aimed at managing the company as a whole (managing the company's assets, managing the planning of the company's activities). Special functions of corporate regulation are related to the management of marketing, finance, production, and personnel.

According to [5] corporate governance should be based on the following basic principles:

1) compliance with the objectives and goals of the enterprise;
2) consistency and consistency in their activities;
3) innovative development;
4) balance and completeness in the choice of means of influence;
5) compliance with public values;
6) motivation of staff activities;

According to [6],[7] in transition economies, corporate governance typically faces the following challenges:

1) Real corporate governance is complex in nature, and there is a problem of lack of qualified personnel for its implementation;
2) Errors and shortcomings in its implementation can be interpreted as the inability to apply corporate governance in a particular enterprise;
3) The external environment of the functioning of a business entity is associated with the presence of many uncertain factors. For example, the imperfection of the legal framework, corruption, which limit the capabilities of the enterprise and prevent its effective operation.
4) Lack of effective control over the activities of business entities by the state;
5) Owners of controlling stakes often ignore the interests of small shareholders, primarily in matters of reorganization of the company and payment of dividends.

In our opinion, the main goal of corporate governance should be to improve the efficiency of the company's functioning, taking into account the interests of investors, creditors, employees and management of the enterprise and society as a whole.

According to [8] compliance with the requirements of international standards helps to protect the interests of investors, society, and corporations, reduces corruption risks, and limits abuses by the authorities and management of the enterprise

The development of countries with economies in transition is largely dependent on the activities of their corporate structures.

In [9] criteria for evaluating the effectiveness of corporate activities are proposed (figure 1):

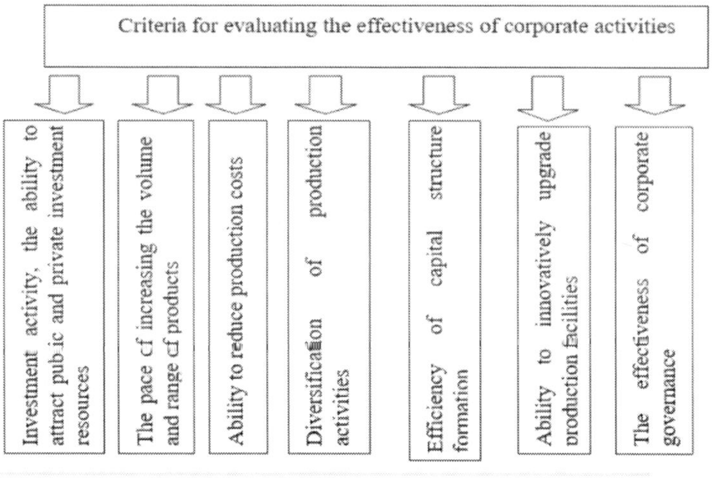

Fig. 1. Criteria for evaluating the effectiveness of corporate activities.

According to [10] management of a corporate enterprise is a system of relationships designed to regulate the procedures and rules for decision-making, as well as the delegation of rights and obligations between individual divisions of the enterprise.

Its effectiveness is achieved by increasing profitability and competitiveness, as well as by establishing principles of effective management and control over its activities.

According to [11],[12] the basic criteria for effective management are indicators that characterize the results of economic activity of the enterprise and the implementation of the principles of corporate governance.

Countries with economies in transition tend to produce products with low added value, and a significant share of GDP in most of them belongs to the agricultural sector of the economy. Although in agriculture there are enterprises of various forms of ownership: personal farms, farms, limited liability companies, agricultural holdings. But it is argohodlings, as a rule, that produce most of the products.

2 Materials and methods

In writing this paper, the authors used an analytical method, with the help of which the studied problems were considered in their unity and development. Taking into account the goals and objectives of the study, the structural and functional method of scientific research was used. This made it possible to study some of the problems associated with improving the efficiency of corporate management of an agricultural enterprise.

3 Results

For strategic planning of the development of economic entities, it is necessary to have long-term guidelines and adapt to the conditions in which they operate.

According to [7] the following components of the external environment of the enterprise can be distinguished: technical and economic, political and legal, foreign economic, demographic, environmental and climatic, and socio-cultural.

In our opinion, if the external environment is stable or its evolution can be predicted, then there is a possibility of effective medium - and long-term planning in the enterprise.

Analyzing the impact of the external environment on the agricultural sector of the Ukrainian economy, it can be noted that the basis of export-oriented agricultural sectors is the cultivation of sunflower and grain crops. The factors affecting the effectiveness of corporate governance are shown in Figure 2.

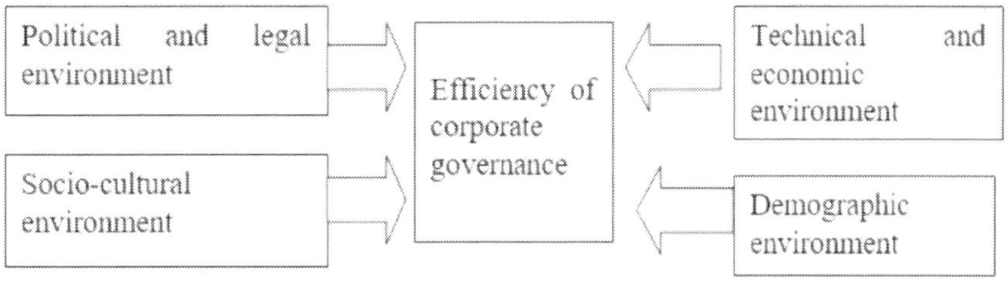

Fig. 2. Factors affecting the effectiveness of corporate governance.

The technical and economic component of the external environment is underdeveloped, its innovative development is at a very low level. The socio-cultural component of the external environment is characterized by poor development of rural areas and low living standards of local residents. The political and legal component is characterized by insufficiently effective state financial policy and corruption risks. The demographic component of the external environment is characterized by insufficient mechanization of production, as well as low qualification of the labor force.

The effectiveness of corporate governance largely depends on how effectively the financial resources of the enterprise are used. Managing the asset structure of an enterprise should harmoniously combine borrowed and own sources of financing, which in turn helps to reduce financial risks and strengthen financial stability.

The financial results of the corporation's business activities are highly dependent on external factors. In our opinion, along with international accounting and reporting standards, it is also necessary to develop and implement additional indicators that characterize the company's activities taking into account local specifics. This, in our opinion, will simplify the corporation's achievement of strategic and tactical goals and facilitate short - and medium-term planning of its activities.

When managing financial flows in corporations, great attention should be paid to ensuring the financial balance of the enterprise and ensuring the required liquidity of its assets.

In our opinion, among the performance indicators of the corporation should be those that link the results of the work of individual divisions and responsibility centers of the enterprise with the financial results of the company as a whole.

4 Discussion

In our opinion, the development strategy of the economic society should be adapted to the changing conditions of the external environment and at the same time combine this with the need to obtain maximum profit from economic activities.

In our opinion, the implementation of structural changes in the company, maintaining stable growth, strengthening the company's position in the agricultural market, and investment activities should be used as components of the corporation's development strategy.

Taking into account the external and internal environment of agricultural enterprises along with an increase in production volumes and increased profits for successful functioning of the Corporation requires upgrading fixed assets, improving the quality of products in accordance with international standards, improving the quality of life of employees, increase in sales of agricultural products and high added value.

5 Conclusions

The role of financial results in the effective management of an agricultural enterprise is quite high. The task of the head of an agricultural enterprise is not only to control the effective work of employees, but also to make decisions aimed at improving the financial stability of the organization.

Any organization exists at the expense of making a profit. And the amount of this profit should be such that not only meet the main obligations, that is, be solvent, but also have a reserve of funds that will be used for the development of the organization.

The agricultural sector has a large amount of costs, for land, keeping pets, caring for plantings, servicing specialized equipment. But at the same time, this sector depends on external factors, for example, weather. Accordingly, adverse weather conditions can reduce the value of the crop, and this in turn will affect the financial results.

A competent manager of an agricultural enterprise, before making a management decision, should conduct a comprehensive analysis of the external and internal environment, identify all possible threats and plan methods of combating them. At all stages of the solution implementation, control is necessary, and when errors are detected, they are immediately eliminated.

Well-thought-out management decisions lead to an increase in the financial stability and profitability of the enterprise. Due to these criteria, the organization develops and strengthens its competitive position in the market, which increases the chances of winning the competition in market conditions. A solvent profitable enterprise is attractive from the point of view of the object of investment, which is very important for an agricultural enterprise. Due to investments, the agricultural enterprise has the opportunity to purchase new high-tech devices used in agricultural production, which will contribute to improving labor productivity and product quality.

References

1. K. Prib, *Practice of crisis management in agricultural enterprises, Ekonomika APK*, 51 (2019)
2. K. Kovács, *Agricultural Enterprises in Hungary* (2018)
3. D.M. Akbasheva, R.A. Goguev, A.R. Salpagarova, A. Arova, L. Urusova, *Accounting at an agricultural enterprise, Scientific development trends and education* (2020)
4. I. Kravchuk, S. Logvinenko, Financial Management of Agricultural Enterprises. Business Inform, 3, 136 (2019)
5. O. Sarapina, T. Pinchuk, T. Shram, Scientific Notes of Taurida National V.I. Vernadsky University. Series: Economy and Management, **70(3)**, 62 (2020)
6. N. Pylypiv, I. Ambroziak, *Agricultural enterprise management system* (2016)
7. V. Voskolupov, Bioeconomics and Agrarian Business **11,** 5 (2020)
8. M. Rudenko, Ekonomìka ta upravlìnnâ APK, 76 (2019)
9. M. Zos-Kior, V. Ilin, M. Kryvobik, Pryazovskyi Economic Herald (2020)
10. V. Tkach, E. Medyukha, N. Zemlyakova, L. Pudeyan, K. Chanturia, E. Moskvitin, IOP Conference Series: Earth and Environmental Science 403, 012134 (2019)
11. N. Nosan, Modern Economics, **22**, 72 (2020)
12. A. Chykurkova, M. Gorichovski, *Management system for agricultural enterprises competitiveness, The economic discourse*, 70 (2019)
13. P. Bryla, *Erratum to: Managing Agricultural Enterprises* (2018)

Regulation of environmental protection measures in the Caspian Sea region

Sergey Yekimov[1,*], *Oleg* Bavykin[2], *Elena* Kuznetsova[3], *Roman* Kucherenko[4], and *Dmitriy* Kucherenko[4]

[1]Publishing House "Education and Science" s.r.o., Olstynska 607/1, 18100 Praha 8, Czech Republic
[2]Moscow Polytechnic University, Moscow, Russian
[3]Kirov State Medical University, Kirov, Russian
[4]"Kuban State Agrarian University named after I.T. Trubilin", Krasnodar, Russian

Abstract. Information obtained as a result of monitoring of water bodies is taken as a basis for legal regulation of economic activities related to environmental pollution. In the context of a globalized economy, the strengthening of environmental protection measures in one of the neighboring countries often leads to the fact that the next plant will be built across the border, and its impact on the environment on a global scale will be the same. In this study, the authors studied the problem of finding new ways to solve the environmental problems of the Caspian Sea. The Caspian Sea belongs to Kazakhstan, the Russian Federation, Turkmenistan, Azerbaijan and Iran. to solve the environmental problems of the Caspian Sea, it is necessary to unify the environmental legislation of these countries, similar to the unification of the environmental legislation of the European Union countries.

1 Introduction

Monitoring of water objects is the collection, processing and delivery of information obtained for the purpose of regular monitoring of hydrological, hydrogeological, hydrogeochemical, sanitary-chemical, microbiological, parasitological, radiological and Toxicological indicators of timely revealing of negative processes, assessment and prediction of waste system recommendations on prevention and elimination of the adverse effects of the proposed action on water resources management. Various types of monitoring of water bodies are considered in [1,2,3,4,5,6], the negative impact of human activity on the state of the Caspian Sea was studied in [7,8,9,10,11,12], in the article [5] recommendations for improving environmental monitoring at the interstate level are proposed. This indicates the relevance of further study of the legal regulation of the ecological state of the Caspian Sea.

According to Article 19 of the Framework Convention for the Protection of the Marine Environment of the Caspian Sea [13], the Contracting Parties shall perform the following duties:

* Corresponding author: 3701313@mail.ru

1. Seek to establish and implement appropriate independent and / or joint monitoring programmes for the marine environment of the Caspian Sea.
2. Agree on the list and criteria of pollutants that are controlled in the Caspian Sea for emissions and production.
3. Independently and / or jointly assess the effectiveness of the assessment of the state of the Caspian Sea and develop measures aimed at limiting, reducing and preventing pollution of the Caspian Sea.
4. Agree on rules for data monitoring and processing programs, as well as their quality assessment.
5. Provide a framework for decision-making and a centralized database that is a common source of information and knowledge for professionals, administrators.

2 Materials and methods

In this paper, the system of monitoring the ecological state of the Caspian Sea and ways of implementing environmental measures were analyzed. Based on the analysis, the conclusion is made about the need for a comprehensive approach that provides for further unification of environmental legislation in the Caspian region.

3 Results

In our opinion, in modern conditions, the problem of developing a more optimal legal regime for the use of natural resources of the Caspian Sea, as well as carrying out measures to protect the environment, is steadily increasing. Modern international law contains a large number of norms that relate to various issues of the use of natural resources and, first of all, the burial and transportation of industrial waste.

The Framework Convention for the Protection of Marine Resources of the Caspian Sea [13] includes all five Caspian countries: Azerbaijan, the Russian Federation, Iran, Turkmenistan and the Republic of Kazakhstan.

All these states produce hydrocarbons from the sea, so they should, in our opinion, jointly solve the problems associated with the release of pollutants into the Caspian Sea, as well as develop measures to control the degree of pollution of the Caspian Sea. Environmental measures taken unilaterally by only one of the states, in our opinion, will not give the proper effect, but will only contribute to an increase in hydrocarbon production in countries where environmental legislation is less strict, which means that the volume of emissions of harmful substances into the Caspian Sea will not decrease.

Therefore, in our opinion, it is necessary to develop various internal interstate acts or regulations on the regulation of individual subjects of the Caspian states regarding the Caspian Sea. Accordingly, based only on the word "framework", it should be understood that this type of convention does not provide a document regulating all aspects of the protection of the environment of the Caspian Sea. we should not forget that the framework nature of the Tehran convention reflects the possibility of forming additional regulatory documents to the convention, as well as the signing of bilateral and multilateral treaties between the Caspian countries. This is the reason for the creation of a legal instrument that will consider the articles of the convention more broadly and in detail, as separately taken. When assessing this phenomenon, not a single regulatory act regulating this activity was identified. That is why, in our opinion, it is important to develop and adopt national regulatory legal acts in each Caspian state, which will be aimed at protecting the environment of the Caspian Sea, which will take into account such aspects as fishing, import and export of goods through the Caspian Sea. Legalize equal production without harming the environmental situation, and clearly distribute measures controlling poaching and shipping. intensive use of vessels for transportation of oil produced in the Caspian Sea region and oil products produced from it causes significant potential risks associated with water pollution.

In addition, due to poaching, the population of sturgeon fish in the Caspian Sea has sharply decreased. Although the Ministry of Natural Resources and Ecology of the Russian Federation is actively fighting this problem, however, in our opinion, this is not enough to solve this problem. We believe that it is necessary to develop and sign international agreements that more closely coordinate the activities of environmental and law enforcement agencies of all countries of the Caspian region.

This should form the overall structure of relations between the Caspian countries, as well as contribute to a more sustainable and stable development of the entire region.

The national legislation of all the Caspian countries is of great importance in regulating the protection and use of natural resources in the Caspian Sea. At the same time, due to corruption and lack of financial resources allocated for environmental protection measures, these laws are not being implemented satisfactorily.

Unified requirements for environmental protection exist in the countries of the European Union, so, in our opinion, it can be very effective to create uniform requirements for economic activities in the Caspian Sea region, including in matters related to the management of industrial waste.

In our opinion, the problem of preventing the consequences of various accidents and emergencies that have a negative impact on the Caspian Sea is also very important. In modern conditions, each of the Caspian littoral countries does not have modern means and infrastructure to eliminate the consequences of such accidents. Given that the development of hydrocarbons on the shelf of the Caspian Sea is growing steadily every year, we believe that all the Caspian countries should develop international legal mechanisms for environmental protection, taking into account the problematic aspects. this will, in our opinion, make a great contribution to the preservation of this natural complex.

4 Discussion

Consideration of the legal regulation of economic activity on the ecological state of the Caspian Sea requires a more detailed study. Theoretical analysis has shown that in the studies of different authors there are different approaches to solving this problem. Most often, we are talking about the recommendation to establish any restrictions on the economic activities of economic entities. These restrictions are often developed without taking into account the interests of enterprises responsible for pollution of the Caspian Sea, and it is often easier for them to pay a fine or look for any other ways than to comply with these restrictions.

In our opinion, it is necessary to involve more widely in the development of environmental legislation not only representatives of regulatory structures, scientists, but also enterprises of the most polluting Caspian Sea, primarily oil producing and refining companies, of all countries.

The final result should be the unification of the environmental legislation of the countries of the Caspian region, as it is done with the environmental legislation in the countries of the European Union.

5 Conclusions

Due to the fact that the Convention [13] provides for the right of each of the Caspian countries to free movement by all means of transport to other oceans and seas, and the process of approving projects for laying pipelines on the seabed, taking into account the verification of environmental safety, is a long and complex process. In our opinion, it is necessary to create uniform requirements for the laying of such communications. This, in our opinion, will contribute not only to the greater preservation of natural resources, but also to a more dynamic development of the region's economy (Figure 1).

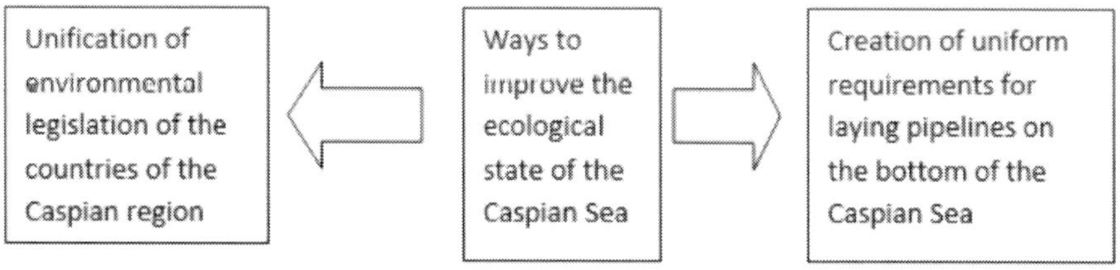

Fig. 1. Ways to improve the ecological state of the Caspian Sea.

In our opinion, the oil companies of the Caspian states should take part in today's consortia to determine the wealth of the Caspian Sea. In the course of field development between two or more areas, there is a need for appropriate production sharing agreements between States. In our opinion, the countries of the region should take a more active part in agreements, including condominiums, on the opportunities of the Caspian Sea, including on shipping, fishing and environmental protection. At the same time, we believe that each state should have agreed on cross-border territorial waters, that is, their fishing zone should be located within one state. The rest of the sea should be open to navigable navigation between States.

Our proposals on the legal status of the Caspian Sea include the experience of using mineral resources and international standards not only for the use of mineral resources, but also the experience of using the wealth of the Caspian basin between the USSR and Iran.

References

1. N. Jafari, Journal of Ecology and the Natural Environment **2**, 25 (2010)
2. E. Lari, B. Abtahi, M. Seyed Hashtroudi, Aquatic Toxicology **170,** 330 (2016)
3. D.N. Katunin, S.N. Yegorov, I.A. Khripunov, E.A. Kravchenko, N.V. Galushkina, G.V. Radovanov, A.V. Azarenko, The main characteristics of hydrological-hydrochemical regime in the lower reaches of the Volga River in 2002. In "Fisheries Researches in the Caspian:Scientific-research works results for 2002", Astrakhan (2003)
4. D.N. Katunin, S. N. Yegorov, I.A. Khripunov, et.al, Main characteristics of hydrologicalhydrochemical regime in the lower reaches of the Volga River in 2003. In "Fisheries Researches in the Caspian: Scientific-research works results for 2003", Astrakhan (2004)
5. A. Korshenko, A.G. Gul, Hdb. Env. Chem. **5,** 109 (2005)
6. A. Korshenko, *A Desk Study Project to determine the fluxes of major contaminants* (2007)
7. Zt. Lobat, *Integrated Environmental Assessment as a Comprehensive Governance Tool for the Caspian Littoral States to Control the Environmental Degradations* (2018)

8. S. Harris-Brandts, D. Gogishvili, Eurasian Geography and Economics **1-25** (2018)
9. N. Sivri, K. Ozcan, E. Başar, F. Ahmadli, *Oil contamination damage and the marine biodiversity: Caspian Sea and Black Sea* (2016)
10. B. Mirnategh, N. Shabanipour, M. Sattari, Seawater, *Sediment and Fish Tissue Heavy Metal Assessment in Southern Coast of Caspian Sea* (2018)
11. H. Eronat, Deu Engineering Faculty Science and Engineering **21**, 47 (2019)
12. T. Härkönen, M. Baimukanov, A. Bignert, et.al, EIA. Caspian Sea Region: Survey of Key Oil and Gas Statistics and Forecasts, July 2006 (2008)
13. Framework Convention protection of the marine environment of the Caspian sea (Tehran, 4 November 2003)

Changes in maximum water holding capacity of chernozem soil caused by soil compaction

Yuriy Bekhovykh*, *Larisa* Bekhovykh, *Aleksey* Lyoevin, and *Evgeniy* Sizov

Altai State Agricultural University, pr. Krasnoarmeyskiy, 98, 656049 Barnaul, Russia

Abstract. The goal of the studies presented in the work is to estimate the changes in the maximum water holding capacity of the soil caused by soil compaction. The object of the research is a loam type of chernozem soil. The research shows that the maximum water holding capacity is a function of the soil density change and indirectly depends on the action of soil compaction factors, including external pressure level, the number of impacts on soil cover, as well as on soil moisture. Soil moisture is indicated as the most important factor in changing the maximum water holding capacity caused by soil compaction. The decrease in the maximum water holding capacity caused by soil compaction is more obvious with an increase in soil moisture, both in the research of the number of impacts and external pressure influence on this indicator. The results of the work show that the change in the maximum water holding capacity caused by a compacting action significantly depends on the initial value of the soil addition density.

1 Introduction

Overconsolidation of the soil cover is one of the factors negatively affecting its fertility and crop yields [1].

Optimal densities are within 1.1-1.3 g/cm^3 for most soil types in Russia [2-4]. On soils with optimal density plants consume moisture more productively, aeration of the root layer is carried out [4], and the yield of cultivated crops increases [4,5].

However, in the conditions of modern agricultural production, there is a tendency to increase the weight of agricultural machinery, which increases the compaction effect on the soil by its undercarriage systems and undercarriage systems of the trailer units used in soil treatment [6-9]. Soil compaction leads to a reduction in the pore space of the soil [10], a deterioration in the water-physical characteristics [11], resulting in a decrease in yield [12]. Soil decomposition processes are slow and often require special energy-intensive measures [13] which ultimately increase the cost of crop cultivation. In this regard, it is rational to carry out crop production taking into account changes in water, air and other soil properties during soil compaction, which will undoubtedly contribute to increasing soil fertility and yield.

The aim of this work is to investigate the potential water content of soil compaction.

* Corresponding author: Phys_asau@rambler.ru

2 Materials and methods

The object of the study is the leached medium loam chernozem of the Priobskoye Plateau.

The objectives are as follows:

1) to identify an indirect relationship between the maximum water holding capacity of the soil and its bilk density;

2) to identify a tendency of change in the maximum water holding capacity of the soil of different moisture conditions in relation to the number of fixed pressures;

3) to identify a correlation between the change in the maximum water holding capacity of the soil of different moisture conditions and the influence of different pressures.

The experimental plots were laid on the fallow field of the "Prigorodnoye" training and experimental farm, located in the forest-steppe zone of the West Siberian Plain, in the north-east of the Priobskoye Plateau. The different pressures on the soil were regulated by the action of a fixed weight load on the compactors of different support areas. An unconsolidated area of black fallow served as a control area. Different soil moisture in the experimental plots was created by artificial moistening.

The determination of the maximum water holding capacity was carried out by a calculation method based on the results of direct measurements of moisture, bulk density and particle density of the soil [14].

3 Results and discussion

The studied chernozem relates to a loam type of leached chernozem soils with a predominance of medium loam in the soil horizons. The chernozem profile is classical with humus horizon A, transitional AB, illuvial horizon B, transitional BC, and carbonate horizon of the parent rock Ck. The surface layer from which the samples were taken was the upper half of the humus arable horizon A (0-29 cm). According to morphological structure, it is moist, black-and-brown with the pronounced lumpy-powder-like structure of aggregates, loose, loam.

According to the granulometric composition, the upper layer of the arable horizon of the studied chernozem relates to the medium loams (Table 1).

Table 1. Granulometry of the arable layer of the studied chernozem on the experimental plot

Soil horizon	Depth, cm	Fraction content (in % content of absolutely dry soil)							Name of the granulometric composition of the soil
		1-0.25 mm	0.25-0.05 mm	0.05-0.01 mm	0.01-0.005 mm	0.005-0.001 mm	<0.001 mm	$\sum$ fractions <0,01 mm	
A(hum)	0-29	7.36	34.68	24.44	6.40	5.24	21.88	33.52	medium loam

As can be seen from Table 1, in the granulometric composition of the studied leached chernozem prevail fractions of medium sand (0.25-0.05 mm), coarse dust (0.05-0.01 mm) and a clay fraction (less than 0.001 mm). The maximum humus content in the surface layer of the chernozem arable horizon did not exceed 3.4%. According to the pH = 6.7, the soil is neutral or close to such.

In order to create a weight load in the study of changes in the maximum water holding capacity of the soil of different moisture from the number of fixed pressure impacts, it was decided to use a load that produced a pressure of 100 kPa by a soil compactor of a certain area. Modern agricultural machinery with close average soil pressure parameters of a single unpaired wheel is most often used in modern agricultural production.

The results of the studies show that maximum water holding capacity is a function of changes in soil density (Fig. 1). The approximation curve of dependence is well described by a logarithmic function and an exponent. The numerical coefficients included in the function of changes in maximum water holding capacity from density reflect the characteristics of the granulometric composition of the studied soil and, supposedly, the content of organic matter in the soil and require further study.

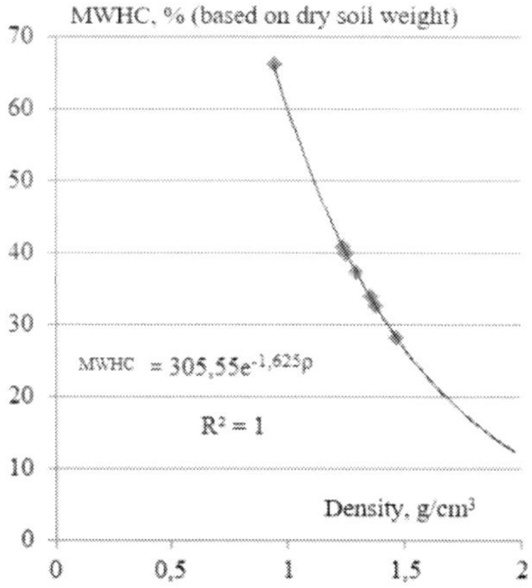

Fig. 1. Dependence of the maximum water holding capacity of the surface layer of the studied chernozem on the bulk density of the soil

The study of changes in the maximum water holding capacity of the soil of different moisture content on the number of impacts of the soil compactor creating a fixed pressure (Table 2) revealed that an increase in soil moisture is a determining factor in reducing the maximum water holding capacity under repeated loads on the soil layer. For example, when the surface layer of the studied chernozem was moistened to values close to 30% of the dry soil mass, even a single compaction with an external pressure of 100 kPa led to an increase in the soil compaction density and a change in the pore space characteristics, which reduced maximum water holding capacity by almost half.

With a decrease in soil moisture, the change in maximum water holding capacity due to external pressure occurs similarly, but the potential moisture content decreases less intensively. According to the graphs (Figure 2), the dependence of the maximum water holding capacity on the number of compactor impacts on the soil shows that the maximum water holding capacity decreases significantly at the first impacts, with the density increasing most intensively [15]. The coefficient of potential moisture content decreases and approaches the abscissa axis asymptotically under multiple impacts of external pressure (Fig. 2). This peculiarity is evident due to the fact that the soil density under the impact of external pressure tends to "saturation" [15].

Table 2. Dependency of density (g/cm^3) and maximum water holding capacity (% of dry soil mass) of the studied chernozem on the number of impacts (N) of the compactor creating a pressure of $100 \cdot 10^3$ Pa on the soil

N	The bulk density of the soil at the different moisture content (g/cm^3) / Maximum water holding capacity (%)			
	14%	18%	24%	28%
0	0. 942/66.2	0.942/66.2	0.942/66.2	0.942/66.2
1	1.236/40.9	1.257/39.5	1.299/37.0	1.350/34.1
5	1.250/40.0	1.299/37.0	1.347/34.2	1.375/32.7
10	1.291/37.4	1.309/36.4	1.391/31.9	1.467/28.2

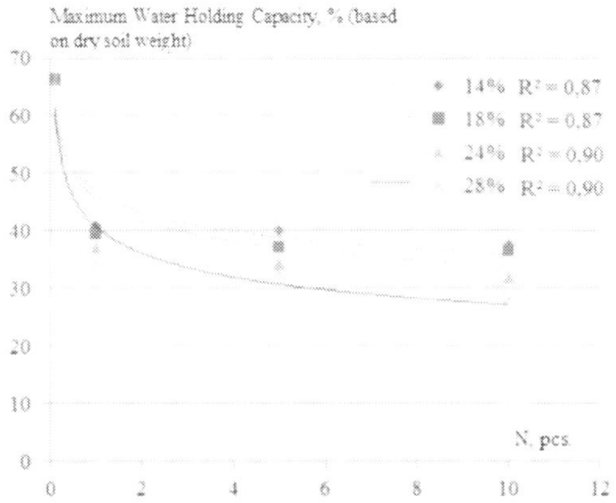

Fig. 2. Dependency of the maximum water holding capacity of chernozem on the number of impacts (N) of external fixed pressure ($100 \cdot 10^3$ Pa)

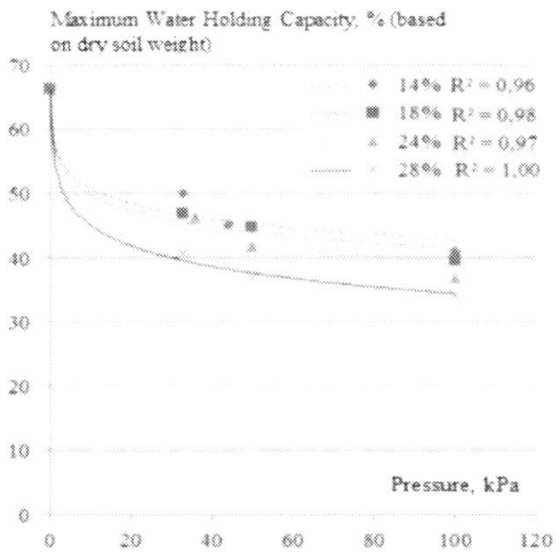

Fig. 3. Dependency of the maximum water holding capacity of chernozem on the value of external pressure

Figure 3 illustrates the graphically represented approximation of the dependence of the maximum water holding capacity of the studied chernozem on different external pressure created by soil compactors with different areas of the compaction surface. The graphically reflected dependence of the maximum water holding capacity on external pressure (Figure 3) is non-linear and tends to asymptotic decrease of the values. The graphs (Fig. 3) also reflect the fact that soil moisture is the most important determinant of the degree of potential moisture content.

The change in the maximum water holding capacity of the soil caused by the compaction impact strongly depends on the initial bulk density of the soil. The more the soil is loosened and the lower the initial density before compaction, the more it increases by a single compaction impact of a fixed pressure, while the pore space decreases more significantly and, as a result, the maximum water holding capacity, which depends on the number and volume of free pores that water can potentially occupy.

4 Conclusions

Based on the research results, the following conclusions can be drawn:

1. the maximum water holding capacity is a function of changes in the soil density and indirectly influenced by soil compaction factors, including on the magnitude of external pressure, on the amount of its impacts on the soil cover, as well as on soil moisture;
2. soil moisture is the most significant factor determining the change in the maximum water holding capacity of the soil caused by compaction;
3. the decrease in the maximum water holding capacity of the soil caused by compaction is more visible with an increase in soil moisture, both when studying the influence of the number of impacts and the magnitude of external pressure on this indicator;
4. the change in the maximum water holding capacity caused by compaction effect significantly depends on the initial soil bulk density;
5. the indicator of the potential water holding capacity (maximum water holding capacity) of the soil asymptotically decreases with multiple impacts of external pressure and with an increase in its value.

References

1. A. G. Bondarev, Eurasian Soil Science, **5** (1990)
2. S. S. Sdobnikov, Agriculture, **2** (2000)
3. I. A. Chudanov, L.F. Ligastaeva, Agriculture, **4** (2000)
4. S. I. Dolgov, I. V. Kuznetsova, I. A. Modina, Problems of soil cultivation: reports of the International meeting dated June 13-15, 1968 (Varna, 1970)
5. Kh. Kh. Ashabokov, L. M. Khazhmetov, Yu. A. Shekihachev, materials of the International (correspondence) scientific-practical conference (2017)
6. N. N. Pogodin, V. V. Kuchko, F. A. Barsukevich, S. V. Shatilo, Melioraciâ, **59** (2008)
7. G. A. Okunev, N. A. Kuznetsov, AIC of Russia, **75(1)** (2016)
8. Z. A. Godzhaev, A. M. Pogozhina, Tractors and agricultural machinery, **5** (2018)
9. G. A. Okunev, N. A. Kuznetsov, A. A. Brazhnikov, Bulletin of the Chelyabinsk State Agroengineering Academy, **69** (2014)
10. Yu. V. Bekhovykh, Bulletin of Altai State Agrarian University, **178** (2019)
11. V. Kh. Katz, S. V. Kuznetsov, Collection of scientific treatises VIM, **66** (1974)
12. V. P. Dyakov, All-Russian scientific and practical conference with international participation (2016)
13. A. G. Bondarev, I. V. Kuznetsova, P. M. Sapozhnikov, Eurasian Soil Science, **4** (1990)
14. A. F. Vadyunina, Z. A. Korchagina, Methods for the study of physical properties of soils (1986)
15. Yu. V. Bekhovykh, Bulletin of Altai State Agrarian University, **177** (2019)

E3S Web of Conferences **262**, 03004 (2021) https://doi.org/10.1051/e3sconf/202126203004
ITEEA 2021

Breeding of comb-wheatgrass (Agropyron pectiniforme Roem. et Schult.) for arid conditions of the steppe zone

S. V. Saprykin[1,*], *V. N* Zolotarev[2], and *I. S.* Ivanov[1]

[1]Voronezh experimental station for perennial grasses-branch of the Federal state budgetary scientific institution "Federal Williams Research Center of Forage Production & Agroecology", Dokuchaev str., 1, Pavlovsk, Voronezh region, Russia
[2]Federal Williams Research Center of Forage Production & Agroecology, Nauchny Gorodok str., 1, Lobnya, Moscow region, Russia

Abstract. In the context of the progressive trend of climate aridization of the main agricultural regions in the forest-steppe and steppe zones, the problem of increasing and stabilizing the high productivity of forage crops is becoming more acute. Perennial grasses are the most affordable and stable source of cheap plant feed, and a means of preserving soil fertility during intensive farming, which has an anthropogenic negative impact on agricultural ecosystems. Perennial grasses play a special role as a biological means of counteracting the degradation of biological, water and physical and agrochemical properties of soils, the processes of their dehumidification and erosion. In arid climatic conditions, the most effective species of perennial bluegrass grasses is the comb-wheatgrass or broad-spiked wheatgrass (Agropyron pectiniforme Roem. et Schult.). This crop is resistant to extremely stressful conditions for cultivation in arid conditions, high production potential for the production of various types of voluminous feed. The limiting factor for the effective economic use of the comb-wheatgrass is the insufficient number of zoned varieties. At the Voronezh Experimental Station for perennial grasses, active breeding work is underway to develop new varieties of comb-wheatgrass. The results of the assessment of the source material in the collection nursery are presented. The best specimens for the creation of complex hybrid populations for the formation of the polycross nursery have been identified.

1 Introduction

In recent years, aridity and drought have become major global issues due to their environmental and social and economic consequences. On the one hand, aridity causes a decrease in the biological and economic productivity of terrestrial ecosystems and poses a serious threat to ecological, biogeochemical, and hydrological processes [1]. Projected climate change is global warming (with concomitant changes in hydrological regimes and

*Corresponding author: gnu@bk.ru

other climatic variables) caused by an increase in the concentration of radiatively active greenhouse gases. Climate models predict that global surface air temperature may increase from 1.8–4.0 C to 4.0–5.8°C in the next few decades [2, 3]. In addition to increasing temperatures and decreasing rainfall, increased frequency, severity and duration of droughts can significantly affect food production. The analysis of the role of climate change trends in crop production revealed the danger of a significant decrease in yield, including due to a reduction in periods of active crop growth, which is probably the result of high temperatures and water stress [4]. The assessment of modeled climate change impacts using statistical and threshold analysis, climate models, and scaling/initialization methods showed a significant shift of all facilities towards a drier climate by the end of the century, and a decrease in herbage productivity on managed (intensive, extensive) pastures [5]. The detrimental impact of climate change on food security can be offset by improved management of crop production, soil and water resources, as well as the introduction of stress-resistant crops and varieties into agricultural production that are able to overcome the negative effects of aridization [6, 7]. The climatic risks of reduced crop production can be reduced through more efficient use of naturally occurring genetic resources and result in the development of new varieties, well adapted to natural changes. Thus, in arid conditions, with a decrease in rainfall by 20-30% in the structure of growth in the production of grain crops, 34% falls on the use of improved drought-resistant varieties [8].

Against the background of the emerging tendency of xerophytization of the species and varietal composition of agricultural crops in the conditions of aridization of the climate for the dry steppe regions of the country, one of the most promising crops for forage production is wheatgrass, a dense bunchy herbaceous loosely bunched polycarpic. In a practical sense, wheatgrasses are the most suitable forage crops that are resistant to climatic and weather stresses, high adaptive potential for cultivation and production of various types of voluminous forages [9]. Wheatgrasses belongs to tribe Triticeae Dum. of the type subfamily of the family Poaceae Barnh., genus Agropyron (wheatgrass), subgenus Agropyron Nevski. According to the data of *The Plant List*, the genus *Agropyron* includes 26 species [10]. In the system proposed by the Federal Research Center N.I.Vavilov All-Russian Institute of Plant Genetic Resources (VIR), there are 24 taxons of the genus *Agropyron* in the flora of Russia: 11 species and 13 subspecies, belonging to the 3 species, of which two have two subspecies, and *A. crictatum* s.l. has 11 subspecies [11].

Currently (as of 2020), 10 varieties of comb-wheatgrass (*Agropyron pectiniforme* Roem. et Schult.), 3 varieties of Siberian wheatgrass (*Agropyron sibiricum* (Willd.) P. Beav.) and 8 varieties of desert wheatgrass (*Agropyron desertorum* (Fisch. ex Link) Schult.) are registered in the State Register of Breeding Achievements approved for use in Russia.

For arid conditions, the most effective crop among perennial bluegrass grasses is comb-wheatgrass (according to GRIN Taxonomy for Plants (2014-05-22) the valid name is *Agropyron cristatum* subsp. *pectinatum* var. *pectinatum* (M. Bieb.) Tzvelev; according to the list of botanical and zoological genera and species provided for by article 1413 of the Civil Code of the Russian Federation of the Ministry of Agriculture of the Russian Federation - *Agropyron pectiniforme* Roem. et Schult.) (Synonym = *A. pectinatum* (Bieb.) Beauv. 1812, Ess. Agrost.: 146. – *Triticum pectinatum* Bieb. – *T. imbricatum* Bieb. – *T. muricatum* Link. – *T. cristatum* (L.) Schreb. var. *hirsutum* Lindem.: 1893. – *Agropyron cristatum* (L.) Beauv. subsp. *pectinatum* (Bieb.) Tzvel.: 1974=*Agropyron pectinatum (M.Bieb.) P.Beauv.* = of *Elymus uralensis subsp. viridiglu European Environment Agency* (European Environment Agency). This species of wheatgrass occupies the most extensive range in the wild [11, 12]. In culture, it also has the highest specific gravity among other wheatgrass species; it also surpasses them in yield [10, 13]. Among the varieties of perennial grasses created at the Voronezh Experimental Station, in 1958 the Pavlovsky 12 comb-wheatgrass was zoned. Pavlovsky 12 variety belongs to the East European steppe ecological and geographical group of wheatgrass [14], and was created on the basis of the wild population of the Don River floodplain. The variety is characterized by high winter hardiness, drought hardiness, unpretentiousness to cultivation conditions. Under favorable conditions, it provides the yield of air-dry mass up to 3.0-4.0 t/ha, seeds - up to 0.6-0.8 t/ha. On poor washed-away, stony and sandy soils, it develops a deep and powerful root system, which is a very much multiple of the weight of the aboveground mass [15]. It has a long productive longevity and high competitiveness, for the 10-12th years of use when sowing the Pavlovsky 12 variety in the composition of multi-species grass mixtures, the grass stand consisted almost entirely of wheatgrass. However, in connection with the long-term reproduction of Pavlovsky 12 and changes in climatic conditions and agrotechnical requirements, farming systems over the past 70 years, the issue of developing new varieties of wheatgrass is relevant.

An important task of wheatgrass breeding currently is the creation of intensive-type wheatgrass varieties, geographically and ecologically differentiated, characterized by high plasticity and resistance to stress factors [15]. To expand the amplitude of the genetic diversity of economically valuable traits, it is necessary to involve the initial material of different geographical origin in the breeding process.

2 Purpose of work

Creation, reproduction and introduction into production of intensive varieties of comb-wheatgrass of a new generation, combining early maturity with high drought resistance, having high fodder and seed yield and other positive signs and properties and suitable for use in arid conditions of the steppe zone.

3 Materials and methods

The studies were carried out in 2017-2020 in the field crop rotation of the Voronezh Experimental Station for perennial grasses, located near the city of Pavlovsk, Voronezh Region, in the steppe zone of the south of the Central Black Earth Region. In the collection nursery of the initial material, specimens of wheatgrass obtained from the A.I. Baraev Research and Production Center of Grain Farming. The climate of the Voronezh region is characterized by continentality, which intensifies from the northwest to the southeast, with warm summers and rather cold winters. In general, the climatic conditions are favorable for the cultivation of most crops. The main limiting factor for the cultivation of agricultural crops is insufficient moisture supply. The total amount of rainfall on average per year is 570 mm in the north and 420 mm in the south of the region. From May to September, rainfall is 240 mm, during the dry period rainfall is 136-179 mm. A characteristic feature is the unevenness of rainfall over the years and seasons. The average annual air temperature varies from +5°C in the north, to +7°C in its southern regions. The lowest temperature drops to -40°, the highest is +40° (the absolute maximum in the south of the region is +42°), the sum of active temperatures ranges from 2600°C to 3000°C. Droughts and dry winds are observed almost every year. Droughts are very dynamic (May, June, July), in the southern regions their duration reaches 70-80 days. Every third year they are intense.

The research was carried out in uncovered single-species crops in breeding and collection nurseries. All records and observations were carried out using guidelines for the study of the world collection of perennial forage plants, breeding of perennial grasses and guidelines for the breeding of perennial grasses. Breeding nurseries were established in a field crop rotation in a wide-row and continuous way. The soil cover of the field plot is ordinary medium-thick, medium-loamy, thin-humus chernozem. The content of P_2O_5 in soil is average, K_2O – increased. Plot area is 5-10 m^2, Pavlovsky 12 variety was used as a standard.

Statistical processing of the experimental data was carried out by the method of analysis of variance based on the method of B.A. Dospekhova (1985) on a PC using Excel programs from the Microsoft Office Word 2007 application package.

4 Results and discussion

In previous years, various breeding methods helped to obtained breeding material, which was tested at different levels of the breeding process. Competitive variety testing was completed in 2019. According to the research carried out, it was found that the best variety of comb-wheatgrass is Pridonsky variety, which was named Nenarokovsky when transferred to the State Commission; it has an excess of the standard for the yield of green and dry matter by 11.4 and 12.1%, respectively, and was transferred to the State Commission for registration in 2020 [15].

The work on creating for the conditions of the Central Black Earth Region a new promising breeding material for wheatgrass with high potential productivity and resistance to the stress effects of abiotic and biotic environmental factors continues. In 2020, a four-year study of the comb-wheatgrass collection provided by the A.I. Baraev Research and Production Center of Grain Farming (Kazakhstan). Phenological observations, cares, the nature of regrowth in spring and after each mowing, the height of forage grass stands (by mowing), accounting for fodder mass and seed yield were carried out in the field at the breeding nursery during the growing season of 2017-2020.

The meteorological conditions of the growing seasons were significantly different, which made it possible to reveal the differences between the studied varieties and the standard. In terms of meteorological conditions, in general, 2017 was relatively favorable for the germination, growth and development of perennial grasses. The winter reserve of productive moisture was the highest over the years of the experiment (Table 1). Sowing was carried out in the 2nd decade of April 2017. The rainfall in April-May (74.7 mm) contributed to the obtaining of good seedlings. Seedlings appeared in early May, tillering began in June, and by the end of the growing season, the crops had formed well-developed grass stands. The condition of the plants before leaving for winter was assessed as good.

Table 1. Hydrothermal conditions during the growing season of wheatgrass.

Year	Indicators	Month					
		April	May	June	July	August	September
2017	Rainfall, mm	29.4	45.3	21.0	19.5	5.4	50.0
	% of the rate	79.4	89.0	34.3	31.9	8.4	135.0
	Air temperature, °C	9.5	19.2	21.7	21.5	24.1	17.2
	Departure from the norm, °C	+2.6	+4.1	+2.7	+2.7	+4.2	+3.6
	Productive moisture reserves in the layer, mm: 0-20 cm.	28.4	15.4	13.5	10.7	0.8	15.4
	0-100 cm	129.8	80.6	85.0	78.8	27.7	81.9
2018	Rainfall, mm	23.3	12.5	18.7	119.3	9.5	48.2
	% of the rate	63.0	24.5	30.6	198.0	15.0	130.0
	Air temperature, °C	10.6	21.5	23.0	24.7	23.6	18.0
	Departure from the norm, °C	+3.7	+6.4	+4.0	+3.7	+3.7	+4.4
	Productive moisture reserves in the layer, mm: 0-20 cm.	21.7	15.3	5.1	0.2	5.4	12.5
	0-100 cm	108.5	92.7	69.0	36.4	71.4	67.6

Table 1. Continued.

2019	Rainfall, mm	17.9	87.6	29.8	114.9	33.5	12.7
	% of the rate	48.3	171.8	48.9	188.0	52.0	34.0
	Air temperature, °C	11.5	21.7	25.9	22.1	22.0	15.6
	Departure from the norm, °C	+4.6	+6.6	+6.9	+1.1	+2.1	+2.0
	Productive moisture reserves in the layer, mm: 0-20 cm.	20.7	11.2	0.7	9.1	13.8	0
	0-100 cm	103.0	84.9	60.6	44.0	78.8	8.6
2020	Rainfall, mm	30.8	44.4	29.5	12.5	8.8	1.0
	% of the rate	83.0	87.0	48.4	20.4	13.7	2.7
	Air temperature, °C	9.8	14.8	24.9	26.6	24.5	18.5
	Departure from the norm, °C	+2.9	-0.3	+5.1	+5.6	+4.6	+4.9
	Productive moisture reserves in the layer, mm: 0-20 cm.	20.9	19.1	9.4	0.5	0	0
	0-100 cm	106.1	89.5	61.0	27.7	5.2	3.5

The winter reserve of productive moisture in 2018-2020 in the meter layer of soil was (103.0-108.5 mm). However, during the growing season, the temperature regime was increased and was characterized by instability, and the rainfall was uneven in its distribution over months and decades. The temperature during the formation of the first cut was higher than the average long-term norms, the excess of which was 3.7-6.4° in April-May in 2018, and 4.6-6.6°C, in April 2020 by +2.9°C, and in May there was a decrease to 0.3°C from the average long-term norms.

The beginning of spring regrowth of plants 2-4 years of age was noted on March 27 - 31. The interphase period "regrowth - flowering" was leveled between the specimens and the standard, its duration was 62–66 days. Evaluation of plant height showed that most of the studied varieties in the first cut, according to the average indicators of a three-year study, formed grass stands with an average height of 77.3–86.0 cm, of which only 3 specimens slightly exceeded the standard 85.0 cm (Table 2).

Table 2. Height of forage grasses of wheatgrass varieties (sowing in 2017, records in 2018-2020).

Name of specimens	first mowing				second mowing		
	Years of life						
	2	3	4	Medium	2	3	Medium
St. Pavlovsky 12 variety	**81.5**	**85.0**	**88.4**	**85.0**	**45.5**	**43.2**	**44.4**
Batyr variety	75.8	75.4	80.8	77.3	46.2	33.0	39.6
Karabalyk 202 variety	75.0	80.6	85.2	80.3	46.0	31.0	38.5
Shortandinsky broad-spiked variety	81.5	85.2	89.2	85.3	40.0	35.0	37.5
Burabay variety	78.5	82.6	78.6	79.9	42.0	31.6	36.8
SGP KL-1565	82.0	82.8	81.4	82.1	38.0	37.2	37.6
SGP KL-1538	80.5	83.8	84.0	82.8	37.0	37.8	37.4
SGP KL-1540	81.5	77.2	81.4	80.0	38.0	36.6	37.3
SGP KL-1560	82.5	83.4	85.8	83.9	38.0	44.2	41.1
SGP KL-1561	81.5	86.2	90.4	86.0	47.0	36.8	41.9
SGP KL-1564	82.0	88.4	85.4	85.3	41.0	33.6	37.3
SGP KL-1562	77.5	77.8	82.4	79.2	38.0	33.8	35.9
SGP KL-1557	81.5	79.6	84.8	81.0	38.0	43.2	40.6
SGP KL-1545	78.5	76.8	84.0	79.8	35.0	35.8	35.4
NSR_{05}	6.18	6.23	6.52	6.31	3.23	3.01	3.17

The second mowing was formed only in 2018 and 2019, the average height for two years in the studied specimens was in the range of 35.4–41.9 cm and did not exceed the corresponding values of 44.4 cm for the standard. The reason for the absence of the second mowing, as noted above, was the uneven distribution of rainfall by months and high temperatures. The amount of rainfall in June, July and August, in 2018 and 2019, contributed to the formation of the second mowing, they fell only in July, respectively, 198 and 188% of the long-term norm. In the severely arid year of 2020, 50.8 mm fell during these months, the shortage of rainfall was 135.2 mm, i.e., each hectare did not receive additional 1,352 tons of water. The uneven distribution of rainfall over the years and high temperatures adversely affected the overall productivity of green mass according to the standard and all specimens (Table 3). The highest yield was obtained in 2 years of plant life. In the third and fourth years of life, the yield of all specimens decreased. Thus, the standard's yield of green mass, respectively, was 20.04, 12.48 and 7.48 t/ha, respectively, i.e., a decrease was noted every year by one third. In our opinion, the yield decline was due to a lack of moisture. This restrained growth processes; during periods of drought, the leaves were observed to dry out, which negatively affected the formation of energy reserves of plants for overwintering and active regrowth next spring. The amount of rainfall and reserves of productive moisture are of great importance for plants, the optimal content of which in spring in a meter layer of soil should be 170-175 mm. The initial period of regrowth and the volume of the yield are positively influenced by the amount of rainfall in the winter period, and during the growing season and their distribution by months and phases of plant development.

Yield is one of the factors that can characterize the resistance of the specimens to biotic and abiotic factors. In the breeding nursery, out of 20 specimens of wheatgrass, 2 specimens were isolated according to the productivity of green mass according to the average three-year data, which exceeded the standard by more than 5%. As noted above, starting from the third year of life, there was a decrease in the yield of all specimens. In the severely arid year of 2020, when only one cut was obtained, the yield of green mass of all specimens is significantly lower than the two previous years. The reason for the absence of the second moving was the uneven distribution of rainfall by months and high temperatures. The amount of rainfall in June, July and August fell 50.8 mm, the shortage of rainfall in 2020 was 135.2 mm, i.e., each hectare did not receive additional 1352 tons of water. The relative humidity of the air has a great influence on the yield of forage and seed productivity, which in 2020 during atmospheric drought and dry winds on some days decreased to 30%, which increased the coefficient of water consumption and evaporation of rainfall from the soil surface.

Table 3. Forage yield of the best specimens in the collection nursery of wheatgrass (sowing in 2017, records in 2018-2020).

Item No.	Specimen	2 year of life		3 year of life		4 year of life		Year average	
		(t/ha)	+/- t	(t/ha)	+/- t	(t/ha)	+/- t	(t/ha)	+/- t
St	**Pavlovsky 12**	**20.04**	**0**	**12.48**	**0**	**7.48**	**0**	**13.31**	**0**
1	Batyr variety	24.10	+4.06	10.00	-2.48	8.50	+1.02	14.20	+0.89
2	Karabalykskiy 202 variety	12.72	-7.32	10.58	-1.90	10.30	+2.82	11.20	-2.11
3	Shortandinskiy broad-spiked variety	14.36	-5.68	11.96	-0.52	10.78	+3.30	12.37	-0.94
4	Burabay variety	15.32	-4.72	10.80	-1.68	7.58	+0.10	11.23	-2.08
5.	SGP KL-1565	17.86	-2.18	11.72	-0.76	7.80	+0.32	12.46	-0.85
6	SGP KL-1538	15.70	-4.34	9.98	-2.50	7.76	+0.28	11.15	-2.16
9	SGP KL-1540	16.86	-3.18	10.46	-2.02	7.72	+0.24	11.68	-1.63
15	SGP KL-1560	17.12	-2.92	11.40	-1.08	9.10	+1.62	12.54	-0.77
16	SGP KL-1561	24.06	+4.02	11.16	-1.32	10.50	+3.02	15.24	+1.93
18	SGP KL-1564	17.90	-2.14	11.02	-1.46	11.12	+3.64	13.35	+0.04
19	SGP KL-1562	21.12	+1.08	10.58	-1.90	9.18	+1.70	13.63	+0.32
20	SGP KL-1557	22.06	+2.02	10.60	-1.88	8.62	+1.14	13.76	+0.45
21	SGP KL-1545	15.90	-4.14	11.84	-0.64	8.56	+1.08	12.10	-1.21
NSR_{05}		1.96		0.87		0.76		1.08	

The most objective assessment of the studied varieties was revealed in 2020, when a more significant decrease in the fodder mass occurred in the standard (Pavlovsky 12 variety). This year, the yield exceeded the standard of 9 specimens by 5 percent or more. It can be assumed that these specimens have a high adaptive potential and increased stress resistance. They are, perhaps, less demanding on soil fertility and their productivity is less dependent on high temperatures and lack of moisture than the standard.

The results of seed yield are shown in Table 4. Depending on the biological characteristics of the specimens and the weather conditions that developed during the years of the study, there were significant fluctuations in seed yields, as well as in fodder mass. The highest yield over the years of research was formed by SGP KL-1538 specimen in the more favorable weather conditions in 2018 (0.84 t/ha), and the lowest in 2020 (0.12 t/ha). The average yield of this specimen over three years of observations was 0.38 t/ha, which is 27.4% higher than the standard yield. The seed yield also varied in other specimens. Among the specimens shown in the table, Batyr variety turned out to be less yielding, the three-year average yield of which was 0.23 t/ha.

Table 4. Seed yield of wheatgrass varieties in the collection nursery, t/ha (sowing 2017, records in 2018-2020).

Name of specimens	2 year of life	3 year of life	4 year insurance	Average of 3 years	% of standard
St. Pavlovsky 12 variety	0.51	0.24	0.15	0.30	100.0
Batyr variety	0.45	0.19	0.06	0.23	77.3
Karabalyk 202 variety	0.73	0.10	0.39	0.29	97.0
Shortandinsky broad-spiked variety	0.75	0.20	0.11	0.35	116.6
Burabay variety	0.63	0.12	0.06	0.27	89.6
SGP KL-1565	0.63	0.09	0.08	0.27	89.4
SGP KL-1538	0.84	0.18	0.12	0.38	127.4
SGP KL-1540	0.43	0.08	0.09	0.25	81.5
SGP KL-1560	0.71	0.15	0.09	0.32	105.5
SGP KL-1561	0.64	0.20	0.16	0.33	110.5
SGP KL-1564	0.72	0.20	0.16	0.36	120.9
SGP KL-1562	0.51	0.13	0.15	0.26	86.8
SGP KL-1557	0.64	0.22	0.13	0.33	109.8
SGP KL-1545	0.53	0.14	0.12	0.26	87.3
SGP KL-1544	0.68	0.19	0.10	0.32	108.1
NSR_{05}	0.062	0.017	0.012	0.030	

5 Conclusion

According to the results of scientific research: experimental data were obtained on the evaluation of a new starting material of the comb-wheatgrass. In the 2017 collection nursery, out of 20 wheatgrass specimens obtained from Kazakhstan, 2 specimens exceeded the standard by more than 5% in terms of green mass according to average three-year data and 9 specimens exceeded the standard in dry 2020.

As can be seen from the tables, starting from the third year of life, there was a decrease in yield in terms of fodder and seed yield of all specimens. Thus, the standard yield of green mass, respectively, by years was 20.04, 12.48 and 7.48 t/ha, i.e., a decrease was noted by one third, in terms of seed productivity of 0.51, 0.24 and 0.15 t/ha. In our opinion, the decrease in yield was due to a lack of moisture. In the severely arid year 2020, when only one cut was obtained, the yield of green mass of all specimens is significantly lower than the two previous years, the most significant decrease in the fodder mass occurred in the standard (Pavlovsky 12 variety). This year, the yield exceeded the standard of 9 specimens by 5 percent or more. It can be assumed that these specimens have a high adaptive potential

 https://doi.org/10.1051/e3sconf/202126203004

and increased stress resistance, which should be continued to be used in research as more drought hardy specimens to improve the available varieties.

Acknowledgments

The authors express gratitude to N. I. Filippova, head of the perennial herb breeding department of the A. I. Baraev Research and Production Center of Grain Farming for the provided specimens from the collection of the center.

References

1. L. Salvati, L. Perini, A. Sabbi, S. Bajocco, Geographical Research, **50(2),** 193 (2012)
2. B. S. Chauhan, G. Mahajan, R. K. Randhawa et al, Advances in agronomy, **123,** 65 (2014)
3. A. Anand, S. Khetarpal, Plant Biology and Biotechnology. Springer, 729 (2015)
4. B. Sen, S. Topcu, M. Türkes et al, Climate Research, **52,** 175 (2012)
5. A. I. Graux, G. Bellocchi, R. Lardy, J. F. Soussana, Agricultural and Forest Meteorology, **170,** 114 (2013)
6. W. D. Dar, C. L. Laxmipathi Gowda, Journal of Crop Improvement, **27(2),** 242 (2013)
7. K. B. Ruiz, S. Biondi, R. Oses et al, Agronomy for sustainable development, **34(2),** 349 (2014)
8. Q. Sohail, H. Naheed, R. Mohammadi, Innovations in Dryland Agriculture. Springer, Cham, 257 (2016)
9. T. Absattar, A. Absattarova, N. Fillipova et al, Molecular breeding, **38(4),** 1 (2018)
10. M. V. Derevyanikova, V. V. Chumakova, V. V. Kravtsov, Science news in the agro-industrial complex. Wheatgrass is a promising crop for cultivation in the southern regions of Russia, **2(11),** 189 (2018)
11. A. V. Bukhteeva, Proceedings on applied botany, genetics and breeding, **169,** 210 (2012)
12. A. N. Danilova, A. A. Sumbembayev, European Researcher, **74(5-1),** 809 (2014)
13. G. K. Bulakhtina, A. V. Kudryashov, N. I. Kudryashova et al, RUDN Journal of Agronomy and Animal Industries, **13(2),** 103 (2018)
14. A. V. Bukhteeva, L. L. Malyshev, N. I. Dzyubenko, A. A. Kochegina, *Genetic Resources of wheatgrass – Agropyron GAERTN,* 268 (2016)
15. S. V. Saprykin, V. N. Zolotarev, I. S. Ivanov, M. G. Ostrikova, IOP Conference Series: Earth and Environmental Science, **663(1),** 012029 (2021)

Complexes of ground beetles (Coleoptera, Carabidae) in agrocenoses of cultivated crops of Krasnodar Territory

Alexandr Bondarenko[1], *Eugeniy* Khomitskiy[2], *Alexandr* Zamotajlov[2*], and *Alexandr* Belyi[2]

[1]Russian Centre of Forest Health, Branch "Centre of Forest Health of Krasnodar Region, Odesskiy Proezd, 4, 350020 Krasnodar, Russia
[2]I.T. Trubilin Kuban State Agrarian University, Kalinina st., 13, 350044 Krasnodar, Russia

Abstract. The paper presents new data on the species composition of ground beetles of row crops in 10 municipalities of the Krasnodar Territory. Pursuant to the research results, 45 species of Carabidae were identified. The chorological analysis and the analysis of the ecological groups noted in the model areas were performed.

1 Introduction

This study continues long-term observations of the local fauna of ground beetles in agrocenoses of the Northwest Caucasus, namely, the study of the structure of the community of ground beetles formed on row crops (sunflower and corn) of the Kuban Region. This kind of research has already been performed both abroad [15-22] and in Russia, in particular, in the central zone of the Krasnodar Territory [1-13], but, at the moment, these observations have been performed most fully, since they cover all the main zones of the region. Obviously, the faunas of ground beetles of various agrocenoses are not constant, their composition is influenced by many factors that form a special microclimate, which is confirmed by the researches performed by the domestic and foreign authors at the agricultural landscapes. We should note that, together with other studies on this topic, these materials allow to track the fauna transformations, associated with the influence of climate change or anthropogenic impact.

2. Materials and methods

The material was collected using modified Barber soil traps [14] in 13 localities (Table 1) from April to October in 2019 and 2020. The traps were 0.5 L plastic cups with an inlet diameter of 95 mm. Herewith, the edges of the glasses were at the same level with the soil surface. A 4% formaldehyde aqueous solution was used as the fixing liquid. In order to

* Corresponding author: a_zamotajlov@mail.ru

prevent the traps from flooding during rain and excessive evaporation of the fixing liquid, plastic covers were installed above them. Means of accounting were placed in a line of 20 pieces, at a distance of 5 m from each other.

Table 1. Characterization and localization of model areas

No	Municipal formation of Krasnodar Region	Crop	Abbreviation
1	Timashevsky district	sunflower	Tim 1
2	Timashevsky district	corn	Tim 2
3	Dinskoy district	sunflower	Din
4	Tbilisky district	corn	Tbil
5	Gulkevichsky district	sunflower	Gulk 1
6	Gulkevichsky district	sunflower	Gulk 2
7	Armavir	corn	Arm
8	Starominsky district	sunflower	Star
9	Kanevskoy district	corn	Kan
10	Kushchevsky district	corn	Kush
11	Pavlovsky district	corn	Pavl
12	Tikhoretsky district	sunflower	Tikh 1
13	Tikhoretsky district	corn	Tikh 2

Based on the results of identification of the material selected from the traps, the species composition of ground beetles in the model areas was determined, and analyzes of the similarity of local faunas, global ranges of species and their landscape-cenotic confinement were performed.

The Jaccard's coefficient of community was used to compare localities by species similarity of ground beetles. The calculation and construction of the dendrogram was performed in the Biodiversity Pro 2.0 software application. For the classification of habitats, as well as ecological groups, the nomenclature provided by the authors earlier was used [4, 5]. The analysis of the landscape-cenotic confinement of ground beetles was performed considering the available data on the occurrence of the species under consideration in natural biocenoses.

3 Results and discussion

Pursuant to the results of research in the fields of row crops, 45 species of ground beetles were identified (Table 2). In each individual model area, there were from 23 to 35 species. The greatest species diversity was in the field of corn in the Tbilisky district, the smallest - in the field of sunflower No. 1 in the Gulkevichsky district.

Table 2. Species composition and occurrence of ground beetles on row crops in the Krasnodar Territory

No	Species	Locality												
		Tim 1	Tim 2	Din	Tbil	Gulk 1	Gulk 2	Arm	Star	Kan	Kush	Pavl	Tikh 1	Tikh 2
1	*Cicindela germanica* Linnaeus, 1758				+				+					
2	*Calosoma auropunctatum* (Herbst, 1784)		+		+			+		+	+	+		+

Table 2. Continued

3	*Carabus cumanus* Fischer von Waldheim, 1823				+	+	+	+						
4	*Carabus planus* Géhin, 1885	+	+	+	+			+			+			
5	*Carabus exaratus* Quensel, 1806	+	+	+	+	+	+	+	+	+	+	+	+	+
6	*Clivina fossor* (Linnaeus, 1758)								+					
7	*Broscus semistriatus* (Dejean, 1828)	+	+	+	+	+	+	+	+	+	+	+	+	+
8	*Poecilus cupreus* (Linnaeus, 1758)	+	+	+	+	+	+	+	+	+	+	+	+	+
9	*Poecilus sericeus* Fischer von Waldheim, 1824	+	+	+	+	+	+	+	+	+	+	+	+	+
10	*Poecilus crenuliger* Chaudoir, 1876	+	+	+	+	+	+	+	+	+	+	+	+	+
11	*Pterostichus niger* (Schaller, 1783)	+	+	+	+	+	+	+	+	+	+	+	+	+
12	*Pterostichus longicollis* (Duftschmid, 1812)	+	+						+			+		
13	*Pterostichus strenuus* (Panzer, 1796)		+		+					+				
14	*Calathus fuscipes* (Goeze, 1777)	+	+	+					+	+	+	+	+	+
15	*Calathus ambiguus* (Paykull, 1790)		+		+			+		+	+	+		+
16	*Calathus melanocephalus* (Linnaeus, 1758)	+	+	+	+			+						
17	*Calathus halensis* (Schaller, 1783)	+	+	+	+	+	+	+	+	+	+	+	+	+
18	*Anchomenus dorsalis* (Pontoppidan, 1763)	+	+	+	+	+	+	+	+	+	+	+	+	+
19	*Synuchus vivalis* (Illiger, 1798)								+				+	
20	*Amara aenea* (DeGeer, 1774)	+	+	+	+	+	+	+	+	+	+	+	+	+
21	*Amara familiaris* (Duftschmid, 1812)				+			+	+					+
22	*Amara ovata* (Fabricius, 1792)								+					
23	*Zabrus tenebrioides* (Goeze, 1777)								+		+		+	+
24	*Anisodactylus binotatus* (Fabricius, 1787)			+	+			+	+	+	+	+		+
25	*Anisodactylus signatus* (Panzer, 1796)	+	+	+	+	+	+	+	+	+	+	+	+	+
26	*Stenolophus mixtus* (Herbst, 1784)	+	+	+	+	+	+	+	+	+	+	+	+	+
27	*Harpalus griseus* (Panzer, 1796)	+	+	+	+	+	+	+	+	+	+	+	+	+
28	*Harpalus rufipes* (DeGeer, 1774)	+	+	+	+	+	+	+	+	+	+	+	+	+

Table 2. Continued

29	*Harpalus attenuatus* Stephens, 1828							+		+				
30	*Harpalus flavescens* (Piller & Mitterpacher, 1783)	+	+	+					+	+		+	+	+
31	*Harpalus tardus (*Panzer, 1796)	+	+	+	+	+	+	+	+	+	+	+	+	+
32	*Harpalus latus (*Linnaeus, 1758)				+				+					+
33	*Harpalus smaragdinus* (Duftschmid, 1812)									+				
34	*Harpalus cupreus* Dejean, 1829	+	+	+	+	+	+	+	+	+	+	+	+	+
35	*Harpalus affinis* (Schrank, 1781)	+	+	+	+	+	+	+	+	+	+	+	+	+
36	*Harpalus distinguendus (*Duftschmid, 1812)	+	+	+	+	+	+	+	+	+	+	+	+	+
37	*Ophonus azureus* (Fabricius, 1775)		+		+	+		+		+	+	+		+
38	*Ophonus diffinis* (Dejean, 1829)				+						+			
39	*Ophonus sabulicola* (Panzer, 1796)			+								+		+
40	*Chlaenius aeneocephalus* Dejean, 1826		+		+		+	+				+		
41	*Microlestes minutulus (*Goeze, 1777)			+	+		+	+		+			+	+
42	*Brachinus alexandrii* F. Battoni, 1984	+	+	+	+	+	+	+	+	+	+	+	+	+
43	*Brachinus crepitans* (Linnaeus, 1758)	+	+	+	+	+	+	+	+	+	+	+	+	+
44	*Brachinus ejaculans* Fischer von Waldheim, 1828	+	+	+	+	+	+	+	+	+	+	+	+	+
45	*Brachinus elegans* Chaudoir, 1842	+	+	+	+	+	+	+	+	+	+	+	+	+

To determine the similarity of the complexes of ground beetles in the studied agrocenoses, a cluster analysis was performed on the basis of the Jaccard coefficient and a dendrogram of similarity was constructed (Fig. 1). The local faunas of the Carabidae in the area under consideration are characterized by a fairly high level of similarity. With a threshold distance of 74%, 4 classes are distinguished, including from 1 to 5 localities. Class 1, which is the most different from all of them (the similarity level is 68%), includes a sunflower field complex in the Starominsky district. The second class is formed by the faunas of the corn fields of Armavir and Tbilisky region. The third class, the closest to the previous one, included the complexes of ground beetles in corn fields in the Kushchevsky, Tikhoretsky, Kanevsky, Pavlovsky and Timashevsky districts. The fourth class is represented by local faunas of sunflower fields in the Gulkevichsky, Tikhoretsky, Dinskoy and Timashevsky districts. Therefore, there is the dependence of the similarity of the local faunas of ground beetles on the cultivated crop, with individual groupings of fields of sunflower and corn. Only the carabid complex of the sunflower field in the Starominsky district differs significantly from all the others. Herewith, the territorial proximity matters for the fields of the same crops. Thus, the greatest similarity of faunas (87%) was observed in sunflower fields in one municipal entity of the Krasnodar Territory - in the Gulkevichsky district (Gulk 1 and Gulk 2).

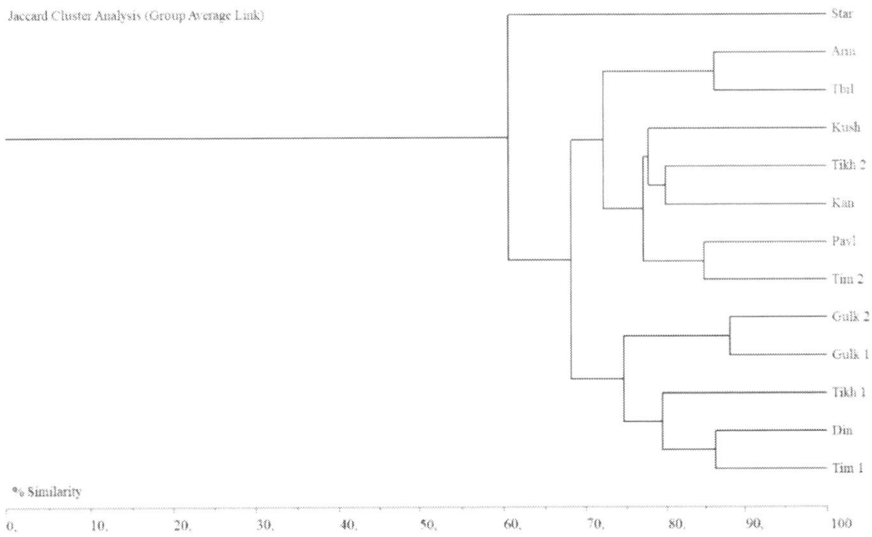

Fig. 1. Dendrogram of the similarity of complexes of ground beetles in the fields of row crops in the Krasnodar Territory. Locality designations are shown in Table 1.

The arealogical analysis of the local faunas of ground beetles of row crops in the area under consideration showed that species with wide polisector and polyzonal ranges prevailed in all localities: Holarctic, Transpalaearctic, Amphipalaearctic, West Palaearctic, Euro-Siberian, etc. Their share ranged from 47% (corn field in Pavlovsky district) to 58% (sunflower field in Starominsky district). In most observation points, a high proportion were European-Mediterranean species - from 13% (field of sunflower No. 2 in Gulkevichsky district) to 24% (field of corn in Kushchevsky district), as well as steppe (Scythian, European-Scythian, partly polisector) - from 13% (sunflower fields in Gulkevichsky and Starominsky districts) to 19% (sunflower and corn fields in Timashevsky district). In general terms, such proportions are typical not only for the carabid complexes under consideration, but also for carabid complexes of completely different agrocenoses in the Krasnodar Territory, including for orchards and vineyards [3].

Pursuant to the results of the analysis of the ecological groups of ground beetles identified in the fields of row crops in the area under consideration, the prevalence of polytopic mesophiles was established. Their share varied from 51.7% (corn field in Kushchevsky district) to 62.5% (sunflower field No. 2 in Gulkevichsky district and corn field in Armavir). The share of steppe mesophiles was also quite high, ranging from 25% (corn field in Armavir) to 34.6% (sunflower field in Tikhoretsky district). The share of other ecological groups was significantly lower, none of them exceeded 7% in any of the localities.

4 Conclusions

Pursuant to the results of the study of carabid complexes in the fields of row crops of the area under consideration, 45 species of ground beetles were identified. There is a fairly high similarity of local faunas. Herewith, the groups characteristic of corn fields and sunflower fields are clearly differentiated. This is probably due to the difference in microclimatic conditions in the area of the soil surface and its upper layer inhabited by most species. Only the fauna of ground beetles in the sunflower field differs significantly from all other groups.

Pursuant to the results of the arealogical analysis, a sufficiently high similarity of the proportions of chorological complexes was established for all model areas. In addition, similar proportions are observed in most other agrocenoses in the Krasnodar Territory. The dominance of species with wide polisector and polyzonal ranges was noted. The European-Mediterranean species also accounted for a significant share. Herewith, in contrast to other agrocenoses, the set of chorological groups in the fields of row crops was much poorer.

Pursuant to the results of the analysis of ecological groups of ground beetles in the model areas, a significant predominance of polytope mesophiles was established, the proportion of which in none of the areas was less than half. Other large group was created by steppe mesophiles. Other ecological groups showed the poor species diversity. The main proportions of the ecological groups of ground beetles were broadly similar and did not depend on the cultivated crop or the territorial distribution of the model plot.

Acknowledgements

The reported studies were partially supported by the RFBR (projects 19-44-230004 r_a and 19-44-233007 r_mol_a).

References

1. A.I. Belyi, L.S. Glushchenko, A.S. Zamotajlov, E.E. Khomitskiy, *Biodiversity, Bioconservation, Biomonitoring*, 23 (2013)
2. A.I. Belyi, A.S. Zamotajlov, E.E. Khomitskiy, I.A. Markova, Proc. Kuban Stat. Agr. Univ., **48**, 35 (2014)
3. A.S. Bondarenko, E.E. Khomitskiy, A.S. Zamotajlov, A.I. Belyi, Proc. Kuban Stat. Agr. Univ., **86**, 37 (2020)
4. A.S. Zamotajlov, A.Yu. Vozzhannikova, A.K. Makaov, Proc. Kuban Stat. Agr. Univ., **20**, 206 (2009)
5. A.S. Zamotajlov, V.N. Orlov, M.V. Nabozhenko, N.V. Okhrimenko, E.A. Khatschikov, M.I. Shapovalov, I.V. Shokhin, Entomological Review, **90**, 333 (2010)
6. A.S. Zamotajlov, R.G. Krivoruchka, Proc. Russian Entomol. Soc., **84**, 42 (2013)
7. A.S. Zamotajlov, E.E. Khomitskiy, A.I. Belyi, Proc. Kuban Stat. Agr. Univ., **52**, 103 (2015)
8. V.Yu. Serdyuk, A.S. Zamotajlov, Proc. Kuban Stat. Agr. Univ., **63**, 90 (2016)
9. V.Yu. Serdyuk, A.S. Zamotajlov, A.S. Bondarenko, Proc. Kuban Stat. Agr. Univ., **75**, 90 (2018)
10. E.E. Khomitskiy, A.S. Zamotajlov, A.I. Belyi, N.B. Nikitskiy, *Biodiversity, Bioconservation, Biomonitoring*, 85 (2015)
11. E.E. Khomitskiy, A.S. Zamotajlov, A.I. Belyi, A.S. Bondarenko, Proc. Kuban Stat. Agr. Univ., **79**, 80 (2019)
12. E.E. Khomitskiy, A.S. Zamotajlov, A.I. Belyi, L.P. Esipenko, I.V. Balakhnina, BIO Web of Conferences, **21**, 00009 (2020)
13. A.S. Zamotajlov, E.E. Khomitskiy, A.I. Belyi, BIO Web of Conferences, **18**, 00030 (2020)
14. H. Barber, J. Elisha Mitchell Sci. Soc., **46**, 259 (1931)
15. F. Talarico, A. Giglio, R. Pizzolotto, P. Brandmayr, Eur. J. Entomol., **113**, 325 (2016)
16. L. Arus, A. Kikas, A. Luik, Žemdirbystė – Agriculture, **99**, 327 (2012)
17. J.C. Lee, D.L. Edwards, BioControl; Dordrecht, **57**, 515 (2012)
18. B. Radawiec, Periodicum Biologorum, **118**, 291 (2016)

19. T. Diekötter, S. Wamser, T. Dörner, V. Wolters, K. Birkhofer, Agricultural and Forest Entomology, **18**, 167 (2016)
20. F. Oberholzer, Th. Frank, Biocontrol Science and Technology, **13**, 99 (2013)
21. S.I. Rondon, A. Pantoja, A. Hagerty, D.A. Horneck, Production Florida Entomologist, **96** (4), 1492 (2013)
22. P.J. Reganold, J.M. Wachter, Nature Plants, **2**, 15221 (2016)

E3S Web of Conferences **262**, 03006 (2021) https://doi.org/10.1051/e3sconf/202126203006
ITEEA 2021

Anthropogenic impact on grain-size distribution and agrophysical properties of soils of cultivated rice lands of Kuban

A.V. Osipov[*], *V.N.* Slyusarev, *V.P.* Vlasenko, and *I.I.* Suminski

Federal State Budgetary Educational Institution of Higher Education "Kuban State Agricultural University named after I.T. Trubilin"

Abstract. This work reviews the relation of the grain-size distribution and impact of the use of rice rotation soils on the main agrophysical indicators of soils that determine the composition and properties of soil cover of modern delta of Kuban. The regularities of dependence between the density of the matrix soil and the grain-size distribution that consist of mineral composition of alluvial soils and rocks have been identified. The trend towards increase in soil density with increase in physical clay and silt content, as well as increase in matrix soil density has been established. No significant differences in agrophysical properties of subsurface soil horizons and underlying formations have been identified. The agrophysical indicators of soils of rice fields and boharic analogues during their agricultural use have been assessed. The hydromorphic soil-forming processes did not lead to considerable changes in agrophysical properties of alluvial formations and soils involved in rice rotation.

1 Introduction

This work is a result of many-year research of agrophysical properties of rice rotation soils. The need in such research is dictated by demands of the agricultural practice. Today it is not known how far are the conversion processes of agrophysical properties of former boharic soils that are currently used in rice rotation. At this, there arise the issues of development of methods for restoration of properties and fertility of the rice field soils.

Three massifs of the rice irrigation systems arranged on the first terrace above flood-plain, ancient delta and modern delta of Kuban river occupying more than 75% of total area of the rice systems that makes about 150 ths. ha are distinguished in the lower reach of Kuban river [1,3].

In this region, the characteristics of the parent rock materials and the soil cover of the rice systems are primarily dependent on geomorphology and relief.

The modern delta of Kuban, as the specific and the youngest geomorphological element of Kuban plains, occupies the area of about 6000 km^2 (600 ths. ha). The main elements of relief of the modern delta of Kuban are the ridge-shaped elevations along the acting and extinguished eriks, plain flat spaces and closed extensive flat depressions (limans) that were flooded for a

[*] Corrensponding author: kubsoil@mail.ru

long period of time during high-water periods. The absolute elevations of the region vary from 2-4 m to zero and even negative points.

The parent rock materials are represented by the modern alluvial formations of various grain-size distribution - from sands to heavy clays. In the western part of the region, the shell deposits are locally spread, they are the bystanders of the recent past when some part of the territory of the modern delta was a shallow gulf of the Se of Azov.

The ground waters in the considered region lay at the interval of 1-3 m, with their mineralization range from 0.5 g/l to 20-40 g-l and more.

Natural vegetation is represented by meadow, primarily, grass herbs in the uplands and by sedges and canes in the depressions. The pseudo-peat layer, the dead semidecomposed cane roots and stems, has formed along the bottom of limans, its thickness varies from 20 cm to 2 m.

The hydromorphous soils belonging to alluvial types have formed in the existing natural conditions: alluvial bog (humus-gley, slimy-peat-gley, slimy-peat), alluvial meadow-bog and alluvial meadow saturated (stratified, normal, black) soils [4,5,9].

The above listed soils are characterized by the presence of one or several buried soil horizons in the thickness of 0-2 m that evidences the intermittency of the soil-forming process and the youth of soils considered.

The elevated relief elements are made up by the deposits with light grain size distribution, primarily, by sand loams, loamy light sands and medium-textured loams with sand interlayers. The clayey deposits and, respectively, the clayey soils are most common in depressions and former limans.

The absence of fractions coarser than 1 mm in the grain size distribution should be noted as a general feature of deposits and soils of the region under consideration.

2 Materials and methods

Considering that the prevailing area of soils involved in rice rotation of the lower reach of Kuban river are located in the modern delta, the agrophysical properties of soils were studied for this geomorphological region.

The thickness of 1.6-2.0 m was researched during study of the agrophysical properties, considering small thickness of humus horizons of 30-60 cm, the researches covered considerable thickness of the parent rock materials. It was found out that no significant differences in agrophysical properties of subsurface soil horizons and underlying formations were observed. It follows that the hydromorphic soil-forming processes did not lead to considerable changes in agrophysical properties of alluvial formations that is related to the youth of the soils under consideration. The duration of the soil-forming process that is repeatedly discontinued by the buried formed soils, fresh alluvial sediments, does not exceed 2-4 thousand years in this region [2].

The basis for the summary were the materials of research of the physical properties of soils carried out in this region in 2008–2020. 110 points were researched according to the unified procedure in the field and laboratory conditions to study the agrophysical properties of soils that made it possible to use the statistical data processing methods and assess their reliability.

3 Results and discussion

The research of the agrophysical properties of soils in different regions of Krasnodar Territory showed that the identified varieties considerably differ by their agrophysical properties that, to a great extent, determines the genesis of the parent rock materials.

Three groups of the parent rock materials and the soils with various agrophysical properties formed on their basis, having the similar grain size distribution, that is the physical clay and physical sand ratio, can be distinguished in the initial approximation in Krasnodar Territory:

- loesslike deposits and formed zonal soils of chernozem type that are characterized by high porosity and fluffy consistency;
- alluvial deposits with hydromorphous soils, with average density and porosity;
- degraded loesslike, deluvial, tertiary marine deposits and saline soils with zonal and intrazonal soils that characterized by high density and low porosity [4,6].

Within the mentioned soil groups, the same varieties of the grain size distribution have rather closely adjacent indicators of the agrophysical properties. With the same genesis of the parent rock materials, depending on the soil-forming and other natural processes (subsidence, compactness, salinization, gleyzation, etc.), the agrophysical properties can be considerably different while having the same grain size distribution [11,12]. The comparison of agrophysical properties of soils of different genesis is given (table 1).

Table 1. Grain size distribution and agrophysical properties of subsurface horizons of various types of soils.

Indicators of soil composition and properties	Ordinary chernozem	Meadow chernozem soils	Alluvial meadow soils
Physical clay content, %	61	61	67
Silt content, %	30	34	26
Silt/dust ratio	1.0	1.3	0.6
Solid phase density, g/cm^3	2.70	2.70	2.74
Bulk density, g/cm^3	1.33	1.57	1.34
Total porosity, %	50.7	41.9	51.1
Aeration porosity, %	15.7	3.7	10.6

The comparison of basic indicators of the agrophysical properties of various types of soils with the same grain size distribution evidences significant difference by the soil bulk density and aeration porosity.

In the modern delta of Kuban, the difference by basic indicators of the agrophysical properties is observed from the steppe zone of the Territory where the parent rock materials are the loesslike deposits that are relatively homogeneous by their grain size distribution, and in the delta of Kuban, the considerable area of the soil cover with the homogeneous grain size distribution makes an exception. Confinedness of the light soils to the uplands and of the heavy soils to the depressions, commonly occurring stratification - the interleave of layers with various grain size distribution that is not smoothed by the relatively short soil-forming process.

Construction of the rice irrigation systems and introduction of rice rotation delineated the soil-forming processes of the rice agrocoenosis and boharic analogues. The alluvial meadow boharic soils began adapting to the steppe conditions, and the soils used in rice rotation of various original genesis began developing in the artificially created meadow-boggy conditions. But the differences in the soil processes actually have not affected the stable indicator - the grain size distribution, no pronounced difference in the grain size distribution of the soils involved in the rice rotation and boharic soils has been identified.

The soils of the rice agrocoenosis of the modern delta are irrigated by Kuban water that carries the silty suspended matters and leaves them in the soil. This would seem to have great impact on the silt content in these soils [10].

After construction of the Krasnodar water reservoir, Kuban water began leaving more than 90% of the carried suspended matters in its bed. In the tail race of the water reservoir, the silt content in Kuban water makes 20–30 g/m^3 in total. But downstream, near Fedorovsky and Tikhovsky hydrosystems feeding the rice irrigation systems of the modern delta, the silt amount in water increases to 50–100 g/m^3 due to bank erosion. With the average irrigation norm for rice of about 12 ths. m^3/ha, about 0.7–1.3 t/ha of silt penetrates the soil.

The silt carried by irrigating waters is actually distributed only in the ploughing horizon of the soil, and its inflow can be calculated in % relative to initial silt content. With the average soil density of about 1.4 g/cm^3, the weight of the ploughing horizon makes: 10000x0.24x1.4 ≈ 3500 t/ha. Its silt content is usually 20-40%, or 700–1400 t/ha. The annual silt inflow by irrigating waters makes 1/1000, or 0.1 % of the initial content. At least 100 years of annual rice-on-rice cultivation will be required in order the heaving of the grain size distribution of the rice field soils is practically noticeable due to the silt inflow by irrigating waters, this heaving process of soils is slow and virtually is not pronounced [13].

The analysis of the research made it possible to establish the leading role in the quantitative indicators of the physical clay composition, namely, its silt fraction content (<0.001mm particles) or the silt/dust ratio (0.01–0.001mm particles). By these indicators, the above described three types of soils differ significantly within the same variety of the grain size distribution. The silt/dust ratio can be a powerful tool in the study and quantification of the water-physical properties in the physical realm. It is noted that the relative silt fraction content is determined not only by the parameters of the water-physical properties, but also by such agrochemical properties as the amount and composition of the absorption capacity, reserves and forms of mobile nutrients, humus content and composition.

The boharic soils and rice rotation soils of the modern delta of Kuban formed on the alluvial deposits showed a regular increase in the silt/dust ratio as the grain size distribution increases, that is, as the physical clay content increases. The statistical processing of the grain size distribution data made it possible to identify average values of the silt/dust ratio for all soil varieties by the grain size distribution (Table 2).

Table 2. Dependence of the silt content and the silt/dust ratio on the physical clay content in soils and alluvial deposits of the modern delta of Kuban.

Soil varieties by grain size distribution	Physical clay content, % (PC)	Silt content, % (s)	Dust content, % (d)	Silt/dust ratio Csd	Silt content, %, of the physical clay content, S
Sandy	< 10	< 1	> 9	< 0.1	< 10
Sandy loam	10-20	1-3	9-17	0.1-0.18	10-15
Light loamy	20-30	3-6	17-24	018-0.23	15-20
Middle loamy	30-45	6-12	24-33	0.25-0.36	20-27
Heavy loamy	45-60	12-21	33-40	0.36-0.50	27-35
Light clay	60-75	21-31	40-44	0.50-0.70	35-42
Middle loamy	75-85	31-41	≤ 44	0.70-0.93	42-48
Heavy clay	> 85	41-56	≤ 44	0.93-1.27	48-55

The given dependence of the silt/dust ratio (Csd) on the physical clay (PC) content has the graphical representation of the exponent and is quite detailed by the empirical equation:

$$Csd=e^{[PC/40-2.2]}, \quad (1)$$

it follows from the formula (1):

$$LnCsd = (PC : 40) - 2{,}2, \quad (2)$$

$$PC = (Ln\ Csd + 2{,}2) \text{ to } 40, \quad (3)$$

where **e** is the base of the natural logarithm;

Ln is the natural logarithm.

The dependence of the silt content in % on the physical clay content has is simpler - the view of arithmetic progression (straight line in the graph), and is expressed as the equation:

$$S = (0.5\ PC + 5), \quad (4)$$

where S is the silt content, %, of the physical clay (PC) content.

It follows from the above:

- each kind of the grain size distribution of the soils being considered corresponds to specific and rather narrow silt/dust ratio;
- the silt content in each kind of the grain size distribution is determined by the physical clay content;
- the quantitative change in the physical clay content corresponds to the qualitative change in its composition, namely, its silt content.

The density indicator of the matrix soil is the most stable over time and is taken as constant in the relevant calculations. In the soils being considered, it varies within a wide range from 2.46 (in grained sands) to 2.78 g/cm^3 (in heavy clays). The research made it possible to identify the conformity of each kind of the grain size distribution and certain density indicators of the matrix soil and their variation within the narrow range (Table 3).

Table 3. Dependence of the matrix density and the bulk density of soils on the physical clay content

Soil varieties by grain size distribution	Physical clay content, % (PC)	Matrix soil density, g/cm^3 (d)	Soil density variation limits, g/cm^3 (d_v)
Sandy	< 10	2.46-2.62	1.08-1.42
Sandy loam	10-20	2.62-2.66	1.16-1.43
Light loamy	20-30	2.66-2.68	1.17-1.43
Middle loamy	30-45	2.68-2.71	1.19-1.44
Heavy loamy	45-60	2.71-2.73	1.22-1.46
Light clay	60-75	2.73-2.75	1.25-1.50
Middle loamy	75-85	2.75-2.76	1.30-1.55
Heavy clay	> 85	2.76-2.78	1.40-1.62

The dependence of the matrix soil density on the physical clay content has the form is graphically represented as exponent. It is described by empirical equations rather detailed:

$$d = e^{(LnPC+36) : 40}, \quad (5)$$

$$40Ln\ d = (LnPC+36), \quad (6)$$

where: Ln is the natural logarithm, e is the base of the natural logarithm.

The dependence of the matrix soil density on the physical clay content, particularly, on the silt/dust ratio is also graphically represented as exponent and described by the empirical equation:

$$d = 2{,}766 + (Ln\ Csd : 16), \quad (7)$$

It follows from the equation (7):

$$Csd = e^{(d - 2.768)/16}, \quad (8)$$

In equations (5–8), the following units are used:
PC – in %, Csd – in unit fractions, d – in g/cm^3.

The mentioned equations are produced on the basis of the summarized definitions of the matrix soil density and grain size distribution, and are of static nature. The deviation of the obtained results of the matrix soil density from those calculated by equations 5 and 7 should be regarded as an anomaly that requires detailed consideration.

The given equations make it possible to calculate the matrix soil density with sufficient accuracy ($P \geq 85\%$) according to the grain size distribution data. This is of practical importance, the grain size distribution analyzes are usually performed in large quantities, density determinations in small volumes.

The physical significance and essence of the identified regularities of dependence between the matrix soil density and the grain-size distribution of soils consist in mineral composition of alluvial soils and rock. The physical clay is a component that includes complex high molecular weight minerals with a density of 2.7-5.2 g/cm^3 - muscovites and biotites, hornblendes and augites, hydrohematites and various iron compounds (hematite, limonite, magnetite), etc. The physical sand mainly includes simpler compounds with a density of 2.4-2.6 g/cm^3 - quartz, kaolinite, orthoclase, microcline, etc. As the physical clay content and silt content in soils increase, the matrix soil density increases proportionally, and vice versa, as the physical sand content increases, the matrix soil density decreases.

The soil bulk density is of the most important characteristics of the agrophysical indicators of soils determining the other indicators - porosity, water capacity, water yield and permeability.

The soil bulk density is not constant, it is subject to seasonal cyclic changes in the boharic conditions, when its minimum values correspond to the spring disequilibrium state of the soil, and its maximum values correspond to the autumn disequilibrium state, and also the unstable water regime leads to significant dynamics of soil density between the indicated seasons [14].

Determination of the density of soils of the same genesis and the grain size distribution with the various natural moisture content gave widely different results. Therefore, during the statistical processing of the density data, we used its values obtained with the moisture content being equal to the lowest water capacity (WC). It should be noted that if the soil had a tendency to swell, this property was manifested to a great extent with the moisture content being equal to the WC. Despite the indicated limitation on moisture content, the variation in the density values of soils with the same grain size distribution was significant. This is related to the study of the agrophysical properties during the entire warm period of the year (from April till October), when there were seasonal changes in the water-physical properties.

Statistical processing of such dynamic values without appropriate ranking is incorrect, and in this case, it is possible statistically to substantiate the limits of variation (min - max) in (Table 3).

Based on the given data, there is a trend towards increase in soil density with increase in physical clay and silt content, as well as increase in matrix soil density. But the listed characteristics are stable over time in contrast to the soil density. That's why, there can be no clear dependence of Y on PC, silt and matrix soil expressed in mathematical terms.

Large bulk of initial data made it possible to identify an equally important regularity: with the relatively homogeneous grain size distribution of soils, a clear tendency to increase in the bulk density in depth is revealed (Table 4).

Table 4. Change in average density along the profile of alluvial meadow soils of the modern delta of Kuban

Averaged thickness of genetic horizons, cm	Clay content, %							
	< 10	10-20	20-30	30-45	45-60	60-75	75-85	> 85
	Bulk density, g/cm^3							
0-23	< 1.15	< 1.16	< 1.22	< 1.19	< 1.22	< 1.25	< 1.30	< 1.40
23-45	1.15-1.24	1.16-1.25	1.22-1.30	1.19-1.28	1.22-1.30	1.25-1.34	1.30-1.40	1.40-1.45
45-70	1.24-1.32	1.25-1.32	1.30-1.37	1.28-1.35	1.30-1.37	1.34-1.41	1.40-1.48	1.45-1.52
70-100	1.32-1.37	1.32-1.38	1.37-1.42	1.35-1.40	1.37-1.42	1.41-1.45	1.48-1.52	1.52-1.57
100-150	1.37-1.41	1.38-1.42	1.42-1.45	1.40-1.43	1.42-1.45	1.45-1.48	1.52-1.54	1.57-1.61
150-200	1.41-1.42	1.42-1.43	1.45-1.46	1.43-1.44	1.45-1.46	1.48-1.50	1.54-1.55	1.61-1.62

According to the Modin's soil compaction estimation scale [7], the density of soils with the sandy, light loamy and middle loamy grain size distribution during the vegetation period (May-September) is estimated as from very loose to medium dense, with the light and middle clay grain size distribution - from loose to dense, with the heavy clay grain size distribution - from medium dense to very dense, and the ploughing horizons of soils with all varieties of the grain size distribution are naturally characterized by the least density values.

4 Conclusion

The main agrophysical indicators and the grain-size distribution of the cultivated rice lands of the modern delta of Kuban differ from the steppe zone of the Territory, this difference depends on the relief elements and the parent rock material composition. Confinedness of the light soils to the uplands and of the heavy soils to the depressions, commonly occurring stratification, as well as the interleave of layers with various grain size distribution.

The introduction of rice rotation after construction of rice irrigation systems delineated the soil-forming processes of the rice agrocoenosis and boharic analogues but did not affect the stable indicator - the grain-size distribution. The regularities of dependence between the density of the matrix soil and the grain-size distribution that depend on the mineral composition of alluvial soils and rocks have been identified. The increase in bulk density of the rice field soils is observed with the increase in physical clay and silt content, as well as increase in matrix soil density.

References

1. E.S. Blazhny, *Soils of Kuban delta and adjacent spaces*, 276 (1971)
2. V.F. Vadyunina, Z.A. Korchagina, *Method for studying the physical properties of soils*, 416 (1986)
3. V.F. Valkov, *Soils of Krasnodar Territory, their use and protection*, 192 (1996)
4. V.P. Vlasenko, Works of the Kuban State Agricultural University, **69**, 118 (2017)
5. V.A. Kovda, *Development of action plan for long-term maintenance of favorable soil-reclamation situation in the Lower Delta of the Kuban*, 341 (1981)

6. M.S. Melnik, IOP Conference Series: Earth and Environmental Science, 32028 (2019)
7. S.A. Modin, *Study of water-physical properties of soils for reclamation construction*, 33-2.1, 02-85 (1982)
8. S.A. Nikolaeva, *Reclamation of flooded soils in Kuban delta and irrigation of soils in the Caucasian plain zone*, 174 (1986)
9. A.V. Osipov, *Change in properties and salt regime of soils of the modern delta of Kuban*, 131 (2016)
10. V.N. Slyusarev, Modern state of soils of the North-West Caucasus, **42**, 99 (2013)
11. N.B. Khitrov, Eurasian Soil Science, **48(7)**, 671 (2015)
12. V.S. Tsovrebov, Soil cover is a national treasure of the people: Collection of articles on the materials of the All-Russian Scientific and Practical Conference devoted to the 50th anniversary of Daghestan Department of Dokuchaev Society, 134 (2012)
13. V.S. Tsovrebov, *Modern resource-saving innovative technologies for agricultural crops cultivation in the North Caucasus Federal District*, 188 (2016)
14. A.Kh. Sheudzhen, *New methods and results of landscape studies in Europe*, Central Asia and Siberia. Monograph in 5 volumes, 118 (2018)

Diagnosis of human-induced degradation of soils of the Azov-Cuban lowland

V.P. Vlasenko*, *A.V.* Osipov, and *V.N.* Slyusarev

Kuban State Agrarian University (Kuban SAU), Krasnodar, Russia

Abstract. The work presents the observational data of soil cover structure dynamics, changes in soil composition and properties under the influence of anthropogenic impact. A decrease of 6.0-7.4% in the share of black soils (Chernozem) over 50 years of observations and a corresponding increase in the area of waterlogged soils associated with closed depressions have been revealed. Due to the high dynamism of soil density with changes in its moisture content, a criterion has been sought to characterize this dependence and a method has been proposed to describe it using dynamic curves. The amplitude of density fluctuations and its dependence on the severity of anthropogenic degradation – 59 -78% – has been established. Correlation analysis confirmed a medium strength relationship (determination coefficient 0.39-0.42) between shrinkage and specific surface area of initial degradation soils and a strong relationship (determination coefficient 0.79-0.86) in meadow-chernozem compacted soils. A strong correlation has been found between the volumetric shrinkage and fines content in the size distribution (determination coefficient 0.84-0.96). A new diagnostic criterion ("T" criterion) for the intensity of hydrometamorphism and, in general, for the physical aspect of anthropogenic soil degradation has been proposed.

1 Introduction

The concept of anthropogenic degradation combines degradation processes caused by technogenesis. There are quite a number of definitions of soil degradation in general and soils in particular – "...soil degradation is understood as the processes and results of changes in their properties and natural regimes, collectively leading to changes in soil functions as an element of the ecological system and reduction of soil fertility" [1-5].

The novelty of the approach proposed in the research is ensured by the comprehensiveness of the issue in hand:

- on the one hand, the in-depth analysis of the dynamics of soil properties degraded due to anthropogenic impact will provide information on the real level of soil fertility, which is the basis for the development of new (promising) farming systems;

- on the hand, the concepts of transition of the agrarian economy to adaptive landscape specific farming system, private methods, particular economic and mathematical models, refined in accordance with the features of the modern market economy, will be a working

* Corresponding author: kirsanovi@mail.ru

tool to assess the productive potential of natural landscapes, environmental and economic evaluation of agricultural systems, and justification of priorities for innovation development of the industry.

The purpose of this research is to identify the dynamics of soil cover structure, composition and properties of soils due to anthropogenic degradation

The objectives are as follows:

- analysis of the land topsoil structure of agriculturally used areas and identification of the main directions of its transformation;
- search for new approaches to the diagnosis of degradation processes;

The relevance of the research is due, on the one hand, to the scale and rate of negative processes (erosion, waterlogging) of up to 1.8 million hectares of land in the Krasnodar region alone; on the other hand, to the lack of current information on the composition and properties of soils;from yet another perspective, to the currently existing imbalance between the recognition of concepts "land, land plot" as a priority over "soil", and the totally unjustified overemphasis on agrochemical characteristics of soils. The physical basis of landscape (agrarian landscape) stability, consisting in soil inalterability, primarily the solid soil phase, which is the material (physical) basis of agrarian landscape stability, remains poorly understood, while the anthropogenic impact on soils is increasing.

The research is aimed at studying the dynamics of soil structure, composition and properties of soils (chernozems) of the Azov-Kuban lowland. Problems related to classification, nomenclature and diagnosis of soils transformed due to hydrometamorphism, erosion, justification of terms and formulation of concepts (waterlogging, degradation, physical degradation, slitogenesis (development of vertic properties in soils), permissible erosion rate, etc.) have been covered in works of a large number of authors [6, 7, 8, 9]. Views on the soil-forming processes in soils hydrometamorphised soils (without the naming of the latter as such) is presented in works by N.B. Khitrova [10, 11]. There is also a fairly wide range of views on the diagnostic criteria for the degradation in soils, and justification for the application of one or another criterion is given [12-14].

The study of research materials on the topic of interest has suggested that human-induced degradation of soils is determined not only and not so much by anthropogenic impact, but by the properties of the soil itself, especially its solid phase. The nature of the dependence of its physical state indicators (density, shrinkage) on humidity varies and can serve as an indicator of the presence or absence of signs of degradation. It is necessary to find indicators characterizing the dynamics of the process of changing the physical state of the soil under anthropogenic impact.

2 Objects and methods

The objects of the research were the soils of the Azov-Kuban plain (Timashevsk district, Krasnodar), subject to anthropogenic degradation processes: alkaline chernozems and meadow-chernozemic compacted soils (Classification ... 1977) or clay-illuvial chernozems and clayed hydrometamorphic (quasigley) and vertic chernozems (as per Russian soil classification, 2004) [15].

The analysis of soil cover structure dynamics was based on materials of soil researches of the land areas of Yeisky, Timashevsky and Dinsky districts and city of Krasnodar (1972-2020), carried out by Kubangiprozem institute and personally by the authors.

The dynamics of soil density with changes in moisture content (drying) was studied in soil samples taken by the ample ring method with a volume of 50 cm3. The soil moisture at the time of sampling is close to the lowest moisture capacity (MC), the replication is three and five times, depending on the horizon. Subsequent drying for 18 days was accompanied by daily weighing and measurement of geometric parameters with a 3D scanner.

The specific surface area of the solid phase was determined using the standard method, a new method was the use of evacuation, which accelerated the saturation process [16].

Volumetric shrinkage was investigated in samples taken for the soil density dynamics analysis using the same pattern. The shrinkage was calculated using the standard method, with moisture parameters of 28-31% at beginning of the experiment and 6-10% (by weight) at the end.

3 Results and discussion

3.1 Changes in the structure of soil cover (SSC) and area of waterlogged lands (WLA)

were identified by analyzing and overlaying the soil maps of the previous rounds of the survey carried out by the Kubangiprozem institute. Our own research suggests that the development of physical soil degradation, especially on arable land, is caused by an increase in the intensity of anthropogenic impact against the background of changing (often cyclical) climatic indicators. This trend is confirmed by the increase in the area of wetlands over the last half century. For example, the percentage of waterlogged lands in Dinskiy and Timashevskiy districts has doubled (from 9.0-15.2 to 18.7-29.7%), and in Krasnodar city it has tripled (from 7.9 to 25.7%).

In our opinion, the increase in WLA is less related to an increase in precipitation, but is mainly due to a change in the structure of soil cover (SSC) of lowland and basin agrarian landscapes towards an increase in the proportion of semihydromorphic chernozem analogues.

For example, in Timashevsk district there has been a 3.5-4.8% decrease in the share of typical chernozems, a complete disappearance of leached chernozems and a 7.2% increase in the share of meadow-chernozemic compacted soils by 6.0% and meadow-chernozemic vertic soils by 7.4% over half a century.

3.2 Diagnosis of anthropogenic degradation

The list of soil properties, and especially parameters that could provide a basis for concluding the presence or absence of anthropogenic degradation is quite large, which is due both to the diversity of its manifestation and to the differences in the approaches applied by the researchers [17]. We will limit to indicators characterizing physical degradation, which is often a consequence of anthropogenic impact, leading to development and being itself the result of hydrometamorphism of soils. Since the essence of hydrometamorphism consists of slitogenesis (almost always) and often, but not always, of gleyification, it is logical to use morphological manifestations, indicators of physical state of soils caused by these processes as diagnostic attributes.

N.B. Khitrov [18], F.I. Kozlowsky [19] et al. The leading diagnostic sign of slitogenesis is morphologically assigned to slickensides. While generally agreeing with this approach, the difficulty of applying this indicator at high (wet) and low (dry) humidity levels must, however, be noted. The use of soil density as a diagnostic criterion also requires the specification of moisture conditions, which is very impractical. There are many suggestions in the literature for the use of indicators characterizing slitogenesis as a process of compaction, swelling and shrinkage.

Despite the diversity of approaches, specific surface area, shrinkage and swelling, and particle convergence are the most common [20]. In our study we used the following system of indicators – fractional makeup (silt and physical clay content), specific surface area of the solid phase, soil density (dynamics in the range 30-5% of moisture weight percentage). The results were subjected to correlation analysis, in order to identify correlations between each other and in relation to the severity of anthropogenic degradation (hydrometamorphism).

3.3 Density of soils

During the observation of soil samples changing in the process of drying it was found that the maximum density at 5% moisture content was 1.7-2.0 g/cm^3 for samples of meadow-chernozemic vertic soils. Density of meadow-chernozemic compacted soils does not exceed 1.5-1.7 g/cm^3. The amplitude of the change in density relative to the original density also varied, ranging from 64% in compacted soils to 75% in vertic soils.

Form of curves (Fig. 1) and, especially, their location in relation to the coordinate axes are individual and are determined primarily (but not exclusively) by a set of physical properties of the soil. The analysis of the curves characterizing the investigated dependence for the arable layer of the studied soils showed a pattern not noted by previous researchers: the slope of the curves to the X-axis increases depending on the severity of hydrometamorphism. In the deeper layers, this process is reflected in an increase in curvature, from a straight line in meadow-chernozemic compacted soils to curves of maximum steepness in meadow-chernozemic vertic soils.

An important process parameter is the slope of the curve of the relationship under study, the physical meaning of which is to characterize the speed of the process.

There is a very significant differentiation in this indicator for soils of different degrees of hydrometamorphism in particular and anthropogenic degradation in general – 55%. While we have established a high degree of correlation between the curve slope angle and the specific surface area of the solid phase of soils, first of all for soils with pronounced signs of hydromorphism, there is a reason to propose this criterion ("T" criterion – in mathematical sense, the tangent of the slope angle to the axis of abscissas) as an indicator of hydrometamorphism and of physical degradation of soils in general. We intend to investigate the possibility of using it as an indicator of the degree of hydrometamorphism.

3.2 The specific surface area and soil shrinkage

The specific surface area (SSA) is considered by many researchers to be the reasonably integral index reflecting the geometry of the solid soil phases. Our research has found a rather significant difference in the specific surface area of soils with different degrees of hydrometamorphism.

The variability of specific surface area is associated with different degree of dispersion of solid phase caused by intensity and duration of hydrometamorphism and, consequently, the index of SSA can serve as a diagnostic criterion of anthropogenic degradation of waterlogged soils in lowland and basin agrarian landscapes.

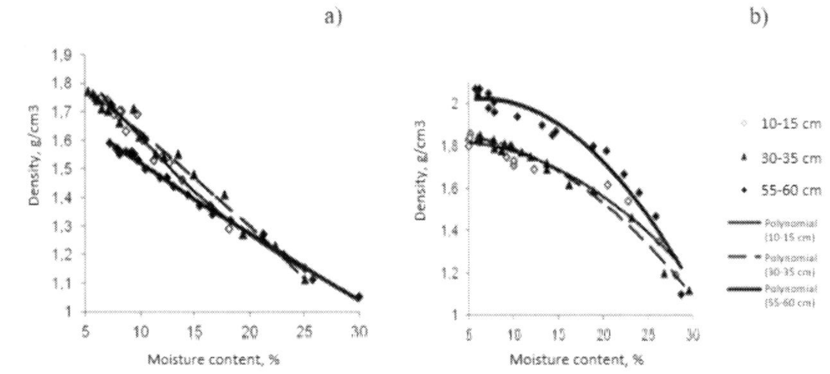

Fig. 1. Dynamics of soil density on drying: a) meadow-chernozemic compacted soils, b) meadow-chernozemic vertic soils.

Shrinkage was determined in the same soil samples in which soil density dynamics was analyzed, which to a certain extent avoided the influence of soil mass heterogeneity, both in the vertical and horizontal directions. The method we used to study shrinkage was no different from conventional methods. The only assumption we made was that the shape (regularity) of the cylindrical specimen remained unchanged during drying.

Based on the data obtained, plots of soil shrinkage (b shr) and moisture content (W) have been prepared (Fig. 2).

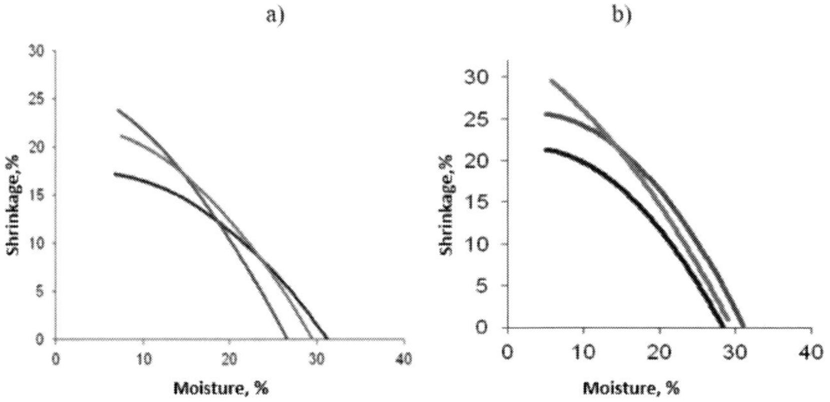

Fig. 2. Progress of soil shrinkage during drying, meadow-chernozemic compacted soils (a), meadow-chernozemic vertic soils (b)
Legend:sample depth
- 10-15 cm
- 30-35 cm
- 55-60 cm

According to the views of Joang and Varkentin [14], which we share, four shrinkage stages are distinguished: structural (I), normal (II), residual (III), shrinkage-free (IV), whose criterion is the different ratio of soil volume to evaporated water.

In our research, stage I shrinkage was absent in all samples, probably due to their high initial humidity (28-30%). In leached chernozems, the shrinkage course corresponds to stage II, when the change in soil volume is due to a decrease in water volume due to evaporation. The shrinkage rate of leached chernozem horizons does not vary and is described by a range of 16-25%.

In hydrometamorphised soils, the shrinkage process is different: initially the process corresponds to the normal (II) and residual (III) stages. The humidity at which this transition takes place varies:

- higher (19%) in vertic soils;
- significantly lower (14%) in the compacted soils.

The transition to stage IV takes place at 12-14% moisture content, varying slightly by horizon. Probably at this stage, occurring quite quickly, there is a strong compaction of the soil and the formation of cracks.

This is due to an increase in the moisture gradient between the surface of the units and their interior, not compensated for by transfer of moisture from the interior of the units to the surface. Stretching forces during shrinkage cause cracks to form. At this stage, the change in moisture content does not exceed 5%, and its (effort) magnitude correlates with the degree of hydrometamorphism. The latter circumstance is associated with more pronounced fissuring in vertic soils than in compacted soils.

In terms of volume shrinkage, the investigated soils differ significantly both in the arable layer and in the profile, ranging from 19% to 30%. The maximum value was observed in meadow-chernozemic vertic soils at a depth of 55-60 cm.

In the upper part of the profile there is a medium correlation dependence of volumetric shrinkage on specific surface in meadow-chernozemic compacted soils – r^2=0.46-0.49 and strong in meadow-chernozemic vertic soils - r^2=0.77-0.81.

The marked decrease in the correlation coefficient in the meadow-chernozemic vertic soils horizon (r^2 =0.77) is probably due to a more complex correlation relationship and requires further investigation into the selection of a more adequate model.

A strong correlation between the volume shrinkage and silt fraction content for all horizons of the soils studied is indicated by a determination coefficient of 0.81-0.94.

4. Conclusions

1. As a result of the research, the direction of SSC dynamics of the Azov-Kuban lowland in the development of hydrometamorphism caused by anthropogenic degradation has been revealed, consisting in an increase in the proportion of semihydromorphic analogues of black soils by 6.0-7.4% over a half-century period of observation.
2. A difference has been found in the amplitude of the density change of soil samples when drying, towards an increase of 11% between soils with different degrees of hydromorphism and compaction.
3. Statistical analysis has allowed to establish a link between the average power of shrinkage of soils and their specific surface area (determination coefficient =0.44-0.49) in meadow-chernozemic compacted and strong (determination coefficient =0.77-0.81) in meadow-chernozem vertic soils. A strong relationship has been found between the volumetric shrinkage and silt content in all soil horizons studied (determination coefficient =0.81-0.94)
4. The possibility of using the indicator "slope of the curve of dependence characterizing the dynamics of soil density during the changes in its moisture content in relation to coordinate axes" (criterion T) has been studied. The high correlation ratio of this index with specific surface of a firm phase of soils at meadow- and meadow-chernozemic compacted and vertic soils of 0.70-0.91 at a level of probability 0.95 allows its use as an indicator of the presence of hydrometamorphism and, in general, of anthropogenic soil degradation in its physical aspect to be considered appropriate.

Acknowledgment

The work was performed with the financial support of the Russian Foundation for Basic Research grant 19-44-230008 and the administration of Krasnodar region.

References

1. A. V. Gordeev, G. A. Romanenko, Problems of agricultural land degradation and productivity recovery in Russia, **68,** 10 (2008)
2. I. Coxhead, R. Øygard, *Land Degradation,* 146 (Cambridge University Press, 2007)
3. P. F. Loiko et al, *Modern agricultural land use in Russia: state, problems and prospects,* 229 (M.: Publishing house RAGS, 2007)
4. I. O. Alyabina, V. A Androkhanov, V. V. Vershinin, S. N. Volkov, N. F. Ganzhara, G. V. Dobrovolsky, A. V. Ivanov, A. L. Ivanov, E. A. Ivanova, L. I. Ilyin et al, *Unified State Register Of Soil Resources Of Russia. Version 1* (2014)
5. A. L. Ivanov, State, rational use and protection of land, **6-1(139),** 22 (2015)
6. O. S. Bezuglova, O. G. Nazarenko, I. N. Iljinskaya, Arid Ecosystems, **10(2),** 93 (2020)
7. N. M. Novikova, O. G. Nazarenko, N. A. Volkova, Ecosystems: ecology and dynamics, **1(2),** 21 (2017)
8. O. S. Bezuglova, K. S. Kazeev, S. I. Kolesnikov, O. G. Nazarenko, In the collection: The current state of chernozems. Proceedings of the international scientific conference, 6 (2013)
9. F. R. Zaidelman, A. S. Nikiforova, L. V. Stepantsova, T. V. Krasina, V. N. Krasin, Eurasian Soil Science, **47(4),** 223 (2014)
10. Yu. N. Vodyanitsky, *Diagnostics of waterlogged mineral soil,* 95 (2008)
11. V. P. Vlasenko, Scientific works, **179** (2012)
12. N. B. Khitrov, L. V. Rogovneva, E. P. Zazovskaya, Eurasian Soil Science, **51(7),** 731 (2018)
13. U. Simsek, F. Mikailsoy, E. Erdel, E.V. Shein, A.G. Bolotov, Eurasian Soil Science, **52(3),** 296 (2019)
14. V.F. Utkaeva, E.B. Skvortsova, V.N. Shchepot'ev, P.M. Sapozhnikov, Eurasian Soil Science, **42(2),** 151 (2009)
15. M. I. Gerasimova, N.B. Khitrov, I. I. Lebedeva, Journal of Soil Science Institute named after V. V. Dokuchaev, **102,** 5 (2020)
16. E.V. Shein, Eurasian Soil Science, **48(7),** 712 (2015)
17. E. V. Shein, T. Sakunkonchak, Y. Y. Milanovskii, D. D. Khaidapova, V. I. Lazarev, Y. Y. Aidiev, Y. Y. Kuznetsov, Eurasian Soil Science, **44(10),** 1097 (2011)
18. N. B. Khitrov, Eurasian Soil Science, **45(9),** 834 (2012)
19. A. S. Sorokin, G. S. Kust, Eurasian Soil Science, **48(1),** 63 (2015)
20. E. V. Shein, Eurasian Soil Science, **48(7),** 712 (2015)

Isolation sources of micro and macro elements among connection rice samples the creation of drought-resistant forms

Yu. K. Goncharova*, *O.A.* Bragina, *E.M.* Kharitonov, *S.V.* Goncharov, and *N.A.* Ochkas

[1]FGBNU "Federal Rice Research Center ", Belozerny, 3, 350921 Krasnodar, RF
[2]Kuban State Agrarian University, 350000 Krasnodar, RF
[3]OOO "Aratai", "Skolkovo", 143026 Moscow, RF

Abstract. Metals content affects not only for nutritional value of plants, but also their resistance to stress. So it was revealed that cationic composition changes hydration of cytoplasm. Hydration of cytoplasm positively correlated with content of alkali metals, a representative of which is potassium, and negatively with amount of alkaline earth metals (calcium, magnesium). Variability of rice collection samples in terms of content of potassium, sodium, calcium, magnesium, ammonium cations has been studied. Sources that are recommended to be used in breeding programs for the creation of drought-resistant forms are highlighted. The potassium content varied in the rice samples from 40 to 90%. In 18 samples, its content did not exceed 70%, in 13 it was above 80%, which makes it possible to recommend them as sources of these trait: Olymp, Ivushka, Dozhdik, Berezka, Rapan, Aurora, Fisht. The study of the relationship between the content of cations showed their presence between the content of potassium and sodium - 0.87, potassium and ammonium - 0.45, calcium and magnesium 0.35.

1 Introduction

Rice is superior in nutritional value to most cereals [1-3]. Both animal and plant organisms need minerals to function. For humans, the most important of them are iron and zinc. Content of minerals in a plant is influenced by both genetic characteristics of sample and external factors, primarily soil on which it grows [4-6]. Rice can be a source of calcium, magnesium, phosphorus, selenium, zinc, manganese, iron, potassium. There is a wide variation in content of micro and macroelements both between subspecies and within a species [7-9].

Content of metals affects not only for nutritional value of plants, but also their resistance to stress. So it was revealed that the cationic composition changes hydration of the cytoplasm. Hydration cytoplasm positively correlated with the content of alkali metals, a representative of which is potassium, and negatively with the amount of alkaline earth metals (calcium, magnesium) [10-12].

* Corresponding author: yuliya_goncharova_20@mail.ru

Colloidal protein particles carrying negative charge, like cations, are surrounded by a hydration shell. Charged particles with a higher mass Mg2 +, Ca2 + are more strongly attracted to negatively charged protein molecules, neutralizing their charge, this causes decrease number of water molecules surrounding the protein. Monovalent charged particles have a weaker effect on hydration, increasing tissue hydration [13-15].

When analyzing breeding material, it should be borne in mind that the concentration of trace elements is up to 10 times higher in the embryo, bran, hulled rice than in polished white rice, which is usually eaten. So magnesium in bran was (5.12%) and in polished rice (0.51%) [16-18].

Adaptability to drought is also influenced by a number of other signs, the efficiency of mineral nutrition, photosynthesis, and tissue hydration. Samples with improved characteristics of drought resistance can be obtained by combining them in one sample [19-21].

2 Materials and methods

Content of cations was determined in 50 varieties Russian breeding and collection samples of FRC. For measurements we used Kapel 105-M and Infralum devices. For the analysis, material was taken from 20 plants of each sample, in duplicate. Preparation of samples to obtain data on content of cations and anions included: selection of plant material (1 gram of sample), its homogenization, fixation with 10% alcohol, holding in a refrigerator for 1 day, centrifugation at 10-15 thousand rpm, selection of supernatant in pure 2 ml. test tube for analysis. The analysis on the Kapel 105-M device was carried out in accordance with the methods for the determination of trace elements. Statistica software was used for statistical processing of the data obtained.

3 Results and Discussions

Figure 1shows chromatogram obtained in MultiChrom program for Windows © 1993-2002 Ampersand Ltd for Flagman variety. Foregrams simultaneously show variability of samples in terms of the content of 5 cations: ammonium, potassium, calcium, sodium, magnesium. The amount of cations was expressed as a percentage per sample volume. Analysis of variance showed the reliability of difference between samples in composition of all studied cations Table 1.

Table 1. Analysis of variance reliability differences in composition of cations.

Signs	Amount squares	Degrees freedom	Middle square	F	p
Ammonium, %	2771,49	47	58,9679	707,615	0,000
Potassium, %	14951,32	47	318,1132	3817,359	0,000
Sodium, %	11780,90	47	250,6574	3007,889	0,000
Magnesium, %	985,44	47	20,9669	251,602	0,000
Calcium, %	49,32	47	1,0494	12,593	0,000

https://doi.org/10.1051/e3sconf/202126203008

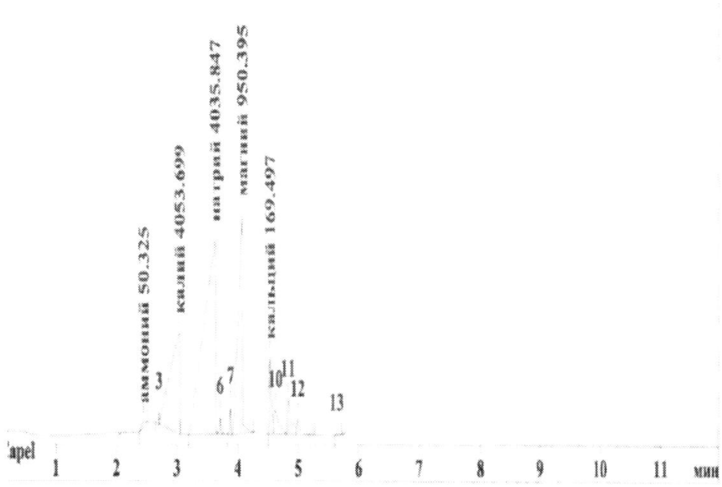

Fig. 1. Chromatogram obtained in MultiChrom program for Windows © 1993-2002 Ampersand Ltd. The number of cations in Flagman rice sample.

The presence of potassium activates many enzymes that regulate protein synthesis. The effect of potassium on respiration and photosynthesis and movement of its products in plants has been established. The increased content of potassium activates the work of stomata, increases turgor. In our studies, the potassium content varied in rice samples from 40 to 90%. In 18 samples, its content did not exceed 70%, in 13 it was above 80, but less than 90%, which makes it possible to recommend them as sources according to the characteristic, these are varieties: Olymp, Ivushka, Dozhdik, Berezka, Rapan, Aurora, Fisht (figure 2).

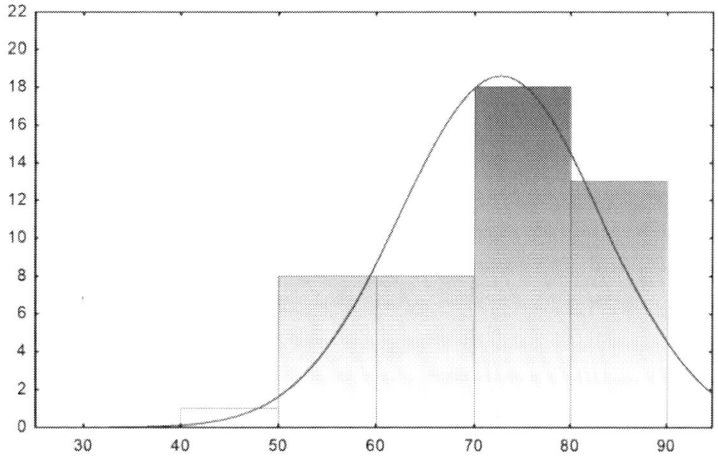

Fig. 2. Distribution samples by groups according to content of potassium cations. Red line is expected normal distribution.

Y-axis is number of samples. X-axis - feature value.

Sodium ions can be toxic. However, some salt tolerant plants (halophytes) contain high concentrations of sodium. Examples of such species: salineros, sea vault, sarsazan, some types of tamarix. Solyanka plants contain up to 6% salts in the cell sap, which reduces

 https://doi.org/10.1051/e3sconf/202126203008

water potential, and ensures flow of water even during salinity and drought. It is possible to avoid the toxicity of salts due to their compartmentalization in vacuoles, which reduces their damage to enzymatic complex of cytoplasm. Sodium content varied in rice samples from 5 - 44%. In 38 samples, its content did not exceed 25%, in 7 samples it was higher than 25, but less than 45%, which allows us to recommend them as sources for this trait: Flagman, Courage, Smuglyanka, Don Nong, Mars (Figure 3).

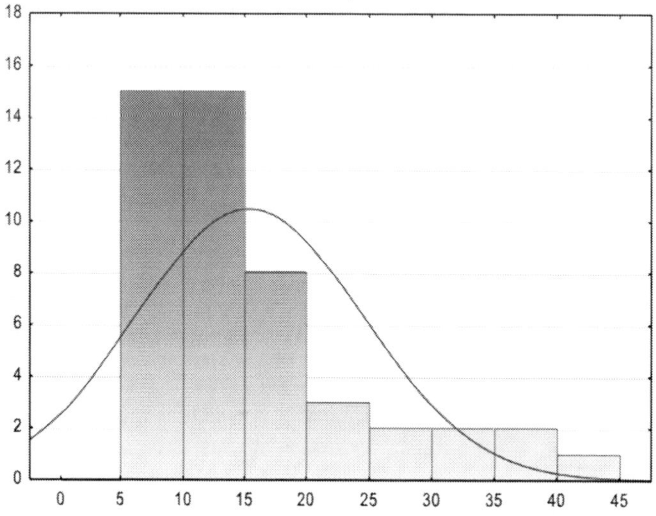

Fig. 3. Distribution samples into groups according to content of sodium cations. Red line is expected normal distribution.

Y-axis is number of samples. X-axis - feature value.

Ca2 + / Mg2 + ratio is of great importance for the life of plants and regulates many metabolic processes. Magnesium content varied in the rice samples from 2 - 14% (Figure 4).

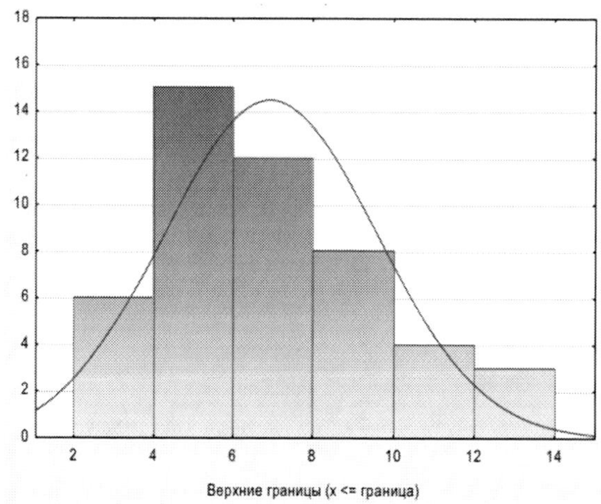

Fig. 4. Distribution samples by groups according to magnesium cations content. Red line is expected normal distribution.
Y-axis is number of samples. X-axis - feature value.

In 40 samples, its content did not exceed 10%, in 10 samples it was above 10%, but less than 14%, which makes it possible to recommend them as sources on this trait: Natasha, Orion, Partner, Gamma.

Calcium is also very important for the body, as it stabilizes the cell membranes, with a lack of it, the root system suffers especially. The calcium content varied in the rice samples from 0.07 - 2.33%. In 36 samples, its content did not exceed 1%, in 7 samples it was higher than 1.5%, which makes it possible to recommend them as sources according to the characteristic, these are varieties: Polevik, Krepysh, Ivushka, Flagman, Courage. The range of variation for studied traits is presented in Table 2.

Table 2. The range variation in cations content in rice samples.

Cation content	N obs.	Average	Minimum	Maxim.	St. Dev.	Std. - errors
Ammonium, %	48	4,16	0,050	18,75	4,43	0,63
Potassium, %	48	72.76	43.78	86,19	10,29	1,48
Sodium, %	48	15,38	5,30	43,58	9,14	1,32
Magnesium, %	48	6,93	3,35	13,86	2,64	0,38
Calcium, %	48	0,76	0,07	2,33	0,59	0,085

Conducted study of correlations between content of cations showed their presence of strong and moderate negative correlations between the content of potassium and sodium - 0.87, potassium and ammonium -0.45, and weak positive correlations between calcium content and magnesium 0.35 table 3.

Table 3. Correlation relationships between content of cations in rice samples.

Trait	Aver age	Magnesiu m, %	Ammon ium, %	Potassiu m, %	Sodium, %	Calcium , %
Ammonium, %	4,33	-0,08	1,00	-0,45	0,06	-0,11
Potassium, %	72,93	-0,06	-0,45	1,00	-0,87	-0,22
Sodium, %	15,55	-0,20	0,06	-0,87	1,00	0,18
Magnesium, %	7,10	1,00	-0,08	-0,06	-0,20	0,35
Calcium, %	0,93	0,35	-0.11	-0,22	0.18	1,00

4 Conclusions

Variability collection rice samples in terms of potassium content , sodium, calcium, magnesium, ammonium cations was studied. Sources that are recommended to be used in breeding programs to create drought-resistant forms were identified.

Potassium content varied in rice samples from 40 to 90%. In 18 samples, its content did not exceed 70%, in 13 it was above 80%, which makes it possible to recommend them as sources of trait: Olympus, Ivushka, Dozhdik, Berezka, Rapan, Aurora, Fisht.

Magnesium content in the rice samples varied from 2 to 14%. In 40 samples it did not exceed 10%. Sources of trait are variety: Natasha, Orion, Partner, Gamma.

Calcium content varied in the rice samples from 0.07 - 2.33%. In 36 samples, its content did not exceed 1%, sources are varietyes: Polevik, Krepysh, Ivushka, Flagman, Courage.

Study relationship between content of cations showed their presence between content of potassium and sodium -0.87, potassium and ammonium -0.45, calcium and magnesium 0.35.

This work was carried out with the financial support of the Russian Science Foundation No. 19-16-00064.

References

1. U. M. Lanfer-Marquez, J. Food Comp Anal, **18**, 4 (2005)
2. M. S. Akter, M. Ahmed, Afr. J. Food Sci, **7**, 8 (2013)
3. B. H. Jenning, C.A. Akoh, J. Food Chem, **114**, 4 (2009)
4. P. Bhattacharjee, R. S. Singhal, P. R. Kulkarni, Inter. J. Food Sci Technol, **37**, 1 (2013)
5. R Bisne, A K. Sarawgi, J. Agric. Res., **33**, 3 (2008)
6. I. Buresova, I. Sedlackova, O. Famera, J. Lipavsky, **56** (2010)
7. C. Diako, E. Sakyi-Dawson, B. Bediako-Amoa, F. K. Saalia, Ann. Food Sci. Technol., **12,** 1 (2011)
8. O. A. T. Ebuehi, A. C. Oyewole, Afr. J. Biote. Chu., **6,** 8 (2008)
9. S. L. Jiang, J. G. Wu, Y. Feng, X. E. Yang, C. H. Shi, J. Agric. Food Chem., **55** (2007)
10. V. C. Mbatchou, S. Dawda, Int. J. Res. Chem. Environ., **3**,1 (2013)
11. A. O. Oko, S. C. Onyekwere, Int. J. Biotechnol. Biochem., **6,** 6 (2010)
12. A. O. Oko, S. I. Ugwu, Afr. J. Biotechnol., **6**, 8 (2011)
13. A. O. Oko, B. E. Ubi, A. A. Efisue, N. Dambaba, Int. J. Agric. Forest., **2**, 2 (2012)
14. R. Sompong, S. Siebenhandl-Ehn, G. Linsberger-Martin, E. Berghofer, Food Chem., **124**, 1 (2011)
15. D. K. Verma, M. Mohan, P. K. Prabhakar, P. P. Srivastav, J. Int. Food Res. J., **22**, 4 (2015)
16. A. Vlachos, I. S. Arvanitoyannis, Critical Reviews Food Sci. Nutr., **48** (2008)
17. K. M. Wang, J. G. Wu, G. Li, D. P. Zhang, Z. W. Yang, C.H. Shi, J. Cereal Sci., **54**, 1 (2011)
18. D. K. Verma, P.P. Srivastav, J. Science Direct Rice Science, **24,** 1 (2017)
19. F. M. Anjum, I. Pasha, M. A. Bugti, M.S. Butt, J. Agri. Sci, **44**, 2 (2007)
20. J. K. Goncharova, S. V. Gontcharov, E. E. Chicharova, Russian J. of Genetics, **54**, 7 (2 018)
21. E. M. Kharitonov, Y. K. Goncharova, E. A. Maliuchenko, Russian J. of Genetics: Applied Research, **7**, 6 (2017)
22. J. K. Goncharova, E. M. Kharitonov, Russian J. of Genetics: Applied Research, **6**, 3 (2016)

Method of increasing yield indicators of rice varieties in process of seed production

N. A. Ochkas[1,3], *Yu. K.* Goncharova[1,3,*], *O. A.* Bragina[1,3], and *K.S.* Krikun[2]

[1]FGBNU "Federal Research Center for Rice", Belozerny, 3, 350921 Krasnodar, RF
[2]FGBOU VO "Kuban State Agrarian University named after I.T. Trubilin ", 350044 Krasnodar, RF
[3]OOO "Aratai", "Skolkovo", 143026 Moscow, RF

Abstract. The paper describes a method for creating new and increasing the yield of existing varieties based on analysis variability elements of rice varieties yield structure under influence of environmental conditions. Seeding rates are used as a source of variability. Following trait are analyzed: seeds field germination, survival plants before harvesting, total tillering of plants, realization of total tillering, graininess of panicle, fertility of spikelets, weight of 1000 grains. Based on analysis, through onnections conjugated elements, the optimal values of elements structure of yield are determined, at which it takes the maximum value. Based on a set of optimal values, theoretical model of studied rice variety is being developed. Then, according to panicles selected in variety, single-row plots are sown for further propagation, rows with parameters of the yield structure elements set same with model of variety are selected, which makes it possible to increase the yield of Mavr rice variety by 39.18 centners / ha or 45.61%.

1 Introduction

The continuing growth of world's population requires the introduction of new breeding technologies into practice, capable of creating in 30 years varieties that are 70% more productive than existing ones [1]. At the same time, it was noted that many crops are close to their yield plateau and further improvement by traditional breeding methods is problematic [2, 3]. Methods associated with the fixation of heterotic effect, the use of genes of wide compatibility, an increase in photosynthetic potential of plants, tissue culture technologies, or induced mutagenesis are considered promising [4, 5, 6].

Recently, the methods of transgenic selection and genomic editing have been rapidly developing; at the same time, their potential danger to humanity has been noted [7-9]. However, many works have shown the potential of traditional breeding, which has not yet been fully realized [10, 11, 12]. In our work, we present a method for creating new and increasing the yield of existing varieties that does not require the introduction of foreign DNA or the effects of mutagens and, therefore, is safe. The method does not require expensive equipment, at the same time its efficiency, according to our data, in some cases exceeds 40%.

* Corresponding author: yuliya_goncharova_20@mail.ru

2 Materials and methods

Seeds of the studied variety are sown in single-row plots with a total area of 3 m^2 with a cassette seeder; plants from the entire plot are manually harvested into sheaves. Modeling of changing environmental conditions, in order to increase the variation of elements of yield structure, is carried out by different rates of sowing rice seeds at the rate of: 1, 3, 5, 7, 9 and 11 million grains per 1 hectare. When shoots appear, they are counted on the plot and recalculated per 1 m^2.

Plant from plot are analyzed in laboratory conditions, where number of plants, shoots and panicles preserved before harvesting is counted, number of spikelets, grains and their weight are determined for 25 plants.

Based on analysis of results, germination rate is calculated (the ratio number of seedlings to the sown seeds); survival (the ratio plants preserved before harvesting to number of seedlings obtained); total bushiness of plants (the ratio of number of shoots to number of plants); realization of plants bushiness (the ratio number of panicles to number of shoots); spikeiness (the ratio of number of spikelets to number of panicles); fertility (ratio of grains number to number of spikelets) and mass of 1000 grains (the ratio of grain mass to their number multiplied by 1000).

The grain yield is calculated by formula:

$$Y = \frac{S * Sg * Sv * Bu * Pb * Nu * F * M}{100}$$

Where: Y - yield, c / ha;
S- is seeding rate of seeds;
Sh - seed germination;
Sv - survival rate of seedlings;
Bu - bushiness of plants;
Pb - realization of bushiness of plants;
Nu - spikeiness of the panicle;
F - spikelet fertility;
M - is mass of 1000 grains;
100 - conversion factor in centners / ha.

Agricultural technology is generally accepted in experience.

Mathematical processing of the results obtained was carried out in the programs Statistica 10.0 and Microsoft Excel. 2010 [14-15].

3 Results and discussion

As an example, to demonstrate the principle of improving yield indicators, we show the calculation on the Mavr rice variety with a colored grain pericarp.

The results of analyzes are entered into the table of initial data, from which it can be concluded that maximum grain yield of rice variety Mavr is 85.91 c / ha is formed at a seeding rate 5 million grains per 1 ha, with an average 72.74 c / ha according to options. Table 1.

Table 1. Initial data of the result of the analysis of yield and elements of its structure of the variety Mavr.

№	Yield structure elements								Productivity, c / ha
	Sowing rate	Germination rate	Survival rate	Toughness	Implementation of tillering	Heading rate	Fertility	Weight 1000 grains	
1	1	0.90	0,67	4,33	1,00	131.50	0,81	24.67	67,99
2	3	0.80	0,42	4.10	0,93	106.80	0,78	25,26	80,07
3	5	0,74	0,49	3,06	0,96	82,70	0,80	24,38	85,91
4	7	0,67	0,47	2,18	0,96	74,90	0,81	26,12	72,45
5	9	0,55	0,55	2,85	0,91	51,60	0,77	25,18	69,79
6	11	0,47	0,39	3,07	0,80	64,40	0,83	22,87	60,23
Average	6,00	0,69	0,50	3,26	0,93	85,32	0,80	24,75	72,74
Minimum	1,00	0,47	0,39	2,18	0,80	51,60	0,77	22,87	60,23
Maximum	11,00	0,90	0,67	4,33	1,00	131,50	0,83	26,12	85,91

When calculating yield through average parameters options elements of its structure according to the above formula is:

$$Y = \frac{S*Sg*Sv*Bu*Pb*Nu*F*M}{100} = \frac{6*0,69*0,5*3,26*0,93*85,32*0,8*24,75}{100} = 106,02c/ha$$

which is 20.11 c / ha higher than the best option by experience. Determination of yield through the maximum values of its elements according to variants of experiment shows: Y = 818.79 c / ha, through the minimum values - Y = 2.9 c / ha. The researcher should not be confused by such figures because they were obtained without taking into account the conjugation of yield structure elements with each other, the elimination of this drawback is achieved when developing a variety model.

3.1 Developing a variety model

To develop a variety model, it is first necessary to determine all relationships between elements of yield structure with each other. For example, to determine relationship between seeding rate of seeds and their germination on graph of the polynomial dependence of the second degree, we determine equation and reliability of approximation coefficient according to data in Table 1, Figure 1.

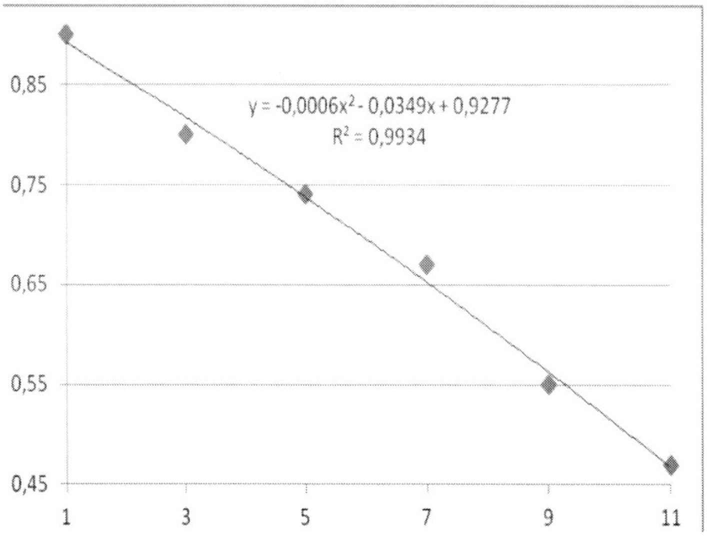

Fig. 1. Relationship of the seeding rate of seeds with their germination, rice varieties Mavr. Y-axis germination rate. X -axis is seeding rate millions seeds per hectare.

Relationship between seeding rate of seeds, rice varieties Mavr, with their germination is described by equation: y = -0.0006x2 - 0.0349x + 0.9277 where: y - field germination of seeds; x- is the seeding rate of seeds, with accuracy of approximation coefficient R^2 = 0.9934, which meets our requirements.

The graph clearly shows that with an increase in seeding rate of seeds, their field germination decreases.

Having determined connections of seeding rate with the rest of elements of yield structure, it was found that they are described by following equations: with survival rate of seedlings y = 0.0022x2 - 0.0406x + 0.636; bushiness of plants y = 0.0406x2 - 0.6438x + 5.1919; realization bushiness of plants y = -0.0022x2 + 0.0121x + 0.9614; spikeiness of panicle y = 0.8513x2 - 17.486x + 149.65; spikelet fertility y = 0.0009x2 - 0.0092x + 0.8131 and 1000 grain weight y = -0.0658x2 + 0.6823x + 23.79.

Substituting sequentially in the obtained equations value of seeding rate within the variation of experiment, we determine the values elements of yield structure and through their product divided by 100 (the formula for determining yield), we establish yield of rice grain Mavr variety at these values, Figure 2.

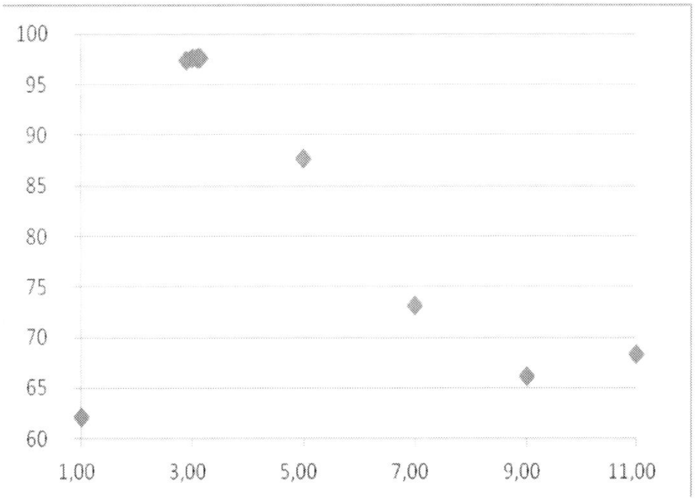

Fig. 2. Productivity of Mavr rice grain at different seeding rates. Y-axis yield. X -axis is the seed rate.

The figure shows that maximum grain yield rice variety Mavr is formed at seeding rate of seeds close to 3 million per 1 ha. Using the trial and error method, we refine this indicator as shown in Table 2.

Table 2. An example of calculating the optimal seeding rate for rice varieties Mavr.

Seeding rate	Yield structure elements							Yield
	Germination rate	Survival rate	Bushiness	Realization of bushiness	Spikeiness	Fertility	Weight 1000 grains	
1,00	0,89	0,60	4,59	0,97	133,02	0,80	24,41	62,088
2,90	0,82	0,54	3,67	0,98	106,10	0,79	25,22	97,390
3,00	0,82	0,53	3,63	0,98	104,85	0,79	25,24	97,559
3,09	0,81	0,53	3,59	0.98	103,75	0,79	25,27	97,624
3,10	0,81	0,53	3,59	0.98	103,62	0,79	25,27	97,627
3,11	0,81	0,53	3,58	0,98	103,50	0,79	25,28	97,628
3,12	**0,81**	**0,53**	**3,58**	**0,98**	**103,38**	**0,79**	**25,28**	**97,629**
3,13	0,81	0,53	3,57	0,98	103,26	0,79	25,28	97,628
3,14	0,81	0,53	3,57	0.98	103,14	0,79	25,28	97,627
5,00	0,74	0,49	2,99	0,97	83,50	0,79	25,56	87,684
7,00	0,65	0,46	2,67	0,94	68,96	0,79	25,34	73,162
9,00	0,57	0,45	2,69	0,89	61,23	0,80	24,60	66,170
11,00	0,47	0,46	3,02	0,83	60.31	0,82	23,33	68,294
Maximum								97,629

As can be seen from the table, the maximum yield 97.629 c / ha is formed at seeding rate 3.12 million seeds per 1 ha, therefore, this seeding rate for Mavr variety is considered optimal.

Next, we establish optimal values remaining elements of yield structure, for example, 0.81 with a yield of 99.44 c / ha is considered optimal for seed germination. To obtain the desired germination, it is necessary to sow with a seeding rate of 3.19 million seeds per 1 ha, while: the survival rate of seedlings will be at level 0.53; bushiness of plants - 3.58; realization bushiness of plants - 0.98; panicle spikeiness 103.68; spikelet fertility 0.79 and 1000 grain weight 25.23 table 3.

Table 3. Productivity of grain of rice variety Mavr when optimizing the elements of its structure.

Structure elements	Seed ing rate	Germinating ability	Survival Bushiness	Bushiness	Implementation of bushiness	Spikeiness	Fertility	Weight of 1000 grains	Yield, centner / ha
Norm seeding	3,12	0,81	0,53	3,58	0,98	103,38	0,79	25,28	97,63
Germination	3,19	0,81	0,53	3,58	0,98	103,68	0,79	25,23	99,44
Survival	**3,20**	**0,81**	0,63	**3,70**	**0,99**	**105,42**	**0,79**	**25,07**	**125,09**
Bushiness	4,26	0,75	0,52	3,95	0,93	101,31	0,80	24,32	119,01
Implementation of bushiness	3,29	0,81	0,57	3,56	0,98	106,15	0,80	24,98	111,99
Colosity	3,36	0,81	0,46	3,88	0,95	104,20	0,80	24,47	94,01
Fertility	5,02	0,74	0,50	3,31	0,95	96,48	0,81	24,73	113,82
Weight of 1000 grains	5,10	0,71	0,53	3,76	0,94	93,63	0,81	24,05	122,53

After analyzing yield of rice grain Mavr variety, when optimizing elements of its structure, it was found that the maximum yield 125.09 c / ha is formed when optimizing the survival rate of seedlings 0.63, which exceeds yield of best option in experiment by 39.18 c / ha or 45.61 %, therefore, values of remaining elements associated with this indicator can be taken as a theoretical model of variety.

Model of Mavr variety: seeding rate 3.2 million seeds per hectare, their germination rate should be at level 0.81; seedling survival - 0.63; bushiness of plants - 3.7; realization bushiness of plants - 0.99; panicle spike - 105.42; spikelet fertility - 0.79; mass of 1000 grains - 25.07; with a yield - 125.09 c / ha.

In other words, at a seeding rate of 3.2 million grains per 1 ha, per 1 m^2, it is necessary to obtain: 259 seedlings of which 163 plants will remain before harvesting, which will form 603 shoots and develop 597 panicles with 105 spikelets each of which will produce 84 grain total weight 2.1 g.

3.2 Improving the yield indicators of the Mavr rice variety:

1. Intravarietal mass breeding panicles is carried out.
2. With a cassette seeder panicles, are sown with single-row plots with a seeding rate 3.2 million per 1 ha.
3. For further work, plots are selected according to the parameters of yield structure elements set by variety model.
4. Next year, seeds from selected plots are sown by plots, with the same seeding rate of 3.2 million per 1 ha, the plant are selected, which are inherited parameters of yields structure elements laid down by model, and then begining to reproduce the improved variety.

4 Conclusions

The selection according to parameters set of model elements of yield structure in the process of seed production allows increasing rice grain yield Mavr variety up to 125 c / ha, which is 39.18 c / ha or 45.61% higher than best variant in experiment.

Acknowledgments

This work was carried out with the financial support of the Russian Science Foundation No. 19-16-00064.

References

1. *Food and agriculture organization of the United Nations*, https:// www.fao.org
2. Yu. K. Goncharova, Russian J. of Developmental Biology, **45** (2014)
3. L. Ge, H. Chen, J.F. Jiang, Y. Zhao, M.L. Xu, Y.Y. Xu, et al., Plant Physiology, **135** (2004)
4. Yu. K. Goncharova, S.V. Gontcharov, E.E. Chicharova, Russian J. of Genetics, **54** (2018)
5. Yu. K. Goncharova, S.A. Vereshchagina, S.V. Gontcharov, Plant Cell Biotechnology and Molecular Biology, **20** (2019)
6. Y. Rhee, R.S. Sekhon, S. Chopra, S. Kaeppler, Genetics, **186,** 3 (2010)
7. M. Lusser, C. Paris, D. Plan, E. Rodriguez-Cerezo, Nature Biotechnology, **30** (2012)
8. H. Zhao, J.D. Wolt, Emerging Topics in Life Sciences, **1**, 2 (2017)
9. A. Whelan, M. Lema, GM Crops & Food, **6**, 4 (2015)
10. M. Lema, *Poster presented at the Int. Symposium on the Biosafety of Genetically Modified Organisms (ISBGMO14)* (2017)
11. H.D. Jones, Nature Plants, **8**, 1 (2015)
12. T. Laaninen, *New plant-breeding techniques*, http://www.europarl.europa.eu
13. V.A. Dragavtsev, J. of Technical Physics, **90**, 10 (2020)
14. A. Kh. Sheudzhen, O. A. Gutorova, E.V. Shein, V. A. Romanenkov, IOP Conference Se ries, **368** (2019)
15. V.I. Lapshin, V.V. Yakovenko, S.N. Shcheglov, BIO Web of Conferences (2020)

Biological effectiveness of the range of disinfectants winter wheat seeds used in Krasnodar Territory

Natalya Anatolyevna Moskaleva[1], *Natalia Nikolaevna* Dmitrenko[1], and Natalya Anatolyevna Sasova[2]

[1]FSBEI HE "Kuban State Agrarian University named after I.T. Trubilina", Kalinina str., 13, 350044 Krasnodar, Russia
[2]Branch of VGBU "Rosselkhoz" in the Krasnodar Territory, Rasshilevskaya str., 329, 350051 Krasnodar, Russia

Abstract. In the conditions of the Krasnodar Territory, the pre-sowing treatment of seeds with drugs against diseases has firmly entered the system of protection of winter wheat. The use of seed dressing agents minimizes the risk of reducing field germination, ensures good overwintering of the crop, while being one of the most environmentally friendly plant protection methods. The idea of the biological efficiency of the seed disinfectants applied in the conditions of the region will allow not only to purposefully select a disinfectant, but also to prevent the resistance of pathogens causing diseases of seeds. An analysis of the range of fungicidal disinfectants used in the Krasnodar Territory in 2019 showed that the range was based on combined disinfectants based on several active substances with a different mechanism and spectrum of action, as well as that the shares of one-component and four-component drugs in it were - 10%, two-component - 20%. Combined dressers based on three active ingredients were the basis of the range used, the share of such preparations was 60%. Analysis of the data obtained allows to conclude that the modern range of dressing agents used in the farms of the Krasnodar Territory is capable of almost completely restraining the development of the pathogenic complex on winter wheat seeds. The maximum biological efficiency at the level of 100% was shown by the preparations in protection against dust and kernel smut, while it was at the level of 96.5% in protection against fusarium root rot.

1 Introduction

In the Russian Federation, a prerequisite for food security and independence is sustainable production of winter wheat grain, which ensures political and economic stability [9]. The Krasnodar Territory takes the leading place in the winter wheat grain harvesting. The soil and climatic conditions of the Krasnodar Territory allow to cultivate this crop in all agroclimatic zones of the region [1,6]. The annual sowing area of this crop in the region is 1.5-1.6 mln ha [2.4]. Crop losses caused by diseases amount to 15 - 35%, of which 60% of losses occur due to poor-quality seed dressing [4,7]. In production, it is important to get not only healthy, but also even sprouts of winter wheat, which is currently almost impossible without pre-sowing seed treatment, since there are practically no healthy lots of seed grain, among which the most dangerous is the Fusarium genus fungi, both in Russia and abroad [10]. It is known that more than 60% of fungal phytopathogens are transmitted through winter wheat seeds [8]. In addition, there can be a large amount of pathogenic microflora of various etiologies [3] on plant residues on the surface of the soil or in its upper layer, therefore, seed dressing is of particular importance. This is a preventive measure against internal and external seed infection. In many farms of the Krasnodar Territory, pre-sowing treatment of seeds with drugs against diseases has become a part of the system of protection of winter wheat. The literature review showed that the use of seed dressing minimizes the risk of reducing field germination, provides a good overwintering of the crop, being at the same time one of the most environmentally safe plant protection methods.

The research was aimed to study the biological efficiency of the seed disinfectant range used in the Krasnodar Territory, considering the pathogenic complex of winter wheat seeds formed during the years of research. Following issues were considered in the course of research:

- weather conditions during the growing season of winter wheat in 2018-2019;

– study of the species composition of the pathogenic complex of winter wheat seeds of the 2019 harvest in the Krasnodar Territory;

– study of the efficiency of the applied range of winter wheat seed disinfectants in the conditions of the Krasnodar Territory

2 Materials and methods

Research on the topic was performed in the research laboratory of the Kuban State Agrarian University at the Department of Phytopathology, Entomology and Plant Protection and in the laboratory of the branch of the Federal State Budgetary Institution "Rosselkhoztsentr" in the Krasnodar Territory. An assessment of the infection with pathogenic fungi of winter wheat grain in the Krasnodar Territory of the 2019 harvest was performed. Winter wheat grain samples were taken from 35 districts of the Krasnodar Territory.

The method of centrifugation and subsequent microscopic examination of wash water were applied to determine kernel smut on seeds. When determining the infection of seeds with micromycete species, the visual method and the method of a wet chamber were used. Determination of the species composition of micromycetes was performed by microscopy.

The weather conditions were quite favorable for the winter wheat overwintering. The first half of May 2018 was warm and dry. Due to the increased temperature, the development of winter crops was accelerated. June was characterized by abnormally hot weather with frequent dry winds. Only in the third ten-day period, precipitation was of a stormy nature and was unevenly distributed over the territory of the region. The first half of May 2019 was warm and dry. Precipitation on some days was in the form of rain, with heavy and very heavy rains (locally). June was abnormally hot and dry, with only occasional thunderstorm rains. Weather conditions in June-early July for harvesting were generally good, but there were short breaks due to rains for 1-3 days. Such cumulative weather conditions influenced the formation of a complex of pathogens during the growing season of winter wheat.

The results of phytoexamination of seeds before sowing seeds in 2018 showed that there were practically no consignments of grain free from seed infection. Therefore, dressing winter wheat seeds before sowing in the conditions of the region is an important technique. The idea of the biological efficiency of the seed disinfectants applied in the conditions of the region will allow not only to purposefully select a disinfectant, but also to prevent the resistance of pathogens causing diseases of seeds.

The biological efficiency of seed dressing preparations is determined by various factors, the most important of which is the toxicity of the active substance for the pest. The modern range of preparations for treating seeds against diseases contains fungicides with a different spectrum of action and action on the harmful organism. Analysis of the range of fungicidal disinfectants used in the Krasnodar Territory in 2019 showed that it was based on combined disinfectants based on several active substances with a different mechanism and spectrum of action (table 1). [5]

Table 1. Biological efficiency of the seed disinfectant range used in the Krasnodar Territory, 2018 - 2019

Disinfectant	Consumption of preparation, l/t	Harmful object	Biological efficiency,%		
			Medium	minimum	maximum
Vibrans Integral, component mixture (175 + 25 + 25 + 10 g/l)	1.5	root rot	93.0	93.0	94.0
Vial Trio, water suspension concentrate (5 + 30 + 120 g/l)	0.8	loose smut and kernel smut	100.0	100.0	100.0
Vitalon, component mixture (400 + 14 g/l)	2.0	loose smut, seeds molding	97.5	95.0	98.0
Grandseal Ultra, component mixture (75 + 45 +20 g/l)	0.5	fusarium rot	85.0	83.0	85.3
Divident Extreme, component mixture (92 + 23 g/l)	0.6	fusarium rot	93.0	92.0	94.0
Lamador Pro, component mixture (100 + 60 + 20 g/l)	0.5	fusarium rot	95.0	93.0	96.0
Maxim Extreme, component mixture (18.7 + 6.25)	1.8	kernel and loose smut	100.0	100.0	100.0
Polaris, ME (100 + 25 + 25 g/l)	1.5	fusarium rot	90.0	88.0	93.0
Polaris, ME (100 + 25 + 25 g/l)	1.5	kernel smut	100.0	100.0	100.0
Celest Max, component mixture (125 + 25 + 15 g/l)	1.5	fusarium rot	85.0	83.0	85.3
Sistiva, component mixture (333 g/l)	1.0	kernel smut	100.0	100.0	100.0
Sistiva, component mixture (333 g/l)	1.0	fusarium rot	95.0	93.0	96.5
Triaktiv, component mixture (100 + 120 + 40 g/l)	0.30	loose and kernel smut	100.0	100.0	100.0

At the same time, the share of one-component and four-component drugs in it was - 10%, two-component - 20%. Combined dressers based on three active ingredients were the basis of the range used, the share of such preparations was 60%.

The biological efficiency of dressing preparations is also determined by such indicators as the degree of seed surface coverage and the concentration of the dressing, agro-climatic conditions, the correct choice of dressing based on the results of phytoexamination (Table 2) [5]

Table 2. Biological efficiency of the applied seed disinfectant range in the Krasnodar Territory, 2019

Disinfectant	Consumption of preparation, l/t	Harmful object	Biological efficiency, %		
			Medium	minimum	maximum
Vibrans Integral, component mixture (175 + 25 + 25 + 10 g/l)	1.5	root rot	93.0	93.0	94.0
Vial Trio, water suspension concentrate (5 + 30 + 120 g/l)	0.8	loose and kernel smut	100.0	100.0	100.0
Vitalon, component mixture (400 + 14 g/l)	2.0	loose smut seed molding	97.5	95.0	98.0
Grandseal Ultra, component mixture (75 + 45 +20 g/l)	0.5	fusarium rot	85.0	83.0	85.3
Divident Extreme, component mixture (92 + 23 g/l)	0.6	fusarium rot	93.0	92.0	94.0
Lamador Pro, component mixture (100 + 60 + 20 g/l)	0.5	fusarium rot	95.0	93.0	96.0
Maxim Extreme, component mixture (18.7 + 6.25)	1.8	kernel and loose smut	100.0	100.0	100.0
Polaris, ME (100 + 25 + 25 g/l)	1.5	fusarium rot	90.0	88.0	93.0
Polaris, ME (100 + 25 + 25 g/l)	1.5	kernel smut	100.0	100.0	100.0
Celest Max, component mixture (125 + 25 + 15 g/l)	1.5	fusarium rot	85.0	83.0	85.3
Sistiva, component mixture (333 g/l)	1.0	kernel smut	100.0	100.0	100.0
Sistiva, component mixture (333 g/l)	1.0	fusarium rot	95.0	93.0	96.5
Triaktiv, component mixture (100 + 120 + 40 g/l)	0.30	loose and kernel smut	100.0	100.0	100.0

The analysis of the data obtained allows us to conclude that the modern range of dressing agents used in the farms of the Krasnodar Territory is capable of almost completely restraining the development of the pathogenic complex on winter wheat seeds. The maximum biological efficiency at the level of 100% was shown by the preparations in the protection against dust and kernel smut, and in the protection against fusarium root rot at the level of 96.5%.

It should be noted that the severity of the pathogenic complex remained at a low level for the last two years due to weather conditions (dry and hot weather prevailed in most of the territory of the region). However, the ubiquitous high infectious stock in the soil and with abundant precipitation during flowering and earing of winter crops increases the possibility of epiphytotic damage to winter wheat; therefore, sowing untreated seeds creates conditions for the accumulation of infection.

Pursuant to the information and analytical center of GBU KK "Kuban Agricultural ICC", if in 2018 out of the required 180.51 th t in fact, 168.02 thousand tons were pickled, which amounted to 93% of grain, then in 2019 this figure decreased and of the required 170.77 thousand tons, 144.76 thousand tons were actually pickled, which amounted to 85% of the sown grain. A decrease in the volume or quality of dressing leads to a deterioration in the phytosanitary situation and the creation of foci of infection, causing contamination of the grain of the new harvest. Thus, pursuant to the phytoexamination, a positive trend was noted in the spread of diseases of seeds of the winter wheat crop in 2019 (Figure 1).

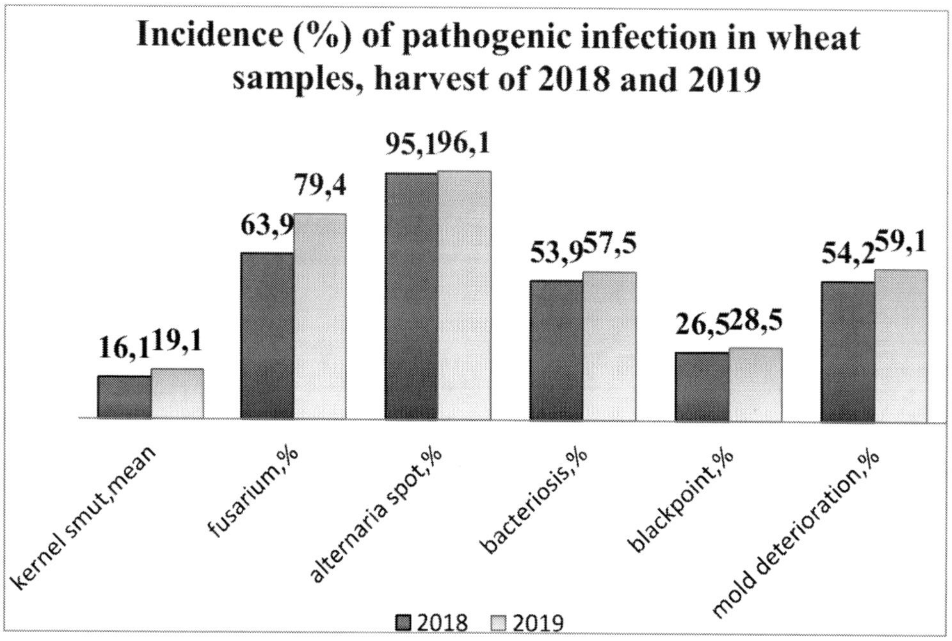

Fig. 1. Dynamics of damage to winter wheat seeds during phytoexamination in 2018 and 2019, Krasnodar Territory (Federal State Budgetary Institution "Rosselkhoztsentr" in Krasnodar Territory).

Thus, it can be argued that seed dressing is an important method in plant protection. A wide range of effective dressing agents is used in the region, 90% of which are combined. However, the full implementation of the protective effect of the drugs and making a decision on the advisability of dressing would require the following: the information about the place of winter wheat in the crop rotation and about the phytosanitary situation in the fields in the current year; forecasting the development and spread of diseases for the next year; the phytopathological examination report on the composition of the pathogenic complex of pathogens and the degree of infection of the seed material.

2 Conclusion

Phytoexamination of winter wheat seeds of the harvest of 2018 and 2019 from 35 districts of the Krasnodar Territory showed that the pathogenic microflora of grain was represented by: (*Alternaria* spp.), Hard smut (*Tilletia tritici*), fusarium *(Fusarium graminearum* etc..*)*, mold (*Aspergillus* spp. and *Penicillium* spp.) and bacterioses (*Pseudomonas* spp., *Xanthomonas* spp.) and black embryo (*Alternaria alternata*).

The results of grain analysis of the harvest in 2018 showed a dominant position in the mycobiota of seeds of representatives of the genus *Alternaria* spp. Alternaria was observed in 97.4% of seed lots. Fungi of this genus had the highest frequency of occurrence on seeds from 55.9 to 72.0%. Infected seeds of kernel smut (*Tilletia tritici*) were noted in 17.2%, lots, with fusarium *(Fusarium graminearum* etc..*)* - 67.5% of lots, mold and bacteriosis - from 50.1% to 53.2% of seed lots and black embryo - 24.2% of seed lots.

In the grain harvest in 2019 in the region, Alternaria was noted in 92.2% of seed lots, kernel smut in 17.4% of lots, Fusarium infected - 72.3%. the weighted average percentage of infected seeds at the edge ranged from 14.5 to 18.3%, which is higher than the 2018 level.

Infection of grain with Fusarium infection increased and was noted in 79.4% of the samples (the frequency of occurrence on average along the edge was 14.2% of the total number of infected seeds, the weighted average percentage of infection in the samples varied from 1.5 to 5.6%). In 2019, the infestation by saprophytic molds and bacteriosis increased in comparison with 2018.

A wide range of effective dressing agents is used in the region, 90% of which are combined. The disinfectants show high biological efficiency against a complex of smut diseases at the level of 99...100 %, root rot, including fusarium infection (biological efficiency - 89 ... 92%).

References

1. A.A. Goncharenko, Bulletin of the Russian Agricultural Academy, **6**, 49 (2005)
2. K.N. Gorpinchenko, Economy of agricultural and processing enterprises, **4**, 46 (2012)
3. M.I. Zazimko, V.Yu. Buzko, P.V. Sidac et al., Plant protection and quarantine, **9**, 19 (2013)
4. A.N. Korobko, The farming system of the Krasnodar Territory on an agrolandscape basis, 352 (2015)
5. Review of the phytosanitary state of agricultural crops in the Krasnodar Territory in 2019 and the forecast of the development of harmful objects in 2020, 108 (2020)
6. E. A. Pikushova, L. A. Shadrina et al., Agroecological monitoring in agriculture of the Krasnodar Territory, **431 (459)**, 82 (2008)
7. Ye.Yu. Toropova, Plant protection and quarantine, **2**, 3 (2018)
8. V.A. Chulkina, Ye.Yu. Toropova, Plant protection and quarantine, **2**, 21 (2007)
9. N.N. Dmitrenko, A.A. Karpenko, V.A. Kurylenko, Proceedings of the Kuban State Agrarian University, **2(77)**, 84 (2019)
10. A.N. Freije, K.A. Wise, Crop Protection, **77**, 139 (2015)

The design of the working equipment for laying a drip tape based on the Grimme GF-75/4 ridge former

N.B. Martynova* and *V.I.* Balabanov

Russian State Agricultural University - MSHA named after K.A. Timiryazev, 127550 Moscow, Russian Federation

Abstract. The article examines statistical data on the amount of precipitation in the spring - summer period and substantiates the need for irrigation measures when growing potatoes in the Moscow region. The rationality of the use of drip irrigation as the most economical and effective method of watering has been proved. With drip irrigation, there is practically no danger of excessive watering, causing the leaching of nutrients from the root layer, an increase in the level of groundwater, and an additional load on the drainage network. The use of drip irrigation will make it possible to maintain an optimal water and air balance during the growing season. The possibility of combining technological operations for laying a drip tape and forming potato ridges is considered. It is found that the resistance of the drip tape laying will not create significant traction resistance, so the working equipment can be mounted on the ridge former. The design of a drip tape stacker based on the Grimme GF-75/4 ridge former was developed and a mechanized drip tape was laid in a potato ridge. It has been proven that the implementation of the drip tape laying and the formation of ridges with one machine will increase the productivity of the work and reduce the energy consumption.

1 Introduction

The amount of precipitation that falls during the growing season in the Moscow region varies widely, high-water years alternate with low-water ones. The difference in the amount of precipitation in the spring - summer period in different years varies 2 - 4 times [1]. Often precipitation is of a stormy nature, the precipitated moisture can not be absorbed into the soil for the lack of time and goes into the runoff. Precipitation during this period is extremely uneven. Days with heavy rains, when their daily fallout can amount to tens of millimeters, alternate with days and even weeks when there is no precipitation [2, 3].

When growing potatoes throughout the growing season, water consumption also changes. In the initial phase of development, the plant has sufficient moisture reserves of the parent root tuber and there is practically no need for additional moisture [4]. However, during this period, the soil is filled with moisture formed in the process of snow melting. In subsequent phases of development, the volume of daily water consumption increases. The maximum

* Corresponding author: nmartinova@rgau-msha.ru

values are observed during the period of flowering and the end of the growth of tops [5, 6]. Further, the daily water consumption decreases slightly (Fig. 1). As the graph shows, in order to obtain a sustainable harvest of a number of agricultural crops, in particular potatoes, additional irrigation is required.

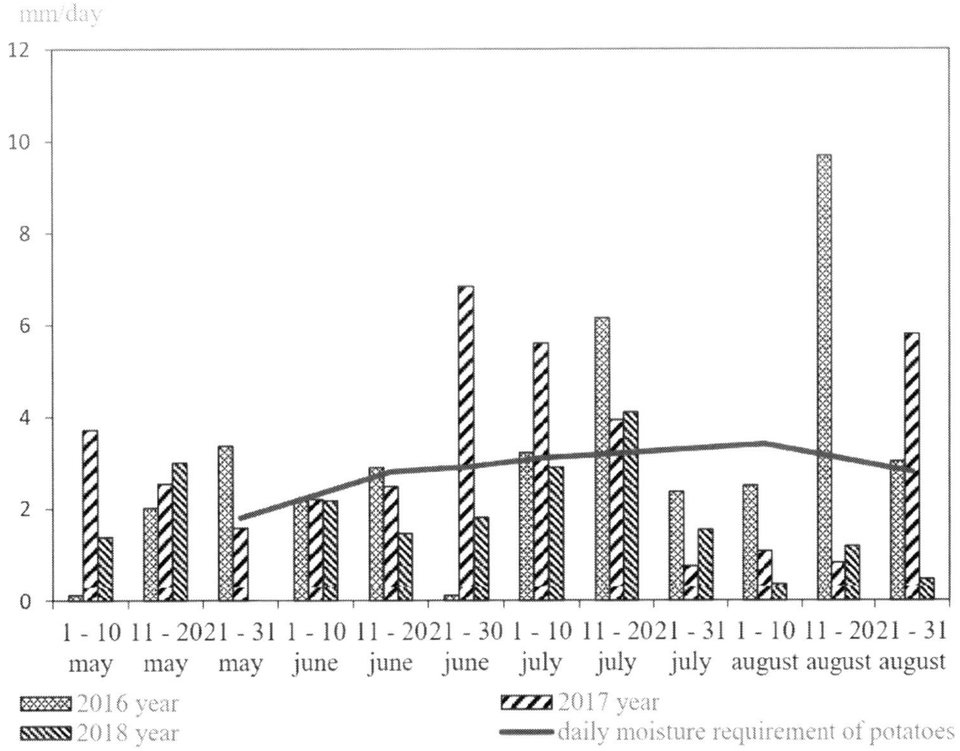

Fig. 1. Dynamics of precipitation in the spring-summer period and the need for moisture in potatoes

Among the irrigation agrarian-meliorative measures, drip irrigation is the most economical. The undoubted advantage of this irrigation method is the delivery of water directly to the root zone [7, 8]. This, in addition to saving water and electricity, practically eliminates excessive watering, and, consequently, leaching of nutrients from the root zone, raising the level of groundwater, and additional load on the drainage network. In addition, in contrast to surface irrigation, during drying, a dense crust does not form, which makes it difficult to aerate the soil. With drip irrigation, the spacing is practically not moistened, this leads to a slowdown in the growth of weeds [9, 10].

The area occupied by drip irrigation in the Russian Federation is about 51 thousand hectares, which is an insignificant percentage of 4633 thousand hectares of the area of all irrigated land. This index is below the world average and is significantly inferior to countries such as the United States or Israel. Pace of drip irrigation implementation is significantly lower than in a number of Asian countries [11].

One of the limiting factors is the low degree of mechanization when laying the drip tape. The equipment designed for mechanized laying of drip tape to the depth of 0.45 m is serially produced in the USA by Andros, Rain Flo, Kennco Manufacturing [12]. However, these tools are expensive and are not designed to put drip tape into a potato ridge. The use of primitive devices for laying the drip tape causes the uneven tension and breaks or bends of the tape.

2 Materials and methods

It is rational to lay the drip tape when growing potatoes in conjunction with the technological operation for the ridges formation, 10 - 14 days after planting. During this period, an increase in daily water consumption of the plant begins, and the combination of operations for ridges formation and laying of a drip tape will increase the work performance, reduce energy consumption [13, 14].

To lay the drip tape into the potato ridge, the equipment was developed for installation on the ridge former (Fig. 2), consisting of a frame on which a guide tube with a colter is fixed to place the drip tape in the ridge at the specified depth. The laying depth is adjusted by vertical movement of the pipe using the holder. On the vertical column of the frame there is an axle with clamping discs for fixing the coil. To prevent the tape from unwinding by inertia in the event of a sudden stop of the machine, a brake device with a tension roller and a spring-loaded pressure shoe is used. During the movement of the machine, the tape is unwound from the coil and enters the guide tube, along which it moves into the potato ridge.

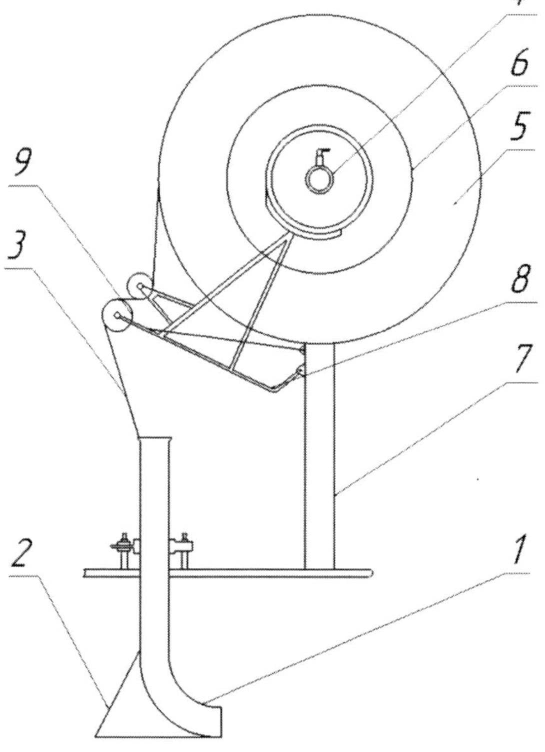

Fig. 2. Construction of drip tape stacker: 1 - guide tube, 2 - opener, 3 - drip tape, 4- shaft, 5 - coil of drip tape, 6 - protective disc with latch, 7 - vertical support post, 8 - braking device, 9 - tensioning roller.

To determine the possibility of installing additional equipment for laying the drip tape on the ridge former, you should calculate the resistance of the drip tape laying, which will be the sum of the soil resistance to development, the friction of the tape in the stacker, overcoming the unbalance of the coil and its friction on the axis [15, 16]:

$$F_{ро} = R_{нг} + F_{лвр} + F_{лдр} + F_{лтр} =$$

$$= \mu \cdot G_p + k_p \cdot h \cdot b + \varepsilon \cdot h \cdot b \cdot v^2 +$$

$$+\left(f_o \cdot G_л \cdot d_o \cdot d_p^{-1} + 2 \cdot G_л \cdot l_э \cdot d_p^{-1}\right) \cdot \left(1 + e^{f_{o} \cdot \beta}\right), \quad (1)$$

Where F_{lv} is the tape pulling resistance, F_{lvr} - friction resistance of the tape in the stacker, kN; F_{ldr} - force to overcome the unbalance of the coil, kN; F_{ltr} - resistance to friction of the coil on the axis, kN; R_{ng} - soil resistance to the development, kN; f_o coefficient of friction of the axle in the bearings; G_l - weight of the tape coil, kN; d_o - axis diameter, m; d_p is the diameter of the coil, m; l_e - displacement of the spool center of gravity from the axis of rotation, m; β - tape wrap angle; μ coefficient of friction of the working tool on the ground; G_p - stacker weight, kN; k_p — resistivity to cutting, kN/m^2; ε - coefficient that takes into account the effect of cutting speed on the resistance to digging; h - development depth, m; b - development width, m.

As calculations have shown, when installing additional equipment on the ridge former, the pulling force will increase by 6 - 8%, which will not affect the performance of the machine, therefore, it is possible to carry out operations to form ridges and lay a drip tape by means of one machine in the same agrarian technical terms (Fig. 3).

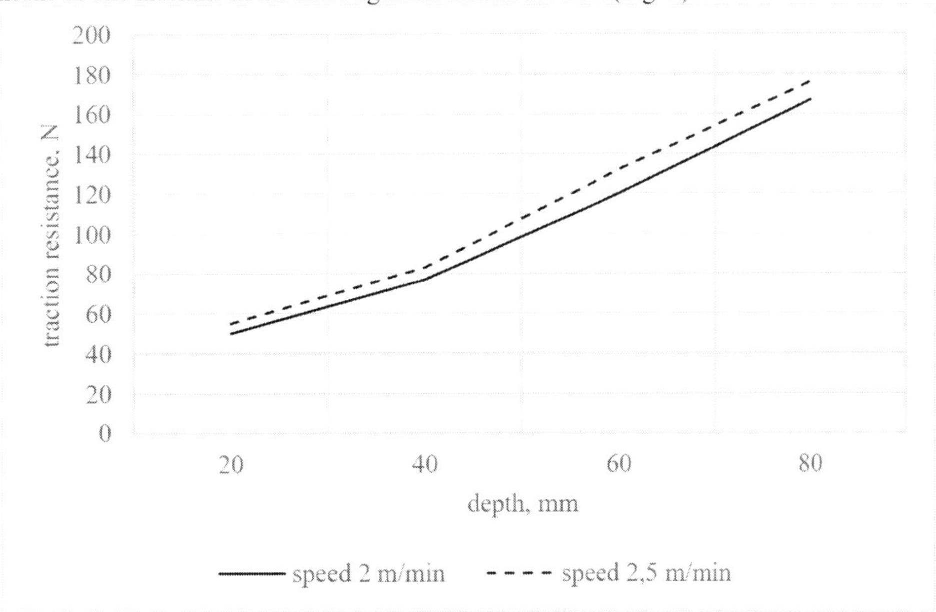

Fig. 3. Depth dependence of the drip tape laying resistance.

3 Results and discussion

According to the research results, the working equipment was mounted on the Grimme GF-75/4 ridge former and the drip tape was laid at the Field Experimental Station of the RGAU-MSHA (Russian State Agricultural University) named after K.A. Timiryazev. During the field research, the conditions for performing the technological operation for laying the drip tape were tried out, the optimal speed mode and the time for performing the operation were determined (Fig. 4).

Fig. 4. Ridge-based drip tape stacker Grimme GF-75/4.

Based on the potato planting scheme, a drip tape with a diameter of 16 mm with a distance between emitters of 300 mm was chosen. The parameters of the developed machine are presented in Table 1.

Table 1. Technical characteristics of the developed machine.

Working body weight	2150 kg
Working body width	3,0 m
Working speed	0,9 – 1,4 m/sec
Performance	0,51 – 0,97 ha/h
Drip tape diameter	16 mm
Dripper water consumption	1,6 L/h
Dripper spacing	300 mm

Watering was carried out in all subsequent phases of the growing season until the beginning of the wilting of the tops. Watering was carried out at a soil moisture content of 70% of the PPV at a constant rate, with the adjustment of the time between waterings (Fig. 5).

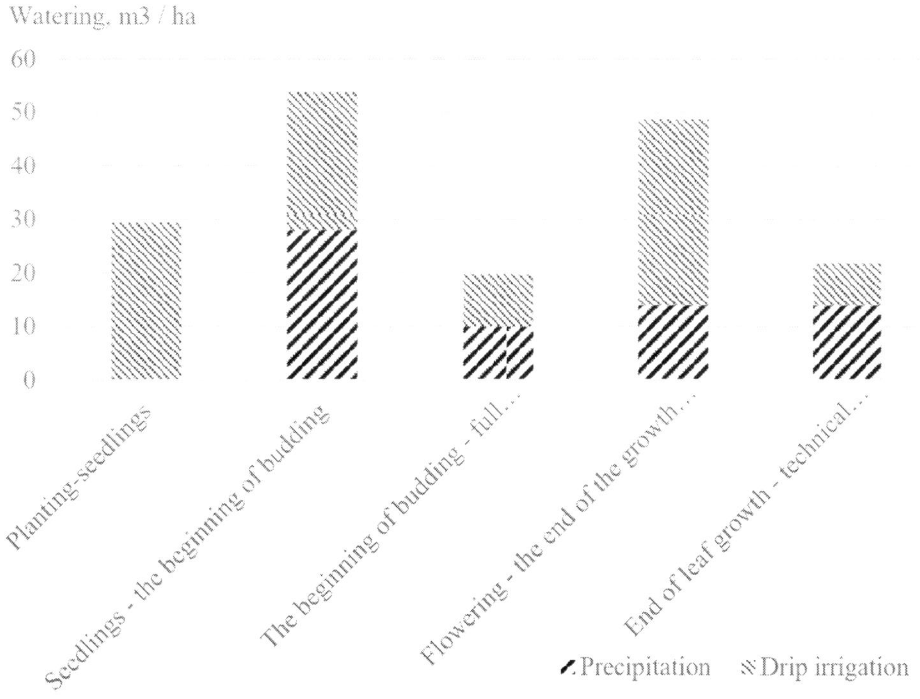

Fig. 5. Water consumption of potatoes at different stages of development.

4 Conclusion

When growing potatoes in the Moscow region, irrigation is required to obtain sustainable yields. Drip irrigation will make it possible to moisturize the root zone without over-watering. Drip irrigation will make it possible to maintain optimal water and air balance during the growing season. It is advisable to lay the drip tape together with the operation for the formation of ridges, as in the early phase of potato development there is no need for additional watering due to the moisture reserves of the parent root tuber. The resistance of the drip tape laying did not exceed 180 N, therefore, the installation of the working equipment for the drip tape laying on the ridge former will not cause a significant increase in traction resistance and will not affect the operational characteristics of the machine. The developed design of the drip tape stacker based on the Grimme GF-75/4 ridge former will increase work productivity and reduce energy costs due to the joint implementation of technological operations.

Acknowledgements

The article was prepared with the support of the Ministry of Education and Science of Russia within the framework of agreement No. 075-15-2021-032 dated March 23, 2021 on the provision of a grant in the form of a subsidy for the creation and development of an engineering center on the basis of an educational institution of higher education and (or) a scientific organization in within the framework of the federal project "Development of infrastructure for research and training" of the national project "Science and Universities".

References

1. N.N. Dubenok, Bolotin D.A., Novikov A.A., Bolotin A.G. Bulletin of the Nizhnevolzhsky agro-university complex: science and higher professional education, **51 (3)**, 81 (2018)
2. O.A. Starovojtova, N.E. Shabanov, The Herald of the Moscow State Agricultural Engineering University named after V.P. Goryachkin, **74 (4)**, 34 (2016)
3. J. Reyes-Cabrera, L. Zotarelli, M.D. Dukes, D.L. Rowland, S.A. Sargent, Agricultural Water Management, **169**, 183 (2016)
4. N.B. Martynova, A.Yu. Korneev, The Herald of the Moscow State Agricultural Engineering University named after V.P. Goryachkin, **84 (2)**, 18 (2018)
5. A.S. Shtahko, V.N. Shkura, Scientific journal of the Russian Research Institute of Melioration Problems, **31 (3)**, 39 (2018)
6. S.M. Vasiljev, V.N. Shkura, A.S. Shtahko, Bulletin of the Nizhnevolzhsky agro-university complex: science and higher professional education, **52 (4)**, 316 (2018)
7. Kh.A. Abdulmazhidov, A.S. Matveyev, The Herald of the Moscow State Agricultural Engineering University named after V.P. Goryachkin, **72(2)**, 40 (2016)
8. A.S. Ovchinnikov, A.A. Buber, Yu.P. Dobrachev, V.V. Borodychev, Bulletin of the Nizhnevolzhsky agro-university complex: science and higher professional education, **54 (4)**, 65 (2018)
9. A.S. Ovchinnikov, V.S. Bocharnikov, M.P. Mesherjakov, O.V. Bocharnikova, Bulletin of the Nizhnevolzhsky agro-university complex: science and higher professional education, **48 (4)**, 10 (2017)
10. N.N. Dubenok, D.A. Bolotin, S.D. Fomin, A.G. Bolotin, Bulletin of the Nizhnevolzhsky agro-university complex: science and higher professional education, **48 (4)**, 22 (2017)
11. M.N. Makani, S.A. Sargent, L. Zotarelli, D.J. Huber, C.A. Sims, Scientia Horticulturae, **197**, 428 (2015)
12. J. Reyes-Cabrera, L. Zotarelli, D.L. Rowland, M.D. Dukes, S.A. Sargent, American Journal of Potato Research, **91 (5)**, 504 (2015)
13. V.N. Krasnoshchekov, D.G. Olgarenko, Environmental Management, **4**, 51 (2016)
14. B.M. Kizjaev, N.B. Martinova, Irrigation and Reclamation, **13(3)**, 62 (2018)
15. E.V. Zhalnin, The Herald of the Moscow State Agricultural Engineering University named after V.P. Goryachkin, **82 (6)**, 10 (2017)
16. V.I. Balabanov, A. Li, The Herald of the Moscow State Agricultural Engineering University named after V.P. Goryachkin, **95(1)**, 10 (2020)

The availability of new potato cross-breeds for industrial processing

Soltan Soslanbekovich Basiev*, *Petr Zaharovich* Kozaev, *Asia Taimurazovna* Doeva, and *Diana Petrovna* Kozaeva

Federal State Budgetary Educational Institution Of Higher Education "Gorsky State Agrarian University", 362040 RNO-Alanya, Vladikavkaz, Russia

Abstract. Under the conditions of the Republic of North Ossetia-Alania, an assessment of the availability of new cross-breeds of their own breeding to industrial processing of French fries according to morphological, biochemical and consumer quality indices has been carried out. It has been specified that the requirements for tubers on morphological grounds for the production of French fries are satisfied by cross-breeds 10.11/53 and 10.11/804, characterized by an elongated form of tuber, flat and smooth surface, shallow depth of the eyes. According to the quality of the finished product we should select the cross-breed 10.11/1787 due to the faintly darkening pulp of the frozen semi-finished product, the attractive appearance of the finished French fries, their good taste and smell. Cross-breeds 10.11/53 and 10.11/804 should be recommended for the cumulative French fries.

1 Introduction

Potato tubers are processed mainly for crispy potatoes, French fries, starch and mashed potatoes. Each type of processing requires tubers of a certain variety. For example, high-starch varieties (13-18% or more) are required for processing for starch and alcohol. For processing for food the varieties with increased content of dry matter and low reducing sugars are more suitable.

The manufacturer should remember that the consumer is very discerning to quality. The processing industry has strict standards for raw materials, as the products produced shall be of high quality and at the same time cost-effective. Potatoes used as raw materials for the processing industry shall meet a number of quality requirements: the size and shape of tubers, damage and defects, the dry matter content and color [1].

The shape of tubers, the condition of their surface, the depth of the eyes have a significant impact on the economic indices of processing. The optimal shape of tubers for the production of fries - at least 50 mm to length. It should be borne in mind that the abrasive cleaning is used in the production lines and when using oblong shape tubers compared to rounded ones there are large wastes. The eyes on the tubers should be superficial, as deep cleaning is required manually, which requires additional workers and

*Corresponding author: basiev_s@mail.ru

causes an increase in waste. During mechanical cleaning, the total losses from the main cleaning and subsequent post-treatment can reach up to 25–45% [2].

An important index of the variety is the color of flesh. For processing, varieties with yellow flesh are preferable. Another important aspect is color uniformity. The uneven color causes the French fries to appear dark at one end. The reason for this is aging (wilting) of tubers and secondary germination.

Some varieties tend to develop "sugar bags" when aging after long-term storage. In cases of strong secondary germination, starch flows from the primary tuber to the secondary germinated tuber. The outflow of starch begins at the bottom of the tuber and can cause liquefaction of flesh. Liquefaction occurs in flesh of the secondary germinated tubers or on top of a long tuber; when cut, it has a watery transparent appearance. Even spongy textures are observed.

Thus, an increase in the production of potato products requires the creation of special varieties. Breeding potatoes for availability for fried potato products industrial processing is associated with certain difficulties due to the need to combine many traits in one genotype. The quality of the finished French fries is of great importance for the content of reducing sugars, on which the color of the resulting product depends, and therefore, breeding work in the direction of creating varieties that do not accumulate them is promising. A large amount of reducing sugars corresponds to the dark color of fried potato products, as well as their bitter taste. Tubers intended for processing should contain 0.25% sugars max. [3, 4].

In Russia, there are no high-starchy potato varieties, and therefore monitoring the starch content of new varieties is one of the most necessary elements of potato selection for - quality. Starch is a mixture of amylose and amylopectin - high molecular polysaccharides, the ratio of which depends on the viscosity, gelatinization, swelling, the ability to gel-forming ability. By increasing the content of amylopectin, the brittleness of fried potato products can be reduced [5].

In breeding for high starch content, the selection effectiveness of the parental forms for crossbreeding and phenotypic selection of the breeding material have been proven. Starch accumulation depends on the duration of the potato growing season, and the shorter it is, the less the genes responsible for starchiness can appear themselves [6].

The creation of high-starchy varieties is a laborious task, which is associated with the peculiarity of the cultivated type of potato, which is not characterized by a large degree of starch accumulation. In this regard, the wild species *S. chacoense Bitt., S. andigenum Juz. et Buk., S. demissum Lindl.* And others are often used in breeding programs. The situation is aggravated by the existence of a negative correlation between yield and starch content.

With an increase in the dry matter content, the industrial processing of potatoes becomes simpler, the quality of the finished product increases, and the fat content decreases in it. The low dry matter content results in watery and soft French fries. However, very high dry matter content can result in a hard consistency in the final product. Tubers prepared for the production of French fries should contain 20-24% dry matter [7, 8].

2 Materials and methods

Potatoes were grown in a mountainous zone (1400 m above sea level, Alagirsky district of North Ossetia - Alania, Kurtat village, the department branch of agriculture, plant growing, selection and seed production of the agronomic faculty of the State Federal-Funded Educational Institution of Higher Professional Training "Gorsky GAU").

The investigation aim was to evaluate the potato cross-beeds of the State Federal-Funded Educational Institution of Higher Professional Training "Gorsky GAU" for availability to industrial processing of French fries.

The investigation objects are potato cross-breeds of our own selection: 10.2/752 (Bars × Adretta), 10.2/771 (Bars × Adretta), 10.11/53 (Roko × Romano), 10.11/804 (Roko × Romano), 10.11/1787 (Roko × Romano), 10.11/1800 (Roko × Romano).

Potato tubers were evaluated according to the requirements of GOST 26832-86 "Fresh potatoes for food processing" [9].

The following calculations and observations were carried out during the experiments [2]:

1. The morphological characteristics of potato tubers were determined (according to 9-grades scale):

- the shape, size and surface of the tubers were assessed visually. The shape factor of a tuber was calculated as the ratio of its length to width. For French fries, varieties with a long and very long shape are preferred, the factor of which varies from 1.7 to 2 or more. The surface of tubers is flat and smooth. The surface quality was assessed according to the following scale: 9 - smooth; 5 - with shallow indentations (up to 3 mm); 3 - with slight warts and cracks (up to 5mm); 1 - with warts and cracks (more than 5 mm).

- the number (by calculating) and the depth of the eyes (by measuring with a special ruler) were determined. To reduce waste during cleaning, tubers should have a shallow depth of eyes, and their number should be small.

- assessed the color of the peel and flesh organoleptically, which is important from the consumer point of view. For the production of French fries, yellow meat varieties are preferred, providing a golden color to the finished product. The flesh color varies within the following shades: white, cream, light yellow, yellow, dark yellow.

2. The biochemical indices of potato tubers were assessed: starch and dry matter content by the gravimetric method. The obtained data of dry matter as a percentage ratio were converted into grades, where 16.0-18.0 %% corresponded to 1-3 grades, 18.1-21.0 %% - 4-6 grades, 21.1-24.0 %% - 7-9 grades. The starchiness of potatoes affects the taste of the finished product, as well as the consumption of energy and oil.

3. Tubers were assessed for availability for making French fries:

- habit, shape and size of the bars organoleptically, where 9 grades - bars of regular shape, uniform in color and size, 5 grades - bars of uneven size and color with spots occupying 15-20% of the surface; 1 grade - broken, spotted, irregularly shaped blocks;

- darkening of the flesh of frozen bars organoleptically: from no darkening (9 grades) to very strong (1 grade);

- the consistency of the bars flesh organoleptically.

3 Results and discussion

To cook French fries, several tubers were peeled by cutting them into 10 × 10 mm pieces. Washed, blanched for 2 minutes, washed again. They were fried until half cooked, cooled, and then frozen in a freezer at a temperature of - 35 ... 40 °C. The flesh darkening of the frozen semi-finished product was noted, after which it was fried until completely cooked.

According to the data obtained (Table 1), from the presented samples, a uniform, unexpressed yellow color of the product with no burning was noted in the 10.11/1787 cross-breed. Cross-breeds 10.11/53 and 10.11/804 when roasted give an uneven, unexpressed yellow color. Other cross-breeds form brown spots.

The flesh consistency of all varieties is approximately the same and is characterized by a slight separation between the crust and core, as well as uneven roasting. The difference in taste and smell of cross-breeds is also small and ranges from 5.0 to 6.4 grades.

Table 1. French fries quality.

Cross-breed	Color, grade						Consistency, grade						Habit, shape, size, grade						Taste, smell, grade					
	1	**2**	**3**	**4**	**5**	**avg**	**1**	**2**	**3**	**4**	**5**	**avg**	**1**	**2**	**3**	**4**	**5**	**avg.**	**1**	**2**	**3**	**4**	**5**	**avg**
10.2/752	3	4	3	5	4	3.8	5	3	3	6	6	4.6	5	6	3	5	6	5	3	7	3	6	6	5
10.2/771	2	3	3	2	4	2.8	7	4	3	5	7	5.2	3	4	3	7	4	4.2	7	6	4	4	5	5.2
10.11/53	6	5	6	4	5	5.2	5	4	6	5	7	5.4	6	5	5	6	6	5.6	7	8	6	5	6	6.4
10.11/804	7	7	4	4	5	5.4	7	5	5	4	5	5.2	8	5	5	7	5	6	8	6	6	6	4	6
10.11/1787	6	7	6	6	7	6.4	7	4	6	6	6	5.8	8	7	6	7	7	7	6	6	7	6	7	6.4
10.11/1800	2	3	2	3	5	3	5	4	3	5	4	4.2	3	3	3	4	4	3.4	7	4	5	4	6	5.2

It is noted that in habit, shape and size, the cross-breed 10.11/1787 is superior to other samples, as evidenced by obtaining bars that are fairly uniform in color and size, of the correct shape. Basically, the investigated cross-breeds, when roasted, formed swellings, thinning, spots.

Table 2. The availability assessment of the finished product of cross-breeds for fries processing.

No.	**Cross-breed**	**Flesh darkening of a frozen semi-finished product, grade**	**Color, grade**	**Flesh consistency, grade**	**Habit, shape, size, grade**	**Taste, smell, grade**	**Dry matter, grade**	**Overall grade**
1	10.2/752	7	3.8	4.6	5	5	3	4.7
2	10.2/771	3	2.8	5.2	4.2	5.2	3	3.9
3	10.11/53	6	5.2	5.4	5.6	6.4	3	5.3
4	10.11/804	5	5.4	5.2	6	6	3	5.1
5	10.11/1787	7	6.4	5.8	7	6.4	3	6.0
6	10.11/1800	5	3	4.2	3.4	5.2	3	4.0

According to the data presented in Table 2, the slightly darkening flesh of the frozen semi-finished product complies with the cross-breeds 10.2/752 and 10.11/53 (7 grades each). Other cross-breeds darken moderately, while the 10.2/771 cross-breed darkens strongly, and therefore the color and habit of the finished product were unattractive. We can also characterize the accumulation degree of reducing sugars in cross-breeds according to this index. So, in cross-breeds 10.2/752 and 10.11/1787 it is minimal which is extremely important in the production of fried potato products.

The cross-breeds formed approximately the same amount of dry matter - within 17.1-18 %% (due to the short growing season of the investigated samples), which complies with 3 grades when assessing the availability for cooking French fries.

So, according to the overall grade of the finished product availability assessment we should select the cross-breed 10.11/1787 due to the faintly darkening flesh of the frozen semi-finished product, the attractive habit of the finished French fries, their good taste and smell.

It has been specified that according to the morphological grounds of tubers, cross-breeds 10.11/53 and 10.11/804 (7.5 and 6.8 degrees, respectively), characterized by an elongated tuber shape, shallow depth of eyes, and a flat smooth surface of tubers, satisfy most of all the requirements for the production of potatoes. Cross-breeds 10.2/771, 10.11/1800, 10.11/1787 comply with these requirements to a weak degree.

Table 3. The availability assessment of the cross-breeds for fries processing.

No.	Cross-breed	Tuber grade	Finished product grade	Average
1	10.2/752	6.0	4.7	5.4
2	10.2/771	4.5	3.9	4.2
3	10.11/53	7.5	5.3	6.4
4	10.11/804	6.8	5.1	6.0
5	10.11/1787	5.3	6.0	5.7
6	10.11/1800	4.8	4.0	4.4

According to the overall grade (Table 3), which is calculated by adding the grades obtained in assessing morphological, biochemical parameters, as well as the quality of the finished product, cross-breeds 10.11/53 and 10.11/804 should be recommended for the production of French fries.

4 Conclusion

1. The requirements for tubers on morphological grounds for the production of French fries are satisfied by cross-breeds 10.11/53 and 10.11/804, characterized by an elongated form of tuber, flat and smooth surface, shallow depth of the eyes,

2. According to the quality of the finished product we should select the cross-breed 10.11/1787 due to the faintly darkening pulp of the frozen semi-finished product, the attractive appearance of the finished French fries, their good taste and smell.

3. Cross-breeds 10.11/53 and 10.11/804 should be recommended for the cumulative French fries.

References

1. G. Kuzdra, Potatoes and vegetables, **6**, 6 (1986)
2. K. A. Pshechenkov, *Methodological guidelines for assessment of potato varieties availability for processing and storage* (2008)
3. L. V. Nezakonova, A. P. Pingol, Potato protection, **1**, 8 (2011)
4. S. Basiev, T. Lazarov, M. Gaplaev, F. Gerieva, I. Shishkhaev, IOP Conference Series: Earth and Environmental Science, **659**, 012085 (2021)
5. G. I. Piskun, L. N. Kozlova, Potato growing: collection of scientific investigation, **52** (2013)
6. V. L. Makhanko, L. N. Kozlova, O. B. Nezakonova, Land husbandry and plant protection, **3**, 62 (2013)
7. D. V. Abrosimov, *The selection principles of parental pairs and selection methods for breeding potatoes for increased starchiness* (2007)
8. N. F. Sintsova, Z. F. Sergeeva, T. A. Osipova, Agricultural science of the Euro-North-East, **3**, 32 (2015)
9. GOST 26832-86. Fresh potatoes for processing into food. Specifications (2010)

Agri-food market of the region: strategic analysis and development prospects

Anzor Gyatov[1,*], *Zarina* Soskiyeva[2], *Madina* Marzhokhova[1], *Anzor* Shardanov[1], and *Oksana* Bagova[1]

[1]Kabardino - Balkarian State Agricultural University named after V.M. Kokov, Lenin Avenue, 1V, 360030 Nalchik, Russia
[2]Gorsky State Agrarian University, Vladikavkaz, Russia

Abstract. The agri-food market is distinguished by a high degree of dependence on external factors, such as climatic and weather conditions, uneven supply and demand for agricultural products due to the seasonality of production, which explains the dynamism and instability of this market segment. [1] Due to the indicated reasons, the process of current and strategic planning, analysis, determination of development prospects, is an integral part of agricultural production. The article studies the state of the agri-food market of the Kabardino-Balkarian Republic on the scale of the North Caucasian Federal District. The dynamics of the indicators of the produced crop and livestock products in the context of the regions of the NCFD are considered and analyzed, the comparative characteristics of the analyzed values are performed. The results of the analytical study made it possible to build a SWOT analysis model, which identifies the strengths and weaknesses of the agri-food market of the Kabardino-Balkarian Republic, as well as the opportunities and threats of its development.

1 Introduction

The regional food market is, on the one hand, a subsystem of the national food market, and on the other hand, it acts as a subsystem of the regional social-and-economic system [2].

The relevance of the study is due to the importance of the agri-food market in the region, not only at the level of the subject, but also in general for the country's macroeconomic development, national food security.

The purpose of this study is to analyze the current state of the agri-food market of the Kabardino-Balkarian Republic in terms of assessing its strengths and weaknesses, as well as opportunities and threats, in order to determine the most favorable prospects for the development of the regional agri-food market.

* Corresponding author: gyatov.anzor@bk.ru

2 Materials and Methods

The information base for the study of the regional agri-food market was the data of the Federal State Statistics Service, as well as literary sources of domestic and foreign authors.

The work found the application of the monographic research method, as well as such methods of cognition as: analysis, synthesis, generalization, the method of economic comparison and the method of SWOT analysis.

3 Results and discussion

To determine the main directions of the strategic development of the regional agri-food market, a comprehensive diagnosis of the state of the agrarian market of the region at the present time is required.

The efficiency and efficiency of agricultural production is largely determined by the provision of the regional agro-industrial complex with a material and technical base [3], table 1 shows the dynamics of changes in indicators characterizing the state of fixed assets in agriculture in the Kabardino-Balkarian Republic.

Table 1. Fixed assets of the Kabardino-Balkarian Republic and the efficiency of their use.

Indicators	Year					2019 to 2015 in %
	2015	2016	2017	2018	2019	
The cost of fixed assets in agriculture, RUR mln	27344	38643	43540	42392	36992	135.3
New fixed assets commissioned, RUR mln	4302.7	10110.2	738.0	1305.1	746.0	17.3
Fixed asset depreciation value, %	30.4	40.6	46.0	47.4	47.3	155.5
Fixed assets renewal ratio,%	9.8	25.7	6.9	11.5	6.7	68.4
Fixed assets liquidation ratio,%	0.4	0.5	0.8	1.1	0.4	100
Return on assets of agriculture	1.43	1.13	1.04	1.16	1.46	102.1
Capital intensity of agriculture	0.70	0.88	0.96	0.86	0.69	98.6

Source: compiled by the author based on the data of [4].

The data in Table 1 indicate the unsatisfactory state of fixed assets of agriculture in the Kabardino-Balkarian Republic, in particular, a high degree of depreciation, a low level of renewal of fixed assets, the exception is the rate of return on assets, which has a slight positive trend over the analyzed period.

The presented information indicates that the material and technical support of agro-industrial production is the "weak" side of the regional agri-food market of the republic. The result of this state of affairs in the industry is: underdevelopment of the infrastructure of the agro-industrial complex, often extensive forms of organization of work, due to a lack of equipment, as well as a reduction in acreage (Table 2).

Table 2. The structure of the cultivated areas of the Kabardino-Balkarian Republic.

Indicators	Year						2019 to 2010 in %
	2010	2015	2016	2017	2018	2019	
Sown area, agricultural crops, th ha	287.9	282.6	282.8	283.7	283.0	281.8	97.8

Table 2. Continued.

The structure of sown areas of agricultural crops in farms of all categories, as a percentage of the total sown area, %							
cereal and leguminous crops	175.3	207.8	207.7	210.7	209.1	212.1	121.0
industrial crops	44.1	26.8	30.0	27.8	31.0	28.5	64.6
potatoes and vegetable and melon crops	30.8	29.1	27.9	28.5	24.7	23.7	76.9
forage crops	37.7	18.9	17.2	16.7	18.2	17.5	46.4

Source: compiled by the author based on the data of [4].

The data in Table 2 indicate, in general, a decrease in sown areas over the past 10 years in the KBR by 2.2%. However, it should be noted that positive dynamics for the analyzed period is noted for grain and leguminous crops - an increase in acreage by 21%. For the rest of agricultural crops, there is a decrease in sown areas: industrial crops by 35.4%, potatoes and vegetable and melon crops - 23.1%, fodder crops - 53.6%. Thus, it can be concluded that cereals and leguminous crops prevail in the region's crop production.

For the most vivid picture of the volume of agricultural products produced on the scale of the KBR, let us compare these indicators for the regions of the North Caucasian Federal District (Table 3).

Table 3. Agricultural products in farms of all categories (in actual prices), RUR mln.

Indicators	Year						2019 to 2010 in %	Damage location in the Russian Federation
	2010	2015	2016	2017	2018	2019		
RF	2462187	4794615	5112356	5109475	5348777	5907955	239.9	-
NCFD	197873	384914	428066	436752	461495	493569.8	249.4	5
The Republic of Dagestan	47410	94478	108473	118960	124371	132923.2	280.4	14
The Republic of Ingushetia	3231	6885	8518	9031	10392	10812.1	334.6	70
Kabardino-Balkar Republic	23588	37920	42424	45710	49385	54133.1	229.5	35
Karachay-Cherkess Republic	15040	26032	26802	27581	29513	33334.8	221.6	53
Republic Northern Ossetia-Alania	16910	23210	22165	22291	24198	29990.5	177.4	54
The Chechen Republic	10893	17549	21123	25997	27778	32420.9	297.6	52
Stavropol Territory	80801	178840	198563	187182	195858	199955.1	247.5	6

Source: compiled by the author based on the data of [5].

The volumes of agricultural products for the analyzed period tended to grow in all regions of the North Caucasus Federal District, for example, in the Republic of Ingushetia, the volume of agricultural production increased more than 3 times, in other regions more than 2 times, with the exception of the Republic of North Ossetia-Alania, where the volume of production agriculture increased 1.77 times. However, if we consider the volumes of agricultural products produced on a national scale, it should be noted that the KBR ranks 35th in the country, yielding in this indicator to the Stavropol Territory (6th place) and the Republic of Dagestan (14th place) from the regions of the NCFDt.

Let us consider the production volumes of the main agricultural crops in the regions of the NCFD, for the analysis and comparison of the indicators of the Kabardino-Balkarian Republic on the scale of the federal district (Table 4).

Table 4. Gross harvest (thousand tons) of major agricultural crops in all categories of farms in the NCFD regions.

Crops	Year				2019 to 2010 in %
	2010	2017	2018	2019	
Grain (in weight after processing), th tons					
NCFD	8456.8	13261.1	11980.3	11424.8	135.1
The Republic of Dagestan	211.1	414.9	359.5	385.0	182.4
The Republic of Ingushetia	43.9	47.4	89.4	40.7	92.7
Kabardino-Balkar Republic	**644.0**	**1157.3**	**1128.1**	**1136.1**	**176.4**
Karachay-Cherkess Republic	143.7	476.1	437.9	493.9	343.7
Republic Northern Ossetia-Alania	399.7	636.2	763.8	791.2	197.9
The Chechen Republic	124.6	422.0	268.4	177.7	142.6
Stavropol Territory	6889.7	10107.2	8933.2	8400.2	121.9
Sunflower seeds, th t					
NCFD	388.0	684.2	594.6	560.2	144.4
The Republic of Dagestan	3.8	9.1	8.3	8.4	221.1
The Republic of Ingushetia	3.4	1.9	3.3	0.2	5.9
Kabardino-Balkar Republic	**43.2**	**28.4**	**32.9**	**26.1**	**60.4-2**
Karachay-Cherkess Republic	7.0	15.5	17.0	15.0	214.3
Republic Northern Ossetia-Alania	1.2	3.6	2.4	1.4	116.7
The Chechen Republic	5.0	24.8	4.4	2.3	46.0
Stavropol Territory	324.4	600.9	526.3	506.9	156.3
Potatoes, th t					
NCFD	1063.2	1065.9	1064.1	1029.8	96.9
The Republic of Dagestan	295.5	357.0	356.3	353.5	119.6
The Republic of Ingushetia	29.5	49.8	37.3	33.0	111.9
Kabardino-Balkar Republic	**200.8**	**184.9**	**182.9**	**196.0**	**97.6**
Karachay-Cherkess Republic	165.6	143.9	123.8	107.9	65.2
Republic Northern Ossetia-Alania	119.4	62.7	99.3	89.2	74.7
The Chechen Republic	21.8	38.5	32.1	32.7	150.0
Stavropol Territory	230.5	229.1	232.5	217.6	94.4
Vegetables, th t					
NCFD	1632.6	2447.6	2405.1	2442.6	149.6
The Republic of Dagestan	973.6	1451.7	1438.5	1432.1	147.1
The Republic of Ingushetia	1.8	2.8	3.9	5.5	305.6
Kabardino-Balkar Republic	**315.0**	**498.9**	**470.4**	**404.4**	**128.4-2**

Table 4. Continued.

Karachay-Cherkess Republic	56.3	79.3	77.2	73.3	130.2
Republic Northern Ossetia-Alania	35.3	19.3	28.4	31.0	87.8
The Chechen Republic	25.2	75.2	75.2	130.3	517.1
Stavropol Territory	225.4	320.5	311.6	365.9	162.3
Fruits and berries, th t					
NCFD	268.4	487.5	557.2	703.8	262.2
The Republic of Dagestan	107.6	157.7	161.1	173.3	161.1
The Republic of Ingushetia	0.8	10.9	20.7	25.6	3200.0
Kabardino-Balkar Republic	**90.4**	**215.2**	**257.2**	**349.5**	**386.6-1**
Karachay-Cherkess Republic	3.4	7.5	10.3	15.0	441.2
Republic Northern Ossetia-Alania	14	18.9	18.8	43.2	308.6
The Chechen Republic	12.0	17.8	20.3	20.9	174.2
Stavropol Territory	39.3	59.6	68.6	76.4	194.4

Source: compiled by the author based on the data of [5].

The production of grain and leguminous crops for the analyzed period in the KBR increased by 76.4% and in 2019 amounted to 1,136.1 thousand tons, some role in this was played by an increase in acreage for these crops by 21%, in this product category in terms of production, the Kabardino-Balkarian Republic from the regions of the North Caucasus Federal District is second only to the Stavropol Territory, where grain production in 2019 amounted to 8400.2 th t. In terms of sunflower production, the leading place in the North Caucasus Federal District belongs to the Stavropol Territory - 506.9 thousand tons in 2019, followed by Kabardino-Balkaria - 26.1 thousand tons, while it should be noted that the reduction in sunflower production for the analyzed period in the KBR was 39, 6%. The largest gross potato harvest in 2019 in the North Caucasus Federal District belongs to the following regions: the Republic of Dagestan - 353.5 th t, the Stavropol Territory - 217.6 th t and the KBR - 196.0 th t. In terms of vegetable production, the Kabardino-Balkarian Republic from the regions of the North Caucasus Federal District is second only to the Republic of Dagestan - 1432.1 th t and 404.4 th t, respectively. The production of fruit and berry crops in the KBR in 2019 amounted to 349.5 th t, for 2010-2019, the production of this product category increased almost 4 times. It should be noted that the production of fruit and berry crops is the most developed branch of agricultural production in the Kabardino-Balkarian Republic, providing a leading position not only among the regions of the North Caucasus Federal District, but also in the Russian Federation, since in the production of fruits and berries the KBR is second only to the Krasnodar Territory in the country, where the gross product collections in 2019 amounted to 497.6 th t. In general, it should be noted that in recent years, the intensification of import substitution, the domestic fruit and berry sector shows positive dynamics and becomes more and more attractive [6].

Table 5. Productivity of the main agricultural crops in all categories of farms in the Kabardino-Balkarian Republic and the North Caucasian Federal District, c/ha.

Crops	Year				2019 to 2010 in %
	2010	2017	2018	2019	
Kabardino-Balkar Republic					
Grain (weight after enhancement)	37.4	56.3	54.1	54.8	146.5

Table 5. Continued.

Sunflower seeds	14.4	16.5	18.8	18.4	127.8
Potatoes	162	204	217	236	145.7
Vegetables	175	259	290	260	148.6
Fruits and berries	119.0	200.1	215.9	249.4	209.6
North Caucasian Federal District					
Grain (weight after enhancement)	32.2	41.9	37.8	35.9	111.5
Sunflower seeds	13.1	17.6	16.5	17.0	129.8
Potatoes	128	170	173	176	137.5
Vegetables	189	269	268	276	146.0
Fruits and berries	60.8	113.0	122.7	143.5	236.1

Source: compiled by the author based on the data of [5].

Comparison of yield indicators for the Kabardino-Balkarian Republic and values for the North Caucasian Federal District (Table 5) testifies in favor of the KBR, since practically for all years and for all categories of crops, yield variations in the republic exceed those for the NCFD. These circumstances are due to a number of factors, in particular, favorable natural and climatic conditions; the use of modern technologies in the organization of agricultural production, for example, the system of intensive gardening; the use of mineral fertilizers and modern plant protection products; the use of quality seeds, high-performance varieties and crops; increasing soil fertility.

Table 6. Dynamics of the number of livestock by regions of the NCFD.

	Livestock (th animals)			Place in RF in 2019	Livestock of sheep and goats (th animals)			Place in RF in 2019
	2017	2018	2019		2017	2018	2019	
NCFD	2126.4	2091.9	2087.6	5	8952.8	8359.5	8314.7	1
The Republic of Dagestan	1004.0	960.5	952.1	2	5339.5	4743.8	4647.1	1
The Republic of Ingushetia	57.0	65.4	67.2	64	152.2	198.3	216.2	20
Kabardino-Balkar Republic	**265.7**	**265.1**	**268.6**	**22**	**364.0**	**375.9**	**383.9**	**13**
Karachay-Cherkess Republic	157.5	157.4	160.7	45	1080.1	1094.4	1137.8	6
Republic Northern Ossetia-Alania	83.4	91.2	92.7	59	58.9	74.8	98.7	34
The Chechen Republic	239.7	251.5	252.8	24	248.3	238.5	265.4	18
Stavropol Territory	319.1	300.7	293.5	20	1709.9	1633.7	1565.6	3

Source: compiled by the author based on the data of [5].

When analyzing the regional agri-food market, along with the indicators of sown areas, gross harvests and crop yields, indicators characterizing the livestock sector of agriculture in the region are of great importance. The basis of animal husbandry is the breeding of cattle, sheep and goats.

The data in Table 6 indicate a positive trend in the Kabardino-Balkarian Republic for the analyzed period, both in the number of cattle - 1.1% and in the number of sheep and goats - 5.5%. In terms of the number of cattle in 2019 in the NCFD, the leading positions belong to the Republic of Dagestan - 952.1 th animals, the Stavropol Territory - 293.5 th animals and the Kabardino-Balkarian Republic - 268.6 th animals. The three leaders in terms of the number of sheep and goats among the regions of the NCFD are as follows: The Republic of Dagestan - 4647.1 th animals, the Stavropol Territory - 1565.6 th animals and the Karachay-Cherkess Republic - 1137.8 th animals, then the KBR - 383.9 th animals. In general, it should be noted, in contrast to crop production, animal husbandry is a weaker competitive position of the Kabardino-Balkarian Republic. Further strengthening of the competitive advantages of the industry requires an increase in the productivity of livestock and poultry (Table 7).

Table 7. Livestock production in the NCFD.

Crops	Year				Place in RF in 2019
	2017	2018	2019	2019 to 2017 in %	
Production of livestock and poultry for slaughter (slaughter weight), th t					
NCFD	655.1	681.9	706.9	107.9	7
The Republic of Dagestan	143.6	148.4	150.8	105.0	23
The Republic of Ingushetia	3.4	3.7	4.0	117.6	76
Kabardino-Balkar Republic	73.9	70.5	72.1	97.6	43
Karachay-Cherkess Republic	29.2	27.9	31.2	106.8	57
Republic Northern Ossetia-Alania	20.9	19.2	23.0	110.0	64
The Chechen Republic	23.7	23.5	24.6	103.8	63
Stavropol Territory	360.4	388.5	401.3	111.3	5
Milk production, th t					
NCFD	2618.4	2627.2	2694.9	102.9	5
The Republic of Dagestan	875.6	892.7	913.3	104.3	7
The Republic of Ingushetia	93.1	98.4	107.4	115.4	66
Kabardino-Balkar Republic	490.5	499.2	514.4	104.9	24
Karachay-Cherkess Republic	185.4	187.0	193.1	104.2	47
Republic Northern Ossetia-Alania	144.5	155.0	167.9	116.2	53
The Chechen Republic	286.9	285.8	290.7	101.3	40
Stavropol Territory	542.5	509.1	507.9	93.6	25

Table 7. Continued.

Egg production, mln pcs.					
NCFD	1587.1	1607.2	1567.7	98.8	7
The Republic of Dagestan	255.0	245.0	246.1	96.5	44
The Republic of Ingushetia	14.2	19.7	23.3	164.1	73
Kabardino-Balkar Republic	229.5	229.8	230.1	100.3	45
Karachay-Cherkess Republic	81.3	83.7	82.5	101.5	65
Republic Northern Ossetia-Alania	70.5	69.2	69.2	98.2	66
The Chechen Republic	146.3	131.4	122.7	83.9	56
Stavropol Territory	790.4	828.3	793.8	100.4	25
Wool production, t					
NCFD	24258	24181	21922	90.4	1
The Republic of Dagestan	15061	15481	13126	87.2	1
The Republic of Ingushetia	332	343	428	128.9	19
Kabardino-Balkar Republic	976	1000	1033	105.8	10
Karachay-Cherkess Republic	1403	1342	1380	98.4	8
Republic Northern Ossetia-Alania	95	104	114	120.0	35
The Chechen Republic	595	575	546	91.8	17
Stavropol Territory	5823	5336	5294	90.9	3

Source: compiled by the author based on the data of [5].

The data in Table 7 indicate a positive trend in the production of livestock products for the analyzed period in the Kabardino-Balkarian Republic: milk production - 4.9%, egg production - 0.3%, wool production - 5.8%, with the exception of livestock and poultry production. for slaughter, where there was a decrease in the indicator by 2.4%.

In general, it should be noted that in terms of crop and livestock production, the Kabardino-Balkarian Republic is one of the top three in the North Caucasian Federal District, along with the Stavropol Territory and the Republic of Dagestan. However, for the objectivity of the conclusions, it is necessary to take into account that the number of people employed in agriculture in the Stavropol Territory exceeds the number of people employed in agriculture in the KBR by 2.5 times; in the Republic of Dagestan - 3.3 times. The sown area of the Stavropol Territory is 11.2 times, and the area of the Republic of Dagestan is 1.2 times higher than the same indicators for the Kabardino-Balkarian Republic. It should also be noted that the total land area of the Stavropol Territory is 6616.0 th ha, the Republic of Dagestan - 5027.0 thousand hectares, and the Kabardino-Balkarian Republic - 1247.0 th ha.

To assess the state of the regional market for agricultural products for strengths and weaknesses, as well as to identify promising areas of development - opportunities and potential threats, the SWOT analysis method can be used [7] (Table 8).

Table 8. SWOT analysis of the development of the agri-food market of the Kabardino-Balkarian Republic.

Strengths:	Weaknesses
- favorable natural and climatic conditions and geographical location for agricultural production; - the market of fruit and berry products occupies a leading position on a national scale; - the use of modern, innovative technologies in production (intensive gardening, modern vegetable and fruit storage facilities); - developed infrastructure network in the region (gasification, electrification, water supply systems; transport and logistics networks); - human resources, characterized by the presence in the region of specialized secondary and higher educational institutions for the training of specialists in the field of the agro-industrial complex.	- reduction of acreage; - insufficient level of technical equipment of agricultural production; - with the exception of certain industries, such as gardening, vegetable growing, agriculture of the republic, it is characterized by a low level of innovation; - low standard of living of the rural population; - underdeveloped specialized agricultural infrastructure, for example, seed production, agrochemistry (often preference is given to imported analogues), breeding, veterinary medicine, etc.; - the system of state support and subsidies does not fully satisfy the needs of agricultural producers; - an underdeveloped risk insurance system that meets modern conditions for organizing agricultural production.
Opportunities.	Threats:
- the regional agri-food market has the potential for import substitution of agricultural products; - development of export opportunities in some sectors of the agro-industrial complex; - involvement of unused land in agricultural production; - formation and development of an effective mechanism of state support for the agricultural sector at the regional level; - development of production of environmentally friendly products, as a result of the growing demand for such products; - the availability of innovative technologies, advanced experience, which makes it possible to increase the efficiency of agricultural production. It is proposed to consider innovative activity as a set of scientific, technical, organizational, financial and commercial measures aimed at commercializing the generated relevant knowledge and technologies [8]; - enhancing the quality of the management system in agriculture; - creation of conditions for enhancing the quality of life, the standard of living of the rural population, in order to curb migration processes in rural areas.	- high dependence of agricultural production on weather conditions, as a result of which the risks of natural disasters, crop failure, etc. - reduction in the volume of state support for the agricultural sector - the agri-food market (like any market) is subject to market fluctuations, both in the domestic and foreign markets; - decrease in soil fertility, depletion of the potential of land, natural, water resources, due to non-observance of scientific recommendations for the rational use of natural resources; - low level of investment attractiveness of the agro-industrial complex for external investors.

4 Conclusions

Analysis of the agri-food market of the Kabardino-Balkarian Republic showed the existing "strengths" and "weaknesses" in the agricultural production system of the republic, as well as revealed the possibilities for growth and development of the agri-food market in the region, revealed the potential for strengthening the competitive position of the agricultural sector in the domestic and foreign markets. However, the implementation of the "opportunities" for the development of the regional agri-food market presented in the SWOT analysis is associated with overcoming "threats" and dangers of various nature, without the resolution of which the effective and progressive development of the agri-food market of the Kabardino-Balkarian Republic is impossible.

Acknowledgments

The study was performed with the financial support of the Russian Foundation for Basic Research, project No. 20-010-00853A.

References

1. N. Yu. Polunina, In the collection: Statistical analysis of the socio-economic development of the constituent entities of the Russian Federation, Materials of the 5th International Scientific and Practical Conference, 249-253 (2018)
2. O. I. Vanyushina, V. N. Minat, Bulletin of Rural Development and Social Policy, **2(18)**, 2-9 (2018)
3. E. V. Salnikova, M. E. Otinova, Economics of Agricultural and Processing Enterprises, **10**, 40-48 (2020).
4. *Official site of the Federal State Statistics Service*, https://stavstat.gks.ru
5. *Regions of Russia, Socio-economic indexes* (2019), https://rosstat.gov.ru
6. A. N. Baydakov, A. V. Nazarenko, O. N. Babkina, Kant, **4(29)**, 213-219 (2018)
7. E. V. Afanasyev, A. A. Bykov, S.M. Golovatyuk, AIC: Economy, management, **5**, 40-46 (2019)
8. N. Kh. Kairova, R. Ye. Shokumova, Z. S. Marzhokhov, REVISTA INCLUSIONES, **7**, 435-451 (2020)

Legal and organizational means of innovative development of the agro-industrial complex in Russia

Alexander Savoskin[1,2,*], Victoria Kalitskaya[1], Maria Vilacheva[1], *Takhir* Toguzaev[3], and *Aslan* Zhemukhov[3]

[1]Ural State University of Economics, 8 March/Narodnaya Volya Str., 62/45, 620144 Ekaterinburg, Russia
[2]South Ural Federal University, Lenin Avenue, 76, 454080 Chelyabinsk, Russia
[3]Kabardino-Balkarian State Agricultural University named after V.M. Kokov, Lenin Avenue, 1V, 360030 Nalchik, Russia

Abstract. The performance of agriculture in Russia remains unsatisfactory, which requires the state to take active measures to support and develop it. In recent years, government policy has been focused on the formation of a high-tech agro-industrial complex, the introduction of digital means and ensuring food security. At the same time, the measures which are taken by the state do not form a single integrated system and do not ensure the achievement of the desired effect. The publication analyzes the current state of legislation in the field of agro-industrial complex support. Particular attention is paid to the Federal Scientific and Technical Program for the Development of Agriculture for 2017 - 2025. Based on the results of the research, conclusions about the need to adopt a comprehensive federal law and strengthen the role of the State Council of the Russian Federation in the development of a national agrarian policy are drawn, measures to improve legislation in the field of supporting the innovative development of agriculture are substantiated, and additional measures of state support for the agro-industrial complex in the Russian Federation are proposed.

1 Introduction

The current state of agriculture in Russia and the economy as a whole does not allow to get around without the active role of the state. New challenges, including the coronavirus pandemic, also actualize the role of the state in supporting and developing the domestic agro-industrial complex, especially in the area of import substitution. At the same time, the analysis of regulations leads to the conclusion that there is no unified concept of support and development of the agro-industrial complex in Russia.

At the same time, the legal regulation of the agro-industrial complex is not only extremely fragmented (which is at least inconvenient), but also does not have sufficient depth, which means that it does not fully take into account the specifics of the agricultural

*Corresponding author: savoskinav@yandex.ru

manufactures and processors (in comparison with other types of production and processing).

The main problem of legal regulation of agricultural support is the lack of comprehensive legislative regulation. So it should be noted that despite the large scale of the 2020 constitutional reform, the country's Basic Law was not supplemented with provisions on agriculture, food security or other similar norms. There is also no specialized federal law that would define the powers of bodies of all levels of public authority in relation to agricultural manufactures and processes of raw materials. Also, the Decree of the President of the Russian Federation No. 474 dated July 21, 2020 "On the national development goals of the Russian Federation for the period up to 2030" does not say anything about agriculture.

However, it would be wrong to say that Russia is not taking measures to support the innovative development of the agro-industrial complex [1,3]. So, in the interests of the development of agriculture, the President of the Russian Federation initiated measures aimed at the implementation of the state scientific and technical policy (we are talking about the Decree of the President of the Russian Federation dated July 21, 2016, No. 350). The New Food Security Doctrine (also approved by the President of the Russian Federation dated January 21, 2020) also focuses on the innovative development of the agro-industrial complex.

The main problem is the fact that today the legal regulation of the development of agriculture is actually entrusted to the Government of the Russian Federation, which, as an executive body, is called upon to implement the agricultural policy in the country, but not to generate it. It is important to take into account that the subjects of the development and implementation of strategic positioning documents have different decision-making mechanisms and different competencies, and their artificial separation is of great managerial importance. As mentioned above, in the Russian Federation, such a division is not traced, and the main generator of the legal field for the development of the agro-industrial complex is the Government of the Russian Federation in conjunction with the Ministry of Agriculture of the Russian Federation which is subordinate to it [6].

This is clearly seen from the plenty and content of acts of the Government of the Russian Federation in the area we are analyzing, and above all, from the materials of the State Program for the Development of Agriculture and Regulation of the Markets of Agricultural Products, Raw Materials and Food (approved by the Decree of the Government of the Russian Federation dated July 14, 2012, No. N717), the Strategy for the development of the agro-industrial and fishery complexes of the Russian Federation for the period up to 2030 (approved by No. by the Government of the Russian Federation dated April 12, 2020), the Strategy for the Development of the Food and Processing Industry of the Russian Federation for the Period up to 2020 (approved by the Government of the Russian Federation dated April 17, 2012), the Strategy for Sustainable Development of Rural Areas of the Russian Federation for the period up to 2030 (approved by the Government of the Russian Federation dated February 2, 2015), the Long-term Strategy for the Development of the Grain Complex of the Russian Federation until 2035 (approved by the Government of the Russian Federation dated August 10, 2019), etc. It is noteworthy that many program documents were adopted in the form of orders of the Government of the Russian Federation, that is, as a general rule, they are not normative legal acts, which directly contradicts their purpose and content as acts of a program nature, which can negatively affect the implementation of innovative technologies in agroecosystems.

At the next level of legal regulation, there are acts directly from the Ministry of Agriculture of the Russian Federation. Among the large number of them, in relation to the development of innovative technologies in agroecosystems, one should highlight: Departmental target program "Ensuring general conditions for the functioning of the branches of the agro-industrial complex" (approved on July 9, 2020) and the Action Plan of the Ministry of Agriculture of the Russian Federation for 2020-2025 (approved on August 24, 2020).

At the present time, the main tool for the innovative development of agriculture and related industries the state has determined state programs and subprograms adopted for their development. Among the ones that have been implemented for a long time, one can, for example, single out the State Program for the Development of the Fisheries Industry (approved by the Government of the Russian Federation dated April 15, 2014), and relatively new is the state program for the Integrated Development of Rural Areas (approved by the Government of the Russian Federation dated May 31, 2019); federal project "Creation of a support system for farmers and the development of rural cooperation" of the national project "Small and medium-sized businesses and support for individual entrepreneurial initiative" (approved by the minutes of the meeting of the project committee dated December 11, 2018); federal project "Export of products of the agro-industrial complex" of the national project "International cooperation and export" (approved by Presidium of the Council under the President of the Russian Federation for Strategic Development and National Projects dated December 24, 2018); Federal Scientific and Technical Program for the Development of Agriculture for 2017 - 2025 (approved by the Government of the Russian Federation dated August 25, 2017).

Analysis shows that documents in the field of development of the agro-industrial complex are adopted not only by state authorities, but also by all kinds of committees, councils or presidiums under them. Which do not belong to the number of power institutions, which raises a big question about the legal force of the documents approved by them. Due to the absence of a single coordinating body, these documents have an extremely fragmented content, are not always interconnected, and therefore are difficult to analyze and evaluate in their totality. That in the future does not allow to objectively judge the real effectiveness of the efforts undertaken.

2 Materials and methods

The publication is based on an analysis of the regulations of the President of the Russian Federation, the Government of the Russian Federation, the Ministry of Agriculture of the Russian Federation and official statistical data (including reports of the authorities and reports of scientific organizations prepared by order of the authorities), presented on the websites of the Ministry of Education and Science of the Russian Federation, the Ministry of Agriculture of the Russian Federation, the official website of the Federal Scientific and Technical Program for the Development of Agriculture for 2017-2025, as well as on materials of open data of federal executive bodies posted in accordance with the Federal Law "On providing access to information on the activities of state bodies and local self-government bodies."

Within this research, an optimal model of legal regulation of the innovative development of the agro-industrial complex in Russia is being built on the basis of the dialectical unity of its legal, general scientific (analysis, synthesis, deduction and induction, abstraction, structural-functional method) and special methods (formal-legal, method of legal construction, formal-logical, systemic, technical and legal analysis, statistical method). After that, a variant of the optimal legal regulation of the development of the agro-industrial complex of Russia was proposed.

 https://doi.org/10.1051/e3sconf/202126203014

3 Results and Discussion

Perhaps the most effective and, as practice shows, a popular tool for implementing an innovative state agricultural policy is the Federal Scientific and Technical Program for the Development of Agriculture for 2017-2025 (hereinafter referred to as the Federal special-purpose research and development programme) and the subprograms adopted in its development "Development of Potato Breeding and Seed Production in the Russian Federation "(Approved by Resolution of the Government of the Russian Federation No. 559 dated May 05, 2018); "Development of selection and seed production of sugar beet in the Russian Federation" (approved by the Decree of the Government of the Russian Federation No. 1615 dated December 21, 2018); "Creation of a domestic competitive cross of meat chickens in order to obtain broilers" (approved by Resolution of the Government of the Russian Federation No. 782 dated May 28, 2020).

At the beginning of 2021, the following industry subprograms of the Federal special-purpose research and development program are at various stages of preparation: "Development of production of feed and feed additives for animals", "Improving the genetic potential of cattle of specialized beef breeds", "Improving the genetic potential of dairy cattle", "Improving the genetic potential of small ruminants", "Breeding and seed production of oilseeds" , "Selection and seed production of vegetable crops", "Development of viticulture, including nursery growing", "Development of selection and processing of grain crops", "Development of nursery and horticulture", "Development of selection and seed production of corn", "Development of selection and seed production of industrial crops", "Development of aquaculture", "Agricultural machinery and equipment", "Development of technologies for the production of pesticides and agrochemicals of biological origin for use in agriculture", "Development of technologies for the production of medicines for veterinary use." These programs have not yet been approved by the Government of the Russian Federation, but in developing them, the key task was to create and introduce competitive domestic technologies in agriculture. The work carried out in the country to introduce new technologies in the domestic agro-industrial complex and to ensure the country's food security directly depends on the quality of the development of these subprograms and the success of their implementation [7].

Taking into account that some of the subprograms of the Federal special-purpose research and development program are already being implemented and official statistical information has appeared on them, we will analyze them for problems.

The subprogram "Development of potato breeding and seed production in the Russian Federation" is aimed at ensuring stable growth in production volumes, as well as the implementation of high-quality seed potatoes of modern competitive domestic varieties based on the use of new high-tech Russian developments and complex scientific and technical projects of a full innovation cycle. To date, 31 companies have taken part in the program and 12 new domestic competitive potato varieties have been created ("Northern Light", "Elena", "Gulliver", "Samba", "Prime", "Carmen", "Indigo", "Triumph", "Bear", "Kumach", "Varyag", "Explosive").

At the same time, the analysis of the implementation of these subprograms revealed the following problems. First, project customers often do not have the opportunity to fully utilize the allocated state support funds. This is due to both the late delivery of budget financing and organizational difficulties in the implementation of measures, in particular, the transfer of the deadlines for the execution of contracts for some obligations. Second, when using the current editions of the subprogram passports, the customers cannot confidently continue the implementation of projects, and the monitoring center and the Program Directorate face difficulties in assessing the results of the subprogram implementation (the solution to this problem is seen in the updating and approval of the previously approved project passports by the Program Council). Third, the participants in the subprogram do not meet the targets. Fourth, the lack of specific deadlines for communicating budget funds to customers and project participants (scientific and educational institutions). Fifth, there is a high probability of not reaching the planned values of some indicators, in particular, in terms of the number of developed plant protection products. Sixth, the lack of demand or low demand for new domestic potato varieties, difficulties in selling the obtained seed material. Seventh, the weather factor is a risk for production and can make adjustments to the implementation of projects and subprograms. Eighth, restriction of transportation and delivery of goods due to the prohibition of the movement of vehicles. Ninth, an increase in the timing of the manufacture of goods and the implementation of activities of the subprogram from suppliers due to the impossibility of timely fulfilment of obligations under agreements, contracts, conventions. Tenth, a change (fluctuation) in the exchange rate, which entails an increase in the cost of some work (for example, the purchase of material and technical values).

Elimination of existing and timely prevention of emerging risks will reduce the degree of negative influence of these factors. In order to update and bring into compliance with the schedule of the work performed, the timing of achieving the planned indexes and indicators, the 21st customer of Complex scientific and technical program determined the need to amend the previously approved project passports. The main issue requiring a prompt decision was the approval by the Presidium of the Council of the Program of new editions of passports. For this purpose, the monitoring center analyzed the submitted new editions of passports and prepared the corresponding conclusions.

The expert group assessed the impact of the changes introduced on the achievement of the results of each project and prepared recommendations for the Program Council: the continuation of the implementation of the Complex scientific and technical program in the new edition of the passports for 18 customers was agreed (LLC "AgroStar", LLC "Zolotaya Niva", LLC "Kolpakov", JSC "Pogarskaya Potato Factory", LLC "Agrofirma Slava Potato Yalchiki ", LLC APK "Lyubovskoye", LLC "FH SeDeK", SEC "Nikolsk", LLC " Zolskiy potato", LLC "SoyuzAgro", LLC" SSK "UralskIy potato", LLC "FAT-AGRO", JSC "Ozyory", LLC "Agricultural enterprise" Dary Malinovki ", LLC TPK "Elita-potato", LLC "AgroInter", CJSC Agricultural enterprise "Michurinets", LLC "Redkinskaya APK"); Refusal to approve the implementation of the Complex scientific and technical program in the new edition of passports for 3 customers (PJSC "Belorechenskoye", LLC "AgroSoyuz Spassk", LLC "Agrofirma" KRiMM ").

Separately, it should be noted that the issue of agreeing on new editions of passports remains very relevant, since when using outdated current editions of passports, customers cannot confidently continue the implementation of projects, and the monitoring center and the Directorate of the Program cannot monitor and control the implementation of the subprogram. Thus, the question of the algorithm and procedure for promptly making changes to the passports of the Complex scientific and technical program remains open.

The subprogram "Development of selection and seed production of sugar beet in the Russian Federation" is implemented with the aim of creating competitive hybrids of domestic selection of sugar beets based on the use of new high-tech Russian developments and the implementation of complex scientific and technical projects of a full innovation cycle, development of a system of sugar beet seed production, ensuring stable growth in volumes of industrial production and sale of high-quality competitive seeds of cost-effective hybrids of sugar beet of domestic selection, the creation of modern means of diagnostics of diseases and quality control of seeds of sugar beet hybrids.

At the same time, since the start of the programme, only two enterprises have taken part in it: Federal state budgetary scientific institution "Pervomayskaya Sugar Beet Selection and Experimental Station" and Federal state budgetary scientific institution "All-Russian

Scientific Research Institute of Sugar Beet and Sugar" named after A.L. Mazlumov", which have developed 3 new hybrids of domestic selection of sugar beet, namely: "Ruby", "Carat" and a PMC 129 diploid hybrid. In connection with the above, the Ministry of Agriculture of the Russian Federation announced in December 2020 an additional competition for the selection of complex scientific and technical projects for the provision of grants in the form of subsidies from the federal budget for the implementation of complex scientific and technical projects in the agro-industrial complex [9].

If we abstract from the nuances of the implementation of specific subprograms, then in the process of implementing the Federal scientific and technical program itself, the following systemic issues and problems arose.

First, there were questions about calculating the budget and, first of all, determining the amount of extra-budgetary funds raised for the implementation of the project. If to follow the regulatory framework, then the amount of such extra-budgetary funds for the implementation of the project should be equal to or exceed the amount of the following federal budget funds allocated for state support of the project: subsidies for the implementation of a state task in terms of project work performed by a scientific or educational organization subordinate to the executors' subprograms (with the exception of federal budget expenditures for the development of educational programs); grants in the form of subsidies from the federal budget for the implementation of the project. It is important to emphasize here that when determining the volume of actually attracted extra-budgetary funds for the implementation of the project, the costs of creating capital construction facilities, the acquisition of specialized agricultural machinery and equipment within the framework of the project are not taken into account in the part compensated from the federal budget. In addition, the funds allocated by the budgets of the constituent entities of the Russian Federation for state support of projects are not taken into account when determining the minimum amount of extra-budgetary funds attracted for the implementation of the project.

Secondly, there were problems with the transfer of the unused part of the grants to the next reporting periods. The difficulty here lies in the fact that only the Ministry of Agriculture of the Russian Federation directly has the right to make a decision on changing the terms of the concluded Agreement, including on the basis of information and proposals sent by the recipient in case of establishing the need to change the amount of the subsidy specified in the Agreement, including reducing its amount, as well as an increase in the presence of unused limits of budgetary obligations and the condition that the recipient provides information containing a financial and economic justification for this change.

Third, only specialized organizations are allowed to participate in the programs, which significantly reduces the number of potential grant participants and weakens the competition.

Fourthly, we have to state the fact that the deadlines for the adoption of new subprograms are being delayed (by the way, more than twenty of them were originally announced). For example, in light of the sharp rise in the cost of sunflower oil, already at the end of 2020, it was planned to adopt a subprogram "Development of Breeding and Seed Production of Oilseeds in the Russian Federation", which would ensure the breeding of competitive domestic varieties and hybrids of sunflower, soybeans and domestic rapeseed. The project involves the implementation of comprehensive domestic scientific and technical innovative developments of the full cycle aimed at the development of the system of seed production of oilseeds, should ensure a stable growth in industrial production and the sale of high-quality seeds of competitive varieties and hybrids of sunflower, soybeans and rapeseed of domestic selection, the creation of modern diagnostic tools quality control of seeds of varieties and hybrids of sunflower, soybeans and rapeseed. At the same time, at the time of going the publication to print (March 1, 2021), the state spends serious material and organizational resources to maintain acceptable prices for sunflower oil for the population, but the subprogram, despite its extreme demand, has not been approved by the Government of the Russian Federation

4 Conclusions

1. In Russia, there is no concept of support and development of agriculture that is legally formalized. Today it conditionally replaces by extremely fragmented and not systematized in terms of content and time of adoption a group of program acts of the head of state, numerous acts of the Government of the Russian Federation and the Ministry of Agriculture of the Russian Federation. Formally, these bodies have the right to adopt such acts, but methodologically, this approach seems to be incorrect. In this regard, it is proposed that the foundations of support and development of agriculture be defined in a separate federal law with the discussion of the draft of such a law in the newly created supreme body of power - the State Council of the Russian Federation, which is precisely intended to determine the main directions of domestic policy of priority areas of economic development (Art. 5 of the Federal Law dated December 8, 2020, No. 394-FZ "On the State Council of the Russian Federation").

2. The existing measures to support the innovative development of the agro-industrial complex have provided a certain positive effect, expressed in the creation of new varieties of seeds, vegetables and poultry. However, in general, the measures taken remain insufficiently effective, primarily due to the imperfection of their legal regulation. Access to support measures is complicated by overly formalized and overloaded legislation, the presence of administrative barriers, and the inflexibility of law enforcement agencies. The solution to this problem is seen in the simplification of legislation on measures to support agricultural manufacturers and processors, the transition to the "one window" principle when registering applications for participation in government programs, as well as allowing the use of unspent funds in the reporting period in the current financial year.

3. It is necessary to introduce or improve additional measures to stimulate the innovative development of the agro-industrial complex. These include concessional lending for agricultural manufacturers, organizations and individual entrepreneurs engaged in innovative production, processing and (or) sale of agricultural products [5]. At the same time, it is important to provide that a preferential short-term or investment loan should be at a rate of no more than 0-2% and it could be obtained from any of the banks authorized by the Ministry of Agriculture of Russia. In this case, credit institutions should be reimbursed for lost income directly from the federal budget.

4. At the present time, measures for the innovative development of the agro-industrial complex are weakly linked to the strategic goals of the development of Russia, reflected in the Decree of the President of the Russian Federation No. 474 dated July 21, 2020 "On the national development goals of the Russian Federation for the period up to 2030". This gap should be filled primarily by including in the development of new or improving existing federal scientific and technical programs, taking into account the achievements of digitalization and information technology [4, 8, 10].

References

1. E. A. Egorov, Zh. A. Shadrina, G. A. Kochian, P. F. Paramonov, Bulletin of Russian Agricultural Science, **5**, 4 (2020)
2. A. L. Ivanov, I. S. Kozubenko, I. Yu. Savin, V. I. Kiryushin, Bulletin of Russian Agricultural Science, **5**, 4 (2018)

3. I. S. Kozubenko, I. Yu. Savin, Bulletin of Russian Agricultural Science, **5**, 9 (2017)
4. A. V. Kolesnikov, O. Domozhirova, Agroindustrial Complex: Economics, Management, **1**, 27 (2020)
5. V. V. Maslova, G. M. Savkina, Agroindustrial Complex: Economics, Management, **2**, 29 (2020)
6. V. A. Panfilov, Bulletin of Russian Agricultural Science, **1**, 4 (2020)
7. V. A. Panfilov, Bulletin of Russian Agricultural Science, **1**, 13 (2019)
8. V. I. Trukhachev, Bulletin of Russian Agricultural Science, **2**, 15 (2019)
9. I. G. Ushachev, A. F. Serkov, V. S. Chekalin, M. V. Kharina, agro-industrial complex: economics, management, **1**, 3 (2021)
10. V. P. Yakushev, Bulletin of Russian Agricultural Science, **2**, 11 (2019)

E3S Web of Conferences **262**, 03015 (2021) https://doi.org/10.1051/e3sconf/202126203015
ITEEA 2021

The mechanism of legal regulation of innovative technologies in the agro-industrial complex of the Eurasian Economic Union: realities and prospects

Veronika Meshcheryagina[1,*], *Maria* Vilacheva[1], *Alexander* Kurdyumov[1], *Madina* Kyarova[2], and *Svetlana* Temmoeva[2]

[1]Ural State University of Economics, Yekaterinburg, Russia
[2]Kabardino-Balkarian State Agricultural University named after V.M. Kokova, Lenin Avenue, 1V, 360030 Nalchik, Russia

Abstract. The article contains a research of the elements of the legal mechanism for regulating public relations in the field of digitalization of the agro-industrial complex of the EAEU, highlights the legal means of integrating the norms of international and domestic law (using the example of the Russian Federation) in the mechanism of regulation of the agro-industrial complex of the EAEU. It is concluded that in the process of implementing the EAEU norms into the domestic legislation of Russia, there is a combination of two independent legal systems, respectively. The elements of the mechanism for the implementation of EAEU norms include transformation (inclusion of international norms in national legislation in amended form) and incorporation (exact textual reproduction of international norms in national legislation), which must be taken into account when regulating the digitalization of the agro-industrial complex. After analyzing the current regulatory framework for regulating innovative technologies in the agro-industrial complex of the EAEU, the authors have to admit its compressed volume and low elaboration from the standpoint of legal technology, as well as propose priority areas for legal regulation of innovative technologies in the agro-industrial complex of the EAEU.

1 Introduction

The processes of digitalization of the agro-industrial complex of the Eurasian Economic Union play a leading role in building modern economic integration of its member countries, in accordance with the Agricultural Development Strategy approved by the Eurasian Economic Commission, where it is indicated, that innovative approaches should become the basis for the further development of food security in the region, a more complete import substitution, the production of high-tech products, an increase in export volumes through

* Corresponding author: metsheryagina@yandex.ru

the use of IT technologies, information databases, smart contacts of states and business entities, etc. [1].

The objectives of the agreed agro-industrial policy set out in the Treaty on the EAEU presuppose the use of mechanisms of interstate interaction, among them are: forecasting in the agro-industrial complex; regulation of the common agricultural market; development of uniform requirements in the field of production and circulation of products; integrated information support of the agro-industrial complex.

Of course, these processes cannot take place without the development of international legal norms - the main regulator of public relations, which should subsequently be introduced into the domestic legislation of the EAEU member states. In legal science, such "introduction" occurs through the implementation procedure - a complex legal phenomenon, which is a set of legal forms and methods of adapting an international norm to domestic legislation.

In this context, it should be noted that the introduction of innovative technologies is of an advanced nature and is not always subject to proper legal regulation, which complicates the implementation processes. The Treaty on the EAEU, which entered into force for the Russian Federation on January 1, 2015, specifically stipulates that the task of harmonizing the legislation of the member states is to establish similar (comparable) rules of legal regulation of the common financial market, provided that the remaining differences will not hinder effective functioning of the common financial market.

The development of bilateral and multilateral economic relations of the Russian Federation, mutually beneficial and equal partnership with foreign states based on respect for the principles of independence and sovereignty, pragmatism, and upholding national priorities are the priority tasks of the foreign policy of the Russian Federation [2]. In this regard, it is very important to pay attention to the theory and practice of the international law impact on the spheres of domestic governance, including the agro-industrial sector of the economy.

This fact leads to the need to conduct research of the existing legal mechanism for regulating public relations in the field of digitalization of the EAEU agro-industrial complex. At the same time, the mechanism of legal regulation will be understood as a set of special legal means, with the help of which the regulation of social relations is performed. Legal means act as tools used in the process of legal regulation [3], they include: norms and principles of law, law enforcement acts, contracts, legal facts, subjective rights, legal obligations, prohibitions, benefits, incentives, penalties, acts of realization of rights and obligations [4].

The goal of the research is to identify the elements of the legal mechanism for regulating public relations in the field of digitalization of the EAEU agro-industrial complex, legal means and ways of interaction between elements.

To achieve this goal, it is necessary to implement the following range of tasks:

– to identify the features of the mechanism of legal regulation of public relations in the field of the agro-industrial complex, inherent to a regional international organization;

– to determine the means of integrating the norms of international and domestic law (using the example of the Russian Federation) in the mechanism of regulation of the agro-industrial complex of the EAEU;

– identify priority areas for improving the mechanism for regulating public relations in the field of digitalization of the EAEU agro-industrial complex.

2 Materials and Methods

To achieve this goal, it is necessary to use both general scientific research methods: analysis, synthesis, deduction and induction, abstraction, structural and functional method, and special methods of jurisprudence and other branches of knowledge: formal legal method, methods of legal design and technical and legal analysis, formal-logical, systemic, statistical methods, etc.

The above mentioned methodological framework will allow us to identify the mechanism of legal regulation of the EAEU agro-industrial complex from the standpoint of the norms of international and domestic law of Russia. Based on the formal legal method, the international norms contained in the acts of the EAEU bodies will be analyzed (international treaties concluded within the EAEU; decisions of the EAEU Interstate Council; decisions of the EAEU Parliamentary Assembly; decisions of the EAEU Commission), judicial practice of the EAEU court and courts of the Russian Federation constituting empirical research base.

3 Results and Discussion

3.1 Features of the mechanism of legal regulation of public relations in the field of the agro-industrial complex, inherent in a regional international organization

Determining the features of the mechanism of legal regulation of any social relations, we inevitably come across the issue of building a system of legislation containing all the necessary legal instruments. However, the complexity of building an appropriate system of legislation within this research is due to the fact that the EAEU does not belong to "classical" international organizations, since it has formed public authorities that are authorized to extend power to private entities; establishes a hierarchy of legal norms within the organization, regulates the procedure for bringing the norms of domestic law to the norms formed by an international organization.

This feature of the EAEU legal system was stated by the EAEU Court in its Decree dated April 08, 2016 in case No. CE-1-2/3-16-KC [5], noting the presence in Art. 6 of the Treaty on the EAEU, a list of sources of EAEU law, which, nevertheless, is not exhaustive (closed), since the Vienna Convention on the Law of International Treaties does not allow limiting the range of sources of international law.

Thus, the hierarchy of the EAEU legal norms is based on the priority of the norms of the Treaty on its creation (Part 1). 3 Art. 6 of the Treaty on the EAEU), lower in legal force will be introduced international treaties within the Union; international agreements of the Union with a third party; decisions and orders of the Supreme Eurasian Economic Council, the Eurasian Intergovernmental Council and the Eurasian Economic Commission, adopted within the framework of their powers.

At the same time, a direct reference is made to the need of developing norms of intra-national legislation on issues related to the Decisions of the Supreme Eurasian Economic Council and the Eurasian Intergovernmental Council.

The system of legislation of the Russian Federation in terms of regulation of the agro-industrial complex is divided into two large levels: legislative (Constitution of the Russian Federation; Laws on amendments to the constitution; Federal constitutional laws; Federal laws; Laws of the Russian Federation) and the level of by-laws (Decrees of the President of the Russian Federation; Resolutions of the Government of the Russian Federation; Orders Ministries and departments of the Russian Federation; acts of local governments, etc.).

Constitution of the Russian Federation (p. 4 Art. 15) establishes the limits of the influence of the EAEU norms on domestic legislation, based on the legal force of the acts by the formulation: "The generally accepted principles and norms of international law and international treaties of the Russian Federation are an integral part of its legal system. If an international treaty of the Russian Federation establishes rules other than those stipulated by law, in this case, the rules of the international treaty shall apply" For a long time, this formulation caused controversial discussions about the priority of international law over the Constitution of the Russian Federation and vice versa. Although, in our opinion, it is methodologically incorrect to pose the question from this angle, since from the standpoint of the reception of international norms, constitutional norms consolidate the mechanism of adaptation and application of an international norm, without putting an ultimatum on the supremacy of international treaties or domestic acts. It seems that it is about this property of implementation that Professor Zh.I. Ovsepyan wrote the following: "Priority of application" rather than "higher legal force" is more acceptable from the standpoint of the requirements of the principle of state sovereignty and, at the same time, respect for the requirements of the Vienna Convention on the Law of Treaties"[6].

An important role in the application of international law on the territory of the Russian Federation is played by the Constitutional Court of the Russian Federation (hereinafter referred to as the Constitutional Court of the Russian Federation), which, in accordance with paragraph 4 of Art. 3 of Federal constitutional law "Concerning the Constitutional Court of the Russian Federation" (Federal constitutional law No. 1, 1994) has the exclusive authority to interpret the norms of the Constitution of the Russian Federation.

Controversial discussions between the Constitutional Court of the Russian Federation and the EAEU court have not yet been observed, however, the preconditions for the emergence of a "conflict" in the interpretation of the EAEU norms and the subsequent execution of the EAEU court decision on the territory of the Russian Federation have already been in history. For example, in the case of exemption from payment of customs duties when importing certain categories of goods into the unified customs territory of the Customs Union, approved by the decision of the Customs Union Commission No. 728 dated July 15, 2011, when the Russian LLC "SeverAvtoProkat" applied to the Eurasian Economic Commission to challenge clause 4 of the Application Procedure, after which, to resolve the dispute, it was decided to send the case to the EurAsEC Court (the successor of which is the EAEU Court).

Disputing the decision of the Customs Union Commission, LLC "SeverAvtoProkat" referred to the peculiarities of the retroactive effect of the law, under which the conditions for the provision and use of tariff preferences in the territory of the Customs Union, determined from July 6, 2010, should no longer fall within the competence of the Customs Union Commission. From that date, such an issue, according to the applicant, could only be resolved by concluding a separate international treaty or by direct settlement in the relevant norm of the Customs Union Code.

In its decision, the EurAsEC Court indicated that due to the specifics of the sources of international law, both contractual legal norms and acts of international supranational organizations have the same legal force. The Constitutional Court of the Russian Federation did not agree with the legal positions of the Board of the EurAsEC Court on the need for identical approaches in the regulation of benefits, giving priority to the norms of the Constitution of the Russian Federation, which, guaranteeing the rights and liberties of an individual and a citizen, presupposes the existence of benefits for business and the inadmissibility of the deterioration of economic entities position in Russia under the pressure of international rights.

Proceeding to other laws regulating the agro-industrial complex, it should be noted the intersectoral nature of the object of legal regulation, inherent in civil law, constitutional law, however, the largest array of normative acts of the agro-industrial complex is concentrated in the field of public administration, that is, inherent in the branch of administrative law and, to a greater extent, by subordinate regulation [7].

Thus, the following features of the mechanism of legal regulation of public relations in the field of agro-industrial complex, inherent in the regional international organization of the EAEU, can be distinguished:

1) the presence of national and supranational legal systems with a specific hierarchy of legal norms;

2) the presence of national and supranational law-making and judicial bodies, the order of interaction of which is not defined at the proper level (for example, in the Russian Federation, the reference point in these issues is the Constitution of the Russian Federation and the Federal constitutional law "On the Constitutional Court of the Russian Federation");

3) a complicated procedure for integrating the EAEU legal norms into domestic legislation, the means of which - transformation and incorporation - will be described in more detail below;

4) when the terms "harmonization of legislation" and "implementation" are related, methodologically, they should be considered as a goal and a process, where the first stage of the implementation mechanism is regulatory activity of the state is aimed at concrete definition the initial norms in international agreements; the second stage - defining the concept of implementation in domestic legislation (general or special reception) in constitutional norms or norms of legislation; the third stage is the activity of the state to concrete definition the provisions of international norms into domestic legislation (transformation and incorporation). The final goal of the outlined legal mechanism is to increase the level of harmonization of the legislation of the EAEU member states.

3.2 Means of integrating the norms of international and domestic law (on the example of the Russian Federation) in the mechanism of regulation of the agro-industrial complex of the EAEU

In the process of implementing the EAEU norms into the domestic legislation of Russia, we are actually faced with the combination of two legal systems, respectively, the elements of the mechanism for the implementation of EAEU norms can be attributed to transformation - the inclusion of international norms in national legislation in amended form; incorporation - accurate textual reproduction of international norms in national legislation.

At the same time, the transformation procedure cannot be arbitrary, in view of the recognition of the norms of international law as an integral part of the Russian legal system. The subject of the transformation of international norms is set out in Art. 15 of the Federal Law "On International Treaties of the Russian Federation". To transform the EAEU norms in the field of agro-industrial complex into Russian legislation, the form of a subordinate normative legal act is most often used - Government Resolutions and orders of various ministries and departments. For example, the Decree of the Government of the Russian Federation "On the distribution of tariff quotas for meat of cattle and poultry in 2021" (No. 2142, 2020) established the distribution of tariff quotas for 2021 in accordance with the decision of the Board of the Eurasian Economic Commission No. 102 dated August 18, 2020 in relation to meat of cattle, fresh or chilled.

Another example: the annual establishment of quotas for certain categories of goods imported into the territory of the EAEU: By the decision of the Board of the Eurasian Economic Commission No. 155 dated September 25, 2018; By the decision of the Board of the Eurasian Economic Commission No. 141 dated August 28, 2018, quotas were established for long-grain rice of Vietnam origin, placed under the release custom procedure for domestic consumption; cattle meat, etc. These norms are disclosed in the Decree of the Government of the Russian Federation No. 1495 dated December 07, 2018; Decree of the Government of the Russian Federation No. 1498 dated December 08, 2018;

Decree of the Government of the Russian Federation 1524 dated December 14, 2018; Order of the Ministry of Economic Development of Russia No. 11 dated January 18, 2019; Order of the Ministry of Economic Development of Russia No. 9 dated January 18, 2019; Order of the Ministry of Economic Development of Russia No. 10 dated January 18, 2019).

In general, from the analysis of the legislation, it should be concluded that the form of transformation of the EAEU norms is not applied as regularly as the incorporation, about which we`ll talk later.

Incorporation is applied when international norms give rise to specific legal relationships, establish general rules within the framework of which member states adapt obligations to their systems.

So, the most striking example of the incorporation of EAEU norms is the development of a standard EAEU Customs Code, formalised in the Appendix to the Treaty on the Customs Code of the Eurasian Economic Union (Signed in Moscow April 11,2017), which became a general law regulating relations in the field of movement of goods, works and services in the EAEU member states.

There are common situations in which a direct reference to the EAEU legal norms is applied in the legal documents of the customs authorities. So, in the Decision of the Council of the Eurasian Economic Commission No. 57 dated July 10, 2020 "On Amendments to the Rules for Determining the Origin of Goods Imported to the Customs Territory of the Eurasian Economic Union (Non-Preferential Rules for Determining the Origin of Goods)", the issues of VAT rates for the export of goods are regulated in detail. Namely - 0%, if supporting documents are submitted to the tax authority (clause 1 of article 72 of the EAEU Treaty, clause 3 of the Protocol on the procedure for levying indirect taxes and the mechanism of control over their payment when exporting and importing goods, performing work, rendering services (Appendix No. 18 to the Treaty on the EAEU)), which are automatically applied by the customs authorities of the Russian Federation.

Another example: the Order of the Government of the Russian Federation N 151-r dated February 02, 2015 "On Approval of the Strategy for sustainable development of rural areas of the Russian Federation for the period up to 2030" already makes a direct reference to the EAEU norms in the field of food clusters.

At the same time, it is necessary to stimulate the development of the constituent entities of the Russian Federation with favorable natural and economic conditions for conducting intensive and competitive agro-industrial production, using interregional and regional innovation and investment projects for the production of certain types of agricultural products, raw materials and food.

To do this, it is necessary to develop an all-Russian scheme for the location of agro-industrial production, on the basis of which it is advisable to define and form specialized production zones for certain types of agricultural products, introduce a management system and implement a rational intra-regional distribution of agricultural production, taking into account the possible development of food and processing industries, the creation of interstate food clusters within the framework of the Eurasian Economic Union.

It is important to note that the incorporation of EAEU norms into Russian legislation involves not only rule-making bodies, but also the judiciary. By way of illustration, one must have to indicate the Determination of the Judicial Collegium for Economic Disputes of the Supreme Court of the Russian Federation (No. 306-ES20-387 dated May 21, 2020,) in case No. A06-2706/2019 on the import of goods of required quality into the Russian Federation.

3.3 Priority areas for improving the mechanism for regulating public relations in the field of digitalization of the EAEU agro-industrial complex

One of the main acts in the field of digitalization of the EAEU agro-industrial complex is the Decision of the Eurasian Intergovernmental Council No. 4 dated October 25, 2017, which approved the procedure for developing initiatives within the EAEU digital agenda, for submitting them to expert platforms. Nevertheless, what is the procedure for the operation of the sites, the format and other issues are not considered in this legal act. In fact, practice is appearing in such a way that expert platforms have turned into scientific events and symposia, and respectively, it is very difficult to analyze the proposed initiatives.

It seems that at the level of this regulatory legal act, it is worth designating expert platforms as separate sections of the Internet portal of general information resources and open data of the EAEU (https://opendata.eaeunion.org/ru-ru/Pages/thematicsection.aspx).

The EAEU Portal of General Information Resources and Open Data (hereinafter referred to as the EAEU PGIROD) today is the only integrated digital resource that allows you to receive information about electronic registers of agricultural products, structures of electronic documents and information, regulatory and reference information, information support on the implementation of interstate electronic services, etc. Given the great importance of this portal, there is no separate EAEU normative act, which would regulate in detail its status, sections, etc. There is only a small article 4.2.5.6. with the name "Requirements for the technological portal of an integrated system" in the Decision of the Board of the Eurasian Economic Commission No. 55 dated April 02, 2019 "Terms of reference for the development of an integrated information system of the Eurasian Economic Union" and the conceptual base [8], which, in addition to general requirements, does not disclose the specifics of the sections portal, and the concepts are characteristic only for the customs segment, beyond which the Portal has already gone.

Nevertheless, a positive practice would be the introduction of personalized user registration with the formation of a personal account, after which business entities could post their own initiatives in a separate section, publicly discuss and vote on the need for adoption.

Specifically, the digitalization of the agro-industrial complex is mentioned in the Decision of the Council of the Eurasian Economic Commission No. 2 dated February 04, 2015 "List of indicative indexes of the development of the agro-industrial complex of the member states of the Eurasian Economic Union", which indicates the list of domestic information resources of the agro-industrial complex (a unified register of varieties of agricultural plants, general information resource of data forecast indicators of the agro-industrial complex of the Member States; a resource of data on the prices of agricultural products and food products produced in the Member States, etc.), which will be posted in an automated form on the EAEU PGIROD after that.

In addition, an information and analytical subsystem of the integrated system is being installed, which should ensure the collection, processing and storage of the following information: on the economic and production indicators of agriculture; on the volume of trade in food products and agricultural raw materials; on average selling prices, average consumer prices and average contract prices for the main types of agricultural products, food and processing industry products and food products, etc.

However, in practice, this is not implemented, since within the framework of the EAEU PGIROD Portal in the "Agro-Industrial Complex" section there are only answers to frequently asked questions in the field of the agro-industrial complex; information on export support for participants in the foreign economic activity of the member states in the field of agro-industrial complex; Information on international trade rules; information on points of entry/exit based on the general list of places of the arrival of goods in the customs territory of the EAEU and places of departure of goods from the customs territory of the EAEU; information on the websites of trade missions of the EAEU member states in third countries.

Obviously, this situation cannot be recognized as admissible and a detailed legal study of the provisions specified in the Decision of the EEC Council No. 2 at the level of the legislation of the Russian Federation is required.

It also seems necessary to integrate into the EAEU PGIROD a link to the Information and analytical system for operational monitoring and assessment of the state and risks of scientific and technical support for the development of agriculture, which was developed on the basis of the Decree of the Government of the Russian Federation No. 996 dated August 25, 2017 "On the approval of the Federal Scientific and Technical Program for the Development of Agriculture for 2017-2025" and provides a wide range of opportunities to analyze the information infrastructure of the expert community in the field of assessing the state and risks of scientific and technological development of agriculture, assessing the intermediate and final results of its implementation.

With regard to the development of new software, today a legal base has been formed, which indicates mathematical formulas and calculation algorithms, which should be converted into electronic form for the convenience of calculations by the subjects of the agro-industrial complex. It would be possible to place these software products in the public domain on the EAEU PGIROD. For example, the relevant calculation methods are in the Recommendation of the Board of the Eurasian Economic Commission No. 11 dated April 24, 2017 "On the Methodology for assessing the effectiveness of measures of state regulation of the agri-food market and support of the agro-industrial complex" when assessing the effectiveness of measures of state regulation of the agri-food market and state support for agriculture. The methodology makes it possible to determine which criterion (indicator, goal) led to a change in the assessment of the effectiveness of government regulation measures, which, in turn, makes it possible to develop recommendations to improve the effectiveness of government regulation measures.

Similar mathematical calculation algorithms are also available in the domestic legislation of the Russian Federation [9], and can be quite successfully implemented in the form of a software product on the Internet portals of the information retrieval system "Export opportunities of Russia" or "The Single portal of the foreign economic information of the Ministry of Economic Development of Russia. The more so because these portals have a direct reference in the regulations [10].

Another promising area of digitalization of the agro-industrial complex, which has not yet found legal consolidation in international treaties and domestic acts, is the integration of information resources between state Internet portals and private business entities introducing new technologies in the production process.

Among them, information from large enterprises that have branches in different constituent entities of Russia, which in the agricultural sector create the maximum added value in agriculture, would be especially valuable [11].

Perhaps the most susceptible to digitalization processes and supported by a regulatory framework are the processes of moving agricultural products across customs borders. The basic Customs Code of the EAEU [12] includes electronic optimization of customs procedures for import and export of agricultural products: remote release of goods, mandatory preliminary notification, remote payment of customs payments, electronic declaration, "single window" procedure, etc.

After analyzing the regulatory framework for regulating innovative technologies in the agro-industrial complex of the EAEU, we have to state its compressed volume and low elaboration from the standpoint of legal writing.

4 Conclusions

Based on the study, it can be concluded that the legal mechanism for regulating public relations in the field of digitalization of the agro-industrial complex of the EAEU is at the junction of international and domestic law and expresses the implementation of international norms through transformation and incorporation.

The EAEU Treaty provides the formation of a single economic space within the territories of the participating countries, similar mechanisms for regulating the economy on the basis of harmonized or unified legal norms, the creation of supranational judicial and law-making bodies. This feature predetermined a special procedure for the implementation of the norms of international treaties of the EAEU, concluded both within the Union and with third countries, into the domestic legislation of the Russian Federation and their impact on the regulation of public relations in the economic sphere, through transformation (not used so regularly) and incorporation (applies in the event than international norms give rise to specific legal relations, establish general rules within the framework of which participating states adapt obligations to their systems).

Having analyzed the regulatory basis for regulating innovative technologies in the agro-industrial complex of the EAEU, we have to state its compressed volume and low elaboration from the standpoint of legal writing. In fact, the only area of digitalization of the agro-industrial complex that has due legal regulation relates to customs legislation.

The currently existing EAEU PGIROD Internet portal can become constructive information and analytical system of the EAEU in the event of additions to the acts of public supranational bodies of the EAEU. Unfortunately, now this portal does not reflect the necessary information about the agro-industrial complex of the participating countries, provided for in the Decision of the Council of the Eurasian Economic Commission No. 2 dated February 04, 2015. It also seems necessary to integrate a link to the Information and analytical system of operational monitoring into the EAEU PGIROD and assessing the state and risks of scientific and technical support of agriculture development.

A good practice would be to introduce personalized user registration with the formation of a personal account, after passing which business entities could post their own initiatives in a separate section, publicly discuss and vote on the necessity for adoption.

References

1. The food security doctrine of the Russian Federation, approved by the Decree of the President of the Russian Federation N 120 dated January 30, 2010 "On the approval of the Doctrine of food security of the Russian Federation"; The strategy for the development of the information society in the Russian Federation for 2017-2030, approved by the Decree of the President of the Russian Federation N 203 dated May 09, 2017 "On the Strategy for the Development of the Information Society in the Russian Federation for 2017-2030", etc.
2. Decree of the Government of the Russian Federation No. 717 dated July 14, 2012 "On the State Program for the Development of Agriculture and Regulation of the Markets of Agricultural Products, Raw Materials and Food"//SPS "ConsultantPlus"
3. Resolution of the Eurasian Economic Union Court dated April 08, 2016 in case No. CE-1-2/3-16-KC <On termination of proceedings in the case upon the application LLC "Remdizel" Production Enterprise (Russian Federation) on recognizing an additional note to positions 8408 10 110 0 - 8408 10 990 0 ТН ВЭД EAEU, contained in the table of the Explanations to the unified nomenclature of goods subject to the foreign trade of the EAEU (volume 6, sections I - XXI, groups 1 - 97), annex to the

Recommendation of the EEC Board dated 03/12/2013 N 4, contrary to Article 5 of the Treaty on the Eurasian Economic Commission dated 11/1/2011, as well as the Decision of the EEC Council dated 07/16/2012 N 54> (Together with the "dissenting opinion of judge Neshataeva T.N. (case No. CE-1-2 / 3-16-KS) ")

4. Order of the Ministry of Agriculture of Russia No. 517 dated August 28, 2020 "On approval of methods for calculating the values of certain indicators of the State Program for the development of agriculture and regulation of markets for agricultural products, raw materials and food", SPS "ConsultantPlus"
5. Order of the Government of the Russian Federation No. 1806-r dated October 18, 2010 "On Approval of the Comprehensive Program for the Participation of the Russian Federation in International Cooperation in the Field of Agriculture, Fisheries and Food Security"//SPS "ConsultantPlus"
6. Decision No. 60 of the Interstate Council of the Eurasian Economic Community "On the Concept of Establishing an Integrated information system for foreign and mutual trade of the customs union and priority measures for its implementation" (Adopted in St. Petersburg dated November 19, 2010); Agreement on the creation, functioning and development of an Integrated Information system for foreign and mutual trade of the Customs Union dated September 21 (2010)
7. Customs Code of the Eurasian Economic Union (Appendix No. 1 to the Agreement on the Customs Code of the Eurasian Economic Union), SPS "ConsultantPlus"
8. Decree of the President of the Russian Federation No. 640 (2016)
9. A. Kurdyumov, Scientific conference on New Industrialization, 81-84, 240 (2019)
10. E. Butsenko, A. Kurdyumov, A. Semin, Mathematics, **8**, 1480 (2020)
11. A. Kurdyumov, E3S Web of Conferences, 176, 05028 (2020)
12. O. Kozhevnikov, Pravoprimenenie, **1(4)**, 5-13 (2017)
13. O. Ruzakova, E. Butsenko, A. Kurdyumov, *Education excellence and innovation management through vision*, 711-719 (2020)
14. A. V. Savoskin, A. V. Kurdyumov, M. A. Zadorina, O. A. Kozhevnikov, V. A. Meshcheryagina, *Participation* (2021)
15. A. V. Savoskin, A. V. Kurdyumov, V. A. Meshcheryagina, M. N. Vilacheva, *Digitalization of the Russian economy as a national goal* (2021)
16. A. D. Dusupova, International Public and Private Law, **5** (2019)
17. J. I. Ovsepyan, Lawyer-Jurist, **15**, 77 (2015)
18. T. Yu. Falkina, *Actual problems of history, politics and law*, 6-8 (2017)

E3S Web of Conferences **262**, 03016 (2021) https://doi.org/10.1051/e3sconf/202126203016
ITEEA 2021

Certain aspects of digital development in AIC

Indira Mikitaeva[1,*], *Eldar* Malkanduev[1], *Pavel* Zhufan[2] and *Saida* Shardan[3]

[1]Kabardino - Balkarian State Agricultural University named after V.M. Kokov, Lenin Avenue, 1V, 360030 Nalchik, Russia
[2]Mendel University in Brno, 1, Zemedelska, 61300, Czech Republic
[3]North Caucasian State Academy, Cherkessk

Abstract. The social and economic development of the territories includes the existence of stable interregional food and commodity linkages, the formation of a food economy, an increase in food resources conducive to the development of the AIC and the entire economic system. Digitalization and the implementation of information technologies ensure high-quality information content of all subdivisions of the AIC, enable optimization of the flow of resources and goods within the complex, leading to facilitate the task of transforming the agro-industrial complex of Russia, further "effective and sustainable agro-industrial production that corresponds to the world level ensuring the country's food security, the saturation of the market with food "[4]. An effective mechanism for the regional innovative economy management is, in our view, the complex establishment of a high-quality information base, the development of information technology networks to transcend traditional technologies. The success in the transition to the modern stage of the digital economy in the AIC prejudges the growing importance of the of the country, its regions and individual industries in the formation of a single information field. The federal policy policy is directed at creation of new information technologies and communication channels throughout Russia and all over the world. The regional policy in the field of information should be based on the interests of the regional users and the general federal interests of a group of industries located in the region. The sectoral policy in the field of information should take into account the interests of one industry, as well as enterprises and organizations within the structure of this industry.

1 Introduction

The problem of digitalization has become a priority at the current stage and its importance in society is constantly increasing. The latest achievements in informatics, computer science and telecommunications have created a new type of high technology - digital technology. The information economy development, that is primarily information technology (IT) services and advanced information and communication technologies (ICT), have become globalized and have become a defining feature of the country's level of development.

* Corresponding author: diseconkbgau@mail.ru

The creation of digital technologies is focused on the economic efficiency and sustainability of the country and its regions. The most developed enterprises in the industry regarding information technology can promote synergies in increasing their efficiency mainly by reducing transaction costs in combining into a common information exchange network.

The relevance of the development and implementation of modern information technologies in the agro-industrial complex of Russia is determined by the strategic role and importance of the AIC for the provision the country with agricultural products, raw materials and food.

2 Materials and Methods

During the research, such methods as analysis and synthesis were used. The research materials are the works of prominent academic economists from around the world and statistical data.

3 Results and Discussion

The study identified that there are challenging tasks of creating information technologies in the sectors of the AIC, which may be summarized as follows: Primarily, there are structural weaknesses in the digital transformation ecosystem, a relatively low level of digital skills, there is no open innovation culture.

The tasks of establishing global information complexes in the sectors of the AIC include:

- to improve federal, sectoral and regional policy in the field of information and to implement modern information technologies;
- to develop information technology networks, information services market and to integrate into the world system of creating new information technologies in the AIC.

Addressing these problems requires AIC staff development in information technologies and telecommunications, in order to implement new information technologies, and to develop consulting and expert services networks. The development of a new information culture is facilitated by the higher education system, engaged in providing the intellectual potential in the field of information, network technologies for the sectors of the AIC. According to Evgenii Sandomirskii, EPAM Systems Senior Business Development Manager, "Today people have become the main asset, and information has become the main resource. It is necessary to equip people with relevant information, to invest in their skills, talents and to provide them with convenient tools for work"[3]. E. Sandomirskii pointed to the McKinsey & Company research, whereby three factors influence the success of a company's digital transformation:

- ease of access to information for all employees in combination with transparent communications;
- independent use of self-service services by employees;
- close cooperation among various units.

The development of training centers that focus on training both administrators and users is required to ensure that any user could master the office suite in short time. Nowadays, not only technical support is important, but also assistance with implementation. The implementation projects, assisting administrators during implementations, from our point of view, specifically requires particular attention.

It should be mentioned that at the current stage, certain successes have been achieved in terms of creating information management systems related to the qualitatively new level of application software development. Software, as an independent and basic use of information technologies in the field of management systems for AIC enterprises, organizations and sectors, covers almost all management fields.

But how to switch to new software so that the effectiveness of the organization does not fall, but grows? Concerns over the inconvenience and unfamiliarity of work are justified. Since there are often difficulties related to established habits of using office applications, on format compatibility. Issues concerning the compatibility with other application software (EDM, BPM, GIS, etc.), enabling the product to be seamlessly implemented into the automated organization's business processes tended to be determined already at the testing stage. Although some of the issues are removed during the initial acquaintance with the product, a more detailed study of the software product requires improvement in the qualifications in authorized training centers.

The foregoing emphasizes the complexity and at the same time the relevance of the tasks for both the users of the programs and their developers.

To make significant improvements in this area, the Russian Government promotes the implementation of disruptive innovation by investing in national broadband infrastructure. In the era of implementation of import substitution programs, the state declared a quest for import substitution of office products, developed recommendations on the transition of public authorities at various levels and state-owned companies to the preferential use of indigenous software, including indigenous office software. The recommendations lists the measures to be included in the transition schedule, formulated indicators of the transition efficiency, the share of national office suites to the total use of office software to be achieved.

Projects for the program implementation are actively developing, the automation loops are expanding, progressively covering new functionality, including those related to the robotization of typical operations,to the use of artificial intelligence algorithms, and multidimensional data visualization (Fig. 1).

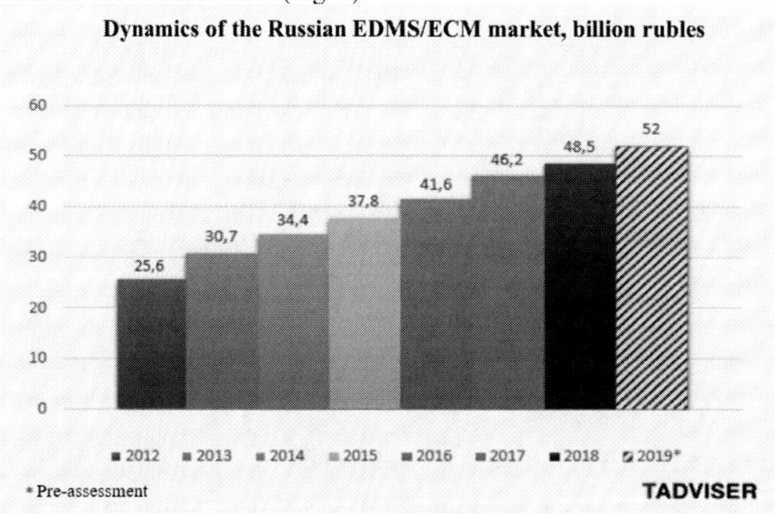

Fig. 1. Dynamics of the Russian EDMS market, billion rubles
*Compiled from TAdviser data

According to TAdviser, the market growth was about 5%, amounting to 48.5 billion rubles. At the end of 2018, the indicators of a number of major players shrank. According to the preliminary results of 2019, the market growth rate of 7% is expected, nevertheless its volume will not exceed 52 billion rubles [7].

The results of 2019 demonstrated that, despite all the government's initiatives, it is too early to talk about mass import substitution. However, the first signs of this process have already emerged. At the end of 2019, the Ministry of Digital Development, Communications and Mass Media carried out central procurement of Russian office programs and antiviruses for a total amount of over 1 billion rubles.

This situation makes actual the position of A. N. Bogatyrev, who stated: "The technological revolution is gradually turning into an intellectual and informational one. Information becomes not only a subject of communication, but also a profitable commodity, certainly, an effective contemporary means of organization and management of social production ... and the social and economic development of society as a whole"[8]. It can be seen that the significance of informatization in all sectors of the economy is constantly growing.

Regarding the agricultural sector, the timely and high-quality spread of digital technologies in organizations and enterprises of the AIC, considering the large extent of the territory of Russia, is economically profitable. The system of information flows, providing high-quality informativity of all sectors of the AIC, makes it possible to optimize the flows of resources and goods inside the complex and, consequently, reinforces the effectiveness of market mechanisms in the reproduction process.

The use of digital systems in the industries of the AIC is very promising due to the following features:

- a wide variety of production and processing of raw materials;
- a significant number of technological processes, many of which are energy- and material-intensive;
- the seasonal nature of the work;
- a large proportion of personnel employed directly in the production of products.

As a consequence of the above features, it is very difficult to analyze the economic and financial activities of organizations, which accentuates the need to use information technologies to obtain a set of information on which it is possible not only to reduce losses in production significantly, but also to improve the use of production facilities and increase the efficiency of production and sales.

The fundamental importance of increasing production and promoting agricultural products is also due to the leading role in the maintenance of the food security of the country and its regions. Considering the strategic and systemic role of the AIC, extensive use of digitalization can significantly enhance the competitiveness of agricultural products and the investment attractiveness of the industry, enhance the linkages "between the key stakeholders of the digital ecosystem, including the public and private sectors, civil society and the scientific and educational community" [9].

Today, a balanced growth of the agro-industrial complex of Russia is impossible out of the information field. The agro-industrial complex of the Kabardino-Balkarian Republic is no exception, the growth of which is associated with the provision of high-quality informativity in all areas and sectors, and specialists - with access to regional, national or international telecommunication networks.

The interrelationship of various production structures through the establishment of stable vertical and horizontal linkages, the use of the potential and principles of partnership between the state and private business, modern information technologies, innovative approaches will allow the agro-industrial complex to reach a competitive level of production development [6].

The information support system has a significant impact on the choice of potential areas of investment by investors, since new capabilities of information and computer technologies provide access to the study, for example, of an investment passport of a particular municipal district.

We have investigated the official website of the Maysky municipal district of the Kabardino-Balkarian Republic, which provides comprehensive information on both the National projects, Federal and regional programs implemented in the territory, and indicators on the state of the economy and social sphere of the municipality (Fig. 2). The section "Investment activity" presents the investment policy of the municipal district, which is to create the most attractive conditions for investment in the development of local production and social infrastructure. The same section contains a list of local laws and secondary legislation governing business and investment activities, the existing forms of their municipal support: The program of social and economic development of the Maysky municipal district for the period 2016-2020, the Municipal program "Development and support of small and medium-sized enterprises in the Maysky municipal district of the Kabardino-Balkarian Republic for the period 2016-2020", the Municipal program "Development of agriculture and regulation of markets for agricultural products, raw materials and food for the period 2013-2020 in the Maysky municipal district", On the support of small and medium-sized enterprises, On the Investment and Entrepreneurship Board to the head of the local administration of the Maysky municipal district and others.

Fig. 2. Fragment of the municipality official website

The list of priority areas for investment in the Maysky municipal district includes: agriculture, processing industry, mechanical engineering, chemical production[5]. A register of investment sites, free and for possible use in the implementation of investment projects in the Municipal District is presented.

To improve the awareness of business entities and consumers of goods, works, services, the section "Promotion of Competition" provides information on the state of competition and activities to promote the development of competition.

On the official website and on the Internet portal about investment activities in the constituent entity of the Russian Federation, in a separate section, a publication of interest on measures and infrastructure to support small and medium-sized enterprises (including a separate subsection for agricultural producers) can be found. Publications are structured by the development priorities of the constituent entity of the Russian Federation and types of activities, indicating the procedure for obtaining mentioned support (including in the form of a scheme) and indicators of the efficiency of the regional program within which this activity is carried out. The amount and detail of information disclosure should reflect the coherence of the investment programs of natural monopolists with the territorial development plans of the entity.

The specified example demonstrates the interest in creating a unified information space for receiving a set of information for the further growth and development of the region.

Due to the digital transformation based on a set of technologies, big data, the Internet of Things, cloud computing, artificial intelligence and a number of others, the most important condition for the dynamic growth of economic sectors of the regions and the country as a whole will be created. In the fair view of Shardan S.K., the involvement of breakthrough technologies has become a vital necessity. An effective mechanism for managing the regional innovative economy according to the author is "the integrated implementation of such functions as the formation of an information base about management objects ..."[2]. Let us only clarify that the increase in cloud services, digital platform significantly reduces the time period of the innovation cycle and thereby causes an increase in the rate of renewal of innovative products and technologies.

4 Conclusions

The market environment, in the context of the rapid development of information technology, places on each participant high demands associated with the increasing amount of information and the increasingly complex system of relationships between market participants and a wide range of stakeholders. In order to meet the requirements, correctly focus and find the right solutions, market participants and business entities need to move beyond traditional technologies [1].

As a result of the study, we have identified some aspects that determine the success in the transition to the modern stage of the digital economy in the AIC.

1. The federal policy aimed at creating a unified information space is highly relevant and timely. It relates to the intensified development of education in the field of informatics, retraining and advanced training of specialists of all levels, the development of scientific research in the field of creating new information technologies and communication channels all over Russia and all over the world.

2. The regional policy in the field of information should be based on the interests of the regional users and the general federal interests of a group of industries located in the region. The regional aspect of digitalization includes the stable interregional food and raw materials linkages, the creation of a food base, and the increase in food resources conducive to development of the AIC and the entire economic system.

In creating a single information space, it is important to take into account regional differences. Each region develops its own strategy for the transition to import substitution, its typical workplace, which includes, besides an office suite, an operating system, an office, e-mail, antivirus, an electronic document management system. In addition, in each region there are government information systems (GIS) which solve specialized problems. For example, they are variable in the timing and sequence of stages of a transition project. One first chooses a certain operating system, installs it, then transfers all GIS to it, and only then purchases an office suite, mail and antivirus, that is, application systems. The second approach is slightly different: it implies to install something that does not depend on the choice of the operating system first. For example, "R7-Office", running on all domestic operating systems. Such an installation, for the record, is technologically safe, since it does not strategically affect the possibility of the subsequent choice of an operating system.

3. The sectoral policy in the field of informatization should proceed from the interests of one sector, taking into account the development of the regional and federal levels of informatization. Being in general relationships with each other, the interests of enterprises and organizations that are part of the structure of this industry are resolved at the sectoral level.

4. Finally, each agricultural organization or processing enterprise, as the primary unit of informatization, solves the problem of implementing new information technologies and telecommunications, comparing its interests with the policy of the industry, region, state.

This approach will obviously result in the provision of the social and economic development of the territories and the implementation of the main strategic national priorities.

These are the main provisions of the trends of digitalization and the implementation of information technologies that contribute to facilitate the task of transforming an important sphere of the economy - the agro-industrial complex of Russia.

The article was prepared with the financial support of the Russian Foundation for Basic Research, project No. 20-010-00838A

References

1. A. I. Altukhov, *The Economic Problems Of Innovative Development Of Grain Subcomplex In Russia* (2015)
2. Saida K. Shardan, Nazgul B. Davletbayeva, Svetlana S. Morozkina, Deschi Sh. Musostova, Anastasia V. Sharopatova, Journal of interdisciplinary research, http://www.magnanimitas.cz/
3. Digital transformation day 2020, https://www.tadviser.ru/index.php
4. Indira R. Mikitaeva, Marina T. Tekueva, Mikhail Kh. Balkizov, Tanzila H. Sozayeva, Journal of Organizational Behavior Research, **3(2)**, 192 (2018)
5. Investment passport of the Maysky municipal district, https://mayadmin-kbr.ru/
6. *National models of economic systems: collective monograph* (2019)
7. Russian EDMS/ECM market, https://www.tadviser.ru/
8. A. N. Bogatyrev, V. A. Panfilov, V. I. Tuzhilkin et al., *System of scientific and engineering support of food and processing industries of agriculture in Russia* (1995)
9. T. H. Sozaeva, A. Yu. Pshigosheva, S.A. Gurfova, I. R. Mikitaeva, *Agrarian territories in the context of the formation of a digital economy: problems and prospects* (2020)

State and prospects of competition, communications and logistics of agro-food products

Hadis Rakhaev[1,*], *Anzor* Gyatov[1], *Zalina* Ivanova[1], *Elvira* Kokova[1], and *Akhmat* Chochaev[2]

[1]Kabardino-Balkarian State Agricultural University named after V.M. Kokov, Lenin Avenue, 1V, 360030 Nalchik, Russia
[2] «Development corporation of the Kabardino-Balkarian Republic», JSC

Abstract. Communication and logistics are one of the most important features of the markets; they permeate and constitute its content. Communication and logistics for commodity producers are also as important attribute as production itself, and sometimes even more. For this reason, commodity producers themselves directly and explicitly, but more often through other institutions (including the government), tend to set up stable exclusive communication systems and logistics. This article analyzes the state of the prospects for the formation of communications and logistics networks for agricultural products. The correlation of existing communications and logistics of agricultural products and other types of products (including finished industrial products of various purposes: from machinery, equipment, chemical products to defense, hydrocarbon, carbon and other raw materials, timber and other goods) is analyzed. The established linkages are revealed. They were quantified, calibrated and classified. The existing principles (comparative and absolute advantages) are reviewed and new principles are formulated (marginal player, marginal linkages), which describe the real situation in the markets of agricultural products more correctly. New criteria for grouping and reformatting existing communication and logistics networks, which make it possible to increase the competitiveness of domestic agricultural products are proposed.

1 Introduction

An observation of the modern and historical practice of the agriculture development shows one trivial case: it is possible to produce (and producing) a good-quality product in a large volume, but not to transform/turn it into a commodity, i.e. not to sell it and in this way all production costs will be a loss. This experience was typical to many countries, from ancient Greece, Egypt, Babylon, and China to the present day. Most often, the basis of an unfinished metamorphosis, when a product does not turn into a commodity, is seen as the low quality of the product and high competition in markets that the manufactured product could not overcome. Therefore, a pair of relations: price-quality acts as a criterion for

* Corresponding author: r3bizengin@mail.ru

overcoming the barrier. But actually, it is much more complicated. Soviet and Russian agriculture, as well as the agriculture of many other countries, shows that often the problem lies not in this respect, but in communications and logistics, i.e. in the networks. The logic of the appearance of the indicated and described effect of not passing the expected metamorphosis: product-commodity, for products of domestic agriculture, is that the transformation of a product into a commodity is related not to the consumer qualities of the product, but with logistics, communications, i.e. so that the manufacturer finds its consumer. This situation causes the emergence of intermediaries, like those agents who find their consumer for each producer and their product for each consumer. Thus, the experience indicates that it is not enough to produce a product and to supply it with the necessary qualities and price, so that it becomes a commodity (i.e. sold), it is necessary to create its own consumer for each product or create your own communications and logistics.

However, it is one thing when talking about a single product, single producer and consumer, and even a certain industry, and completely different thing when talking about a branch of the national economy and, moreover, agriculture, which produces tens, hundreds of thousands of items of products that are necessary turn into commodities.

Clearly, it is necessary to proceed from the features of these products when forming a communication and logistics system for agricultural products and raw materials. First thing is that they, as a rule, belong to the so-called perishable products. Therefore, the manufacturer has to sell them as soon as possible. There are two forms of marketing: direct or to the final consumer and processor or to the intermediate consumer. At the same time, the manufacturer faces the issue of cost, i.e. in the first case, he earns more than in the second. That is why the manufacturer seeks to find the final consumer of his products himself. Secondly, many of them cannot be kept in the traditional storage, but require special storage and transportation conditions. Thirdly, they correlate (and even complement) with ecology, physiology, and other traits, i.e. can have a negative impact on the environment (carry environmental harm), as well as on the physiology of people and animals. They can be carriers of diseases, pests, etc. Therefore, special measures are required not only for storage and transportation, but also consumption. There are also other features of these products that distinguish them from products of industry, construction, transport and services. However there are also qualities that set them apart for the so-called better side from their competitors. Primarily, they are as versatile as many raw materials. Secondly, they are universal in consumption (use), i.e. do not need any special knowledge, skills and means. And in this regard, they are kind of primitive. Thirdly, they act as the basis for the biological existence of humans and animals. In addition to the listed features, there are also many properties described in any good merchandise science textbook.

The listed features of agricultural products and raw materials certainly must be reflected in the formation of communications and logistics, which, in our view, was the main mistake of the Soviet (and then repeated and seems to continue also in the Russian) system of communications and logistics of agricultural products and raw materials, but, apparently, they are not decisive in the transformation of a product into a commodity. The analysis revealed several fundamental problems in the existing system. The first problem is the lack of consistency in the formation of communications and logistics of agricultural products and raw materials. The second problem is the lack of consistency in the formation of such a system. The third problem is the lack of a principled approach and mechanism in the formation of communications and logistics networks, the use (and following) of stereotypes and tradition, the absence of heuristic solutions. In this paper we are going to consider on these problems, as well as on how to form effective communication and logistics systems.

2 Materials and Methods

The empirical base of the study is official statistics on the production and foreign trade of agro-food and other products published in statistical collections of the FSSS (Rosstat), as well as the results of research by domestic and foreign authors published in the public press: monographs, articles, educational and other manuals. The theoretical and methodological basis of the research was formed by the works of A. Smith, D. Ricardo, J. St. Mill, A. Marshall, M.I. Tugan-Baranovsky, N.D. Konratyev, V.V. Leontiev, P. Samuelson, B. Olin, M. Porter, P. Lindert and other foreign and domestic academic economists, and the principles, categories and concepts formulated in their works. The basic concepts (category and principle) are the Ricardian principle of comparative advantage, in the interpretation of Olin-Heckscher-Samuelson, which is complemented by communications and networks. According to the above, the principle is formulated as follows: the competitive preference, if other things being equal, has not the product with the lowest cost of production, but the product that can be delivered (provided) to the consumer more quickly. Thus, the competitive advantages (preferences) do not consist in production, but in existing (used) communications and logistics. The theoretical and methodological basis of communications and logistics has been developed in various fields of practice and theory: from natural science, technology to humanitarian disciplines. They also receive significant attention in economic and organizational research. In this regard, the epochal works of A. Bogdanov, L. von Bertalanffy [1, 2, 3] formulating methodological basis, as well as a large number of works by contemporaries that can be identified as theoretical and practical in this area [4, 5, 6, 7, 8, 9] should be mentioned.

3 Results and Discussion

The major features of any market are competition, foreign markets, communications and logistics. Currently, it is very difficult for agricultural producers to enter not only foreign markets, but even domestic ones without their own exclusive communications and logistics. In the USSR exclusive logistics systems were formed for agricultural products on world markets. However, mainly the products of domestic agriculture were sold to the countries of the socialist community, as well as African and Latin American countries. Moreover, the prices for products turned out to be symbolic, since the products for export were sold under certain agreements. As for the food and agricultural markets of developed countries: France, Great Britain and Germany in Europe, the USA and Canada in America, etc., they were essentially closed for our agricultural products [10, 11, 12]. Moreover, we imported large quantities of grain, animal oil, vegetable oil and many food products from the USA, Canada, France and Greece. Large amounts of citrus and tropical crops (bananas, coffee, coconut, etc.) were imported from African and Latin American countries. Large amounts of sugar were imported from Cuba, tea from India, vegetables from Hungary and Bulgaria. This was explained by the fact that we did not have our own communications and our own logistics in the sale of food and agricultural products. At present, when the competition in the world agro-food markets has intensified, it is difficult, or even impossible, to sell one's own agricultural products without appropriate sales channels. As for the implementation into communications and logistics of other countries, which is often carried out for some types of products, it results in losses and is inefficient on its own, since we act mainly as a "guest" and not a "owner" of communications. Therefore, it is not we who dictate the terms, but we but we who are dictated by the conditions for the sale of products. Experience shows that the solution of this fundamental problem is impossible without the support of the government. Moreover, this is a worldwide practice that is used by all developed countries all over the world [11, 13, 14, 15]. However, one of the promising directions associated

 https://doi.org/10.1051/e3sconf/202126203017

with the integration of flows of agricultural products, with flows of arms and other military products, as well as raw materials (gas, oil, coal, etc.) could be used for us. Agricultural products could be implemented into the existing supply chains and communications of the above military and raw materials (not only hydrocarbons, but also other ore and nonmetallic materials). Such a variant is rather cheaper for the Russian government than to form new agricultural logistics and communication chains and corridors. Complex, systemic nature in relation to communications and logistics networks is required.

Thus, the listed problems constitute a complex of conditions hindering the development of agriculture. Therefore, the growth that has been observed in domestic (including in most regions) agriculture in recent years was formed, firstly, by a limited set of industries (mainly grain production, pork production, poultry production, and some other segments), secondly, the price factor was a major factor in its provision (price increases led to an increase in price trajectory in the overall growth trend), thirdly, by means of food embargo measures established and implemented since 2014 by decrees of the President of the Russian Federation and Resolutions of the Government of the Russian Federation [16, 17]. These measures ensured a kind of "greenhouse" development of national agriculture, protecting it from foreign competitors. (During this period, the so-called "national champions" were born as large vertically integrated national agro-food companies: Miratorg Agribusiness Holding, RIF Trade House, Agrocomplex named after N.I. Tkachev, RUSAGRO Group, Krasnodarzernoprodukt Group (KZP), Prioskolie, Stavropolsky broiler LLC, Agrokombinat Yuzhny JSC, Agro-Invest LLC. The latter can be compared with those that existed in the 90s and early 2000s [18]). However, according some experts, the situation in certain sub-sectors and segments will probably change radically with the opening of borders and the liberalization of foreign relations.

Where do the material and analysis result in? Primarily, the development of domestic agriculture requires the formation of efficient communications and logistics networks, i.e. the creation of a network effect in the industry, a specific effect of the presence of networks in the reproductive process of the industry (product, production). By its very nature, the network effect is multiplicative, i.e. defined somewhere (on one link or product) it stimulates activity in other chains and products. Today, it is clear that access to external markets (as well as to domestic markets) is not possible without own (exclusive) communications and logistics [19, 20, 21, 22]. But the establishment of such communications and logistics proves to be very expensive and complex, both because of the high level of competition and, in fact, because of the lack of free markets, and because of our companies' lack of experience in establishing such communications and logistics networks. Of course it is not necessary to give up on this, since the struggle for agro-food markets in the world will only increase and if peculiar «windows of opportunities» appear, it is necessary to be ready to take up the released traffic and post them for own purpose. (Such experience exists. An example is the grain market, in which Russia, since 2014, has become the dominant player and global actor [23, 24, 25]. But it should not be forgotten that Russia has been coming to this role since the second half of the 90s and the entire first decade and half of the second 2000s [26, 27]. There is a market opportunity for poultry [28] and pork [29, 30, 31], which are still poorly developed by domestic agricultural industry. But we should not forget about the power and greatness of Russia and two or three segments with which we can enter world markets. Resource potential of Russian agriculture allows to create dominance in more than a dozen commodity segments

The experience of the grain industry indicates the costs that had to be incurred to obtain its communications and logistics (traffic) only in the grain market. But there is a less expensive and more efficient way to build own communications and logistics networks in the area of agro-food. It refers to the entry (integration, cooperation) of the agro-food sector with the defense sector, i.e. the integration and cooperation of the export of food and agricultural raw materials into existing (established and effective) communications and weapons logistics networks.

However, weapons and agro-food are, from the commercial point of view, diametrically opposed products, and therefore incompatible at the first glance. Moreover, communications and logistical networks of armaments represent specific networks in terms of both technological and organizational and institutional characteristics. But this, in our view, is a superficial view and superficial perception of the two products and their communications. In fact, there is a high correlation between them in a number of areas: consumer, country, territory, transport and other communications, etc. Furthermore, there is a need to integrate and cooperate agro-food communications and logistics networks in communications and arms logistics. Besides, there is a strong correlation between these products.

So the question is: how to do this technologically and institutionally? In our view, the integration of agro-food products into the communications and logistic networks of MIC products should take place sequentially and in several stages.

The first stage examines (separately) the communications and logistics of MIC and AIC products, identifying major trends, problems, counterparties, countries, territories, and other peculiarities. As a result, communication and logistics maps that show traffic are built, as well as their formalization and quantification. In the second stage, it is possible to integrate networks and traffic into each other and form a single logistic network. The correlation between the flows of MIC and APC products is evaluated, possibilities of their overlap and formation of common traffic are estimated. In the third stage, new communications and logistics networks are modeled by combining correlating traffic and communications. In the fourth stage, a mechanism for the joint promotion of MIC and AIC products is being developed. At this stage, weapons (MIC products) are first taken over, i.e. communications and logistics are provided to agro-food products, but in the next stage agro-food products will provide their communications and logistics networks with weapons. In the new communications, the products of the AIC (MIC) are first offered as a «load» to the products of the MIC (AIC), but then as a complementary (i.e. supplementary) product.

4 Conclusions

A complex of analytical and monograph studies indicated that: 1) the competitiveness of national agricultural products depends less on the cost of their production as well as on the quality, as on efficient communication and logistics. Whereas previously the latter were a kind of supplement to consumer competitiveness, with the growing number of agents offering their products, they become a determinant of the competitiveness architecture; 2) Russian agriculture loses in the sale of its products largely due to the lack of efficient exclusive communications and logistics and the use of traditional communications, logistics and traffic, established partly even in the 18-19th centuries. This leads to the conclusion that, in order to promote own agricultural products effectively, it is necessary to develop own modern communications and logistics networks; 3) the formation of modern communications and logistics networks is very expensive and therefore often beyond the capacity of individual states (for example, the EU has a long-standing strategy of joint communications and logistics networks, as have the US and Canada, as well as other countries and territories. Even China has developed and implemented a strategy "one way - one destiny", thus connecting national logistics networks to its communications). For Russia, EurAsEC, BRICS and local alliances are optimal from this point of view. Mainly EurAsEC; 4) at the same time, there is a high competition within the EurAsEC in the field of agricultural products with Kazakhstan, due to long-standing historical traditions: they grow almost identical agricultural products: grain, livestock, etc. This situation requires the search for new communications and logical networks that might not compete with an EurAsEc and strategic partner, within traditional communications and logistics networks, and, on the contrary, to contribute active conquest of external markets, both jointly and separately. For this purpose, it is proposed to link the marketing of agricultural products to communications, logistics networks and the traffics of military products. Primarily, there is a high correlation between arms sales and agricultural products, (it has been indicated that the marketing of agricultural products was most correlated with armament, high technology, and the least with primary commodities: oil, gas, coal, etc. This indicates a complementary (non-competitive) character between the named products); secondly, there is a high correlation between certain types of arms and agricultural products; thirdly, there is a high correlation between country and territorial traffic, fourthly, there is, by contrast, a high but negative, as well as a low negative, low positive correlation between MIC products and AIC products in certain country and territorial traffic. Moreover, the latter is often found where MIC products are very important, i.e. they take up either all or most of the traffic. As an addition, the products of raw materials or processing industries are used. On these traffic, the products of the national APC tend to be marginalized. We believe that it is precisely because of the latter that the above-mentioned variations in correlation occur. It is probably the filling of these traffic by implementing AIC products into the established MIC communications and logistics networks that will, firstly, increase the volume and structure of agricultural products sold, secondly, the latter should also support sales of MIC products.

References

1. A. A. Bogdanov, *Tectology: General Organizational Science* (1989)
2. A. L. Takhtadzhyan, *Principia tectologica. Principles of Organization and Transformation of Complex Systems: An Evolutionary Approach* (2001)
3. L. von Bertalanffy, *Studies in General Systems Theory: A Collection of Translations* (1969)
4. D. Bowersox, D. Kloss, *Logistics: An Integrated Supply Chain* (2017)
5. E. I. Nikitina, *International Logistics* (2018)
6. V. A. Konovalenko, M. Yu. Konovalenko, *Foundations of Integrated Communications* (2015)
7. D. A. Novikov, *Network Structures and Organizational Systems* (2003)
8. V. N. Burkov, N. A. Kuznetsov, D. A. Novikov, Automation and Telemechanics, **12**, 96-115 (2002)
9. B. Z. Milner, *Organization Theory* (2002)
10. *Agricultural Sector of the United States at the end of the 20th Century* (1997)
11. N. Andreeva, MEMO, **7**, 91-96 (2009)
12. I. Buzdalov, Society and Economy, **1-2**, 136-153 (2015)
13. V. Demyanenko, D. Rylko, MEMO, **8** (1998)
14. N. D. Avarsky, Kh. N. Hasanova, V. V. Taran, *Theoretical and practical aspects of commodity circulation in the agro-food market (foreign and domestic experience)* (2014)
15. A. I. Altukhov, *Directions and principles of the development of interregional and interstate food and raw materials relations* (2000)

16. Decree of the President of the Russian Federation on August 6, 2014 No. 560 "On the application of certain special economic measures to ensure the security of the Russian Federation"
17. Order of the Government of the Russian Federation No. 778 dated August 7, 2014 "About measures for implementation of Presidential decrees of the Russian Federation "On the application of certain special economic measures to ensure the security of the Russian Federation"
18. D. Rylko, R. Jolly, *Organizational changes in Russian agriculture: the emergence and consequences of the activities of "new agricultural operators"* (2005)
19. V. P. Korovkin, A. N. Nezadorov, *Foreign economic activity in the agro-industrial complex* (2003)
20. V. I. Nazarenko, *Russia and foreign countries. Agricultural Policy Models* (2008)
21. M. Porter, *International Competition* (1996)
22. M. Tracy, *Agriculture and Food in Developed Economies: Introduction to Theory, Practice and Policy* (1995)
23. V. Demchuk, Latifundist.com (April 15, 2020)
24. S. Kulikov, Rossiyskaya Gazeta (September 7, 2016)
25. Grain crops market in Russia – 2021, Indicators and projections, https://tebiz.ru/
26. A. V. Gordeev, V. A. Butkovsky, Russia - Grain Power (2004)
27. Export of grain from Russia: history and modernity, https://vvs-info.ru/
28. V. Fisinin, Animal husbandry of Russia, **3**, 2-5 (2014)
29. O. A. Kholodov, M. A. Kholodova, I. F. Gorlov, O. P. Shakhbazova, N. I. Mosolov, A. V. Glushenko, D. A. Mosolova, *Analysis of prospects for the development of dairy cattle breeding in Russia: forecasts and scenarios*
30. A. L. Zolkin, E. V. Matvienko, M. V. Shavanov, *Innovative technologies in agricultural crops breeding and seed farming*
31. V. A. Labinov, Animal husbandry of Russia, **1**, 4-5 (2014)
32. National Union of Pig Breeders, http://www.nssrf.ru/documents.php

Path of sustainable development of agricultural ecosystems in the context of digitalization

Tanzilya Sozaeva[1,*], *Svetlana* Gurfova[1], *Dinka* Marius-Sorin[2], and *Alfina* Pshigosheva[1]

[1]Kabardino - Balkarian State Agricultural University named after V.M. Kokov, Lenin Avenue, 1V, 360030 Nalchik, Russia
[2]Transylvania University of Braşov, Bulevardul Eroilor 29, 500036 Braşov, Romania

Abstract. Digitalization provides new opportunities for determining the path of sustainable development of agricultural and ecological systems. The article attempts to consider the problems related to the emerging digital divide and the possibility of their leveling accompanying this process both at the international and national levels. Digital transformation is taking place not only in the technology sector and digital companies: its scale extends to all areas of the world economy. It was revealed that digital transformation, accompanied by the development of digital platforms, provides a more complete interaction of agricultural ecosystems with the digital world. Digital platforms have been identified to facilitate the integration of business, social and technological processes that form digital service ecosystems. The main reason for the expansion of the digital segment of agricultural ecosystems is the increase in the transactional sector. The creation of IT platforms based on the digitalization of production and technological processes, jobs; logistics, collaboration of agribusiness entities and public-private partners, consumers, competitors, investors, authorities has been proposed.

1 Introduction

The problem of digitalization of agricultural ecosystems and sustainable development of rural areas is one of the key issues of modern world science. Digital transformation is accompanied by the development of digital platforms. In this regard, agricultural ecosystems are collaborating with the digital world. The result is the definition of a new path for sustainable development. However, at the same time, digital gaps arise, the need to overcome them by states, which is becoming a priority area of policy in developing countries. Digitalization is taking place not only in the technology sector and digital companies, its scale extends to all areas of the world economy. The digital segment of agricultural and ecological systems is constantly expanding. The main reason for this is the increase in the transactional sector, which accounts for more than 70% of the national GDP of developed countries. A diversified economy presupposes the presence of large volumes of unique data circulating both within the country and abroad, as a result of which

* Corresponding author: sozaytanzilya@yandex.ru

information traffic in the national economy grows. It should be noted the higher efficiency of the digital economy in the market, where there are a larger number of participants and there is a high level of penetration of information and communication technologies (ICT).

2 Materials and Methods

The materials were the works of foreign and domestic scientists in the field of the theory of agricultural ecosystem management, statistical data of the Federal State Statistics Service, reports of the Ministry of Agriculture of the Russian Federation, as well as reports of the UN, FAO, OECD. The system analysis and synthesis were used as the main research methods, which made it possible to assess the current state of the use of ICT and identify the main directions in solving digitalization problems.

3 Results and Discussion

In the context of digital globalization, there is a need for a positive synergistic effect on the sustainable development of agricultural ecosystems. Digital transformation of agriculture is part of the nationwide task of the integrated development of agricultural territories. In this regard, it is necessary to locate and specialize in agricultural and industrial production based on a multi-level integrated information space and modern digital technologies. Digital transformation is driven by four trends: mobile technologies, business intelligence, cloud computing, and social media. The entire agri-food chain and its links will change under the influence of digitalization. It will be possible to manage the resources of any element of the system using optimization, an individual approach, rationality and predictability. The system will operate in real time through hyper-connectivity, data-driven. Value chains will be characterized by the coordination of optimal management models for agricultural land, plants and animals. The result of the functioning of digital agriculture will be the creation of efficient agri-food systems adapted to climate change. Furthermore, the level of food security, profitability and stability of enterprises and organizations of the agricultural and industrial complex (AIC) will increase.

Note, that the introduction of digitalization in various spheres of society causes heated disputes among scientists. Thus, the philosophical understanding of this process is today one of the most relevant areas of study of the problems and prospects of IT technologies, considered in the context of NBIC convergence [1, 12, 13].

Digitalization processes are based on the formation of infrastructure, ecosystem and culture of ICT users. The agricultural and industrial complex is positioned as one of the most dynamic and promising points of ICT application. Thus, in 2018, the volume of exports of services using digital technologies amounted to 50% of world exports of services, and the size of the digital economy, according to estimates, was in the range of 4.5-15.5% of global GDP. Nearly 40% of the value added in the global ICT sector comes from the United States and China. These two countries account for 75% of patents with blockchain technology, 50% for the Internet of Things, over 75% of the cloud computing market, and 90% of the market value of the largest 70 digital platforms [2, 3].

The domestic market of digital technologies in the agricultural and industrial complex in 2020 is estimated at RUB 360 billion, and by 2026 it should grow 5 times. In 2019, the highest costs for the implementation and use of digital technologies in the Russian Federation were observed in the Central Federal District (RUB 1,717,059.7 million). (Table 1).

Table 1. Costs for the implementation and use of digital technologies in 2019, RUB million.

Russian Federation and its constituent entities	Total	Including	
		Internal expenses	External expenses
Russian Federation (RF)	2316831.4	1828652.5	488178.9
Central Federal District (CFD)	1717059.7	1365467.4	351592.3
Northwestern Federal District (NWFD)	127356.0	95509.9	31846.1
Southern Federal District (SFD)	46607.0	35725.8	10881.2
North Caucasian Federal District (NCFD)	13802.3	12405.5	1396.8
Volga Federal District (VFD)	176901.6	141721.1	35180.5
Ural Federal District (UFD)	106103.4	76775.0	29328.4
Siberian Federal District (SFD)	81955.8	62394.1	19561.7
Far Eastern Federal District (FEFD)	47045.6	38653.7	8391.9

The source: [7]

The influence of social and economic transformations and geopolitical factors on the agricultural and industrial complex led to an increase in the indicators of the agricultural sector of the domestic economy, which was not a high-tech industry and did not show a significant increase in labor performance and efficiency. Digital transformation is most in demand today in crop production, processing of agricultural products, as well as their collection, storage and logistics.

Since 2010, organizations of the constituent entities of the Russian Federation have seen a significant increase in the use of local area networks and servers, and since 2015 - "cloud" services (Fig. 1-4).

Modern economic development of agricultural territories is impossible without state regulation of digitalization processes. In the Federal Scientific and Technical Program for the Development of Agriculture for 2017-2025 and the National Technological Initiative (NTI), great importance is attached to IT issues in agriculture. This is reflected in the FoodNet and AeroNet Roadmap, the concept of smart agriculture and priority projects totaling up to RUB 3.3 billion. The departmental project "Digital Agriculture" of the Ministry of Agriculture of the Russian Federation contained 5 units that ensure the adoption of intelligent digital solutions in order to increase labor productivity and sustainable development of agricultural territories. The process of digitalization of the Russian economy lags significantly behind developed countries. Thus, according to AB InBev Efes, the share of IT specialists in the total number of employees in the agricultural sector in the USA, Germany and Great Britain, respectively, is 4.3; 4.5 and 4.1%, while in Russia this indicator is at the level of 2.4% [4].

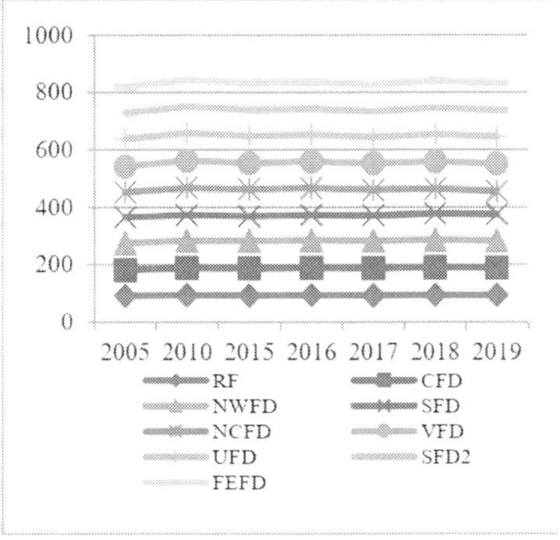

Fig. 1. Personal computers,% [7].

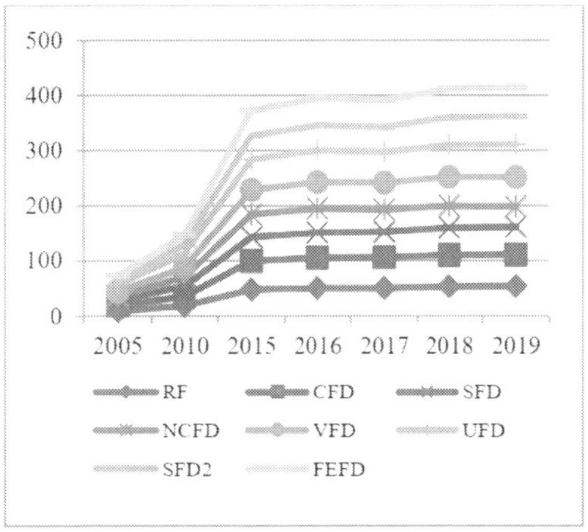

Fig. 2. Servers,% [7].

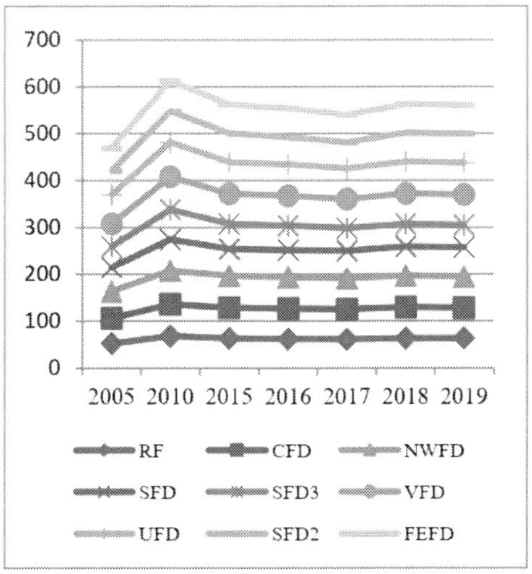

Fig. 3. Local area networks,% [7].

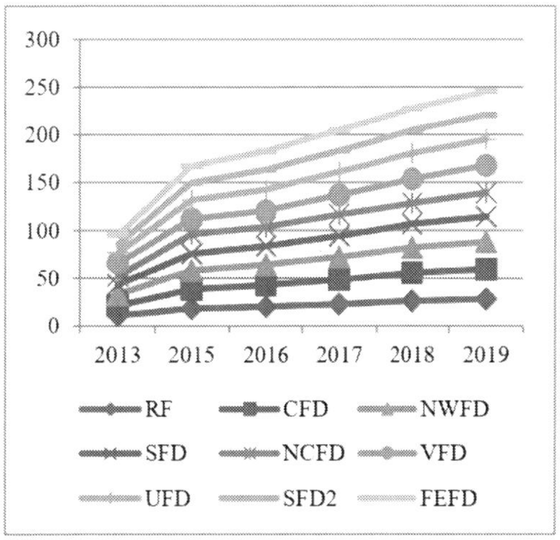

Fig. 4. Cloud services,% [7].

In the context of defining a path for sustainable development, digital agriculture can provide economic, social and cultural, as well as environmental benefits: the former - by increasing productivity, spending money efficiently and wisely using market opportunities; the latter - by expanding communication and increasing inclusiveness; still others - by optimizing the use of resources and adapting to changing climatic conditions.

The digitalization process has led to a change in the social and economic paradigm of social development. Digital transformation using advanced IT technologies has reshaped the picture of competition, blurred boundaries, changed business models [5, 10, 15]. To unleash the full potential of agricultural digitalization, governments should consider creating and developing an enabling regulatory environment. A high level of administrative capacity is required to develop and effectively manage e-Government projects. In closing the digital divide, governments must initiate the participation of farmers, investors in the agri-food sector, and encourage start-ups to digitize smallholder farmers.

Reaping the potential benefits that are expected from the digitalization of agricultural ecosystems, in full, presupposes a holistic, systems approach. In this regard, the agricultural production systems, the rural economy and the management of natural resources must change dramatically. The formation of the digital economy in the agricultural sector is aimed at the transparency of the business environment, attracting investment in the development of digital literacy, the introduction of IT technologies in the agricultural and industrial complex, expanding the relationship of stakeholders in the digital ecosystem, including the public and private sectors, production and the scientific and educational community [8]. Thus, in 2020, as part of the digitalization of Russian agriculture, a single agricultural ecosystem was created: the strategic partnership Digital Agro (Uralchem subsidiary), Agrosignal and Cognitive Pilot, which are part of the SberX ecosystem. The partnership was formed to create a unique product that will combine full cycle agro-ERP and unmanned vehicle management technologies. The expected result is a software and hardware complex that will be able to control equipment, collect, process and transform telemetry data in an unmanned mode, as well as conduct all economic activities of the organization using the capabilities of a single digital field. The total volume of products, services and facilities sold by the parties to the agreement will amount to up to RUB 20 billion per year by 2023 It is planned to introduce these technologies in 30% of the agricultural and industrial complex market in the amount of over 20 million hectares. The agricultural ecosystem includes a number of elements. In particular, we are talking about the interface; algorithms by which Big Data is analyzed and processed; mobile applications for users from Digital Agro; Cognitive Agro Pilot software and hardware complex, a system that allows autonomous control of agricultural machinery using artificial intelligence (AI); the digital platform "Agrosignal", which is the core of the "Digital Agro" ecosystem (it will serve a base for three technologies integrated into a single agro-ERP). The agricultural ecosystem assumes the possibility of solving all the practical problems facing the economy in one interface; application of accumulated agronomic knowledge and modern technologies in the field of soil science; rational use of equipment and increased control over the implementation of agricultural work. The users can control crop rotation using historical data, constantly updated information from sensors installed on agricultural machinery, knowledge of the soil and climate in their farm, using the Digital Agro application on a tablet. With the help of numbers, you can control all processes and data, set precise tasks for machine operators and monitor their implementation in real time.

Digital transformation of processes on farms will help reduce risks and increase the potential yield of agricultural crops during seasonal work, optimally reduce their cost as a result of reducing losses of a non-production and production nature. The services and solutions of Digital Agro are focused on reducing costs up to 20%, thanks to the consolidated perception of information about the farm and the ability to manage it in real time. As a result of the application of Cognitive Agro Pilot technologies, it is expected to reduce crop losses by 3% and increase the harvesting speed by 25%. The introduction of all technologies of the agricultural ecosystem will provide a cumulative effect in the form of an additional reduction by an average of 13% in direct and indirect costs.

Three problems are identified that hinder the development of the digital transformation of agriculture in Russia. The first is the lack of competencies possessed by owners and managers of agricultural formations; the second is in an extremely small number of organizations that could actually do something in this direction; the third is that, as a rule, many proposals are unsuitable for the majority of small and medium-sized farmers that produce most of all agricultural products.

In this regard, solving the problem of digitalization of agriculture requires taking into account a number of factors:

1) lack of systemic and official data, especially their digitalization in urban and rural areas; incomplete information on government support and regulatory support for digitalization. This prevents understanding of the very concept of digital transformation of agriculture;

2) significant gap in the introduction of digital technologies into the agricultural sector of developed and developing countries, both in global companies and in local and family farms; the introduction of modern agricultural technologies is determined by the availability of financial resources and the level of education;

3) economies of scale: the larger the enterprise, the easier it is to implement digital technologies. In other words, large farms have advantages over small ones. This leads to the emergence of inequality between them and, therefore, inequality between developed and developing countries.

Thus, the digital inequality of agricultural territories due to different levels of access to digital infrastructure does not allow the use of a single set of measures for regions and agricultural formations.

The priority areas in the field of digitalization are:

- collection of more complete information reflecting the applied digital technologies and the process of digital transformations by regions, including separately urban and rural, and population groups;
- creating sustainable business models that allow small farms to be involved in the digital transformation of agricultural territories;
- development of an index reflecting the development of digital agriculture, taking into account the existence of basic and accompanying conditions for the development of digitalization, as well as the potential impact of the process on the economy, ecology and society.

The "Index of Readiness for Implementation of Digital Technologies in the Agricultural Sector" is considered as a basis. It was developed in 2015 by experts from the FAO Regional Office for Europe and Central Asia. The stages of the implementation of the process of digital transformation of agricultural territories have also been determined.

Digitalization affects the increase in the efficiency of economic processes, the growth of competitive advantages, the reduction of unemployment, the rating of countries in the world markets, the development of digital payment systems and electronic money, contributes to the formation of a synergistic effect of competition in commodity markets [5, 6, 8]. The positive is the possibility of digital technologies in solving problems of performing large-scale operations; growth in labor productivity; providing additional benefits to consumers and increasing participation in social and political life. However, there are also negative consequences, in particular, the suppression of the individual by digital addiction and cybersecurity. Moreover, the benefits of digitalization are uneven due to limited access to the Internet; there is an increase in the polarization of the labor market; the influence of natural monopolies is increasing. In this regard, it is impossible to consider digitalization unequivocally. Positive and negative characteristics in the aggregate suggest multivariance of the final results, caused by changes.

It is important to note that the digitalization of the economy will have a significant impact on the transformation processes of labor migration in the future. It caused a change in the behavior of economic agents: a special social group of people appeared, which is engaged in intellectual work and is not tied to a specific place of residence. Digital globalization has caused transformational processes in the migration space and labor relations, including by increasing special interest in highly qualified workers employed in promising sectors of the economy. As a result, the share of highly qualified specialists in the structure of international migration has grown significantly.

To implement digital transformations with the required level and scale, especially in agriculture with its industry specificity, the following conditions must be met:

- fact of the existence of the technologies themselves, characterized by a certain quality, sufficiency of quantity, readiness to work (connectedness), availability, including financial, is the same for all users - both legal entities and individuals;

- formation of the ecosystem of the digital economy, by which we mean a set of economic agents and conditions ("habitat"), with the possibility of their effective functioning, thanks to the comprehensive connections between them, on the basis of which they interact with each other, exchange resources and transform them, carry out relationships with each other;
- available infrastructure combined with its efficient use;
- corresponding "digital" culture. To fulfill their needs and interests, the users must have basic digital skills, be digitally literate and competent.

Obviously, the place of residence (type of settlement), a city or a village, has a significant impact on the digitalization process as a whole, including due to the "cumulative effect" of factors such as income level, education level, lack of infrastructure, remoteness from communications, etc. Therefore, the digital skills of the urban population when working, e.g., with a word processor and spreadsheets, as well as when transferring files from a computer to peripheral devices and vice versa, are 10-15% higher than that of rural users. Such conclusions are valid for almost all indicators reflecting the digitalization of agriculture. A relative exception is the electronic display of state and municipal services to the public. Here the gap between urban and rural areas is reduced to about 10% [9].

The main determinant of economic development, both at the national and regional levels, is the use of digital technologies in order to reduce production costs, meet the needs of the market, since the volume of information is growing squared faster than the volume of production [10, 11, 14]. The priority task of the digital transformation of the agricultural economics is the integration of objective information flows of agricultural producers and government data into the digital agriculture platform in order to provide accurate recommendations to market participants, using artificial intelligence and activating innovative processes. Pilot projects will ensure, through big data, the digital transformation of the agricultural sector of the economy with the formation of digital platforms. Currently, only 10-15% of Russian farmers use digital platforms. With regard to drones and sensors that allow collecting information about the weather, controlling the operation of equipment, monitoring crops and other key agricultural processes, the share of coverage is higher here, and, therefore, the picture looks more optimistic.

4 Conclusions

It is important to note that it is necessary to create IT platforms based on the digitalization of production and technological processes, jobs; logistics, collaboration of agribusiness entities with public-private partners, consumers, competitors, investors and authorities. The basic requirements for the creation of digital platforms have been developed for the following groups:

1) high availability (in terms of economic entities of the agricultural and industrial complex with advanced technologies);

2) relative availability (in terms of enterprises of agricultural service, food and processing industries, trade, poultry farms, stable and profitable functioning);

3) adaptable (in terms of medium and large agricultural enterprises that have sufficient material, financial and labor resources in order to use modern information technologies);

4) not capable of independent adaptation (we are talking about unprofitable enterprises, small peasant (farmer) and other households).

Regional pilot projects will allow, using big data, to digitize agriculture based on digital technologies, and to activate them in everyday life. It is advisable to develop new tools that allow the state, business and society to interact on the principles of inclusiveness, to expand the participation of society in determining science and technology policy. Moreover, it is very important how the rural population will perceive advanced technological solutions, which is closely related to the level of requirements for digital competencies and their dynamics.

The digital economy presupposes a new stage of introducing the achievements of scientific and technological progress in all sectors and spheres of social and economic activity, as well as in the foreign and domestic policy of the state. The created platform solutions for digital agriculture have a goal: to form a modular platform that will be able to globally predict supply and demand. In this case, various types of information will be used: from satellites, soil and climatic, from tax, customs, statistical authorities and others. The information will have established formats and will be provided online. For efficient agriculture, the transition to management based on planning and risk forecasting and economic security, interfaces are required that allow connecting logistics and transport systems and feedback data processing systems ("counter-manufacturer") in order to maximize consumer preferences. Furthermore, the role of non-profit associations (unions) in the scenario of digital transformation of agricultural ecosystems, the implementation of innovative solutions, which will lead to the formation of a private-state digital agricultural and industrial complex management system, is growing.

Acknowledgment

The study was conducted with the financial support of the Russian Foundation for Basic Research within the scientific project No. 20-010-00838 A.

References

1. J. Canton, *Managing Nano-Bio-Info-Cogno Innovations: Converging Technologies in Society*, 33-45 (2006)
2. Digital Economy Report. 2019. UNCTAD (2019)
3. European Commission. 2019. 2nd Survey of Schools: ICT in Education (2019)
4. S. A. Gurfova, Economics and Entrepreneurship, **3 (116)**
5. Indira R. Mikitaeva, Marina T. Tekueva, Mikhail Kh. Balkizov, Tanzilya Kh. Sozaeva, Orgutsel davranis arastirmalari dergisi, **3(2)**, 189-197 (2018)
6. Kh. M., Rakhaev, T. Kh., Sozaeva, S. Kh. Sulumov, *Spatial Polarization of National Economy: Current State and Development Priorities*, 50-55 (2019)
7. Regions of Russia. Social and economic indicators, Rosstat (2020)
8. T. Kh. Sozaeva, A. Y. Pshigosheva, A. M. Tramova, T. Kh. Toguzaev, International Journal of Recent Technology and Engineering, **8(3)**, 8286-8292 (2019)
9. T. H. Sozaeva, A. Yu. Pshigosheva, S.A. Gurfova, I. R. Mikitaeva, *Agrarian territories in the context of the formation of a digital economy: problems and prospects* (2020)
10. S. A. Tumenova, T. Kh. Sozaeva, Knowledge Economy: Noospheric Context of Development European Proceedings of Social and Behavioural, 434, 3226-3236 (2019)
11. S. A. Tumenova, M. M. Mustaev, F. A. Mambetova, European Proceedings of Social and Behavioural Sciences, 92, 339, 2561-2568 (2020)
12. A. V. Frolov, Innovations, **7(177)**, 63-73 (2013)
13. J. A. Schumpeter, *The Theory of Economic Development* (1982)
14. I. L. Litvinenko, Z. S. Zhangorazova, Y. B. Mindlin, R. A. Shichiyakh, N. Y. Veselova, G. V. Petruk, International Journal of Applied Business and Economic Research, **13** (2017)

15. Kh. Uyanayeva, N. Kokov, T. Toguzayev, M. Kuchukov, F. Alibius, 3rd International Conference on Agribusiness, Environmental Engineering and Biotechnologies, AGRITECH-III 2020, 2 (2020)

Methodological approaches to the analysis of the functioning of agricultural ecosystems in the context of anthropogenic environmental changes

Laura Khalishkhova*, *Anzhela* Temrokova, *Inga* Guchapsheva, *Oksana* Bagova, and *Madina* Marzhokhova

Kabardino - Balkarian State Agricultural University named after V.M. Kokov, Lenin Avenue, 1V, 360030 Nalchik, Russia

Abstract. The study provides methodological approaches to the analysis of the functioning of agricultural ecosystems in the context of anthropogenic environmental changes, based on the assessment of rationalization and efficient use of natural capital through energy consumption for the production of final products, and describes the agricultural ecosystem operating on the principles of market economy. A set of general methodological principles for analyzing the functioning of an agricultural ecosystem has been proposed. An algorithm for achieving optimal ecological and economic parameters of the functioning of the components of the agricultural ecosystem is presented, including a set of step-by-step actions, as well as multi-functional dependences of the reproductive capabilities of the main components of the agricultural ecosystem. A complex of ecological and economic parameters of the reproductive process of the agricultural ecosystem, used for its optimization, has been developed. The interpretation of terms and concepts, generalized in the course of the research within the framework of the presented approach, is carried out. A new principle of agricultural ecosystem organization is highlighted, considering it as a social, ecological and economic system.

1 Introduction

Intensification of human production activity, causing an increase in technogenic pressure on agricultural ecosystems, leads to a disruption of biological equilibrium, which manifests itself in a significant depletion of agrocenoses due to the death of useful microflora components, disorganization of microbiological processes in soil biota, production of phytotaxic elements, an increase in soil fatigue rates, transformation of the nature of infection plants, negative transformations in the immune status of cultivated plants. Self-limitation of the rate of continuous growth of productivity is the main reaction of the agricultural ecosystem in the established manifestations.

* Corresponding author: L a u r a@inbox.ru

The thesis put forward actualizes the research devoted to the formation of methodological approaches to the analysis of the functioning of agricultural ecosystems in the conditions of anthropogenic changes in the environment and the identification of functional relationships of the structural components of agricultural ecosystems with substantiation of the optimal parameters of their functioning. The solution of the tasks set will provide the requested level of environmental and economic efficiency, taking into account the limits of technogenic impacts on agricultural ecosystems, and will create conditions for their functional sustainability, as well as develop approaches to managing the agricultural ecosystem as an integral self-organizing structure functioning under the principle of operational closed condition.

2 Materials and Methods

In the process of functioning, the agricultural ecosystem concentrates substances and energy, transforming them into a new form, replenishing the expended elements in the form in which the dynamic optimum is maintained. The more nature contributes to the functioning of the agricultural ecosystem, the more efficient and cost-effective it looks. Moreover, the assessment of the efficiency of the functioning of the facilities, including agricultural ecosystems, is carried out within the framework of a generally accepted approach based on the theoretical provisions of a market economy: the ratio of the cost of the final product and its market price, the ratio of income and the cost of fixed assets. However, in our opinion, this approach presupposes some distortion of the actual production costs. In market environment, the products that have the lowest cost with the same market price are more competitive. Furthermore, the cost reduction is due to the performance of the natural resources involved in production processes. Thus, the improvement of economic parameters, and moreover, the progressive dynamics of social and economic development, presuppose a non-linear growth of the consumed natural energy.

In the framework of the modern economic approach, natural resources are not endowed with value, since no human labor is invested in their creation. The cost arises only at the moment the resource is processed into a new substance, or energy. Thus, the development of machinery and technology provides a reduction in the cost of a unit of production. We do not fully and completely share these views, since the cost of the final product does not take into account the cost of natural resources.

In the existing system of economic coordinates, labor is a meaningful, purposeful activity that is aimed at transforming natural components to meet the needs of the population. Therefore, labor is considered only that activity that is aimed at the removal of these natural components and their processing. However, in the conditions of industrialization and the gradual depletion of natural capital, another type of labor activity arises, associated with the restoration of damage to the environment as a result of its incorrect exploitation. Labor activity in this case does not contribute to the growth of incomes in the implementation of the economic process, which means that the cost of products increases if we translate the calculations into the energy calculus plane. The authors of the developed approach cite important parameters for the development of the social economic system in their works: the maintenance of society in energy terms costs an average of 270 thousand kcal/day per person, and an increase in gross output in agriculture by 1% is achieved by increasing involvement energy capacities by 3%, and the production of 1 ton of useful products is accompanied by the formation of 99 tons of waste [1, 2].

Thus, the energy required for the development of the social and economic system is formed as the sum of the costs of energy created by man and the energy produced by the natural system:

$$E = E_{man} + E_{natur}$$

An objective measure for assessing the rationalization and efficient use of natural capital is, among other things, the measure of energy consumption for production [3]. In our opinion, this is the only reliable measure of the profitability and economic efficiency of an agricultural ecosystem.

The set of general methodological principles for analyzing the functioning of the agricultural ecosystem is:

- the principle of self-organization and orderliness of the expenditure of matter, information and energy, ensuring the formation of the integrity and stability of the agricultural ecosystem, characterized by fractal reversibility (systems that transform their structure only to the extent that external conditions are transformed [4]);
- the principle of consistency of the goals of the development of the natural and social system, which ensures directional movement towards sustainable and harmonious functioning;
- the principle of "we work for ourselves, we work for others", which focuses on taking into account the contribution of nature to meeting the needs of the social and economic system (the contribution of nature to the development of the social and economic system in comparison with the contribution of the social and economic system itself);
- the principle of managing the agricultural ecosystem as a self-organizing structure, characterized by an integral structure, functioning according to the rules of conceptual and operational isolation and self-regulation due to permanently decreasing external energy and energy generated by the agricultural ecosystem itself;
- the principle of energy efficiency of the functioning of the agricultural ecosystem, aimed at the study of energy flows and the study of energy consumption, the growth of target output parameters, without environmental degradation [5].

3 Results and Discussion

This study outlines methodological approaches to the analysis of the functioning of agricultural ecosystems in the context of anthropogenic environmental changes and provides a description of the agricultural ecosystem operating on the principles of market economy. Within the framework of the presented approach, it becomes necessary to interpret the applied terminology and concepts generalized in the course of the study, including the functional stability of the agricultural ecosystem, the complex of ecological and economic parameters of the reproductive process of the agricultural ecosystem, the invariant of the structure of the agricultural ecosystem, the conceptual and operational closure of the agricultural ecosystem, self-organization and self-regulation of the agricultural ecosystem of development of the attractive goal of the agricultural ecosystem.

The functional stability is a significant sign of agricultural ecosystems, characterizing its resulting efficiency and is a dynamic optimum in the parameters and system functions of the agricultural ecosystem components, which is based on verified and empirically substantiated regulatory and design solutions [6, 7].

The functional stability of the agricultural ecosystem to the effects of biotic and abiotic factors is largely determined by the anthropogenic component and is due to the observance of the consistency of the interrelationships of the components that set the state "anthropogenic load - the ability to self-reproduction". Establishment of dependencies between the reproductive abilities of the components of the agricultural ecosystem should take place within the framework of the rules for the functioning of complex natural and technogenic systems on the basis of a set of economic, ecological, social, and biological laws. The methods used to identify the relationships between the reproductive abilities of the agricultural ecosystem include empirical and statistical, correlation and factor analysis. Within the framework of the indicated methods, biogeochemical, dynamic, balance and nonlinear models are used that establish multi-functional dependencies. The use of these models in the analysis makes it possible to obtain a fairly accurate statistical estimate with the determination of the degree of influence of anthropogenic impact on the reproductive capacity of the agricultural ecosystem component. The calculations make it possible to identify and substantiate the directions for reducing technogenic pressure and form a toolkit for ensuring the sustainable functioning of the agricultural ecosystem.

Egorov E.A. verified multi-functional dependencies of the reproductive capabilities of the main components for agricultural ecosystems with the placement of fruit plantations. Thus, exceeding the standard parameter of soil compaction by 1% or more affects the decrease in the yield of fruit crops by 0.5%; exceeding the permissible pesticide load by 1% or more reduces the reproductive potential of plantings by 2.7%; increasing the use of the biological method (replacing chemical pesticides with biologically active supplements) reduces the residues of insecticides and fungicides in the soil by 1.62%, increases the yield by 0.6%, reduces the cost of protective measures by 3.4%, on production in general by 2.5%, which leads to an overall increase in profitability by 2.4% [7].

In modern conditions, agricultural ecosystems should be tuned to reduce the technogenic component and reduce the consumption of resources used in the production process. The key criteria for optimizing the ecological and economic state of the agricultural ecosystem and its components are the limiting minimization of the negative anthropogenic impact on the environment, the maximization of the productive potential of the agricultural ecosystem, and the maximization of the yield per unit of consumed technogenic resources.

The main methods for achieving the indicated criterion values are a selection of plant varieties adapted to the corresponding agricultural ecosystem, rational placement of fruit plantations, and the use of biomass. The indicated methodological approaches allow the development of instructions and regulations for production activities in the agricultural ecosystem.

The algorithm for achieving optimal ecological and economic parameters of the functioning of the components of the agricultural ecosystem includes a set of step-by-step actions:

- formulation of a local task within the agricultural ecosystem;
- development of a system of assessment indicators;
- identification of interdependencies and interactions between the components of the agricultural ecosystem;
- formation of a system of criteria for optimizing the ecological and economic state of the agricultural ecosystem;
- modeling of processes and calculation justification of the ecological and economic parameters of the system.

The functional stability of the agricultural ecosystem is associated with the achievement of a dynamic optimum in the system functions and parameters of its biological components for the realization of the production potential, which is achieved primarily on the basis of adaptability, i.e., the wide involvement of biological resources in reproduction processes.

Zhuchenko A.A. considers the construction of sustainable agricultural ecosystems based on an increase in the species and genotypic diversity of cultivated plants, their adaptive distribution; genetic determination of the ability of agricultural ecosystems to utilize the natural and anthropogenic resources of the environment with the greatest efficiency in the process of life as the main factors of biologization and ecologization of intensification processes in crop production [8].

In his studies, Shchukin S.V. considers the choice of species and breeds, taking into account the produced crop products, as well as with local agricultural and climatic conditions; calculation of the optimal number of animals depending on the existing resource base and cultivated crops; ensuring the functioning of the agricultural ecosystem with the minimization of the used fossil energy and the leveling of the harm caused to the environment among the main factors of biologization and ecologization of animal husbandry [9].

Table 1 shows the generalized parameters of the ecological and economic state of the agricultural ecosystem and the reproduction process, the observance of which ensures functional stability and a given level of efficiency. Modeling the functional stability of the agricultural ecosystem allows you to obtain a reliable statistical assessment of the systemic relationships that are formed within the agricultural ecosystem, and formulate regulations for the impact on its reproductive potential with a calculated justification for reducing technogenic pressure in a set of areas, as well as develop tools for achieving sustainability parameters.

Table 1. Complex of ecological and economic parameters of the reproductive process of the agricultural ecosystem, used for its optimization.

Stage of the reproductive process	System component	Estimated indicator
Agricultural ecosystem		
Formation of means of production (distribution chain)	biocenosis	
	plants (producers) animals (consumers)	stress fluctuation of the realizability of the production potential of system components
		acceptable levels of pesticides
		preservation of biologically inherent yield and productivity
	soil microbiota (destructors)	Physical and chemical and particle size distribution of the soil; density; moisture supply and mineralization; humus content; the volume of microbial biomass; the ratio of symbiotes and pathogens; pesticide residues
	agrocenosis	
	agrotechnical regulations	implementing the potential of yield and productivity
		indicators of excess of the average economic yield and productivity
		break-even point indicators
		relative indicators of the functionality of the agricultural ecosystem
	infrastructure systems	relative performance indicators of the use of infrastructure systems
Social and economic system		
Implementation of the production process (production area)	production and technological activities	need for infrastructure, human and natural resources
		return on capital investment; income to capital investment ratio
		identification of optimal proportions in the ratio: yield/productivity - costs - income
		ratio of fixed and variable costs; ratio of income and costs; ratio of income and working capital
		integral indicator of resource efficiency
Market system		
Conversion of produced value into monetary value. (distribution chain)	commodity, economic and marketing activities	indicators of product standardization; product quality indicators
		indicators of product competitiveness
		indicators of profitability of production and sales

Identification of the optimal range of indicators of the ecological and economic efficiency of the agricultural ecosystem highlighted in the table creates the basis for increasing its capabilities, taking into account the combination of factors of the external and internal environment and ensuring the growth of qualitative and quantitative factors, which will form a solid basis for sustainable development.

The invariant of the structure of the agricultural ecosystem consists of a set of elements of the social and economic system and ecosystem, including the components of the natural environment, society and the infrastructural basis of a certain territory, taking into account its area and ecological capacity. The presence of these elements creates the basis for its stability and ability to self-reproduction, determines the formation of coordinated multifunctional interconnections of the components that set the state of "anthropogenic load - self-reproduction". The establishment of dependencies between the reproductive components of the agricultural ecosystem occurs within the framework of the rules for the functioning of complex natural and technogenic systems on the basis of a set of economic, environmental, social, and biological laws. The invariant structure of the agricultural ecosystem is characterized by the ratio of matter and energy flows along the formed food chains: "plant - cattle - man" and "plant - man". Thus, specialization of the agricultural ecosystem arises. The society determines the choice of primary and secondary biological products, forms a set of plant and animal species, as well as methods of growing them, which will ensure the highest productivity and yield. The state of the agricultural ecosystem resources, i.e., soil properties, biodiversity, hydrological and hydrochemical regimes of the agricultural ecosystem are largely determined by the parameters of agricultural ecosystem exploitation and the level of technogenic pressure. As a result of the management of the agroexystem by the society, the "conservative" circulation of matter and energy is accelerated, and it goes into an abiotic state. The agricultural ecosystem suppressed and transformed the mechanisms of self-regulation inherent in natural systems, which leads to a decrease in resistance. The intensity and direction of the processes of exchange of matter and energy between cultivated crops and raised animals determines the productivity of the agricultural ecosystem.

Conceptual and operational closure of the agricultural ecosystem is separation from the environment in the sense of the independence of the results of recursive evolution and structural accommodation from the influences of the external environment. This is a system property determined by a set of processes, which is clearly visible only within the system. In this case, we are not talking about the isolation of the agricultural ecosystem, but only about the operational closure, i.e., its self-sufficiency. The notion of "structural fit" helps explain the relationship of the agricultural ecosystem to the environment, and how the system is adapted to it. Put in other words, the mutual consistency of the agricultural ecosystem and the environment determines the mechanism of coexistence of systems and forms a space of opportunities for each of them. Structural correspondences do not affect or limit the self-determination of the agricultural ecosystem. Furthermore, they, being reflected in the forms and operations of the agricultural ecosystem, will make it possible to form a structure that will be ready for resonant processes. In this case, we are talking about identity achieved by structural correspondence, that is, a completely different composition of elements of the agricultural ecosystem and the environment, with deep structural coherence, allows them to act as a form of something one. Structural conformities limit the agricultural ecosystem's capabilities, but leave it free at such an optimal level that it can behave "unadapted". The adaptation of the agricultural ecosystem to the surrounding world occurs by compensating for the internal excess of its capabilities, reacting with various limitations of these capabilities. By the closed nature of the agricultural ecosystem, we mean "operational closed condition", not thermodynamic one. The operations carried out by the agricultural ecosystem in the process of functioning do not allow it to go beyond the limits of functioning; on the contrary, they ensure its closure. Otherwise, the agricultural ecosystem would not be able to isolate itself and form the boundaries of the complexity of the external world, and, therefore, would not be able to arise and exist. Closed condition, at the same time, concerns only the operations of the agricultural ecosystem, acting as a condition for its openness and presupposes purposeful development towards a certain spontaneously formed given state.

Self-organization and self-regulation as properties of an agricultural ecosystem are a symbiosis of its structure and processes in which a complex dynamic system is formed, improved and reproduced due to the supply of external energy. This is an immanent property of a complex system that tends to a state of dynamic optimum. In this case, we mean the imbalance of the inflow and outflow of matter and energy, which initiates the presence of a positive "balance", which acts as the basis for the self-development of the agricultural ecosystem. The alienated energy and substance of the agricultural ecosystem concentrates and spends on self-improvement and strengthening of the organizational structure. We call this property self-organization of the agricultural ecosystem. From the standpoint of self-organization, agricultural ecosystems are social and ecological and economic formations that, resulting from the mutual expediency of these systems, achieve maximum stability and productivity.

The systems are considered self-organizing because they are formed without an estimated goal. Their goal is attractive in the sense that, being formed, the system entails a process, since it acts as the most effective opportunity for the production of products. We call an attractive goal a goal that arises under the influence of interrelated events at the initial moment, which predetermine the development of the process towards a given goal, all other things being equal. Put in other words, the task is not to identify the optimal transition to the desired goal, but to organize the state of the system according to the goal, that is, to create in the present an attractor of the future that initiates the present. Self-organizing holistic agricultural ecosystems are endowed with their own attractor, that is, such a state that they form together with the environment, and which they can achieve if all the initial conditions of the internal and external environment remain unchanged throughout the entire period of movement of the system towards achieving its goal. However, under natural conditions in agricultural ecosystems, on the way to achieving an attractive goal, random or non-random events occur that change the attractor as such. Thus, the attractive target becomes wandering or "strange", changing its coordinates in space and time. It acts as a limit, in the process of approaching which the development of the agricultural ecosystem stagnates, the processes taking place inside it stabilize, and the system reaches the level of sustainable development and dynamic optimum. The paradox is that an attractive goal allows you to form the prospects for the development of an agricultural ecosystem in accordance with its future state. The state has not yet been reached, but it is determined by the future. In a methodological sense, positioning an attractor as a goal of developing an agricultural ecosystem turns out to be quite effective. Attractive goals are critical for all types of agricultural ecosystems.

4 Conclusions

Within the framework of the existing methodology for the functioning of agricultural ecosystems in the management structure, there are no elements that, upon reaching a certain predetermined level of productivity, would compensate for the harm caused to the environment from further growth of its performance [10]. Once this level of productivity is reached, the agricultural ecosystem should switch to intensive methods that increase resilience and productivity. The activity of the social and economic system should be anti-entropic in relation to the ecological system in which it is located. There is a given state of the agricultural ecosystem, which it must achieve after a specified time, including the qualitative and quantitative parameters of the products produced, as well as the structure, forms and amount of consumed energy, information and mass of matter. Social and economic systems are able to identify their given state. Thus, the agricultural ecosystem can be based on new principles of organization and be considered as a social, ecological and economic system, provoking the inter-penetration of systems, which presupposes the improvement of their condition. The status of the agricultural ecosystem and the output characteristics are changing. In addition to cost-effectiveness, sustainability and productivity of the system arises. The part of the ecosystem that is part of the agricultural ecosystem should be transformed into the plane of market relations and operate on the principles of "supply - demand - price", with the determination of prices in energy and monetary terms. The governing body of the agricultural ecosystem must implement the function of comparing the given state of the agricultural ecosystem with the one achieved at the current moment. A given state is formed for specific goals and is, in principle, achievable. The ratio of the given and current state allows you to establish the action of negative feedbacks and assess the performance of the agricultural ecosystem, as it moves towards a given state. Energy analysis and the results of energy assessment of production activities due to the functioning of the agricultural ecosystem allows one to compare agricultural ecosystems of various specializations, to quantify the share of energy introduced by the society and nature. Within the framework of the proposed methodological approach, the productivity and efficiency of the agricultural ecosystem are determined by the energy contribution of the ecosystem. We consider this circumstance to be a strategic guideline for the development of agricultural ecosystems.

Acknowledgments

The study was supported by the Russian Foundation for Basic Research, project No. 20-010-00935A

References

1. T. Odum Howard, *Environmental accounting: EMERGY and environmental decision making* (1996).
2. A. S. Mindrin, *Energy and economic assessment of agricultural products* (1997)
3. V. S. Chesnokov, Globalization Studies, **2**, 181-187 (2010)
4. A. V. Pozdnyakov, The Earth's Surface, Layered Relief and the Rate of Relief Formation: Proceedings of the Irkutsk Geomorphological Seminar, 223-225 (2007)
5. O.V. Feldman, E. A. Denisenko, D. O. Logofet, *Problems of the Environment and Natural Resources*, 66-81 (1998)
6. E. A. Egorov, Scientific works of State Scientific Institution "North Caucasian Research Institute of Horticulture and Viticulture", 2, 7-21 (2016)
7. E. A. Egorov, Zh. A. Shadrina, G. A. Kochian, Scientific works of State Scientific Institution "North Caucasian Research Institute of Horticulture and Viticulture", 2, 9-14 (2013)
8. A. A. Zhuchenko, Economics of Agricultural and Processing Enterprises, **8**, 1-3 (2011)
9. S.V. Schukin, *Greening agriculture (transferring traditional agriculture to organic) / Series of training manuals "RUDECO Retraining of personnel in the field of rural development and ecology"* (2012)
10. L. Z. Khalishkhova, A. Kh. Temrokova, M. A. Marzhokhova, O. I. Bagova, Z. Sh. Babaeva, Revista Inclusiones, **7** (2020)

Prospects for national agriculture in the new reality

Madina Eneeva[1], *Madina* Gazaeva[1], *Takhir* Toguzaev[1], *Magamed* Israilov[1], and *Zalim* Kaisinov[2]

[1]Kabardino-Balkarian State Agricultural University named after V.M. Kokov, Lenin Avenue, 1V, 360030 Nalchik, Russia
[2]Regional Fund "Center for Entrepreneurship Support of the KBR"

Abstract. The term "new reality" entered the lexicon of economists in the last decade. It means fundamental changes in the economy, which are predominantly economic in nature, also expanded in engineering and technology, sociology, politics and psychology. In 2020, the "new reality" supplemented with Covid-19. It seems that the reason for its strong influence on the economy is that Covid-19 is becoming a kind of economic factor related not even with the development of pharmacology, epidemiology, etc., but with the fact that it touched one of the most dynamically developing spheres of the economy - services. Agriculture has also faced the "new reality" in a new century. But in contrast to the economy, its "new reality" is formed by qualitatively different factors: climate, weather, social polarization of societies, increased consumption and decrease of the matrix of basic factors of agricultural production. In this regard, a study of the prospects for national agriculture merits attention.

1 Introduction

Modern achievements in agriculture are as follows: precision farming, growing vegetables, mushrooms and berries on the principles of hydroponics, intensive horticulture, robotic dairy farms, as well as cattle feeding complexes, poultry farms, pig farms, etc., actually represent a new wave (a kind of re-) industrialization [1]. These are all industrial technologies. Consequently, we seem to return to the 30s of the last century. Obviously, the above pattern will give a certain growth, however, firstly, it will not be so high, secondly, it will not be so long, and thirdly, it does not exclude new crises, etc., but above all, fourthly, it shows limited efficiency (high total costs with low results). For example, intensive orchards (and horticulture in general), as well as vegetable growing together with similar technologies in animal husbandry, have their own limited (time) period, so-called life cycle (LC). Moreover, according to some analysts, "this miracle of bioengineering cannot live on its own", since it requires active mineral feeding, chemical protection and huge water consumption. The situation in beef cattle production is as follows: poultry production ranks first in terms of profitability (currently existing technologies allow to fatten a broiler in 42 days, to increase the egg-laying capacity and also change the poultry appearance (to decrease density and size of feather, etc.), as well as pigs (pig is brought to the required commercial condition in 6 months). The longest growing process is retained by cattle (it takes 18 months to bring the bull to a lean condition). Under the most favorable conditions, no further than the mid-30s and early 40s. Then comes their natural extinction (this should be kept in mind when investing). Using and investing in such technologies we limit ourselves in the trajectory of development, i.e. we plan our backlog in advance. And that's what matters.

2 Materials and Methods

The theoretical and methodological basis of the study is worked out and represents a theory of systems. At the same time, there are some peculiarities related to food and agricultural products.

2.1 Methods

The study is based on a set of analytical and descriptive methods to identify, formalize and quantify current economic, organizational, technological and institutional ties between various sectors and AIC segment. Among the analytical methods, the emphasis is on the correlation-regression and index methods to quantify the relationship, identify proportions between different sectors and segments. Among descriptive methods, attention is paid to the description of obtained dependencies, identified trends and proportions. The main purpose is to interpret the obtained estimations.

The empirical study is based on the official statistics, systematized in the statistical collections of the FSSS of Russia (Rosstat), its territorial branch for the KBR (TB FSSS Kabardino-Balkariyastat). In addition to the mentioned Rosstat data, the results of works of foreign and domestic researchers are used as an empirical base.

3 Results and Discussion

The majority of experts currently accept that the main problem of the development of agriculture in the country is associated with several factors: firstly, seed material (currently imported seed material is used for most agricultural crops; the domestic material share is generally not more than 10%; moreover, it is poorly adapted to soil local conditions, agricultural landscape, etc.), secondly, technologies (most technologies are resource and energy intensive; the resource and energy consumption per unit of the produced product of domestic agriculture exceeds foreign ones by several times. Besides, the main technologies are either completely foreign or on foreign licenses and patents), thirdly, ecology (the use of intensive technologies has a great impact on soil, water and the surrounding system as a whole. They are frequently used without regard to environmental impact and environmental assessment. Especially negative effects cause the use in areas with ..., which include the territories of the far north, tundra, mountains, waterways), fourthly, the low processing level of the produced product. Similar problems are observed in animal breeding. Primarily, low development of the meat market and meat and dairy industry in Russia as a whole results in high production cost of both beef meat and milk [2]. Experts believe that the reason for the high prime cost of beef and milk is "in the economic inefficiency of the projects chosen by our businessmen". According to experts, the main idea of "cheap meat" is the cooperative model, in which "small family farms integrate and cooperate with larger partners" using a system of labor division. Small family farms raise newborn calves up to 8-12 months. And then the larger partners purchase young bulls with further fattening and slaughter. In the USA, 7,000 thousand farmers maintain 15 million livestock animals.

Meanwhile in Russia, the main meat livestock is concentrated in agricultural holdings, where the cost of calf management is 1.5-2 times higher due to additional costs of managing a huge farm". For this reason, the National Union of Beef Producers (NUBP) promotes the idea of integrating farmers and larger businesses. "This model should be based on families, each contains 150-200 heads. A promising farming business begins with this amount, although both small farmers and private household plots can participate in such cooperation." They believe that by integrating farmers with larger companies, beef farming will grow faster. In our view, the problem (and reason) of the low development of domestic livestock and dairy production has been identified. But we believe that the proposed solution is incorrect, i.e. leading to a dead end. Basically, an evolutionary path of development and problem solution is proposed. But first, let us provide a few comments on this proposal.

At the outset, in our view, it is irrelevant to compare the North American model of organization and management of cattle breeding with the Russian one. Dating back to the pre-Soviet times, the Russian model assumed both landlord farms and farmsteads, and if the former grew into collective (collective-farm), state-owned farms, then the second grew into personal subsidiary farms (private farms or modern peasant farms). This is a historical practice. It seems impossible to change it while remaining within the traditional farming model. Industrialization just only corrected this model, but did not change it due to the significant conditions in Russia: large territories, lack of private ownership of land, etc. But it seems to be a purely technical or even ideological incorrectness, which could be ignored as meaningless, as domestic practice since the 90s and even the end of the 80s has shown that the farmer's way of developing national agriculture for Russia is both dead-end and impossible. And it is not only about psychology, but also about practice. As a matter of fact, in the same source it is reported that "half of the cattle are still in personal subsidiary plots (PSP). Nevertheless, there are less people willing to keep their own livestock. And the increase in beef production by farmers and at agricultural enterprises (for ten years the share of herd of peasant farm enterprises and agricultural firms increased by only 10%) does not compensate the decrease in the output of private household plots. A more significant mistake, in our view, is the proposal to form cooperative and integration ties in the organization of the cattle-breeding segment of the national agriculture. Such model has a so-called attractive character, but actually it causes (and will produce) a slowdown associated with a shortage of investment. This situation is caused by the current economic growth in agriculture, which is associated with industrialization, i.e. has a structural character. The very model of such growth (and development) was first described by J. Schumpeter [3] back in the last century, it is kept on constant investment and innovation, which are of a limiting nature. The point is that increasing investment is required to ensure continued growth; wherein, the growth rate of the investments should not be lower, and the further from the beginning, the higher than the economic growth itself. This paradox of industrial or structural growth is reflected in the PRC, as well as other countries, including our own history [4, 5, 6, 7]. There is one more aspect in this attractive project, which can be illustrated, for example, by the agriculture in Kabardino-Balkaria, one of the most advanced and developed in Russia. The point is that most of the investments are directed to limited sectors and segments that are very attractive (i.e., they give the maximum return since the absence of the competitors in such niche, i.e., there is an absolute advantage). For example, in the KBR there is a crop production, and within it there is a segment related to: a) horticulture (intensive horticulture is the main consumer of investments inside this segment, its share in the total investment in agriculture is bigger than the half); b) vegetable growing (within which, also there is a dominant segment - growing tomatoes in greenhouses); c) viticulture, (with a high share of investments). The fourth sector is animal husbandry, within which the dominant positions are taken by: poultry and cattle breeding (cattle).

Other sectors and segments of the agro-food market: fish farming, berry farming, beekeeping, as well as sheep breeding, goat breeding, horse breeding, etc., occupy a rather small share in the investment system. Due to the listed sectors and investments in them, the agriculture of the KBR is among the leading in the technological and technical equipment of industries in Russia [8, 9]. Particularly, we proceed from the existing criteria, for example, the volume of grain production per capita exceeds 1000 kg, the republic has achieved a kind of food security, has a high level of provision of the population with vegetables, fruits, as well as meat, etc. The share of the products of these agricultural sectors in the food export of the KBR is large [10]. However these agricultural sectors (and even segments) continue to receive an active flow of investments, direct and indirect, due to various kinds of preferences. At the same time, experts do not notice a decrease in return per investment unit. According to the laws of economics, it should be expected, on the one side, an outflow of investments, and on the other, a change in the system of preferences [11]. And if the first is stimulated by the appearance of new sectors and the overflow of capital into them, then the second is stimulated by the first and by the compression of the total volume of investments [12]. Meanwhile to maintain this sector in the existing parameters: employment, share in GRP, share in different levels of incomes to budgets of different levels, etc., it will be necessary to increase the volume of investments in it. In this context, the most important criteria to predict the situation in this sector of agriculture is the assessment of the life cycle (LC) of the created sector and the industries concentrated in it. According to approximate estimates (since exact estimates related to foreign analogs differ, and there are no such data about domestic production), the length of the life cycle of the industrial technologies is 10-15 years. But this period is not final. Due to the acceleration of time in the conditions of new scientific, technical and industrial revolutions it seems to be reduced. There is a reason to reduce this period to 8-12 years. So what that mean?

Primarily, the current "advanced" technologies in the field of domestic agriculture, described above in a maximum of 15 years, will no longer be associated as such not only in the world, since world agriculture has already reorganized to new technologies in accordance with the criteria of the new industrial, scientific and technological revolution, but even inside the country, since the new technologies are being formed, for example, in the Moscow region, Kuban and a number of other regions of the country. Conversely, this means that already in the middle of the past decade we should have redistributed investment resources from these sectors and these industries to new ones related to the new industrial revolution and the new scientific and technological revolution. But contrary to the logic of development, we continued to carry out massive investments in ... past technologies. It appears, that currently the investments are holding our development stronger than the chains with which Prometheus was chained to the rock. Now our agriculture have to live according to another laws - the laws of dwindling investments, i.e. we are obliged to support the existing production and agricultural sectors until the investments will be exhausted. There is no other way, since we are obliged to pay for investments, i.e. pay interest on loans, support all production, technological, technical, etc. infrastructure (i.e. refrigerators, sorting rooms, the production of posts, nets to containers and caretaker specialists, etc.). Thus, we are in the position of a vicious circle due to the fact that we were involved in the restoration of the industrial matrix, which once (back in the 30s) was successfully carried out, but cannot break out of this industrial paradigm of development. And until we stay within its borders, we will multiply our backlog.

It raises several questions. The first is what to do? The second is where to develop, what is the prospect to strive, which should be formed today? The answer to the first question is to accelerate the search for new mechanisms and models for the transition to innovative agriculture. This task is not easy. The fact is that currently in our agriculture there are at least three different types of agriculture, set into economic structures: a lot of atomized peasant households (so-called personal subsidiary plots), peasant farm enterprises and an advanced associated industrial, which concentrated in large agricultural enterprises (VIS and agricultural holdings). Depending on the region, the share of each structure in territorial agriculture is different. Moreover, in different sectors and segments of the agro-food market, this share is also different. Large enterprises ignore not only small regions, but also many sectors and segments. And it's understandable. Their activities are based on obtaining income, profits - super income and super profits, most small regions cannot allow them to expand and to receive their super profits. But even large enterprises are reluctant to innovate and switch to new technologies until their investments have been recouped. As a result, agriculture is held hostage to these principles.

There are two main directions for solving this problem. The first is radical and revolutionary. It is related to the "unfastening" of the existing ballast, which is represented by the first two structures. But the third structure is not advanced, too. It only represents an expanded and industrialized form of the previous ones. Besides, it cannot exist without the two previous structures, since there are sectors and segments of the agro-food market that have no interest for him. The second direction is evolutionary. It is based on cooperation and integration of the first two structures into the third. This is the usual path of development of agriculture, with several subdirections, which were tested in the USSR in the 30s and then in the 50s, when industry and advanced agricultural mode (state farms and agro-industrial associations) transferred their experience to collective farms, agricultural cooperatives. Another subdirection is associated with the organizational, financial and technological integration of small peasant farms (households, still mistakenly called personal household plots) to large enterprises (VIS, holdings, etc.). The last direction presupposes a sort of satellite character of relations, formed on the basis of the transfer of some functions to small farms; thus, a kind of belt of small enterprises is forming around large enterprises, which are engaged in the completion and processing of particular industries. In this regard enterprises have not so much an economic function as a social one. This form of relations presupposes that someday small farms (especially peasant farmsteads) will die out, but as long as they exist, it is necessary to provide them with the possibility of existence or residence. Consequently, we are talking about clearing the territory for themselves, which we can currently see. But this, in our view, does not solve the problem.

Agriculture remains the kind of agriculture it was a thousand years ago. So, in this way, we do not solve the global problem - food production without traditional agricultural technologies. And this is the fundamental task set by the current industrial, scientific and technological revolutions. The answer to the second question lies in the major format of the rejection of the so-called traditional agriculture, in which the basic element is the land and the so-called natural environment or the conditions for the transition to precision fermentation technology. There are two mechanisms according to this direction. The first is related to precision technologies that express the accuracy of the problem, i.e. it is needed to know in advance what we want to receive at the exit. Another (second) mechanism is related to fermentation, which means the technological processing of raw materials with the help of special substances to the finished product. As a fermentation system, bioreactors are used, which are a complex of biomaterials/substances (from molecules and ferments to ...) [13]. In general, precision fermentation technology means the creation of food products with the pre-planned properties (taste, food, etc.) in bioreactors. It means that for food production there is no need for land and so-called natural conditions: solar radiation, temperature, humidity, irrigation, land and other known conditions involved in the production of agricultural products. All these processes for the production of the final product take place in a bioreactor - closed autoclaves that can be placed ... in the kitchens of citizens. However, this technology needs several elements: 1) bioreactors, i.e. a set of biologically active substances (bacteria) with specific properties, 2) raw materials from which the final products will be produced by bacteria (for example, presscake, etc.), 3) energy to feed the bioreactor.

In this regard, today it is necessary to reform the matrix of agricultural development from industrialization to a new organization, which will be based on precision fermentation.

4 Conclusions

Analysis of the state of agriculture leads to the following conclusions: 1) the reindustrialization of national agriculture, which began in the last decade, will lead in the growth of agriculture. However, this growth will turn out to be nondurable and unstable, depending on the conjuncture, and above all, fading and expensive. Increasing costs will be required to provide growth. Eventually, growth will be absorbed by rising costs. Production will reach absolute inefficiency; 2) changes in climate and weather will have the growing negative impact on agriculture. There is a reason to meet the transformation of the economy from taking to protecting, i.e. instead of removing resources from the environment, to spend more on preserving and restoring the natural environment. As a result, we will get an inverted economy. What is needed with respect to the described tendencies: 1) To differentiate national agriculture into several modes (segments): a) weather-dependent (a segment that has a strong dependence on the state of climate and weather and with increasing climate and weather changes will demonstrate a deterioration in its performance. This segment currently occupies the largest volume both in terms of concentration of resources and in terms of production. However, there will be an increase in costs not for the production of products, but for the elimination of the negative impact of climate and weather. Moreover, these costs will grow exponentially and at some point it will become unprofitable to engage in the production of products, because the result will be several times lower than costs. Such circumstances occasionally occurs in the history of agriculture. Recently, it has covered the so-called water-intensive products and their production [14, 15, 16]. In addition, it is necessary to provide a clear zoning of agriculture. Another direction is the use of crops and technologies adapted to changing climatic and weather conditions), b) organic or ecological (a special segment engaged in the production of environmentally friendly products. Its capabilities will remain within the existing capabilities. At the same time, the market capacity of these products is expanding and will continue to grow due to the growth of incomes of the population, on the one side, and the reduction in the production of environmentally friendly products), c) precision (a segment that is related to the development of precision fermentation. This is the least developed and essentially new segment of the current national agriculture. Therefore, an accelerated formation of this sector (segment) of national agriculture is required. But this isn't about traditional precision agriculture based on the use of modern technologies for planning traditional agriculture [17, 18, 19], but of the creation of a new sector of national agriculture, designed to produce food (and agricultural) products directly, bypassing the stage of traditional agriculture), d) (segment associated with the production of bacteriophages. This isn't about the creation of traditional agricultural products and raw materials for food, etc., but the production of bacteria for various industries: from food to pharmaceuticals, hygiene, etc.), e) (a segment associated with the production of intellectual products that do not have material carriers, i.e. formulas, programs, and other products); 2) to ensure continuity and interconnection between segments, similarly to communicating vessels, due to the reciprocal mechanism of overcoming the transfer of agriculture [20].

References

1. V. A. Kushkhova, M. Sh. Gazaeva, A. V. Gyatov, Z. M. Ivanova and M. N. Eneeva, IOP Conference Series: Earth and Environmental Science (2019)
2. A. Kholodov, M. A. Kholodova, I. F. Gorlov, O. P. Shakhbazova, N. I. Mosolov, A. V. Glushenko and D. A. Mosolova, IOP Conference Series: Earth and Environmental Science (2019)
3. J. Schumpeter, *The Theory of Economic Development* (1982)
4. I. V. Stalin, *Economic Problems of Socialism in the USSR* (1952)
5. R. K. Allen, *Capital Accumulation, Soft Budget Constraints and Soviet Industrialization* (1997)
6. R. Allen, *From Farm to Factory: A New Interpretation of the Soviet Industrial Revolution* (2013)
7. G. I. Khanin, Free Thought-XXI, **5**, 72-94 (2002)
8. H. M. Rakhaev et al., *Agriculture of Kabardino-Balkaria: state, potential, problems, prospects for modernization* (2016)
9. V. A. Kushkhova, E. S. Bakkuev, Z. M. Ivanova, A. M. Tramova and Z. M. Khochueva, IOP Conference Series: Earth and Environmental Science, 42007 (2019)
10. A. V. Gyatov, S. A. Bogatskaya, M. H. Zhetteeva, Agrarian Bulletin of the Urals, **4**, 91-100 (2020).
11. N. G. Baryshnikov, E. A. Cherdantseva, *Reproduction in Agriculture: Priorities and Prospects* (2012)
12. K. Marx, Capital: The Process of Capitalist Production, **2** (1978)
13. Yu. A. Ramazanov, *Biotechnology: State and Development Prospects*, 287-289 (2015)
14. *Global Climate Change and Risk Forecast in Agriculture in Russia* (2009)
15. V. P. Yakushev, E. E. Zhukovsky, Bulletin of RAAS, **2** (2010)
16. H. M. Rakhaev, A. V. Gyatov, Agrarian Bulletin of the Urals, **12(179)**, 27-32 (2018)
17. T. M. Batte, W. M. Arnholt, Computers and Electronics in Agriculture, **38**, 125-139 (2003)
18. *Precision Agriculture* (2009);
19. Machine Technologies for the Production of Products in the System of Precision Farming and Animal Husbandry: Materials of the 3rd Scientific and Practical Conference (2005)
20. F. M. Tallas, M. R. Uyanaev, Economics and Entrepreneurship, **4-2**, 1085-1088 (2015)

Bioengineering systems for protection and improvement of urbanized areas of coastal and recreational zones

Saligadzhi Kurbanov[1,*], *Akhmed* Sozaev[1], *Astemir* Shogenov[1], and *Abduvali Karshiev*[2]

[1]Kabardino-Balkarian State Agricultural University named after V. M. Kokov, Lenin Avenue, 1V, 360030 Nalchik, Russia
[2]Tashkent State Agrarian University, Universitetskaya, 2, 100140 Kibray district, Tashkent region, Republic of Uzbekistan

Abstract. The problems of protection and environmental management of disturbed and urbanized areas of coastal and recreational zones are urgent for the southern regions of Russia. Exogenous natural processes, intensively developing in the basins of small rivers of mountain zones, pose dangerous situations for populated areas and tourist centers. The applied methods of controlling exogenous processes require ecologization and adaptation to the natural environment. Engineering systems are required to solve the complex problem of reliable protection of urbanized and recreational areas and restoration of disturbed land areas. Bioengineering systems, interconnected structures, which effectively perform engineering and biological functions of protecting territories and restoring disturbed lands have to be developed. For scientific substantiation of proposed bioengineering systems, theoretical and experimental research methods were used, as well as scientific and technical achievements in the fight against exogenous natural processes. The main study results are new constructive and technological solutions for environmental bioengineering systems in disturbed areas of coastal and recreational zones. In compliance with the regulatory requirements for environmental technologies, protective and regulatory structures and slope fastenings have been developed, forming bioengineering systems. Theoretical foundations and recommendations for the design and construction of bioengineering systems for river bed regulation, protection and restoration of disturbed lands in protected areas have been developed. For further studies, scientific, technical and technological foundations for the development of bioengineering protective systems from the proposed structures have been developed. In the future, technical conditions and projects of standard structures of protective and regulatory structures will be developed.

1 Introduction

The North Caucasus and the South of Russia are known for their wonderful natural conditions and amazing natural monuments. At the same time, the entire infrastructure of

* Corresponding author: Kurbanov-salih@rambler.ru

regions is located in the basins of small rivers. The total length of areas exposed to hazardous effects of natural exogenous processes reaches several hundred kilometers. The annual damage reaches hundreds of billions of rubles. There is a serious issue of protection and environmental management of coastal and recreational zones. This issue is relevant for many countries of the world [1, 2, 3]. Exogenous natural processes, intensively developing in the basins of small rivers of mountain zones, pose dangerous situations for populated areas and tourist centers.

The applied methods of controlling exogenous processes require ecologization and adaptation to the natural environment. Special engineering systems are required to solve the complex problem of reliable protection of urbanized and recreational areas and restoration of disturbed land areas. That is, they should simultaneously perform both engineering and biological functions for protection and restoration of water protection and recreational zones [4, 5].

Tourism development and urbanization are capturing new territories, including water conservation and recreational ones. There was a need for environmental regulation and environmental management of disturbed areas of coastal and recreational zones [6, 7, 8].

The practice of recent decades shows the technogenic nature of the methods used to control exogenous processes, negatively affecting nature. The need emerged for effective and biopositive designs of protective and regulatory structures that form protective bioengineering systems. For this, the authors have developed a protection concept of coastal and recreational zones, where the main emphasis is on the development of bioengineering systems [9, 10]. Scientific foundations and recommendations for the design and construction of bioengineering systems for river bed regulation, protection and restoration of disturbed lands in protected areas have been developed.

Protective and regulatory structures are used to regulate river reaches and protect coastal and eroded lands. They are widespread and extended structures that have a great impact on the river and coastal zone ecosystems [11, 14, 15]. Engineering solutions have been developed for the construction of protective and regulatory structures and fastenings of a biopositive design based on the use of local materials, which is positively taken by the natural environment, and form bioengineering systems [9, 12, 13, 16-24].

Experimental studies of biopositive designs on rivers prove their reliability and environmental efficiency.

2 Materials and Methods

Today, protective and regulatory structures and fastenings are built without taking into account the alluvial and hydrobiological regimes of small rivers, as well as soil and landscape conditions of mountain recreational zones, their remoteness from transport communications and industrial zones. There are no scientifically based methods, regulatory and technical recommendations for the design and construction of effective and reliable structures for bioengineering systems used to regulate channels and protect coastal and recreational zones.

There are methods to protect the natural environment [6] based on natural mechanisms of self-regeneration of environmental elements (rivers, lakes, reservoirs, soil layer, etc.). The use of bioengineering systems and technologies to accelerate natural processes for the restoration of disturbed lands is not considered. A unified theory and evidence-based recommendations are required. At present, the authors are developing theoretical foundations and conditions for development of bioengineering systems. According to the Ecological Doctrine of the Russian Federation, the issues of environmentally friendly methods for engineering protection of territories are considered as a priority [13, 14].

To develop complex bioengineering systems, the authors have prepared a list of effective and biopositive designs and fastenings from their own innovative developments and scientific achievements of recent years [16-24]. On the basis of many years of research and developments on the stated topic, the authors have received more than 30 patents and published more than 25 scientific articles. Table 1 shows brief descriptions and diagrams of some structures and their elements (protected as inventions and useful models) for development of bioengineering systems for the protection and restoration of coastal, eroded and other disturbed land areas in recreational zones.

More detailed field and analytical studies were performed on areas of small rivers in urbanized and tourist zones:

— reliability and environmental efficiency of the regulation and protective structures and fastenings of small rivers have been studied;

— the alluvial regime of rivers, zones of fish migration and spawning have been studied;

— the flood regimes of small rivers and regulation methods have been studied;

— the efficiency of anti-erosion structures and fastenings have been studied;

— the ecological state of tourist base areas and the arrangement of tourist trails have been studied.

The study results showed the need to use biopositive materials and products as engineering anti-erosion measures. The proposed (table 1) bank protection, channel regulation and anti-erosion structures ensure the development of bioengineering systems for protection and restoration of disturbed lands.

There is also a number of similar author's developments on protective and regulatory structures, confirmed by patents for inventions and prepared for application for new patents. Structures of and their construction technologies are biopositive. They provide favorable conditions for fish spawning and germination of green plants in coastal and recreational zones.

2 Results and Discussion

The study results and current environmental problems of disturbed lands, as well as the existing experience of controlling natural exogenous processes, make it possible to assess the need for development and implementation of bioengineering systems. Over the past 20 years, some of the new proposed structures, made from local natural materials, have been built and tested in natural conditions on the North Caucasus rivers. The study has shown high reliability and efficiency of these structures. They are overgrown, merged with the environment and continue to perform engineering functions, having transformed in bioengineering (nature-like) structures. Material costs for construction and maintenance are more than 2 times less than traditional reinforced concrete structures.

The preparation of the main technical solutions for the manufacture of prefabricated elements of structures from local materials - sand, vegetable soil, reeds, etc. The use of prefabricated products and structures made of natural materials provides an accelerated restoration of disturbed and eroded land areas. Prefabricated products (wiepped fascine, heavy fascine, flexible mat) serve as the basic elements of bioengineering structures. Using these standard products and elements, various structures are assembled and erected, forming protective and restorative bioengineering systems. These systems will simultaneously perform both engineering and biological tasks.

The most significant results of studies and development include theoretical foundations and technical conditions for the design and construction of bioengineering systems for land restoration and protection. The technical parameters, main criteria, constructive and

https://doi.org/10.1051/e3sconf/202126203021

technological connections for the development of bioengineering systems have been determined.

The development and implementation of bioengineering systems ensure high efficiency of conducted flood and erosion control measures to protect and restore disturbed areas. Over time, they turn into biopositive designs, overgrown with shrubs and grass, which help the development of natural processes of restoration of eroded lands.

Table 1. Innovative technical solutions for the development of bioengineering systems.

Names and numbers of patents, diagrams of structures and devices	Summary and purpose of technical solutions
1. Heavy fascine of biopositive design and its construction method Patent No. 2396391 E02B 3/12 dated 08/10/2010; Patent No. 2369687 E02B 3/12 dated 10/102009 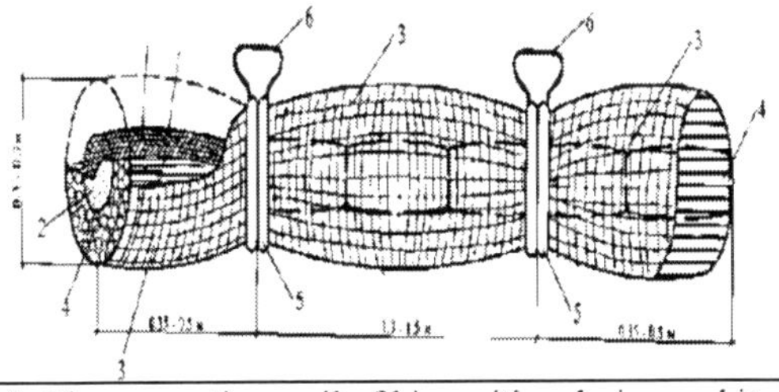	It is made of bags filled with plant soil and packed in a shell of wiepped fascines. Then it is wrapped in a gabion net and tied. It is designed to create protective structures, restore slopes and eroded lands under conditions: $i < 0{,}01$, $\upsilon_i = 3 - 4$ m/s, Fr = 0.8…1.5. (i - channel slope; υ_i - average water flow rate, m/s; Fr - Froude number)
2. The supporting wall of biopositive design and its construction method Patent No. 2399717 E02B 3/12 dated 09/20/2009 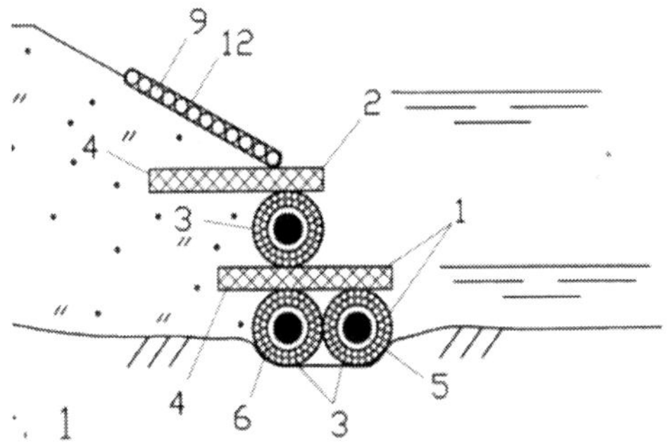	The supporting wall is two-stage and is made of 3 heavy fascines and 4 gabion mats. On the slope above the wall, an attachment is arranged from 12 mats made of 9 wiepped fascines. It is recommended for use against collapse and for restoration of coastal slopes under conditions: i = 0.005...0.01, $\upsilon_i = 2 - 4$ m/s, Fr = 0.5…1.5.
3. The dam dike of biopositive design and its construction method Patent No. 2336388 E02 3/06. Bull. No. 29, 2008 Patent No. 2319805, 2008 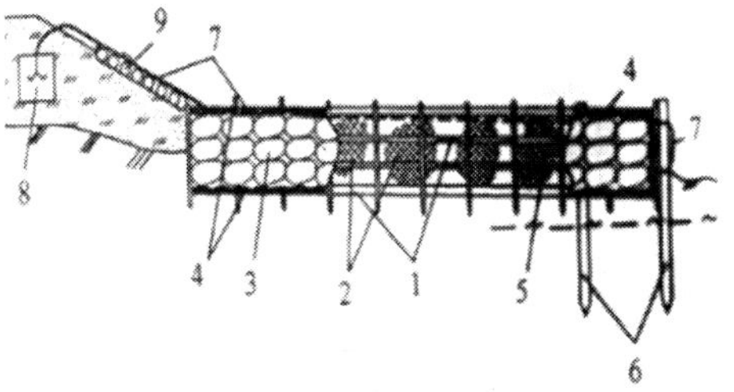	Wooden spatial crib structure 1 is filled with gabion mats 2. The head of dam dikes is reinforced with wooden piles 6. The structure is anchored 8. It is recommended for channel regulation and coastline stabilization under conditions: i = 0.0005...0.002, $\upsilon_i = 0.5 - 2$ m/s and Fr < 0.5. Experimental studies were performed at the Terek and Zolka rivers.

Table 1. Continued.

4. The slope fastening of biopositive design and construction method Patent No. 2351708 E02B 3/12 dated 04/10/2009 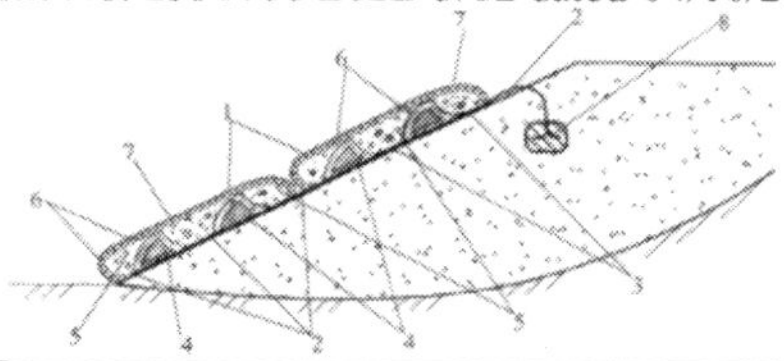	Fastening of interconnected mats 1 and anchored with wires 2 to concrete cubes 8. Mats in a shell of wiepped reed fascines 6 are filled with local soil 5 and have inclusions in the form of pockets with fertile soil 4 and plant seeds. It is recommended as an erosion control structure.
5. The slope fastening of biopositive design fascines and its construction method Patent No. 2399718 E02B 3/12 E02D 17/20; ate 09/20 2010; Patent No. 2398930, 2010 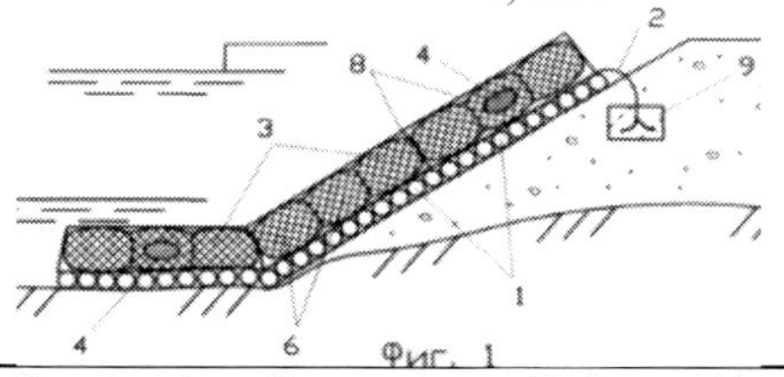	The fastening layer-by-layer includes a number of reed fascines 1, a coarse mesh 8, a number of heavy fascines 3 and on top of a coarse mesh 8. The meshes are interconnected and tied to anchors 9 with wire. It is designed to protect and restore coastal slopes from erosion when: i = 0.0005...0.015, υ_i = 0.7 – 4 m/s, Fr = 0.5…1.5.
6. Construction of supporting walls from gabions Patent for Invention No. 2336389, 2008 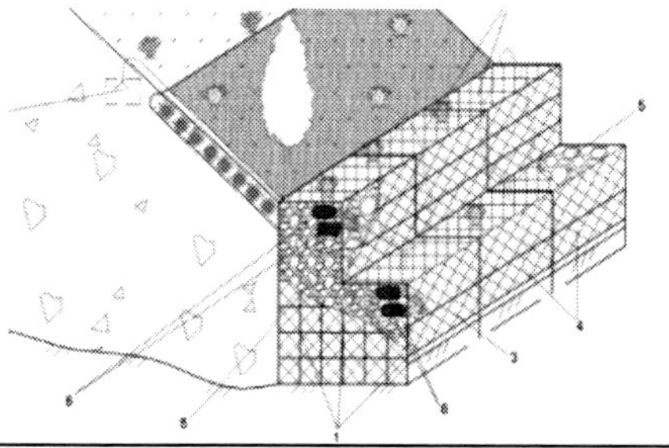	A two-stage supporting wall made of gabions 3, reinforced with reinforcing cage 1. Gabions have 5 stone filling and 6 inclusions from bags with plant soil. It is recommended to use it against erosion and collapse of slopes under conditions: i = 0.005...0.015, υ_i = 2 – 4.5 m/s, Fr = 0.5…2.0. Experimental studies were performed at the Cherek and Nalchik rivers
7. Erosion control works of biopositive design and its construction method Patent for Invention No. 2449078, 2012 Patent for Invention No. 2451129, 2012 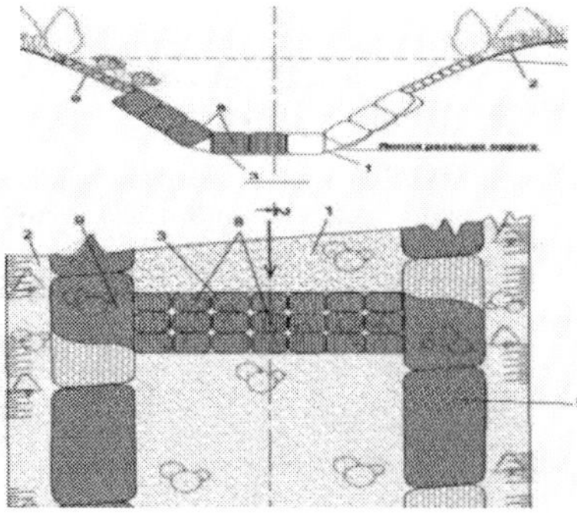	Cross rapids 8 are arranged at the bottom of a ravine of heavy fascines 3 at a certain distance. Above the rapids, the coastal slopes are reinforced with mats 9 made of wiepped fascines 5 and a gabion mesh. It prevents the development of ravines and slope erosion.

Table 1. Continued.

8. Coastal fastening made of fascines and gabion mats of biopositive design and its construction method Patent for Invention No. 2565264, 2015 Patent for Invention No. 2569828, 2015 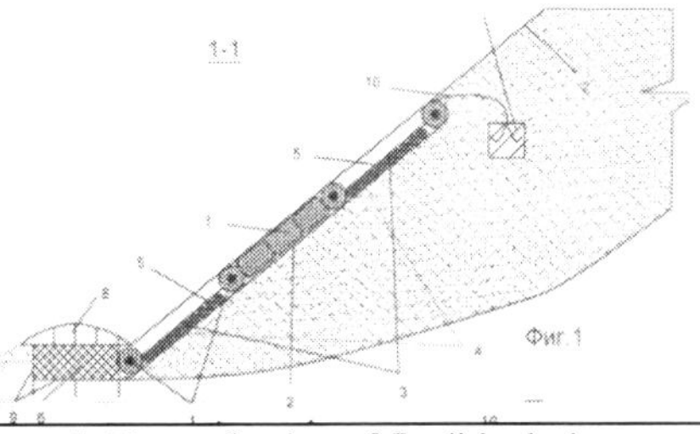	The fastening is made of interconnected heavy fascines, laid in longitudinal and transverse rows to form square cells. The cells in one layer are packed with wiepped fascines in dense rows and covered with a gabion mesh. Under the slope, gabion mats of a prismatic shape are arranged, covered with segmental-shaped reinforcing lattice frames and attached to anchors using cables or wire. It is recommended to protect river banks from erosion and destruction.
9. Heavy fascine of flexible design Utility model patent No. 183818, 2019 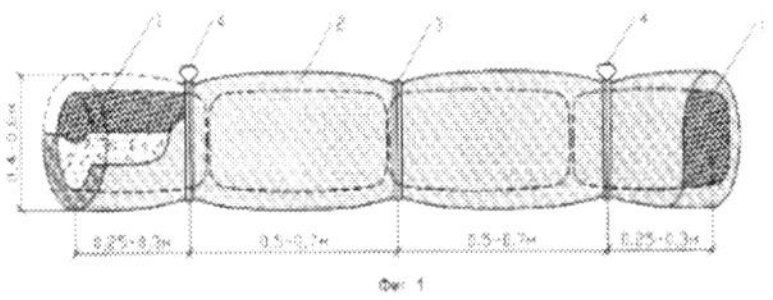	The fascines are made of bags filled with local soil mixed with dry vegetation, wrapped in two layers of geomat, tied in several places with wire. They can be laid in rows on a prepared bank slope, tied together, anchored and secured.
10. Supporting walls are used to protect the coastal areas of rivers from collapse and erosion and their construction method Patent for Invention No. 2645334, 2018 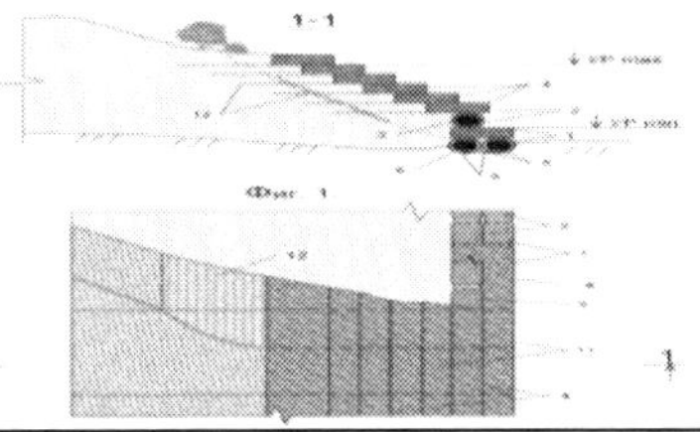	The coastal anchorage contains two types of supporting walls, one on top of the other. The first wall is two-stage 1 and 2, made of heavy fascines 3 and gabion mats 4. The second supporting wall is a stepped reinforced soil structure. It is made of 4 gabion mats with drains and reinforced soil mass. It is recommended to protect coastal zones from collapse and erosion.

4 Conclusions

1. Technical conditions and recommendations for implementation of bioengineering systems into the practice of water management and environmental construction have been developed.

2. Based on study results, the efficiency and reliability of proposed bioengineering structures for foothill and flat areas of small rivers have been confirmed.

3. More effective methods for the manufacture of biopositive products from natural materials and the accelerated construction of protective and regulatory structures have been proposed.

4. The technological connections and conditions for the layout and binding of structures and fastenings for the foothill and mountain zones of small rivers have been determined and refined.

5. New effective ways of fastening coastal slopes, protection and restoration of disturbed lands in accordance with environmental technologies have been confirmed by patents.

The development and implementation of bioengineering systems into the practice of water management and environmental construction will effectively solve the problems of engineering protection and environmental management of disturbed land areas.

In the future, based on the results of the implementation of new constructive and technological solutions, the main technical parameters, criteria and constructive and technological connections for the development of bioengineering systems of protective and regulatory structures on protected and restored land areas will be found and determined.

In conclusion, based on study results, scientifically grounded technical conditions will be developed for the regulation of sections of small rivers, the design and construction of environmentally effective structures for protective, regulatory and anti-erosion structures.

References

1. D. L. Armand, *Agricultural Erosion and Control* (2009)
2. J. M. Anderson, *Ecology and Environmental Sciences: Biosphere, Ecosystems, Man* (1985)
3. V. P. Bondarev, *Geomorphological Analysis and Forecast of Gullying* (2010)
4. T. K. Bykovskaya, T. A. Paramonova, Bulletin of Moscow State University, **17 (2)**, 8-14 (1998)
5. K. S. Losev, I. I. Mazur, R. G. Mamin, *Natural Resources, Nature Reserves and International Environmental Problems* (2009)
6. S.V. Krivitskiy, Bulletin of Moscow State Construction University, **4**, 285-290 (2009)
7. B. Nebel, *Environmental Science: The Way the World Works* (1993)
8. Yu. Odum, *Ecology* (1986)
9. S. O. Kurbanov, A. A. Sozaev, Ecology and Industry of Russia, **24 (8)**, 34-39
10. S. O. Kurbanov, A. A. Sozaev, T. M. Chapaev, Proceedings of the int. scientific-practical conference, 268-278 (2020)
11. S. O. Kurbanov, S. M. Zhemgurazov, Zh. H. Nastaeva, Polythematic Network Electron ic Scientific J. of the Kuban State Agrarian University, **149 (5)** (2019)
12. S. O. Kurbanov, A. A. Sozaev, Polythematic Network Electronic Scientific J. of the Kuban State Agrarian University, **118**, 916-936 (2016)
13. S. O. Kurbanov, E. A. Kushaeva, M. M. Khasanov, Electronic scientific J. "Engineerin g Bulletin of the Don", **4** (2017)
14. S. O. Kurbanov, A. A. Sozaev, "Applied Scientific News - 2014": Materials X Int. Scientific-Practical Conf., 40-43 (2014)
15. S. O. Kurbanov, A. A. Sozaev, A. B. Balkizov, T. M. Chapaev, Science Education Practice: proceedings of the Int. University Science Forum, 117-130 (2020)
16. S. O. Kurbanov, A. A. Sozaev, M. M. Shakhmurzov, Method of Erosion Control Const ruction of Biopositive Design, **1**, 8
17. S. O. Kurbanov, K. S. Kurbanov, The method of manufacturing heavy fascines of biopositive construction, **8**, 8
18. S. O. Kurbanov, A. A. Sozaev, Gabion mat of biopositive design, **28**, 7

19. S. O. Kurbanov, F. T. Dudarova, The construction method of a slope fastening of a bipositive design, **9**, 6
20. S. O. Kurbanov, F. A. Srukhova, The construction method of an anti-erosion biopositiv e design for gully restoration, **29**, 7
21. S. O. Kurbanov, A. A. Sozaev, D. V. Dulaeva, The construction method of a coastal fastening from bipositive fascines, **33**, 6
22. S. O. Kurbanov, S. M. Zhemgurazov, M. M. Khasanov, Gabion mat of flexible design, **11**, 6
23. S. O. Kurbanov, A. A. Sozaev, F. T. Dudarova, Construction of supporting walls for riverbank protection against collapse and erosion, **6**, 6
24. S. O. Kurbanov, A. A. Sozaev, E. A. Kushaeva, Heavy fascine of flexible design, **28**, 6

Yield efficiency for apple trees depending on intensive orchard systems in the Kabardino-Balkarian Republic

Abdulabek Rasulov[1], *Beslan* Beslaneev[1,*], *Muchazir* Kalmykov[1], and *Akrom* Ishnazarov[2]

[1]Kabardino-Balkarian State Agricultural University named after V.M. Kokov, Lenin Avenue, 1V, 360030 Nalchik, Russia
[2]Tashkent State University of Economics, Islam Karimov Str., 49, 100066 Tashkent, Republic of Uzbekistan

Abstract. The study investigated effects of intensive orchard density of different apple cultivars on the yield on the M9 and NK stocks, in comparison with planting on the MM106 stock with the same spindle-shaped crown pruning in the foothills of the KBR. In the first 3 variants, the orchard is on espalier and on drip irrigation, in the 4th and 5th variants - without espalier, irrigated along the furrows. With a non-spaller system and without drip irrigation, the average yield of the Idared, Golden Delicious, Prikubanskoye varieties was 35-37 t/ha, and in the espalier orchid on a M9 dwarf rootstock and on drip irrigation with a planting density of 3170 trees/ha the average yield of the Golden Bee, Granny Smith and Gala varieties was at around 40-43 t/ha. With a superdense planting of Jonagold and Golden Bee trees varieties (6660 trees/ha), the average yield was 66.2-68.7 t/ha. The production of a high-density orchard of the Red Delicious Sandidge variety planted in 2012 with a planting scheme 3,5×0,5 m (5700 trees/ha) was 30 t/ha in the 3rd year, 50 t/ha in the 4th year, and 55 t/ha in the 5th year.

1 Introduction

The innovative development of horticulture and greenhouse vegetable production by introducing modern European technologies, which have greatly progressed over the past two to three decades has recently become one of the main directions of development of the agro-industrial complex of the Kabardino-Balkarian Republic.

"Since the last decade of the 20th century, a new impetus was given to the horticulture development in the world, as a result of a new technology developed in Holland and based on a number of innovative methods, such as increasing the density of planting dwarf apple trees up to 3 thousand trees per 1 ha, the use of crowned seedlings with formed fruit buds at planting, the use of drip irrigation, foliar spray of trees with fertilizers containing not only the main nutrients N, P, K, Ca, Mg, but also microelements and a number of other innovations"[1-4]. "Such orchards enter the fruiting period already in the 2nd year after

* Corresponding author: beslaneev@mail.ru

planting and promptly giving a yield of up to 10-15 t/ha, and in the 4th-5th year the yield reaches 40-50 t/ha" [11]. "The implementation of this technology allowed China to increase its apple production in 15 years from 5-6 million tons to 25 million tons per year. The other countries have succeeded in reducing the total area of orchards by 2-3 times, while remaining the total production of fruits at the same amount" [1-2].

In Russia, the implementation of the new technology into horticulture started 15-20 years later than in Europe, with the exception of Sad Gigant JSC in the Krasnodar Territory [1, 5]. In the Kabardino-Balkarian Republic, the first superintensive orchards on European technology began to be planted in 2009-2010, in other republics of the North Caucasus 5-6 years later [6]. The North Caucasus has the most favorable natural and climatic conditions for the development of intensive horticulture on European technology. The same planting systems, technological methods, varieties as in European countries are used here.

In recent years, in the Kabardino-Balkarian Republic, orchards of two types are predominantly planted: a superintensive orchard of dwarf trees on an M9 stock on espalier and on drip irrigation with a planting density of 3 thousand or more trees per 1 ha, and the second type is an intensive orchard of semi-dwarf trees on MM106 and NK2 stocks without espalier and drip irrigation with a planting density of 850 - 1000 trees per 1 ha.

"Areas of new plantings in Kabardino-Balkaria for the period 2010-2019 amounted to approximately 10 thousand hectares, the ratio of orchards of the first and second types is approximately equal. The total area of orchards in the republic is 20 thousand hectares, by 2025 the production of fruits will reach 500 thousand tons (in 2020 they reached 330 thousand tons). The capacity of fruit storages is 150 thousand tons. The choice of these types of orchards is also related to the fact that since 2015 the state has been subsidizing 80% of all costs for planting orchards with a density of more than 800 trees per 1 ha" [6, 7].

It should be noted that in the North Caucasus there are still not enough fruit tree nurseries to grow seedlings with a branched crown, which are necessary for the planting of superintensive orchards, so the seedlings have to be imported from the European Union countries. Predominantly farmers cultivate planting material according to the old technology, that is, producing of one-year-old trees without side branches, which are used for planting orchards of the second type.

The cultivation features of seedlings of a new type are the following. It is necessary to use free from viruses stocks and cuttings for scions [12]; in the production of one-year-old seedlings with formed crown (having side branches), the technology provides for a decrease in planting density from 50 to 30-35 thousand stocks per 1 ha; to carry out high budding and meticulous care (irrigation, fertilization); to carry out 2-fold treatment of plant tops using a manual spray bottle of the Arbolin (a mixture of Gibberellin A3 with benzylaminopurine or benzyladenine) during the period of the most active growth of grafted trees (June) [3], in order to temporarily suspend the apical growth and, thereby, stimulate the awakening to the growth of lateral shoots. Manual nipping of apical leaflets also gives some awakening of lateral buds, however, there is no simultaneous fruit buds setting, that means the acceleration of tree flowering is not observed [7, 8].

In addition to one-year-old seedlings with lateral branches (with formed crown with fruit buds), it is possible to cultivate the same seedlings from winter grafting plantings. Winter graft trees in the first year usually reach a height of 100-120 cm, that means they do not reach the 1st grade standard, so they are left for the 2nd year. According to the new technology, before the beginning of the second growing season, they are cut off at a level of 60 cm, the growing lateral shoots are removed, leaving only the uppermost one, and when its height reaches about 60 cm, the top is sprayed with Arbolin to stimulate the rise of side branches and the setting of fruit buds. Arbolin (Promolin) is not registered in our country; its replacement, Applin, has recently appeared, but its effectiveness on seedlings has not yet been studied.

The advantage of such orchards is an early entry into fruiting (in the 2nd year after planting), a significant reduction in the share of manual labor in the most laborious processes, pruning and harvesting. The advantage is also reflected in obtaining highly competitive products, not worse than foreign ones on quality [1-3]. Such orchards can recoup investment by the age of 4, and this is 2-3 times earlier than traditional orchards on medium-sized stocks [3, 5].

So what are the differences between an orchard laid out on Eurotechnology and an orchard laid out on the old technology? The main difference is the quality of the planting material. Foreign planting material is grafted on an M9 dwarf stock, it is free from viruses, quite high - up to 170-180 cm, the body diameter is 17-18 mm, in the zone of the 1st tier on the trunk there are about 7-10 lateral branches (at a height of 60 to 100 cm). Grafting (or budding) is carried out at a height of 25-30 cm above the root collar. This increases the influence of the dwarf stock on the growth of the aboveground part of the fruit tree.

In turn, the trees are planted so that the grafting site is about 15 cm above the soil surface. After planting, the branches and the main brunch are cut slightly - by 15-20 cm, i.e. preserve the entire aerial part of the tree. The planting scheme is usually 3.5×0.9 m (3170 trees/ha). Such an orchard just after planting looks like a 3-year-old, due to a significant number of overgrowing branches and twiglets of different lengths.

Planting is carried out not very deep, trees can bend and fall, so an espalier is needed to immediately tie tree trunks to it. The sometimes installed temporary espalier is made of wooden stakes, to which a double twine is pulled, and it is unreliable in a rainstorm with a squally wind. It is advisable to install an espalier made of reinforced concrete posts and several rows of wire before planting an orchard. The height of the support should be 4.5 m, given that an anti-hail net is being pulled over the tree crowns (the production of poles and nets is already being carried out in the KBR).

Shallow planting of trees and the absence of post-planting pruning of seedlings made it necessary to constant irrigation of trees. For this purpose, primarily a sprayer tank with two hoses, irrigating two rows at once is used. Later on, drip irrigation is set up as soon as possible.

The formation of the trees crown in the orchard is carried out according to the "slender spindle" type, which means the presence of a central trunk and numerous lateral branches that have a horizontal position. The diameter of the crown in the lower part is up to 1.0-1.1 m, in the middle part is 0.7-0.8 m, in the upper part is 0.3-0.4 m, in this zone, the presence of long branches is not allowed [2].

In the summer months (July - August) of the first year after planting, it is important to deflect branches longer than 40 cm to a horizontal position and tie them up. In winter or early spring, all branches with acute angles of discharge that could not be rejected during the growing season are removed. At the same time a thorn of about 1 cm is left, when cutting out the branches and the cut is left horizontal so that the shoot that appears later from the hemp appeared from below, and not from above, and had a horizontal position.

In the 2nd and 3rd vegetation, similar summer operations are carried out. In winter, together with competitors, branches with the thickness of 2/3 of the main brunch are removed; the main brunch should be noticeably thicker than the side branches and dominate over them.

In addition to the works described above, on trees at the age of 4 and more years in the winter, regular thinning of the crown is carried out, 1/3 of the longest and most fertile branches are removed, ensuring that all 3-4-year-old fertile branches are gradually replaced in 3 years.

The soil between the rows is kept according to the lawn system, which is gradually formed from natural grass in 2-3 years with 3-4-fold annual mowing of growing grasses with a frontal Italian mower. The near-stem strips are kept clean, treated with herbicides (Roundup and its analogues) twice per vegetation.

Mineral supplements (fertilizers) in tinned soil conditions are more efficient to apply together with irrigation (fertigation). Foliar feeding is also carried out on the leaves with chemicals containing fertilizers with macro and microelements (Aquarin, Ravstorin, etc.).

The latest highly effective chemicals are used as means of protection against diseases and pests. Spraying them is usually carried out at least every 7-10 days, and in case of heavy rainfall, treatments are carried out even more often. The total number of treatments of the orchard with means of protection reaches 20 or more per season.

The purpose of our work is to monitor the growth and fruiting of new intensive orchards in the Kabardino-Balkaria conditions, as well as to identify the key features of the technology of this orchard type, which provide the targeted harvest.

In the process of fruit production intensification, according to the financial resources of commodity producers, various systems and types of apple tree orchards are created. In countries with developed horticulture, agricultural producers have completely switched to a high-intensive orchard type with a planting density of 3-3.5 thousand low-growing trees per 1 ha on a dwarf M9 stock with a yield of about 50 t/ha [1, 2, 7, 8]. There are similar orchards in Kabardino-Balkaria on an area of up to 10 thousand ha. Due to the current economic situation, medium-intensive orchards with the placement of about 850-1000 semi-dwarf trees per 1 ha on NK2, MM106 and M26 stocks, with a targeted yield of about 30-35 t/ha are being laid on an equal basis with high-intensive orchards. Unlike the first type orchards, they do not require supporting structures and drip irrigation, greatly reducing the cost of planting and arranging the orchard [6]. In some cases, on a limited area of 0.2-0.3 ha of household farms, super-dense plantings (5500-6500 trees per 1 ha) are laid on the M9 stock, with espalier and drip irrigation [11].

2 Materials and Methods

Observations and registrations were carried out according to generally accepted methods [10].

The impact of the planting density on the yield on the M9 stock (3170 pieces); 5700 and 6660 trees per 1 ha, 2500 trees on the SK4 stock, was investigated in comparison with planting on the MM106 stock, 870 trees per 1 ha in the conditions of the foothills of the KBR. Crown formation in all plantings is spindle-shaped. In the first 3 variants, the orchard is on an espailer and on drip irrigation, and in the 4th and 5th variants - without an espailer, irrigated along the furrows. Different varieties of apple trees were studied.

The purpose of the research was to give a comparative assessment of the orchard productivity in plantings of different types, as well as to determine the indicators of biological productivity to characterize the production process.

Plantings of apple trees are of different planting densities (scheme and year of planting, stock, studied varieties are presented in the table). The orchards are located in the foothill area on land owned by peasant farm enterprises, located in the suburb of Nalchik. The soils are leached chernozeums with an average supply of basic nutrients. In the years under review, the temperature and humidity conditions were within the normal range, the average annual temperature was 9.3°C, precipitation was 600-630 mm per year.

Plantations on MM106 and NK4 stocks are cultivated without support and drip irrigation, but irrigation is carried out during dry periods by overflow (on MM106) or with a hose (on NK4). In these orchards were used domestic seedlings without branching.

Espalier and drip irrigation are installed in the plantations on the M9 stock. Seedings with branching on the M9 stock were brought from Italy. In all plantings, the crown is spindle-shaped, the tree height is 3.0 m, with the exception of the most dense planting, where the height of the trees reaches 3.5-3.8 m. A high agricultural background, which is necessary for a modern intensive orchards has been achieved in orchard plantings.

3 Results and Discussion

From the information presented in the table, it can be seen that the yield of the orchard differs significantly depending on the type of planting.

When comparing two orchards cultivated on a unsupported structure (the first two variants), it can be seen that the Idared variety in an orchard with 870 trees (MM106 stock) has a yield of 4.9 t/ha, or 12.3% more. The average difference for the two varieties is only 2.8 t/ha and is within the experimental error. Consequently, both types of orchards have the same productivity in the range of 37-35 t/ha and can be estimated as similar. As indicated in the table, orchards on espalier and drip irrigation (3rd variant) with 3 thousand trees per 1 ha, which is most widely used, in general, has a higher yield in the range of 43.0-47.0 t/ha.

There is definitely a fluctuation in yields from year to year in all orchards, which requires more careful setting of standards (removal of excess sets) in years with abundant fruit setting, which is not always carried out basing on the estimated yield [7].

The highest yield was recorded in the orchard with superdense planting (1.5×1.0 m) 66.2-68.7 t/ha, which is 26.6 t/ha, or 39.4% more than the third variant. Given this experience, such type of planting with 6.0-6.5 thousand trees per 1 ha has the right to life and is used in agricultural households. However, the rows must be increased to 2.0 m (with a planting scheme 2.0×0.7-0.8 m), it is recommended to use varieties with moderate growth (Gala, Red Delicious Sandidge, etc.), although ordinary highly productive varieties (such as Golden Delicious, etc.) have shown themselves quite well.

In order to reduce the cost of purchasing seedlings, it is recommended to plant stocks in a permanent place and carry out budding at the location of the future orchard. A gasoline engine mounted on a trolley is used for spraying trees in narrow raws. A vertical spray boom with fittings from a tractor sprayer is also welded onto the base of the trolley.

Table 1. The productivity of orchards depending on the stock, planting system and tree planting density.

Tree planting scheme, stock	Trees amount, thousand per 1 ha	Apple tree variety	Yield by years (t/ha)		
			2019	2020	Ave rage
Orchard system - unsupported					
5×2.3m, MM106	0.87	Idared	38.0	41.4	39.7
		Golden Delicious	43.2	30.4	36.8
4×1m, NK4	2.50	Idared	33.2	36.4	34.8
		Prikubanskoye	35.0	37.0	36.0
Orchard system - espalier					
3.5×0.9, M9	3.17	Golden Bee	42.5	47.0	44.8
		Granny Smith	40.0	46.0	43.0
		Gala	44.4	49.6	47.0
3.5×0.5, M9	5.70	Red Delicious Sandidge	55.6	50.5	53.0
1.5×1.0, M9	6.66	Golden Bee	67.4	70.0	68.7
		Jonagold	60.5	72.0	66.2
LCD_{05}			5.0	4.2	4.6

In an orchard with a low-growing variety Red Delicious Sandij and a planting scheme 3.5×0.5 m (5700 trees/ha) in an orchard on an espalier and drip irrigation, it was identified that in the 3rd year the yield of the orchard reaches 30 t/ha, and in 4th year and further 50-55 t/ha (Table 1). In this case to obtain a yield of 50 t/ha it is enough to have 45-50 fruits on a tree. With a distance between trees of 50 cm, it is carried out extensive pruning, all growths longer than 30 cm are cut on a stump, the tree is column-shaped, spur-type varieties are used.

4 Conclusions

Consequently, in the conditions of the Kabardino-Balkaria foothills, unsupported orchards on MM106 stocks (870 trees/ha) and NK4 (2500 trees/ha) with formed spindle-shaped crowns and with the use of productive Idared, Golden Delicious, Prikubanskoye varieties provide an average yield of 35-37 t/ha for two years.

The espalier orchard of the Golden Delicious, Granny Smith and Gala varieties on a dwarf M9 stock and drip irrigation with a planting density of 3170 trees/ha has a yield in the range of 40-43 t/ha.

The productivity of a high-density orchard with 5.7 thousand trees/ha of the Red Delicious Sandidge variety is 30 t/ha in the 3rd year, and 50-55 t/ha for the 4th-5th years.

The superdense planting of trees (6660 trees/ha) of Jonagold and Golden Bee varieties provides an average yield of 66.2-68.7 t/ha for two years.

All mentioned types of orchards and planting systems are quite effective, have the right to exist and are used depending on the production and economic conditions of farms.

References

1. V. A. Gudkovskii, A. Klad, Horticulture and Viticulture, **4**, 2-6 (2001)
2. I. V. Mukhanin, *Modern system for creating and cultivating intensive apple orchards*, http://asprus.ru
3. I. V. Mukhanin, L. V. Grigorieva, *A highly profitable intensive orchard with a "compa ct spindle" formation*, http://asprus.ru
4. A. N. Fisenko, E. A. Egorov, V. P. Popova, Weak Gardening, **1**, 60-64 (1999)
5. A. A. Solomakhin, *Gardening and Nursery*, http://asprus.ru
6. A. R. Rasulov, A. K. Ezaov, T. H. Pshikhachev, Z. M. Shakhmurzov, *Cultivation of Intensive Apple Orchards in Kabardino-Balkaria (Recommendations)* (2012)
7. A. R. Rasulov, A. S. Sarbashev, A. H. Balov, Problems of the development of the agro-industrial complex of the region, 1, 101-104 (2019)
8. V. A. Alferov, *Optimization of technological and economic parameters of the structure of agrocoenosis and regulations for the cultivation of fruit crops and grapes*, 237-242
9. P. V. Kondratenko, A. M. Silaev, V. V. Torop, Horticulture and Viticulture, **3**, 14-16 (2008)
10. *Program and methodology for the variety study of fruit, berry and nut crops* (1999)
11. A. R. Rasulov, M. M. Kalmykov, B. B. Beslaneev, Collection of scientific works based on the results of the VIII International Scientific and Practical Conference dedicated to the memory of Honored Scientist of the Russian Federation and KBR, Professor B.Kh. Zherukov, 80-83 (2020)

12. A. R. Rasulov, B. B. Beslaneev Priority directions of innovative development of agriculture: Materials of the All-Russian Scientific and Practical Conference, 60-63 (2020)

Production potential of the Kabardino-Balkarian Republic pasture ecosystems

Aida Tamahina[1,*], and *Urfa Turan Ogly* Turabov[2]

[1]Kabardino-Balkarian State Agricultural University named after V.M.Kokov, Lenin Avenue, 1V, 360030 Nalchik, Russia
[2]Azerbaijan State Agricultural University, AZ2000, Ataturk Avenue, 450, Ganja, Azerbaijan

Abstract. The pasture digression of meadows followed by soil deflation is one of the pressing environmental problems. This problem is typical for mountain pastures that are constantly in economic circulation. The article presents the results of a geobotanical survey of the Zolsky pastures on the territory of the Kabardino-Balkarian Republic in 2018-2020. The results of a geobotanical survey show that the pasture phytocenoses are characterized by high floristic diversity due to the heterogeneity of edaphic and orographic factors. The flora of the pastures is represented by postwood moist sedgy-and-tussock-grass, mesophilic woodreed-and-agrostidinic grass, cereal forb, forbs cereal mesophilic and subalpine meadows, low sedgy meadow steppes. The average yield for the pasture period varies from 7.2 to 16.6 centners/ha of dry eaten mass. The consequence of prolonged pasturage and excessive pasture load was a decrease in alpha and beta diversity, the formation of low-productive secondary plant communities of non-food, poisonous and weed grasses, the destruction of sod and soil outcropping. Restoration of degraded pasture ecosystems is possible on the basis of ecological intensification, which provides for the regulation of pasture loads, adherence to grazing terms, phytomelioration using perennial grasses, and short-term isolation of pastures from grazing. This will prevent erosion processes, increase biodiversity, productivity, forage value of grass stand and stability of pasture ecosystems.

1 Introduction

Pasture ecosystems perform the production (production of feed for livestock), environment-forming (air and water purification, soil protection from erosion, formation of soil bioproductivity, regulation of biological processes and regulation of the number of living organisms), information (genetic resources, information on the structure and functioning of natural ecosystems) and recreational services. Due to the fact that the main ecosystem service of pastures is a production service, the assessment of the volume of ecosystem services of pasture ecosystems in Russia is carried out by the share of natural fodder eaten by livestock, the average amount of fodder eaten by livestock, the number of fodder units that natural pastures produce per year per 1 hectare. area of the region (on average, 29 centners of feed units per year). In general, the used volume of the productive ecosystem

* Corresponding author: aida17032007@yandex.ru

services of the majority of pasture ecosystems in Russia is many times less than their actual volume. The share of natural fodder eaten by livestock in most of the territory of Russia does not exceed 4%. However, in the mountainous regions of Russia, the degree of pasture use is close to 50%, and the share of natural fodder eaten by livestock reaches 19% [1].

Despite the enormous ecosystem and economic importance, the current state of pasture ecosystems is of serious concern due to their intensive and irrational exploitation. One of the urgent environmental problems is the degradation of pasture ecosystems, which manifests itself in a decrease in the productivity and quality of herbage, changes in the composition of flora under the influence of pasture, technical, recreational digression and pesticide pollution [2, 3]. An increase in the intensity of grazing in pasture meadows leads to a simplification of the structure and a decrease in species diversity, an increase in the role of digression-resistant species of forbs and soil, a decrease in the balance of N and P in the soil, its compaction and desiccation [4, 5, 6].

This problem is clearly manifested on the territory of the Kabardino-Balkarian Republic (KBR), where the meadows in economic circulation are constantly used as hayfields and pastures. The area of pastures and hayfields in the KBR is 311 and 58.7 th ha, respectively. The distant pastures, where annually up to 75% of the total livestock of the republic are kept, are of particular value. Ineffective and disorderly use of natural forage lands in the republic has led to a sharp decline in returns and an increase in environmental stress. This is especially true of mountain hayfields and pastures, the main part of which is degraded, subject to erosion, littered with harmful, poisonous and inedible grasses [7, 8].

In connection with the above, the purpose of the study was to assess the current state of the KBR pasture ecosystems.

2 Materials and Methods

The main research method was the geobotanical survey of the Zolsky pastures (Zolsky district of the KBR) on an area of 950 hectares in 2018-2020. The description of phytocenoses was performed accounting areas (AA) of 100 m^2. The association was named pursuant to the dominant species. The projective cover of the species was assessed as a percentage of the AA total area. The grass stand was disassembled into economic groups (cereals, sedges, legumes, forbs, poisonous grasses) with the selection of well and satisfactory eaten ones. When describing edaphic conditions, the type and method of soil moistening were noted. The economic characteristics of the pastures were given according to the participation of useful and harmful plants in the herbage, the state of the herbage in relation to its trampedness and economic productivity. The level of pasture digression was assessed on a scale: 1 - weak, 2 - moderate, 3 - medium, 4 - strong. To monitor alpha and beta diversity, forms of geobotanical descriptions for individual communities were used. Alpha diversity was assessed based on the total number of species in the community based on the data of route counts and descriptions of sample plots, the beta diversity was assessed using Whittaker measure (β_w). Statistical processing included determination of the correlation dependence (r) between the studied parameters of pasture phytocenoses and cluster analysis.

3 Results and Discussion

Zolsky pastures are located in the mountainous zone of the KBR (Malye Kuraty tract) at the junction of the forest and subalpine belts. The vegetation formed on mountain-meadow chernozem-like, mountain meadow-boggy soils, on plateaus and gentle slopes, on steep slopes of the eastern, southeastern and western exposures in a cool climate, is very diverse.

The flora of the pastures is represented by post-forest moist sedge-pike, mesophilic reed-field grass, cereal-forb, forb-cereal mesophilic, subalpine meadows and low sedge meadow steppes. In the surveyed area, 11 types of phytocenoses were identified (Table 1).

Table 1. The main types of pasture phytocenoses in the Zolsky district.

Item No.	Phytocenosis type	Soil type	Productivity for the pasture period, kg/ha of dry eaten mass	Hydration
Mountain belt				
1	Sedge-pike, pike-sedge	Mountain meadow-boggy	16.6	atmospheric, soil and singer
2	Postwood mesophilic cereals, cereal-sedge-forb	Mountain meadow chernozem	16.8	atmospheric
3	Forbs-field-grass	Mountain meadow chernozem-like gleyic heavy loamy	12.1	atmospheric
4	Heavily trampened cereal-weedy-forb	Mountain meadow chernozem-like leached gleyic heavy loamy	10.0	atmospheric
5	Forbs and cereals	Mountain meadow chernozem-like friable stern loamy	11.0	atmospheric
6	Cereal-herb-low-sedgy	Mountain-meadow chernozem-like friable soddy heavy loamy surface-gravel-stony	13.2	atmospheric
7	Forbs-field-grass, reed-field grass-field	Mountain meadow chernozem-like friable stern loamy	7.2	atmospheric
8	Grass-forb with low sedge	Mountain meadow chernozem-like friable stern loamy	6.6	atmospheric
9	Forb-low-sedge	Mountain meadow chernozem	7.3	atmospheric
Alpine belt				
10	Forbs and cereals	Mountain meadow chernozem-like gleyic heavy loamy	10.6	atmospheric
11	Forb-cereal meadow red-tailed fescue grass	Mountain meadow chernozem-like friable soddy heavy loamy	15.4	atmospheric

To assess alpha diversity, 11 most characteristic geobotanical descriptions were selected, in which 85 species were registered.

Sedge-pike and pike-sedge type of pastures is found on gentle slopes of northern exposure, plateaus, in depressions between slopes. Humidification is atmospheric, ground and drip. The grass cover of these meadows is quite dense. UP-1 is dominated by *Deschampsia cespitosa*, *Carex pallescens*, *C. leporina*, *C. melanostachya* and *Luzula multiflora*. Minor types are *Festuca pratensis*, *Veratrum lobelianum*, *Astrantia maxima*, *Gladiolus communis*, *Filipendula vulgaris*, *Geranium sanguineum*. *Carex acuta*, *Juncus gerardii* are abundant in bushy areas, and related species are *Filipendula ulmaria*, *Anthriscus nemorosa* and *Trollius ranunculinus* Scarce are *Poa longifolia*, *Calamagrostis phragmitoides*. The projective cover is 65-90%, the height of the grass stand varies from 40 to 60 cm.

Postwood mesophilic meadows are represented by forbs-field grass and reed-field grass types of pastures on gentle and steep slopes of western and northern exposures. The dominants of phytocenoses on AA-2 are *Agrostis planifolia* and *Calamagrostis phragmitoides*, and the codominants are *Anthriscus nemorosa*, *Bromopsis variegata* and *Phleum phleoides*. *Rhinanthus serotinus*, *Gladiolus communis*, *Geranium sanguineum*, *Filipendula ulmaria*, *Valeriana officinalis*, *Dactylorhiza urvilleana*, *Veratrum lobelianum*, *Brachypodium pinnatum*, *Cirsium obvallatum*, *Galium ruthenicum* take a significant part in the composition of the herbage. Legumes are represented by *Trifolium hybridum*, *T. pratense*, *T. ambiguum*, *Vicia angustifolia*, *Lotus caucasicus*, *Medicago falcata*. The projective cover is 90-100%, the height of the herbage is 40-60 cm. Infestation with poisonous and harmful herbs (*Veratrum lobelianum*, *Ranunculus montanus*, *Cirsium obvallatum*) reaches 5%.

Grain-weedy-forb modification of pastures is noted on the plateau and gentle slopes of the northern exposure. The site is heavily trampened. *Deschampsia cespitosa*, *Poa pratensis*, *Alchemilla xanthochlora*, *Ranunculus polyanthemos* and *R. montanus* grow on AA-3. Poisonous, harmful and weeds are abundant - *Veratrum lobelianum*, *Cirsium obvallatum*, *Rumex confertus*, *Ranunculus montanus*, *Urtica dioica*. The total projective cover is 55-60%. Height of grass canopy is 50-60 cm.

Postwood steppe forbs-cereal pastures are represented by forbs-meadow red-tailed fescue grass and herb-gramineous types. The herb-meadow red-tailed fescue grass pasture type is found on the gentle slope of the northeastern exposure. The basis of the AA-4 herbage is *Festuca pratensis*, *Agrostis planifolia*, *Alchemilla xanthochlora*, *Leontodon hispidus*, *Carum carvi*, *Cirsium obvallatum*. *Veratrum lobelianum* is abundant in the lower part of the slope. The projective cover is 85-90%. Height of grass canopy is 10-30 cm.

The herb-cereal type of pastures is noted on the gentle slopes of the northern, northwestern exposure and the steep slopes of the eastern exposure. The dominants on AA-5 are *Agrostis planifolia*, *Bromopsis variegata*, *Phleum phleoides*, *Briza media*, *Festuca ovina*. Minor types are *Brachypodium pinnatum*, *Poa pratensis*, *Koeleria caucasica*, *Carex pallescens*. The most abundant legumes are *Trifolium hirtum* and *Lotus caucasicus*. Wild grasses are *Filipendula vulgaris*, *Astrantia trifida*, *Achillea millefolium*, *Veratrum lobelianum*. The projective cover is 100%, the height of the herbage is 35-40 cm.

In the herbage of the steep-slope forb-cereal pasture (AA-6) dominate *Bromopsis erecta*, *Koeleria luerssenii*, *Helictotrichon pubescens*, *Carex humilis*. Among legumes, *Anthyllis macrocephala*, *Trifolium ambiguum*, *T. montanum*, *Vicia angustifolia*, *V. varia* are present in significant quantities in the herbage. Forbs are represented by steppe (*Galium verum*, *Thymus serpyllum*, *T. marschallianus*) and meadow-steppe specie (*Filipendula vulgaris*, *Origanum vulgare*, *Psephellus leucophyllus*, *Salvia verticillata*, *Plantago atrata*). The projective cover is 90-95%, the height of the herbage is 15-40 cm.

The cereal-forb type of pastures is found on the steep slopes of the western exposure. On AA-7, *Bromopsis erecta*, *Brachypodium pinnatum*, *Helictotrichon pubescens*, *Briza media*, *Festuca ovina* dominate among cereals, the codominant is *Carex humilis*. The most abundant forbs are *Astrantia trifida*, *Pyrethrum coccineum*, *Filipendula vulgaris*, *Leontodon hispidus*, *Origanum vulgare*. There were small amounts of *Scabiosa bipinnata*, *Potentilla erecta*, *Anthemis melanoloma*, *Geranium sanguineum*, *Ranunculus caucasicus*. Legumes are represented by *Trifolium ambiguum*, *T. montanum*, *T. hirtum*, *Onobrychis biebersteinii*, *Vicia angustifolia*, *Lotus caucasicus*. The projective cover is 85%, including legumes - 10%, the height of the herbage is 15-40 cm.

Grass-forb-low-sedge and forb-low sedge types of pastures are found on steep and gentle slopes of the southern, southeastern exposure. The dryness of the habitat is due to the runoff of atmospheric precipitation from the slopes. The dominants in the herbage of gentle slopes (AA-8) are *Carex humilis*, *Festuca pratensis*, *Bromopsis erecta*, *Origanum vulgare*, *Alchemilla xanthochlora* and *Trifolium ambiguum*. On steep slopes, grasses (*Bromopsis erecta*, *Koeleria luerssenii*, *Festuca ovina*) and herbs (*Psephellus leucophyllus*, *Plantago atrata*, *Thymus nummularius*) dominate in the herbage. Legumes include *Anthyllis macrocephala* and *Lotus caucasicus*. The projective cover is 70-80%, the height of the herbage is 15-30 cm. Contamination (more than 3%) with the poisonous *Euphorbia iberica* was noted.

The vegetation of the subalpine belt is represented by mesophilic and steppe grass-grass pastures on gentle slopes of northern and northeastern exposure. The dominants of the mesophilic herb-grass pasture (AA-9) are *Bromopsis erecta*, *Calamagrostis phragmitoides*, *Agrostis planifolia*, *Astrantia trifida*, *Betonica macrantha*, *Inula orientalis*, *Alchemilla xanthochlora* amd *Veratrum lobelianum*. Associated species are *Myosotis alpestris*, *Anemonastrum fasciculatum*, *Trollius ranunculinus*, *Carex tristis*, *C. pallescens*. Legumes, in particular *Trifolium ambiguum* and *T. hirtum*, play a significant role in the composition of the herbage. The projective cover is 90%, the height of the herbage is 45-50 cm.

Bromopsis erecta, *B. variegata*, *Koeleria caucasica*, *Agrostis planifolia* dominate in the herbage of the steppe forb-cereal type (AA-10). The minor species are *Festuca ovina*, *Carex humilis* and *C. tristis*. *Festuca woronowii* occurs only once. Legumes are represented by *Trifolium montanum*, *T. ambiguum*, *T. hirtum* and *Vicia angustifolia*. The most abundant forbs are *Betonica macrantha*, *Lomelosia caucasica*, *Inula orientalis*, *Leontodon hispidus*, *Astrantia trifida*. *Veratrum lobelianum* was noted in the hollows (about 4% of the projective cover). *Sedum acre*, *Thymus marschallianus*, *Festuca valesiaca* are found at rock outcrops. The projective cover reaches 100%, the height of the herbage is 50 cm.

On average, the annual stocks of pasture fodder amount to 11.53 thousand c/ha of dry eaten mass. Most of the surveyed Zolsky pastures (about 70%) are characterized by poor and medium quality forage. At the same time, in terms of species composition and relief (slopes with a steepness less than 25°), all pastures are suitable for grazing cattle.

On the territory of the Zolsky pastures, small areas of little grazed vegetation (reference areas) have been preserved on the slopes of the western, northeastern and southeastern exposure. Straight-edged, mixed-herb-low-sedge and herb-variegated associations grow here. The alpha diversity of the reference site (AA-11) is formed by 32 species. The projective cover is 100%, the height of the herbage is 60-65 cm. The results of monitoring of pasture phytocenoses of the Zolsky pastures are presented in Table 2.

Table 2. Parameters of the surveyed pasture phytocenoses.

UP	Level of pasture digression, points	Species richness	Average projective cover, %	Average height of herbage, cm	Productivity of aboveground phytomass, kg/m^2
1	4	18	78	40	0.14
2	2	22	97	50	0.17
3	4	10	58	55	0.10
4	3	8	87	20	0.07
5	4	15	98	38	0.12
6	2	17	93	28	0.14
7	2	22	85	30	0.13
8	4	14	75	20	0.13
9	2	24	90	48	0.11
10	2	21	100	50	0.15
11	1	32	100	63	0.22

The following gradations of correlation were revealed between the studied parameters: moderate (r = 0.3-0.5) between the level of pasture digression and the average height of the herbage, the average projective cover and the productivity of the aboveground phytomass; average (r = 0.51-0.70) between the level of pasture digression, species richness and productivity of aboveground phytomass, as well as between species richness, average projective cover and average height of herbage; high (r = 0.71-0.90) between species richness and productivity of aboveground phytomass, average projective cover and level of pasture digression. The cluster analysis shows that the main factor of the productivity and biodiversity of pasture ecosystems is the level of pasture digression (Fig. 1).

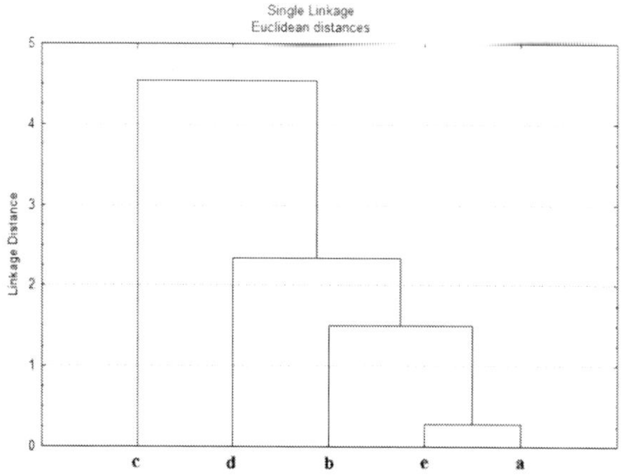

Fig. 1. Dendrogram of clustering parameters of pasture phytocenoses: a - level of pasture digression; b - species richness; c - average projective cover; d is the average height of the grass stand; e - productivity of aboveground phytomass.

The beta-diversity of the Zolsky pastures is generally characterized by a β_w value of 3.63. In communities with a moderate level of pasture digression, β_w decreases to 2.72, and with a moderate and strong level - to 2.40. The beta diversity of the reference plot and phytocenoses with moderate pasture digression is higher (βw = 3.26) compared to the same indicator for the reference plot and phytocenoses with strong pasture digression (βw = 2.74). This indicates an increase in the number of common species, a decrease in biodiversity and resistance of plant communities with an increase in pasture load, regardless of the type of phytocenosis, spatial and soil heterogeneity.

Due to the duration of grazing and excessive pasture load in the ecotopes of the Zolsky pastures, negative trends are observed:

1) an increase in forbs in the composition of forb-reed and forb-variegated rangelands, the formation of low-productive secondary clumps of *Alchemilla xanthochlora*, destruction of sod and outcropping of the soil, the development of non-eating ruderal plants on the released substrate; on the plateaus and gentle slopes of the northern exposure, replacing grass stands with good fodder plants with weeds and poisonous vegetation;

2) the formation of a network of paths and gullies in places of excessive grazing, soil swelling with erosion of sod on steep slopes and the formation of landslide and erosional landscapes.

In this regard, we should note that in the KBR, soils on an area of more than 290 th ha (23% of the total area of the republic), including 99 th ha of pastures, are subject to intense water erosion. The water erosion of land leads to a sharp loss of humus (annually up to 6.1 t/ha). Over a ten-year period, the humus content in the topsoil of the Zolsky district decreased from 7.2 to 6.4% [9].

A high pasture load on natural pastures is a powerful destructive ecological factor affecting the populations of valuable forage legumes and grasses, which requires the search for new approaches, methods and means to increase the efficiency and sustainability of natural forage lands. In areas subject to significant anthropogenic pressure, an important condition for the conservation of flora biodiversity is the implementation of various tactics and life strategies by plants [10]. The population dynamics of meadow plants is determined not only by grazing, fertilization and mowing, but also by the heterogeneity of the landscape and the fragmentation of the habitat. A regional species pool in a heterogeneous landscape is characterized by a higher species richness, as a result of which it can be the main source of the species richness of pasture phytocenoses, depending on the ability of species to spread, their abundance, seed bank and their germination.

To assess the species composition of meadow plants, an integral approach is required, based on identifying the relationship between the ability of plants to disperse and the productivity of aboveground phytomass, landscape heterogeneity, and disturbances caused by agricultural practices. Taking these factors into account will improve the efficiency of agroecological schemes for the restoration or conservation of biodiversity of pasture ecosystems [11]. The key task of the rational use of pasture ecosystems is the search for ecologically and economically efficient management regimes to maintain the ecological functionality and biodiversity of plant communities [12]. Therefore, to predict the botanical, functional composition and ecosystem services of pasture meadows, it is necessary to take into account the interaction between managerial, edaphic and climatic factors [13].

An important measure for the restoration of degraded pasture ecosystems is the regulation of pasture loads and compliance with the timing of grazing [14]. Grazing of pastures in the summer-autumn period compared to the spring-summer period should be less intense and short-lived. This is due to the formation of a powerful root system by plants in the fall and the enrichment of the soil with nitrogen during the decomposition of overgrown grass (autumn regrowth of grass is equivalent to the introduction of 60–120 kg/ha of nitrogen into the soil). The grazing regime in the middle of summer is advisable on pastures with good grass regrowth in spring [15].

Important methods of pasture restoration and prevention of erosion processes are phytomelioration using perennial grasses, in particular, the *Trifolium* and *Arrhenatherum* species, which ensure the reconstruction of destroyed sod, high density and an optimal set of valuable plants in the herbage [16, 17], as well as short-term isolation of pastures from grazing, allowing to increase the productivity of herbage, species diversity of legumes and cereal components and reduce erosion processes [18, 19]. The nature reserve status excludes the practice of a purely consumerist attitude towards natural forage lands. Therefore, the use of the reserved regime of individual areas of pastures should be determined by the productive capacities of vegetation, soil cover and the state of the landscape as a whole. The listed activities underlie the ecological intensification of pasture meadow cultivation, which presupposes landscape approaches based on the rational use of the natural functional capabilities of pasture ecosystems and the potential for their sustainability [20].

4 Conclusion

The results of a geobotanical survey of pasture ecosystems using the example of the Zolsky pastures of the Kabardino-Balkarian Republic indicate their significant floristic diversity due to the heterogeneity of edaphic and orographic factors. The flora of the pastures is represented by postwood moist sedgy-and-tussock-grass, mesophilic woodreed-and-agrostidinic grass, cereal forb, forbs cereal mesophilic and subalpine meadows, low sedgy meadow steppes. Their average yield for the pasture period varies from 7.2 to 16.6 centners/ha of dry eaten mass. A negative tendency due to excessive pasture load is the formation of low-productive secondary plant communities of non-food, poisonous and weed grasses, destruction of sod and soil outcropping, and a decrease in floristic biodiversity. To restore degraded pasture ecosystems, ecological intensification is promising, providing for the regulation of pasture loads, adherence to grazing terms, phytomelioration using perennial grasses, and short-term isolation of pastures from grazing. This will prevent erosion processes, increase biodiversity, productivity, forage value of grass stand and stability of pasture ecosystems.

References

1. E. N. Bukvaryova, D. G. Zamolodchikov, *Ecosystem services of Russia: Prototype of the national report. Services of land ecosystems* (2016)
2. O. G. Bembeeva, Bulletin of Institute for Integrated Research of Arid Territories, **2(37)** (2018)
3. N. V. Tyutyuma, G. K. Bulakhtina, Izvestia of the Lower Volga Agro-University Complex: Science and higher education, **3(39)** (2015)
4. P. M. S. Muratchayeva, Works of Institute of geology of the Dagestan Scientific Center of RAS, **65** (2015)
5. A. D. Sambuu, A. B. Dapylday, N. G. Khomushku, Bulletin of science and practice, **4(9)** (2018)
6. A. A. Titlyanova, Soil Science, **12** (2007)
7. A. Ya. Tamakhina, A. A. Gadieva, A. Ch. Kagermazova, Bulletin of KrasSAU, **8** (2013)
8. N. L. Tsepkova, M. N. Fisun, *Mountain pastures of Kabardino-Balkaria* (2005)
9. A. Ya. Glushko, D.S. Frolko, Terra economicus, **10 (3)** (2012)
10. L. Bondarieva, et al., Ukrainian J. of Ecology, **9(2)** (2019)
11. E. Gaujour, et al., Agron. Sustain. Dev., **32** (2012)
12. J. Liira, M. Issak, U. Jogar, M. Mandaja and M. Zobel, Annales Botanici Fennici, **46(5)** (2009)
13. A. Michaud, et al., The J. of Agricultural Science, **150(2)** (2012)
14. G. Gasanov, et al., Arid ecosystems, **1(54)** (2013)
15. K. G. Magomedov, R. K. Kamilov, G. D. Kagirov, Scientific news, **4** (2016)
16. Yu. I. Sukharev, V. V. Borodychev, E. B. Dedova, S. A. Sangadzhiyeva, Environmenta l engineering, **5** (2011)
17. K. G. Magomedov, N. V. Berbekova, Advances in current natural sciences, **8** (2016)
18. G. K. Zvereva, Siberian ecological magazine, **16(5)** (2009)
19. N. V. Elesova, Ukrainian J. of Ecology, **3** (2019)
20. P. Tittonell, Curr. Opin. Environ. Sustain, **8** (2014)

Sowing time influence on the sweet corn productivity in the mountainous ecologically clean zone of Kabardino-Balkaria

Zalim-Geri Shibzukhov[1,*], *Alim* Kishev[1], *Zalina* Shibzukhova[1], *Murat* Eziev[1], and *Ilahun* Akbar[2]

[1]Kabardino-Balkarian State Agricultural University named after V.M. Kokov, Lenin Avenue, 1V, 360030 Nalchik, Russia
[2]Xinjiang Agricultural University, No. 311 Nongda East Road, 830052 Urumqi, People's Republic of China

Abstract. With a properly selected sowing time, the highest productivity of sweet corns is achieved. In the conditions of the mountainous zone of Kabardino-Balkaria, the timing of sowing sweet corn has not been studied, therefore, determining the optimal sowing time is one of the most important issues in the successful cultivation and obtaining high yields of sweet corn. In our studies, the sowing time influenced the survival rate of sweet corn plants. Optimal sowing favors the dynamic growth and development of plants, the root system, allows better assimilation of moisture in the soil, thus increasing the growth of green mass and the formation of larger and higher quality grains. Based on the analysis of data on the structure of yields, in all respects, the second sowing period (late April - early May) is ahead of other options, especially later sowing dates. The amount of products obtained depends on the weight of one earcorn. The number of marketable earcorns drops sharply at late sowing dates. The sharp decline in the productivity of sweet corn plants is explained by the fact that in the late sowing period, the soil moisture level is at a low level due to the rise in high temperatures.

1 Introduction

"Especially valuable for food use represents a grain of sweet corn in the phase of milky or milky-wax ripeness. In this regard, it is mainly grown for the canning industry, as well as for use in food in fresh, frozen and dried form" [1, 2, 3, 4].

"The cultivation of sweet corn is of great agronomic importance. Subject to high agricultural technology, it helps to clear the fields from weeds and is a good predecessor for grain crops, in particular for wheat" [7].

"Sweet corn is an economically profitable crop for farms. Its short growing season, especially of early ripening varieties, allows two crops to be grown in the same field during irrigation. It should also be noted that to obtain one ton of grain of edible corn using a

*Corresponding author: konf07@mail.ru

technological complex of machines and herbicides, significantly less costs are required compared to other vegetable crops" [5, 7, 9, 11].

In mountainous areas, there have been no studies to determine the optimal timing for sowing sweet corn. The area occupied by this culture was so insignificant that no one has ever asked this question.

Therefore, to resolve this issue, we chose the topic of scientific work and set the following goals and objectives for its solution.

The purpose of the research: to identify the optimal time for sowing sweet corn in the mountainous zone of Kabardino-Balkaria for the highest yield of grain with the highest quality characteristics.

To achieve the goal, the following tasks were set:

- to determine the optimal sowing time in the mountainous zone of Kabardino-Balkaria;
- to study the effect of sowing dates on the productivity of sweet corn.

The scientific novelty of the work lies in the fact that for the first time in the mountainous zone of Kabardino-Balkaria in an ecologically clean area, the sowing dates for sweet corn plants were studied.

The practical value of the work. Recommendations based on the analysis of the data obtained will clarify the timing of sowing sweet corn and will be useful to agricultural producers.

2 Research methodology

The object of research is sweet corn of the Megaton F1 hybrid.

Agrotechnics on a test field are recommended for conventional corn in the area. The predecessor of sweet corn is winter rye.

The plot area was 100 м^2, the replication was fourfold.

Location of the study: mountain zone of Kabardino-Balkaria Baksan region, village Upper Kurkuzhin is located at an altitude of 850 meters above sea level.

Soil - typical black soil.

To solve the set tasks, studies were performed in the form of phenological observations, where the phases and dynamics of growth and development of sweet corn, plant density, plant survival, mass of the green part of plants, etc. were noted.

The yield of sweet corn was carried out on a plot, took by hand at 60-70% grain moisture. The data obtained were recalculated per hectare.

Mathematical processing of the obtained data was carried out according to the method of B.A. Dospekhova.

To solve the problems, the following sowing dates were determined for testing, based on the recommendation of literary sources for the cultivation of ordinary corn

1. I sowing date mid-April (April 15-25) - control;
2. II sowing date end of April - beginning of May (April 25 - May 05);
3. III sowing date mid-May (May 15-25);
4. IV sowing date end of May - beginning of June (May 25 - June 05);
5. V sowing date mid-June (June 15-25).

3 Results and discussion

With a properly selected sowing time, the highest productivity of sweet corns is achieved. In the conditions of the mountainous zone of Kabardino-Balkaria, the timing of sweet corn sowing has not been studied, therefore, determining the optimal sowing time is one of the most important issues in the successful cultivation and obtaining high yields of sweet corn.

In our studies, the sowing time influenced the survival rate of sweet corn plants.

"Optimal sowing favors the dynamic growth and development of plants, the root system, allows better assimilation of moisture in the soil, thus increasing the growth of green mass and the formation of larger and higher quality grains" [6, 8, 10].

An important indicator of crops is plant survival after germination and before harvesting. The future grain yield depends on the number of plants [12, 13, 14]. We found that the survival rate of sweet corn plants during the growing season depended on the sowing time (Table 1).

Table 1. Influence of sowing dates on plant survival, on average for 2019-2020

Options	Quantity of plants (thousand pcs / ha)		Survival (%)
	after germination	before harvesting	
I sowing time (k)	65.5	62.4	95
II sowing time (k)	65.7	63.8	97
III sowing time (k)	65.4	62.2	95
IV sowing time (k)	63.6	58.7	92
V sowing time (k)	60.3	54.2	89

As can be seen from the table, late sowing periods sharply reduce plant survival. The largest percentage of surviving plants was found at the 2nd sowing period, which amounted to 97% (2% higher than the control variant). The difference in the control variant with the lowest survival rate is 6%.

Phenological observations of plant growth occupy an important place among studies. Based on the data obtained, the vegetation period of the plant is determined. According to our observations, at the beginning of the growing season, the growth phases in all variants of the experiment proceeded in the same way. Shoots appeared on the 10-15th day. Further studies have shown that the influence of the sowing time is clearly traced to the onset of the growth phases.

The results of a study on the effect of sowing time on the growth of green mass in sweet corn plants are shown in Table 2.

Table 2. Influence of irrigation regimes on the growth of green mass in sweet corn plants, t/ha.

Options	Day from the beginning of the growing season			
	10	30	60	80
I sowing time (k)	3.4	37.2	64.6	74.4
II sowing time (k)	3.6	40.4	66.2	76.8
III sowing time (k)	3.4	37.5	64.2	74.6
IV sowing time (k)	3.2	35.4	62.1	70.7
V sowing time (k)	2.8	28.5	56.3	63.4

Analyzing the data in Table 2, it can be seen that the greatest increase in the green mass of plants of 76.8 t/ha was revealed in the second sowing period. At late sowing dates, the plants sharply reduced the productivity of green mass. The 5th sowing period turned out to be poor for the accumulation of green mass and differed from the control variant by 11 t/ha.

In our experiments, the sowing time significantly affects the grain yield (Table 3). As can be seen from the table, the yield and yield structure strongly depended on the sowing time.

Table 3. Influence of irrigation regimes on the yield and structure of the yield of sweet corn crops.

Options	Number of marketable earcorns c 1 m^2 (pcs.)	A number of granules on the earcorn (pcs.)	Weight of one earcorn (g)	Grain weight from 1 m^2 (g)	Weight of 1000 grains g	Weight of 1000 grains (t/ha)
I sowing time (k)	5.8	594	396	894	295	9.1
II sowing time (k)	6.3	607	408	997	311	10.3

Table 1. Continued

III sowing time (k)	5.7	584	389	845	287	8.8
IV sowing time (k)	4.9	570	371	789	270	7.5
V sowing time (k)	3.7	488	268	422	212	4.1
LCD_{05}						1.5

On the control option, the yield in grain was 9.1 t/ha, which is higher than other options except for the 2nd sowing date. On the 2nd sowing period, the yield was 10.3 t/ha.

Based on the analysis of the data on the structure of the yield, in all respects the second sowing period is ahead of the other options, especially the later sowing dates. The amount of products obtained depends on the weight of one earcorn. The number of marketable earcorns drops sharply at late sowing dates. The sharp decline in the productivity of sweet corn plants is explained by the fact that in the late sowing period, the soil moisture level is at a low level due to the rise in high temperatures.

4 Conclusions and recommendations for production

In the mountainous zone of Kabardino-Balkaria, it is necessary to sow sweet corn at an early date.

Optimal dates for sowing sweet corn in a mountainous zone mid-end of April

The yield of sweet corn is highly dependent on the timing of sowing. So, a delay in sowing by 10-15 days, the grain yield can decrease by 12-15%.

For the successful cultivation of sweet corn in the mountainous zone of Kabardino-Balkaria, we recommend sowing at the end of April.

References

1. I. M. Khanieva, Yu. M. Shogenov, Z.G. S. Shibzukhov, Materials of the international scientific and practical conference of the research centre "Povolzhskaya scientific corporation", 162 (2017)
2. Yu. M. Shogenov, Z. Gatazhokov, M. H. Khaniev, Grain farming, **3-4**, 37 (2007)
3. Yu. M. Shogenov, Z. A. Ivanova, M. B. Khokonova, F. H. Nagudova, Science and Education - XXI Century, **2013**, 41 (2013)
4. Yu. M. Shogenov, Z. S. Shibzukhov, S. B. Elmesov, T. S. Vindugov, Materials of the international scientific-practical conference dedicated to the year of ecology in Russia, 344 (2017)
5. A. Ezov, Z.-G. Shibzukhov, B. Beslaneev, Z. Shibzukhova, M. Khantsev, International Scientific and Practical Conference «Development of the Agro-Industrial Complex in the Context of Robotization and Digitalization of Production in Russia and Abroad» (2020)
6. K. Nazranov, E. Didanova, Z.-G. Shibzukhov, M. Orzalieva, B. Nazranov, International Scientific and Practical Conference «Development of the Agro-Industrial Complex in the Context of Robotization and Digitalization of Production in Russia and Abroad» (2020)
7. M. S. Sidakova, E. N. Didanova, A. Y. Kishev, R. Z. Abdulkhalikov, International Scientific and Practical Conference «Development of the Agro-Industrial Complex in the Context of Robotization and Digitalization of Production in Russia and Abroad» (2020)

8. Z.H. Khan, S.K. Khalil, Ali, M. Ikramullahc, A. Iqbal, B. Islam, K. Ali, W. Ali, M. Ahmad, S.G. Afridi, F. Shah, Fresenius Environmental Bulletin, 6568 (2019)
9. R. Mohi-Ud-din, N. A. Dar, L. Singh, Saba-Shafia, R. Jan, K. A. Khan, Vegetos, 459 (2017)
10. B. K. Mehta, F. Hossain, V. Muthusamy, R. U. Zunjare, J. C. Sekhar, H. S. Gupta, Indian Journal of Agricultural Sciences, 1543 (2017)
11. A. A. N. Franco, P. S. V. Filho, C. A. Scapim, R. S. Okumura, O. J. Marques, A. Y. Numoto, Australian Journal of Crop Science, 831 (2016)
12. K. Z. Berbekov, I. M. Khanieva, M. S. Sidakova, Materials of the I International Scientific and Practical Conference, 28 (2018)
13. L. Kh. Azubekov, M. Kh. Khaniev, I.M. Khanieva, Yu. M. Shogenov, International Scientific Research, **3(32),** 158 (2017)
14. A. M. Temirzhanov, L. A. Andretsova, Materials of the International Scientific and Practical Conference, 36 (2020)

The productivity of the vetch and oat mix in crop rotation when using various fertilizer systems in the Vologda region

Olga Vasilievna Chukhina[1,*], *Larisa Alexandrovna* Korelskaya[2], and *Nikolai Sergeevich* Demidov[1]

[1]Federal State Budgetary Educational Institution of Higher Education "Vologda State Dairy Academy named after N.V. Vereshchagin" (Federal State Budgetary Educational Institution of Higher Education "Vologda State Dairy Academy"), Vologda Dairy, Russia
[2]North West Research Institute of Dairy and MeadowPaste Estate named after A.S. Emelianov is a separate unit of the Federal State Budget Institution of Science "Vologda Research Center of the Russian Academy of Sciences",) Vologda Dairy, Russia

Abstract. In 2017-2019, it was revealed that fertilizers provide 36 - 51% of the annual crops green mass yield, the collection of raw protein increases by 1.5 - 1.6 times. When N12 was introduced against the background of phosphorus and potassium fertilizers, its content corresponded to 12.50%, and when the N75 was introduced, it increased to 13.6 - 13.7%. Takeaway for N, P, K increased with the use of $N_{75}P_{35}K_{130-160}$ respectively by 1.5 - 1.6 times, by 40 and 80-90% compared to control. The content of NO_3 in green mass increased with the use of fertilizers, amounted to 72 - 86% of the maximum allowable concentration. The doses of $N_{75}P_{35}K_{130-160}$ fertilizers are 32 - 34 ton/ha of green mass in a vetch and oat mix and the more negative balance of the fertilizer elements investigated.

1 Introduction

Dairy farming is developed in the Vologda region, so providing high-yielding cows with a balanced diet is one of the most important tasks of forage production.

To provide livestock farming with high protein feed on a large area in agricultural enterprises of North-West Russia cultivated perennial bean-grain herbs, in crop rotations mainly clover timothy mix. Annual crops such as pea and oat and vetch and oat mixes are of no small importance, which are distinguished by a high content of nutrients, including essential amino acids, are of great importance for the biologization of agriculture that are necessary in the structure of sown areas.

Fertilizers are given great importance when cultivating a vetch and oat mix. Scientific doses of fertilizer improve the quality and quantity of agricultural products, the accumulation of nitrates at the same time does not exceed the maximum allowable value [1, 2, 3].

* Corresponding author: ochukhina@mail.ru

Investigations were conducted to inspect the effect of minimal and calculated doses of fertilizer on the productivity of the green mass of the vetch and oat mix under the conditions of the European North of Russia in the Vologda region.

2 Materials and methods

The site of the field stationary experiment, located on the experimental field of the Federal State Budgetary Educational Institution of Higher Education "Vologda State Dairy Academy", during 2017 - 2019, in 4 several-fold repetition. Experimental plots are broken rectangular form with a total area of 140 m^2, with the accounting area - more than 25m^2, with a systematic location. The soil of the experimental site has the following characteristics: sod-podsolic; according to the granulometric composition – medium clay-loam; complies with the average level of cultivation. Experimental plots were located in a four-year field crop rotation [3, 4, 5]. The experience examined the varieties of vetch and oat, made in the State Register of breeding achievements, admitted for use in the North West zone of the Russian Federation [3, 4].

The experiment explored options: control (no fertilizer) (1 var.), $N_{12}P_{16}K_{16}$ (2 option), $N_{75}P_{35}K_{130}$ (3 option), $N_{75}P_{35}K_{160}$ (4 option), $N_{50}P_{20}K_{100}$ + afteraction 40 ton/ha of the decomposed dung (5 option).

At 3rd, 4th, 5th options of fertilizer doses for application were determined by the planned recommended balance ratios (Kb) according to Yu.P. Zhukov. It was planned as the ratio of fertilizer elements removal by the planned yield of culture in 25 ton/ha to the dose of fertilizers applied in the converted percentages [6]. The actual Kb shows the balance of the fertilizer elements in the soil and was as the ratio of the actual removal of the fertilizer elements to the dose of their input.

During the experiment, fertilizers containing the main elements of N, P, K were applied according to conventional cultivation technology, in the main fertilizer P and K, and NH_4NO_3 - under pre-planting cultivation, when sowing $N_{12}P_{16}K_{16}$ [3, 4, 5, 7, 8].

A solid method was used to account for the output of products from the unit area, then the green mass of crops led to humidity of 75%, complying with the standard, to establish the ratio between crops (vetch and oat) sampled bundles.

To determine the content of element N in a green mass of cultures the method of Kjeldahl was used, raw protein - GOST 13496.4-93 (conversion factor 6.25), movable phosphorus - on a photocolorimeter, exchanged potassium - on a fiery photometer.

Dispersion analysis – in Excel with counting automation [9].

As the previous investigations of the authors on this issue have shown: complete calculation systems of fertilizers provide 53% of the green mass yield of annual forage crops, significantly - by 11 – 14 GJ/ha [7] and more (by 26.3-31.8 GJ/ha) [8] increase the collection of exchange energy by the green mass of the vetch and oat mix. The fact that fertilizers increase the yield and nutritional value of annual crops has also been established as a result of other investigations [10 - 13].

During the growing season of the vetch and oat mix, the hydrothermal ratio (HTR) in June 2018 was below the multi-year value. The high HTR was observed in 2017, exceeding the multi-year values in June and July more than twice. And 2019 was characterized by a lower temperature regime and excess moisture, especially in July (Figure 1).

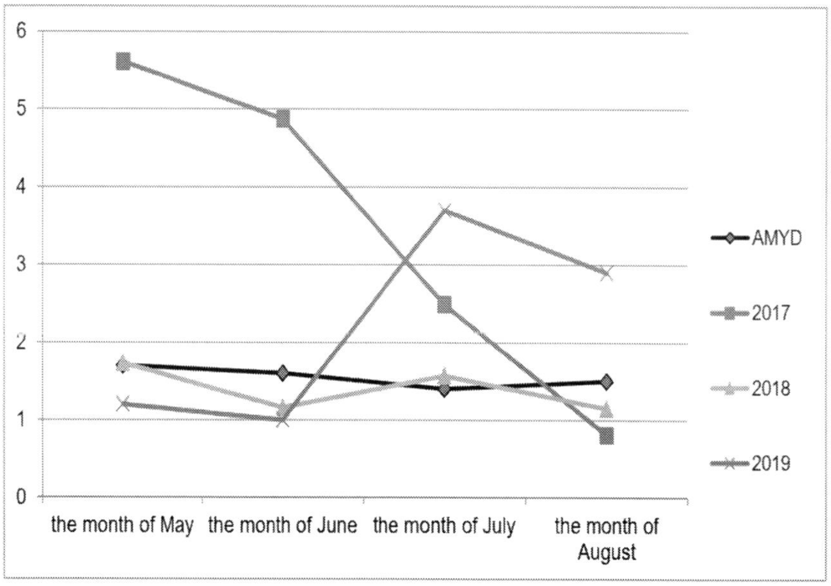

Fig. 1. The HTR by months of 2017-2019 investigations in comparison with AMYD (average multi-year data).

3 Results and discussion

In 2017-2019, when the productivity of annual forage crops was investigated, conditions were good for the growth and development of vetch and oat plants under the mix conditions, therefore, the planned level of crop productivity was achieved and even exceeded (Figure 2).

The most favorable conditions turned out to be in 2017, when excess moisture, optimal air temperature served favorably and ensured crop productivity, which, when $N_{75}P_{35}K_{130-160}$ was applied, exceeded both the other two investigated years and the average value by 4.1 3.0 ton/ha.

Doses of $N_{75}P_{35}K_{130-160}$ provided a significant increase in yield compared not only with the control (1 option), but also with fertilization only when sowing at a dose of $N_{12}P_{16}$ K_{16} (2 option).

During the investigation years, the inspected calculated doses of fertilizers differed insignificantly. Apparently, an increase in the dose of potash fertilizers from 130 to 160 kg of d.a./ha did not affect the index.

The investigated doses increased the yield of green mass up to 25.7 34.1 ton/ha, as shown in other references [3, 4, 5, 7, 8].

 https://doi.org/10.1051/e3sconf/202126203025

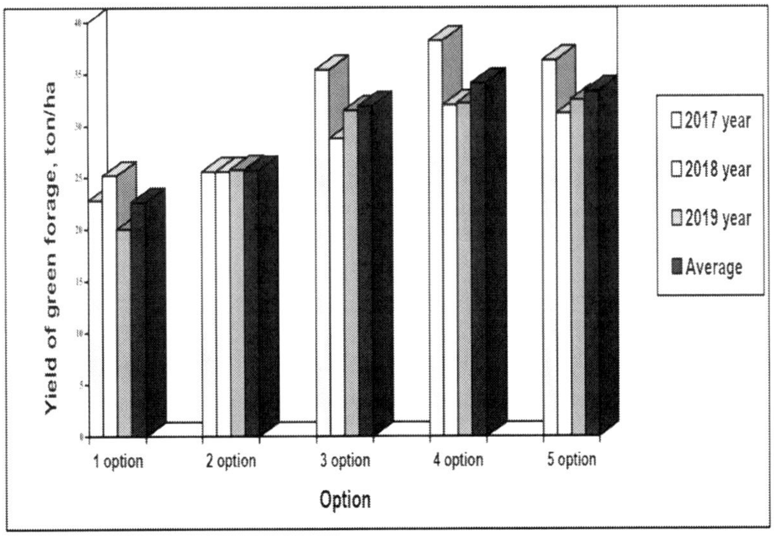

Fig. 2. Productivity of green mass obtained as a result of investigations for 2017-2019, ton/ha.

So, the productivity of cropped fallow (predecessors of winter rye) was $N_{75}P_{35}K_{130-160}$ 31. 9 - 34.1 ton/ha when used.

The content of "raw" protein changed slightly depending on the weather during the investigation years. Its lowest value was observed in 2018, and the highest was in 2019. Spring vetch, as a legume crop, used part of the nitrogen due to nitrogen fixation from the atmosphere. This amount accounts for up to 40% of the total nitrogen removal by the crop. Nitrogen nutrition and crops - a component of this mix, oats is improved. It is believed that it is required to significantly reduce the dose of nitrogen fertilization for this crop. And when applying N75, inhibition of the culture can be observed at the first stages of growth and development, although under our conditions the negative impact of high doses of nitrogen fertilizers was not observed on the development of crops in the mix [4, 5, 6, 7, 11] (Figure 3).

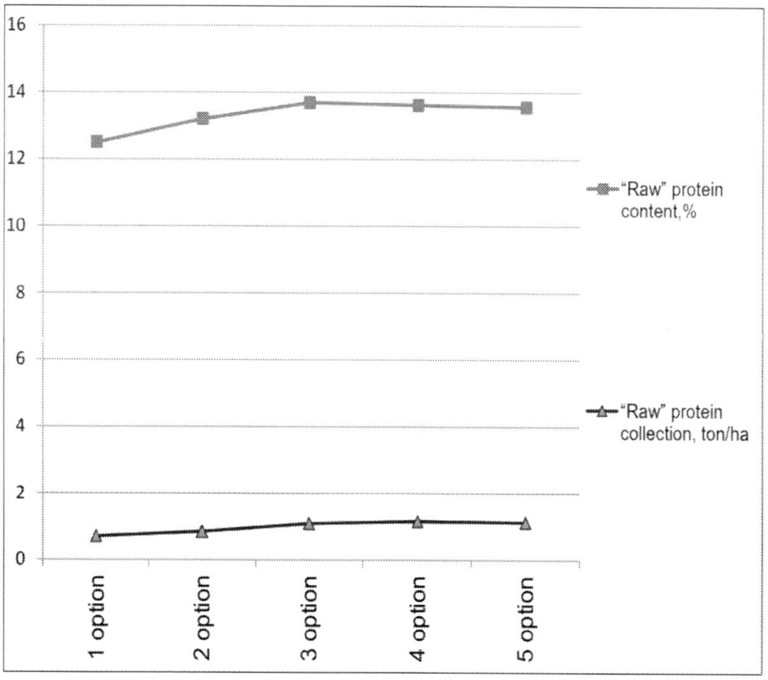

Fig. 3. Content in percentage and collection of "raw" protein, ton/ha.

The manuscript notes that with an increase in the doses of fertilizers applied, there is a tendency to an increase in the content of "raw" protein. So, when applying N12, its content complied with 12.50%, and when applying N75, it increased to 13.56-13.69%. Moreover, different fertilization systems, mineral and organomineral (options 3 and 5) differed only by 0.13% in absolute value and by almost 1% in the relative value of the trait. Apparently, increasing doses of nitrogen fertilizers are more important for growth and increase in protein content.

Protein collection changed significantly during fertilization compared with the control option, increased by 1.2 times in option 2 and 1.5-1.6 times when using the calculated doses (Figure 3), as in previous investigations [4, 9, 10]. Obviously, as the yield increases, the collection of "raw" protein also increases.

So, when applying N12, the content of "raw" protein complied with 12.50%, and when applying N75, it increased to 13.56-13.69%.

The content of NO_3 increased in the green mass of the annual crops mix when fertilizers were used: the application of $N_{12}P_{16}K_{16}$ during sowing increased by 30%; full doses of fertilizers (3-5 options) increased the content by 23 - 47%. The application of $N_{50}P_{20}K_{100}$ against the background of the afteraction 40 ton/ha of organic fertilizers caused an increase in nitrates in comparison with the equivalent investigated mineral fertilization system by 14 mg/kg. The content of nitrates at doses of fertilizers $N_{75}P_{35}K_{130\text{-}160}$ amounted to 72 - 86 percent of the maximum allowable concentration (MAC). In the experiment, with the application of nitrogen, phosphorus and potash fertilizers in the optimal ratio of elements for the culture, the content of nitrates exceeding the MAC is not observed (Figure 4).

So, the content of NO_3 in the green mass of the annual crops mix increased by 30 - 47% when fertilizers were applied.

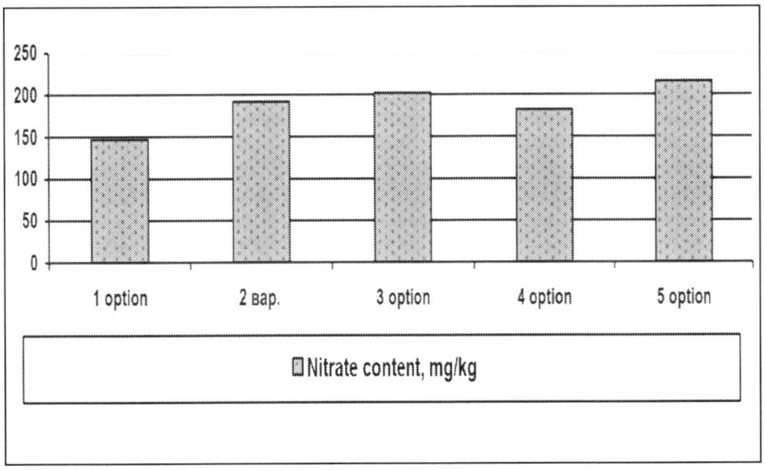

Fig. 4. NO_3 content in vetch and oat mix for 2017-2019, mg/kg.

The economic removal of nitrogen, phosphorus and potassium by the crop increased with the use of fertilizers. Nitrogen removal when applying $N_{75}P_{35}K_{130-160}$ exceeded control by 40-43 kg. Moreover, the organomineral fertilization system of the crop (option 5) did not actually differ from the mineral fertilization systems, compare 118 and 119 kg. The removal of phosphorus with the yield did not differ, it was 65 kg against the background of fertilizers. In terms of potassium removal, the organomineral and equivalent mineral systems did not differ; they were slightly inferior to the variant with the maximum dose of potassium, only by 12-13 kg. Apparently, potassium is the limiting element in the soil under such conditions (Figure 5).

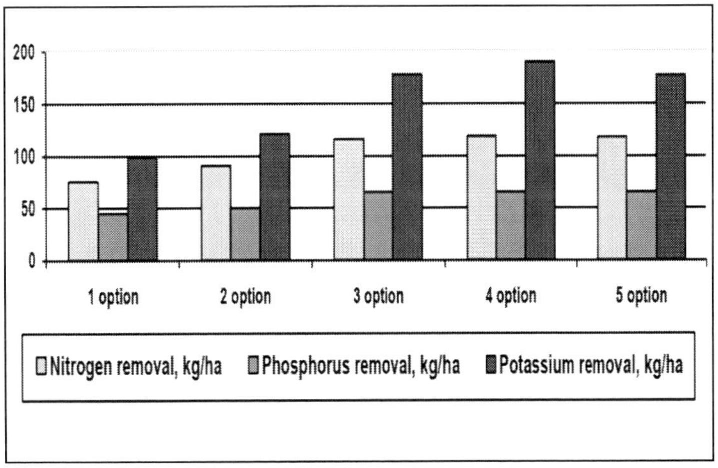

Fig. 5. Fertilizer elements removal in the experiment, on average for 2017 - 2019, kg/ha.

Fertilizer doses in 3–5 variants increased N removal by 50–60%, P_2O_5 — by 40%, and K_2O — by 80–90% compared to the control.

It should be assumed that the actual balance factors of fertilizer elements will also increase in numerical terms with an increase in the element removal, while the balance will decrease.

Actual balance factors for nitrogen amounted to 155-158%, exceeding the planned level by 35-38% in absolute value. They were significantly higher in phosphorus, by 85%, in potassium, they exceeded the planned value by 36 - 39% (Figure 6).

This is due to the fact that the removal significantly exceeded the planned value due to the high crop yield, which exceeded the planned level by 40%.

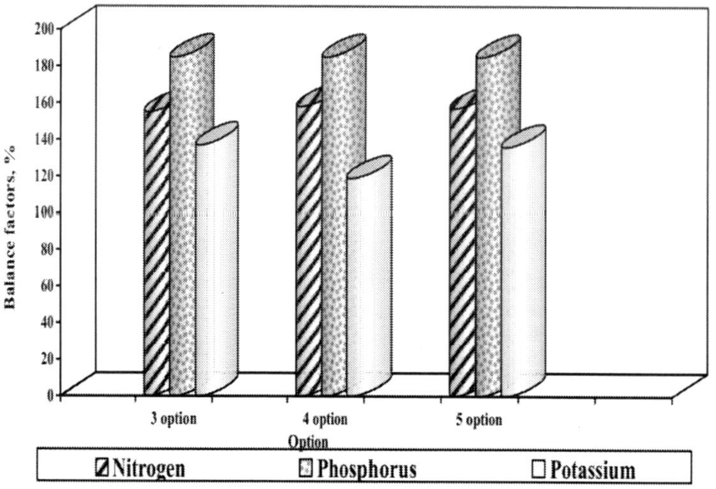

Fig. 6. Balance factors of fertilizer elements used obtained on investigation results, for 2017-2019, %.

So, the actual balance factors for nitrogen amounted to 155-158%, exceeding the planned level by 35-38% in absolute value, for phosphorus they were significantly higher than the planned ones, by 85%, for potassium they exceeded the planned value by 36-39%.

Consequently, the applied doses of fertilizer $N_{75}P_{35}K_{130\text{-}160\ \text{fertilizers}}$ when obtaining 34 ton/ha of green mass in a vetch and oat mix do not provide zero balance on fertilizer elements. A negative balance of all the fertilizer elements is detected, i.e. elements removal is much higher than the input with fertilizers.

Thus, the application of fertilizers at the rate of $N_{75}P_{35}K_{130\text{-}160}$ increases the yield of green mass by 36 – 51%, the collection of raw protein - by 1.5-1.6 times compared to the control. Accordingly, the yield of plant residues after annual forage crops increases, which contributes to a greater return and accumulation of organic matter in the soil in comparison with other inspected options.

References

1. B. A. Yagodin, *Agrochemistry*, 584 (2016)
2. V. A. Pilipchak, Feed production, **5**, 15–17 (2004)
3. O.V. Chukhina, Fertility, **1(106)**, 22-25 (2019)
4. O. V. Chukhina, Dairy Bulletin, **4(36)**, 141-154 (2019)
5. E. I. Kulikova, *The influence of different doses of fertilizers on crop productivity and fertility of sod-podsolic soil in crop rotation: dissertation*, 162 (2012)
6. Yu. P Zhukov, Land husbandry, **1**, 40-42 (1988)
7. O.V. Chukhina, *Nutritional value of vetch and oat mix when using fertilizers in crop rotation*, 201-206 (2019)
8. O. Chukhina, IOP Conference Series: Earth and Environmental Science, 6th Int. Conf. on Agriproducts Processing and Farming, 012035 (2020)
9. B. A. Dospekhov, *Field experiment technique*, 351 (1985)

10. O. P. Kamneva, On the complex use of lime and mineral fertilizers in the cultivation of vetch and oat mix for green fodder, 1(45), 116 – 121 (2016)
11. T. P. Sabirova, Vladimirsky Agricolist, **4(86)**, 33-37 (2018)
12. A. N. Fadeeva, Feed production, **2**, 14-16 (2001)
13. P. Hubert, S. Hugo, W. des Korner, Legume construction in Deutschland, RAPS, **12(1)**, 44-46 (1994)

Technological solutions for the cultivation of potatoes in the organic farming agroecosystem

D.A. Maksimov, *V.B.* Minin, *A.N.* Perekopskiy, and *A.M.* Zakharov

Federal State Budgetary Scientific Institution 'Federal Scientific Agroengineering Center VIM', branch in Saint Petersburg, Filtrovskoje highway, 3 p.o. Tiarlevo, 196625 Saint Petersburg, Russia

Abstract. One of the main tasks of organic production of agricultural products is to reduce the load on agroecosystems while simultaneously obtaining a yield indicator of high consumer quality, as with intensive production technologies, and this can be attained only through the fuller use of the biological potential of a crop, species, variety. In the production of agricultural products using biologized machine technologies, they interact with the environment: the atmosphere, hydrosphere and soil. The research is aimed to the creation of adaptive biologized machine technologies that ensure the efficient use of natural resources in the production of agricultural products with a minimum impact on the environment. The research object is the potato agrocenosis in the crop rotation of tilled and field crops. The crop rotation includes potatoes, beets, perennial legumes and cereals. In the experiment, potatoes of the Udacha variety (superelite), zoned for the Leningrad region, were cultivated. Particular attention was paid to the assessment of possible environmental risks in the implementation of the selected technological solutions for growing potatoes, as well as the fight against weeds, since the use of synthetic herbicides, fungicides, insecticides, etc. radically contradicts GOST 33980 - 2016 "Organic products. Rules for production, processing, labeling and sale".

1 Introduction

In the context of the intensification of the crisis in the domestic economy and the growing tension in the political situation in the world, the problem of the country's food security is becoming more urgent. Russia's current high dependence on food imports poses an immediate threat to the country's sovereignty and requires vigorous measures to replace food imports with nationally produced products.

In this regard, recently, the accelerated import substitution of food has become one of the main directions of the state agrarian policy aimed at achieving food independence of the country [1].

One of the promising areas that can contribute to the development of import substitution in the Russian Federation is the simultaneous development of potentially competitive, environmentally friendly areas of agricultural production. One of these areas is the organic production of agricultural products, both animal husbandry and crop production, also including organic production of seeds, animal feed, and so on, to complete a complete production cycle and complete self-sufficiency of the country's population. [2-4].

The creation and implementation of machine technologies in agricultural production without proper monitoring and forecasting of the consequences of their impact on the environment in the long term cannot guarantee not only the country's food security, but also creates a threat of irreversible negative consequences.

The research is aimed to the creation of adaptive biologized machine technologies that ensure the efficient use of natural resources in the production of agricultural products with a minimum impact on the environment.

2 Materials and methods

The object of research is potato agrocenosis in organic crop rotation, where row crops and field crops are involved.

The organic crop rotation is six-field and includes fields:

1-potato variety "Udacha";

2-beetroot "Two-seeded TSKHA";

3-coverless sowing of clover and timothy;

4-perennial grasses 1 year;

5-perennial grasses 2 year;

6-perennial grasses 3 years with sowing, after plowing winter rye for green fertilization [5-7].

Work in the crop rotation of organic production continued (the beginning of the experience was laid in 2016). In the field season of 2020, an experiment was performed in which three groups of factors were studied in the framework of the use of biologized technology of potato cultivation:

– the level of mineral nutrition, due to the introduction of organic fertilizer BIAGUM, produced by IAEP-branch of FSBRI FNATS VIM;

– the effect of additional biological nitrogen provided by the introduced microorganisms - nitrogen fixers (Flavobacterin);

– the effect of biofungicides, biological plant protection products, the development of the FSBRI VIZR.

Since the principles of organic production provide for the restoration of natural soil fertility and the use of new innovative, nature-like technologies that do not damage the world around them, but exist in harmony with it and allow restoring the balance between the biosphere and technosphere disturbed by humans, key attention was paid to assessing possible environmental risks when applying the selected technological solutions in the crop rotation of organic production [8, 9].

In the experiment, potatoes of the Udacha variety (superelite), zoned for the Leningrad region, were cultivated, its predecessor was white mustard green manure. Organic fertilizer BIAGUM [10] was applied in two versions, 4.3 and 8.6 t/ha (if translated into nitrogen, then 80.0 and 160.0 kg/ha).

The experiment scheme is presented in Table 1.

Table 1. Scheme of options for planting potatoes

Experiment No.	Amount of fertilizer	Biological product
1	None	None
2	None	Kartofin
3	None	Flavobacterin
4	80.0 kg of nitrogen	None
5	80.0 kg of nitrogen	Kartofin
6	80.0 kg of nitrogen	Flavobacterin
7	80.0 kg nitrogen + 80.0 kg nitrogen in top dressing	Kartofin

8	80.0 kg nitrogen + 80.0 kg nitrogen in top dressing	Kartofin + Fitolar G
9	160.0 kg of nitrogen	Kartofin

In the experiment with potatoes, the biological preparation "Kartofin", developed by the specialists of the FSBRI VIZR, was used; it was created on the basis of the collection producer strain B. subtilis I-5 12/23, by spraying at a rate of 3 l/ha. In one of the variants, a multifunctional multi-bioconversion microbiological preparation Fitolar ShV, G based on the collection producer strain T. asperellum T-36, developed at the FSBRI VIZR, was also used for introduction into the soil along with planting potatoes [11, 12].

In two variants of the experiment, a complex microbiological preparation with a nitrogen-fixing function developed by VNIISKhMB, Flavobacterin, is used.

Two options were compared, one with plant protection with biological devices, and the other without protection with biological preparations (without organic fertilizer Biagum and with the introduction of organic fertilizer Biagum with a dosage of 80 kg of nitrogen per hectare) [13, 14].

Plot area - 5.6 × 11 = 61.6 m^2. The repetition rate was fourfold, the location of the plots was randomized.

On May 26, organic fertilizer Biagum was transported to the experimental landfill. The application of organic fertilizer and cutting of the ridges for planting potatoes with the simultaneous incorporation of fertilizer into the ridges was performed on the same day, these two technological operations were performed one after the other in order to prevent nitrogen from escaping into the atmosphere. The planting of potatoes with simultaneous treatment with the Kartofen preparation was performed by a two-row planter Lida - 201 on May 27, 2020, and on June 9, soil samples were taken from individual plots [15].

An analysis of the data obtained shows that the total potato yield in 2020 is slightly lower than in 2019, due to the prevailing weather conditions (slightly less warm May and increased precipitation in June - August.

3 Results and discussion

The yield of standard tubers in the experiment was obtained in the range from 18.47 to 31.77 t/ha, depending on the option, table 2 and figure 1.

Table 2. Yield of standard potato tubers of the Udacha variety and the share of small potatoes in the biological yield.

Experiment No.	Dose, nitrogen kg/ha	Biological product	Yield (t/ha)	Share of small potatoes in, %
1	0	None	20.25	2.07
2	80	None	25.88	1.68
3	0	Flavobacterin	22.87	1.36
4	80	Flavobacterin	27.33	1.38
5	0	Kartofin	18.47	2.4
6	80	Kartofin	27.51	1.66
7	80	Kartofin + Fitolar, G	28.20	1.84
7	160	Kartofin	31.77	0.83
HCP_{05}			2.07	

Without the use of compost and plant protection products, the yield was 20.25 t/ha. The use of compost made it possible to raise it to 25.88 t/ha, that is, by more than 5 tons.

Flavobacterin, which promotes both the development of soil nitrogen fixation and the enhancement of the general condition of potato plants, provided a significant increase in yield of 2.62 t/ha even without compost. The use of compost, together with Flavobacterin, increased the yield by another 4.46 t to 27.33 t/ha.

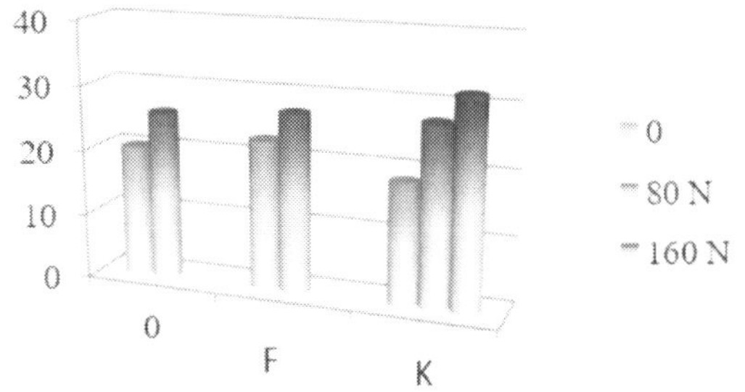

Fig. 1. Impact of biological products and fertilizers on yield.

The potato yield in the experiment with the biofungicide Kartofin, which was not provided with compost, did not differ significantly from the control variant.

The use of compost together with Kartofin gave a yield increase of 9.04 t/ha, however, in the variant where Fitolar G was additionally used, the increase was already 9.73 t/ha.

The maximum yield of 31.77 t/ha was achieved with the introduction of an increased dose of compost (160 kg N/ha) together with Kartofin.

It should be noted that the lowest content of small potatoes was in the variants with Flavobacterin and in the variant with an increased dose of compost and Kartofin. This phenomenon is explained by the formation under these variants of the most favorable conditions for the full development of potato tubers.

The use of biological products Kartofin, SK and Fitolar G significantly reduced the development of fungal diseases on potato plants during the entire observation period. The biological efficiency for the development of a complex of fungal diseases ranged from 81.2 - 95.3% and only before mowing the tops decreased to 43.8 - 67.3%.

It was found that biological products Kartofin, SK and Fitolar G had a positive effect on the infestation of tubers, reducing this indicator by 22 - 36% in comparison with the control.

The dry matter content in the tubers ranged from 18.82 to 22.34%, table 3. It should be noted that the lowest dry matter content was recorded in variants without the use of biological products. This indicates that biological products promoted more efficient vital activity of potato plants and better biosynthesis. As expected, a lower dry matter content is also observed in the variant with the maximum compost dose, here is also the maximum concentration of nitrates. The increased content of nitrates promoted the active accumulation of water.

Table 3. Content of dry matter and nitrates in potato tubers.

Experiment No.	Fertilizer quantity, kg nitrogen/ha	Biological product	Dry matter content, %	Content of nitrates, mg/kg
1	0	None	19.24	46
2	80	None	19.80	61
3	0	Flavobacterin	20.49	210
4	80	Flavobacterin	21.40	178
5	0	Kartofin	22.34	125
6	80	Kartofin	20.05	169
7	80	Kartofin + Fitolar, G	20.72	119
8	160	Kartofin	18.82	226

Content of nitrates in potatoes of all options is lower than MAC. When using biological products, the content of nitrates in potato tubers increased, which reflects the increased mineralization activity of microflora in these variants of the experiment, and potato plants used this additional amount of nitrogen.

In the biologized technology used, gentle soil cultivation in the spring (cultivation and immediately cutting of ridges) was performed at the optimum time. A significant number of earthworms in the soil of the experiment during the growing season served as an indicator of the favorable soil conditions. Only biological pesticides were used, which also did not harm natural biodiversity.

The only environmental threat was the accumulation of nitrates in the soil during processing and composting.

The compost was applied to the soil on the twentieth of May, a few days before the potatoes were planted. With the first dose of compost, about 1.5 g of compost got into each kilogram of soil, and with it 1.5 mg of mineral forms of nitrogen and about 4 mg of total nitrogen. Table 4 shows the results of calculating the rate of accumulation of forms of mineral nitrogen in the soil of the experiment for the period from the end of May to 10 June.

Table 4. The impact of organic fertilization on the rate of accumulation of forms of mineral nitrogen in the soil, kg nitrogen/ha per day

Fertilizer quantity, kg nitrogen/ha	Year of research	
	2019	2020
None	2.42	1.48
80	6.58	5.18
160	9.20	9.38

It should be noted that starting from the twenties of May, when the first soil samples were taken (before composting), the average daily temperature was above 10 degrees. This contributed to the development of soil microflora and the active development of the mineralization process. In the control variant, 1.48 - 2.42 kg/ha forms of mineral nitrogen (predominantly nitrate) were mineralized and formed in the soil. The composting has significantly intensified this process. It should be noted that with a dose of compost, a relatively small amount of nitrogen-containing compounds (about 4 mg/kg of total nitrogen) was introduced into the soil, however, in variants with compost, 5.18 mg of nitrate and ammonium nitrogen were formed daily in 2020 and 6.58 mg - in 2019. Thus, the compost activated the subsoil microflora and provided additional mineralization of soil nitrogen, significantly exceeding the nitrogen content in the compost itself.

Table 5 shows the results of the analysis of soil samples taken during the growing season of 2020.

Table 5. Nitrates in soil, experiment in 2020

Fertilizer amount, nitrogen/ha	Steel	22.05.20	9.06.20	29.06.20	21.07.2020	25.10.20
None	0	6±3	29	283	12	83
80	0		165	398	27	138
None	Flavobacterin		75	205	6	72
80	Flavobacterin		211	334	40	116
None	Kartofin		75	288	12	90
80	Kartofin		152	368	11	111
80	Kartofin + Fitolar		102	403	6	165
160	Kartofin		245	464	23	83

Noteworthy is the high accumulation of nitrate nitrogen by the end of June in the control variant (283 mg/kg). This is due to the high content of organic matter in the soil (8 - 10%), which provides such active mineralization and an increase in the content of nitrates, even without composting, but with regular treatments. As expected, both compost and biological products contributed to an even greater activation of mineralization and a further increase in the content of nitrates in the soil. However, these nitrate stocks are not particularly hazardous as they are the developing potato plants almost all and swallowed. By the twenty-first of July, no more than 40 mg of nitrates per kg of soil remained in the soil, but the absorption of nitrogen continued.

Particular attention should be paid to the accumulation of significant amounts of nitrates and all mineral nitrogen in the autumn period (by the end of October). This phenomenon is associated, on the one hand, with active soil cultivation during the harvesting of potatoes, which contributed to the activation of mineralization processes, and on the other hand, with the established rather warm weather.

In the second experiment, a variety testing of potato varieties was performed (table 6).

Table 6. Yield indicator in the experiment on variety suitability for organic production conditions

Cultivar	Yield (t/ha)	Small-size fraction share in, %
League	15.47	3.63
Ryabinushka	19.03	3.08
Nevsky	22.50	1.50
Yelizaveta	15.11	0.94
$HCP_{0.95}$	1.486	

Under rather harsh conditions (without using compost) the highest yield was shown by the variety Nevsky, the smallest Liga and Elizaveta. Apparently, varieties Liga and Ryabinushka could not realize their potential due to insufficient nutrient supply. This is evidenced by the increased share of small potatoes in the biological yield. Nevertheless, all four varieties have shown a fairly high productivity under test conditions.

Thus, pursuant to the 2020 research results, the efficiency of the developed version of the technology for the production of organic potatoes is confirmed. The combined use of compost at a dose of 4.3 t/ha and biological products made it possible to achieve the yield of standard tubers of 27.3 - 28.2 t/ha with their low incidence of fungal diseases. With an increase in the compost dose to 8.6 t/ha, the yield increased to 31.77 t/ha, but the dry matter content in the tubers decreased and the nitrate content increased, while biological products significantly activated the accumulation of mineral forms of nitrogen in the soil, which contributed to the enhancement of nitrogen nutrition of potato plants. Significant accumulation of nitrates in the soil in the autumn period was noted, which can pose a threat of nitrogen losses, primarily nitrates, as a result of leaching. Therefore, it is required to propose methods of planting winter stubble crops after harvesting row crops that will eliminate these threats.

4 Conclusion

Regularities of the formation of potato productivity from the prevailing weather and climatic conditions, the supply of nutrients and plant protection products, the selected cultivation methods were obtained. The combined use of organic fertilizer BIAGUM, produced in the department of engineering ecology of the IAEP branch of the FSBRI FNATS VIM and modern biological devices used for plant protection, which were developed by specialists of the FSBRI VIZR, has proven itself, which made it possible to obtain a good yield indicator for organic production, it amounted to 32 tons/ha of potatoes of high consumer quality. Herewith, it was confirmed that by applying current technological methods, technical means in organic production technologies, it is possible to obtain a yield close to the yield indicators of intensive technologies. Herewith, organic production of agricultural products, in contrast to intensive and high-intensity, can significantly reduce the load on agroecosystems, by eliminating the use of synthetic herbicides, fungicides, insecticides, synthetic growth regulators, etc., and maintaining a healthy state soil and an increase in its fertility and biological activity is achieved through the use of special crop rotations, including the cultivation of legumes and other green manure crops.

References

1. S. Hyun, S. M. Yang, J. Kim, K. S. Kim, J. H. Shin, S. M. Lee, B.-W. Lee, R. M. Beresford, D. H. Fleisher, Computers and Electronics in Agriculture, **181,** № 105936 (2021)
2. O. Gorb, I. Yasnolob, T. Chayka, O. Zoria, T. Dugar, P. Shvedenko, O. Kalian, Y. Lypii, N. Protsiuk, Journal of Environmental Management and Tourism, **11,** 1588 (2020)
3. A. P. Glinushkin, V. I. Startsev, L. V. Startseva, IOP Conference Series: Earth and Environmental Science, **459,** 062069 (2020)
4. S. O. Bastaubayeva, R. K. Sagitov, International Journal of Pharmaceutical Research, **12,** 2156 (2020)
5. H. S. Wulanningtyas, Y. Gong, P. Li, N. Sakagami, J. Nishiwaki, M. A. Komatsuzaki, Soil and Tillage Research, **205,** 104749 (2021)
6. X. Hou, R. Li, Transactions of the Chinese Society for Agricultural Machinery, 51, 262 (2020)
7. C. A. Kwiatkowski, E. Harasim, Agronomy, **10,** 1045 (2020)
8. I. M. Fomin, E. E. Oreshin, G. A. Loginov, A. M. Zakharov, In the collection book: Ecology and agricultural technologies: agro-engineering solutions. materials of the 7th International Scientific and Practical Conference, 141 (2011)
9. G. A. Loginov, I. M. Fomin, E. E. Oreshin, A. M. Zakharov, Technologies and technical means of mechanized production of crop and livestock products, **82,** 51 (2010)
10. V. B. Minin, V. D. Popov, D. A. Maksimov, A. A. Ustroev, S. P. Melnikov, E. Papushin, Agronomy Research, **18,** 1359 (2020)
11. J. Jin, J. D. Wood, A. Franks, R. Armstrong, C.Tang, Soil Biology and Biochemistry, **144,** 107780 (2020)
12. Yu. A. Titova, I. L. Krasnobayeva, Technologies and technical means of mechanized production of crop and livestock products, **2(99),** 164 (2019)
13. N. V. Romanovsky, A.N. Perekopsky, Technologies and technical means of mechanized production of crop and livestock products, **98,** 101 (2019)
14. A. N. Perekopsky, A. M. Zakharov, Journal of Advanced Research in Technical Science, **18**, 61 (2020)
15. T. Gnatiuk, Biosystems Diversity, **26,** 71

E3S Web of Conferences **262**, 03027 (2021) https://doi.org/10.1051/e3sconf/202126203027
ITEEA 2021

Environmental investment as a tool for ensuring economic security of business entities: accounting aspects

Alexey Popov*

Ural State University of Economics, 8 Marta, 62, 620144 Ekaterinburg, Russia

Abstract. The paper is devoted to the issues of companies' investing environmental assets to ensure their economic security. The role of environmental investments in strengthening the potential of an economic entity is justified, which has a positive impact on the environmental security of regions and the state as a whole. The necessity of analyzing a group of environmental indicators of economic security of an economic entity is formulated. The author considers features and methods of evaluating the effectiveness of environmental investment projects and analyzes the system of legal regulation of forming accounting information used to evaluate the above effectiveness. The classification of environmental objects is given, their structure, principles of formation, recognition and evaluation being considered. The author summarizes promising areas of improving economic science in order to form more representative data to make well-grounded investment decisions, which will have a positive impact on the performance of economic entities, investment strategy and ensure sustainable development.

1 Introduction

In modern economic conditions characterized by development and modernization of industrial production, integration of industries, digitalization of economic processes and universal globalization, issues of environmental safety are becoming increasingly relevant since significant environmental problems related to global warming and climate change, loss of the biodiversity, desertification of territories and other adverse economic factors affecting the environment, growing environmental damage from emergencies and man-made disasters, air pollution, water pollution (underground water resources included), as well as biological resources of the world oceans affect the interests of the state, industry and citizens.

The current environmental situation in the Russian Federation reveals signs of a significant level of anthropogenic impact on the environment, the one caused by results of the past activities of enterprises included,

As noted in the "Fundamentals of State Policy in Environmental Development of the Russian Federation till 2030", approved by the President of the Russian Federation on April

* Corresponding author: prepodpopov@yandex.ru

30, 2012, more than 54% of the urban population in more than forty regions are subject to the negative impact of air pollution, which exceeds the established standards. The volume of untreated industrial and domestic wastes dumped into open sources remaining significant and a negative trend associated with a decrease in the quality of soils and land prevailing, we observe losses in soil fertility, a decrease in the quality of agricultural land, deterioration in fertility and yields, and as a result, their removal from the fund of agricultural land, as well as a trend of desertification of a number of territories of more than 27 subjects of the Russian Federation on an area of more than 100 million hectares and increased amount of waste which is not involved in the subsequent economic turnover and dumped in the territories in violation of the requirements of environmental safety. Doctor of Economics, Professor Arkhipov writes that "from the perspective of the neo-industrial paradigm, GDP growth in modern Russia caused by inflation of petrodollars and the raw materials orientation of this economy leads to the depletion of natural resources and pollution of the environment and destroys the unity of biological and economic systems, extractive and manufacturing sectors, short-term and long-term public interests, accumulation and consumption." [1]

On the basis of the indicated prerequisites, the issues of investing in environmental facilities have proved to be of paramount importance at the present stage. These issues are discussed in detail in the works of Russian scientists-researchers Arkhipov A. I.[1], Khafilyev R. R.[2], Erokhina V. N.[3], Borlakova A. K.[4], Egorova M. S.[5] , etc.

Professor Magaril (DSc) notes that "the actual problem is that when conducting an economic analysis of an investment project, the data obtained at the stage of environmental analysis are underutilized or are not used at all. In other words, environmental factors are not economically assessed and are therefore excluded from the analysis."[2]

In accordance with the urgency of the problems, the purpose of this work is to reveal the features of taking into account environmental factors when evaluating the effectiveness of investment projects in order to strengthen the economic security of business entities.

When describing the above features, we can't but mention the contribution of Professor Medvedeva to the development of methodological recommendations for the implementation of environmental and economic assessment of the effectiveness of projects of planned economic activities. She defines environmental and economic efficiency as " "an indicator characterizing the ratio of the overall economic benefits to losses from the project, including external environmental effects and related social and economic consequences affecting the interests of the population and future generations as a result of the implementation of the project." [6] The methodology proposed by this researcher implies the integration of the economic components (costs and benefits) of environmental activities into the program into the program of analysis of the effectiveness of investment projects.

A number of foreign researchers, in particular Newt S. [7], Cortazar G. [8], Lingart J. B. [9] and others address this problem and their development is of special scientific interest. In particular, Bonifant B.S. notes: "In the past, industrial enterprises opposed the introduction of environmental standards, fearing significant restrictions and high costs for compliance with environmental requirements. Today, however, many companies are moving towards improved business, increased productivity, and innovative technologies, while gaining additional competitive advantage from environmental investments. "[10]

The implementation of environmental investments is directly related to strengthening economic security of economic entities, which is justified by Belik I.S. [11], Kesyan Z.A. [12], Medvedev P.V. [13], Buletova N.E. [14], Vlasenko M.N. [15], etc.

Batova V.N. notes that "the problem of ensuring the economic security of entrepreneurship cannot be considered isolated from environmental aspects. This relationship should be based on a comprehensive analysis of the entire set of social, economic and environmental indicators in society. The urgency of these tasks is largely due to the fact that until now the environmental component has not been an independent element in the system of economic security. "[16]

Vysotskaya T.R. justifies in her work such an instrument of economic security as environmental liability insurance, which will reduce environmental risks if environmental damage is to be compensated. [17]

Thus, on the basis of the synthesis of the research results, the conclusion is made about the prevailing role of financial, personnel and social indicators of the economic security of the enterprise used to analyze and assess the level of the latter. However, we think these groups of indicators are not enough for a comprehensive assessment of the economic security of economic entities. In addition to traditional financial, human and social indicators, we suggest including environmental indicators, such as the volume of environmental investments made in total investments, the profitability of environmental investments, the amount of environmental damage and other indicators.

2 Materials and Methods

The research is based on universal general scientific methods of analysis, generalization, analogy, modeling, empirical methods of description and interpretation, as well as theoretical methods of formal and dialectical logic. The methodological basis for the research was the works of domestic and foreign scientists, normative legal acts in force at the time of writing of this work, international financial reporting standards, as well as draft regulatory acts to be implemented.

3 Results and Discussion

The main regulatory document of the Russian Federation used to assess the effectiveness of investments is the "Methodological Recommendations for Assessing the Effectiveness of Investment Projects," approved by the Ministry of Economy of the Russian Federation, the Ministry of Finance of the Russian Federation, Gosstroi of the Russian Federation on 21.06.1999 N VK 477. The specified recommendations:

1) unify the terminology and provide the indicators of the effectiveness of investment programs;

2) systematize and unify the rules and requirements that pre-project and project data must meet;

3) provide a methodology for calculating the performance indicators of investment projects in accordance with international standards and regulations;

4) justify the need to compare and critically analyze various options for technical, organizational and financial solutions to carry out a particular investment project;

5) take into account the specifics of launching some special projects that require an innovative approach and non-standard methods of evaluating effectiveness.

Table 1 shows options of the proposed methodology used to evaluate effectiveness of investment projects.

Table 1. Range of Investment Project Performance Assessment Methodology

Features of the methodology for evaluating the effectiveness of investment projects

https://doi.org/10.1051/e3sconf/202126203027

Table 1. Continued

Assessment of financial feasibility of the project	Assessment of the effectiveness of participation of companies in the project	Decision-making on state support of the investment project	Comparison of alternative investment projects and selection of the most effective	Assessment of the economic consequences of selecting one or more projects	Preparation of conclusions for state, industry and other types of expertise investment justifications, feasibility studies, projects and business plans	Economic monitoring and decision-making on changes in financing

When doing investment analysis, one can use various methods of evaluating investments including:

1. Discounted cash flow methods, which include:
1.1 Net Present Value (NPV);
1.2 Internal Rate of Return (IRR);
1.3 Modified Internal Rate of Return;
1.4 Profitability (RR-Rate of Refund);
1.5 Discounted Payback Period (DPP).
2. Traditional assessment methods:
2.1 Accounting rate of Return (ARR);
2.2 Payback Period.

In order to assess the effectiveness of an investment project, environmental factors included, it is necessary to determine accurately the cash flows associated with investing in environmental facilities, cash flows associated with choosing alternative, less environmentally friendly options, and integrate the above data into the overall performance assessment system.

The data on the enterprise's environmental costs are generated in the accounting system, which provides complete and reliable information on the property used for environmental protection, environmental obligations and costs. In accordance with the new business conditions, economic entities are actively introducing an environmental accounting system, as a subsystem of accounting. The issues of formation of environmental accounting are discussed in the works of a number of domestic economists, such as Kozhukhova O. S. [18], Ilyicheva E. V. [19], Amanzholova B. A. [20], Saenko K. S. [21], etc. The dissertations on these problems defended by Cheprakova T. N. [22], Murueva E. K [23], and others deserve attention. We can't but mention the fact that the fundamental principles of environmental accounting in international practice were formulated in the previous century by Odum H. T.[24] and Akhnad I. J.[25], and later developed by Schaltegger S. [26]

The current four-level system of legal regulation of accounting at the first (the highest) and the second (the level of federal and industry standards of accounting) levels does not contain special rules of formation of information about the company's environmental activities in accounting.

The third-level legal document regulating the formation of financial information on environmental investments and environmental costs in accounting and in financial statements is the letter of the Ministry of Finance of the Russian Federation N PZ-7/2011 "On Accounting, Formation and Disclosure of Information on the Environmental Activities of a Company in Accounts ".

This letter presents the following classification of environmental accounting items, as shown in table 2.

Table 2. Classification of environmental accounting

Items of environmental accounting			
Capital investments (investments in environmental assets)	Current expenses on environmental measures (environmental costs)	Estimated (environmental) obligations	Sources of financing environmental activities
- construction of facilities to protect air from pollution by substances and prevent climate changes (devices for separating gas impurities and dust from industrial gas; automatic air pollution control systems; power generation plants based on alternative sources (wind and solar energy, biogas, thermal waters, etc.);water protection and other wastewater collection and treatment facilities; - installing facilities for decontamination, composting, processing of production and consumption wastes; - construction of anti-erosion hydro-technical, anti-mudslide, anti-landslide, anti-avalanche, anti-dump, including coastal ones, facilities for terracing steep slopes, recultivation of lands, installing equipment for monitoring soil and groundwater pollution, other actions of protection and rehabilitation of lands, surface and underground water; - building facilities protecting environment from noise, vibration and other physical effects (screens, shields, partitions, embankments, hedges); - measures to ensure environmental radiation safety aimed at reducing or preventing the impact of radioactive substances on the environment, rehabilitation of contaminated areas, neutralization and disposal of medium and high radioactive waste, etc. - research and development of environmental equipment, installations, structures, plants and facilities, advanced environmental protection technology, means of protecting environment from negative anthropogenic impacts	- maintenance charges for and repair of environmental-related property, plant and equipment; - costs of raw materials, materials, fuel and electricity used to operate environmental facilities; - upkeep of personnel serving environmental facilities; - rent (leasing) payments, insurance payments of environmental equipment and facilities - costs of collection, storage (disposal) and processing(neutralization), destruction, disposal of production and consumption waste on their own , - costs of independent control over the harmful impact on the environment and monitoring activities, scientific and technical research, management of environmental activities in the company; - costs of current measures to preserve and restore the quality of the environment that has been disrupted as a result of previous economic activities - compulsory insurance of civil liability of the organization operating a dangerous production facilities, - payment for negative impact on the environment; - other ongoing measures to reduce the harmful impact on the environment and prevent climate changes.	obligations to restore the disturbed ecological system in connection with the acquisition, construction and manufacture of potentially environmentally hazardous property, plant and equipment, in particular obligations to recover the costs that the company is expected to incur in the performance of the estimated obligation to dismantle and dispose of the property, plant and equipment and restore the natural resources on the site occupied	- companies' own sources of financing environmental activities - part of the retained earnings of the company aimed at environmental activities - state assistance (subventions, subsidies) provided to organizations for environmental activities (environmental control; collection, disposal of waste and wastewater treatment; protection of flora and fauna and their habitat; applied scientific research in the field of environmental protection; other issues in the field of environmental protection) - loans and credits for environmental activities (borrowed sources)

We would like to consider the basic principles of evaluation of the above items of accounting and their impact on the generation of cash flows in order to assess effectiveness of environmental investment projects.

The International Financial Reporting Standard (IFRS 16) "Fixed Assets" states that acquisition of fixed assets may be for safety or environmental purposes. Despite the fact that the purchase of these assets does not directly generate economic benefits (income) from their operation, it is often necessary for the economic entity to receive income from the use of its other assets. Such assets meet the criteria for identification as part of assets, as they allow enterprises to generate revenues from related assets that exceed the benefits potentially available in the event of a refusal to acquire such assets. For example, chemical industry can introduce new chemical technologies that ensure compliance with environmental regulations and regulations for manufacture and storage of hazardous chemicals, and therefore the associated modernization of production facilities is recognized as an asset, since without it the economic entity cannot produce and sell chemical products. Accordingly, it is advisable to distinguish this category of assets into a separate subgroup of fixed assets, called "environmental assets."

When assessing effectiveness of environmental investments, one should take into consideration the amount of costs for their implementation, excluding VAT and other reimbursable taxes. It is formed currently according to the accounting rules of accounting standard FAS 6/01 "Fixed Assets" in domestic practice, and according to the above IFRS 16 in international practice. As of today the process of reforming domestic accounting in accordance with international standards is very intensive, the program for developing federal accounting standards is in operation, annually being improved (at the moment - the program approved by Order of the Ministry of Finance of Russia of 05.06.2019 N 83n), which prescribes the date of entry into force of the new Federal Accounting Standard (FAS)"Fixed Assets" from January 1, 2022. Therefore, in this paper we will rely on international standards and cover the components of the costs included in the initial cost of an environmental asset under the rules of IFRS 16.

The initial cost of environmental assets includes:

1) the price of its purchase, including customs duties, fees and other payments and non-refundable taxes on the purchase, minus trade discounts and adjustments;

2) all costs directly related to the transportation of the asset to the place of intended operation and bringing it into the condition necessary for use in accordance with the intentions of the management of the business entity;

3)the initial analytically generated value of the costs of dismantling and decommissioning an item of fixed assets and restoring the site occupied by it, for which the company assumes responsibility either when acquiring this item or as a result of its use for a certain period for purposes other than the production of inventories during this period.

The environmental assets directly related to the process of acquisition, construction and manufacture of the object also include:

- personnel remuneration costs arising directly from the construction or acquisition of an item of property, plant and equipment;
- site preparation costs;
- initial expenses of delivery and handling operations;
- installation costs;
- the cost of verifying the proper functioning of the asset, remaining after deducting the net earnings from the sale of products produced during the delivery of the asset to the intended location and bringing it to the required condition (for example, samples obtained when testing equipment);
- fees for professional services rendered and other costs.

These costs form initial cash outflows in order to calculate the net present value and other indicators for assessing the effectiveness of investments.

The current costs for environmental measures (environmental costs), the list of which is given in Table 2, are shown by the company in accordance with the current federal standard FAS 10/99 "Organization costs" as part of expenses for ordinary activities, or as part of other expenses, depending on the type, direction of expenditure and its compliance with the subject of the company's activities in the sum of recognized amounts. For example, the costs of environmental monitoring, testing and verification of products for compliance with environmental standards are unambiguously related to the costs included in the cost of products, work, and services. The costs of paying fines for violation of environmental legislation in accordance with paragraph 11 of FAS 10/99 are considered as other results of activity and are related to 91 "Other incomes and expenses." The specified current expenses are borne by the business entity in the process of implementation of the investment project, accordingly, cash flows on these operations are also subject to consideration in the performance assessment.

Environmental obligations as an integral part of the valuation obligations are formed in accordance with the rules of the federal accounting standard 8/2010 in the amount reflecting the most accurate monetary estimate of the costs required for the settlement of the formed obligation, for example, disassembly, dismantling, removal, disposal of equipment and other fixed assets, restoration of the initial properties of the occupied area, payment of environmental fees for the disposal of harmful and hazardous waste, etc.

When assessing the effectiveness of an investment project cash outflows under these obligations are made at the stage of its completion, therefore, when forming an estimated obligation, the estimated costs are subject to discounting using the discount rate used by the business entity.

When making environmental investments at the expense of their own sources of financing (retained profit), additional cash flows are not generated, accordingly, this fourth object is not taken into account in assessing the effectiveness of the environmental project, a positive cash flow being formed due to the funds of state assistance (subsidies received, subventions). The latter increases the net present value and this flow should be taken into account in calculations.

And when making investments in environmental assets at the expense of borrowed funds, it is necessary to follow the current FAS 15/08 "Accounting expenses on loans " to calculate efficiency since this standard prescribes the rules for showing basic expenses on loans, i.e. payable interest, as well as additional costs. At the same time, the interest to be paid is subject to capitalization (inclusion in the value of the investment asset), if the environmental asset is recognized as investment during the period of acquisition, construction and manufacture, while fulfilling a number of conditions specified in paragraph 9 of FAS 15/08, and otherwise relate to the structure of other expenses and reduce the financial result of the current period. It is therefore very important to identify the relationship of cash outflows to the project in order to estimate the cash flows used to calculate the effectiveness of the investment project.

4 Conclusion

Thus, the considered rules for assessing assets of environmental accounting, which form the cash flows necessary for calculating the effectiveness of investing environmental projects, will allow economic entities to choose the optimal project, carry it out and, thereby, strengthen their economic security indicators. Further research in this sphere should include the issues of reforming accounting and evaluation of environmental accounting assets in accordance with the Federal Accounting Standards development program since the international rules, on which the draft standards are based, need to be adapted to the domestic rules of accounting and conducting economic activities.

References

1. A.I. Arkhipov, L.A. Kormishkina, L.P. Koroleva, Bulletin of the Institute of Economics of the Russian Academy of Sciences, **3** (2018)
2. R. R. Khalfiev, E. R. Magaril, Bulletin of USTU-UPI. Series Economics and Management, **2**, 81 (2009)
3. V. N. Erokhina, T. V. Lesina, Science, **7(4)** (2015)
4. A. K. Borlakova, SRRM, **6(75)** (2012)
5. M. S. Egorova, Vectors of well-being: economy and society, **1(1)** (2011)
6. O. E. Medvedeva, *Methodological recommendations for the implementation of environmental and economic assessment of the effectiveness of projects of planned economic activities*, 96 (2004)
7. C. Nehrt, Strategic Management Journal, **17(7)**, 535-547 (1996)
8. G. Cortazar, E. S. Schwartz, M. Salinas, Management Science, **44(8)**, 1059-1070 (1998)
9. J. B. Linhard, Journal of Safety Research, **36(3)**, 257-260 (2005)
10. B. C. Bonifant, M. B. Arnold, F. J. Long, Business Horizons, **38(4)**, 37-48 (1995)
11. I. S. Belik, *Environmental and economic security* (2013)
12. Z. A. Kesyan, Bulletin of the Adyghe State University, Economics, 4 (2010)
13. P. V. Medvedev, Problems of market economy, **1**, 47-53 (2015)
14. N. Ye. Buletova, *Ecological and economic safety: nature, content and problems of diagnostics in the regions of Russia* (2013)
15. M. N. Vlasenko, Yu. N. Shedko, Business strategies, **2** (2017)
16. V. N. Batova, Yu. V. Ramzaeva, Development of ecological production in small business as a factor in improving the quality of life, Materials of the scientific-practical conference, 130-132 (2014)
17. T. R. Vysotskaya, Russian economic Internet journal, **2**, 69-77 (2012)
18. O. S. Kozhukhova, Management of economic systems: electronic scientific journal, **3**, 112 (2012)
19. E. V. Ilyicheva, Fundamental research, **1**, 66-67 (2009)
20. B. A. Amanzholova, N. V. Fribus, International accounting, **9(351)** (2015)
21. K. S. Saenko, Audit and financial analysis, **5**, 048-058 (2007)
22. T. N. Cheprakova, *Accounting ecological accounting in the management system of agro-industrial complex organizations* (2006)
23. E. K. Murueva, *Environmental aspects of accounting (on the example of the forest sector of the economy)* (2007)
24. H. T. Odum, *Environmental accounting: emergy and environmental decision making*, 707 (1996)
25. Y. J. Ahnad, E. Lutz, The World Bank Symposium / The World Bank, 118 (1989)
26. S. Schaltegger, R. Burritt, *Contemporary environmental accounting: issues, concepts and practice* (2017)

Digitalization of crop production as a factor in strengthening the economic security of the region

Alexey Pestryakov, *Nadezhda* Sbrodova, and *Alisa* Titovets

Ural State University of Economics, 620144, Yekaterinburg, Russia

Abstract. Digitalization determines the development of all processes in the economic and public areas, including the issue of strengthening economic security. The article discusses the problem of digitalization of one of the types of economic activity of the region. The relationship between the development of agriculture and economic security of the region is presented. The authors identify the degree of damage and threshold values in the estimation of the economic security of the region and approbate the original method in the Sverdlovsk region. The article discusses the aspects of the existing level of economic security of the Sverdlovsk region and identifies the problems. The paper provides recommendations for solving these problems and makes calculations of benefits from the introduction of certain Russian digital technologies into operation.

1 Introduction

The year of 2020 stimulated activation of various digitalization processes. Many business entities, regardless of the type of activity, as well as the authorities of different levels, introduced digital technologies into their operations. It should be noted that the use of digital technologies has passed from the category of additional advantage in competitiveness into the category of mandatory elements of economic security. Thus, as part of the Economic Security Strategy of the Russian Federation until 2030, it was determined that poor development of the technologies of the digital economy is a challenge to economic security and it threatens successful development [1].

In traditional view, economic security characterizes the state of economic processes from a qualitative perspective; it determines the ability of the economy to provide all economic entities with the necessary conditions and resources [2]. In the Economic Security Strategy of the Russian Federation up to 2030, it is clarified that economic security is not just a quality of economic processes, but it is rather security of the economy and its resistance to threats (external and internal), which, at the same time, ensures freedom, unity and sustainability in development [1].

With regard to the fact that the Russian Federation is a federal state, the general situation in the country is determined by the conditions in the regions, including the field of economic security. The economic security of the region is the state of the regional economy, "generating growth of regional competitiveness and resilience to the effects of internal and external threats" [3]. T. V. Uskova and I. A. Kondakov under the economic security of the subject of the Russian Federation understand "a set of conditions and factors characterizing the stability of the economy, the sustainability and graduality in its development, the degree of independence and its integration with the economy of the country, as well as the ability of regional government bodies to create mechanisms for the implementation and protection of interests of economic entities, maintaining the economic stability of the territorial community "[4].

One of the priority tasks of ensuring economic security at any level is to forecast challenges and threats. The most important global challenge today is digitalization. Digitalization is a change associated with the use of digital technologies in all aspects of human society [5]. It can also be designated that digitalization is the adoption or expansion of the use of digital or information technologies by the organization, a branch, country, etc. The effective functioning of the economy has become dependent on the digital medium [6, 7].

One of the threats for economic security connected with digitalization is a growing technological lag and increase of financial dependence on the state of those activities that, by virtue of conservatism, cannot start transition to new technologies. This fact has been mentioned by V.B. Bethelin, Academician of the Russian Academy of Science, who wrote that "it is impossible to form the necessary future "on the basis of borrowing technologies and components" [8].

One of these activities is agriculture, particularly crop production, which is especially vulnerable. The relevance of this research is determined by the following problems: to the absence of a competent digitalization process [9], poor communication between the consumer and producers of rural equipment, conservative attitude of workers and a number of concomitant reasons associated with the problem of self-sufficiency of the region

The interconnection between economic security and agricultural activities has been proven by many researchers: agriculture plays a significant role in the country's economy, so the activities of agricultural entities affect economic security, and modern conditions require comprehensive approaches to the "recovery" of the food sector [10,11,12].

The experts argue that there is a serious impact of agriculture on economic security of a country as a whole and a separate region and/or enterprise. Summing up, ensuring self-sufficiency of production inside the country and the region it is possible to bring the security to a qualitatively new level of development.

2 Materials and methods

It is necessary to assess the impact of crop production on economic security with the help of indicators of the economic security of the region. The basis for such analysis is the Decree of the President of the Russian Federation of May 13, 2017 No. 208 "On the Strategy of Economic Security of the Russian Federation for the Period until 2030" [1].

Considering the degree of influence of crop production on the economic security of the region, it is possible to reveal the following pattern, which is reflected in the specific indicators presented in Table 1. Interpretation of these indicators may help to determine the positive and negative trends in strengthening the economic security of the region.

Table 1. Relation to crop production and threshold values of indicators for economic security of the region assessment

Name of economic security indicator	Threshold value	Degree of relation to crop production, from 1 to 3

Table 1. Continued

Gross regional product per capita	min. 100%	3
Unemployment level	max. 8%	2
Proportion in imports of foods	max. 30%	3
Proportion in manufacturing	min. 70%	2
Wear-and-tear of the assets	max. 60%	3
Proportion of investment in Gross regional product	min. 25%	1-2
Proportion of expenditures for R&D in Gross regional product	min. 2%	1
Balanced financial result		2
Production index		2
Correlation between the foods produced and consumed in the region	min. 1	3
Income per capita		2

We have made an expert assessment of the degree of influence of crop production on a 3-point scale. Evaluation of these indicators makes it possible to assess the contribution of each type of economic activity to the economic security of the region, which will allow in the future to adjust the activities in order to strengthen economic security.

3 Results and discussion

It is necessary to describe the impact of crop production and the indicators for the economic security of the region assessment on the example of the Sverdlovsk region (table 2).

Table 2. Indicators of economic security of the region in crop production in the Sverdlovsk region in 2018 [13,14]

Name of economic security indicator	Threshold value	Real value
Gross regional product per capita	min. 100%	97%
Unemployment level	max. 8%	1,98% in rural areas /1,17% in cities
Income per capita	As the average in the region or higher	25199 roubles per month (2017)
Proportion in imports of foods	max. 30%	14%-60% depending on the type of product
Proportion in manufacturing	min. 70%	82,76%
Depreciation of fixed assets	max. 60%	57,50%
Share of investments in GRP	min. 25%	31,66%
Balanced financial result of the crop production industry	more than 0	72 mln. rub.
Industry production index	min. 100%	99%
The ratio of produced food products in the territory region to consumed	min. 1	0,3-0,6
Share of R&D expenditures in GRP	min. 2%	1,30%

These data demonstrate that the state of crop production in the Sverdlovsk region is unsatisfactory in terms of a number of significant indicators, which has a great impact on the economic security of the region, creating risks that may affect the future. In other words, if at the moment the situation does not seem particularly dangerous, in 10-15 years (provided that there are no corresponding regulatory measures) it is likely that the region may become dependent on imported agricultural products.

Thanks to this research methodology, the relevance of the problem is revealed, which is associated with a number of reasons.

1. Lack of production of the necessary food products for domestic consumption in the region;
2. Volatility of production with a relatively stable level of consumption;
3. The need to introduce new technologies, since the existing technologies do not maintain the crop yield at the proper level;
4. The need for an integrated development of the territory, since this situation is caused, among other things, by a decrease of the resource base for industrial production.

In the Sverdlovsk region, the share of crop production in agriculture is only 36.3%, while cattle breeding is 63.7% [13]. Such disproportion indicates a weak diversification of the economy, as a result, dependence on one type of manufactured product, which creates risks for the economic security of the region.

Despite the fact that for the Sverdlovsk Region, as well as for the entire Ural Federal District, agriculture is not a priority area of activity, an important issue for any region is the complexity of the development of the territory and self-sufficiency in mass crops. Rigid specialization is dangerous and always creates the basis for the emergence of risks for the economic security of the region.

It is important to take into account that the economic security of cattle breeding is also directly affected, since more than 50% of all areas in the region are used specifically for the fodder base. As a consequence, crop production security creates a safety bag for cattle breeding.

Ensuring the economic security of the Sverdlovsk region through the development of crop production implies the participation of 3 categories of entities: regional government bodies, manufacturers of machinery and equipment for agriculture and manufacturers of agricultural products (Fig. 1).

Fig. 1. The influence of entities on economic security of the region
Where: Государство – State; Производители сельскохозяйственной продукции – manufacturers of agricultural products; Производители машин и оборудования для сх - manufacturers of machinery and equipment for agriculture.

Almost all processes in agriculture become more automated every year. The more up-to-date the tools are, the better and more efficient the results will be. At the same time, the technical and technological backwardness of the agricultural industry is an important problem affecting import substitution process; it also prevents increase of the efficiency of agricultural enterprises. The income of agricultural producers is insufficient to modernize the equipment.

A comprehensive analysis of the Russian agricultural machinery market by experts identifies the following important aspects:

– An import-oriented model has been formed in the Russian agricultural machinery market;

– The structure of the agricultural machinery market in 2018 was characterized by a 13.4-fold increase in the volume of imports compared to domestic production, and the trade balance was negative and amounted to -3 million units;

– China became the leader in import supplies at the end of 2018 (over 38%);

– Belarus buys most of the products of Russian exporters (over 72%).

If the import of agricultural equipment is systematically reduced and support for domestic goods is provided, then manufacturers of agricultural machinery and equipment will be able to:

– increase profits from the sale of equipment and components;

– strengthen their own performance efficiency;

– enhance the prestige of the activity;

– increase the number of jobs.

Thus, thanks to the effective interaction between the subjects, it is possible to make a step towards import substitution and strengthen the economic security of the region, while reaching the strategic goals of the state and the region as well.

The main methods for achieving these goals are: increase of investments in agriculture, effective land use and land reclamation, elimination of industrial stagnation, stimulation of domestic mechanical engineering to address the needs of agriculture and use of modern technologies.

And since one of the subjects is the state, we studied the activities of the authorities. Thus, on the territory of the Sverdlovsk region, 49 different programs are being implemented for the development, support, and introduction of areal agriculture. At the same time, it is important to note that this indicator is more associated with the cattle breeding sector than the crop sector.

According to the action plan of the state program of the Sverdlovsk region "Development of the agro-industrial complex and the consumer market of the Sverdlovsk region until 2024", the total funding is 6,436,604.5 thousand rubles. At the same time, the total financing of measures for the development of agriculture is 61.7%, development of cattle breeding is 31%, crop production development is only 7.3%, which includes support for the purchase of seeds for growing forage crops, the provision of unrelated support to agricultural commodity production in the field of crop production and increasing soil fertility.

By identifying these three main subjects of relations, it is possible to achieve the economic security of the region, taking into account modern trends in the most optimal way and reaching the required level of self-sufficiency of the region.

Currently, there are the following reasons for the violation of relations between the subjects shown in Figure 1 (1,2,3):

– lack of interest of manufacturers of agricultural machinery in the modernization and provision of the necessary units of equipment for agriculture;

– low investment activity due to the opinion that modernization of agriculture is unprofitable;

– government support for unprofitable enterprises;

– dependence on imports;

– low wages and low prestige of work in the industry.

In order to restore the broken ties between the subjects, it is proposed to purchase and integrate a number of innovations, paying particular attention to Russian technologies (Table 3), since this directly affects the economic security of the region.

Table 3. Economic calculations of integration of some technologies

Technology	Name of producer	Name of service	Cost of service	Total cost	Calculation
Control system for feeding, fertilizing and planting based on data from sensors located in the field	Crop control system «Magrotech» (Russia)	Calculation of minimal fertilizing	From 30000rub per field	30000rub	Saves up to 24*112*3=**8064rub** per one fertilizing component, where 24 rub is the cost of one kg of nitrate fertilizer, 112 is the amount of saved kilograms, 3 – seasons when the system is working (spring, summer, autumn)
		Agricultural consulting	From 330rub per hectare	330*50 =16500rub	
		Online access	From 150000rub a year	15000rub	
	Service "MyDataPlant" by the company Kleffmann (Germany)	Online access	140 to 560 rubles per hectare	7000-28000rub	
Complex systems of farm management	Cropio(USA)	Online access	60-300rub a year per hectare (without the cost of installing)	3000-15000rub	Total: **45000-12000rub,** the system increases production by 3, so the number of staff reduces by 3 as well, the average salary of combine operator is 15000 – 40000₽, 8 people usually works on one field.
	Company "InfoBiS" (Russia)	«Agrosygnal» System	Subscription is 400-700 rub per month	4800-8400rub	
Meteorological stations and sensors located in the field collecting environmental data	Sensors by «Pynco» (USA)	1 sensor	22500-32000rub		In one region, 34 companies were affected by overwatering. The damage to each of them amounted to more than 2 million rubles, while compensation was only about 1 million rubles.
		2 sensors	45000-65000rub		
	Sensors by «AllMeteo» (USA)	Wind sensor	6400-32000		
		Humidity/temperature sensor	16000rub		
	Meteorological station by «Dives» (USA)	Meteorological station	55000rub		

So, there are 288 farms in the Sverdlovsk region in the sphere of crop production, the equipment will cost approximately 100,000 rubles, that is, 28,800,000 rubles will be needed for all farms in the region, and this is only 6.13% of the funds allocated for the development of crop production. Therefore, redistribution of the rest of the funds will provide enterprises with basic technologies.

When introducing such a process as digitalization in the crop production industry, there may be some problems. Therefore, we have compiled a map of risks (table 4) that can be faced in the current realities or in the near future.

Table 4. Map of risks.

	Risks	Occurrence probability	Amount of loss	Impact	Party
CURRENT	Employee's conservatism	1	1	2	Agriculture, state
	The belief that digital transformation is unprofitable	2	1	3	Agriculture
	Insufficient stimulation of domestic machine building	2	2	4	State, machine building
	Irrational use of resources	2	3	5	Agriculture
	Insufficient financing of new technologies purchase	2	3	5	State, machine building
POTENTIAL	Bureaucracy(slow process of money acquisition)	2	2	4	State
	Poor IT knowledge of specialists working in agriculture	2	2	4	Agriculture, machine building
	Absence of technological solutions	2	3	5	State, machine building, agriculture
	Weak responsibility of recipients of funds	2	3	5	Agriculture, machine building
	Loss of connections with international partners	2	3	5	State, machine building, agriculture
	Sales: poorly developed logistics, lack of granaries, lack of information.	3	3	6	Agriculture

Assessment of the impact of risks has shown that negative trends in the digitalization of agriculture are quite obvious. Despite the fact that in the nearest future there are risks of possible losses, these losses can be avoided with the help of careful planning and good communication between the three main subjects of relations.

4 Conclusions

Summing up the results of the study, we analyzed crop production from a different position, not typical of the state. Comprehensive development of the region, economic security and focus on results are the factors that were taken into account in this research. As a result, crop production has an impact on the economic security of the region, which is proved by the list of indicators that we demonstrated earlier. This means that the authorities are interested in improving these indicators and should develop programs for the digitalization of agriculture. Moreover, we reviewed the successful experience of technologies implementation in other regions of the Russian Federation and calculated the average cost of technologies in order to understand the necessary threshold for launching digital development of farms. Thus, we have identified one of the most probable and promising methods for the development of crop production, not only to increase indicators in each area, but also to strengthen the economic security of the region as a whole.

References

1. On the Strategy of the Economic Security of the Russian Federation for the Period up to 2030, Decree of the President of the Russian Federation dated May 13, 2017 No. 208, http://www.consultant.ru
2. V. P. Bykov, O. A. Dembovskaya, E. M. Lebedko, Economics and efficiency of production organization, **5**, 60–63 (2006)
3. Yu. A. Fridman, Bulletin of the Kuban State Technical University, **1**, 122-125 (2015)
4. T. V. Uskova, I. A. Kondakov, Economic and social changes: facts, trends, forecast, **2 (14)**, 37-50 (2011)
5. E. Stolterman, A. C. Fors, *Information Technology and the Good Life. in Information Systems Research: Relevant Theory and Informed Practice* (2004)
6. R. M. Nizhegorodtsev, *Information Economy*, 1, 163 (2002)
7. A. Berisha-Shaqiri, M. Berisha-Namani, Mediterranean Journal of Social Sciences, **6(6)** (2009)
8. V. B. Betelin, Digital Economy: Imposed Priorities and Real Challenges. State Audit, **¾,** 22-25 (2017)
9. J. Pfeiffer, A. Gabriel, M. Gandorfer, Agric Hum Values **38**, 107-128 (2021)
10. V. Surovtsev, Yu. Nikulina, Economist, **6**, 79-84 (2010)
11. M. I. Korolev, Economics, 283 (2011)
12. V. V. Gubin, Vestnik NGIEI, **5(48)**, 32-38 (2015)
13. Statistical collection: Agriculture in Russia, https://rosstat.gov.ru
14. Statistical collection, Regions of Russia. Socio-economic indicators, https://rosstat.gov.ru

Rural and agroindustrial tourism as a factor of socio-cultural and economic development of the territory

A. A. Oshkordina*, and *E. G.* Radygina

Ural State University of Economics, 620144 Yekaterinburg, Russia

Abstract. The paper discusses current issues and problems of organizing and implementing event and tourist activities as an effective tool for popularizing and attracting attention to rural areas, which, in the context of the COVID 19 preventive campaign around the world, is a relevant factor for many categories of the population. The authors have identified the positive and negative factors of tourist activities organization, affecting the socio-cultural and economic level of rural development. The main directions of organization, implementation and development of rural and agroindustrial tourism on the territory of a separate subject of the Russian Federation have been determined. Most important, in the current difficult conditions of the tourist industry development, in many territories of Russia domestic tourism is becoming a priority, focusing, among other things, on the creation of new products of rural and agroindustrial tourism.

1 Introduction

In current socio-economic conditions, the multiplier effect of tourism can make a significant contribution to the economy development of rural areas as a factor of additional financial resources; it promotes the formation of the territory social infrastructure, as well as facilitates the popularization and increase of the cultural, cognitive and patriotic consciousness of the population. Today, event tourism in rural areas does not have a massive character; the exception is the ubiquitous dacha tourism, which showcases a high level of popularity in Russia. It should be remembered that the events of the past year have seriously reoriented the activities of travel companies to create new tourist products, primarily domestic ones, most of which belong to rural tourism. Therefore, in our opinion, the popularization and attractiveness of rural tourism in modern conditions are relevant and require regulatory and financial support from the government. In addition, today there is no clear understanding of the terminological apparatus and the concept of "rural tourism", which greatly complicates the process of collecting, processing and evaluating information and statistical data in the organization and implementation of event, excursion and tourist activities in rural areas.

Rural tourism is not only a certain tribute to the widespread fashion, but also a kind of necessity for the urban population living in an unfavorable ecological and psycho-

* Corresponding author: al2111la@yandex.ru

emotional environment of the modern and dynamically changing rhythm of life. A holiday in the countryside aims to combine enjoying the nature and learning historical, cultural and ethnographic aspects of the heritage, such as customs, crafts, national cuisine, clothing, etc. of the population living in the territory. In addition, rural tourism can also develop a research focus in the form of collecting folklore, learning local languages and dialects, collecting plant and mineral raw materials and substances, etc. According to I. V. Smirnova, the population, tired of the hustle and bustle of a big city, seeks to relax physically, mentally, to enjoy the beauty of nature, to touch the cultural heritage of historically significant places, which are not as popular as those in large tourist centers [1].

This study aims to identify the main directions of rural tourism development as a factor of increasing socio-economic efficiency of rural areas of the Sverdlovsk region.

Thus, the territory of our country, including the Sverdlovsk region, has a powerful potential of natural, landscape, climatic, cultural, historical, ethnic and economic resources. This potential ensures implementation of various types of tourist activities in rural areas. At the same time, it should be noted that the implementation of ideas and activities in rural areas can be possible only in environmentally friendly territories that are not touched by industrial production, which pollutes the environment of rural areas.

2 Materials and methods

The study was conducted on the statistical, accounting, reporting, and information data on the activities of travel industry organizations and enterprises located in the Sverdlovsk region. The study also included research data from the related works by domestic and foreign scientists in the field of rural and agroindustrial tourism. While processing and systematizing the information and analytical data we applied such methods of study as grouping, detailing, synthesis, historical and logical research method, as well as methods of economic analysis.

3 Results and discussion

A group of Russian scientists T.A. Volkova, V.V. Minenkova, D.V. Maksimov. rightly claim the existence of prototypes of rural and agroindustrial tourism in the pre-Soviet and Soviet period, in contrast to many young scientists, who define rural tourism as a relatively young type of tourist activity, though quite a promising one [2]. So, agrotourism dates back to the times when travelers using horse-drawn carts had to stop for the night at an inn in the countryside. Later, in the 17th century, in Russia, summer vacations at dachas (country houses) gained wide popularity with the nobles, and that can be considered as a closer predecessor of modern agrotourism [3]. Since the beginning of the Soviet Union, almost the entire population, from schoolchildren to those ending their work life, including students of secondary vocational and higher educational institutions, were engaged in the system of agroindustrial activities through involvement in agricultural work. During the existence of the Soviet Union, labor and recreation camps were organized for schoolchildren and students with a round-the-clock stay in rural areas. Employees in all branches and subsectors of the national economy took an active part in agricultural work of the nearby collective and state farms. Thus, the involvement of almost the entire working-age population of the Soviet Union aged 12 to 60 did not merely enable to obtain economic effects for agricultural enterprises; it realized an educational and cognitive function for the young people, while fostering diligence, mutual assistance, curiosity, solidarity, love for their native land and Motherland, etc.

Currently, in many countries of the world, such as Switzerland, France, Italy, Spain, Ireland, Scandinavian countries, rural tourism is undergoing rapid development, both from the perspective of domestic and inbound tourism, while attracting a significant number of domestic and foreign visitors from all over the world. According to expert assumptions and calculations, the annual profit of rural tourism in these countries is more than a hundred million dollars.

Russia ranks 5th in the world in terms of agricultural land [4]. According to Rosstat (Russian Statistics), the rural population on January 1, 2019 was 37.5 million people, or about 25% of the entire population of the Russian Federation [5]. According to the All-Russian Agricultural Census that took place in 2016, the number of peasant (farm) households decreased compared to 2006 and was 136.6 against 253.1 thousand, and the number of individual entrepreneurs in the countryside increased from 32 to 38 thousand. At the same time, the average land area per farm for the specified period increased from 102.6 to 268.9 hectares, and for individual entrepreneurs - from 106.2 to 149.0 hectares [6]. Yet, the official statistics indicate the inconsistency and ambiguity of economic processes and patterns in the development of agriculture in Russia (Table 1).

Table 1. Socio-economic criteria for rural areas development in Russia [7]

Criteria	2014	2015	2016	2017	2018
Average annual number of employees, people	4921	4900	4890	4481	4346
Number of workplaces	21759	21868	21902	20923	21191
Average wage, roubles	17194	19238	21268	23529	25820
Number of profitable farms,%	73,6	77,0	77,7	75,6	73,8
Number of unprofitable farms,%	26,4	23,0	22,3	24,4	26,2

The analysis of Table 1 indicates a decrease in the average annual number of agricultural workers in 2018 by almost 12% compared to the base year of 2014, which can also indicate positive developments in this economy sector associated with the introduction and use of innovative technical tools and technologies, as well as with the labor automation in agriculture, which reduces the amount of manual labor. Unfortunately, along with the implementation of automation tools in large agroindustrial farms, a decrease in the average annual number of employees in this economy sector can be associated with an outflow of labor resources from sparsely populated areas, which in turn leads to a high demand for qualified personnel in rural areas. At the same time, the reduction in the average annual number of workers by 12% led to the reduction in the number of workplaces in agriculture by only 3%, which once again confirms the possible fact of the widespread replacement of heavy manual labor with highly efficient automated labor. Yet, the level of wages in agriculture during the analyzed time period has a tendency of annual growth by an average of 9-11%, though this is two times lower than the average wage in the country. A more detailed cross-section of agricultural workers' wages in different regions of the Russian Federation proves that the wage level varies. The number of profitable farms varies approximately within the same limits and averages 75%; about 25% of farms suffer losses. In our opinion, it is the attraction of tourist flows for the purpose of visiting the countryside that can create conditions for increasing the profitability of farms.

According to S. K. Volkov, over the past decade rural tourism has begun to actively develop in Russia, namely in the Altai Territory, Kaliningrad, Leningrad and Pskov Regions, Krasnodar Territory, and the Republic of Bashkortostan. Tourism experts do not get tired of predicting the unprecedented scale of rural tourism development, for which Russia has enormous resources [8]. Thus, it is necessary to stress an increase in the volume of tourist flows within rural and agroindustrial tourism throughout the whole territory of the Russian Federation. At the same time, it is necessary to note the consumers' dynamically changing preferences and demands; it requires attracting investments in order to form recreational and socio-cultural infrastructure in rural areas. Thus, the implementation of event and tourist activities in rural areas solves a number of socio-economic problems: raising the socio-economic level of rural development by increasing the number of tourist flows; saving and increasing jobs by raising the interest of the local population in the revival of national culture and their ancestors' identity; developing recreational and social infrastructure in rural areas; preserving natural resources; creating small enterprises for souvenir manufacturing and crafts development in sparsely populated areas.

Yet, despite the dynamically developing rural tourism on the territory of the Russian Federation in recent years, the industry is now experiencing serious problems that require immediate solutions. Thus, according to I. L. Polyakova and M. P. Grigorieva, the current state of rural tourism in Russia and its development trends make it possible to identify a number of factors that hinder its further development:

- insufficient support for rural tourism as a separate branch of the economy;
- undeveloped legislative and regulatory framework;
- lack of a comprehensive system for stimulating rural tourism;
- poorly developed tourist infrastructure, including the transport system;
- low living standard of the rural population of Russia's regions;
- insufficiently effective mechanism for managing the rural tourism sector;
- unfavorable climate conditions;
- terrorism, unfavorable crime situation;
- underdeveloped rural infrastructure;
- weak promotion of the agricultural product in the domestic and international markets;
- lack of a recognizable brand;
- low level of investment in rural tourism;
- lack of support programs for rural areas [9].

N. A. Mironova describes the development of agrotourism in Russia as follows: "Unfortunately, at the moment, the development of agrotourism in Russia is a declared slogan, without any progressive movements in this direction. With assistance of the state, an association for agrotourism promotion has even been launched, which declares a large number of objectives on its own website, but judging by the activity on the portal and the nearly total absence of news updates, the real activity of the association is either at an extremely low level, or completely absent" [10]. O. K. Slinkova states the following priority measures that can give an impetus to agro-tourism development in the Russian Federation: developing a legislative framework to regulate functioning of this sector of tourism; providing support for agrotourist farms in the form of preferential loans for the tourist infrastructure development; developing and implementing regional programs to support agrotourism[11].

Thus, rural and agrotourism development is directly proportional to the state assistance in the field of legal, organizational and managerial support for tourist businesses with the possibility of attracting industrial capital. And thus, the actualization of the activities, which aim at organizing and implementing new tourist products and events in rural areas, is caused by their powerful impact on the level of socio-economic and demographic development of rural areas.

The territory of the Sverdlovsk region was formed on January 17, 1934 and is located on the border of Europe and Asia at the intersection of transcontinental flows of raw materials, goods, financial, labor and information resources. The Sverdlovsk region today is a large economically developed territory of Russia with a high level of business, cultural and social activity, one of the most promising subjects of the Russian Federation. According to the official statistics, the permanent population of the region at the beginning of 2021 is 4 million 290 thousand people, of which 85% is urban, and only 15% is rural.

There are 47 cities, 26 working settlements and urban-type settlements, 1804 rural settlements on the territory of the region [12].

Despite the harsh climatic conditions, agriculture is developing rapidly in the Sverdlovsk region. So, at the beginning of 2021, 390 agricultural organizations, more than 970 peasant and farm households and entrepreneurs are engaged in agriculture, about 440 thousand people have subsidiary farms. According to the latest data, more than 818.0 thousand hectares in the Sverdlovsk region are allocated for sown areas for agricultural crops; more than 24 thousand people are officially employed at agricultural enterprises. Thus, according to the main socio-economic development criteria, the Sverdlovsk Region is among the top ten regions of the Russian Federation. In addition, the socio-cultural infrastructure formed on the territory of the region enables many rural areas to implement event, sports and tourist activities.

Table 2. Implementation of tourist events in agricultural territories of the Sverdlovsk region in 2021

Indicators	Number of events	Specific gravity, %
Total event activities	350	100
Yekaterinburg	102	29
Municipalities	191	55
Rural areas	57	16

Given the data of the Tourism Development Center of the Sverdlovsk region on the implementation of tourist events, the following conclusions can be drawn: annually, more than 350 events of various types take place on the territory of the Sverdlovsk region; about a third of all events are held in the administrative center of the region – the city of Yekaterinburg; municipalities of the region show a high level of activity in the implementation of event tourism, the share of which is more than half of the total number of events declared; only 16% falls on the implementation of activities in rural areas. In addition, it is necessary to reflect the fact that 28 of the activities presented in Table 2 are carried out in seven rural settlements, i.e. more than half of the tourist activities in rural areas are organized by nearly the same settlements that have long been involved in tourist flow organization and are increasingly participating in these processes every year. If we talk about the degree of active participation in the events of all rural areas of the region, we have to state a sad fact: only about 30 rural settlements of the region out of 1804 take at least some part in the organization and implementation of tourist activity, which primarily indicates a low level of interest of both the rural population and the authorities of rural administrations and agricultural enterprises. At the same time, in recent years, the government of the Sverdlovsk Region has provided event organizers in sparsely populated towns and settlements with a unique opportunity to apply for financial, organizational and legal support free of charge.

As an example, we would like to mention a unique young project in the village of Marmorskoye - "Day of Marble Music" within the "Marble Mile" («Mramornaya milya») project, which was successfully launched in 2019 with the support of the "Idea" («Ideya») Social Activity Fund, the head of the village of Marmorskoye Vera Nufer and the board of the "Ural Marble" («Ural'skiy mramor») enterprise. For the second time, in a spent marble quarry, at the bottom of an ancient sea, musicians performed music pieces of various composers and eras. According to the results of the "Russian Event Awards" competition, this project was recognized as the best regional tourist project, and the Sverdlovsk region took the 1st place in the rating of the All-Russian Award in the field of event tourism [13].

Another unique project, known for over 40 years and attracting the attention of tourists not only from the Sverdlovsk region, but also from other regions - "Nizhnyaya Sinyachikha and the Museum-Reserve of Wooden Architecture". It is noteworthy that this project

https://doi.org/10.1051/e3sconf/202126203029

implementation was initiated by a single person, Ivan Danilovich Samoilov. He went to the realization of his idea for many years, starting in 1967. Today, the Nizhnyaya Sinyachikha Museum-Reserve of Wooden Architecture is the only museum of this kind on the territory of the Sverdlovsk Region. In the Middle Urals, it can be compared to the museum in Khokhlovka, Perm Region [14]. Every month, Nizhnyaya Sinyachikha is visited by 100 to 500 tourists, attracting additional funds, forming the social and cultural infrastructure of the territory, and creating new jobs for the population. Today, the museum-reserve serves as a basis for children's and youth's "circles" and folklore groups, which pursue the goal of preserving the cultural heritage and traditional national crafts, while fostering patriotism and love for the native land and fatherland among the younger generation. It is in Nizhnyaya Sinyachikha that from 6 to 10 events take place annually, attracting visitors from various regions of our country and from abroad. Thus, Nizhnyaya Sinyachikha is proof of possible organization and implementation of projects when there is personal interest and activity of the local population.

Nizhnie Tavolgi, an old Ural village with a population of about 350 people, has been famous for its pottery since ancient times. Together with Verkhniye Tavolgi, the village has recently organized a center of folk crafts. A modern tourist infrastructure has been created, providing a comfortable stay on the territory of these settlements. There have been developed tourist routes, which vary in duration and thematic component: 1) a hike along the path of power with a visit to the springs of living and dead water and Ermak rock; 2) a trip to the Old Believers' skete "Bogodan"; 3) a visit to the cult center of the Neolithic period – Shaitan lake. The presence of a tourist social infrastructure facilitates organizing festivals of a regional and international scale, such as the tea eco-festival "When Tavolga Blossoms" («Kogda tsvetet Tavolga»), the Festival of arts and crafts "Secrets of the Gemstone Ring" («Tayny samotsvetnogo kol'tsa»), etc. Thus, the territory is actively developing and creating additional jobs due to implementation of tourist projects and products.

4 Conclusions

In the current socio-economic conditions of rural areas development, organization and implementation of rural tourism can play an important part. At the same time, the share of agrotourism in the Russian tourist market does not exceed 2% [15]. The availability of a huge variety of natural, climate and geographic, cultural and ethnic, and historical riches of Russia's rural areas, including the Sverdlovsk region, makes it possible to organize and implement tourist products with a high degree of competitiveness, which in turn will ensure solving various problems of socio-economic and cultural nature in rural areas. At the same time, rural tourism today has not received a mass character on the territory of Russia, but has the so-called point distribution and directly depends on the activity and initiatives of the local population and rural administrations. Yet, the implementation of rural tourism ideas and projects requires support from the state authorities.

Thus, to meet the challenge and develop agrotourism in Russia, in our opinion, it is necessary to:

- develop legal documents regulating tourist activity at all stages of its implementation;
- develop regional target programs that aim to provide financial support for current projects and ideas of rural tourism organization;
- attract financial resources from private organizations and investors on a preferential or gratuitous basis;
- raise the population awareness of the possibilities to participate in various event activities of rural tourism by means of mass media;

- update the development of short-term weekend tourist products with the use of event tourism elements (hiking, making products of folk art with your own hands, engaging in agricultural work, collecting medicinal herbs, making agricultural products, etc.);

- popularize thematic corporate events, weddings, banquets, children's parties, etc. on the territory of agricultural enterprises.

The complex of developed measures, in our opinion, is to increase and facilitate agrotourism popularization in rural areas. In current conditions, agrotourism should be perceived as an effective economic tool for rural areas development from various perspectives.

References

1. I. V. Smirnova, Bulletin of the Mari State University, **1(13)**, 87-90 (2014)
2. T. A. Volkova, V. V. Minenkova, D. V. Maksimov, A. A. Ponamorenko, Scientific journal of KubSAU, **132(08)** (2017)
3. V. V. Minenkova, D. V. Maksimov, T. A. Volkova, Yu. I. Karpova, Global scientific potential, **8(41)**, 7-13 (2014)
4. Site of the Federal State Statistics Service, http://www.gks.ru
5. Agriculture in Russia, 2019: Statistical collection, Rosstat, 91 (2019)
6. All-Russian agricultural census of 2016, Preliminary results: Statistical Bulletin, 7 (2016)
7. Agriculture in Russia, 2019, Statistical collection, Rosstat, 91 (2019)
8. S. K. Volkov, Economics, entrepreneurship and law, **6(17)** (2012)
9. I. L. Polyakova, M. P. Grigorieva, Service in Russia and abroad, **11(5)**, 31-43 (2017)
10. N. A. Mironova, Moscow Economic Journal, **5** (2020)
11. O. K. Slinkova, Scientific result, Business and service technologies, **3(3)**, https://cyberleninka.ru/
12. Official website of the government of the Sverdlovsk region, http://midural.ru/
13. Information portal "ProPolevskoy", https://propolevskoy.ru/
14. Pavel Raspopov, Lower Sinyachikha and the Museum-Reserve of Wooden Architecture, https://uraloved.ru/
15. What hinders the development of agritourism in Russia? http://www.agropages.ru/

Trends and risks of developing local consumer markets in rural areas

Elena Dvoryadkina, *Larissa* Khacheva, and *Natalia* Istomina

Ural State University of Economics, Ekaterinburg, Russia

Abstract. The paper considers trends and risks of development of local consumer markets of rural territories, measures of state programs aiming at sustainable developing consumer markets. The concepts of municipalities of rural territories and consumer markets of rural territories are given. The main indicators of the consumer market, such as turnover of retail trade and catering, as well as indicators of the consumer service sector, are analyzed.

1 Introduction

Providing for socio-economic development of rural territories is traditionally among priorities of public policy, one of the effective mechanisms for the implementation of the latter being government and national programs.

The national program is a document of strategic planning, containing a package of planned measures interconnected in terms of tasks, dates, executing officers and resources, and state policy instruments ensuring within the framework of the implementation of key state functions the achievement of priorities and goals of public policy in the field of socio-economic development and national security of the Russian Federation [1]. The national program is the most important means of implementing domestic public policy focused on developing social and economic spheres of life, and it should be aimed at carrying out large scientific and investment projects to solve complex problems within the bounds of federal, regional and local authorities. While working on the annual federal budget the government adopts dozens of programs to be financed.

The main documents of public strategy and programming in the context of the objectives of the present research are the following : "Strategy for the Sustainable Development of Rural Territories of the Russian Federation till 2030" adopted by the Government of the Russian Federation dated February 2, 2015 No. 151-r, the National Programs "Integrated Development of Rural Territories" adopted by the Government of the Russian Federation dated May 31, 2019 No. 696, "Development of Agriculture and Regulation of Markets of Agricultural Products, Raw Materials and Food" adopted by the Government of the Russian Federation dated July 14, 2012 No. 717. [2-4].

Local consumer markets in rural areas are therefore an integral component of rural socio-economic development.

2 Materials and Methods

The analysis of modern processes and trends in the development of rural areas is done in the works of such domestic scientists as Nefedova T. G.[5], Treivish A. I. [6], Gazizov R. M. [7], Darmograi O. V. [8], Dvoryadkina E. B., Belousova E. A. [9], Ilinykh A.V. [10], Kleymenov D. S. [11], Kovtun B. A., Papelo V. N., Yarmanov V. V. [12] and others. Local consumer markets of municipalities are studied from different viewpoints by such scientists as Kovaleva I. V., Kucherenko T. V. [13], Nikiforova O. P., Antokhonova I. V. [14], Chernysheva E. V. [15], Razorvin I. V., Usova N. V., Bardasova N. V. [16] and others.

In order to analyze trends in the development of local consumer markets of municipalities of the Sverdlovsk region, we use the statistic data provided by Rosstat (Current Statistical Survey). Being typical rural territories, the above municipalities act as a testing ground for the study of local consumer markets. The concept of 'rural territories' is a collective one since it may embrace various elements of administrative-territorial or municipal-territorial structure. E.B. Dvoryadkina and E.A. Belousova defined the concept of 'rural-type municipality'. "This municipality, as a kind of territorial structure of local self-government, has a special economic and legal status related to the laws of functioning of rural economy, and, unlike rural territories, it can be unambiguously identified as an object of statistical service in order to optimize interaction in the system of population-local self-government- production complex "[17]. Rural-type municipalities thus include municipal areas and rural settlements.

3 Results and Discussion

To fulfill the national program "Development of Agriculture and Regulation of Agricultural Product, Raw Material and Food Markets for 2013-2020," the Government of the Sverdlovsk region adopted a similar target program "Development of the Agro-industrial Complex and Consumer Market of the Sverdlovsk Region till 2025" [18]

The regulatory framework of the program includes the following legal acts: Federal Law of 29.12.2006. No. 264-FZ "On the Development of Agriculture", the Doctrine of Food Security of the Russian Federation approved by the Russian President (the Decree of 30.01.2010), "The Concept of Long-term Social and Economic Development of the Russian Federation till 2020" dated 17.11.2008 No. 1662-r, "The Methodical Instructions of the Ministry of Economic Development of the Russian Federation on Development and Implementation of State Programs of the Russian Federation" adopted by the Ministry of Economic Development of the Russian Federation dated 22.11.2013 No. 690 [19-22].

The proposed measures, carried out within the framework of the national program "Development of the Agro-industrial and Consumer Market of the Sverdlovsk Region till 2025," are based on the following principles:

- stability. The new program retains effective directions and measures to support agriculture;
- systemacy. The program covers economic, social and environmental aspects of development of the above branch and combines support measures with market regulation and rural development;
- co-financing. The implementation of the program measures involves co-financing of agriculture from the federal and regional budgets;
- government and private partnership. It involves joint efforts of the government and business to achieve the objectives of the program. [23]

The goals and objectives of the public program of the Sverdlovsk region "Development of the Agro-industrial Complex and the Consumer Market of the Sverdlovsk Region till

2025" correspond to the goals and objectives of the main strategic documents: "Strategies for the Development of the Food and Processing Industry of the Russian Federation for the period up to 2020", adopted by the Government of the Russian Federation, dated 17.04.2012 N 559-r, "Strategies for the Sustainable Development of Rural Territories of the Russian Federation for the Period up to 2030" adopted by the Government of the Russian Federation, dated 02.02.2015 N 151-r, "Strategies for the Socio-economic Development of the Sverdlovsk region for 2016 – 2030" adopted by the Government of the Sverdlovsk Region with the law dated December 21, 2015 N 151-OZ "On the Strategy for the Socio-Economic Development of the Sverdlovsk Region for 2016-2030." [24, 25]

One of the goals of subprogram "Development of the Consumer Market of the Sverdlovsk region" is to ensure effective activities of state authorities in the field of the agro-industrial complex and the consumer market. To achieve this goal one should

- improve the quality and safety of food products in the consumer market of the Sverdlovsk region;
- increase consumer awareness and literacy on issues of quality assurance, food safety and protection of consumers;
- diversify, improve quality and raise competitiveness of food produced in the Sverdlovsk region;
- develop multi-format trade;
- ensure integrated development of rural areas.

As part of the municipal structure of the Sverdlovsk region, 94 municipalities were formed within the boundaries of administrative-territorial units by January 1, 2020. They include 68 urban districts and 5 municipal districts, which consist of 5 urban settlements and 16 rural settlements. The paper gives an analysis of the consumer market of rural territories based on the above municipalities.

The methodology includes a SWOT analysis and an analysis of the most important indicators of the consumer market of rural areas in dynamics.

Table 1 presents a SWOT analysis of the consumer market to identify strengths and weaknesses of the development of the consumer market in rural areas, as well as opportunities and threats.

Table 1. SWOT Analysis of the Consumer Market (compiled by the author)

Strengths (S)	Weaknesses (W)
1. Advantageous economic and geographical position of the Sverdlovsk region for the development of the industry.	1. Low solvency of part of the population, especially in areas remote from the regional center.
2. A steady positive trend in the development of the consumer market industry in the Sverdlovsk region.	2. Lack of qualified specialists, high turnover of personnel, weak motivation of staff to be in the profession, low level of wages.
3. High saturation of the market with food of local production and processing, as well as agricultural products.	3. Presence of hard-to-reach remote areas.
4. Regional support programs for catering enterprises providing services to socially vulnerable segments of the population of the Sverdlovsk region (from the regional budget).	4. Lack of state regulation of the catering industry, weak legislative framework regulating activities in the domestic market of the Russian Federation.

Table 1. Continued

5. Sufficient provision of the population with stationary retail space and seats in enterprises of the public network.	5. A weak material and technical base of small and medium-sized catering enterprises, lack of modern resource-saving equipment.
Opportunities (O)	Threats (T)
1. Development of a system of quality control, safety control of raw materials, food products and services.	1. Territorial asymmetry in the development of municipalities, especially remote and rural territories.
2. State support (methodological, organizational, financial ones) of public catering enterprises providing services to socially unprotected segments of the population of the Sverdlovsk region.	2. The dependence of social enterprises in concluding contracts for the supply of products and the provision of services on the fundamental criterion of "price," which significantly affects the quality of goods (services).
3. Further development and improvement of the network of consumer market enterprises.	3. Penetration into the market of counterfeit, low-quality products
4. Development of transport infrastructure.	4. Increasing rural outflows to cities or more socio-economically developed areas.

The SWOT analysis reveals a number of weaknesses and threats in the consumer market of rural territories of the Sverdlovsk region. Rural territories of the Sverdlovsk region are quite complex for the development of the consumer market. However this analysis shows that the development of multi-format infrastructure of the consumer market will eliminate the unevenness of the industry in municipal areas, the special manifestation of which is recorded in remote and inaccessible settlements.

3.1. Analysis of consumer market indicators of municipal districts of Sverdlovsk region

3.1.1. Analysis of Retailing and Catering

Table 2. Retail Turnover (without small businesses), in thousand rubles [26]

Municipalities	2017	2018	2019
Bajkalovskij	224281	312414,2	357248,7
Kamyshlovskij	113697	306928	290109
Nizhneserginskij	1840347,6	2416770,3	2401225,6
Slobodo-Turinsk	404097,7	448305,5	513577,7
Taborinskij	101024,4	112076,3	128394,4
Total	**2683447,7**	**3596494,3**	**3690555,4**

Table 3. The Total Volume of All Food Products Sold within the Boundaries of the Municipal District during the Financial Year, in thousand rubles. [26]]

Municipalities	2017	2018	2019
Bajkalovskij	437759,4	460635,7	500466,9
Kamyshlovskij	571249,8	567406	569232,5
Nizhneserginskij	1588716,5	1643208,4	1805920,6
Slobodo-Turinsk	389868,9	410203,3	469169,5
Taborinskij	46287,4	53854,9	65985,1
Total	**3033882**	**3135308,3**	**3410774,6**

The analysis of such indicators as 'retail turnover (without small businesses)' and 'the total volume of all food products sold within the boundaries of the municipal district during the financial year' shows positive dynamics, which indicates the sustainable development of the consumer market.

Table 4. Retailing and Catering Facilities [26]

Facilities	Number of facilities per year		
	2017	2018	2019
stores	718	698	677
pavilions	75	77	74
stalls and kiosks	21	19	17
chemist's	21	21	20
branch pharmacies	20	21	22
canteens	31	32	32
refectories	84	84	85
hypermarkets			1
supermarkets	10	8	8
specialized grocery stores.	96	95	93
specialized non-food stores	134	128	131
stores discounters	84	82	80
restaurants and cafes	23	29	30
minimarkets	345	340	318
department stores			
other shops	59	55	55

The number of retail facilities has slightly decreased, which is caused by the outflow of rural residents to cities and other areas. The number of catering facilities has increased, which proves that the public catering services are in demand.

Table 5. Turnover of Catering Services, in thousand rubles [26]

Municipalities	2017	2018	2019
Bajkalovskij	14687,8	15557,4	18254,6
Kamyshlovskij	12287,7	14348,5	12890,7
Nizhneserginskij	17359	21300,8	29002
Slobodo-Turinsk	12007,3	11944,8	12581,4
Taborinskij	43,6		
Total	**56385,4**	**63151,5**	**72728,7**

The indicator 'public catering turnover (without small businesses)' shows positive dynamics, which indicates sustainable development of the consumer market and confirms the conclusion of Table 5.

Table 6. Hard Drinks Sales to the Population, in thousand rubles [26]

Municipalities	2017	2018	2019
Bajkalovskij	51939,4	47485,1	41134,4
Kamyshlovskij	15744,6	28226,2	25229,43
Nizhneserginskij	128377,52	124477,36	115089,03

Table 6. Continued

Slobodo-Turinsk	43985,07	31453,2	31305,98
Taborinskij	43985,07	31453,2	31305,98
Total	**284031,66**	**263095,06**	**244064,82**

The indicator 'hard drinks sales to the population' has significantly reduced, which may indicate the effectiveness of the government social policy.

3.1.2 Analysis of Consumer Services

Table 7. Consumer Service Establishments [26]

Consumer Service Establishments	**Number of facilities per year**		
	2017	**2018**	**2019**
Total	162	173	177
shoe repair	6	4	3
repair of fur products	15	11	12
repair of electronic equipment	2	5	6
service centers	32	35	39
furniture manufacturing	13	16	16
dry cleaner's			0
laundries			0
housing repairs	4	6	7
baths, showers and saunas	6	6	7
hairdresser's/barber's	31	32	32
photographer's	15	10	12
undertaker's	19	15	14
Other services	19	33	29

Table 8. Number of Chairs in the Hairdresser's/Barber's [26]

Municipalities	**2017**	**2018**	**2019**
Bajkalovskij	14	15	12
Kamyshlovskij	9	12	12
Nizhneserginskij	24	24	25
Slobodo-Turinsk	8	8	8
Taborinskij	1	1	1
Total	**56**	**60**	**58**

Such indicators as 'amount of consumer service facilities' and 'amount of chairs in hairdresser's/ barber's' show a slight decrease in the number of facilities.

3.1.3 Analysis of per Capita Indicators

Table 9. Per Capita Indicators

Indicators	**2017**	**2018**	**2019**
Retail turnover per capita(without small businesses)	26,83	36,15	37,3
Total food commodities per capita (without small businesses	30,34	31,52	34,47

Table 9. Continued

Trading rooms per capita	0,94	0,98	0,97
Service rooms in catering facilities per capita	0,124	0,128	0,132

Per capita indicators in rural areas are lower than the ones in urban areas. This proves a lower living standard of rural residents.

4 Conclusion

The consumer market of rural areas is an integral part of rural economy. Its sustainable development ensures developing rural areas, improving quality of life of rural residents, and contributes to the development of the agro-industrial complex. Retail and catering industries show positive dynamics. This fact indicates a stable development of the consumer market. The Sverdlovsk region shows the following trends in developing consumer services, typical of the Russian Federation on the whole:

- a gradual decrease in demand for a number of traditional consumer services, which are supplanted by the increased technological equipment of the household. The analysis of the consumer market in terms of consumer services in rural areas of the Sverdlovsk region shows that the structure of consumer services is still dominated by consumer services of the traditional type. The consumer markets, aiming at the high-income segment of services based on the desire to improve the quality of life, mobility and rational use of free time, are not expanding in rural areas. This also indicates the existing problems in the field of consumer services, such as: a poorly developed infrastructure of consumer services, a shortage of a fixed network of consumer services in rural areas, a low level of development of mobile forms of service provision; an undeveloped system of inter-municipal relations of consumer service organizations in carrying out joint projects that contribute to the development of the consumer services market; a lack of qualified personnel in the field of consumer services, the lack of prestige of consumer service professions.

References

1. State programs of the Russian Federation, Official site of the Standing Committee of the Union State. M., (2010) https://www.postkomsg.com/
2. Order of the Government of the Russian Federation dated 02.02.2015 No. 151-r "On approval of the Strategy for sustainable development of rural areas of the Russian Federation for the period up to 2030", Official website of the Government of Russia, http://static.government.ru/
3. Decree of the Government of the Russian Federation dated May 31, 2019 No. 696 "On approval of the state program of the Russian Federation" Integrated development of rural areas "and amendments to some acts of the Government of the Russian Federation", Official Internet portal of legal information, http://publication.pravo.gov.ru/
4. Decree of the Government of the Russian Federation of July 14, 2012 No. 717 "On approval of the State program for the development of agriculture and regulation of markets for agricultural products, raw materials and food for 2013–2020", Official website of the Government of Russia. M., http://static.government.ru/
5. T.G. Nefedova, Spatial Economics, **15(4)**, 36-56. (2019)
6. A. I. Treivish, Questions of geography, **141**, 51-71 (2016)

7. R.M. Gazizov, Bulletin of the Krasnoyarsk State Agrarian University, 11, 348–351 (2013)
8. O. V. Darmogray, *Improving the management of the socio-economic development of rural municipalities*, 24 (2010)
9. E. B. Dvoryadkina, E.A. Belousova, Regional economy: theory and practice, **15(12)**, 2255–2272 (2017)
10. A.V. Ilinykh, Bulletin of the Chelyabinsk State University, **25(380)**, 44–48 (2015)
11. D.S. Kleymenov, *Improving the management of the development of rural areas*, 23 (2016)
12. B. A. Kovtun, V. N. Papelo, V. V. Yarmanov, *Comprehensive development of rural areas: factors, institutions and mechanisms of advanced development*, 400 (2019)
13. I.V. Kovaleva, T.V. Kucherenko, International Journal of Humanities and Natural Sciences, **4-2**, 173-176 (2019)
14. O.P. Nikiforova, I. V. Antokhonova, *Local food markets in new conditions: research methodology and methods of regulation*, 164 (2016)
15. E.V. Chernysheva, *Loyalty of consumers of products of local enterprises of the agro-industrial complex in local markets: formation in the organizational and cultural context of the regions*, 88 (2015)
16. I.V. Razorvin, N.V. Usova, N.V. Bardasova, Agrarian Bulletin of the Urals, **6**, 93-98 (2019)
17. E. B. Dvoryadkina, E.A. Belousova, Bulletin of SUSU, Series "Economics and Management", **12(1)**, 48 (2018)
18. Decree of the Government of the Sverdlovsk region of 23.10.2013 No. 1285-PP "On approval of the state program of the Sverdlovsk region" Development of the agro-industrial complex and the consumer market of the Sverdlovsk region until 2025, Electronic fund of legal and normative-technical documentation, http://docs.cntd.ru/
19. Federal Law of December 29, 2006 No. No. 264-FZ "On the development of agriculture", ConsultantPlus, http://www.consultant.ru/
20. Decree of the President of the Russian Federation of January 30, 2010 No. 120 "On the approval of the Doctrine of food security of the Russian Federation", Official site of the President of Russia, http://kremlin.ru/
21. Order of the Government of the Russian Federation of November 17, 2008 No. 1662-r "Concept of long-term socio-economic development of the Russian Federation for the period up to 2020", Official website of the Government of Russia, http://static.government.ru/
22. Order of the Ministry of Economic Development of the Russian Federation dated November 20, 2013 No. 690 "On Approval of Guidelines for the Development and Implementation of State Programs of the Russian Federation", Official Internet Portal of Legal Information, http://pravo.gov.ru/
23. A.A. Izmalkov, Bulletin of the Voronezh State Agrarian University, **1-2**, 211-217 (2014)
24. Order of the Government of the Russian Federation of 17.04.2012 No. 559-r "Strategy for the development of the food and processing industry of the Russian Federation for the period until 2020", Official website of the Government of Russia, http://government.ru/

https://doi.org/10.1051/e3sconf/202126203030

25. Law of the Sverdlovsk region dated 21.12.2015 No. 151-OZ "On the strategy of socio-economic development of the Sverdlovsk region for 2016-2030", Electronic fund of legal and normative-technical documentation, http://docs.cntd.ru/
26. Database of indicators of municipalities of the Sverdlovsk region, Federal State Statistics Service, https://gks.ru/

Rethinking the residential urban environment in Iraq - literature review

*Al-Shaibani Alaa Abdali Hanoon**, and *Dhumad Kadhim Faris Dhumad*

Baghdad University, Iraq

Abstract. Is a new study to understand the residential architectural environment in Iraq through survey of a series of international, regional and local research and studies related to the residential architectural environment (1991-2019), which approximates 20 studies. the general problem the absence of a clear theoretical scenario that describes the architectural residential environment for the circumstances experienced by the Iraq now. Therefore, the aim of study to give an accurate description and definition of vocabulary and concepts through Theoretical conception to achieve a broader understanding of the residential architectural environment in Iraq (definition of the residential living environment, Housing satisfaction standards and description of the problem of informal housing). This is done by introduction of a three-stage model research, and classification of three items (Housing Satisfaction, Urban Context, Architectural Formation).The results of this study are a set of factors are related to spatial objective reasons for different architectural residential environments are similar the Iraq circumstances. and the research put its observations in a table divided into three axes, to archived aim.

1 Introduction

Housing structure and related problems have intensified in Iraq since 2003 due to rapidly rising housing needs, non-compliance with laws and regulations, and ineffective government programs and policies in the face of growing cities, reflecting declining living standards in many cities. cities, Since the creation of the Iraqi state for over 100 years, many housing plans and programs aimed at solving housing problems and access to a living environment meet the objective needs of the population, but most of these strategies and programs have not been implemented due to the lack of an integrated approach in aspects. This study presents a new concept for understanding the residential architectural environment in Iraq by reformulating and identifying the components of the living environment in them and identifying the most important problems and defining the principles of composition, and then Develop solutions that contribute to the provision of decent housing. Therefore, the importance of the study lies in the study of the basic vocabulary and factors that influenced the formation of the living environment in the Middle East, Middle East, Eastern Europe and the West, as well as in Asia such as China, India and Malaysia. What constitutes a model of strategic vision, built on an objective basis in the process of redefining the components of the living environment, and how to reflect the achievement of comfort in a residential area based on the definition of the components of the principles of residential architecture applied in the Middle East and in Iraq in particular W, in (Iraq).

Studies of the first group, Classification of residential architectural environment in Iraq.

concerned the concept of residential architectural environment [1], [2], [3], [4], [5], [6], [7], [8], [9], [10]. and focused on the meanings and form of residential architecture. Accordingly, they presented a set of issues related to socio-cultural factors and housing satisfaction standards that directly influence the formation of a residential architectural environment. where one could develop a conceptual formula entitled "Problems of Formation", which would later become the basis for the development of a conceptual framework to paraphrase the residential architectural environment. It was like that. in the table (1)

Group Two Studies, Housing Satisfaction Standards or Socially Acceptable Housing.

are devoted to housing policies, programs and standards [11], [12], [13], [14], [15], [16], [17], [18].aimed at solving the housing crisis in general. They referred to housing programs that could reduce the size of housing problems in different countries and establish housing criteria that can be adopted in addressing the challenges of the residential architectural environment in Iraq. It was like that. in the table (2)

Third group studies, Informal housing.

focused on informal and casual housing [19], [20], [21], [22], [23], [24], [25]. These are the most common housing problems in Iraq. At present, due to informal and casual housing, the urban environment does not provide a decent life for both individuals and families and society as a whole. The Researcher was able to find a new formula of knowledge (perspectives of problems) to form a conceptual framework in the future to develop appropriate solutions to describe problems, as well as through research in table (3).

1.4 A summary of the concepts and terminology that make up the residential architectural environment. Research - Model and Strategy Formulation for Redeveloping the Architectural Environment in Iraq By defining and understanding the components of this environment, it is based on concepts and factors, Seven factors have been identified (Social and Cultural Factors, Economic, Environmental and Climate, Geographic, Political and Security, Subjective or psychological planning) and six concepts (Concept of architectural emergence, Housing problems, Housing standards, Housing satisfaction and decent housing, Programs and strategies, Identity and transformation) that shape the living environment in a step towards solving the research problem by redefining the residential architectural environment and understanding its characteristics.

2 Ways to solve the identified problems

The study also shows that the problem under study will go through three main stages in order to solve it based on the classification. These classifications have been ordered according to previous studies and a diagram which is shown in Figure 1 below These classifications are considered as the core for solving the research problem, studying the architectural habitat in Iraq, clarifying the activities taking place in the residential environment, identifying them and further making planning and urban planning decisions after familiarizing themselves with the existing problems through the approved standards and achieving the main problem in the residential environment ... The study identified three categories of management studied. The researchers prepared tables indicating the most important concepts and factors related to the residential architectural environment, which were explained by previous studies and the reasons for their appearance, as well as the observations of the researchers as part of the action plan to analyze the selected studies. As shown in (Table 1,2,3). 1. Residential architectural environment. 2. Criteria for satisfaction with housing or socially acceptable housing. 3. Informal housing.

Table 1. Showing summary of research related to Architectural formation and the environment

Index	Study	Related Factors	Causes of the Phenomenon	Related Concepts	Research Notes
A1	(Mariam Yakub, 2017) "Open Source Architecture: Redefining Residential Architecture in Islamabad".	Related to climatic and cultural factors that constitute the residential environment and provide housing options	The need to address the shortage of units and control the residential environment according to specific architectural standards for design	Related to the concept of residential architectural formation	There is no objective definition of the residential environment, although the four options for residents are defined.
A2	(Tatiana Vadimovna, 2017) " ADAPTATION OF THE LIVING ENVIRONMENT OF MASS micro districts TYPICAL DEVELOPMENT TO MODERN SOCIAL ECONOMIC PROCESSES".	Social and environmental factors	The need for the environment social housing instead of industrial residential environment	Concept of residential formation and residential standards	It is necessary to identify the research items which constitute the backbone of the residential architecture.
A3	(Zhanna, 1993) "Conditions for the formation of low-rise residential buildings for high-density urban development (on the example of Iraq) b"	Climate and social factors	Formation of low rise residential buildings for high density development	Related to the concept of problems of architectural formation	The study did not address residential criteria in determining the levels of residential buildings.
A4	(A. GHAFFARIANHOSEINI etc., 2014) "The Essence of Malay Vernacular Houses: Analysis of the Socio-Cultural".	Social, cultural and climatic factors	The nature of the population's association with daily behavior	Related to the concept of residential architectural formation	Adopting the study as a way of life and a work methodology to assess the assessment of different environments
A5	(MAZDAK ARMSTRONG, 2017) "evolution of Residential Building in Iran based on Organization of space"	Related to the historical and culturalfactors of a society	Transition to the concept of mass construction and loss of fertility and social values	Concept of residential formation and residential standards	The resulting urban context has eroded the true spirit of architecture in Iran and has been reflected in the social relations of the inhabitants.
A6	(Jiboye Est, 2014) Significance of house-type as determinant of residential quality in Osogbo, Southwest Nigeria	Social factors and factors related to the quality factor of the same person	The desire of the inhabitants is to provide them with adequate housing through various traditional and contemporary houses	Related to the individualization of the formation and the re-formulation of the residential environment and the standards of residential satisfaction	The concept of residential quality is inherent to the concept of evaluation and through it, housing satisfaction or socially acceptable housing is achieved.

Table 1. Continued

A7	(Mohammed Thajil, 2005)"THE SOCIAL INFLUENCE OF THE CHANGMENT OF THE HOUSE FORM (BAGHDAD CITY)"	Social and aesthetic factors	Because of social impact, the shape of housing changes.	Related to the concept of architectural formation and formation standards	Changing the shape of housing is associated with architectural and design characteristics linked to social changes and economic level.
A8	(Aysu Akalin, 2009) "Architecture and engineering students' evaluations of house facades: Preference, complexity and impressiveness".	Social and environmental factors/psychological and subjective factors	Reflecting the behavior of individual identity by the inhabitants themselves, where they change the facades of their homes.	Related to the concept of problems of architectural formation and future prospects of formation	The preference is a general urban context, as well as individual behavior that arises in the inhabitants and crystallizes into reality in changing the facades of houses, individually, reflecting their individual identity.
A9	(Omar Al-Hafeet, 2018) "A systematic assessment of architectural approaches to solving the Iraqi housing problem."	Housing satisfaction criteria and subjective factors	Lack of a preferred architectural style for residents	Architectural design concept and housing satisfaction	The aim of the study is socially acceptable housing as the most important factor in the residential architectural environment
A10	(Zhdanova Irina Viktorovna, 2013) "Architectural methods for assessing and improving the consumer properties of the living environment".	Social and aesthetic factors Planning factors	Development of a system for assessing the quality of architectural and spatial solutions of the living environment	Housing standards Housing satisfaction and decent housing	Improving living conditions is associated with improving the functional characteristics of residential buildings through the creation of new functional relationships

Table 2. Summary Research on Housing Standards, Programs and Policies

Index	**Study**	**Related Factors**	**Causes of the Phenomenon**	**Related Concepts**	**Research Notes**
B1	(Tiwari Est, 2016) "Housing Markets and Housing Policies in India"	Economic factors as well as housing policies and programs	Residential growth, living conditions, lack of financial resources and poor program implementation	Related to the concept of problem prospects	Make decent housing a constitutional right and address programs and policies with objective programs
B2	(Onyemaechi Est, 2016) "Motives and Motivation for Implementation of Public Private Partnerships (PPPs) in Housing Provision in Nigeria"	Economic factors and political factors	Addressing the acute shortage of housing units and the desire for partnership between the public and private sectors	Related to the concept of problem prospects and concept of partnership	The relationship between the public and private sectors is one of the most important elements of facing the housing crisis in an approved residential environment. The partnership is based on the desire of both parties to develop the residential environment and achieve profits.

Table 2. Continued

B3	(Huda Ahmed, 2015) "National housing policies in Iraq"	Economic factors, political factors and housing standards	Because of the failure to implement the policies adopted to resolve the housing crisis in Iraq	Related to the individual housing problems as well as policies and programs	Iraqi policies and programs must be based on a future vision based on the reality of the country.
B4	(Khalaf A. Ali. est,2017) "Planning for Affordable Housing Units in Dohuk / Iraq, Survey"	Economic and financial factors as well as factors related to housing indicators such as family income	Because of economic factors and investment trends as well as laws that do not meet ambition	Related to the concept of problem prospects	Encouraging investment is one of the best ways to solve affordable housing as investment laws change according to elected environments
B5	(Lien, Laura Est 2016) "Assessment and analysis of housing accessibility: adapting the environmental component of the housing enabler to United States applications"	Social and accessibility factors	Because of the problem of disability in how to reach homes	Related to the concept of housing satisfaction and housing standards	The study adopted specific measurement measures to achieve the principle of arrival, and quantitative measurement, which is one of the best methods adopted to show the results of a particular experiment.
B6	(Christian A. L et al., 2016) "Housing Policy in Switzerland, UK and USA".	Political and security factors	Housing policy lessons in a developed country are one of the most important drivers for housing policy development in the Middle East.	Programs and strategies	Using the experience of developed countries in finding solutions to housing problems in Iraq, taking into account social standards to achieve this goal
B7	(Xin Liu et al., 2016) "Exploring the Problems of Quality Housing Design in China."	Economic and planning factors	Understand the challenges of improving the quality of housing design in China.	Housing problems and housing standards, programs and strategies	Examining the most important elements for improving design quality and identifying variables and causes
B8	(Rachel M .N. et al., 2019) "Assessing Resilient Aspects in the Housing Sector to Overcome Housing Stress in Northern Iraq."	Economic factors and planning factors	This study aims to assess the sustainability aspects (environmental, economic and social) of the housing sector.	Housing Problems, Housing Standards, Programs and Strategies.	Achieving a balance between social and economic standards in residential buildings in accordance with the adaptation criteria

Table 3. Summarizing research on informal housing

Index	Study	Related Factors	Causes of the Phenomenon	Related Concepts	Research Notes
C1	(Adekoyejo Est, 2017) "Emergence of Informal Housing: Implications for Development of Low-Cost Housing Delivery Strategies in Abeokuta, Nigeria"	Economic factors and building strategies	High property prices in city centers	Related to the concept of formation problems and decent housing standards	Testing the housing strategies in the study area and diagnosing the objective solutions
C2	(Nabil Taha, 2010) " The Informal Areas and the ways Treated it with Planning Standard for Residential Areas"	Environmental factors, planning factors and standards-related factors	Increasing the size of families and economic conditions and the desire to own a housing unit	Related to housing problems as well as to adopted programs and policies	Dealing with informal houses as a needing and a housing balance
C3	(Hanan Mahmoud, 2008) "The phenomenon of random housing in some Arab cities, an analytical study of the causes of emergence and planning scenarios of the possibilities of treatment"	Economic factors, social factors and the way resources are managed	Unprogrammed planning and non-realistic solutions to existing problems	Related to housing problems, as well as to adopted programs and policies	An experienced Arab's experiences represent a basis for dealing with informal housing, taking into account the specific environment
C4	(Haidar Attia, 2014) " The Spatial Analysis of Random Residence Phenomenon Reality in Al-Kufa City "	Social factors and criteria for planning	The acute shortage of housing, absence of laws, its application, and commercial speculation	Related to the concept of problem prospects and the concept of standards	The theme of random neighborhoods is no less important than other official neighborhoods in terms of physical characteristics and geographical location
C5	(Miguel Amado Est ,2018) 'TRANSFORMATION OF INFORMAL SETTLEMENTS: CASE OF CAPE VERT'	Standards and principles of urban planning	The lack of clarity of land ownership is one of the reasons for the increase in informal areas	Related to future prospects to resolve the problem	The design of a program that helps in identifying, distributing and clarifying land ownership is one of the reasons for the increase in informal areas
C6	(Yue Zhang,2017) The credibility of slums: Informal housing and urban governance in India	Economic factors as well as principles of urban planning and policies	Informal neighborhoods have proven their credibility through economic, political and housing opportunities.	Programs, policies and standards	Formalizing informal neighborhoods through government programs to provide credibility

Table 3. Continued

C7	(Paula Meth ,2017) INFORMAL HOUSING, GENDER, CRIME AND VIOLENCE: THE ROLE OF DESIGN IN URBAN SOUTH AFRICA	Planning, design and security, political and social factors	Excessive permeability of residential structures is the biggest problem in generating violence and crime.	The concept of residential problems residential programs and residential standards	It addresses the treatment of crime and violence through housing programs, redesign and resettlement in informal neighborhoods.

2.1 Analyze and discuss the results

The factors and concepts that make up the residential architectural environment are identified, determined by seven factors and six concepts, Table (4) This was done using previous studies that show the extent of its impact and its relationship with the residential environment. The process of formulating the results was done by knowing the degree of relationship of each concept and each factor with the living environment, where the results and available knowledge were interrupted by three groups of studies that were accepted in the study, namely (studies related to the formation of a residential architectural environment, studies related to housing standards, problems, policies and research related to informal housing), it was found that there is a need to reformulate and define the residential architectural environment, knowing the degree of connection (between the residential architectural environment and factors and specific concepts), This relationship is divided into three levels (1-strong correlation with degree 2, 2-average correlation with degree 1.3-weak correlation with degree 0). This explains the levels of concepts and factors involved in redefining and articulating a living environment capable of providing a better quality of life for its inhabitants.

Table 4. The most important factors and concepts related to the residential architectural environment.

Influencing Factors
Social and cultural factors
Economic forces
Environmental and climatic factors
Geographic factors
Political and security factors
Subjective or psychological factors
Planning factors

Related concepts
Concept Architectural formation
Housing problems
Residential standards
Housing satisfaction and decent housing
Programs and strategies
Identity and transformation

2.2. Methods for achieving the satisfaction of residents with elements of the living environment

Research is a model and strategic research for reformulating the architectural environment based on the vocabulary of this environment, determined by that time by the preparation of theoretical foundations based on two concepts of knowledge related to the subject of our research (problems of architectural configuration) and (shaping the prospects for possible solutions). If research shows that the process of connecting to any habitat depends on the nature of the environment chosen in terms of the quality of daily life and the most appropriate housing schemes, and the other party determines the incentives and housing standards based

 https://doi.org/10.1051/e3sconf/202126204001

on the availability of economic opportunities to meet housing needs, thereby providing the population with more opportunities (Fig. 1).

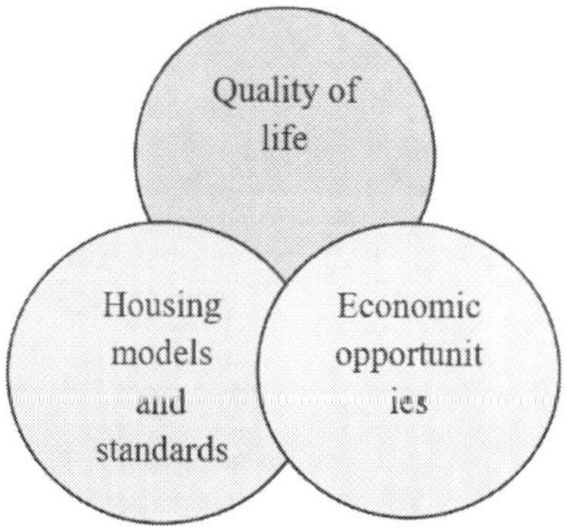

Fig. 1. Scheme for determining criteria for satisfaction with housing

2.3. Redefining Residential Architecture in Iraq

A residential architectural environment can be defined as a social activity for individuals or a society with residential elements that make up an environment based on social norms and traditions, in accordance with modern conditions and requirements, as well as constant changes in intellectual products that are formed in the human mind ... And thus, reflecting on the behavior of people on a daily basis and based on this research, a new definition of the architectural environment can be developed: as an urban context that depends on the socio-cultural economic context of specific environmental conditions and in accordance with standards and models. Housing is based on systems and laws and programs that are sustainable taking into account the requirements of modern life, the population. And thus, access to socially acceptable housing, providing the concept of decent housing or housing satisfaction. (Fig. 2).

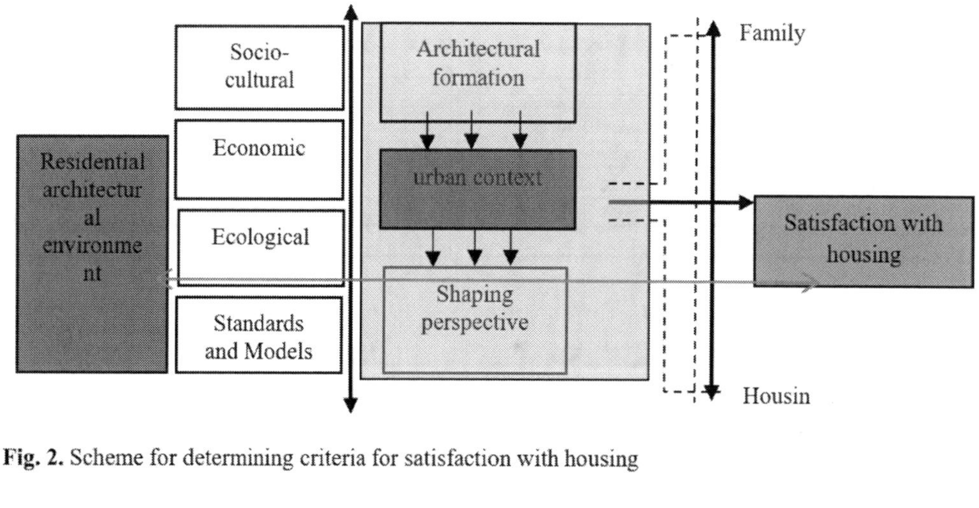

Fig. 2. Scheme for determining criteria for satisfaction with housing

3 Conclusion

1. The main stages and patterns of development of the architecture of the residential environment in Iraq have been identified (1. the formation of the residential architectural

environment. 2. The criteria for satisfaction with housing or socially acceptable housing. 3. Informal housing). By rethinking the architectural environment in Iraq, which includes a set of factors that make a living environment acceptable to Iraqis by conducting and analyzing global, regional

2. Examination of the existing literature has shown that the research available in Iraq is scarce and of low quality, since it does not use modern research methodology that could cover the living environment in different aspects and from different points of view. The most important factors that are related to the problems of the living environment are reflected in the realities of the country, as well as in informal or casual housing, which requires close examination.

3. Previous research develops social patterns after working with architectural environments in which the researcher can choose the best way to solve the problem of the elite living environment from certain aspects of the research. This is consistent with behavior, customs and social traditions and understanding of other aspects of tension, divided into economic, political and natural aspects of security.

4. The study presents a summary of cognitive trends from previous works through their main observations and analysis, which in the future can become a guideline in the preparation of a theoretical framework that determines the main research problem. The study identified factors that influenced the architectural environment in the public sphere, which are divided according to objective directions, which can be understood by three basic concepts (Satisfaction with housing, urban context, architectural formation). They are related to objective reasons for the various residential environments considered in the study. These equations and factors created architectural concepts and research formulas that can be adopted to understand the broader residential architectural environment. It is very important to determine the criteria on the basis of which decisions are formed for specific forms of the living environment, which would reflect the life of the population.

References

1. M. Yaqub, Open Source Architecture: Redefining Residential Architecture in Islamabad, School of Architecture, Art, and Historic Preservation Theses and Projects, 28 (2017)
2. O. Al-hafith, B. K. Satish, S. Bradbury and P. D. Wilde, Front. Archit. Res., **24**, 3 (2018)
3. T. V. Zaitseva, *Adaptation of the residential environment of micro-districts of mass typical building to modern socio-economic processes*, 127 (2017)
4. Zh. B. Shahin, *Conditions for the formation of low-rise residential buildings for high-density urban development (on the example of Iraq)*, 23 (1993)
5. A. Ghaffarianhoseibi, U. Berardi, N. D. Dahland, *The Essence of Malay Vernacular Houses : Analysis of the Socio-Cultural*, 156 (2014)
6. M. Irani, P. Armstrong, A. Rastegar, Asian Cult. Hist., **9(2)**, 46 (2017)
7. A. D. Jiboye, Front. Archit. Res., **3(1)**, 20 (2014)
8. A. Akalin, K. Yildirim, C. Wilson, O. Kilicoglu, J. Environ. Psychol., **29(1)**, 124 (2009)
9. Al-fadillh m. Th., *The social influence of the changment of the house form (Baghdad city)* (2005)
10. I. V. Zhdanova, *Architectural methods for assessing and improving the consumer properties of the housing environment*, 29 (2013)
11. P. Tiwari, J. Rao, *Housing Markets and Housing Policies in India* (2016)
12. P. Onyemaechi, M. Samy, Mediterr. J. Soc. Sci., **7(2)**, 149 (2016)
13. A. J.Huda, *National Housing Policies in Iraq- A Comparative Study* (2015)

14. K. A. Ali, L. M. Raswol, ZANCO J. Pure Appl. Sci., **28**, 342 (2017)
15. L. L. Lien, C. D. Steggell, B. Slaug, S. Iwarsson, J. Hous. Built Environ., **31(3)**, 565 (2016)
16. C. Q. L. Xue, Habitat Int., **57**, 242 (2016)
17. Rahel, M. Amin, S. Salah, Contemp. Urban Aff., **3(1)**, 67 (2019)
18. C. A. L. Hilber, O. Schöni, The Housing Challenge in Emerging Asia: Options and Solutions., 210 (Japan, 2016)
19. B. Adekoyejo, A. Olatunde, Int. J. Innov. Res. Dev., **6(3)** (2017)
20. M. Amado, F. Poggi, A. Martins, N. Vieira, A. R. Amado, Creat. Commons CC BY, **18**, 1 (2018)
21. H. M. Al-juboury, *The Phenomenon of Informal Settlements in Some of Arabic Cities* (2008)
22. N. Taha, *The Informal Areas and the ways Treated it with Planning Standard for Residential Areas* (2010)
23. Al-Garaawi Hayder Atiya Abd Nasir, *The Spatial Analysis of Random Residence phenomenon Reality in Al-Kufa City* (2014)
24. Y. Zhang, *Land Use Policy The credibility of slums : Informal housing and urban governance in India*, 1 (2017)
25. P. Meth, Brit. J. Criminol, **57**, 402 (2017)

Financial instruments of "green" economy development in Russia

M.V. Dubrova[1], *L.V.* Gusarova[1], *N.N.* Zhylina[2], *M.R.* Shamsutdinova[2], *V.G.* Ignatyev[2]

[1]Financial University under the Government of the Russian Federation
[2]Kazan Cooperative Institute, Branch of Russian University of Cooperation

Abstract. Environmental problems are getting worse every year. The root cause of green finance is the climate change, pollution and the depletion of natural capital that threaten Russia's competitiveness and productivity. Green financing means providing finance while enhancing the resource efficiency and reducing the impact on the environment and the global climate. Balanced adherence to financial, economic and environmental criteria is the main foundation. The new emerging financial instruments of a green economy do not have technical fundamental differences, but qualitatively form new conditions for the responsible investment process. The Russian green finance market emerged in 2018. The past period can be characterized by both a quantitative and a qualitative breakthrough in the system of green finance in Russia. At the same time, the issues of developing a mechanism of state support for the activities of non-profit organizations in the field of green finance, the participation of non-profit organizations in the implementation of environmental projects through the development of green financing instruments require consideration.

1 Introduction

Green financing means providing finance while enhancing the resource efficiency and reducing the impact on the environment and the global climate. Green finance instruments include the green loans and the green bonds. Green banks and green funds are engaged in green financing.

In practice, the "green" bonds are the bonds of issuers engaged in the field of alternative energy, environmentally friendly modes of transport, creation of smart buildings and others aimed at projects on ecology and sustainable development. Today, almost 50 countries around the world use similar types of financing.

For securities to be recognized as "green", which means the possibility to entertain the public support and a number of preferences, they should pass certification for compliance with certain criteria, which is carried out by consulting, engineering, legal and auditing companies, as well as rating agencies.

Such bonds are circulated on special stock exchanges at reduced rates. They are of particular interest to large financial institutions. Very often, governments and central banks subsidize the interest rate on green projects. Pension funds are most willing to invest in such bonds.

2 Materials and methods

The research utilized the general scientific research methods, e.g. comparison, analysis and synthesis, grouping, comparison, graphical method of analysis, trend and statistical analysis. The logical, historical and complex approaches to the study of theoretical materials were used as well.

The research methodology includes the examination, generalization, assessment of the accumulated knowledge and experience of domestic and foreign scientists.

A critical review of regulations governing the issuance and circulation of financial instruments made it possible to identify problems in the development of a "green" economy, to identify the need to clarify and legislatively consolidate key concepts, features of the issuance and circulation of "green" bonds, to develop directions of state support for the development of the "green" market financial instruments.

3 Results and discussion

Russia faces the problem of the lack of a large number of participants in the largest organizations in the field of responsible investment, which does not allow it to contribute to the development of solutions to global problems. This problem should be solved by joint efforts of both the government and market participants.

The dynamics of environmental investments in the Russian Federation over the past decade (Fig. 1) allows us to conclude that the market for responsible investment is developing, the "green economy" is becoming a vector of innovative projects.

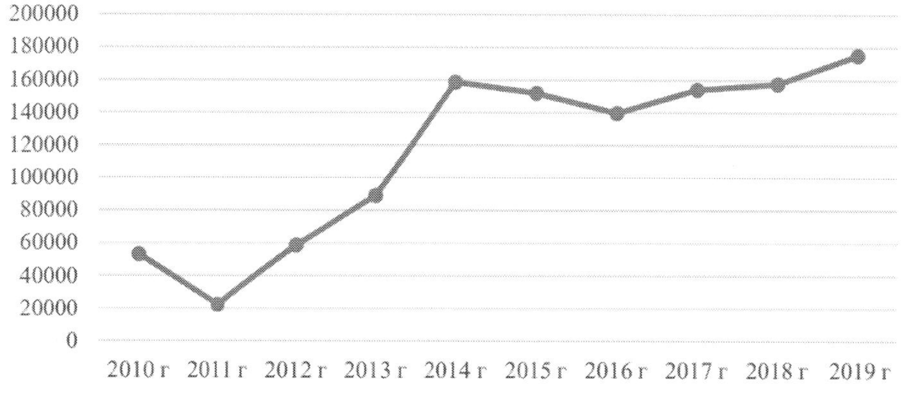

Fig. 1. The fixed assets investments aimed at environmental protection and rational use of natural resources, RUR mln

The dynamics of investments in fixed assets aimed at environmental protection and rational use of natural resources allows us to conclude that a system of financing environmental projects has been formed, which reinforces the previously stated thesis about the development of a "green economy" and "green finance" in Russia.

The first full-fledged support platform for environmental projects "Sustainable Development Sector" in the Russian financial market is an example of the green finance idea implementation. At the beginning of 2020, five issuers in Russia have placed seven issues of green bonds for a total of 7.55 RUR bln and 500 EUR mln.

Currently, a new ecological culture of life is being formed all over the world, while green projects are developing, which, in addition to environmental friendliness, include universal approaches and acquire an ideological sound. Russia sees an increase in financing in the field of the green economy and investment practice in green projects, which indicates the inclusion of a global ecosystem in the global ecological framework. National projects in Russia correlate with the UN Sustainable Development Goals, are complex in nature, balanced by three sustainable development components and compliant with the economic, social and environmental factors of the country's development. Trends in green finance in Russia are important for changes in the eco-financial sphere and the global financial market. The Russian economy is close to the ten largest world economies, while the Russian financial market is not even included in the first hundred of financial markets. Pursuant to the World Competitiveness Index, the Russian financial market is lower than the Turkmen one. Therefore, the development of the eco-financial sector will make it possible to remedy the actual situation and allow to attract additional financial resources, which are less expensive than in Russia, for the Russian green projects.

Financial resources are needed to save the environment, and it is important to optimize the cost of attracting them. In addition, the investors should be shown the benefits of investing in green projects, which can be, for example, image-building. In this regard, there is a tendency to attract institutional investors who, owing to green investments, could improve their image.

It is beneficial for Russia to use green finance, since ignoring the environmental problems associated with economic growth jeopardizes the results already attained. Such lack of action may cause significant economic, social and environmental consequences for the country. This damages not only the Russian economy, but also the health of citizens. The long-term functioning of "green finance" is also associated with many unresolved economic problems.

Two roles are distinguished in the green project relations framework: institutions and organizations of the public and private sectors. They stimulate the development of a green financial market and promote efficient investment and resource efficiency. In the private sector, banks play a leading role in attracting institutional investors.

One of the priority tasks of enhancing innovative financing in Russia is the formation and development of a market for "green" financial instruments, which includes bond loans, loans and other instruments for financing the environmental projects. The list of these instruments is described in the Action Plan of the Ministry of Economic Development of the Russian Federation for the period from 2019 to 2024, pursuant to which national rules and standards for the verification of green financial instruments are being actively developed, contributing to the development of the green bond market, and, consequently, the development of a green economy. It is necessary to increase the rate of promotion of "green" financial instruments to ensure the formation of a competitive national financial market in Russia. The lag in the development of the market for "green" financial instruments may impact the implementation of environmental projects, since investment processes associated with sustainable development projects require a transparent financial mechanism. Consequently, the development of the market for green financial instruments in Russia requires the formation of a regulatory framework that facilitates the verification of these instruments and their involvement in the investment process.

The legal consolidation of the status of "green" bonds, the consolidation of the issue procedure and the features of the issue, the requirements for information disclosure will contribute to the verification of such financial instruments and the accelerated development of the green bond market.

According to experts, the Russian green bond market capacity exceeds 1 trln RUB, therefore, at the moment the potential of this instrument is underestimated, which means it is highly attractive for both issuers and investors. Currently, the official "green" bond status is assigned to the limited list of financial instruments enshrined in the relevant register, including the "Resource Saving of the Khanty-Mansi Autonomous Okrug" bonds (construction and operation of a landfill for placement, neutralization and processing of solid municipal waste); Bank Centrinvest (energy efficient projects, renewable energy sources and environmentally friendly transport); JSC Commercial Real Estate FPK Garantinvest (modernization of shopping and multifunctional centers and the green technology implementation), JSC Russian Railways, which issued the green Eurobonds aimed at foreign investors (financing and refinancing the purchase of passenger electric trains). The list of green bonds is not limited to the listed issuers, but many issues do not have the official status of green instruments, despite financing projects related to environmental protection.

The noticeable event in the development of the market of "green" financial instruments in Russia was the accession of Moscow Exchange to the "Sustainable Stock Exchanges" (SSE) global initiative in April 2019. Joining the SSE and cooperation with the Russian Union of Industrialists and Entrepreneurs (RUIE) - "Responsibility and Openness" resulted in the daily calculation of two sustainable development indices. The first index is calculated on the basis of the prices of shares of 22 issuers that are leaders in information disclosure in the field of ESG, and the second index - "Sustainable Development Vector", is calculated on the basis of the prices of shares of 15 issuers that showed the best dynamics of indicators in the field of sustainability and corporate social responsibility.

The growth of the green bond market is impossible without the implementation of a mechanism of state support and a system of additional benefits for market participants, both for issuers and for investors.

The green bond market state support mechanism may include the following measures:

1) Legislating the green bond principles and standards. The participation of the state in unification of green financial instruments makes it possible to ensure market transparency, proving the reliability of the instrument and monitoring the implementation of environmental projects for the financing of which investments are formed.

2) Regulation of liquidity and activity of the financial market through the issuance of "green" bonds by the state (issuance of bonds by development institutions, subjects and municipalities to maintain the priority and strategic investment projects for sustainable development).

3) Ensuring risk management by scaling green projects, issuing collateralized green bonds, providing institutional investors with access to the bond market; securitization of loans at the design development stage.

4) Determination of the financial incentive (incentives for environmental taxes and fees, government guarantees, budget subsidies, direct government funding at the stage of preparation, project design, government insurance, etc.) provision conditions.

5) Provision of guaranteed demand for financial instruments through public procurement with a priority of environmental components.

It seems necessary to develop and deploy the mechanisms for the creation and development of a green cluster in the economy on the basis of public-private partnership, which will allow in the long term to obtain the effect of a double benefit - financial "recovery" of the economy and environmental benefits.

A new stage in the development of the modern world financial system is the globalization of "green" finance, which is an objective process and coincides in time with global reforms. The peculiarity of the analyzed process is the formation of a global market for "green" financial instruments, the innovative nature of the initiatives and national practices of countries with developing markets.

4 Conclusion

The recommended measures for the development of the entire sphere of the "green economy" in Russia were presented, but they are primarily aimed at supporting the non-profit organizations' private initiatives, which are primarily interested in the fact that their activities became known not only in local level, but also at the level of the constituent entity of the

Russian Federation and the federal level. Without the proper level of financial support from the state and private investment, non-profit organizations are doomed to develop ineffective and inefficient innovative projects.

Undoubtedly, nobody canceled the innovative environmental project grant support mechanism. However, as the foreign countries' promising experience shows, the "green economy" has much more financial resources and instruments than a conventional economy. The creation of only a green investment bank, by analogy with Great Britain, will allow a significant amount of financial resources to be directed to solving the most pressing and complex environmental problems.

References

1. V.V. Arkhipova, HSE Economic Journal, **21(2)**, 312 (2017)
2. Activity of NPO within year of ecology of 2017, https://mediarepost.ru/
3. M. L. Dorofeyev, Banking, **4**, 23 (2020)
4. T.M. Kozhevnikova, S.G. Ter-Akopov, Social and economic phenomena and processes, **3 (049)**, 78 (2013)
5. A.V. Kucherov, O.V. Shibileva, Young scientist, **4**, 561 (2014)
6. K.G. Musailova, World trend in "green" economy and prospect of "green" finance in Russia, 39 (2018)
7. Center of Competence and Green Expertise of the National Association of Concessionaires and Long-Term Investors in Infrastructure. Annual report "Green Finance in Russia 2020" (2020)
8. V.A. Tsvetkov, A.G. Gurinovich, V.A. Vernikov, I.Yu. Fedorova, M.V. Dubrova, A.Z. Namitulina, Abashilov Shekhmagomedovich Khadzhi-Murad, International Journal Of Engineering And Technology, **10(2)**, 1258 (2019)
9. ESG factors in investment. – MIRBIS, (2019), https://www.pwc.ru/

E3S Web of Conferences **262**, 04003 (2021) https://doi.org/10.1051/e3sconf/202126204003
ITEEA 2021

Ways of the lead-bearing slag waste utilization

Gaukhar Zahievna Turebekova[1], *Saule Ospandiyarovna* Akhmetova[2,*], and *Zarina Ilesovna* Bagova[1]

[1]M.Auezov South Kazakhstan University, Shymkent, Republic of Kazakhstan
[2]Almaty Technological University, Almaty, Republic of Kazakhstan

Abstract. The article presents the results of research on the influence of lead production slag on the environment and ecology of the city. Due to the open storage of slags, an excess of maximum permissible levels (MPL) of lead was established: near the plant, the concentration of lead in the soil exceeds 3000 mg/kg, while the MPL is 3.2 mg/kg. The data of analyzes obtained by methods of scanning electron microscopy and X-ray microanalysis, performed on a scanning electron microscope (SEM), as well as the results of X-ray fluorescence semi-quantitative analysis, performed on an X-ray fluorescence wave-dispersive spectrometer, have been presented. The results of thermal analysis of samples in air in the 20-1000 °C temperature range are presented.It was revealed that lead slag contains a fairly high amount of toxic metal compounds. It has been established that the average content of toxic metals in slags formed as a result of the activity of a lead plant is: lead - 2, 38%; zinc - 9.81%; copper - 0.97%; iron - 25.31%; silicon oxide - 24.62%; calcium oxide - 16.21%; potassium oxide - 1.42%; sulfur - 1.35% of the total sample weight.The offered technology of a selective method for the extraction of non-ferrous metals allows to enhance the ecological state of the environment and reduce their negative impact on human health by means of the disposal of the lead production slags.

1 Introduction

As a result of the activities of the Shymkent Lead Plant, about 2 million tons of man-made waste in the form of lead-containing slags have accumulated. Lead production slags contain a large amount of toxic compounds of heavy metals such as lead, zinc and cadmium, which are the dangerous environmental pollution sources [1-4].

Even after the closure of the plant, the remaining waste in the form of slag is a source of environmental pollution of soil, groundwater and air. The environmental expertise revealed a huge excess of the maximum permissible concentration (MPC) of lead in the soil near the former lead plant: currently, at the site of the plant and near the concentration of lead is more than 3000 mg/kg in the soil with the MPC 3.2 mg/kg due to open storage of slags, which cause irreparable damage to the vitals of living organisms and the health of the population of nearby areas (densely populated microdistricts of Shymkent Kazygurt,

* Corresponding author: sunrise_kz@mail.ru

Sauleand other nearby streets of the Abaysky district), as well as hinder the labor safety and industrial sanitation to various business facilities located in this area [5-7].

Lead and zinc compounds are dangerous to a human due to their significant toxicity and ability to accumulate in the body. Lead poisoning ranks first among the professional intoxication of workers in the field of production. For instance, wastewater containing zinc compounds is not suitable for irrigation of fields, the negative effect of zinc compounds on microorganisms and soil microfauna reduces soil fertility, although zinc plays a certain role in plant metabolism. Pursuant to research data and analyzes of the content of lead compounds in plants growing near slag storage, an excess of the MPL from 1.83 to 8.13 times was recorded [8-10]. The environmental damage caused by the long-term storage of slag is seriously dangerous when the slag is in an acidic environment (soil pH <4). Contamination of the soil of Shymkent with compounds of lead and other metals contained in waste slags causes the ingress of harmful chemical compounds into the human body. Lead compounds impact the human nervous system, cause the decrease in intelligence, changes in physical activity, coordination of hearing, impact the cardiovascular system and have a significant impact on life safety [11]. Lead intoxication ranks first among the professional intoxications, and there is a tendency to its increase. About 40% of workers affected by lead exposure are women. Lead is especially dangerous for the breastfeeding women, because it has the ability to cross the placenta and accumulate in breast milk. Unfavorable meteorological conditions during this period of the year also contribute to the accumulation of lead in the lower atmosphere layers. Lead enters the body through the gastrointestinal tract and respiratory system, and then spreads through the blood throughout the body. Herewith, the danger of the lead dust inhalation is much more dangerous than the presence of lead in food. Lead compounds also accumulate in bones, partially replacing calcium in phosphate. Getting into human soft tissues - muscles, liver, kidneys, brain and lymph nodes [12], the lead causes a disease called plumbism. Like many other heavy metals, lead (in the form of ions) blocks the activity of certain enzymes. The activity of enzymes decreases 100 times with an increase in the concentration of lead in the blood 10 times - from 10 to 100 micrograms per 100 ml of blood. At the same time, anemia develops, the hematopoietic system, kidneys and brain are affected, and intelligence decreases. This impacts the health of the population and especially children, who are most susceptible to lead poisoning. Lead can easily enter the body with drinking water if it comes into contact with metal: in the presence of carbon dioxide, soluble bicarbonate slowly passes into the solution. It is enough that there is only one milligram of lead in one liter of water, and drinking such water becomes very dangerous. Most often, lead comes from the atmosphere into the soil in the form of oxides, where it gradually dissolves, passing into hydroxides, carbonates, or another form of cations [13]. Food is the main source from which lead enters the human body. In addition to this, the important role is played by the inhaled air plays, as well as the lead dust that children ingest. About 30-50% of the inhaled dust is retained in the lungs, and a significant proportion of it is absorbed by the blood. Absorption in the gastrointestinal tract of the adults is 5-10%, with up to 50% in case of children. Deficiency of calcium and vitamin D increases the gastrointestinal tract lead absorption rate. On average, the human body absorbs 26-42 μg of lead per day. About 90% of the total amount of lead in the human body is found in bones, with 60-70% in case of children [14]. In addition to lead, the lead slag contains zinc compounds able to significantly impact the life safety and the environment. Wastewater containing zinc is not suitable for irrigation of fields, therefore, the increased content of zinc impacts the microorganisms and microfauna of the soil, significantly reducing its fertility. Many manifestations of zinc intoxication are based on the competitive relationship of zinc with a number of other metals. For instance, residents of nearby areas showed a significant decrease in the total level of calcium in the blood serum. Excessive intake of zinc into the body was accompanied by a drop in the calcium content not only in the blood, but also in the bones, at the same time the absorption of phosphorus was impaired; as a result, osteoporosis developed. The toxicity of zinc oxide is explained by its catalytic activity, which poses a mutagenic and oncogenic hazard, disrupting the human life safety [15]. Thus, due to the great harm to the health of the population of the neighboring regions of Shymkent, the problem of disposal of lead production slags is very acute and relevant.

2 Materials and methods

The study of the material composition was performed on bulk slag material, externally black, with a particle size ranging within 2-6 mm. In the course of the study, a heavy fraction was isolated from the sample, from which polished artificial polished sections (briquettes) were made. The polished sections were analyzed under a LEICA DM 2500P microscope. In addition, the sample was examined under a microscope in immersion liquids, as a result of which samples were taken for further research.

Scanning electron microscopy of the slag was performed on a JEOL-6490 LV microscope (JEOL, Japan) in order to measure the scale factor of the video image by obtaining the surface of an object with a high spatial resolution, as well as elemental analysis and elemental mapping at low magnifications.

The X-ray fluorescence semi-quantitative analysis of the slag was studied on an Axios X-ray fluorescence wave-dispersive spectrometer (PANalytical BV, Netherlands), and the determination of the elemental composition of the substance was performed by the X-ray fluorescence analysis (XRF) method.

In addition, a differential thermal analysis of slag samples was performed on a Q-1000/D derivatograph, system of F. Paulik, J. Paulik and L. Erdey provided by "MOM" (Hungary, Budapest). The survey was performed in air, in the temperature range of 20-1000 °C (293-1273 K), the heating mode was dynamic (dT/dt = 10), the reference substance was calcined Al_2O_3, the sample weight was 500 mg with a scale division of the sample weight - 500 μV. The studies were performed with the following parameters: the sensitivity of the balance is 100 mg, the sensitivity of the other measuring systems of the device: DTA = 250 μV, DTG = 500 μV, TG = 500 μV T = 500 μV. The applied method is based on recording changes in the thermochemical and physical parameters of a substance that can be caused during its heating. The thermochemical state of the sample is described by the corresponding curves: temperature (T), differential thermoanalytical (DTA), thermogravimetric TG and differential thermogravimetric (DTG). It is noted that the differential thermogravimetric (DTG) dependence is a derivative of the TG-function. The optimum thermochemical parameters obtained during the high-temperature treatment of the test system allowed to reveal the nature of the destruction of thermally active components. The powder sample composition was identified by the morphologies of the thermal curves and the obtained numerical values of the intensities of endo- and exothermic effects, using the thermogravimetric readings of the TG lines associated with them. The analysis results were compared with the data given in the atlases of thermal curves of minerals and rocks, and compared with the descriptions of the thermal behavior of monomineral samples presented in reference sources and accumulated in the data bank of the laboratory of physical methods of analysis of the Institute of Metallurgy and Refinement of the National Academy of Sciences of the Republic of Kazakhstan, Almaty.

3 Results and discussion

3.1 Discussion of the lead production slag analysis results

Pursuant to the X-ray diffractometric slag sample analysis results, the formation of the following useful components was revealed: copper in the form of sulfides, complex compounds of oxides of lead, zinc, iron and copper, which are in the form of inclusions in an amorphous host matrix of complex composition. The lead production slag laboratory study results showed a fairly high content of non-ferrous metal compounds. This allows to state that the process for their extraction and disposal of toxic waste of lead production is technically and economically feasible. The research results are shown in Figure 1.

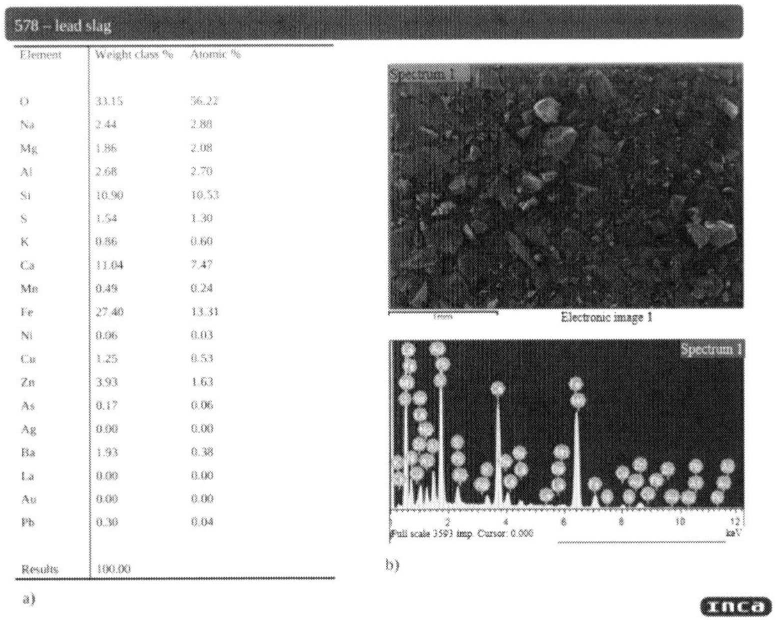

Fig. 1. Scanning electron microscopy and X-ray microanalysis of lead-containing slag; a) the quantitative content of elements; b) morphological-qualitative analysis

The chemical method of analysis, as well as the results of X-ray fluorescence analysis of the slag, shows that the average content of toxic metals in the slags of lead production is in (%,%): lead - 2.38; zinc - 9.81; copper - 0.97; iron - 25.31; silicon oxide - 24.62; calcium oxide - 16.21; potassium oxide - 1.42; sulfur - 1.35 and moisture up to 3% of the total sample weights, the data of which are given in table 1.

Table 1. The content of the components of the lead production slag, formed during the raw material processing

Sample No.	Content of components,%							
	Pb	Zn	Cu	Fe	SiO_2	CaO	K_2O.	S
1	2.0	9.6	1.1	26.1	25.4	14.0	1.2	1.3
2	2.1	9.4	1.0	26.0	18.3	15.2	1.0	1.2
3	2.2	9.6	0.92	25.0	20.2	16.4	1.2	1.18
4	2.5	9.7	0.96	25.3	25.2	16.37	1.4	1.7
5	2.55	9.8	0.98	30.6	24.9	16.4	1.8	1.52
6	2.6	11.2	0.9	28.9	24.2	16.2	1.68	1.36
7	2.7	9.6	0.93	25.1	25.4	16.35	1.75	1.52
average value	2.38	9.84	0.97	26.71	24.62	16.21	1.42	1.35

The analysis of Table 1 in terms of the indicators of the components indicates that the content of lead and zinc, as the main elements of non-ferrous metals, ranges (in%) for lead within 2-2.7, and for zinc - within 9.5-11.3, which allows to extract them into target products, and the rest of the components can be used in the production of mineral fertilizers as trace elements or in the production of stone products.

A slag sample in dynamic heating mode on (DTA-, DTG- and TG-) curves in different temperature ranges left a series of effects caused by endo- and exothermic reactions, which are shown in Figure 2, and their quantitative values are shown in Table 2.

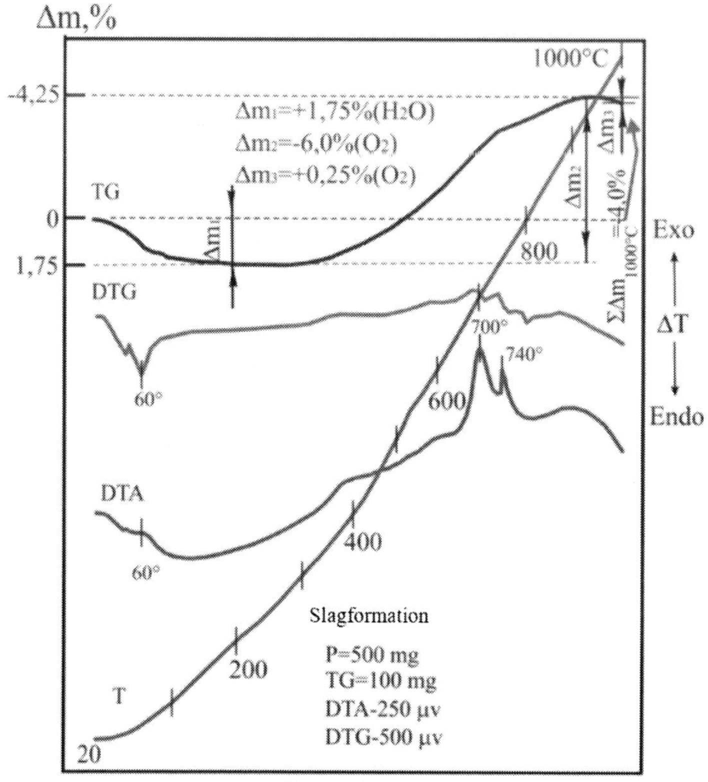

Fig. 2. Derivatogram of a lead-containing slag sample

Table 2. Sequence of the volatile component weight loss quantitative values

Weight loss sequence	Weight loss, %	Volatile components	Decomposition temperature interval, °C
Δm_1	1.75	H_2O	20-200
Δm_2	-6.0	$-O_2$	280-930
Δm_3	0.25	O_2	930-1000
$\Delta m_{1000°C}$	4.0	H_2O, $-O_2$, O_2	20-1000

Note: in Figure 2 and Table 3, the minus (-) sign at the Δm_2 value means the introduction of atmospheric oxygen into the system. This sign is the opposite of the sign applied to the weight loss parameter.

In the intervals of low temperatures of 20-200 °C (293-473K), an endothermic effect with a weight loss Δm_1 equal to 1.75% of the sample weight is noted in the system under study, as shown in Figure 2 and Table 2. Many powder samples contain surface water, which is removed to the atmosphere within the specified temperature range. In this case, the main part of the evaporated molecular water can be attributed to the dehydration of particles of powder slag adsorbed by H_2O. After the sample dehydration process, the

https://doi.org/10.1051/e3sconf/202126204003

enthalpy of the system, in the range of 200-280 °C (473-553 K), practically does not change, which is caused by the absence of weight loss. It should be noted that in the range of 280-930 °C (553-1203K), the thermogravimetric curve (TG) is steadily shifting upwards - towards an increase in the sample mass, which is caused by the introduction of atmospheric oxygen into the system and the formation of oxides of a new transformation. The increase in mass is accompanied by a rise in the DTA-curve line, which, within 640-800 °C (913-1073 K), recorded clearly visible exothermic peaks at 700 (973 K) and 740 °C (1013 K). The processes that caused the introduction of heat into the system are associated with the oxidation of the ferrous components of the sample. Oxygen-enriched iron oxides are usually formed within the indicated temperature ranges. These high temperature peaks are caused by the transition of iron oxide from a lower acidity level to a higher level. At a higher temperature of 930 °C (1203 K), the increase in sample mass reaches its limit ($\Delta m2 = -6\%$). Further heating of the sample to 1000 °C (1273 K) leads to a decrease in its weight by 0.25%.

Thus, based on the results of differential thermal analysis, it is possible to select the main optimal temperature-time modes of the process of roasting and extraction of lead and zinc, as well as the technological design of the synthesis of non-ferrous metals from technogenic waste of lead production.

2.2 Selection of technology

As stated above, the main waste products of JSCYuzhpolimetall are calcium arsenate and granulated slag. Calcium arsenate is a lead production by-product. It is a mixture of basic calcium salts of orthoarsenic acid. Calcium arsenate is removed and stored in a specialized secure landfill, since it is the hazard class II products. Granulated dump slag is the end product of lead production and after granulation with granulating water is fed into sump, and then sent to the dump by motor transport. Dust, which is also a waste, is generated in the production of sinter during the electric cleaning of sulfur dioxide sintering machines in dry electrostatic precipitators of the dust and gas collection shop. Lead dusts are fire and explosive and are classified as hazard class 1 substances.

Based on the chemical composition of slag waste, a technology for selective extraction of lead and zinc compounds was selected. Scientists have been working on the problem of selecting and developing a technology for extracting non-ferrous metals from lead-zinc production for more than one year. This problem is due to the fact that since the rich deposits of non-ferrous metals were exhausted, the state's economic development required more and more consumption of non-ferrous metals. In the process of research and industrial testing, we have modernized the traditional technology for the selective extraction of lead and zinc compounds, that is, the well-known [16-25] technological processes for processing lead-containing slag by the Waelz method and refining using the Waelz kiln were reconstructed: the slag was fed from two ends (cold and hot) Waelz kiln. For the recovery of lead and zinc from lead slag, a Waelz kiln sublimation process has been developed. In the process of blowing a mixture of slag and coke, it is bubbling, the resulting gases, together with metal vapors, are sent to the dust collection stage, where sublimates are captured in a dust collection chamber, cooler and sleeve filters. Purified gases after dust and gas cleaning systems are released into the atmosphere.

To obtain lead oxide and zinc, slag and coal and limestone are fed into the component supply zone. At the same time, high-pressure compressed air is fed into the Waelz kiln. In the course of chemical reactions, lead-zinc slag sublimes to the formation of lead oxide and zinc by blowing in a certain amount of a mixture of pulverized slag and carbon-containing material from the hot end of the rotating Waelz kiln. The carbon-containing material plays the role of a reducing agent and volatilizes as CO_2 during oxidation. And in the form of residues, technogenic mineral formations are formed with the content of silicon, calcium and iron.

After the release of gaseous lead and zinc, the gas flow is directed to the cooling pipes, where the main reaction takes place:

$$PbO+CO=Pb+CO_2 (1)$$

$$ZnO+CO=Zn+CO_2 (2)$$

Pursuant to Figure 3, the technological process of the toxic metal extraction method consists in the following: hot slag (1), class less than 10 mm, by means of a belt conveyor (2) is fed simultaneously with coal (4) of class less than 10 mm, which by means of a belt conveyor (4) is fed in certain ratios to the mixing process (5) of hot slag with coal. The mixed mixture is sent to the Waelz kiln (7), where compressed air containing a mixture of pulverized raw materials from slag dust and carbonaceous material of class less 1 mm (6) is blown. In the Waelz kiln, the process of sublimation of lead oxide and zinc takes place under the influence of high temperature. Carbon of carbon-containing material, fines of coal, metallurgical coke, etc., used in the technological process as a reducing agent leads to the formation of lead, zinc and carbon dioxide. Lead and zinc are then oxidized pursuant to the reaction PPb+ O_2=PbO, Zn+ O_2=ZnO and re-enters the dust collection chamber.

Then the exhaust gases are fed to the cooling process, where zinc oxide and lead dust is cooled in the cooling chimney (9-10). The smoke temperature after cooling the system is reduced to 150 °C. The smoke, passing through the sleeve filter, completely traps zinc oxide and lead dust. The dust cleaning system efficiency is 99.8%. Lead and zinc oxide collected in a sleeve filter are fed into big bags (11-12). Furnace gas containing such harmful gases as sulfur dioxide is subjected to a double alkalization process pursuant to the reaction:

$$NaHCO_3+SO_2 \rightarrow NaHSO_3+CO_2 \quad (3)$$

$$CaO+NaHSO_3+O_2+CO_2 \rightarrow CaSO_4 \cdot 2H_2O+ NaHCO_3 \quad (4)$$

The sulfur dioxide absorption efficiency is about 95%.

For a more complete and efficient recovery of lead and zinc from man-made slags of lead production, sublimation processes have been developed in traditional rotary drum kilns (Waelz kilns) by blowing in the dust of a class less than 0.1 mm and the process of bubbling a mixture of slag and coke with a fraction of 0-5 mm. Together with metal vapors, the gases that are formed during this process are sent to dust collection. Sublimates are captured in a dust collector, a cooling chamber and sleeve filters, and the treated gases are discharged into the atmosphere (13), as shown in Figure 3.

Fig. 3. Technological scheme of the toxic metal extraction method

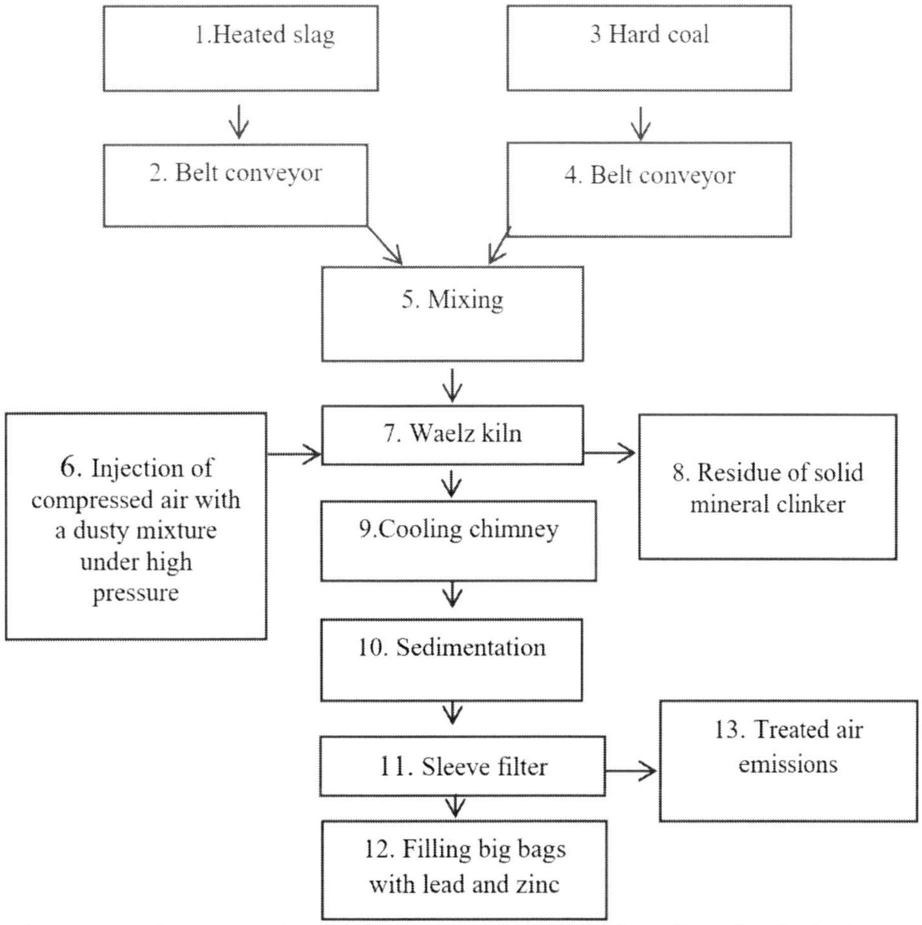

Pursuant to the proposed technology, a system of injection of a pulverized mixture of lead slag and coke or pulverized coal of less than 1 mm class from the hot end of the kiln is used, and the bulk of the initial charge components of less than 10 mm class is fed from the cold feed end. This considers the peculiarities of the kiln operation and the synthesis of non-ferrous metals, enhances the recovery rate and production efficiency of lead and zinc oxides. The proposed technology is distinguished by the use of lead production slag fed into a rotary drum kiln in two oppositely directed flows. The dispersion of the composition is less than 10 mm, containing slag, coke fines and internal overburden of coal mining from the cold end of the rotary drum kiln and the pulverized part with a dispersion of up to 1 mm together with a carbon-containing material with a dispersion of up to 1 mm from the hot end of the rotary kiln allow to intensify the synthesis of zinc and lead oxides and reduce energy consumption by up to 10-15%.The proposed technology utilizes the lead production slag, which is a sintered granular form material with a size ranging within 2-6 mm, containing a small amount of slag up to 10 mm in size, with a bulk density of about 2 $t/м^3$ and an angle of repose of 35 degrees.In addition to lead slag, the technological line utilizes the carbon-containing material, the chemical composition of which is shown in Table 3.In order to increase the environmental and economic efficiency of the technology, we propose feeding a mixture of pulverized slag material into the Waelz kiln from the hot end and pulverized carbon-containing materials from the hot end of the kiln, and the bulk of the initial charge components from the cold feed end.

Table 3. Carbon-containing material chemical composition

	Component content (%,%)					
Material name	C free	SiO_2	Al_2O_3	Fe_2O_3	$S_{общ.}$	Moisture
Coal mining screening	35-65	27-32	2.5-6.5	2.6-2.9	up to 0,01	up to 0,01
Coke fines	85	2-4	2-3	1-2	up to 0,01	up to 0,01

The lead oxide, zinc oxide and copper concentrate are produced by feeding slag from lead production and coke into the Waelz kiln, while high pressure compressed air is fed into the Waelz kiln. This process gives a reaction: lead-zinc slag sublimes to form lead oxide and zinc, coal is burned and volatilized as CO_2. The remainder contains a technogenic mineral formation containing silicon, calcium and iron. After the metallic lead, zinc and copper volatilization, the gas flow is directed to cooling and sedimentation of the separated lead, zinc and copper oxide dust.

To reduce the level of impact on the environment, before being discharged into the atmosphere through a high-rise stack, the gases removed from the Waelz kiln go through a successive stage of dry and wet cleaning.

Construction of a processing lead slag plant will provide for:

- the mitigation of the impact of the dumps (waste heaps) (the historical formations) on the environment;
- the exclusion of the irrational use of lands occupied by slag dumps;
- the partial satisfaction of the market demand for raw materials for the production of non-ferrous metals: lead, zinc and copper

The proposed safe technology for the disposal of lead-containing slag waste by extracting toxic metals eliminates local pollution and the potential impact on the safety of historical lead slag, as well as eliminates the long-term occupation of large amounts of land with this slag and ensures the implementation of a circular green economy.

3 Conclusion

1. The results of thermal, X-ray fluorescence semi-quantitative analyzes, scanning electron microscopy and X-ray microanalysis of lead slag showed that the lead production slag contains toxic metals that are dangerous sources of environmental pollution;
2. It has been determined that the average content of toxic metals in slags formed during the raw material processing is: lead - 2. 38; zinc – 9.81; copper – 0.97; iron – 25.31; silicon oxide – 24.62; calcium oxide – 16.21; potassium oxide – 1.42; sulfurs – 1.35% of the total weight of test;
3. The results of chemical analysis showed that the composition of lead production slag contains a sufficiently high amount of non-ferrous metal compounds: the qualitative composition and content of non-ferrous metals in lead slags allows to make the process of utilization of toxic waste from lead production technically and economically feasible;
4. A new technology for a more complete and selective extraction of lead and zinc oxides from slag waste of lead production has been developed;
5. Application of the proposed technology allows to reduce the fuel and energy costs while reducing the amount of dust and gas emissions by 10-15%;
6. The use of a new technology for the selective method of extracting non-ferrous metals will improve the ecological state of the environment and reduce the negative impact on human health through the disposal of toxic slags from lead production;

7. The developed technology allows to efficiently process waste man-made slags of lead production with the production of lead and zinc oxides, as well as for the production of mineral fertilizers containing microelements and as building materials. Herewith, a significant contribution is made to the development of a system for the rational use of natural and secondary resources.

References

1. Whowill rid Shymkent of 2 million tons of lead sludge?,https://otyrar.kz/
2. Methodological guidelines for calculating emissions (emissions) of pollutants into the atmosphere during mechanical processing of metals (in terms of specific emissions) RND 211.2.02.06-2004. Approved by order of the Minister of Environmental Protection dated 20.12.2004. No. 328-r.
3. S.Yu.Yusfin, L.I.Leontyev, P.I.Chernousov, Industry and environment, 469 (2009)
4. A.V. Tarasov, A.D. Besser, V.I. Maltsev, V.S. Sorokina, Metallurgical processing of secondary lead material, 224 (2003)
5. Technological regulations "Construction of a metallurgical plant for the production of zinc oxide, lead oxide and copper concentrate at the address Ordabasy district, Badamsky rural district, 029 quarter, section 1563, Turkestan region, cadaster 19-293-023-1563", license No. 17010101 dated 05.06.2017
6. Hygienic standards for atmospheric air in urban and rural areas, approved by order of the Minister of National Economy of the Republic of Kazakhstan No. 168 dated February 28, 2015
7. Sanitary Rules "Sanitary and Epidemiological Requirements for Establishing a Sanitary Protection Zone of Production Facilities", approved by order of the Minister of National Economy of the Republic of Kazakhstan dated March 20, 2015 No. 237
8. GOST 17.2.3.02-78 Nature protection. Atmosphere. Regulations for establishing permissible emissions of noxious pollutants from industrial enterprises.
9. GOST 17.2.301-86 Nature protection.
10. GOST 17.2.1.01-76 Nature protection. Atmosphere. Classification of emissions by composition.
11. Lead waste water treatment, https://ru-ecology.info/
12. N.F.Izmerov, Lead and health. Hygienic and medical-and-biological monitoring, 256 (2000)
13. Ye.O. Muradova. Life safety (2006)
14. M. Duisebaev, Zh. Rakhmanberdieva, B. Amiraliev,Poisk, **3 (1)**, 37 (2010)
15. M.K.Dyusebaev, R.K.Niyazbekova, D.M.Zhantasova, Zh.N.Rakhmanberdieva, M.E.Tolegen, Bulletin of the Almaty University of Energy and Communications,**3/2 (10)**, 62 (2010)
16. V.M. Garin, Industrial ecology (2005)
17. Sh.M.Moldabekov, K.T.Zhantasov, V.K.Bishimbaev, K.D.Aibalaeva, M.K.Zhantasov, Zh.M.Altybaev, Innovative patent No. 79381 - A method for producing complex-mixed mineral fertilizers. Committee on Intellectual Property Rights of the Ministry of Justice of the Republic of Kazakhstan. Publ. on 07.11.2011 Bulletin No. 11

18. K.T.Zhantasov, K.D.Aibalaeva, L.Kh.Frangulidi, M.R.Barlybaev, D.S.Berzhanov, B.N.Yurchenko, M.K.Zhantasov, Technological equipment for the production of yellow phosphorus, 444 (2014)
19. K.T.Zhantasov, Kh.T.Zhanov, G.A.Murzagaraev, N.A.Miroshnikov, Development of the technology for pelletizing phosphorite fines in a rotary kiln. Thesis report (1979)
20. K.T.Zhantasov, N.A.Mironov, V.A.Ershov, A.M.Kunaev, O.T.Tyumbaev et al. Industrial tests of the pelletizing of phosphate-siliceous fines in a rotary kiln (1981)
21. N.A. Miroshnikov, E.U. Zhumartbaev, A.M. Kunaev, K.T. Zhantasov, O.T. Tyumbaev Chemical Raw Materials Research Committee, 1, 48 (1993)
22. K.T.Zhantasov, Development and implementation of low-waste and energy-saving technologies in the production of yellow phosphorus: Abstract of a thesis for a doctorate in technical sciences: 05.17.01 - Technology of inorganic substances.-Shymkent,45 (1998)
23. K.T.Zhantasov, Series of chemical technology and environmental engineering, 92 (1996)
24. M.Mäkelä, I.Välimäki, R.Pöykiö, H.Nurmesniemi, O.Dahl, Int. J. Environ. Sci. Technol. (2013)
25. Z.Youcai, Z.Chenglong, Pollution Control and Resource Reuse for Alkaline Hydrometallurgy of Amphoteric Metal Hazardous Wastes, **18**

About the asymmetry structure of the leaf blade Common plantain

Sergey G. Baranov[1,*], *Igor E.* Zykov[2], *Tatyana S.* Biryukova[2], *Lubov* V. Fedorova[3], and *Alla M.* Antsyshkina[3]

[1]Vladimir State University named after A.G. and N.G. Stoletov, Gor'kogo, 79, 600005 Russian Federation
[2]State Humanitarian Technological University, 22, Zelenaya Street, 142611 Orekhovo-Zuevo, Russian Federation
[3]I.M. Sechenov First Moscow State Medical University, 8-2 Trubetskaya str., 119991 Moscow, Russian Federation

Abstract. Using the method of geometric morphometrics the contour of the image of the leaf blade of the Common plantain (*Plantago major*) was aligned along the axis of symmetry to answer the question of whether the asymmetry and shape were influenced by environmental pollution by vehicles and climatic conditions. Procrustes distances showed that fluctuating asymmetry was higher in roadside populations in 2019. In the control populations, a mixture of two types of asymmetry, fluctuating and directional, was obtained. In 2020 with high precipitation (47% more than in 2019), asymmetry was represented by higher directional asymmetry (p <0.0001), although the overall asymmetry remained the same. The nonparametric Kruskal-Wallis test showed the influence on the shape only climatic environments of the year (p <0.001). The geographical location of the populations and the combined effect of the factors year and the place of leaves gathering did not affect the shape of the leaf blade. In 2020, no data were obtained on the excess of asymmetry in roadside populations compared to the control, therefore authors conclude about the weak bioindicative properties of the Common plantain in response to traffic pollution.

1 Introduction

The high morphological plasticity of plants is a serious cause interferes bioindication on fluctuating asymmetry index. Developmental stability, which is asessed on the basis of undirected, fluctuational deviation from bilateral symmetry, is considered as a genetic and phenotypic property [1,2,3]. Some researchers believe that fluctuating asymmetry (FA) is associated with plastic variability and adaptation, that is, FA is included in the adaptation arsenal, but does not serve as a measure of stress. Fluctuating asymmetry refers to a special type of phenotypic variation that depends on genetic properties.To determine the effect: dose - stress - FA, it is important to take into account the stress load gradient and the choice

* Corresponding author: bar.serg58@gmail.com

of suitable control, as well as careful measurement and checking various types of measurement errors [4]. The studies in which the value of FA is associated with stress exposure from the environment are carried out with various plant species, but they do not always show unequivocal positive results. A large number of publications on the indicative properties of silver birch arouse unremitting interest in the field of fluctuating asymmetry. For example, the influence of the elevation of the relief and the geographical position of populations with climatic features on FA is reported [5-10]. The obstacles are the different genetic status of the studied populations, i.e. different trajectories canalized ontogenetic path and unexplored effects hormesis and paradoxical effect with inadequate reaction to toxins [11]. The development of the methods of geometric morphometrics makes it possible to determine the difference in the coordinates of the left and right halves of the leaf blade. For this, the configuration of the leaf blade samples is aligned along the symmetry axis, and the FA value is determined in two way analysis of variance (Procrustes ANOVA). The proposed work took into account the advantage of this analysis for study the fluctuating asymmetry and shape of the leaf blades Common plantain (Plantago major), a common ruderal plant that can serve as an indicator of environmental pollution. Studies with the measurement of dimensional characteristics have shown that the plantain leaves often possess directional asymmetry [12-13], and the high plasticity of the leaf shape depended on the salinity of the soil and the geographic characteristics of the population area [14]. Exceeding a certain threshold of the FA value indicates a deviation in developmental stability. The relationship between plastic variability and developmental instability has not been sufficiently studied and probably based on the principle of feedback [15-16]. Therefore, a comparison of the shape and asymmetry, using the coordinates of the labels applied to the image of the leaf blades, can help in answering the question the relationship between two types of variability. In the proposed paper, we studied the shape and asymmetry of the leaf plate depending on the pollution of the air along the highways for 2 years with different volume of rainfall in three sites separated by distance 100 km along the same geographical latitude.

2 Materials and Methods

2.1 Sites and sampling

Three geographical points were selected, located on the same latitude with a difference of c. one degree and on a distance about 100 km from each other. One in Vladimirskaya (56°15′81.74″N ; 40°46′58.51″E) and two in the Moscow region: Orekhovo-Zuevo (55°47'48.4"N ; 39°00'46.8"E) and the Izmailovo region in Moscow (55°46'46.9"N ; 37°44'01.7"E).

At each sampling site, the experimental one covered square 50-100m × 1km along the road. The control zone with the same square lied not less than 400 m from the experimental zone. The sampling areas had fairly uniform physicochemical properties, which could indicate a close trajectory of ontogenetic development. Both zones did not differ in illumination and represented an open area occupied by common ruderal urban vegetation (quinoa, chicory, bluegrass, wheatgrass); the projective cover of plantain was 30-50%. The leaf blades with a half leaf width of 6.0 ± 0.1 cm, two to three from each of 50 plant, were harvested in August-September 2019-2020 and photographed twice. From each site, 100-150 leaf blades were sampled, i.e. a total sum was about 2,000 leaf blades. The calculation of vehicle emissions was carried out according to the standart methodology by the number of vehicles passing along the highways per unit of time. The total emission, taking into account all pollutants on the highways, was 0.002g/sec (Orechovo and Moscow) and 0.003 g/sec (Vladimir, standart error ±0.0001).

3 Measuring and statistics

We considered both the plant and the leaf blade as a conventional experimental unit. The first two labels applied were not paired and formed an axis of symmetry; the other 25 pair represented the homologous paired traits. Labeling was carried out twice on each image in the JPEG format. The software used (TPS and MorphoJ), as well as recommendations for their use, are freely available on the website of the morphogeometric group https://sbmorphometrics.org. The essence of Procrustes generalized analysis of variance (GPA) is in the constructing a consensus shape, removing the difference in the size of leaf blade samples and in evaluating the difference in the variance of paired labels XY coordinates [17-19]. First, the coordinates of the consensus centroid were determined. Then the difference in shape (classifier 'individual'), the value of FA ('individual × side'), and directional asymmetry, DA ('side') was statistically determined on the mean F- Goodall ratio, correspondingly $F_{I\times S}$ and F_S.

On the basis of vector coordinates, the covariance matrices of two types were created – symmetric matrices and asymmetry matrices. The canonical variate analysis (CVA) was carried out on the data XY coordinates. The symmetry matrix showed the difference in coordinates in the samples among the same landmarks and was used to test the shape of the leaf blade. The asymmetry matrix showed the difference in homologous landmarks between left and right sides and revealed differences in bilateral asymmetry. The difference assumed the distances between the centers of the sets of coordinates, the so-called Procrustes distances. An alternative method was the Kruskal-Wallis analysis. The statistical evaluation used the significance level $\alpha = 95\%$. Permuting multiplication of samples was carried out in 10,000 rounds. The measurement error was calculated in the percentage of the MS value of fluctuating asymmetry.

4 Results and Discussion

4.1 Bilaterall asymmetry

Procrustes analysis of the total amount of leaves for 2 years showed a high content of fluctuating asymmetry in the whole set of leaves, in control, and in leaves sampled near the road (Table 1).

Table 1. Result of GPA for: a) total, b) control and: c) experiment.

Effect	*SS*	*MS*	*df*	*F*
a) Individual (all leaves)	2.55	0.02	150	21.3^{***}
Side	0.03	0.00	50	0.68^{ns}
Ind x side	0.12	0.00	150	23.53^{***}
Residual	19.04	0.00	561200	
b) Individual (control)	1.24	0.02	50	13.58^{***}
Side	0.01	0.00	50	0.1^{ns}
Ind × side	0.09	0.00	50	52.05^{***}
Residual	10.24	0.00	293000	
c) Individual (road)	1.24	0.02	50	68.29^{***}
Side	0.03	0.00	50	1.57^{ns}
Ind × side	0.02	0.00	50	11.09^{***}
Residual	8.81	0.00	268200	

Notes: *** – $p < 0.0001$; ns – $p > 0.05$– of square; *df* – degree of fredom; *MS* – mean squre; *F* – criterion of Goodall

In all three cases, the consensus size did not differ (p >0.05). Classifier "individual" showed a difference in shape (p <0.0001). The control and experimental leaves showed the same size of consensus p-value and significant fluctuating asymmetry without directional asymmetry (the "side" factor is not statistically significant). The value of F-Goodall ratio was higher in control, than in experiment (correspondingly: 52,05 и 11,09). The analysis of populations carried out separately in 2019 and in 2020 showed the next results: in 2019, FA value slightly prevailed in the roadside population, in all cases there was the presence of directional asymmetry (Table 2).

Table 2. Two kids of asymmetry in 2019-2020.

Road					**Control**				
2019									
Effect	*SS*	*MS*	*df*	*F*	*Effect*	*SS*	*MS*	*df*	*F*
Side	0.0001	0.0001	50	1.99**	Side	0.04	0.001	50	8.93***
Ind × side	0.0060	0.0001	100	2.0***	Ind × side	0.01	0.001	100	2.49***
Residual	4.3676	0.0001	144500		Residual	5.96	0.001	170100	
	2020								
Side	0.041	0.001	50	7.49**	Side	0.116	0.001	100	16.09**
Ind × side	0.011	0.000	100	3.28**	Ind × side	0.062	0.001	50	17.31**
Residual	4.116	0.000	123300		Residual	3.920	0.000	122500	

Notes: *** – $p < 0.0001$; ** – $p < 0.001$

In 2020, compared to 2019, the content of directional asymmetry was 3.8 times higher near the road, and 1.8 times higher in the control (see values $F_S = 7.49$ and $F_S = 17.31$; p <0.0001). The error measure (residuals) was from 0.1% (control) to 9.01% (road). Accordingly, the FA value in plants near the road was lower.

Procrustes distances in 2020 between the set of coordinates of control leaves and leaves near the road did not differ (Procrustes distance 0.003; p > 0.05). In 2019, the difference was significant (Procrustes distance 0.005; p <0.0001). Thus, the rainy growing season in 2020 reduced the differences in asymmetry. The asymmetry differed in the control-experiment pair in the summer of 2019, with average precipitation value. The comparison of the two years of follow-up showed that there were no differences in overall asymmetry. This was demonstrated by the diagram of the canonical analysis of variance for the first component CV1 (Fig. 1A). The difference in asymmetry between years was higher (0.008; p <0.0001) than the difference between the total control and all leaves by the road (0.003; p = 0,002 Fig. 1B).

Fig. 1. Difference in asymmetry component of shape variation (CV1). A: red color – 2020; cyan color – 2019; Procrustes distances among groups: 0.008 (p <0.0001). B: control (red) and road (cyan); Procrustes distances among groups: 0.003 (p = 0.002).

GPA for total set of leaves in 2019 revealed the FA free of DA ($F_{I \times S}$ = 10.4; p <0.0001; F_S was not significant; p >0.05). In 2020 there were both type of asymmetry: ($F_{I \times S}$ = 3.27; p <0.0001; F_S = 18.04; p <0.0001). Thus, in a wet year, an increased directional asymmetry was observed, in contrast to a temperate climatic year, where fluctuating asymmetry occupied big dole of the asymmetry. The larger samples neutralized the presence of directional asymmetry. This was true for both years of study. The sample size with n = 50, even with a fourfold increase in replicas, was clearly insufficient to obtain a representative result. So the plantain leaf blades had a high fluctuating asymmetry, tested only at the maximum sample sizes, which should be taken into account when evaluating the FA and developmental stability.

4.2 Shape of leaf blade

Non-parametric Kruskal-Wallis analysis showed the impact of the only factor 'year' on the shape of the leaf blades (p <0.001), a sampling site had no effect on the shape. The differences in Procrustes distances between the centers of the control and experimental sets in 2020 were less high than within the same samples, which indicated accented variability in the morphology of leaf blades in a rainy summer. The symmetric component of shape variation varied pronouncedly from year to year in CVA (Fig. 2).

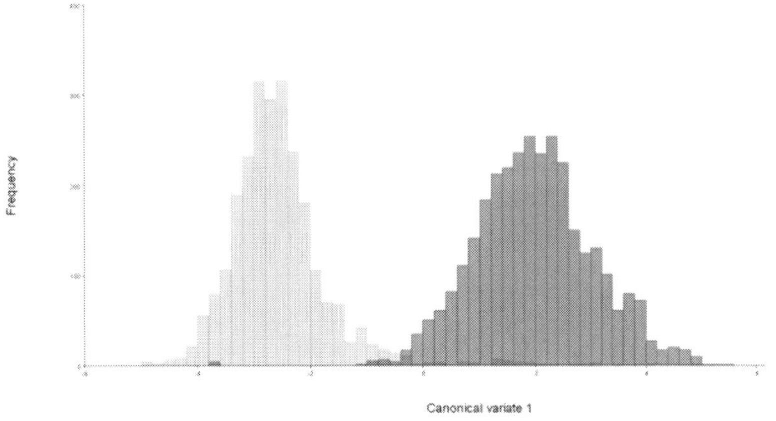

Fig. 2. Symmetric component of shape variation CV1: left –2020, right – 2019. Procrustes distances among groups: 0.042; p <0.0001.

In the summer of 2020 the shape of the blade looked oval, and in 2019 the shape was close to the ovoid form with a narrowed proximal. Thus, if in 2019 plantain acted as a reliable bioindicator (75% of positive results), then in 2020 the results did not allow us to classify this plant as an indicator of developmental stability. This is confirmed by the results obtained on metrical bilateraly symmetrucal traits [13]. Note that bioindication as a property is not so often encountered in recent works, especially concerning the indication of developmental stability by the FA index. The transit of asymmetry from FA to DA, or vice versa, deserves attention as the interaction of genom and phenotype. The opinion about directional asymmetry as a bioindication property for assessing developmental stability remains controversial, although it is known that DA is emerged under stress environment, implying a genotype effect on plant morphology [20]. The high morphological plasticity of this species is the serious reason it could not be considered a reliable indicator with 90-100% positive results.

A weak correlation was obtained between the two components of the shape (r = 0.13-0.14; p <0.001), i.e. shape has been associated with asymmetric variability. Directional asymmetry, as latent, expressed within high degree of freedom df only at small sample sizes and could depend on the frequency distribution of the sample. For example, the control leaf pool showed directional asymmetry at all level of classifier (individual, leaf and image). Testing of experimental leaves showed statistically significant FA ($F_{I\times S}$ = 5.05; p <0.0001) with statistically insignificant directional asymmetry. Subsequent testing at the image level revealed directional asymmetry (F_S = 2.4; p <0.0001). Such a "hidden" DA was revealed when testing the asymmetry of woody plants [21-22]. The further research will be reffered to the study of the heterogeneity of the samples distribution. The multivariate factor analysis of shape and asymmetry over several years of research as data accumulates is also important.

5 Conclusions

a) Plantain leaf blades had a pronounced fluctuating asymmetry, which was determined with a significant volume of leaves (more then 150 leaf blades).
b) Large plantain, like other herbaceous perennial plants (for example, strawberry species), has dubious bioindication properties on the fluctuating asymmetry index, because climatic factors, for example, the amount of precipitation, have a stronger effect than man-made pollutants.
c) During 2 years of observation, the structure of asymmetric variability, available at high sample sizes, changed from FA to directional asymmetry under high humidity environment. The shape of the leaf plate of Common plantain varied significantly correlating to the total asymmetry which contained two kinds varying components – FA and DA.

References

1. S. Baranov, I. Vinokurov, L. Fedorova, Gene Expression and Phenotypic Traits 2019 Dec **13** (2019)
2. S.V. Dongen, J. Evol. Biol., **19(6)** (2006)
3. J.H. Graham, J.M. Emlen, D.C. Freeman, Nonlinear dynamics and developmental instability (2003)
4. N.S. Fox, J.J. Veneracion, J.L. Blois, Ecol. Evol., **10.7** (2020)

5. E.V. Koroteeva, D.V. Veselkin, N.B. Kuyantseva, O.E. Chashchina, In Doklady Biological Sciences, **460(1)** (2015)
6. V. Zverev, A.D. Lama, M.V. Kozlov, Ecological indicators, **1(84)** (2018)
7. E. Shadrina, N. Turmukhametova, V. Soldatova, I. Korotchenko, and G. Pervyshina, Symmetry, **12(8)** (2020)
8. V.M. Zakharov, E.G. Shadrina, N.V. Turmukhametova, E.N. Ivantsova, E.A. Shikalova, V.Y. Soldatova, I.E. Trofimov, Biol. Bulletin, **47** (2020)
9. D. Miljković, M. Stefanović, S. Orlović, M.S. Neđić, L. Kesić, S. Stojnić. Alpine Botany, **129(2)** (2019)
10. J. Brunet and M. L. Van Etten, International Journal of Plant Sciences, **180**:9 (2019)
11. E.A. Erofeeva, B.N. Yakimov, Symmetry, **12(5)** (2020)
12. M. Veličković D. Tošković. Periodicum biologorum, **117(3)** (2015)
13. M.Velickovic, T. Savic, Plant species biology, **27(1)** (2012)
14. M.J. Smekens, P.H. van Tienderen, Acta Oecologica, **22(4)** (2001)
15. R. M. Doret, J.A. Draghi, and M.C. Whitlock, Evol. Lett., **4(6)** (2020)
16. C. P. Klingenberg, Frontiers in Ecology and Evolution **7: 56** (2019)
17. N.S. Fox, J.J. Veneracion, J.L. Blois, Ecology and evolution, **10(7)** (2020)
18. V.Viscosi, P. Fortini, D. E. Slice, A. Loy, C. Blasi, Plant Biosystems, **143(3)** (2009)
19. S.G. Baranov, I.E. Zykov, T.S. Biryukova, L.V. Fedorova. IOP Conference Series: Earth and Environmental Science, **666(4)** (2021)
20. M.V.Veličković, and D. Jakovčev-Todorović, Genetika, **38(2)** (2006)
21. S.G. Baranov, Emerging Science Journal, **2(4)** (2018)
22. S.G. Baranov, Computer Science On-line Conference (2018)

Biotechnological preparation of sewage and livestock manure from livestock farms for the needs of horticulture

Anatoly Bondarenko[1,*], *Liudmila* Kachanova[2], *Yuri* Kolosov[1], *Agnessa* Shakhmurzova[3], and *Anatoly* Bittirov[3]

[1]State Federal-Funded Educational Institution of Higher Professional Training "Don State Agrarian University", Krivoshlykova 24, rural settlement Persianovskiy, 346493 Rostov region, Russia
[2]State Federal-Funded Educational Institution of Higher Professional Training "Don State Technical University", Gagarin Square 1, 344000 Rostov-on-Don, Russia
[3]Kabardino-Balkarian State Agricultural University named after V.M. Kokov, Lenin Avenue, 1V, 360030 Nalchik, Russia

Abstract. Ensuring food security of the state provides for obtaining high yields of agricultural crops and the development of livestock facilities. The aim of the investigation is to develop an accelerated biotechnological preparation of sewage and livestock manure from livestock farms for irrigation of pastures and fertilization of fields for forage crops. Closed-loop systems for products output on pig farms and cattle farms using biotechnology for the preparation of livestock manure for irrigation of fields in the production of forage crops have been developed. The biotechnology of accelerated preparation of sewage and liquid livestock manure is based on the principles of their mechanical separation into fractions with the subsequent preparation of liquid concentrated organic fertilizers using nutraceuticals, which speed up the processes of their preparation in a much shorter time than using traditional technologies. A closed-loop system for products output on sheep and pig farms and cattle farms using biotechnology is proposed. The basis of biotechnology in the processing of solid livestock manure is the principle of accelerated components composting using nutraceuticals. The implementation of accelerated preparation system of sewage and livestock manure for irrigation of pastures and fertilization of fields for forage crops contributed to an increase in their yield up to 20.54%. At the same time, the milk yield per forage cow increased by 8.47-14.20%; average daily weight gain - by 13.30–47.73%.

1 Introduction

To ensure the food security of the state, a well-coordinated work of the agro-industrial complex and its basic industries: horticulture and animal husbandry is required. In 2019, a record harvest of grain crops was gathered in the Russian Federation in the amount of 119

[*] Corresponding author: bondanmih@rambler.ru

million tons, in 2020 - 122.0 million tons. In the Rostov region in 2020, 11 million tons of early grain crops and more than 1 million tons of corn for grain were harvested.

Obtaining high yields is accompanied by the removal of significant amounts of nutrients from the soil and, as a result, its depletion.

In recent years, the system of organic farming has been developing, which, along with obtaining ecologically clean products, is based on the principle of preserving the soil fertility of agricultural lands. More than 60 million tons of humus are mineralized in the arable soils of the Russian Federation annually, and about 27 million tons are replenished. To replenish it, it is required to use organic fertilizers, which have a positive effect on the balance of organic substance in the soil. Currently, there is a difficult situation with the provision of farmland with organic fertilizers.

Over the past 30 years, the output of livestock manure in the Rostov region has decreased from 32 million tons to 3.2-3.5 million tons, which is clearly not enough to restore soil resources. This problem can be solved by mobilizing all resources of organic raw materials, first of all, livestock manure sewage and livestock manure from animal enterprises.

For the effective management of animal husbandry, it is necessary to have its own forage base. For this purpose, restoration of pastures and fields for forage crops is required on the basis of the development and implementation of accelerated biotechnology for the preparation of sewage and livestock manure as organic fertilizers. In the Rostov region, pastures occupy 2,497 thousand hectares (24.8% of the total area). Hay meadow occupy 90.2 thousand hectares (0.9% of the total area) [1].

In this regard, the problem of the effective functioning of animal enterprises with a closed technological cycle on their own forage base is relevant, it is of scientific and practical interest.

The aim of the investigation is to develop an accelerated biotechnological preparation of sewage and livestock manure from livestock farms for irrigation of pastures and fertilization of fields for forage crops.

2 Materials and Methods

The analysis and synthesis of livestock manure preparation systems for use is based on a systematic approach to the study of the issues under consideration and is determined by the internal and external relations of the animal husbandry and horticulture sectors. System analysis is a process of use, at which the parameters of the system operation are determined from the known parameters of the system and the environment.

System synthesis is a process of its investigation, at which, according to the required indicators of the system operation and the existing restrictions on resources in known environmental conditions, it is necessary to determine the parameters of the system (promising biotechnologies for the preparation of sewage and livestock manure). Practical methods of system investigation are based on the analysis, synthesis and observance of their unity at all stages of research [2-4].

On the basis of the above mentioned, systems of a closed production cycle have been developed on pig farms, cattle farms and sheep farms. The particular importance for the steady operation of animal enterprises are the operations of feeding, removing livestock manure, its accumulating, storing and processing into high-quality organic fertilizers [5].

Depending on the technology of keeping animals on farms and complexes, they produce solid (bedding) manure (LM), semi-liquid manure (SLM) and liquid manure (LM). Fresh, all these types of livestock manure have a large amount of pathogenic microflora and pose a danger to flora and fauna in places of their accumulation and storage.

Livestock manure produced at animal and poultry enterprises in the legislative framework of the Russian Federation is defined as waste of IV and V hazard classes [6]. According to the Russian Ministry of Natural Resources livestock manure, dung and other organic substances and materials formed in animal husbandry as a result of keeping farm animals and which are waste products of the latter ones, after disinfection in accordance with sanitary rules, as well as organic fertilizers produced on their basis, can be sold as a product, complying with the following conditions:

– in the presence of statutory, design, technological and other documents, which make it possible to classify livestock manure, dung, other organic substances formed at enterprises as a result of economic activities, to products;

– in the presence of developed and approved technical conditions for fertilizers based on livestock manure, dung, other organic substances, which take into account state and industrial standards, sanitary and hygienic standards and regulations, environmental requirements, as well as in the presence of process procedure for livestock manure, dung, organic substances and materials;

– when reflecting in the economic, accounting records of the enterprise, operations with the specified livestock manure, dung, other organic substances and materials as operations with products.

Therefore, if the above requirements are met, livestock manure produced in animal enterprises can be considered a product. Implementation of the requirements for transferring waste (livestock manure) into a product (organic fertilizer) requires improvement of the technical means for recycling organic waste [7, 8].

Particular attention in the production of livestock products abroad is given to reducing the emission of ammonia and other gases into the atmosphere. In the Netherlands, pollution reductions are being tested on small pilot farms [9]. Similar experiments are being carried out in Denmark, where supplementary systems for the use of protein for cattle feed and their effects on ammonia production are further investigated [10]. Much attention has been paid to the research of multi-stage cleaners to reduce emissions of harmful substances on pig farms [11]. Research is under way in European countries on groups of microorganisms that reduce ammonia emissions at poultry farms [12]. The most effective direction is the recycling of organic livestock and poultry waste into organic fertilizers both independently and in conjunction with organic crop waste (straw, non-grain portion of other agricultural crops) [5].

Scientists in the 70s and 80s of the twentieth century actively developed technologies for processing all types of livestock manure into organic fertilizers and technical means for their implementation. In the case of large animal enterprises producing liquid manure and dung, mechanical separation technologies have been developed, followed by solid and liquid organic fertilizers. In the processing of semi-liquid manure, the preference was given to the technologies of compost production using various moisture absorbers (straw, peat, etc.). Bedding manure, as a rule, was processed by traditional technology: accumulation, drilling, periodic drilling, applying in the fields (Figure 1) [13].

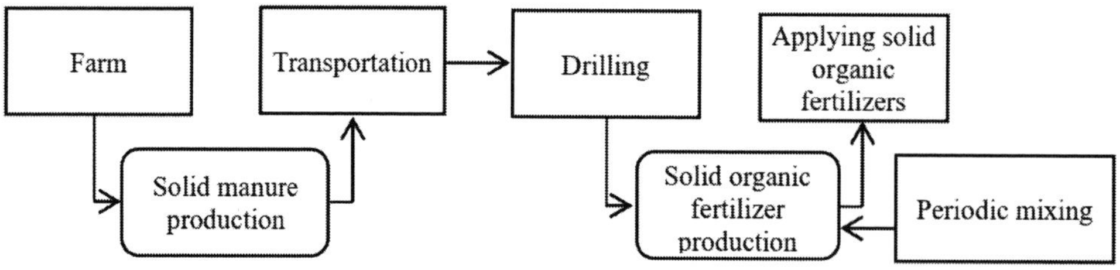

Fig. 1. Solid manure processing scheme

Recycling semi-liquid manure (W = 86-92%) provides the operation of its transporting, accumulation, compost production using a moisture absorber - straw, or applying in the field (after soaking in the storage sediments for up to 12 months) as a semi-liquid organic fertilizer (SLOF) (Fig. 2).

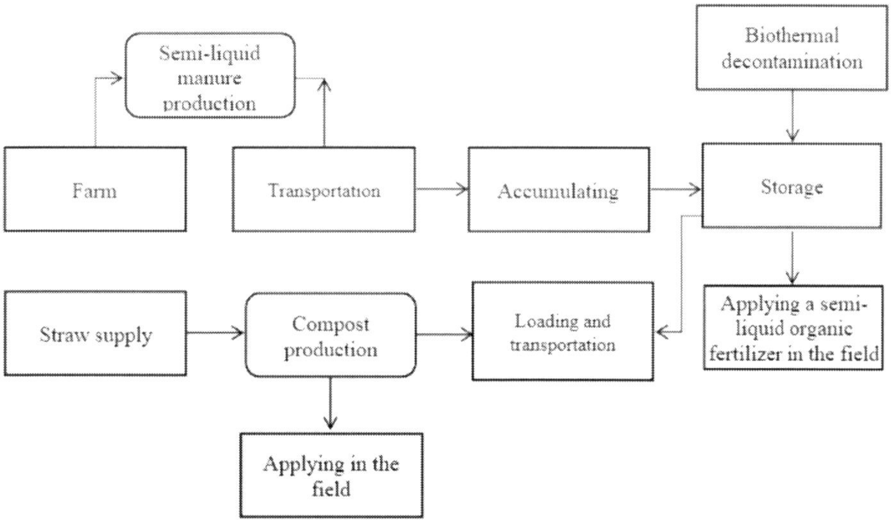

Fig. 2. Solid manure processing technology scheme

Liquid manure produced at animal enterprises is processed under two options.

Option 1. Production of liquid organic fertilizers (LOF) by soaking up to 12 months in storage ponds.

Option 2. Mechanical division of liquid manure into solid (TF) and liquid (FF) fractions with separate processing of them into solid (SOF) and liquid (LOF) organic fertilizers (Figure 3).

Under option 1, liquid manure accumulates in the farm sediment and is pumped by pipelines into two-section storage ponds. Natural decontamination occurs within 12 months, after which liquid organic fertilizer is applied in the fields by means of specialized machines for fertilizing feed and grain crops. The drawbacks of this technology are: a high risk of environmental pollution, so, dung storage facilities often do not have bottom and side wall insulating coating; significant costs of mixing and loading of the LOF; high doses of application - from 400 to 700 ^{3}/ha.

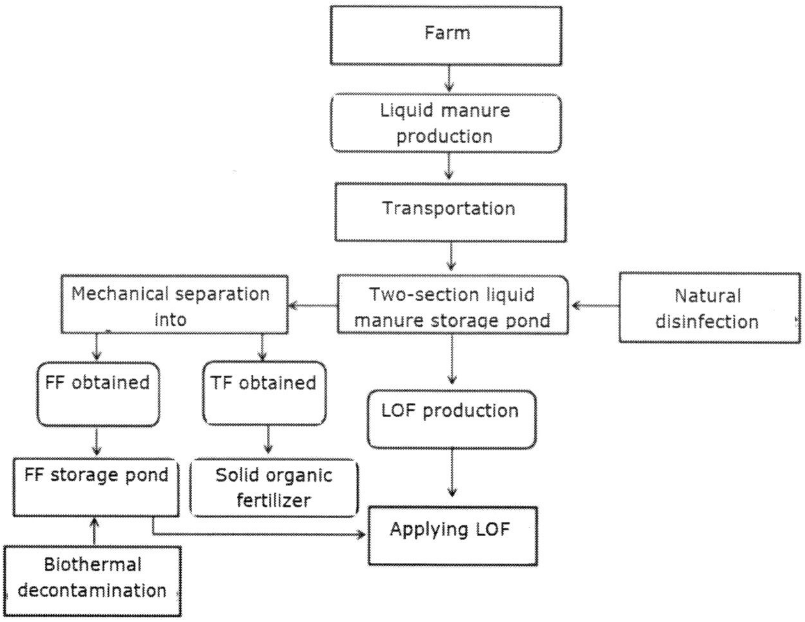

Fig. 3. Liquid manure processing technology scheme

Option 2 is the most acceptable, which involves separating liquid manure into solid and liquid fractions and separating them from integration into solid and liquid organic fertilizers.

The analysis of the liquid manure separation system for the fraction showed the following. Developed earlier systems of machines for processing liquid pig manure at large pig farms are not effective at small farms, because they have high productivity, relatively large dimensions, require additional labor costs, etc.

For small and medium-sized farms, it is appropriate to develop technical devices that are organically related to the main production or are a natural extension of it. In any case, it should be simple, compact, reliable in operation technical means, designed to handle small daily volumes of manure.

3 Results and Discussion

The analysis showed that existing systems for preparing sewage for irrigation of pastures and livestock manure for fertilizing fields did not satisfy agricultural producers for lack of effective technologies and technical means to implement them.

The essence of the closed cycle of the farm operation lies in the fact that the cattle population is provided with vital options - feed and watering, produced on its own base, which is ensured by waste-free production that guarantees environmental safety and economic stability of the enterprise.

Figure 4 shows that the amount of produced pig farm (PPF) depends on the cattle population. The main product is meat, with effectively technological processes for the steady production associated, first of all, with the creation of a microclimate, feeding, watering and cleaning livestock manure.

The microclimate systems of animals and their watering are sufficiently performed and are selected depending on the number of cattle animals and the volume of premises where they are located [8, 9, 14].

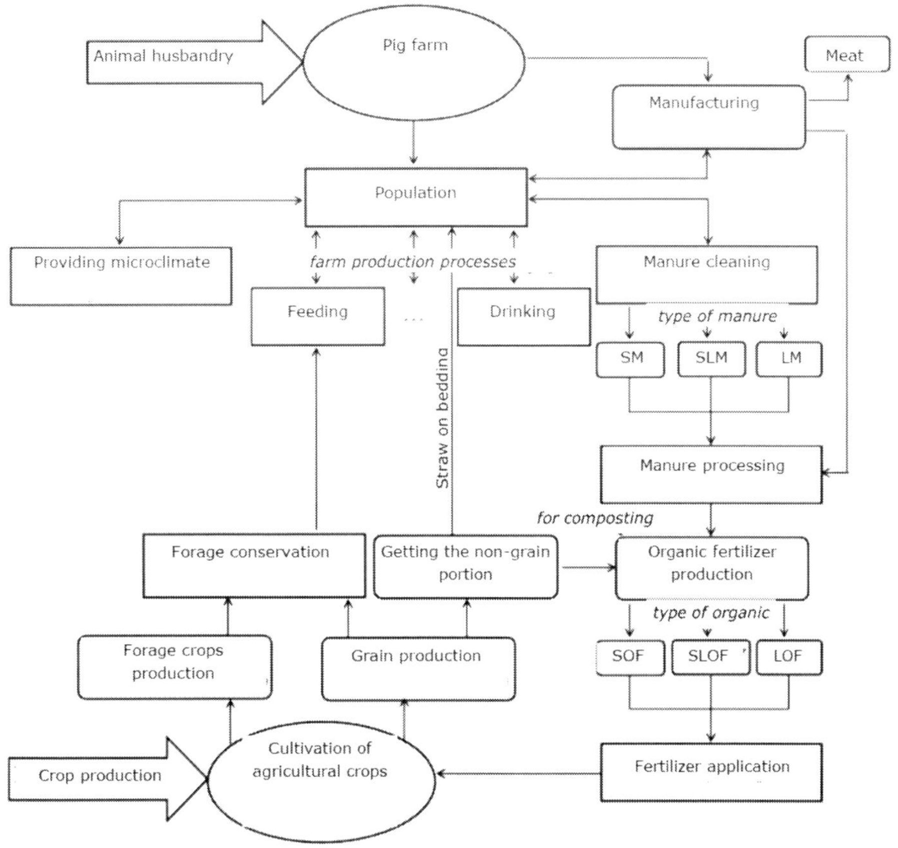

Fig. 4. Closed-loop systems for products output on produced pig farms using biotechnology for the preparation of livestock manure for irrigation of fields in the production of forage crops

An important factor influencing the profitability of livestock production is the availability of a feed base. The use of its own feed base in the closed cycle of livestock production allows not only to increase profitability, but also to improve the ecological situation at the places of livestock enterprise and to preserve the soil fertility of agricultural land.

Figures 5 and 6 show the closed cycle systems for the production of cattle farms and sheep farms. The basic principles of these systems are similar to the system represented in Figure 4.

In the closed cycle system of the animal enterprise, an important task is to develop accelerated biotechnology of sewage and livestock manure preparation as organic fertilizers for the cultivation and production of forage crops. This is especially relevant for the processing of liquid manure and dung sewage.

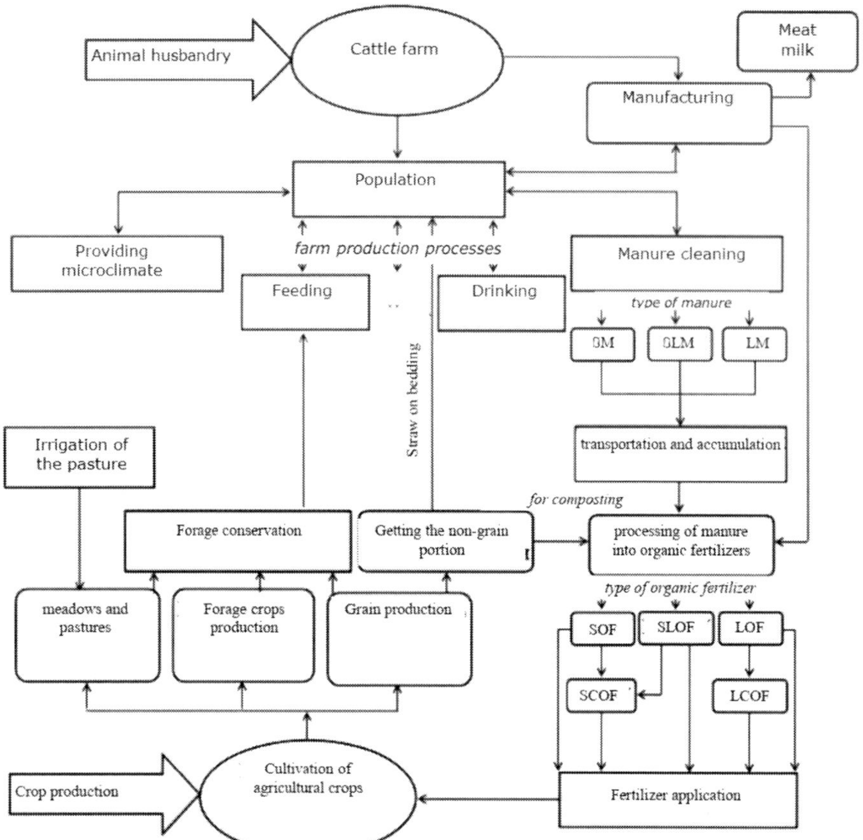

Fig. 5. A closed-cycle production system for cattle farms using sewage and livestock manure preparation biotechnology for irrigation and pasture fertilizing

Research on the development and implementation of agricultural waste management biotechnology is performed in the Russian Federation and abroad.

The solid, semi-liquid and liquid manure produced on the farms is to be recycled (Fig. 4, 5).

Technology of manure processing through dehydration and further sterilization, according to some experts, is very energy-intensive. Thermal treatment of liquid or solid fraction at high temperatures causes loss of food items for plants. In addition, the main requirements for technologies of processing livestock waste and obtaining organic fertilizers are the preservation of their biological activity and the maximum content of nitrogen, phosphorus and other elements.

One possible way to dispose of livestock waste is biological processing using micro and macroorganisms, allowing the rapid and efficient processing of a significant amount of livestock manure and dung.

A promising way of biologically disposing of livestock waste is to cultivate microorganisms. For the fermentation of livestock manure mainly used filamentous fungi (solid-phase cultivation), and bacteria, yeast and fungi are deeply cultivated on dung sewage. The cultivation of bacterial crops on waste from livestock complexes has not been widespread due to the limited use of bacteria for forage.

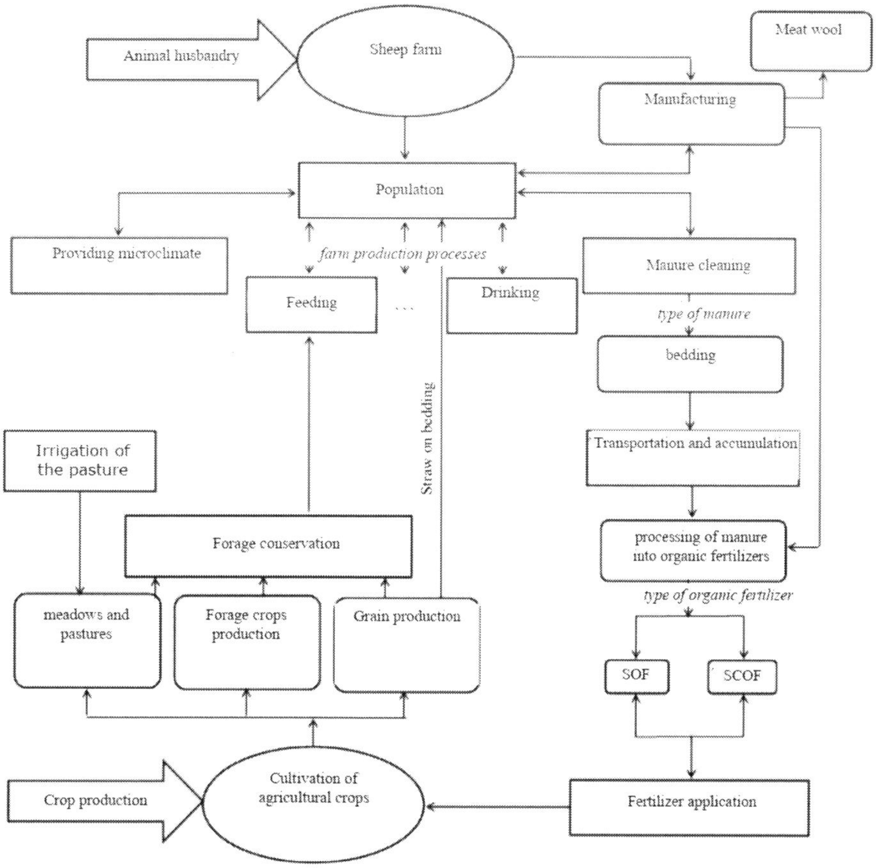

Fig. 6. A closed-cycle production system for produced sheep farms using sewage and livestock manure preparation biotechnology for irrigation and pasture fertilizing

There is a wide range of microorganisms capable of consuming secondary agricultural and processing products with microbial biomass. The most promising are fast-growing microorganisms that can absorb unhydrized agricultural waste. Filamentous fungi and yeast comply most with these requirements. It is known that concentrated sewage is essentially a ready breeding ground for these microorganisms, as they contain all the necessary components, including vitamins and trace elements [15-20]. In addition, the rational use of livestock sewage allows to obtain an additional yield, in monetary terms estimated equivalent to the use of 650,000 tons of nitrogen, 300,000 tons of phosphorus and 600,000 tons of potash fertilizers throughout the Russian Federation.

In general, the best way to dispose of waste for economic and environmental reasons is to use it as a secondary raw material. The above-mentioned methods of waste bioconversion can produce, sometimes unexpectedly, very significant results with beneficial implications for the development of biotechnology and energy. Unfortunately, most of them are implemented very poorly.

Scientists of the Don Agricultural University developed biotechnology of accelerated processing of different consistency organic waste using nutraceuticals. Adding up to 5% of nutraceuticals to the recyclable volume of organic waste (livestock manure) makes it possible to get in a short time (from 7 to 12 days) high-quality concentrated organic fertilizers: solid (SCOF), liquid (LCOF) (figure 5, 6), which are applied in the field with doses of up to 4 t/ha. This biotechnology allows for the quality preparation of sewage for irrigation of pastures and livestock manure for fertilizing fields both in the production of feed and for other crop production needs.

The system of high-quality irrigation of pastures and fertilization of fields for feed crops has been tested in a number of agricultural organizations in the Rostov region. The system is used in the production activities of the Agricultural Production Cooperative-collective farm named after S.G. Shaumian, Miasnikov district, Agricultural Production Cooperative (collective farm) "Rodina" of the Matveyev-Kurgan district, CJSC "Kolkhoz Sovietinsky" in Neklinovsky district, Rostov region.

For example, in the Agricultural Production Cooperative-collective farm named after S.G. Shaumian, Miasnikov district was one of the first in the region to test the high-quality system of pastures irrigation and fertilizing the fields for feed crops. The production activities analysis of the Agricultural Production Cooperative-collective farm named after S.G. Shaumian over a 5-year period revealed an increase in the main performance indicators of the horticulture and animal husbandry industries. Yields of forage crops increased by 180 c/ha, gross milk production increased by 1684 tons, yields per 1 forage cow increased by 771 kg, the average daily weight gain increased by 251 grams. In general, sales of products for the period under review increased by 358 million rubles, net profit increased by 107.8 million rubles. The profitability of the sold products reached 24% in 2019, which is 5 percentage points higher than in 2013. The enterprise's profitability in 2019 was 81%, which is 48 percentage points higher than in 2013.

4 Conclusions

Based on the production investigations and financial activities of enterprises they revealed: the introduction of a high-quality system of pastures irrigation and fertilizing fields for forage crops had a positive impact on the results of production activities. In the Agricultural Production Cooperative (collective farm) the yield of forage crops, yields and weights of animals increased. Receipts from sales of enterprise agricultural products is growing steadily. Net profit varies somewhat, which is due to the increase in production and sales costs. The net profit of organizations also tends to increase.

Enterprises implementing a system of pastures high-quality irrigation and fertilizing fields for forage crops use the latest scientifically sound technologies, which contributes to the efficiency of the process and labor discipline. It is worth noting that the use of pastures high-quality irrigation system and fertilizing fields for forage crops contributes to the social orientation of enterprises, as new jobs are being created and the environmental situation in the agricultural sites is improving.

References

1. V. N. Vasilenko, *Zonal farming systems of the Rostov region for 2013-2020* (2013)
2. E. N. Krylatykh, V. Z. Mazloyev, V. V. Rau, A. S. Strokov, *National Economy: ensuring Food Security in the face of Integration and Globalization* (2015)
3. E. I. Lipkovich, APK: Economics and Governance, **5**, 12-20 (2014)
4. A. M. Gataulin, Russian Agriculture Economy, **1**,, 87-90 (2009)
5. A. M. Bondarenko, L. S. Kachanova, *Technologies and Techniques for the Production and Application of Organic Fertilizers* (2016)

6. A. Yu. Briukhanov, D. A. Maximov and others, *Recommendations for organizing and conducting industrial environmental control of the processing and use of livestock manure (dung)* (2012)
7. A. I. Bogachev, M. G. Poluhina, Agricultural and Energy Supply, **3(7)** (2015)
8. I. Yu. Svinariov, I. N. Mikhailova, KubGAU Science Journal, **91(07)**, 1147-1156 (2013)
9. J. Oenema, and ethers, NJAS – Wageningen Journal of Life Sciences, **58**, 39-48 (2011)
10. Van Duinkerken, G.V.C., Journal of Dairy Science, **149**, 351-367 (2011)
11. Zhao Y., Effectiveness of multi-stage scrubbers in reducing emissions of air pollutants from pig houses Transactions of the ASABE, **54**, 285-293 (2011)
12. Aubert C., Comference paper for les 9emes Joumees de la Recherche Avicole, Tours, France, 116-120 (2011)
13. Yu. E. Domashenko, S.M. Vasilyev, Polythematic network electronic scientific journal of the Kuban State Agrarian University, **106(02)**, 568-579 (2015)
14. I. V. Atanov, A. M. Bondarenko, D. I. Gritsay, *Reconstruction and technical re-equipment of farms and cattle complexes: a study guide* (2017)
15. R. G. Gosmanov, A. H. Volkov, A. K. Galiullin, A. I. Ibragimova, *Sanitary Microbiology* (2010)
16. K.-S. Chen, Environmental Technology, **33(4)**, 487-495 (2012)
17. F. Bichai, Water Research, **46(18)**, 6040-6050 (2012)
18. Alley, M. International Fertilizer Industry Association, 61 (2009)
19. H.-W. Gripentrog, K. Persson, Managing variability - tools, methods and software: Third European conference on precision agriculture, **2**, 671-676 (2001)
20. I. Kristaponyte, Agronomy Research, Estonia, **3(1)**, 45-54 (2005)
21. T. I. Gulyaeva, V. I. Savkin, E. Y. Kalinicheva, O. V. Sidorenko, E. V. Burayeva, Journal of Environmental Management and Tourism, **9(8)**, 1789-1798 (2018)
22. V. V. Prokhorova, E. N. Klochko, O. N. Kolomyts, A. V. Gladilin, International review of management and marketing, **6**, 159-164 (2016)

Biotical cycle in single-crop sowing and mixed agrophytocenosis of forage crops

Aida Tamahina[1,*], and *Urfa Turan Ogly* Turabov[2]

[1]Kabardino-Balkarian State Agricultural University named after V.M. Kokov, Lenin Avenue, 1V, 360030 Nalchik, Russia
[2]Azerbaijan State Agricultural University, AZ2000, Ataturk Avenue, 450, Ganja, Azerbaijan

Abstract. The biotic cycle is based on the assessment of chemical elements in the production, degradation processes, during the deposition and resynthesis of organic compounds. The biotic cycle of agrophytocenoses, in contrast to natural ecosystems, is characterized by a significant imbalance of macro- and microelements. One way of solving this problem is to approximate the composition and structure of agrophytocenoses to natural plant communities. The objective of the research was to investigate the biotic cycle of elements in single- and mixed-seeded crops of *Galega orientalis* Lam., *Inula helenium* L., *Symphytum asperum* Lepech., *urtica dioica* L. The research was conducted in the mountainous zone of the Kabardino-Balkarian Republic in 2015-2019 on grey forest soil. In mixed agrophytocenoses, the biotic balance becomes less negative compared to monoculture crops. The excess of consumption over the return of N, Ca, K, P to the soil averaged 3.96; 40.94; 334.02; 9.46 kg/ha, respectively, over 5 years. The inclusion of *Galega orientalis* in a mixture with nonlegurninous forage grasses increased the soil cultivation index from 0.87 to 0.90. The results show the high potential of a polyculture system based on the combined cultivation of legumes and non-legumes in increasing the sustainability of farming.

1 Introduction

The biotic cycle is based on the assessment of chemical elements in the production, degradation processes, during the deposition and resynthesis of organic compounds. The nitrogen cycle, being the most complex of the element cycles, has its own microbiological cycle of transfer from the atmosphere to the soil, transformation from organic to mineral forms, and release from the soil to the atmosphere. There are 3 sub-cycles of nitrogen in herbaceous biogeocenoses: microbial with a turnover of 5-6 cycles per year, vegetative with a turnover of 1 year, and humus with a turnover of 400-500 years. Common nitrogen stocks in herbal biogeocenoses range from 4000 to 17000 kg/ha, with 96% of this amount in slightly mineralized organic matter, the remaining part is accumulated in mortmass, phytomass and in mineral form. The balance of nutrients in natural plant communities in different years can be either positive, or negative. If the ecosystem is in a terminal state, a

* Corresponding author: aida17032007@yandex.ru

zero balance is usually established within 5-6 years. If the ecosystem is in slow succession, then a zero balance is established within one or two years [1].

The nature of the biotic cycle varies according to the cropping patterns (monoculture, mixed crop, narrow specialization crop rotations) and soil management strategies (tillage, no-till) [2]. In agrophytocenoses, the mineralization of organic matter is not compensated by the humification of plant residues, as their entry into the soil is reduced compared to their loss. Nitrogen, P, K and Ca losses are due to their removal with the crop. In addition, the release of these elements from the mineralizing plant residues exceeds their uptake by the plants. On grassland, the N value increases in mortmass and decreases in phytomass and soil organic matter. When grasslands are fertilized, the N value increases in green phytomass and in soil organic matter. Seeded grasslands with a high percentage of legumes receive about 50 kg/ha of nitrogen per year from the atmosphere through symbiotic nitrogen bonding and up to 2 kg/ha through non-symbiotic nitrogen bonding [1].

One possible way of solving the problem of the negative biotic balance of forage agrophytocenoses is to bring their composition and structure closer to natural plant communities. Formation of optimized mixed agrophytocenoses is based on the selection of species that complement each other fluctuationally, seasonally, successively, longitudinally and functionally [3, 4]. An important principle in the formation of sustainable, self-renewing agrophytocenoses is the differentiation of ecological niches and the ecological complementarity of plants of different adaptive strategies. Due to the fact that an important role in ensuring the stability of agrocenoses is played by the amount of return of substances removed out with the harvest of fodder mass, mixed agrophytocenoses should approach the natural systems and ecological equilibrium (sustaining) in the degree of closeness (balanceness) of circulation of substances [5, 6].

Information on biotic cycling and nutrient balance in mixed crops is scarce and mainly concerns grass-and-legume mixtures. In view of the above, the objective of the study was to investigate the biotic cycle of elements in single-species and mixed crops of non-conventional forage grasses of high nutritional, biological and energetic value.

2 Materials and Methods

The objects of the research were the parameters of the biotic cycle of single-species crops and mixed agrophytocenosis of fodder galega (*Galega orientalis* Lam.), alant (*Inula helenium* L.), prickly comfrey (*Symphytum asperum* Lepech.), stinging nettle (*Urtica dioica* L.). The research was carried out in the mountainous area of the Kabardino-Balkarian Republic (950 m above sea level). Seeding was performed in 2013. The wide-row seeding method was used with row spacing of 70 cm and seed application rate of 1 mln. pcs./Ha (*Galega orientalis*, *urtica dioica*), 0.25 mln. pcs./Ha (*Inula helenium*), 0.1 mln. pcs./Ha (*Symphytum asperum*). Plot area of 20 m^2. For the purpose of creation of a mixed agrophytocoenosis in 2013, segments of roots and rootstock of the studied species were planted with a plant spacing of 70 cm and a row spacing of 70 cm. Systematic arrangement. Planting was done in alternating strips of 5 m in length. No agro-technical activities were carried out on the sites, standing grass crop was used for haymaking (2 times during the growing season). The biotic cycle parameters in the monoculture and in the mixture were recorded in 2015-2019. The green phytomass was cut at soil level, then litter was collected from the site. Underground phytomass was counted using the 1 dm^3 monolith method to a depth of 40 cm. Replication: 4-times. The plant matter samples were dried to air-dry weight and weighed. The rate of mineralization of root and crop residues was analyzed by embedding a mixed sample in the soil and periodically checking the mass loss of the sample. The content of mobile P and K compounds (according to Chirikov), exchangeable Ca (by complexometric method) and pH were determined in the soil samples. The soil cultivation index (Ic) was determined as the arithmetic mean of the relative indices of agrochemical properties (pH, phosphorus, potassium, humus) [7]. The contents of N (Kjeldahl), Ca (complexometric method), P (photometric method), K (flame emission spectroscopy) were determined in the above-ground and below-ground phytomass. The uptake of elements from the soil was estimated based on their content in the above-ground and below-ground mass. The return of elements to the soil was assessed by symbiotic nitrogen fixation, the amount of elements leached by precipitation from above-ground organs (N, P), released by decomposition of underground organs and litter, released by lifetime from underground organs (Ca, P, K).

3 Results and Discussion

The study area is moderately humid. There is 500-800 mm of rainfall during the growing season, with 50% of all precipitation occurring between May and July. The amount of temperature does not exceed 1800°C. The grey forest soil of the experimental site is medium-deep (humus layer thickness 25-28 cm), medium loamy, slightly rocky. The physical clay content in the soil is 50-60%, with a coarse dust fraction (34%) predominating and silty fraction of 10-15%. The density of the plowing soil horizon is 1.26 g/cm^3 and there is significant compaction of the soil horizon at a depth of 35-45 cm. The content of humus in the soil is average, the reaction of the soil solution is slightly acidic (pH_{KCl}=6.7) the content of mobile forms of phosphorus and potassium is elevated, Ic=0.87 (Table 1).

Table 1. Agrochemical indexes of the experimental plot soil.

Sampling depth (cm)	Humus (%)	P_2O_5 (mg/kg of soil)	K_2O (mg/kg of soil)	Ca exchangeable (mg/100 g of soil)
0-10	4.2	82.0	105.0	360.0
19-28	2.3	63.0	98.0	332.0
30-40	1.5	47.0	82.0	304.0

The starting point for the calculation of the biotic balance was the estimation of the main withdrawal of nutrients – accumulation of dry matter in above-ground (2 cuttings per season) and below-ground phytomass (Figure 1).

The general trend for all variants is an annual increase in below-ground phytomass. Compared to the original value, the weight of roots and rootstock has increased in the crops of *Galega orientalis*, *Inula helenium*, *Symphytum asperum*, *Urtica dioica* and mixture by 1,97; 2,43; 2,12; 3,70 and 2,25 times respectively. The average annual specific weight of underground phytomass in total biomass of dry matter was 55.86% for *Galega orientalis*, 37.0% for *Inula helenium*, 56.06% for *Symphytum asperum*, 27.57% for *Urtica dioica*, and 46.21% for mixed agrophytocoenosis.

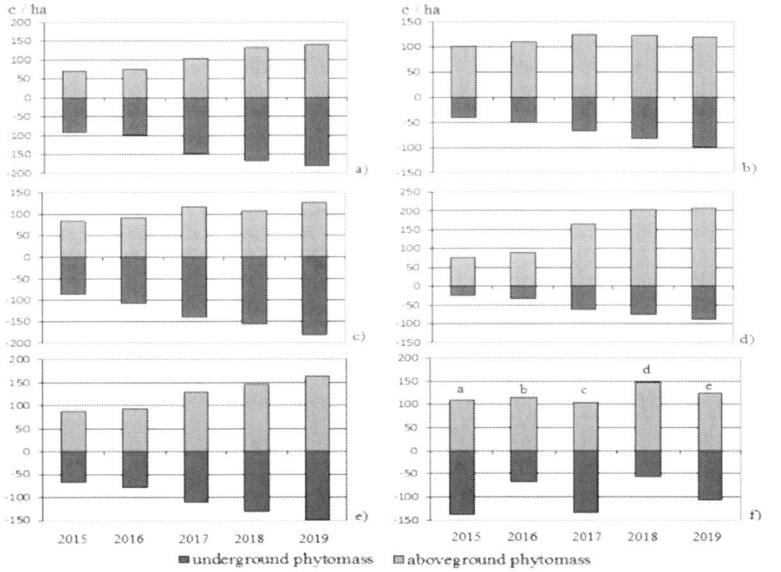

Fig. 1. Dynamics of above-ground and below-ground phytomass, dry weight, cwt/ha, in single-species crops.

Galega orientalis (a), *Inula helenium* (b), *Symphytum asperum* (c), *Urtica dioica* (d), mixtures (e) and the annual average of the variants (f).

The nitrogen balance in single-species crops of *Galega orientalis* is positive, due to symbiotic nitrogen bonding. In the remaining grasses, the difference between N intake and consumption was negative in all years of the study (Table 2).

Table 2. N value of crops.

Years	Yield, kg/ha				Input, kg/ha			
	Galega orientalis	Inula helenium	Symphytum asperum	Urtica dioica	Galega orientalis	Inula helenium	Symphytum asperum	Urtica dioica
2015	214.6	209.2	200.5	225.2	260.2	171.3	164.6	183.8
2016	277.8	290.3	246.3	256.9	384.1	234.6	200.3	208.6
2017	416.4	344.6	318.6	371.2	453.2	266.9	256.7	297.8
2018	463.1	358.4	310.9	382.0	480.0	287.9	250.7	306.1
2019	514.8	362.1	366.5	405.3	545.4	290.6	294.0	324.3
∑	1886.7	1564.6	1442.8	1640.6	2122.9	1251.3	1166.3	1320.6

Ca balance is positive only in the *Galega orientalis crop.* The greatest difference between the input and yield of Ca was observed for *Urtica dioica* (Table 3).

Table 3. Ca balance in crops.

Years	Yield, kg/ha				Input, kg/ha			
	Galega orientalis	Inula helenium	Symphytum asperum	Urtica dioica	Galega orientalis	Inula helenium	Symphytum asperum	Urtica dioica
2015	46.4	45.4	35.2	450.3	34.5	31.9	29.2	387.1
2016	60.2	50.6	41.0	518.6	47.0	38.1	33.8	443.4
2017	71.9	52.4	40.6	611.8	64.6	40.3	33.4	520.2
2018	86.1	55.3	48.4	760.0	75.4	43.7	39.8	642.4
2019	97.0	56.5	50.7	812.2	81.6	45.2	41.8	685.4
∑	361.6	260.2	215.9	3152.9	303.1	199.2	178.0	2678.5

K and P in all crops are characterized by a negative balance due to their significant accumulation in the underground mass and their removal with the hay crop (Tables 4, 5).

Table 4. K balance in crops.

Years	Yield, kg/ha				Input, kg/ha			
	Galega orientalis	Inula helenium	Symphytum asperum	Urtica dioica	Galega orientalis	Inula helenium	Symphytum asperum	Urtica dioica
2015	106.8	654.4	620.8	434.2	82.6	312.9	298.7	220.2
2016	134.0	771.1	719.7	490.2	94.0	361.9	340.3	243.8
2017	162.7	795.6	802.1	521.5	106.1	372.1	374.9	256.9
2018	185.9	826.5	909.3	645.2	115.9	385.2	419.8	308.9
2019	189.1	840.0	926.2	679.1	117.2	390.8	427.0	323.1
$\sum$	778.5	3887.6	3978.1	2770.2	515.8	1822.9	1860.7	1352.9

Table 5. P balance in crops.

Years	Yield, kg/ha				Input, kg/ha			
	Galega orientalis	Inula helenium	Symphytum asperum	Urtica dioica	Galega orientalis	Inula helenium	Symphytum asperum	Urtica dioica
2015	60.4	80.0	68.1	84.5	53.6	69.8	69.0	73.5
2016	78.5	86.3	74.5	90.1	68.7	75.0	65.3	78.2
2017	98.0	95.5	83.3	114.3	84.5	82.7	72.5	98.2
2018	118.0	102.0	80.6	130.0	101.3	88.1	70.3	111.3
2019	132.7	110.8	95.2	140.2	113.5	95.3	82.4	119.7
$\sum$	487.6	474.6	401.7	559.1	421.6	410.9	359.5	480.9

In mixed agrophytocenoses, the biotic balance becomes less negative compared to monoculture crops (Table 6). On average, the excess of yield over input for N, Ca, K and P over 5 years of the study was 3.96, 40.94, 334.02 and 9.46 kg/ha, respectively (Figure 2).

The results show an increase in the biotic cycling of chemical elements in mixed agrophytocenoses compared to single-species crops of *Galega orientalis*, *Inula helenium*, *Symphytum asperum* and *Urtica dioica*. In our opinion, the principle of chemical element imbalance compensation should be considered when optimizing the placement of grasses in a polyculture. The reduction of element imbalances in *Inula helenium*, *Symphytum asperum*, *Urtica dioica* crops is ensured by their co-growth with a legume component (*Galega orientalis*).

Table 6. N, Ca, K, P balance in a mixed agrophytocenosis.

Years	Yield, kg/ha				Input, kg/ha			
	N	Ca	K	P.	N	Ca	K	P.
2015	310.2	272.5	537.9	75.1	288.3	256.6	301.5	69.4
2016	331.9	380.4	610.5	86.3	295.2	347.8	332.0	78.6
2017	365.5	433.8	691.9	99.5	391.4	392.1	366.3	89.5
2018	464.5	474.2	800.0	109.9	488.6	425.7	411.6	98.2
2019	513.0	577.0	891.3	112.7	501.8	511.0	450.1	100.5
$\sum$	1985.1	2137.9	3531.6	483.5	1965.3	1933.2	1861.5	436.2

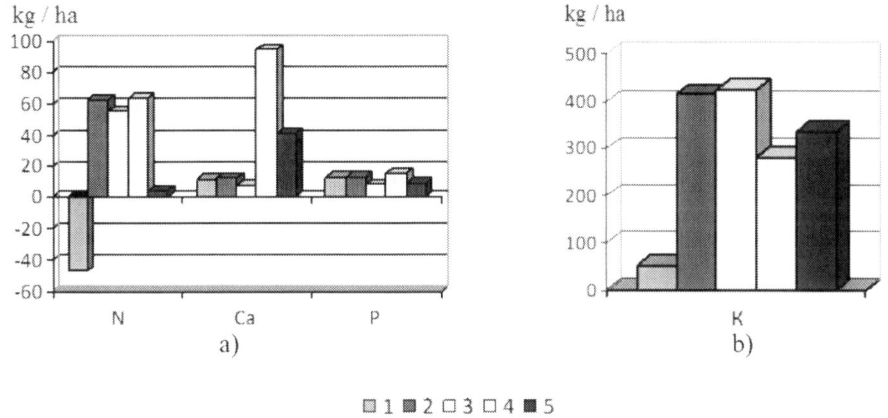

Fig. 2. Excess consumption of N, Ca, P (a), K (b) by *Galega orientalis* (1), *Inula helenium* (2), *Symphytum asperum* (3), *Urtica dioica* (4) and mixture (5) over the return of elements to soil. Average over the 5 years of the study, kg/ha.

In the mixed agrophytocenosis, *Galega orientalis*, *Inula helenium*, *Symphytum asperum* and *Urtica dioica* differ significantly in the requirements for individual of mineral elements, more effective use of resources by reducing the inter-species competition and regional differentiation of ecological niches, which leads to increased productivity and stability of the system.

Compared to non-legume grasses, the *Galega orientalis* agrophytocenosis is characterized by the greatest balanceness of chemical elements, which contributes to improved soil fertility. In 2019, the Ic for humus increased by 0.05 due to a 0.23% increase in humus content. The pH shift towards alcalinity (from 6.7 to 6.9) is due to the release of nitrogen in the form of ammonium during the decomposition of the dead plant mass of *Galega orientalis*. After 5 years, an increase in the content of mobile forms of phosphorus (from 82 to 112 mg/kg) and potassium (from 105 to 135 mg/kg of soil) was noted. As a result, soil Ic increased by 0.05 (Table 7).

The inclusion of *Galega orientalis* in mixtures with non-legume fodder grasses improves soil agrochemical parameters (increasing humus content by 0.02%, pH by 0.08) and increases Ic to 0.90. Improving the balanceness of the biotic cycle brings the mixed agrophytocenosis closer to natural ecosystems.

Table 7. Changes in soil agrochemical parameters in *Galega orientalis* agrophytocenosis.

Indicator	Indicator value					Ic	
	2015	2016	2017	2018	2019	2015	2019
Humus, %	4.20	4.20	4.23	4.30	4.43	0.62	0.67
pH_{KCl}	6.7	6.7	6.8	6.8	6.9	1.00	1.00
P_2O_5	82.0	92.0	100	110.0	112.0	0.87	1.00
K_2O	105.0	112.0	120	128.0	135.0	1.00	1.00
Ic						0.87	0.92

The compulsory inclusion of legume grasses in mixtures with other crops has been confirmed by numerous studies. The main advantages of legume cultivation are saving energy by reducing the need for nitrogen fertilizers, release of organic matter to the soil with an optimum C:N ratio (20-40:1), facilitating nutrient circulation in the soil, restoring inaccessible forms of soil phosphorus, retaining and ensuring the availability of soil water

[8-12]. The deep root systems of a number of legume species facilitate the exchange of nutrients by root exudates and their absorption/recirculation, as well as the infiltration of water into deeper soil layers. The impact of legumes on soil carbon sequestration is more pronounced for forage, green manure and cover crops, which return organic C and N to the soil and release hydrogen gas, which promotes the development of nodules of symbiotic nitrogen-bonding bacteria in the rhizosphere [13-21]. Some legume crops are able to mobilize the bonded forms of soil phosphorus by secreting organic acids (citrate and malate) and other P-mobilizing compounds from their roots [22].

A multicultural system based on the co-cropping of legumes and non-legumes is one aspect of sustainable farming. The main advantage of the polyculture system is the ecologisation of crop production, based on nitrogen bonding and nitrogen transportation from the legumes to non-legumes, soil protection from water and wind erosion, stabilization of soil temperature, water accumulation in the soil profile, improved physical characteristics, increased soil fertility and biological activity [23, 24].

4 Conclusion

The biotic cycle is based on the assessment of chemical elements in the production, degradation processes, during the deposition and resynthesis of organic compounds. Natural ecosystems are characterized by a zero balance of inputs and removal of chemical elements. In the biotic cycle of agrophytocenoses there is a significant imbalance of macro- and microelements. One way of solving this problem is to approximate the composition and structure of agrophytocenoses to natural plant communities. As a result of evaluation of biotic cycle of elements in single-species and mixed crops of *Galega orientalis* Lam., *Inula helenium* L., *Symphytum asperum* Lepech., *Urtica dioica* L., the relatively high balanceness of biotic cycles of nitrogen, phosphorus, potassium, calcium and improved soil agrochemical parameters in *Galega orientalis* agrophytocenosis was established. The biotic balance in the mixed agrophytocenosis was less negative compared to the single-species crop. The excess of consumption over the return of N, Ca, K, P to the soil over 5 years averaged 3.96; 40.94; 334.02; 9.46 kg/ha, respectively. The inclusion of *Galega orientalis* in mixture with non-legume fodder grasses improved soil agrochemical parameters (increasing humus content by 0.02%, pH by 0.08) and increase of soil cultivation index to from 0.87 to 0.90. Improving the balanceness of the biotic cycle brings the mixed agrophytocenosis closer to natural ecosystems. The results show the high potential of a polyculture system based on the combined cultivation of legumes and non-legumes in increasing the sustainability of farming.

References

1. A. A. Titlyanova, Soil Science, **12,** 1422-1430 (2007)
2. F. Negash, et al., Adv. Crop Sci. Tech., **6(1)**, 326 (2017)
3. V. Krautzer., W. Graiss, E. M. Poetsch, *Integrating Efficient Grassland Farming and Biodiversity*, 186-189 (2005)
4. Z. Sh. Shamsutdinov, E. Z. Shamsutdinova, J. of Russian Academy of Agricultural Sciences, **3**, 37-38 (2007)
5. A. Ya. Tamakhina, Forage Production, **10**, 2-4 (2009)
6. A. Ya. Tamakhina, Russian J. of Ecology, **40(4)**, 261-266 (2009)
7. N. Gopp, et al., Soils and the Environment, **1 (1)**, 32-44 (2018)
8. J. F. Angus, J. A. Kirkegaard, J. R. Hunt, M. H. Ryan, L. Ohlander, M.B. Peoples, Crop Pasture Sci., **66**, 523-552 (2015)

9. J. L. Hernanz, V. Sanchez-Giron, L. Navarrete, Agric Ecosyst Environ, **133**, 114-122 (2009)
10. M. Peoples, et al., Symbiosis, **48**, 1-17 (2009)
11. S. Preissel, M. Reckling, N. Schläfke, P. Zander, Field Crop Res., **175**, 64-79 (2015)
12. J. Shen, et al., Plant Physiol., **156**, 997-1005 (2011)
13. A. Bichel, M. Oelbermann, P. Voroney, L. Echarte, Carbon Manag., **7**, 1-10 (2016)
14. F. Stagnari, A. Maggio, A. Galieni, M. Pisante, Chem. Biol. Technol. Agric., **4(2)**, 1-13 (2017)
15. C. Carranca, M. O. Torres, M. Madeira, Agron Sustain Dev., **35**, 1095-1102 (2015)
16. E. Hajduk, S.Właśniewski, E. Szpunar-Krok, Soil Sci. Ann., **66**, 52-56 (2015)
17. S. K. Kakraliya, et al., *Legumes for Soil Health and Sustainable Management*, 277-314 (2018)
18. M. Latati, A. Bargaz, B. Belarbi, M. Lazali, S. Benlahrech, S. Tellah, Eur. J. Agron, **72**, 80-90 (2016)
19. R. L. Lemke, Z. Zhong, C. A. Campbell, R. P. Zentner, Agron J., **99**, 1719-1725 (2007)
20. Y. Wang, P. Marschner, F. Zhang, Plant Soil., **354**, 283-298 (2012)
21. Y. Yu, L. Xue, L.Yang, Agron Sustain Dev., **34**, 633-640 (2014)
22. P. J. Hocking, Adv. Agron, **74**, 63-97 (2001)
23. K. Adamczewska-Sowińska, J. Sowińsk, *Soil Health Restoration and Management*, 279-319 (2020)
24. V. N. Prokhorov, Bulletin of Polessky State University, Series in Natural Sciences, **1**, 6-10 (2009)

Application of biomonitoring technology in environmental engineering based on model groups of amphibiontic organisms (insecta: odonata)

Sergey Kozminov, *Khasan* Ketenchiev, *Anzor* Paritov, *Elena* Krapivina, and *Bela* Sunsheva

Kabardino-Balkarian State University named after H.M. Berbekov, st. Chernyshevsky, 173, 360004 Nalchik, Russia

Abstract. The paper presents the results of the analysis of the use of amphibiontic organisms as bioindicators of the ecological environment, as model groups in ecological engineering. The solution to the conceptual problem of ecological engineering in the conservation of biological diversity is to take an inventory of natural, urbanized and agroecosystems. Among amphibiotic insects, the order Odonata stands out, with morphological plasticity, biological progress, adults and dragonfly larvae actively colonize aquatic and terrestrial ecosystems of various types. When moving to habitat in mountain biotopes, the linear body dimensions of dragonfly larvae of the species L. depressa L., S. meridionale Selys, C. puella L. decrease, which indicates unfavorable, contrastingly changing habitat conditions in mountain ecosystems. The taxonomic composition of dragonflies in lowland water bodies is represented by the suborders Anizoptera (7 species), Zygoptera (5 species), Caloptera (1 species). In the foothill water bodies - Anizoptera (5 species), Zygoptera (4 species), Caloptera (1 species), in mountain water - Anizoptera (4 species), Zygoptera (2 species). Most of the dragonfly taxa are characteristic of the plains and foothills with eutrophied biotopes characteristic of polysaprobic reservoirs with heavy pollution. In oligosaprobic biotopes, with little or no organic pollution in the water, dragonflies Ae. cyanea Müller and C. splendens Harris. The confinement of species to certain habitats, characterized in varying degrees of pollution by organic substances and their oxidation products, makes it possible to use dragonflies as bioindicators and organize background monitoring of ecosystems in the Central Caucasus.

1 Introduction

At present, the problem of preserving the species diversity of living organisms and the rational use of natural resources is quite acute. Protection of the surrounding world should be based on the integration of various approaches, on the basis of a comprehensive study of the interaction of all levels of organization of wildlife [1]. A solution is possible with the adoption of measures to ensure the long-term functioning and stability of all types of ecosystems, disturbed and undisturbed by anthropogenic impact. The impact of negative technogenic processes on biological diversity, sustainable development and functioning of ecosystems, defines environmental safety as the protection of the environment [2]. In this regard, environmental engineering is designed to solve a number of environmental and biological problems to ensure the preservation of taxonomic diversity, to reduce various types of pollution on ecosystems.

As a system of scientific and engineering activities, environmental engineering solves the problems of ensuring environmental quality control measures for the environment of aquatic and terrestrial ecosystems, the distribution and preservation of habitats of various groups of living organisms, environmental monitoring and the long-term existence of all types of natural biotopes with their plant communities and animal populations.

One of the conceptual tasks of ecological engineering is the conservation of biological diversity, based on fundamental research on the inventory of natural, urbanized and agroecosystems that ensure the transformation and circulation of matter and energy from producers to the highest trophic levels. In this aspect, ecological monitoring seems to be important, providing constant monitoring of the functioning of ecosystems with a recreational load, as well as existing natural, or reserves, providing an inflow of biological resources for reproduction used by ecological engineering. Biological control of the quality of the environment can be one of the key methods of environmental engineering, the solution of which is the search for model groups of living organisms that are sensitive to changes in the ecological environment and their components, which are bioindicators of pollution. On the basis of environmental monitoring, it is possible to resolve issues of environmental safety, as well as to comply with the norms and principles of rational use of natural resources [3].

Based on biomonitoring, it is possible to fundamentally improve the tracking scheme for the state of ecosystems of any type. In this regard, the most interesting is the group of amphibiontic organisms, as it is wholly dependent in its development, both on the aquatic and terrestrial-air environment. Among all amphibiontic organisms, the most interesting are amphibiontic insects, which in many water bodies, their ecosystems and biotopes dominate in terms of quantitative indicators, biomass and number of species [4]. Among all amphibiotic insects, the order of dragonflies *Odonata* stands out, known for its morphological plasticity, biological progress, high mobility of adults, and habitation in intrazonal biotopes, whose representatives occupy the niche of active predators [5, 6]. Adults and larvae of dragonflies actively colonize aquatic and terrestrial, both natural and urbanized ecosystems of various types, actively transform matter and energy from lower trophic levels to higher ones. It is also important that dragonflies play an important role in the destruction of blood-sucking insects, larvae and imago of mosquitoes, which are carriers of pathogens of various diseases [7].

The imaginal and preimaginal phases of dragonflies, as the most important component of living nature, are interconnected with both biotic and abiotic factors, which are determined by relationships with other living organisms and the inorganic environment [8, 9] and developed on model species and their populations [10]. Most taxa of dragonflies are quite sensitive to changes in the physicochemical composition of the habitat, drainage of reservoirs, regulation of river flow and deterioration of water quality, which makes it possible to use the potential bioindicator role of dragonflies, as well as to compile lists of rare and endangered species in aquatic and terrestrial ecosystems [11, 12].

Biological control of water quality in the Central Caucasus is a problem, the solution of which is to preserve the biodiversity of aquatic and terrestrial ecosystems of the unique mountainous and historically formed region. The anthropogenic effect on natural ecosystems has caused significant damage to the fauna of various groups of animals, including many rare, endemic and relict taxa of the Caucasus. Dragonflies, leading an amphibiotic lifestyle, being one of the mass components of aquatic and near-aquatic biocenoses, interact with both organic and inorganic environments, are exposed to the nature and varying degrees of pollution of natural ecosystems with organic and inorganic substances and their compounds, products of their decay as a result of oxidative-restorative reactions.

Based on the most important relevance of understanding the bioindicator role of dragonflies in environmental monitoring of ecosystems disturbed and undisturbed by anthropogenic impact, application in environmental engineering, we set a goal for the study: to analyze the distribution and composition of the dragonfly community in a comparative aspect, to determine the taxonomic composition, some ecological, biological and adaptive features , to determine the possibility of different taxa for use in background monitoring, in bioindication of the qualitative composition of the natural environment.

2 Materials and Methods

This work is based on the materials obtained earlier [3, 13-17] and additionally processed with the specification of the original materials and literature sources [18-20]. The study of the distribution of dragonflies by the altitudinal-belt and horizontal vector, the determination of the types of environmental indicators was carried out on various terrestrial, near-water and water plains (up to 500 m above sea level), foothill (500 – 1000 m above sea level) and mountain (1000 – 2500 m above sea level) ecosystems of the Central Caucasus of various origins in the Kabardino-Balkarian Republic. When carrying out field work, which was carried out during the growing season, the route method (100 m) was used, as well as harvesting of areas of 10 – 15 m^2. Collection, species identification and office processing were carried out according to generally accepted methods of entomological research [21-24]. To clarify the original materials, study the ecological and biological features and check the taxonomic affiliation in laboratory conditions, dragonfly larvae (about 3.5 thousand specimens in total) were kept, which were additionally processed.

3 Results and Discussion

When determining the types of bioindicators, it is necessary to take into account not only high morphological plasticity, but also many biological features, such as seasonal variability, summer time, habitats, leading to the differentiation of the species composition. Dragonflies form communities under the influence of physical and geographical conditions, which is associated with the biological characteristics of the species. The influence of biological features primarily affects the penetration of species of southern origin into the mountains, the northern ones can live only in the mountains, and if they descend into the foothills, then only to an insignificant extent [25]. Adults of dragonflies are found in different geographic zones with a sharp decrease in the number of species, while species of northern origin are susceptible to extinction, which illustrates various stages of the general process [26]. At the same time, under the influence of physical and geographical conditions, there is an adaptive variability of the morphology of dragonflies, the color of the body and their parts, the features of the structure of the wings [27, 28]. The distribution of dragonfly larvae entirely depends on the dynamics of water temperature, the abundance and composition of the food supply, the depth and flow of the reservoir. The location of reservoirs with sufficient depth in the lowering of the relief, the overgrowing of the banks with shrubs, contributes to the experience of the unfavorable conditions of the winter period by the larvae, which were subordinate to the heating of the water and the natural food base. [29]. According to the nature of habitats, the preimaginal stage is divided into 4 groups:

inhabitants of exclusively stagnant water bodies, mainly stagnant, mainly flowing and flowing water bodies [30].

When studying the composition of the species composition, distribution, composition of communities, ecological, biological and adaptive characteristics of various taxa to aquatic and terrestrial habitats, it is necessary to clearly define the boundaries of the species range, which is the result of the action of various factors, their complex and the confluence of historical events and evolution. Most species of dragonflies strive to expand the living space that they occupy, interact with the environment according to the rule of limiting factors that move away from the optimum ecological habitat of the species, limiting its existence in new ecological conditions. The limiting factors affecting dragonflies are very diverse, they include both abiotic and biotic and anthropogenic ones. The quick response of dragonfly larvae and imago to these factors is determined and depends on their hereditary characteristics, which determine the optimum needs of the species, its rate of response to changes in environmental factors and their complexes.

When analyzing the distribution, ecological and biological characteristics of dragonflies, it is necessary to take into account the peculiarities of the amphibiontic way of life, which determines a number of restrictions for full development and metamorphosis. At the same time, dragonfly larvae are more dependent on changes in environmental conditions than adults [18], which affects the data obtained on the distribution, composition and structure of the dragonfly community, their discrepancy in taxonomic, quantitative and qualitative indicators.

The basis of the dragonfly fauna of the Central Caucasus is formed by the autochthonous core of taxa, which was formed by historical interacting factors, such as: a warm climate favorable for the habitation and development of dragonflies; a sufficiently optimal humidity regime for the passage of an amphibiotic lifestyle; mountainous relief, which created the whole range of local water and near-water stations with various ecological conditions.

Changes in the ecological habitat that exert strong pressure on dragonflies have both negative and positive effects. The negative impact is determined not only in the destruction of water bodies necessary for the development of the preimaginal phase, but also in their pollution, destruction of breeding sites and limitation of the full development cycle to metamorphosis. The positive side of anthropogenic impact is the creation of new reservoirs in artificial and agroecosystems, which are suitable for the development of larvae, metamorphosis and emergence of adults, which subsequently colonize near-water and terrestrial ecosystems. Therefore, the development and rather rapid colonization of artificial reservoirs by representatives of dragonflies indicates the possibility of faunistic movements and a sharp change in the historically established community of dragonflies. The assimilation of artificial ecosystems of various types and purposes by dragonflies occurs not only due to taxa characteristic of the territory, but also due to species southern in origin that do not previously populate natural ecosystems, changing the species composition, but in general the biological diversity of dragonflies. This determines the fundamental foundations of studies of faunistic movements, species diversity, adaptive, ecological and biological characteristics, the search for species of bioindicators of the natural environment, both among adults and larvae of dragonflies.

The preimaginal phase of dragonflies prefers to develop in small reservoirs or biotopes, which are characterized by strong ecological vulnerability and even complete extinction. A negative factor for dragonflies is the habitation of the preimaginal and imaginal stages of development in different biotopes, a long development cycle, and anthropogenic pollution, leading to the degradation of aquatic ecosystems. A number of factors affect the imago of dragonflies: death from pesticides when processing agricultural systems, road transport when flying over highways, as well as a large catch by collectors.

The main abiotic factor that determines and restrains the spread of dragonflies in ecosystems is the temperature regime of water bodies, which are habitats where the preimaginal phase develops up to the emergence of adults. The distribution of dragonflies along the altitude vector, within each altitudinal belt, is heterogeneous, determined by the ecological, biological and adaptive characteristics of various species, their developmental biology and the ability to interact with various abiotic factors. Dragonflies, as ectothermic organisms, are entirely dependent on external energy sources, and their adaptive and regulatory capabilities are limited [31]. Therefore, the taxonomic composition, distribution and structure of the dragonfly community depends on temperature and its dynamics for a long time, which also affects the biomass of the natural food base, the rate of metabolism and ontogenesis of individual preimaginal stages. Dragonflies are quite resistant to various temperatures, which is determined by avoiding low temperatures, since they spend the winter in water bodies. In this case, the emergence of adults in the north occurs later than in the southern regions, and in a harsh climate, a transition to habitat in biotopes with stagnant water, which warms up better, is noted [18]. Although the distribution of dragonflies is entirely dependent on temperature, this dependence is much less than that of other land dwellers, so the temperature factor has to be given close attention.

Analysis of the temperature regime of water bodies in the altitudinal-zonal aspect shows a certain discrepancy both in the temperature dynamics of aquatic ecosystems and a shift in the maximum temperature achievement towards the summer period, which affects the decrease in the quantitative and qualitative indicators of the natural food supply of dragonfly larvae. In the Central Caucasus, the shift in the beginning of the growing season towards the summer period from the plain to the mountainous zone (1 – 2 months) leads to a decrease in the temperatures of aquatic ecosystems of the mountains (5 – 10 °C) [32]. When dragonfly larvae move into the mountain zone, either an increase in the number of individual stages occurs, with a decrease in the development time, or an increase in the duration of development of stages, with a general decrease in the size of the body and their parts. We noted a decrease in body size in the larvae of dragonflies of the species *Libellula depressa* L., *Sympetrum meridionale* Selys, *Coenagrion puella* L., when moving to habitat in water bodies of the mountain zone, which indicates rather unfavorable, contrastingly changing habitat conditions in mountain ecosystems, as well as adaptive plasticity of these taxa of southern origin. Species of northern origin, such as *Aeshna cyanea* Müller, which create isolated areas [18], develop preimaginal in low-temperature mountain water bodies, although actively mobile imago occurs in the foothill zone, or can fly into the plain.

Characterization of the natural food base is important for determining the taxonomic composition and community of dragonflies. Representatives of different suborders, although they are predators, have their own local food spectra. The species composition of the suborder *Anisoptera* consists of fairly large forms, the natural food supply of which in mountain conditions is sufficient in quantity and quality for the development of the preimaginal and imaginal stages. The food base of representatives of the suborder *Zygoptera* is based on zooplankton, which sharply depletes its quantitative and qualitative indicators in foothill and mountain water bodies. At the same time, the younger preimaginal stages feed exclusively on zooplankton with a low individual biomass, whose representatives, under conditions of a short growing season and low water temperatures, deplete the taxonomic composition and have a local occurrence with specific physicochemical conditions of water composition [32]. It should be noted that the presence of near-water and aquatic vegetation has a direct impact on the taxonomic and quantitative composition of dragonflies *Zygoptera*, whose larvae and adults use vegetation as a substrate for metamorphosis, or shelter and rest [21, 33]. The qualitative composition of vegetation has little effect on dragonflies [18]. Dragonflies of the suborder *Anizoptera*, among which there are many forms of larvae leading a near-bottom lifestyle, are not very susceptible to the nature of vegetation change. However, the structure of the soil in terms of such indicators as the inclusion of silt and detritus in the composition, the formation of its complex, is formed as a result of the presence of near-water and aquatic vegetation. The composition of the natural food base is determined by quantitative indicators and biomass of vegetation, their influence on the composition of nutrients in water, the development of zooplankton and other forms of invertebrates, benthic organisms, which are food objects for dragonfly larvae and the successful passage of the entire life cycle.

The distribution of dragonflies along the altitude and horizontal vector, their occupation of habitats convenient for habitation, is determined by the river network with permanent inland and floodplain water bodies. It is the basin of any river with its complex hydrological system that creates an arena of life for dragonflies, ensures the development and movement of individuals of the species within it. The mineral composition of the water has little effect on dragonflies, the preimaginal stage can tolerate its small size and are not tied to it [18]. The rather weak mineralization of the water bodies of the Central Caucasus is determined by the complex belonging to the hydrocarbonate class of the calcium group, which has little effect on the distribution of dragonflies and their ecological and biological features.

Continental water bodies of any type are divided into groups according to the nature of pollution by organic substances, products of their oxidation and decay [34, 35]. A quantitative and qualitative analysis of water bodies by the altitude vector of the Central Caucasus shows a significant discrepancy both in the availability of convenient aquatic biotopes for dragonflies to inhabit, and in the nature of saprobity, or pollution by organic substances. Water bodies of the plain, to a lesser extent of the foothills, consist of polysaprobic and mesosaprobic biotopes. Rare oligosaprobic biotopes are represented inhomogeneously and are fed from spring waters, or are formed as a result of ice melting. The aquatic ecosystems of the mountains mainly consist of oligosaprobic biotopes. Rare polysaprobic and mesosaprobic biotopes are various shallow bays, backwaters, or temporary puddles. The discrepancy in the gradient of the saprobity of water bodies in the plain, foothill and mountain zones is associated with the descent of oxygen-rich glacial waters into the plain zone, the accumulation of various organic substances on its way and the warming up of water. The distribution of the taxonomic composition of dragonflies showed that dragonfly larvae belonging to the suborders *Anizoptera* (7 species), *Zygoptera* (5 species), *Caloptera* (1 species) live in lowland water bodies, *Anizoptera* (5 species), *Zygoptera* (4 species), *Caloptera* (1 species) in foothill water bodies, in the mountains – *Anizoptera* (4 species), *Zygoptera* (2 species). Most of the taxa are characteristic of the plains and foothills; their habitats are associated with eutrophied biotopes characteristic of polysaprobic water bodies with heavy pollution. For them, the most characteristic indicator species are: *Libellula depressa* L., *Sympetrum meridionale* Selys, Orthetrum cancellatum L., *O. albistylum* Selys, *Crocothemis erythraea* Brulle, *Coenagrion puella* L. For mesosaprobic biotopes with moderate pollution, the following are typical: *Lestes sponsa* Hansemann, *L. dryas* Kirby, *C. pulchellum* Van der Linden, *Platycnemis pennipes* Pallas, *Anax imperator* Leach, *Erythromma najas* Hansemann. Oligosaprobic biotopes with little or no organic pollution in the water are characterized by the larvae of the dragonflies *Aeschna cyanea* Müller and *Calopteryx splendens* Harris.

The confinement of dragonflies to certain biotopes, differing in the nature of saprobity, as well as in the altitude vector, can be associated not only with the ecological characteristics of various taxa to experience unfavorable conditions and develop in different physicochemical composition of the aquatic environment, but also be associated with biological processes and conditions of their course. One of the most important phenomena characterizing the dragonfly community and its taxonomic composition is metamorphosis, emergence and expansion of imago carrying a reproductive function, which determines the possibility of dragonfly spreading and the development of new habitats and territories. he main external environmental factor that determines the beginning and termination of emergence, summer of adults, and its period is the average daily air temperature [20], as well as with temperature dynamics during the growing season and the achievement of optimal values. Comparison of the time of metamorphosis and the temperature of water and air showed that the main emergence of adults is associated with a shift from the end of April to June from the plain to the mountains, when the water warms up to 16 – 18 °C. At the same time, on the plain and foothills, the maximum air temperature was in the range of 21 – 23 °C, and in the mountainous zone – 18 – 20 °C [32]. The hatching of adults is associated with the individual ecological and biological characteristics of various taxa, the ability to adapt to the temperature dynamics of different water bodies and terrestrial ecosystems and their origin. Metamorphosis and emergence is associated with the individual ecological and biological characteristics of various taxa, their origin and historical development. The formation of the faunistic complex, as well as the composition of populations and communities of dragonflies, is deeply associated with ecological and biological adaptations not only to survival in changing habitat conditions, but also to the possibility of settling new habitats, both along the altitude and horizontal vectors.

4 Conclusion

The formation of the larval population, communities of adult dragonflies, the distribution of taxa in various terrestrial and aquatic biotopes is determined by the interaction of physicochemical complexes, the factor of the habitat and the ecological and biological characteristics of different species. Cognition of such interaction will determine the ways and possibilities of using the preimaginal and imaginal phases of development as a model group for ecological monitoring of ecosystems. The taxonomic composition of dragonfly populations clearly determines the pollution of the environment, natural ecosystems, agroecosystems and their stations. The confinement of certain species to certain habitats, characterized in varying degrees of pollution by organic substances and products of their oxidation, makes it possible to use dragonflies of 14 species as bioindicators and organize background monitoring of ecosystems in the Central Caucasus. As amphibiontic organisms of dragonflies clearly capture changes in the aquatic and ground-air environment, are associated with abiotic factors and their dynamics. Given the wide territorial distribution of agroecosystems, it is possible to use them as a reserve of rare and protected species (*A. imperator* Leach, *L. depressa* L., *C. puella* L., *L. sponsa* Hansemann, *L. dryas* Kirby), endangered [36, 37]. Taking into account the interaction of habitat conditions and biology of larvae, as well as adults of dragonflies, it is possible to determine a complex of favorable and unfavorable factors for the development of dragonflies. Each of the factors interacts with each other and does not determine the composition of populations, natural communities, and the biology of individual taxa. They dynamically act through all life cycles on the biodiversity of dragonflies in ecosystems, determine the possibility of development of dragonflies not only in their usual habitats, but also to populate newly formed ones.

References

1. A. V. Yablokov, S. A. Ostroumov, *Wildlife Conservation: Problems and Prospects* (1983)
2. A. B. Savinov, V. A. Basurov, Bulletin of the Nizhny Novgorod University. N.I. Lobachevsky, Biology, **3(1)** (2014)

3. E. A. Krapivina, S. G. Kozminov *The use of model groups of living organisms in ecological monitoring of ecosystems in the western part of the Central Caucasus* (2013)
4. S. M. Golubkov, *Functional ecology of amphibiotic insect larvae* (2000)
5. B. B. Rodendorf, *Infraclass Libellulones* (1980)
6. B. N. Shvanvich, *General entomology* course (1949)
7. B. F. Belyshev, A. Yu. Kharitonov, S. N. Borisov, *Fauna and ecology of* dragonflies (1989)
8. C. L. Pierce, P. H. Crowley, Ecology, **66**, 5 (1985)
9. C. Blois, A. Cloares, Z. Tierpsychol, **68**, 4 (1985)
10. V. M. Titar, S. S. Kulchitsky, *Population monitoring of the zone of anthropogenic impact on the example of two species of dragonflies* (1989)
11. F. L. Carle, Odonatologica **8**, 4 (1978)
12. N. Gaithier, Rhone-Alpes nature, **24** (1990)
13. Kh. A. Ketenchiev, S. G. Kozminov, Z. Kh. Gemueva, A. A. Alakulova, South of Russia: ecology, development, **12**, 4 (2017)
14. Kh. A. Ketenchiev, A. Yu. Kharitonov *Dragonflies of the Mediterranean* (1999)
15. S. G. Kozminov, Kh. A. Ketenchiev *Dragonfly larvae as bioindicators of aquatic ecosystems* (1999)
16. Kh. A. Ketenchiev, A. Yu. Kharitonov, S. G. Kozminov, T. A. Avtaeva, Sh. A. Kushalieva, *Dragonflies of the Mediterranean faunal subregion of the Paleo-subtropical region of the Boreal kingdom* (2020)
17. S. G. Kozminov, Z. A. Kubatieva, Grozny natural science bulletin **5**, 1 (19) (2020)
18. B. F. Belyshev, A. Yu. Kharitonov, *Geography of dragonflies (Odonata) of the Boreal faunal kingdom* (1981)
19. B. F. Belyshev, A. Yu. Kharitonov *Geography of dragonflies (Odonata) of the Meridian faunal kingdom* (1983)
20. B. F. Belyshev, A. Yu. Kharitonov, S. N. Borisov, *Fauna and ecology of dragonflies* (1989)
21. A. N. Popova, *Dragonfly larvae of the fauna of the USSR (Odonata)* (1953)
22. B. F. Belyshev, A. Yu. Kharitonov, *Keys to dragonflies by their wings (genera of the Boreal faunistic kingdom and adjacent lands, species of fauna of the USSR)* (1977)
23. Kh. A. Ketenchiev, A. Yu. Kharitonov, *Keys to the Caucasus dragonflies* (1998)
24. K. K. Fasulati, *Field study of terrestrial invertebrates* (1971)
25. A. N. Bartenev, Bulletin of the State Museum of Georgia, **2** (1925)
26. A. N. Bartenev, Proceedings of the North Caucasian Association of Scientific Research of Innovators, **72**, 14 (1930)
27. A. N. Bartenev, Works of the North Caucasian Hydrobiological Station, **3**, 1 (1930)
28. A. N. Bartenev, Works of the North Caucasian Hydrobiological Station, **3**, 3 (1930)
29. A. Yu. Kharitonov, Ecology, **3** (1975)
30. E. S. Shalopenok, *Distribution features of dragonfly larvae in water bodies of different types* (1989)
31. M. Bigon, J. Harper, K. Townsend, *Ecology. Individuals, populations and communities* (1989)

32. S. G. Kozminov, *Distribution and ecology features of dragonfly (Odonata) larvae in the Central Caucasus* (2019)
33. W. I. Lucas, *The aquatis (naiad) Stage of the British dragonfiies (Paraneuroptera)* (1930)
34. E. N. Pavlovsky, S. G. Lepneva, *Essays on the life of freshwater* animals (1948)
35. L. I. Kotova, L. P. Ryzhkov, A. V. Polina, *Biological control of water* quality (1989)
36. Red Book of the RSFSR (animals) (1983)
37. Red Book of the Kabardino-Balkarian Republic (2000)

Evaluation of the ecological efficiency of soil cultivation machines with innovative dynamic working bodies

Nozim I. Dzhabborov, *Aleksandr V.* Dobrinov*, and *Galina A.* Semenova

Federal Scientific Agroengineering Center VIM, branch in Saint Petersburg, Saint Petersburg, Russia

Abstract. The research results are related to the problem of enhancing the environmental safety of technical means in crop production. A research object is a tillage machine with standard and innovative dynamic working bodies. The research is aimed to a comparative ecological assessment of the efficiency of using typical and dynamic working bodies for surface tillage. The scientific novelty is represented by the regularities of changes in resistivity per unit of active frontal area of the investigated working bodies, inertia force and total kinetic energy, per hectare fuel consumption, fuel economy and reduction of emissions of toxic components at different speed modes of operation of the tillage unit. During the research, the methods of experimental research in the field, analysis and generalization of experimental data were used. It has been established that the resistivity per unit of active frontal area of a dynamic tillage working body with an energy-storage transmission mechanism within the range of speed variation from 2.22 to 3.33 m/s decreases by 33-41% in comparison with a typical working body. With an increase in the speed of the tillage unit with standard working bodies from 2.22 to 3.33 m/s, the fuel consumption increases from 5.12 to 9.21 kg/ha. Within these limits of the speed of the tillage unit with dynamic tillage working bodies and energy-accumulating transmission mechanisms, the per hectare fuel consumption increases from 4.81 to 8.30 kg. The use of dynamic working bodies on the tillage machine allows to save from 0.310 to 0.910 kg of fuel per 1 ha. Herewith, the decrease in the level of emissions of toxic components is 6-10%, which ensures a decrease in the environmental load of soil-cultivating units during the cultivation of crop products.

1 Introduction

According to the latest statistics of the Federal Service for State Registration, Cadastre and Cartography (Rosreestr), the total arable land in Russia is more than 116.8 million hectares. Based on this circumstance, the size of emissions of toxic components from the use of energy facilities calculated during the agricultural year on a national scale can vary within 2.44 - 3.30 million tons.

* Corresponding author: a.v.dobrinov@yandex.ru

Research analysis shows that the amount of toxic components formed during the combustion of 1 kg of diesel fuel averages 57 g, including carbon monoxide - 21 g; hydrocarbons - 4g; nitrogen oxides - 18g; sulfur dioxide - 8g; aldehydes - 1 g and soot - 5 [1-3].

An assessment of the fulfillment of agroecological requirements by tillage machines during their operation at high speeds is given in [4]. The article notes that soil cultivation is currently performed mainly at increased speeds from 10 to 15 km/h, while using combined multi-operational Russian and foreign machines, many of which are not adapted to local climatic conditions for a number of design parameters and technological modes. Excessive tillage causes a deterioration in the structure of the soil, while the formation of erosion-dangerous particles is activated.

The article [5] describes methodological approaches to the development of systems for precision and adaptive landscape farming. The issues of a combination of technological operations, a decrease in the degree of swampy soils and rugged relief are considered.

Modern approaches to the use of energy - and environmentally efficient machine technologies and technical means are outlined in [6].

Evaluation of emissions of harmful substances in the field of technical service and repair of agricultural machinery is also important in its operation. For example, in [7], the results of assessing the concentration of harmful substances in the air in the working area of the galvanic section are given when hardening parts of harvesting machines with chromium plating.

The authors of [8] considered the possibilities of enhancing and enhancing the indicators for assessing the environmental safety of technical means during the procurement of green pressed fodder.

The studies of the authors of the article [9] were devoted to the assessment of the use of a self-regulating system for heating diesel fuel to reduce emissions of toxic components when burning diesel fuel by motor vehicles.

In the works of the authors of this work [10-14], ways of reducing the environmental load of agricultural machinery on the environment are also outlined, agroecological principles for the formation of a zonal system of soil cultivation are formulated, innovative soil-cultivating working bodies with dynamic properties are created, the use of which significantly reduces fuel consumption per unit of processed soil. area.

The analysis of research shows that the methods and means of ensuring the environmental safety of technical means in various sectors of the national economy, developed by scientists, turned out to be very effective for the development of the theory and practice of solving environmental problems.

At the present stage of development of agricultural production, further enhancement of environmental technologies and technical means is required to reduce emissions of harmful substances from mobile sources.

In this regard, the development of methods and ways to reduce the environmental load of agricultural machinery on the environment, in particular, based on the creation of machines for soil cultivation, seems relevant.

One of the effective ways to solve this problem is the development of innovative tillage working bodies and, accordingly, machines with dynamic properties that reduce fuel consumption per hectare and enhance the quality of soil cultivation.

At the IAEP, a branch of the Federal State Budgetary Scientific Institution FNATS VIM, we have developed and created experimental samples of innovative dynamic tillage working tools. Comparative experimental studies of typical and dynamic tillage working bodies and machines have been performed.

This article presents the results of a comparative environmental assessment of the efficiency of using dynamic and typical tillage working bodies and machines.

2 Materials and Methods

The purpose of the experimental studies was to obtain experimental data for a comparative assessment of the ecological efficiency of using dynamic working bodies with an energy-storage transmission device and standard typical working bodies for surface tillage, including loosening with the simultaneous cutting of weeds.

The experiments were performed on the territory of the research and production test site of the institute in the summer-autumn period of 2020.

Figure 1 shows the tillage working bodies for surface tillage.

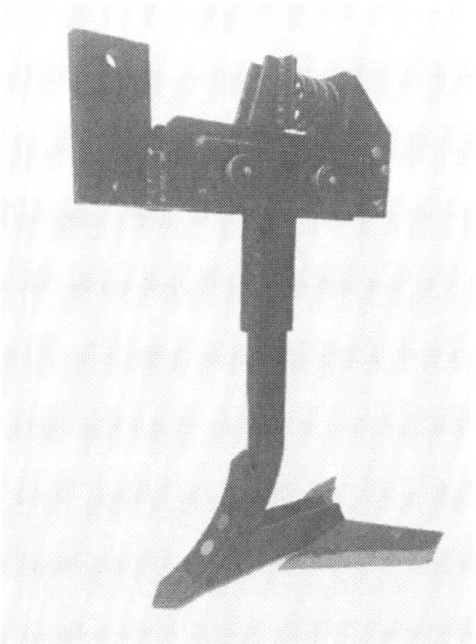

a

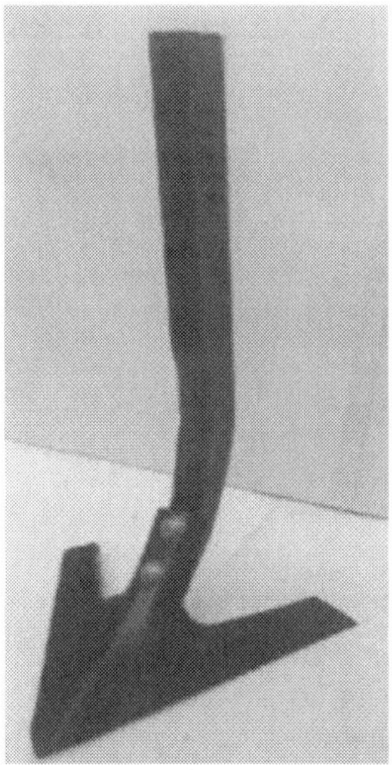

b

Fig. 1. General view of a dynamic (a) with an energy-storage transmission device and a typical (b) tillage working bodies

The experiments were performed under certain soil conditions:
- soil type - soddy-medium podzolic;
- relief, degrees - 1-2;
- ridge of the surface of the field, cm - 3-4;
- soil hardness before processing in a layer of 5-20 cm - 0.85 - 1.0 MPa;
- soil moisture in the layer 0 - 10 cm - 13.5%, in the layer 10 - 20 cm - 16.8%.

The experimental data were processed pursuant to the method described in [15].

3 Results and Discussion

Based on the generalization of the experimental data, graphical dependences of the resistivity per unit of active frontal area of a typical and dynamic tillage working bodies on the speed of movement were constructed (Fig. 2).

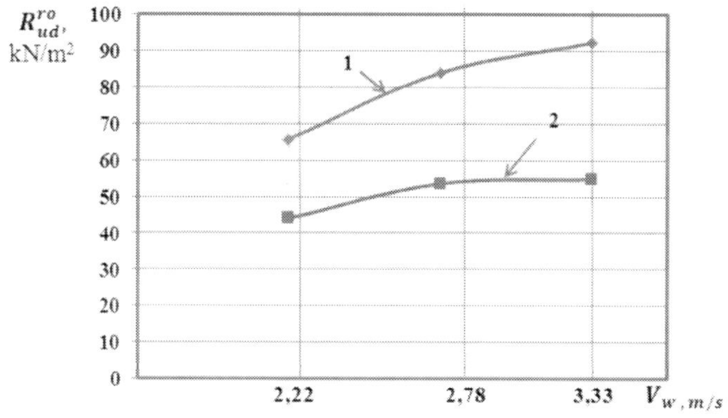

Fig. 2. Dependences of resistivity R_{ud}^{ro}per unit of active frontal area of typical (1) and dynamic (2) tillage working bodies on the speed of movement V_w

The regularities of changes in the resistivity of tillage working bodies, described by the following empirical dependencies, have been revealed:

- typical working body:

$$R_{ud}^{ro} = -15{,}64566V_w^2 + 110{,}86047V_w - 103{,}24814; \quad (1)$$

- dynamic working body:

$$R_{ud}^{ro} = -13{,}39148V_w^2 + 84{,}10741V_w - 76{,}68787. \quad (2)$$

It was found that with an increase in the speed V_w of movement from 2.22 to 3.33 m/s, the specific resistance R_{ud}^{ro} of a typical tillage working body increases from 65.75 to 92.42 kN/m².

Within the same limits of speed change, the resistivity R_{ud}^{ro} of a dynamic tillage working tool ranges from 44.03 to 54.89 kN/m².

Analysis of the R_{ud}^{ro} values shows that the specific resistance of a dynamic tillage working body, within the range of speed V_w variation from 2.22 to 3.33 m/s, decreases by 33-41%, in comparison with a typical working body.

Figure 3 shows the graphical dependence of the total inertia force on the vibration acceleration of dynamic tillage working bodies.

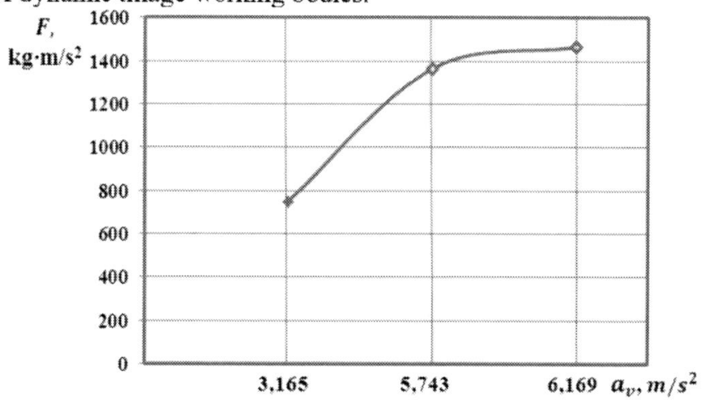

Fig. 3. Dependence of the total inertia force F on the vibration acceleration a_v of dynamic tillage working bodies

Using the Lagrange interpolation formula, based on the obtained experimental values, an empirical dependence was established that describes the regularity of the change in the total inertia force of dynamic tillage working bodies:

$$F = -0{,}00544a_v^2 + 237{,}64933a_v - 0{,}10161. \quad (3)$$

It has been experimentally established that with an increase in the speed of the soil-cultivating unit from 2.22 to 3.33 m/s, the vibration acceleration a_v of dynamic soil-cultivating working bodies increases from 3.16 to 6.16 m². Within the range of vibration acceleration a_v variation from 3.16 to 6.16 m², the total inertia force F of the working bodies increases from 752.0 to 1465.75 kg m/s².

Figure 4 shows the dependence of the total kinetic energy of oscillation on the vibration speed of dynamic tillage working bodies.

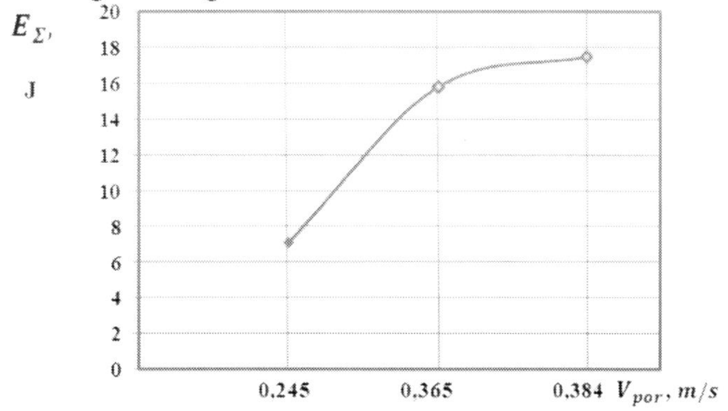

Fig. 4. Dependence of the total kinetic energy of oscillation E_Σ on the vibration speed V_{por} of dynamic tillage working bodies

The interpolation of the generalized experimental data made it possible to reveal the regularity of the change in the total kinetic energy of the vibration from the vibration velocity of the dynamic working bodies:

$$E_\Sigma = 115{,}61593V_{por}^2 + 1{,}98262V_{por} - 0{,}29759. \quad (4)$$

It was found that when the vibration velocity V_{por} changes from 0.24 to 0.38 m/s, the total kinetic energy E_Σ increases from 7.12 to 17.51 J. This indicates that at higher speeds, dynamic working bodies with an energy-storage transmission device generate more energy, which is used to create a dynamic impact on the cultivated layer of soil and enhance the quality of its processing.

Figure 5 shows the dependences of the specific fuel consumption per unit of the cultivated area of a tillage tool with dynamic and typical working bodies on the speed of its movement.

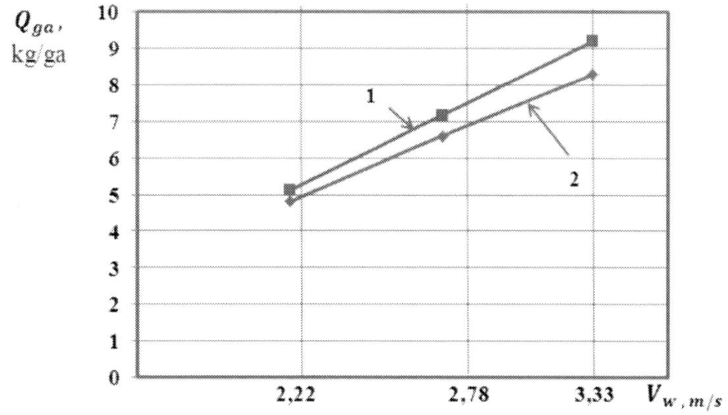

Fig. 5. Dependences of specific fuel consumption Q_{ga} per unit of the cultivated area of a tillage tool with dynamic (2) and standard (1) working bodies on the speed V_w of its movement.

With an increase in the speed of the tillage unit with standard working bodies from 2.22 to 3.33 m/s, the fuel consumption Q_{ga} increases from 5.12 to 9.21 kg/ha. Within these limits of the speed of the tillage unit with dynamic tillage working implements, the per hectare fuel consumption increases from 4.81 to 8.30 kg.

Regularities of changes in specific fuel consumption per unit of treated area are revealed, which are described by empirical dependencies:

- a tillage unit with typical working bodies:

$$Q_{ga} = 0{,}01111V_w^2 + 3{,}62300V_w - 2{,}97783; \quad (5)$$

- soil cultivating unit with dynamic working bodies:

$$Q_{ga} = -0{,}12753V_w^2 + 3{,}85194V_w - 3{,}11278. \quad (6)$$

Figure 6 shows the graphical dependence of the amount of fuel consumption saving ΔQ_{ga} per 1 hectare of the cultivated area by a tillage machine with dynamic working bodies in comparison with typical working bodies on the speed of the machine.

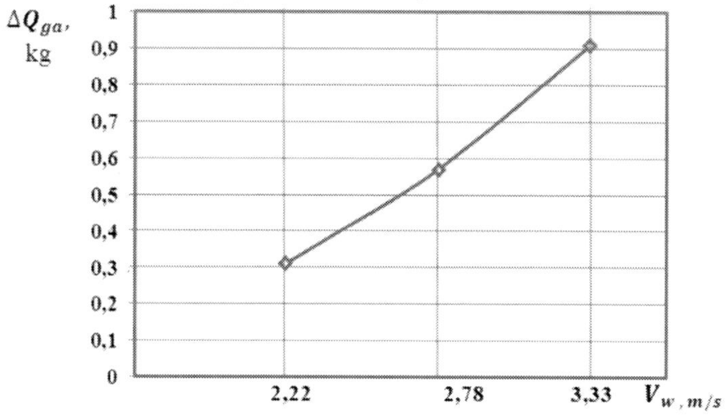

Fig. 6. Dependence of the amount of fuel consumption saving ΔQ_{ga} per 1 hectare of cultivated area by a tillage unit with dynamic working bodies in comparison with typical working bodies on the speed V_w of the unit

As can be seen from Figure 6, within the range of movement speed of $V_w = 2{,}22 - 3{,}33$ m/s, in comparison with typical working bodies, the use of dynamic working bodies in a tillage machine allows to save from 0.310 to 0.910 kg of fuel per ha.

The regularity of the change in the value of the fuel consumption saving ΔQ_{ga} on the speed V_w of the tillage unit is revealed, which is described by the following empirical dependence:

$$\Delta Q_{ga} = 0{,}13865V_w^2 - 0{,}22894V_w + 0{,}13495. \quad (7)$$

As noted above, when 1 kg of diesel fuel is burned, an average of 57 g of toxic components are emitted into the atmosphere.

The graphical (Figure 6) and empirical (formula 7) dependencies obtained in the process of processing and generalization of experimental data made it possible to establish the degree of reduction in the ecological load of the tillage machine with dynamic working bodies in comparison with the use of standard working bodies.

Figure 7 shows graphical dependences of the reduction in the amount of emissions of toxic components ΔM_{tk} (in grams,%) during the operation of a machine-tractor tillage unit with dynamic working bodies on the speed $V_w$of its movement.

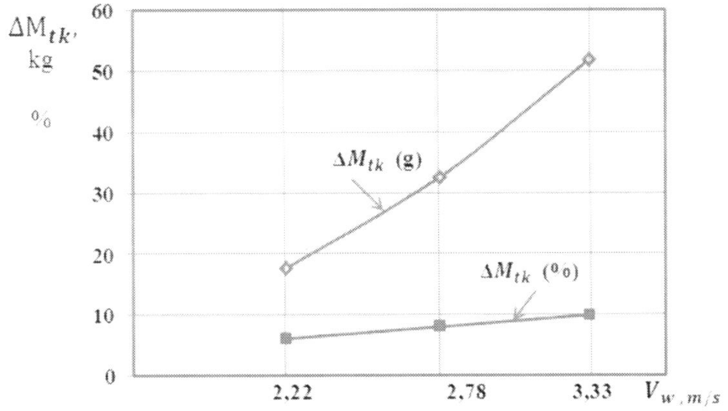

Fig. 7. Dependences of the amount of emissions of toxic components ΔM_{tk} of a machine-tractor tillage machine with dynamic working bodies on V_w the speed of movement

Machine-tractor tillage machines with dynamic working bodies, in comparison with standard implements, correspond to the modern principles of ecological technologies, as they ensure the reduction of emissions of toxic components into the atmosphere. Experimental data indicate that when the speed of the soil cultivating unit with dynamic working bodies changes in comparison with typical working bodies from 2.22 to 3.33 m/s, the emissions of toxic components when processing 1 hectare of area decrease from 17.67 (by 6, 05%) to 51.87 g (9.88%).

The revealed patterns of change in the magnitude of the decrease in the amount of emissions of toxic components into the atmosphere from the speed of movement of the tillage unit with dynamic working bodies are described by empirical dependencies:

$$\Delta M_{tk} = 7{,}87031V_w^2 - 12{,}86938V_w + 7{,}45202, \text{(g)} \quad (8)$$

$$\Delta M_{tk} = 0{,}13718V_w^2 + 2{,}68909V_w - 0{,}59586, (\%) \quad (9)$$

As our research has shown, the use of dynamic working bodies with energy-storage transmitting devices on soil-cultivating implements for pre-sowing soil cultivation makes it possible to enhance its quality and achieve a reduction in fuel consumption when processing one hectare of arable land by 6-10%, which ultimately ensures a decrease in environmental load of MTA crop production.

4 Conclusion

Tillage machines and working bodies used today in modern agricultural production do not fully meet the requirements of environmental safety. The enhancement of tillage machines, in particular their working bodies, should go along the path of enhancing their technological indicators and design parameters, on the basis of giving them the properties of dynamism and adaptability to various soil conditions.

The conducted research results confirm that the use of dynamic working bodies on soil-cultivating implements, in comparison with standard working bodies, is of no small importance for reducing emissions of toxic components into the atmosphere. Thus, within the range of change in the speed of the soil-cultivating unit, from 2.22 to 3.33 m/s with dynamic working bodies in comparison with typical working bodies, emissions of toxic components when processing 1 hectare of area decrease from 17.67 (by 6.05%) to 51.87 grams (9.88%).

References

1. N. N. Krupenio, *Calculation of pollutant emissions from vehicles running on gasoline and diesel fuel, as well as payments for these emissions*, 38 (2006)
2. Online magazine about cars, https://7gear.ru/
3. The content of toxic emissions in the exhaust gases of engines, https://studexpo.ru/
4. V.N. Zvolinsky, *Ecology and Agricultural Machinery*, 2, 69-75 (2009)
5. D.A. Ivanov, N.G. Kovalyov, O.N. Antsiferova, *Ecology and Agricultural Machinery*, 2, 13-18 (2007)
6. A. Yu. Izmailov, Ya. P. Lobachevsky, O. A. Sizov, Agricultural machines and technologies, **4**, 8 (2013)
7. A. E. Krupin, The international technical-economic journal, **2**, 112-117 (2017)
8. S. Ivanov, S. Gakh, I. Skoniyechny, *Ecology and agricultural technology*, 2, 99-105 (2009)
9. A. M. Shuvalov, P. A. Telegin, T. N. Dementyeva, V. S. Ikonnikov, V. F. Kalinin, K. A. Nabatov, *Ecology and agricultural equipment*, 2, 244-248 (2009)
10. N. I. Dzhabborov, A. V. Dobrinov, A. M. Dementyev, Materials of the 7th international scientific and practical conference "Ecology and agricultural technologies: agroengineering solutions", 1, 43-47 (2011)
11. N. I. Jabborov, A. V. Dobrinov, D. S. Fedkin, Regional Ecology, **5**, 23-27 (2015)
12. N. I. Dzhabborov, V. A. Eviev, A. V. Sergeev, G. A. Semenova, Technologies and technical means of mechanized production of crop and livestock products, **4(97)**, 56-65 (2018)
13. N. I. Dzhabborov, A. V. Sergeev, G. A. Semenova, Technologies and technical means of mechanized production of crop and livestock products, **1(98)**, 107-115 (2019)
14. N. I. Dzhabborov, A. V. Sergeev, A. M. Zakharov, A. N. Perekopsky, IOP Conference Series: Materials Science and Engineering. The collection of conference materials, Voronezh State University of Engineering Technologies, 012048 (2020)
15. A. M. Valge, N. I Jabborov, V. A. Eviev, *Fundamentals of statistical processing of Experimental Data in the Conduct of researches on mechanization of agricultural production with examples on STATGRAPHICS and EXCEL*, 140 (2015)

Aquatic ecosystems of the lower reaches of the Zarafshan River. Diversity and ecological groups of molluscs

Khusniddin Baymuradov[1,*], *Tozagul* Zhabborova[2], *Iroda* Tuinazarova[3], *Bekzod* Otakulov[1], *Azamat* Egamkulov[1]

[1]Samarkand State University
[2]Karshi State University
[3]Jizzakh Polytechnic Institute

Abstract. Aquatic ecosystems of the lower reaches of the Zarafshan River revealed the distribution of 49 species and 2 subspecies of molluscs. They are distributed over aquatic ecosystems as follows: in the lower reaches of the Zarafshan River, 32 species and 2 subspecies are distributed, in the Navbakhorsky fish farm - 22 species and in the Khatyrchinsky fish farm - 15 species, in the Nurbuloksaysky fish farm - 22 species and in the Maidonsoysky - 18 species of mollusks. Their belonging to pelolymnophilic, peloreophilic, rheophilic, crenophilic, phytophilic, madicol, limno-crenophilic and telmatophilic ecological groups was considered.

1 Topic relevance

The study of the fauna of aquatic ecosystems in the lower reaches of the Zarafshan River creates important prerequisites for a comprehensive solution of theoretical and practical problems [1,2,4]. The diversity of the environment in aquatic ecosystems has created conditions for the distribution of many rare, endemic and relic animals [3,5,6]. At present, one of the urgent problems is the study of the diversity of mollusks and ecological groups of aquatic ecosystems in the lower reaches of the Zarafshan River.

2 Research object and methods

Below is a comprehensive analysis of the distribution, ecological groups of mollusks, their habitat and the expansion of their ranges [7,8,9]. These mollusks were studied by V.I.Zhadin (1938, 1952), Ya.I. Starobogatov, Z.I. Izatullaev (1984), Z.I. Izatullaev, Kh.T. Baimurodov (2009) and we used the system developed by Z. Izatullaev for Central Asian molluscs.

3 Research results

* Corresponding author: boymurodov1971@mail.ru

Aquatic ecosystems in the lower reaches of the Zarafshan River, hydrographic conditions are complex. There are such sources of water as large and small rivers, lakes, springs and reservoirs, ponds. They differ from each other in a variety of hydrological, physicochemical and hydrobiological regimes. Each type of water is inhabited by molluscs, and each of them has its own ecological groups. It was found that 51 species and 2 subspecies of mollusks inhabit the aquatic ecosystems of the lower Amu Darya (Table 1).

Table 1. Aquatic ecosystems of the lower reaches of the Zarafshan River. Diversity and ecological groups of molluscs

No.	Family and species	Lower course of the Zarafshan river	Navbakhorsky fish farm	Khatyrchimsky fish farm	Sources	Springs	Ecological groups
	Bivalvia class Unionidae family Sinanodonta genus						
1	*Sinanodonta gibba*	+	+	-	-	-	Peloreophile
2	*Sinanodonta orbicularis*	+	-	+	-	-	Peloreophile
3	*Sinanodonta puerorum*	+	-	+	-	-	Peloreophile
	Colletopterum genus						
4	*Colletopterum bactrianum*	+	-	+	-	-	Rheophile
5	*Colletopterum cyreum sogdianum*	+	-	+	-	-	Rheophile
6	*Colletopterum ponderosum volgense*	+	-	+	-	-	Pelolimnophile
	Corbiculidae family Corbicula genus						
7	*Corbicula cor*	+	+	+	-	-	Peloreophile
8	*Corbicula fluminalis*	+	+	-	-	-	Peloreophile
9	*Corbicula purpurea*	+	-	+	-	-	Peloreophile
	Corbiculina sort						
10	*Corbiculina tibetensis*	+	-	+	-	-	Peloreophile
11	*Corbiculina ferghanensis*	-	+	+	-	-	Peloreophile
	Sphaeridae family Musculium genus						
12	*M.hungarica*	+	-	-	+	-	Pelolimnophile
	Euglesidae family Euglesa genus						
13	*Euglesa hissarica*	-	-	-	+	-	Pelolimnophile
14	*Euglesa heldreichi*	-	-	-	-	+	Peloreophile
15	*Euglesa turkestanica*	-	-	-	+	-	Pelolimnophile
16	*Euglesa obliquata*	-	-	-	-	+	Pelolimnophile
17	*Euglesa turanica*	-	-	-	-	+	Pelolimnophile
	Pisidiidae family Odhneripisidium genus						
18	*Odhneripisidium behningi*	-	-	-	+	+	Crenophile
	Kuiperipisidium genus						
19	*Kuiperipisidium terekense*	-	-	-	+	+	Crenophile
20	*Kuiperipisidium issykkulense*	-	-	-	+	-	Crenophile

21	*Kuiperipisidium sogdianum*	-	-	-	+	-	Crenophile
22	*Kuiperipisidium polytimeticum*	-	-	-	+	-	Crenophile
	Gastopoda class Neritidae family Theodoxus genus						
23	*Theodoxus.pallasi*	+	-	-	+	-	Peloreophile
	Valvatidae family Cincinna genus						
24	*Cincinna.pamirensis*	+	-	-	-	-	Pelolimnophile
25	*Cincinna.piscinalis*	+	-	-	-	-	Pelolimnophile
	Belgrandellidae family Bucharamnicola genus						
26	*Bucharamnicola.vucharica*	+	+	-	+	-	Crenophile
	Martensamnicola genus						
27	*Martensamnicola. Brevicula*		+	-	+	-	Crenophile
28	*Martensamnicola.hissarica*	+	+	-	+	-	Crenophile
	Horatiidae family Sogdamnicola genus						
29	*Sogdamnicola. Pallida*		+	-	+	-	Crenophile
30	*Sogdamnicola.shadini*	+	-	-	+	-	Crenophile
	Lymnaeidae family Lymnaea genus						
31	*Lymnaea. Stagnalis*	+	+	-	-	-	Phytophile
32	*Lymnaea.impura*	+		-	-	-	Phytophile
33	*Lymnaea.oblonga*	+	+	-	-	+	Madikol
34	*Lymnaea.goupili*		+	-	+	+	Madikol
35	*Lymnaea.thiesseae*	+	+	-	-	+	Madikol
36	*Lymnaea.truncatula*	+	+	+	+	+	Madikol
37	*Lymnaea.subangulata*		+	-	-	+	Madikol
38	*Lymnaea.almaatina*	+	+	-	+	-	Limno-crenophile
39	*Lymnaea.bowelli*	+	+	-	+	-	Limno-crenophile
40	*Lymnaea.tenera*	+	+	+	-	+	Phytophile
41	*Lymnaea.rectilabrum*	+	-	+	-	+	Phytophile
42	*Lymnaea.psilia*	+	+	+	-	+	Phytophile
43	*Lymnaea.fontinalis*	+	+	+	-	+	Phytophile
44	*Lymnaea.lagotis*	+	-	+	-	-	Telmatophile
	Phisidae family Costatella genus						
45	*Costatella. Acuta*	+	+	-	+	-	Phytophile
	Planorbidae family Planorbis genus						
46	*Planorbis. Planorbis*	+	+	-	-	+	Phytophile
47	*Planorbis.tangitarensis*	+	-	-	-	+	Phytophile
	Anisus genus						
48	*Anisus. Acronicus*	-	+	-	+	+	Phytophile
49	*Anisus.albus*	-	-	-	-	+	Phytophile
50	*Anisus.centralis*	+	-	-	+	-	Pelolimnophile
51	*Anisus.albopersicus*	+	-	-	+	-	Phytophile
Total species		**34**	**22**	**15**	**22**	**18**	

Note: + species present, - species not present

In the middle of the XX century, the anthropogenic destruction of zoogeographic barriers on land caused the penetration of invasive species into the fauna of most territories, in particular, into the hydrofauna, which remained unchanged for a long time.

Aquatic ecosystems of the lower reaches of the Zarafshan River revealed the distribution of 49 species and 2 subspecies of molluscs. They are distributed over aquatic ecosystems as follows: in the lower reaches of the Zarafshan River, 32 species and 2 subspecies are distributed, in the Navbakhorsky fish farm - 22 species and in the Khatyrchinsky fish farm - 15 species, in the Nurbuloksaysky fish farm - 22 species and in the Maidonsoysky - 18 species of mollusks. Their belonging to pelolymnophilic, peloreophilic, rheophilic, crenophilic, phytophilic, madicol, limno-crenophilic and telmatophilic ecological groups was considered.

The shell size of the mollusk species does not directly affect their distribution over the river sections. Representatives of the family Unionidae, which includes large species, and the family Corbiculidae, which includes small species, mean that the accumulation of rivers in the middle reaches is important for the hydrological regime of rivers in the distribution of bivalve molluscs along river beds.

References

1. V.V. Bogatov, *Abstacts of the cinference Mollusks of the Eastern Asia and Adjacents Seas. Vladivostok*, 6-12 (2014)
2. V. V. Bogatov, Ya. I. Starobogatov, Ruthinica, **4(2)**, 147–150 (2004)
3. P. Bouchet, *Inventorying the molluscan fauna of theworld: howfar togon*, 180 (2007)
4. Kh. T. Baymurodov, Advanced studies in science: theory and practice, The Collection of Scholarly Papers Materials of the International Scientific Conference, 239-242 (2016)
5. Kh. T. Baymurodov, Journal of Environmental Science and Engineering, **2**, 692–696 (2013)
6. Kh. T. Baymurodov, Oecologia Montana International journal, 10-12 (2013)
7. Kh. T. Baymurodov, Research publications: nature, ecology and national economy, **7(27)**, 43-52 (2015)
8. M. V. Chertoprud, Monitoring of stream pollution on Structure of macrobenthos, 17 (1999)
9. V. I. Yurishinets, Bull. of Zoology, **1**, 79-84 (2001)

Eco-innovative application of phyto-raw materials in the technology of long-term storage of unrefined sunflower oil

Roza Timakova[1,*], and *Ruslan* Iliukhin[2]

[1]USUE, 620144 Yekaterinburg, Russian Federation
[2]Saint- Petersburg state University of telecommunications. M. A. Bonch-Bruevich, 192232 Saint-Petersburg, Russian Federation

Abstract. Along with the development of agroecosystems, environmental friendliness of agricultural and industrial technologies for long-term storage of agricultural products is being formed as a result of cognitive technologies development, including innovations in biotechnologies. Achieving the biological safety of bulk unrefined sunflower oil is based on the addition of an unconventional antioxidant phyto-raw material to it - stinging nettle, which slows down oxidative processes. In order to prevent oil contamination, the optimal dose of ionizing radiation for the nettle treatment is established - up to 1.5 kGr, which ensures microbiological safety and antioxidant activity (AOA) at the level of 0.763 + 0.016 mMeq, comparable to AOA in the untreated nettle samples after 3 months. After 52 days of storage, the acid number was in the range of 0.99-1.15 mg KOH / g, the peroxide number was 5.09-5.16 mmol active oxygen / kg. Adding 0.5-1.0 wt. % nettle powder to the oil ensures the compliance of organoleptic parameters with the established requirements - transparency, taste and smell. Adding 1.0 wt. % nettle powder to the oil slows down the filtration process of the resulting suspension. The practical relevance of the study is determined by the verification of the research results to prolong the oil shelf life by adding environmentally friendly phyto-raw materials to it.

1 Introduction

One of the national goals in accordance with the Strategy for the Development of the Agroindustrial and Fisheries Complexes of the Russian Federation for the period up to 2030 is creating an agroindustrial complex that develops on the basis of modern technologies for the production, processing and storage of agricultural raw materials and food products and determines two vectors of development: on the one hand, agrotechnology itself for the production of raw materials, on the other hand - their complex processing.

For the development of environmentally friendly crop production, the improvement of agroecosystems is of importance: reclamation of soil as an important biological resource [1]

* Corresponding author: trt64@mail.ru

and ensuring the soil qualitative characteristics, for example, when it is saturated with nitrogen due to changes in the soil biota [2,3]; forming optimal configurations of ridge furrows to improve soil water supply [4] and others. The transformation of agroecosystems based on digital design solutions will help reduce the anthropogenic and technological load on agricultural areas, reduce technological risks and achieve biological safety.

In the medium term, the role of a new technology type - cognitive technologies, determined by the complex of information technologies, biotechnologies and nanotechnologies development [5, 6], is increasing. Digitalization of the agro-industrial sector with the development of the "Mercuriy" ("Mercury") information system, the launch of the StockMeat.com industry program, the introduction of RFID tags for agricultural products, the development of agricultural technology platforms such as "Technology of food and processing industry of the agro-industrial complex - healthy food products" and the "Eurasian agricultural technology platform" and the use of the broadband mobile LPWAN communication to organize large-scale digital agricultural solutions (smart field, smart herd, smart farm, etc.) - increases its investment attractiveness [7, 8].

Within the framework of the strategic direction "Agricultural production" of the "Technology of food and processing industry of the agro-industrial complex - healthy food products" technological platform, one of the promising areas is technologies for long-term storage of agricultural products, including those based on molecular biological methods [9], which can be considered both in terms of physical storage methods that do not affect the chemical composition of the product itself, and in terms of gentle biotechnologies. The choice of innovative technology can be determined by the "circular economy, which, acting as a practical basis for the implementation of a green economy, offers effective business models to ensure more environmentally friendly resource use" [10]. The choice of the processing technology for food raw materials is determined by the goals of processing and the type of food raw materials.

The Russian Federation is the largest exporter of sunflower oil; this predetermines the importance of preserving edible oilseed raw materials. The key preservation factors for vegetable oil during storage are air temperature, darkening of storage areas and lack of oxygen. For the human body, unrefined oils are most useful; yet, the presence of biologically active substances makes unrefined vegetable oils unstable during storage [11]. In the consumer market, a stable niche is occupied by unrefined sunflower oil, which in contrast to refined oil has a shorter shelf life. Unrefined sunflower oil is characterized by a rich taste and smell, it is high in phospholipids, vitamins E, A and K and free fatty acids, which are easily oxidized during storage to hydroperoxides, aldehydes, ketones and alcohols, which reduce the oil nutritional value. In order to provide a prolonged shelf life of the product, biotechnological methods can be applied. The latter are based on the use of phyto-raw materials and serve to slow down oxidative processes due to the content of polyphenols. The use of biotechnological methods can be considered as a promising trend in the greening of technological processes.

According to [12-14], to achieve oxidative stability of vegetable oil and thus, to prolong its shelf life, the most effective way to be used is adding natural antioxidants (AO) to the finished product, such as curcumin for corn oil, green tea extracts for peanut butter, pomegranate peel for coconut oil, sea buckthorn extract, rosemary extracts, meadowsweet, blueberry, capsicum powder, mint and clove for linseed oil, encapsulated extracts of citrus, broccoli and rosemary and olive tree leaves, as well as micellized forms of antioxidants (tocopherin, ascorbic acid) for sunflower oil [15-22].

The research aims to increase the shelf life of bulk unrefined sunflower oil by adding to it the powder of stinging nettle (Urtica dioica) of the Rosaceae family.

2 Materials and Methods

We took the following raw materials to form the objects of study: bulk unrefined premium sunflower oil and nettle powder of the Nettle genus, the Nettle family, the Rosaceae order. Nettle refers to medicinal and technical raw materials, or phyto-raw materials, with an antioxidant orientation due to the content of flavonoids, essential oils, ascorbic acid (vitamin C) and carotenoids.

6 groups of nettle powder samples were prepared, which, to ensure microbiological purity, were treated with ionizing radiation by a stream of accelerated electrons with doses from 0.5 to 3.0 kGr with a step of 0.5 kGr. The choice of radiation doses is determined by the requirements of GOST 33271-2015 "Dry spices, herbs and vegetable seasonings. Guidance on exposure to pathogens and other microorganisms". According to the requirements of Technical Regulations of the Customs Union 024/2011 "Technical regulations for fat and oil products", microbiological indicators for vegetable oil are not stated.

Antioxidant activity (AOA) in the nettle powder was identified without specifying individual antioxidants by the potentiometric method with the use of an MPA-1 analyzer according to the standard procedure [23].

Then 3 groups of samples were formed: 1st control group (samples of bulk unrefined premium sunflower oil), 2nd experimental group (samples of bulk unrefined premium sunflower oil with the added 0.5 wt% nettle powder), 3rd experimental group (samples of bulk unrefined premium sunflower oil with the added 1.0 wt.% nettle powder).

The oxidative damage indicators of the vegetable oil samples, i.e. acid and peroxide numbers were determined by the titrimetric method; microbiological indicators of the nettle powder were determined in accordance with standard analysis methods. The research was conducted in five replicates. The research results were processed by the method of variation statistics.

3 Results and Discussion

At the first stage, the nettle powder was explored. We found out that the number of mesophilic aerobic and facultative anaerobic microorganisms (NMAaFAnM) after the nettle powder treatment with doses of 0.5; 1.0; 1.5; 2.0; 2.5 and 3.0 kGr was $4 \cdot 10^4$, $1 \cdot 10^4$, $7 \cdot 10^3$, $3 \cdot 10^3$, $8 \cdot 10^2$ and $3 \cdot 10^2$ CFU / g, respectively, which correspond to the requirements of the Technical Regulations of the Customs Union 021/2011 "On food products safety". All the nettle powder samples were free of coliform bacteria, E. coli and mold.

Evaluation of antioxidant potential of the nettle powder samples showed that when the radiation dose was increased, AOA decreased with a high coefficient of approximation equal to 0.99. Thus, in the nettle samples treated with a dose of 3 kGr, AOA was 0.677 ± 0.022 mM-eq, which is 22.3% less than in the nettle samples that were not treated with ionizing radiation (Fig. 1).

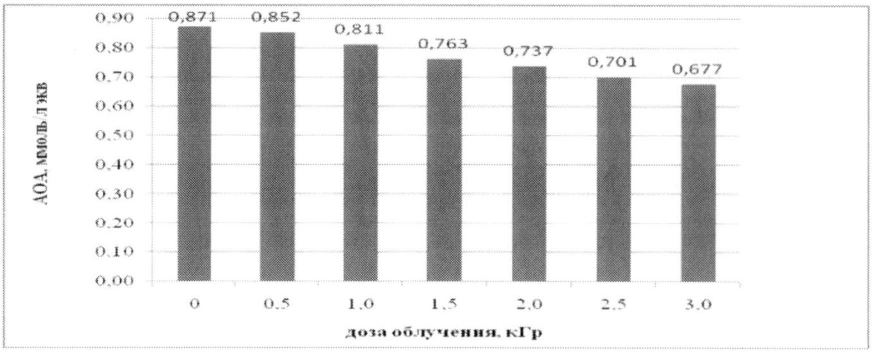

Fig. 1. Changes in antioxidant activity (AOA) in the nettle powder
Where: Доза облучения, кГр - Radiation dose, kGr ; АОА, ммоль/л экв - AOA, mmol / l eq .

The formula for the linear dependence of AOA on the radiation dose is presented as follows:

$$y = - 0.034 x + 0.91 \quad (1)$$

On the basis of the results obtained, the optimal dose of ionizing radiation was established for the use of nettle powder as a phyto-raw material with increased antioxidant activity – up to 1.5 kGr; AOA decreases compared with untreated samples by 12.4% and is 0.763 + 0.016 mM-eq, which is comparable to the decrease in the AOA of untreated samples during storage for up to 3 months.

At the second stage, to obtain a homogeneous suspension, the samples of bulk unrefined sunflower oil were mixed with the nettle powder treated with ionizing radiation of 1.5 kGr in a mixer at a rotation speed of 170-180 rpm and an oil temperature of + 15-18 ° C for 12-15 min. In the experimental samples of the 2nd group, an insignificant suspension of the powder was observed, in the experimental samples of the 3rd group, the suspension was thicker.

After storage of control and experimental samples for 45 days in accordance with the requirements of the regulatory documents and 52 days in accordance with the reserve ratio for non-perishable food products, organoleptic indicators and oxidative damage indicators were analyzed.

As a result of the organoleptic assessment, it was found that the control samples of sunflower oil when stored for up to 45 days met the regulatory requirements, after 52 days of storage, a grid appeared over the sediment and there was a slight smell of oxidation. In the experimental samples of unrefined oil, when stored for up to 45 days and 52 days, the oil was transparent with slight turbidity, the smell and taste were characteristic of unrefined sunflower oil, without any foreign smell or taste.

During storage of the experimental oil samples, suspended particles of nettle powder and phospholipids settle to the bottom, which makes it possible to use such oil after filtration for the production of refined oil. However, in the 3rd experimental group, the filtration process slows down due to a large amount of immiscible residue.

The results obtained for the acid number and peroxide number are presented in Table 1.

Table 1. Values of the acid number and peroxide number in the experimental and control samples of bulk unrefined premium sunflower oil during storage

Shelf life, days	Acid number, mg KOH / g			Peroxide number, mmol act. oxygen / kg		
	1 group	2 group	3 group	1 group	2 group	3 group
0	0,54±0,02	0,40±0,02	0,36±0,03	3,22±0,05	2,88±0,03	2,73±0,03
45	1,23±0,08	0,74±0,04	0,68±0,04	5,79±0,11	4,12±0,06	4,01±0,02
52	1,46±0,02	1,15±0,05	0,99±0,04	6,14±0,08	5,16±0,09	5,03±0,05

Thus, as a result of the undertaken experiment we found out that in all samples of vegetable oil, the oxidative damage indicators during storage for up to 45 days and 52 days corresponded to the requirements of the regulatory documents. At the same time, adding 0.5-1.0 wt. % nettle powder in bulk unrefined sunflower oil resulted in lower acid number in the 2nd experimental group – up to 1.15 mg KOH / g and in the 3rd experimental group - up to 0.99 mg KOH / g, which is lower by 21, 2% and 32,2% compared to the indicators of the control oil samples. The corresponding results were obtained for the peroxide number, which is 16.0% and 18.1% lower than the values of the peroxide number in the 2nd and 3rd experimental groups, respectively, which can be explained by the positive effect of phyto-raw materials on the oil oxidative stability.

3 Conclusion

On the basis of the results obtained, the following conclusions can be made. Adding nettle powder, treated with a radiation dose of 1.5 kGy, in an amount of 0.5-1.0 wt. % to bulk unrefined sunflower oil leads to a slowdown of oxidative processes in the oil in comparison with the control samples due to antimicrobial properties and high AOA in the nettle powder, equal to 0.677 ± 0.022 mM-eq. Despite the lower quantitative indicators of acid and peroxide numbers when adding 1.0 wt. % nettle powder to the oil, it is advisable to use 0.5 wt. % nettle powder for subsequent high-quality filtration of unrefined oil when processed into refined oil. The practical application of the conducted research is determined by the repeatability of the results obtained.

References

1. R.L. Soto, M. Martinez-Mena, M.C. Radilla, J. de Vente, Agr. ecosyst. & environ., **306** (2020)
2. Z. Abail, I. K. Whalen, Pedosph., **31(3)**, 405 (2021)
3. D. Vasu, G. Tiwari, S. Sahoo, B. Dach, A. Jangir, R.P. Sharma, R. Naitam, P. Tiwary, K. Karthikeyan, P. Chandran, Cat., **198** (2020)
4. G. Zhang, F. Mo, F. Shah, W. Meng, Y. Liao, J. Han, Agricul. wat. manag., **245** (2020)
5. N.V. Manokhina, Nov. univer., **2(36)**, 52 (2014)
6. Ye.V. Lutsenko, V.N. Laptev, Politem. set. el. nauchn. zhurn. Kub. gos. ar. un., 164, 128 (2020)
7. S. Ognivtsev, Mezhdun. s.-kh. zhurn., **2(368)**, 77 (2019)
8. Ye.N. Trifonova, Mezhdunar. (och.-zaoch.) nauch.-prakt. konf., 243 (2019)
9. Platforma APK, Strateg. progr. (2021)
10. N. Batova, P. Sachek, I. Tochitskaya, BEROC Gr. Ec. Pol. Pap. Ser., 1, 14 (2018)
11. R.T. Timakova, YU.V. Il'yukhina, Tekhn. pishch. i pererab. prom. APK – prod. zdor. pit., 3, 45 (2020)
12. E. N. Frankel, Food Chem., **57**, 51 (1996)
13. N. V. Yanishlieva, Eur.J. Lipid Sci. and Techn., **103(11)**, 752 (2001)
14. N.V. Makarova, M.S. Voronina, Inn. i prod. bezop., **3(25)**, 82 (2019)
15. M. Asnaashari, R. Farahmandfar, R. Esmaeilzadehkenari, Int. J. Adv. Sci. Eng. and Technol., **5 (3)**, 38 (2017).
16. B. Aydeniz, E. Yilmaz, Food Technol. Biotechnol., **54(1)**, 21 (2016)

17. D. Bopitiya, T. Madhujith, Trop. Agr. Res., **25(3)**, 298 (2014)
18. L.V. Tereshchuk, O.V. Sharmanova, Mezhdunar. nauch.-prakt. konf., 96 (2020)
19. A.V. Bašilov, Vies. Nac. akad. navuk Bielarusi. Sier. ahrar. navuk., **1**, 110 (2009)
20. J.-H. Ahn, Y.-P. Kim, H.-S. Kim, Food Contr., **23**, 528 (2012)
21. P. A. González-Fuentes, M. C. Zuñiga, C. A. Olea-Azar et al., Cien. Inv. Agr., **44(3)**, 262 (2017)
22. A.P. Nechayev, A.V. Samoylov, V.V. Bessonov, Yu.V. Nikolayeva, V.V. Tarasova, O.V. Pilipenko, Vopr. pit, **89(5)**, 101 (2020)
23. Kh.Z. Braynina (RU), A.V. Ivanova (RU), Pat. № RU 2235998C2, RF (2004)

AUTHOR INDEX

AUTHOR INDEX

AUTHOR INDEX

AUTHOR INDEX